Rossiter Worthington Raymond

Mineral resources of the states and territories west of the Rocky Mountains

Rossiter Worthington Raymond

Mineral resources of the states and territories west of the Rocky Mountains

ISBN/EAN: 9783741177408

Manufactured in Europe, USA, Canada, Australia, Japa

Cover: Foto ©Klaus-Uwe Gerhardt /pixelio.de

Manufactured and distributed by brebook publishing software
(www.brebook.com)

Rossiter Worthington Raymond

Mineral resources of the states and territories west of the Rocky Mountains

MINES AND MINING

IN THE STATES AND TERRITORIES

WEST OF THE ROCKY MOUNTAINS;

BEING THE FOURTH ANNUAL REPORT OF

ROSSITER W. RAYMOND,

UNITED STATES COMMISSIONER OF MINING STATISTICS.

WASHINGTON:
GOVERNMENT PRINTING OFFICE.
1873.

LETTER

FROM

THE SECRETARY OF THE TREASURY,

TRANSMITTING

The report of the Commissioner of Mining Statistics, on mines and mining west of the Rocky Mountains, for the year ending December 31, 1871.

TREASURY DEPARTMENT, *March* 20, 1872.

DEAR SIR: I have the honor to transmit herewith the report of Rossiter W. Raymond, Commissioner of Mining Statistics, on mines and mining west of the Rocky Mountains, for the year ending December 31, 1871.

I am, very respectfully,

GEO. S. BOUTWELL,
Secretary.

Hon. JAMES G. BLAINE,
Speaker of the House of Representatives.

CONTENTS.

INTRODUCTORY.

WASHINGTON, *March* 20, 1872.

SIR: I have the honor to transmit herewith my report on mines and mining in the States and Territories of California, Nevada, Oregon, Idaho, Montana, Utah, Arizona, New Mexico, Colorado, and Wyoming, giving a general review of the history of this industry in each district during the year 1871, and of its condition and prospects, with such comments and suggestions as seemed likely to be of use to miners, metallurgists, capitalists, and legislators.

It is pleasant to know that the series of reports of which this constitutes the sixth, (two having been prepared by my predecessor and four by myself,) has been recognized at home and abroad as highly important and valuable, constituting not only a repository of much current information, but a display of the natural resources of the country, and a history of American progress which no other means could supply. Those portions of the reports which discuss the geological, metallurgical, and mechanical problems involved in American mining have been widely studied, quoted, and discussed, and have done much, if I may credit the testimony which has reached me from many quarters, to increase the economy and success of the extraction and reduction of ores. One evidence of this fact is found in the numerous inquiries upon practical points addressed to me by letter, or in personal visits from persons engaged in mining, milling, and smelting throughout the West. The additional labor of correspondence thus thrown upon me will be cheerfully discharged, so far as time and strength permit. While this part of my work has greatly increased, the ordinary duty of collecting, by personal inspection or correspondence, the statistical and technical information from the mining districts required for my annual reports, has also grown to double what it was four years ago. Large numbers of new districts have been opened and made productive since the completion of the Pacific railroads, and the difficulty of obtaining correct information from them is enhanced by the complexity of their communications and financial connections.

The means at my disposal have always been inadequate to the thorough performance of this work as I would like to do it. Ten thousand dollars is scarcely sufficient to provide for the salary of a single assistant, and the necessary traveling expenses. A careful estimate, made in detail, shows that twenty-five thousand dollars would be required to carry out a comprehensive plan, including the employment of resident agents in all the leading districts, and the payment to them of small sums, sufficient to cover their actual expenses in visiting the mines and preparing their consolidated returns. No such agent would receive more than $250, and in most cases the sum would be from $100 to $150. This,has always been my ideal; I cannot realize it with voluntary correspondents, because they are not open to direction or criticism from me. The amount of money I was able to spend last year, for work performed in the collection of statistics, aside from that of myself and one deputy, was only about $500. The result is, that Oregon, Idaho, New Mexico, and parts of other States and Territories, are but slightly treated in this volume, and would have been necessarily ignored altogether, had not a few personal and professional friends very generously prepared notes upon such districts as came within their immediate observation.

The number of gentlemen in all parts of the country who have assisted freely in the collection of materials for this report, is very great. I trust that Congress, if an edition is ordered for general distribution, will put it in my power to acknowledge their kindness by sending them copies of the volume. It has been heretofore a serious annoyance and hindrance to me, that I could not furnish any copies of the mining reports to scientific men at home or abroad, or even to those who had shared in their preparation, except by the unwelcome, and precarious means of soliciting such copies from the Department, or from members of Congress, who are overrun with other applications. The Commissioners of Agriculture and of the General Land Office are not crippled in this way in the distribution of their reports, and I respectfully urge that the industry I represent and the work I perform do not deserve to be thus slighted.

I beg leave to return thanks to the following gentlemen for their assistance, remarking, however, that this list does not include all who have rendered courteous service. The names of many others will be found in connection with the information they have furnished, in the course of this report.

The chapter on California was prepared almost entirely by Mr. W. A. Skidmore, my resident agent in that State, to whose industry and intelligence I am greatly indebted. Mr. Skidmore's acknowledgments (in which I cordially join) for services rendered to him by various citizens of California in the prosecution of his work, will be found in the chapter referred to.

For valuable statistical and scientific material from Nevada, I am under obligation to Messrs. O. H. Hahn, of Eureka; Alexis Janin, of Meadow Valley; B. N. Lilienthal and A. J. Brown, of White Pine; J. W. Hussey, of Elko; the county assessors of the various counties; and the agents of Wells, Fargo & Company.

In Idaho, Messrs. J. M. Adams, of Silver City, and Richard Hurley, of Warren's Camp, have kindly furnished information.

In Oregon, I desire to mention with thanks Messrs.' E. W. Reynolds and W. H. Packwood, of Baker City, and Mr. W. V. Rinehart, of Cañon City.

In Montana, the list of those who extended hospitality and assistance to my deputy and myself is too long for recital. I can only mention generally the agents of the express and stage lines; the editors of newspapers; Hon. James Cavanaugh, late territorial Delegate; Hon. William Clagett, his successor; Governor Potts and his staff; Mr. W. M. Roberts, chief engineer of the Northern Pacific Railroad, and his assistants; the proprietors of mines, mills, and furnaces, and the miners and merchants of all the districts visited. The names of many of these gentlemen will be found in the chapter on Montana.

My acknowledgments to leading citizens of Utah must be equally general. I am indebted for special information with regard to this Territory to Professor W. P. Blake, who examined for me several districts, and a number of other gentlemen to whom credit is given in the appropriate chapter.

In other Territories, I would name particularly Messrs. John Wasson, surveyor general, and Hon. Richard McCormick, Delegate, of Arizona; Surveyor General Willison, and Messrs. Hilgert, Bloomfield, Goulding, and Morehead, of New Mexico; Messrs. Schirmer and Jones, of Denver, Messrs. G. W. Baker and A. von Schulz, of Central City, and Mr. A. Wolters, of Georgetown, Colorado; Messrs. Thomas Wardell, of Rock Springs, and Charles Deuel, of Evanston, Wyoming; and Mr. T. E. Sickels, superintendent of the Union Pacific Railroad, Omaha.

I have had occasion this year more than ever to realize the extraordinary ability and industry of Mr. A. Eilers, who has continued to act as my deputy, and whose thorough scientific training and wide experience of American as well as European mining and metallurgy have been of the greatest assistance in my work. We were both in the field during a part of the year, and traveled about twenty thousand miles in the discharge of official duty.

The recognition expressed in former reports of the courtesies extended by transportation companies should be here renewed. I am under obligations to the Chicago, Rock Island and Pacific, the Union Pacific, and the Central Pacific Railroad Companies, and to Gilmer and Salisbury, proprietors of the Montana stage line, for facilities of travel which considerably enlarged the area I was able to visit with the limited means at my disposal. Wells, Fargo & Company, and all their agents throughout the country were, as usual, most generous in their courtesy and active assistance.

The general condition and prospects of our western mining industry are set forth with so much fullness in the following pages that I will not prolong this letter by a discussion of them. The amount of the estimated bullion product will be found in the appendix. The nature of the increase it exhibits is still more gratifying than its amount, since it shows clearly that the decline of production from superficial mining is more than compensated by a steady augmentation from deeper and more permanent sources.

I have the honor to be, yours respectfully,

R. W. RAYMOND,
United States Commissioner of Mining Statistics.

Hon. GEORGE S. BOUTWELL,
Secretary of the Treasury.

PART I.

CONDITION OF THE MINING INDUSTRY.

CHAPTER I.

CALIFORNIA.

The State of California has a length of nearly seven hundred miles, and a breadth of over two hundred miles, its area being estimated at 154,000 square miles. The State is divided into forty-nine counties, and in twenty-four of these counties mining is carried on to a greater or less extent.

The principal mining region lies on the western slope of the Sierra Nevada, near the central portion of the State, but several important districts are found in the extreme northern and southern parts of the State, as well as on the eastern slopes of the Sierras.

The collection of information and statistics relative to the mining interest over such an extensive area is a task attended with great difficulty and expense, and its satisfactory performance would require a much larger expenditure of money than it has been within the power of the Commissioner to devote to any one State or Territory from the small appropriation made by Congress for this purpose, as the scope of his duties embraces all the extensive mineral region of the country west of the Rocky Mountains, extending from the borders of British Columbia to the Mexican boundary, and embracing the large Territories of Montana, Idaho, Utah, Colorado, and Arizona, in all of which exploration is being carried on and important discoveries are being made yearly, requiring attention and investigation.

As it is impossible, for these reasons, to personally visit and examine more than a portion of the State each year, an effort has been made to obtain information from those districts not visited through the medium of correspondence. This has not proved as successful as was hoped, through the want of interest felt by the mining community in the labors of the Mining Commissioner and his agents. This indifference is to be attributed to the want of knowledge as to the objects of the commission rather than to an unwillingness to impart information. Strange as it may seem in a country where public documents are printed and circulated by the Government with such profusion as in the United States, it happens, with respect to the reports of the Mining Commissioner, that but few copies have reached the class of persons most interested in their contents; and the various editions, both public and private, are already virtually out of print. During the travels of the writer in various parts of the mining region of California, for the past three years, he has not seen more than a dozen copies of the various reports of the Mining Commissioner, though there is a continual demand for them by persons interested in mining. Could these reports be distributed more liberally and judiciously the results would be beneficial to the mining interest, not only by the diffusion of the information contained in them, but it would materially assist in the collection of valuable statistics and information for the future. The recipients of these reports have ever manifested a willingness to aid in the collection of information, and much valuable data have been obtained through their observation and investigation in their respective districts.

In 1869, a set of blanks, five in number, embracing under appropriate headings the class of information desired, were sent by mail and express to persons engaged in mining in various parts of the State, accompanied by a letter explaining their objects and uses. Two hundred and thirty-four sets of these blanks were dispatched, and these elicited but thirty-one replies, though in some cases the returns were comprehensive and valuable.

In 1870, the deep placers of Nevada, Yuba, and Placer Counties were visited and described as thoroughly as limited time and means would permit. Mr. C. Luckhardt visited San Diego County and Inyo County, and much valuable information was gleaned from the returns of the United States census marshals.

During the present year a hasty trip was made through the southern mines, the results of which appear in this report, and letters and circulars soliciting information were sent to such mining counties as could not be visited by an agent. The following is a copy of a circular of which two hundred copies were dispatched to various addresses in different parts of the State, and to the secretaries of mining companies having offices in our principal towns:

OFFICE U. S. COMMISSIONER MINING STATISTICS,
37 Park Row, New York.

DEAR SIR: Being desirous of collecting for my forthcoming Annual Report on Mining Statistics (for 1871) as much trustworthy information as possible concerning the progress and condition of the mining industry in California, and being unable, with the limited means at my disposal, to examine, either personally or through agents, all the districts in the State, I take the liberty of asking your assistance, so far as your district is concerned, assuring you that anything you may be willing to do in this matter will be considered not only as a contribution to the welfare of the mining interest generally, but also as an official and personal favor to myself.

I desire particularly the following information:

An outline of the history of your district for the past year, with notices of any improvements or important works, (such as mills, mining ditches, bed-rock tunnels, &c.,) either in progress or in contemplation.

A general statement of the condition and prospects of quartz, placer, and other mines in your district.

The (estimated) product of bullion for your district (or mine) for the year 1871.

The names, and brief notices, of the principal mines—quartz and placer—in your district.

A list of the stamp-mills in your district, with number of stamps, &c.

A statement of the ruling rates of wages in your district.

In districts or counties where the principal interest is gravel or hydraulic mining, the estimated area of ground (in acres) now being worked, and the extent of the auriferous deposits; an estimate of the yield of hydraulic ground, per cubic yard; and a statement of the yield and expense of milling per cubic yard (where stamps are used) in cement and gravel claims.

Superintendents and secretaries of mining companies, and proprietors of mining ground, are urgently solicited to furnish such details of the operations of the companies they represent, for the past year, as may with propriety be made public.

Unless you otherwise direct, your courtesy will be acknowledged in the report. Persons furnishing information, desiring a copy of my annual report, will be supplied, as soon after its issuance as practicable, on application to my agent in San Francisco.

Please send an early reply to this letter, stating whether I may depend upon you in the matter, addressing your reply to my assistant, Mr. W. A. Skidmore, Box 1513, San Francisco, Cal., to whom also your notes (in any form that may suit your convenience, whether rough and hasty, or written out in full) should be sent as early as practicable.

Very respectfully, your obedient servant,

R. W. RAYMOND,
U. S. Commissioner Mining Statistics.

No more than twenty-four replies were received to this circular, and of these more than half were of no value. Without exception, the foreign companies operating in California—mostly in gravel-mining—have neglected to answer the circular or communicate any information, though in several cases a second circular was sent, accompanied with a letter. Our own large companies have been equally negligent as a general rule. Of the circulars addressed to mining secretaries, but one received attention, though in many cases I have procured the annual reports of the respective companies, when they have been printed.

On the other hand, some replies have been prepared with care, after patient investigation, and contain much valuable data. Of this class is the information furnished by Mr. J. Rathgeb, of San Andreas, Calaveras County, and by Mr. W. M. Eddy, of French Corral, Nevada County. The latter gentleman, at my request, instituted a series of experiments, extending over a period of several months, with a view of testing the yield of gravel, and the expense of treatment by mill process, at French Corral. I am also under obligations to Messrs. Cronise and Crossman, and to Dr. Henry De Groot, of San Francisco; also to Mr. Lyman Ackley, of Smartsville, Yuba County, and Mr. E. N. Strout, of Placerville, El Dorado County.

Condition of the mining interest.—The business of mining for the precious metals has for the past two years made very marked advances throughout all parts of the Pacific slope. Never within a like period has this pursuit so strengthened itself in public confidence, or undergone so great territorial expansion. At first regarded with distrust and suspicion, it has achieved the favor of capitalists to a large degree, and can now enlist their aid more readily than almost any other interest. Mining, instead of being proscribed, as it was at one time, is now not only recognized as a legitimate pursuit, but is fairly regarded as entitled to precedence over most others. Purged of its follies, and with many of its errors corrected by experience, it opens now not only a more profitable but a safer field for investment than any other of our leading industries. In no other department of business on the Pacific coast have the gains for several years past been so liberal or so certain as in this, nor does any other hold out such flattering prospects for large profits in the future. Many of the mineral developments made of late have been enormous, and successful Never has the business of prospecting been pushed so far, nor been attended with such happy results, as during the past few years. From Mexico to Alaska important discoveries are announced, while the opening up of one rich district seems only to point to another still further on in the distant interior. Stimulated by the aid of modern enterprise, silver-mining is being quickened into a new life in our sister republic of Mexico on the south, while the latest accounts from the far north speak of valuable mineral discoveries having been made in our recently acquired possessions in that quarter.

With all these rich discoveries and such a large measure of success, our mining communities have for several years past been comparatively free from those unwholesome excitements that formerly resulted in so much mischief and suffering. Exploration has indeed been exceedingly active, but, having been prosecuted in widely remote and opposite directions, and having been nearly everywhere attended with fortunate results, the discoveries made in one place have, to some extent, neutralized the effects of those made in another, and thus tended to preserve the public mind in a state of equilibrium and prevent a sudden migration toward any particular locality. The hard experience of our mining population has also, it is to be hoped, had something to do in restraining them, during a period signalized by so many important events, from being drawn into one of those precipitate movements that have so often heretofore hurried them away to distant and unremunerative fields of labor. It is also the case, that, while mining enterprise has been greatly stimulated of late, it has not, to such an extent as formerly, been misdirected or overdone. Very rarely have recent undertakings involving large expenditure been entered upon without calculations based on reliable data having first been made, and the whole committed for execution to careful and experienced management. It is not the custom now to build expensive mills or reduction-works without something having been first ascertained as to the capacity of the mine whence ores are expected to be obtained for supplying them. Indeed, a good degree of development is insisted upon as a condition-precedent to considering the question of erecting reduction-works, while in most of our large mines extensive deposits are held in reserve, and exploratory labors maintained well in advance of extraction. In strong contrast with the unfortunate results of the large enterprises set on foot a few years since, are those generally reached by companies who have more recently engaged in quartz and placer mining on a large scale. Then failure was the rule, and success the exception, a condition of things which is now almost reversed, more especially in the State of California. Parties operating in mining stocks continue, as they ever will do, to meet with disaster. Corners are made and the venturesome and unwary are caught, as they ever will be, by their more vigilant and astute adversaries. Reckless ventures and wild speculation are as rife as ever, bringing fortune to some and discomfiture to others, while certain mines are managed, as before, in the interest of selfish rings and combinations. But all this has, it is needless to say, no necessary connection with legitimate mining, which should in no sense suffer disparagement from practices so detrimental to its best interests and so foreign to its objects.

Causes of past failure.—A practical miner of many years' experience in California and Nevada, in investigating the causes of disaster to so many of our mining operations in years past, says failures have often been attributed to the high rates of labor, but a critical investigation would show that it was to be attributed more frequently to other causes, which he thus enumerates: 1st. Too extensive as well as too expensive machinery on too small developments of positive property. 2d. Too much expenditure in corporate associations for speculative purposes, and too little for practical mining. 3d. Too small a percentage of metal saved after heavy expenditure for extracting and raising ores.

In seeking for the cause of many early failures much stress has been laid on the defective and careless manipulation of our gold ores—errors which, now that attention has been so strongly called to them by the press of the State, are rapidly being remedied by experiment and invention. On this subject I quote the following extracts from a series of articles on the " Wastage of the Precious Metals," written for the Mining and Scientific Press by Almarin B. Paul, of San Francisco:

The fact that a high percentage of the precious metals is lost in the manipulating of ores by the present modes of working, no one for a moment questions; but when it comes to any special data, but little has been presented to the public. Some assert their loss to be only a trifle, while others, who more closely investigate, know it to be greater than it should be. Having closely investigated the subject for the past two years, I find the average loss, especially in California, so great, that really I think, if there is not more care taken in the husbanding of our riches, when extracted from the earth, that the Government should take measures to do so.

There is an idea that all gold is readily amalgamated, and therefore it is not necessary to be so particular; consequently there is an unwarrantable degree of carelessness. I have learned, by practical working in both gold and silver, that a higher percentage of silver is more readily obtained by the known system of working for silver than the percentage of gold by its most advanced system, showing that gold milling is far behind silver working, although, as before remarked, gold is considered so " readily amalgamated." Yet to adopt the systems for gold that are used in silver affords no especial relief.

That my readers may have some data, as a corner-stone, to build their ideas upon, before I go too far in my general observations, I will give a few tests of the many which I have made in the last two years, and intersperse with them, as additional evidence, tests of other parties. And here I would call the reader's especial attention to the fact of a goodly percentage of silver in all of our California ores; and I will also remark that the assay of tailings does not even show what percentage of silver the ores may contain, as some may be in the form of chlorides which move off in the water. But to the tests of our gold mining.

Test No. 1. Average yield of ore in mill, $18.60. Wastage after complete washing, including concentrating—silver, $3.14; gold, $10.04; total, $13.18.

Test No. 2. Same mill-tailings 350 feet from mill: silver, $3.93; gold, $5.02; total, $8.98; showing that a percentage secreted itself in its passage down stream.

Test No. 3. Average yield of 150 tons, $3.50. Assays of tailings carefully sampled: silver, $6.28; gold, $13.55; total, $18.83. Silver, $6.28; gold, $8.79; total, $15.07.

The above bad results were occasioned by the extreme fineness of the gold. And even the above does not show the full wastage. To corroborate this I will give some admirable tests made to get at the question of "float gold."

A friend of mine, having somewhat similar ideas to my own, concluded to test the question of float gold as well as he could at the time, and embraced the opportunity of cleaning up the slum from a water-tank for supplying the battery, where the water was used over and over again in consequence of its scarcity. The ores were worked after the usual wet method for gold ores. The water and pulp wore first passed through a sluice to tailing-bed, 190 feet. The tailings being deposited, the water was drawn off at the top, flowing into a well, where it was raised and passed through a sluice 120 feet to tank at battery. This is the tank cleaned up.

The residue was amalgamated in a tub quite rudely, but with a large body of mercury and chemicals. The result was $33 in silver and $56 in gold, making a total of $89 per ton: It will be observed that there were two chances for the metals to precipitate previous to reaching this tank: first, in the tailing reservoir, and second, in the well.

This "float"-metal question is further established by a system of tests made by Mr. G. McDougal, of Grass Valley, who very kindly allowed me to extract the same from his books of tests. And here let me say that these tests are made from water flowing from mills at a point three-fourths of a mile below the mills.

	Cents.
1st test of 20 gallons of water gave	1.10
2d test of 20 gallons of water gave	2.13
3d test of 20 gallons of water gave	.95
4th test of 20 gallons of water gave	.83
5th test of 20 gallons of water gave	1.02
6th test of 20 gallons of water gave	1.13
7th test of 20 gallons of water gave	.97
8th test of 20 gallons of water gave	3.12
9th test of 20 gallons of water gave	1.07
10th test of 20 gallons of water gave	.63
11th test of 20 gallons of water gave	1.01
12th test of 20 gallons of water gave	.90
Average	1.18

It was estimated that 576,000 gallons of this "muddy water" flowed by every 24 hours, which, according to these tests, contain $339.84. Let us carry this calculation a little further.

The average amount of ore worked in 24 hours was given as 58 tons. This shows that $5.85 per ton "floats," which probably is at least 20 per cent. of the yield. Let us run this loss a little further. Suppose the two mills run 250 days in each year, which is not unreasonable, and we have a yearly loss in "float gold" alone, to say nothing of loss by imperfect pulverization and general wastage, of $84,960 from two single mills!

Extend the test as far as you may, on a smaller or larger scale, and wastage stares one badly in the face at every turn.

I made a test of 50 pounds of tailings for a party who took them *a mile below his mill*, and the return was 55 per cent. of what was his average working. I also made a test of three-fourths of a ton, and the result showed the loss in the mill-working to be 63 per cent. From what attention I have given the subject in actual labor, as well as collecting all the data attainable from others, I know that the loss as a whole is fully 50 per cent., and, in the majority of mills, all of 60 per cent. of what the ore contains.

From these data of loss we must come to the conclusion that gold mining, not only in California, but elsewhere on the Pacific slope, as all are operating on about the same system, is not up to the point it should be. In fact, so imperfect is it, that it has been said that our gold mining enterprises, as a whole, may be set down as a failure, when the question of *profit* in all is considered.

One step in advance would be, taking more care. There is too much sloshing about in our gold mining. There is enough in silver, but no comparison between the working of the two metals. This plan of seeing how much can be pounded up and rushed through every 24 hours, is a false, wasteful, and ruinous system.

The profit will be found in *how well* and how cheap it can be done. It is in the right direction, certainly, to reduce ores expeditiously and cheaply, but not to as expeditiously wash *everything* away, having an eye more to pounding up the rock than to taking up the metal.

That our gold ores are so readily amalgamated, is one of the ruinous ideas extant. The majority of California miners are, in fact, but little experienced in all the troublesome accompaniments of even gold ores, considering that if the rock does not pay, it cannot certainly contain it. All, however, admit it to be difficult to extract the gold from iron sulphurets, forgetting that even a small percentage of lead, copper, arsenic, or antimony, which is to be found in nearly all the gold ores of California, vitiates the mercury in a little while, rendering it quite inefficient in collecting even the gold that otherwise from gravity might be taken up. It is too universal to consider that it is only necessary to rig up a set of stamps, apply the power, and let them rip away, smashing rocks, to wash over blankets and copper plates; and all is done with a stream of water to wash the sands off, forgetting that it is equally as potent to wash off the smaller particles of gold.

Some will say, it is all well to talk about loss of metal, but how can we prove it, and where is the remedy?

I will tell you how to prove it, but each must work out his own remedy. For my part, I have worked out the loss by what I consider the remedy—dry amalgamation; but our subject now is loss, not remedy. To awaken the mind for improvements, and be interested in a remedy, miners must first realize their loss. I contend there are several ways of working our gold ores better than the one now universally used in California. If you want to get a clear comprehension of your loss, take, say, 5 tons or more, not less. Reduce the ore *dry* through say No. 20 wire-cloth screens; mix all thoroughly, then spread it out upon a floor about two inches thick. Lay it out in 12-inch squares, take a smaller quantity from 'each square, take samples thus obtained, and again mix them. Again spread out, say, one inch thick, laid out into 4-inch squares, taking a smaller portion from each. Reduce this sample to powder; if too much for average assays, sample again as before. Get 3 or 5 assays from reliable assayers, average the assays. Work your ore by your mill process; compare the results with assays; and in nine cases out of ten every one will find ho possesses more riches than he thought he had. Any other system of testing is unreliable. Pieces of rock can be had to assay more or less, as you want. To get at the value of your mine, the testing of *tons* by this mode is the only safe one.

Mr. Paul further calls the attention of our miners to the existence of a large percentage of silver associated with the gold-bearing quartz:

That there has been still too great a wastage of precious metal, all admit; not only is this in our gold in gold mining, but in the *silver associated with the gold*. This fact is not generally understood, as California miners are not accustomed to getting silver with the gold, a thing precluded by their present mode of working. An investigation will disclose the fact that nearly all the gold ores of California contain no meager percentage of silver. The same may be said of the "gold ores" of other States and Territories. By way of illustration the following assays are given:

	Gold.	Silver.	Total.
Ore from Mariposa County	$12 06	$4 90	$16 96
Sulphurets from Washington Mine	81 40	32 06	116 46
Blanket washings...	4 13	6 77	10 90

The yield of Quartz Mountain ran 9 per cent. silver. The ores of all the counties of California carry silver, and my experiments show they run from 3 to 50 per cent. of the yield.

The closer the concentration from batteries, the higher is the percentage of silver. It is time we were investigating more closely, and outgrowing this rushing system of mining, and, instead of sluicing our silver and gold down streams, seeking modes of working that will produce less wastage. As far as California is concerned, I am satisfied that not more than 40 per cent. of her gold is extracted. The fact is, as before expressed, we are not working for gold or silver, but to crush rock.

Mr. Paul's opinions on the condition of the gold are given in the following extracts:

Our present general system of gold mining is based upon the idea that gold is mainly coarse, while examination will show that the high percentage is in atoms finer than flour itself. In my experiments gold has been taken up so fine that in distilled water it would not precipitate in less than from five to ten minutes. Can you save gold of this kind by running water down stream? Again, can you obtain the gold of this fineness, without minute reduction? Therein lies the secret of high assays before working, and small returns after.

Gold in its matrix, according to the highest authorities, is in a metallic condition. Such being the case, the first requisite is minute reduction, to the fineness of the gold itself, in order to release it. Gold in quartz of gravity enough to resist the pressure of any stream of water is the exception, and this is the aggregating of finer particles, the primary simple condition, in my opinion, being flour or powder of gold. It is the flour of gold we must seek to obtain, to get the wealth of our ores.

In commenting on the disasters that have so often overtaken mining enterprises on this coast, too much stress has sometimes been laid upon a presumed gross mismanagement, or a willful purpose to defraud. To these much of the failure and loss heretofore sustained is undoubtedly due; yet it should be recollected that mining for the precious metals was a business with which our people were wholly unacquainted at the outset. When we embarked in this pursuit, one naturally beset with obstacles and full of inherent difficulty, we had everything to prepare and everything to learn. Not only so, but we had to make our first trials under circumstances that rendered the mining and metallurgical knowledge gained elsewhere of little avail. The experience obtained in the Old World and the rules laid down in the books could not be applied here to advantage. The high prices of labor, the want of material, the character of the ores, the climate, and, in short, all the conditions, were so unlike those to which the experts of other countries had been accustomed, that they, of all others, proved the most inefficient and helpless. In attempting to adapt themselves to surroundings so new and strange, none found themselves at so great a loss or blundered more widely than they. Of all failures, those of our scientific men from abroad were at the first the most signal. Our sins were, therefore, at first, mostly the sins of ignorance, and our errors, the errors of judgment. If, in a few instances, deliberate frauds have been committed, it is no more than has happened in the conduct of many other kinds of business not accounted specially difficult or hazardous. Or if it should appear that there has been some extravagant and even culpable expenditure incurred on mining account, it may yet be claimed that we have introduced many valuable improvements into this business, and advanced it with an energy that more than atones for these wasteful and unwarranted outlays. It is now generally conceded that we have, in the matter of mechanical contrivance, if not also in metallurgical skill, advanced this industry in all its branches far beyond the point where we found it, and even beyond its present status in almost any other country. In the adaptation of means for washing auriferous earth, in the use of hydraulic power, in our ore-crushing machinery, our roasting-furnaces, our concentrators and amalgamators, and in much of our other milling-apparatus and appliances, we can point to improvements that leave us without a rival elsewhere. Both as regards the exploitation of the mines and the beneficiating of the ores, we can justly claim to have reached as great perfection as any other nation in the world, having contributed our full quota towards the wonderful advancement that this branch of mining has undergone during the past quarter of a century.

Present and prospective production.—With so much progress made and so large an aggregate of improvements effected, a large and profitable production of bullion has ensued, leading to heavy investments in mining properties on both home and foreign account, and to an unwonted activity in every department of this industry. The yield of the precious metals for the entire coast has, of course, been greatly curtailed in consequence of the restricted amount of rain that has fallen for the past two years; the effect of which has proved more disastrous to the mining interest of California than of any other portion of the Pacific slope, as we have here many quartz mills dependent on water for their propulsive power, while placer washing constitutes, in favorable seasons, our most prolific source of gold production. But despite this serious interference the California yield for the current year is probably $20,000,000, a sum that would, with the ordinary supply of water, have been increased fully one-fourth. Owing to the introduction of improved machinery and modes of washing, as well as to the employment of more efficient metallurgical processes and labor-saving contrivances, the profit margin is being steadily enlarged in mining operations even while working a lower grade of material; much auriferous gravel being now washed and ore milled with satisfactory gains that would not a few years ago have paid working expenses. Hence the maintenance of such a high rate of production in the face of a drought almost unexampled for its long continuance and severity. During the prevalence of the drought our hydraulic and placer miners have been engaged in the extraction and accumulation of dirt, and in running deep-drainage tunnels, until they have now large quantities on hand, ready for washing as soon as the rainy season will permit. As we have now had three dry years in succession, we may safely count on the incoming winter being a wet one, affording the waiting miner all the water that his needs require. Should this prove to be the case, a heavy yield of gold will be speedily gathered, insuring for next year a larger product than has been realized in California for several years past.* Indeed, a heavy annual increase of bullion in this State may be calculated upon for an indefinite period to come, in view of the impulse lately given to mining enterprise through the liberal investments of capital and other assistance brought to its aid. In the construction of capacious hydraulic works, looking to vastly increased supplies of water, in the extensive opening up of the old river-channels and gravel-banks and in the erection of many large mills and reduction works, to say nothing of the favorable developments being everywhere made in our mines, we have ample assurance of the largely augmented production that awaits us in the future. There is good authority for believing that with the usual supply of water the yield of the California mines alone would be increased $500,000 per month. With all the water available for that purpose introduced into the mines at the head of the State Creek Basin, Sierra County, it is computed that two million dollars could be annually taken from that locality more than it affords at present. With the Von Schmidt ditch completed according to the original plan referred to elsewhere· in this report, the gold-producing capacity of the country to be supplied by it, already one of the most prolific in the State, would at once be more than doubled. In El Dorado County the California Water Company are constructing a system of ditches and reservoirs which will furnish from 30,000 to 45,000 inches of water, miners' measurement, and open a large section of gravel country hitherto untouched or worked only on a small scale; while at Parks Bar in

---

* Since writing the above my anticipations have been more than realized, as the rainfall for December has been almost unprecedented, giving promise of a year of abundance to both the miner and the agriculturist.

Yuba County, and at Indian Bar, (La Grange,) Stanislaus County, extensive water-ditches will supply large tracts of undeveloped ground of great promise. In like manner, many other great hydraulic works projected will, when they come to be finished, contribute, more or less, towards swelling the aggregate bullion product of the State; and as several of them are already well advanced, with the prospect of being carried to an early completion, we may hope to soon enjoy the benefits of their co-operation in stimulating our mines to a more free production.

Investments of capital.—During the past two or three years investments on mining account have been liberal beyond precedent, a great number of valuable properties having been purchased, the most of them with a view to the early inauguration of practical operations upon them. Much of this capital has been drawn from abroad, the English public having been very active in promoting mining enterprises throughout all parts of our Pacific States and Territories. California, Nevada, and Utah have come in for the greater share of these investments, though some of limited extent have been made in Idaho and Montana. The aggregate amount of foreign investments made during this time is variously estimated as ranging between fifteen and twenty millions of dollars. Many of these investments have been made with care, and are likely to prove both safe and profitable. In purchasing properties of this kind much circumspection has latterly been observed, only such being negotiated for as have been somewhat developed, and these being accepted only on the approval of competent experts, based on thorough and careful examination. It has occasionally happened that shares in these mines have been forced up in the market abroad to figures that there was nothing in their condition or prospects to justify. It cannot reasonably be expected that the stocks of mines, however good, should appreciate three or four hundred per cent. in the course of a few months and be able to maintain themselves permanently at this advance. English promoters have made the further mistake of exacting an inordinate commission for their services, necessitating the holders of mines to stock them at extravagant figures. But with all these errors and mishaps, many of the investments made in our mines on foreign account cannot be considered otherwise than fortunate, encouraging the hope that they will be continued and even largely increased in the future.

Our mines have also of late grown much in favor with home investors, it having become evident to all that we cannot longer neglect their development if we expect to enjoy a rapid increase of material wealth hereafter. The experience of the past two years has demonstrated the fallacy of relying upon our agricultural resources alone if we hope to insure for our people an uninterrupted prosperity; while the fact that able-bodied men can at all times command from $2.50 to $3.50 per day in the mines, shows the impossibility of our being able to build up for the present any very extended mechanical or manufacturing industries on this coast. With these wages prevailing, it is obvious that we cannot successfully compete with the low-priced labor of the Eastern States, except in making a few bulky articles, into the manufacture of which our own raw material largely enters. Satisfied of this, there is a growing disposition on the part of the working classes to engage in mining pursuits, causing domestic capital to incline strongly in the same direction.

Recent operations and improvements.—The great extent and the costly character of many of the projects undertaken constitute a notable feature in the late history of mining enterprise on this side the continent.

Nearly all operations and improvements are now planned and carried forward on a large scale, it having become apparent that by this method great gains and economies could be effected. In fact, the altered conditions under which the business of mining must now be prosecuted, have made this necessary. The superficial placers in which men without capital could once earn fair wages, have now become in most places so much impoverished that they can no longer be worked with profit, without a large expenditure first incurred in fitting them up with the requisite apparatus and supplying them with water. With these furnished, they can still be worked with remunerative results, and very often with large gains. There also remains a great extent of rich deposits deeply buried in the beds of the pliocene rivers, with vast banks of auriferous gravel, some of them several hundred feet deep, all of which, though useless to the man without means, can, with the aid of capital, be made to yield up their treasures with profit. To work this class of mines to advantage it becomes necessary that a large area of ground should be secured; to which end the titles to individual claims are first extinguished when they are aggregated in large masses, for which a United States patent is, in most cases, afterward procured. Some of these mining estates now cover an area of several hundred, and occasionally as much as a thousand acres, (in one case, at North Bloomfield, Nevada County, 1,535 acres,) insuring a sufficiency of ground for carrying on extended operations for many years to come. In like manner, where vein-mining is to be engaged in, it has become the custom, at least with the more sagacious and provident class of operators, to make sure of several different lodes, or of a larger portion of some one lode than was formerly the practice. So, too, in the erection of quartz-mills and reduction-works, in the projection of ditches and reservoirs, in rigging up hydraulic apparatus, and in exploring the ancient river-channels, everything is conducted on a scale that causes most former works of the kind to sink into seemingly dwarfish proportions. These formidable and expensive undertakings, while they give promise of early additions to the product of our mines, indicate, at the same time, the confidence reposed in their permanency by a class of men distinguished for practical wisdom. Of the quartz-crushing mills lately built, the majority have been of large capacity, many of them carrying from twenty to forty, and in a few cases as high as sixty stamps. The erection of several sixty-stamp mills is now in contemplation, and it seems probable that those hereafter constructed will generally be of large size. In the districts where the ores require reduction by smelting, larger furnaces will be substituted for the rude and limited works used for pioneer purposes, these having in some instances already been replaced by others of more perfect patterns and increased capacity. We have already in San Francisco one of the most complete and capacious smelting establishments in the United States, and this is now undergoing extensive enlargement and improvement. A recent number of the California Mail Bag says of this enterprise:

Very few persons in our midst have any correct idea of the magnitude of the Selby Smelting Works, the extent of their capacity, or the positive benefit they are to this State and city. For some time large quantities of certain ores were shipped from Nevada and other interior mines to Newark, New Jersey, for reduction; and the bullion so obtained never again found its way back to our markets, being exchanged for various articles of merchandise which interfered with the transactions of our own business community. It was soon discovered, however, that such shipments of ore were unprofitable, and that infinitely greater facilities existed in the immediate neighborhood of the mines, with a corresponding economy of expense for freight, interest, and insurance, and much quicker returns of extracted bullion. So rapid has been the development of this comparatively new industry, that although the Selby Smelting Works contained twenty-five large furnaces, which were kept in constant operation, at the commencement of the present year it has been found necessary to increase the number and enlarge the capacity by making costly additions to the establishment. There are four fifty-vara lots to be filled in up to deep water, and these will be covered completely with the requisite buildings, the work being already under active headway. A large coffer-dam is to be sunk at the outer extremity of the works in deep water, and excavations will be made until the bed-rock of the bay is reached, when a huge chimney will be erected on it, and uprear its lofty, smoking top from out the waters far above all surrounding objects. The works are already the most extensive in the United States, but with the additions will probably equal any known. At present something like 25,000 tons of pig-lead are annually imported into the United States; but during the current year the Selby Works will supply nearly one-third of the whole amount, and it is confidently expected that in five years more they will furnish enough to stop all requirement from abroad. No less than 350 tons are under constant treatment, and this large amount will be increased to 500 tons before the close of the present year. The lead product of these works is admitted to be equal to any other in the Union; and large orders have been received from New York, for the purpose of converting the metal into white-lead superior to any now made. The working force consists of 130 men, divided into two gangs, 65 being employed all day and 65 all night. These people receive liberal wages, and have steady employment the year round. The difference in freight between Nevada and Newark and Nevada and San Francisco is something like $34 per ton in favor of the last-named place, while considerable gain is realized in the greater quickness with which returns are made, involving a saving of interest and insurance. Apart from the general benefit conferred by the existence of so great an industry and the home development of domestic resources, a direct advantage is obtained by our mercantile and trading community, which reflects upon all other classes.

Some of the hydraulic workings in this State, though carried on mostly by single individuals or small companies, are really enterprises of great magnitude, apart from the costly ditches built for conducting water upon them. Our system of water-works, with their extensive canals, high reservoirs, dams, flumes, and iron aqueducts, has long constituted a leading feature in the internal improvements of California.

Principal hydraulic enterprises.—Standing at the head of this category we have the gigantic scheme of Colonel J. W. Von Schmidt, proposing to take the water from Lake Tahoe, and, carrying it through the Sierra Nevada by a tunnel nearly four miles in length, conduct it thence into the mining regions lying adjacent to Bear River and the north and middle forks of the American and their several tributaries, bringing a portion of it on to San Francisco, distant nearly three hundred miles from the point of diversion on the Truckee River, the outlet of Lake Tahoe. Sacramento, Vallejo, Oakland, and other cities along the route of the main aqueduct, are also to be supplied with water. The construction of this work, with its many proposed ramifications, will require two or three years, and involve an expenditure of several million dollars. That portion of the scheme looking to the gathering up of the waters west of the Sierra has already been actively entered upon, the locating surveys have been finished, the dam across the Truckee is in process of erection, and nearly all other preliminary labor completed. The projector of this enterprise has himself invented an ingenious and what promises to prove an effective drilling-machine, described briefly in the report for 1871. One of these is now being built with a view to its early employment on the great tunnel through the Sierra. From his well-known skill and energy, and his fertility of resource, it is believed that Colonel Von Schmidt will soon commence the work of penetrating the mountain, and that he will push the whole to a speedy and successful conclusion.

Next in order, and second only in importance to the great work of Colonel Von Schmidt, is the enterprise now being carried forward by the California Water Company, an association lately organized in San Francisco, with a capital of $10,000,000, for the purpose of engaging in the

business of mining, dealing in agricultural and timber lands, and of
furnishing water for mining and domestic uses at such points as it
may be required along the route of their canals. The following
extract from the Quarterly Mining Review, contained in a late issue of
the San Francisco Commercial Herald, describes the operations, the
character of the property secured, and the objects proposed to be
accomplished by this company.

"They have," says the Herald, "located, by right of discovery,
obtained by purchase and grant, and taken possession of twenty-four
lakes, varying in extent from one hundred to two thousand acres each,
situated in the Sierra Nevada Mountains, in El Dorado County, at an
elevation of from three to five thousand feet above the sea. These
lakes are fed by streams from mountain peaks and gorges of perpetual
snow, and are capable of furnishing a constant stream of 40,000 inches
of water of the purest kind. From these lakes run the Big and Little
Rubicon Rivers and other streams tributary to the Middle Fork of the
American River, which have also been claimed and taken possession of,
according to law. These are all very favorably situated in the high
Sierras, and constitute one of the most convenient, perfect, and valuable water-rights in California. The facilities for constructing dams at
the outlets of these lakes and on the streams are great, admitting of
these natural reservoirs being increased to an unlimited extent, while
the reservoirs that are to be constructed in connection with the work,
and which are to be filled with the waste water of winter, will furnish
an ample supply for an entire year.

"The company has also obtained possession of several thousand acres
of valuable mining ground in the region to be supplied by water, and of
a large area of timber and agricultural lands in the same section, being
the choice selections of near two hundred thousand acres of Government
and railroad lands within the limits of the field of operations; and all
of which can be purchased at from $1 to $2.50 per acre. The country
is heavily timbered with pine, spruce, cedar, oak, &c. The soil is generally fertile, and is well adapted to cultivation of the various products
suited to the different stages of elevation, the lower portion embracing
some of the best fruit and wine-producing country in the State.

"The region to be supplied with water comprises an area of from five
to twenty miles in width and fifty in length, containing numerous towns
and mining camps, creating a demand for from ten to twenty thousand
inches, at a price of ten cents per inch for ten hours' use, independent
of the large amount needed by the company in working its own mines.

"Aside from the amount used in the mines, a vast quantity would be
required for irrigation. Nearly all the cities mentioned would, it is calculated, take more or less water, Sacramento alone being counted upon
for a net annual revenue of $50,000.

"The distance from the lakes in the Sierra Nevada to Sacramento is
about seventy-five miles, and the total ditch and length of pipe to conduct the water to the various mining camps and the city would cost,
probably, $250,000. With it no further extended the sale of water
would return the whole outlay in a very few years, with the chances of
constantly increasing sales in the future. But the great amount of
water owned by the company will justify far more extended works. It
is sufficient in quantity, and in such commanding position, as to supply
the greater portion of the mining and agricultural wants of El Dorado,
Amador, and Sacramento Counties, and, crossing the Sacramento River,
can, if necessary, be carried to the cities of Vallejo, Oakland, San José,
and San Francisco."

Apart from these more extensive and costly projects, a number of works of less magnitude have been commenced, and some of them well advanced for supplying the mines with water in different parts of the State. A ditch of considerable magnitude was completed last summer on the Klamath, carrying water upon a number of high bars along that stream, and giving profitable employment to several hundred men who were before idle. A San Francisco company are pushing forward a large ditch in Tuolumne County, designed to furnish water to the extensive gravel deposits about La Grange, very little of which could heretofore be worked for want of water. Several important enterprises of this kind are also in progress in Yuba, Butte, and adjoining counties, besides numerous works, some of them of immense capacity, intended to carry water upon the great plains of the interior for irrigating purposes.

Already most of the large sources of water supply, of which there are many in the lakes and in the streams having their spring in the Sierra Nevada Mountains, have been utilized and it will not be long before all of them will be made practically available. When this shall come to pass there will be presented such a system of aqueducts leading down the slopes of the Great Snowy Range and across the foot-hills and the broad valleys at its base, as will have no parallel elsewhere in the world. Even now our hydraulic works are a credit to our young State, many millions of dollars having been expended and some masterly engineering displayed in their construction. As forming a part of this system, it will ultimately become necessary to build many large reservoirs, at points naturally adapted to the purpose, for collecting the surplus waters afforded at high stages of the mountain streams, and thence distributing them as required by the wants of the miners during the dry season. These artificial repositories can, in many localities where there are already small lakes and natural basins, be easily constructed, nothing more being required than building a dam at their outlets. In this manner an immense body of water can, in many cases, be easily retained during the season of floods, equalizing the supply and keeping it up throughout the entire year. In view of future demands for water these franchises must come in time to yield large revenues and rank among the most desirable properties in the State.

Encouraging developments, new discoveries, &c.—During the past year, work has been recommenced on many of the mines in this State upon which it had, for various reasons, been discontinued. In nearly all these cases the results since obtained have been satisfactory, showing that the former stoppage was not caused by want of merit in the mines.

All through the interior counties old and often-abandoned claims have been re-opened and idle mills started up and operated with success, raising the presumption that many others now neglected would, in like manner, warrant a resumption of operations upon them.

A large class of mining claims, too, that had for years struggled along, barely able to sustain themselves, are now, with cheapened material, more reliable labor, and improved modes of working, yielding fair and often munificent returns. Among other properties that have long been slumbering, but are soon to be resuscitated, is the famous Union copper mine, in Calaveras County, which, with the advantages of railroad transportation, it is believed can now be worked with profit.

Among what may justly be ranked as new discoveries is a description of gold-bearing deposits denominated "Seam Diggings," and which, according to a statement recently published in the Commercial Herald, have been brought into notice through the late explorations prosecuted in the service of the California Water Company, a considerable extent of them having been found along the line of their new works on the divide between the South and Middle Forks of the American River, in El Dorado County. The formation, here covering a large portion of two entire townships traversed by the company's canal, consists of a soft, easily disintegrated slate, permeated by innumerable small seams of quartz. Many of these seams are rich in free gold, and having become thoroughly decomposed through chemical and elemental action, the metal has been liberated from its matrix, and they, together with the inclosing slates, can now be readily torn down and washed away by the hydraulic process. Some of these slates have already been successfully worked in this manner, the only hinderance to their extensive development having been the want of water. With the quantity that the California Company's ditch will be able to supply, they will hereafter be worked on an extensive scale, and, no doubt, with large profit; and as similar deposits are known to exist in other portions of the State, and so many new water projects are now under way, there is little question but they will add materially to the future product of our California placers.

In the course of the explorations recently carried on in our old riverbeds and other of our deep-lying placers, it has, in many places, been found that, beneath what was generally mistaken for the bed-rock, there is another, and sometimes more than one rich, gold-bearing layer of gravel, adding a novel and valuable feature to these deposits. So important have these lower strata proved that they will cause more careful examinations to be made of the apparent bottoms of these auriferous channels hereafter.

Large yields, strikes, nuggets, etc.—In the month of August, 1871, a piece of nearly pure gold, weighing 64 pounds, was reported by the local press to have been taken from a claim on St. Charles Hill, near the head of Fiddle Creek, Sierra County. This report lacks confirmation, and the story is generally considered exaggerated if not apocryphal.

In September a nugget worth $6,000 was taken from the claim of Bunker & Co., in the State of Oregon.

A man named Fields in October last took from his claim on Kanaka Creek, Nevada County, a piece of quartz gold, weighing 96 ounces, and valued at $1,500. The same man took from his claim in one day 18 ounces.

A party of Chinamen mining in Placer County, in dealing with a trader near their camp the past summer, frequently paid for goods obtained of him with chunks of pure gold, evidently cut off from a larger piece, concerning the size of which or the locality where it was obtained nothing could be learned; though from the reticence of the parties having it in possession it was generally supposed they had pilfered it from some white man's claim.

Occasionally some exceedingly rich strikes have been made during the past year both in vein and placer mining, the clean-ups in the mills and hydraulic washings having in some instances never been surpassed. At several points chispas of large value have been picked up in supposed worked-out claims and heaps of tailings.

Have we diamonds in California ?—The question of the existence of precious stones in California in paying quantities is one which is still in doubt. In various parts of the State agates, carnelian, and the stones of lesser value abound. Near Mokelumne Hill, in Calaveras County, in a strata of an ancient channel, opals are found in large quantities, but of inferior quality. In El Dorado County diamonds have been found at intervals, but no systematic search has been made for them. Mr. W. A. Goodyear, one of the assistants of the State Geological Survey, thus writes to the Placerville Democrat concerning the existence of diamonds in the gravel beds of El Dorado County:

One other point which may be noted as being of some little interest to the miners, as a matter of curiosity if nothing more, although it is no new thing, is the occasional finding of diamonds in the auriferous gravel. From all that I have been able to learn, it appears that not less than ten or twelve diamonds have probably been found within four or five miles of this town. And I have no doubt that many more have been picked up and looked at and thrown away, the finders not knowing what they were. During my stay in El Dorado County I have seen and recognized two of these diamonds, both of which were in the hands of people who did not know what they were, but who had simply saved them as little curiosities on account of their appearance and peculiar shape. For the benefit of those who are not familiar with the stone, it may be stated here that this peculiar *shape* of the diamond is one of the easiest and most characteristic features by which it may be recognized. The commonest shape of the diamond in this country is that of a solid or crystal having *twenty-four triangular faces*. And another remarkable and easily distinguished peculiarity is that these faces, instead of being perfectly flat, as is generally the case with the faces of quartz and other crystals, are very often *curved*, the center of each face being a little higher than the surface toward the edges. The diamond, moreover, is *extremely hard*, and scratches quartz with the greatest case. If, therefore, any one finds a little white or yellowish white crystal with twenty-four of these curved triangular faces, and if on trying it carefully with a crystal of pure quartz, he finds that it easily scratches the quartz without showing the least abrasion itself, he may be tolerably sure that he has a diamond. I would not recommend any one to institute a special hunt for diamonds, since at best they are not remarkably plenty. But it does no harm and takes no time to keep one eye upon the contents of the pan while engaged in cleaning up sluices, batteries, &c., in working the gravel, and though it may not pay to hunt for diamonds, yet it always pays to pick them up when you do happen to see them.

The State Geological Survey.—The legislature of 1860 passed an act appointing Prof. J. D. Whitney State geologist, and authorizing him to make an accurate and complete geological survey of the State. Prof. Whitney was engaged at the time his commission reached him in making a survey of the lead and copper region of the northwestern States under authority of the legislature of Wisconsin. Having accomplished this duty he departed for the new scenes of his labor and arrived in California during the latter part of the year 1860, since which time he has been engaged, with the aid of accomplished and energetic assistants, in making a thorough geological examination of the State, interrupted, however, from time to time, by reason of the failure of the legislature to continue the necessary appropriations for the work.

The result of his labors up to the present time has been the publication of a volume on general geology embodying the results of a preliminary reconnaissance of the State, published in 1865; two volumes on paleontology; and several maps illustrating the geology and topography of the central portions of the State, showing the area and extent of the auriferous deposit, the course of the pliocene rivers, and the course and flow of the lava streams which in the higher Sierra have filled the ancient rivers. Two more volumes on economical geology and mining are now prepared and ready for press, and their early issue will depend on the action of the legislature of 1871-'72, which will be called upon to make the necessary appropriations for their publication.

The following extracts are from the biennial report of Prof. Whitney to the legislature of 1871-'72:

Immediately after being placed in possession of means for continuing the survey, I began to make preparations for a more detailed survey and examination of the most important mining region of the State, namely, the western slope of the Sierra Nevada, between Mariposa and Plumas Counties. There seems to be abundant reason why this part of the State should be worked up with great care, both from a geological and a topographical point of view; and, as it is impossible to have a detailed knowledge of the geology except on a basis of accurate geography, I long since commenced gathering our materials into shape with the idea eventually of publishing a map of the mining counties. Such a map has been partly prepared, and will be submitted to the legislature, in order that they may have the means of judging for themselves whether the completion and publication of this piece of our work may be advisable. The already warmly expressed wishes of many persons interested in mining throughout the State seem to me to point very clearly in that direction. The map, as now prepared, is about six and a half feet by four in dimensions, and covers an area of about nine thousand square miles, extending from Stanislaus County on the south to Plumas County on the north. I propose, however, to extend it in both directions, so that it may be published in four sheets, each reaching across the Sierra from foot-hills to summit, and so arranged that the sheets may be had separately, or the whole combined together into one map.

Immediately after the resumption of the survey, the services of Mr. Amos Bowman were engaged, and he was directed to begin a minutely detailed exploration of the mining belt of the Sierra, with the especial object in view, however, of collecting all the data necessary for a full report on the gravel deposits worked by the hydraulic process. Mr. Bowman took the field in April, 1870, Mr. W. H. Pettee joining him in July, and these gentlemen spent nearly all the remainder of the year in working out the geology and mapping the detailed topography between the Yuba and American Rivers. In the mean time Mr. W. A. Goodyear commenced on the detailed survey of the mining belt of the Sierra lying to the south of the North Fork of the American River, and continued in the field until driven out by the inclemency of the season, having reached a point as far south as Sutter Creek, Amador County.

In the prosecution of this work a large amount of valuable information has already been collected, both of a geographical and geological character. It is confidently expected that the final elaboration of all our materials, with the accompanying maps and sections, will exhibit the phenomena of the gravel deposits of the Sierra, in regard to which there has been so much discussion and such a multiplicity of opinions, in a new light, and that many difficulties which have hitherto perplexed the miners will be solved. It is my desire that this work should be continuously and vigorously prosecuted, until a full and detailed map of the whole western slope of the Sierra has been prepared, accompanied by a corresponding report. This, it is expected, can be accomplished during the next season, if the pecuniary means are forthcoming, the map and report being completed during the following winter.

The question how and when the geological portion of the survey reports shall be published is one that has been much considered; but in the prevailing uncertainty in regard to the moods of the legislature, and the amount which is to be expended on the survey, it has been found impossible to answer it. Some persons have been so unreasonable as to expect a geological map of the State before a geographical one was or could be got ready. As fast as the sheets of the Central California map are engraved we shall color the geology upon them, and we shall do the same with the map of the whole State, which will probably be colored in time to be exhibited to the next legislature, during the coming winter, (1871–72.) As soon as the maps are ready, it will be time to prepare for the publication of the geological volume to accompany them. The engraving of the sections necessary for this volume should also be begun as soon as possible. It is my impression that the geological part of our work can, by condensation and the use of small type, be compressed into two volumes, with an accompanying atlas of maps, sections, and other illustrations. It also seems, at present, as if the best division of the two volumes would be to allot the general geology to one and the economical geology to the other. In that case the first one would contain a systematic description of the geology of California, arranged in chronological order of the formations; while the other would be devoted to a discussion of the character, mode of occurrence, and value of the useful mineral and metallic combinations which they contain. One volume would be for the use of the scientific and general public; the other for those especially engaged in developing our mineral resources.

The labors of the State Geological Survey have now reached a point where many practical advantages may be expected to follow the publication of the results of their investigation relative to the gravel deposits, their extent and mode of occurrence, and the causes of their distribution. In this investigation, the history of past and present work in the mining region has been carefully prepared, and is chiefly valuable as an aid in estimating the extent and value of the undeveloped ground which yet remains. The survey has already more than paid for the sums of money expended in its prosecution, by revealing to the world the almost inexhaustible extent of our mineral wealth. Capital has thereby been attracted, and universal confidence in our resources prevails. It is to be hoped that this useful work may not be stopped at this time, when we are about to reap the reward of the patient and careful labors of Prof. Whitney and his assistants.

Meteorology of California.—The meteorology of California has an important bearing on the general prosperity of the State, as well in its relations to the mining interest as to that of our agricultural prosperity. In future the subject will have a still greater interest on account of its relations to gold production, in consequence of the numerous great undertakings projected or under way for controlling the waters of the State and economizing them for mining and agricultural purposes, and the extensive development of our unworked gravel deposits, which so largely depend for success on copious and regular rains.

The subject has received careful attention for years past on the part of a few scientific gentlemen, who have, from time to time, published the results of their observations for the benefit of the community. Among these are Dr. T. M. Logan, of Sacramento, and Mr. Thomas Tennent, of San Francisco, who, for over twenty years, have kept accurate record of the rain-fall at those places. Recently the Government has established branches of its Signal Service Department at various points on the Pacific coast, and their records and observations will, in future, prove of great utility to the miner and the farmer, as well as a benefit to science at large.

I select for publication the table of Dr. Logan, which I prefer because his observations are made in the central portions of the State, and in closer proximity to the mining region than those of other observers.

Rain table for Sacramento, prepared by T. M. Logan, M. D., arranged according to the seasons, showing the amount in inches of rain each month, during twenty-three years, and for each rainy season; also the mean quantity for every month, and the mean annual amount of rain.

Months	1849.	1850.	1851.	1852.	1853.	1854.	1855.	1856.	1857.	1858.	1859.	1860.
September	0.250	0.000	1.000	0.003	0.000	1.810	Sprinkle.	Sprinkle.	0.000	Sprinkle.	0.025	0.063
October	1.500	0.000	0.180	0.000	0.005	0.050	0.730	0.185	0.655	3.010	0.000	0
November	2.250	Sprinkle.	2.140	0.000	1.500	0.650	0.750	0.651	0.446	0.147	0.493	0.914
December	12.500	Sprinkle.	7.070	13.410	1.540	1.150	2.000	2.360	6.652	4.329	6.834	0.181
January	4.500	0.650	0.530	3.000	3.250	3.670	4.019	1.375	2.444	2.984	2.310	2.068
February	3.500	0.350	0.190	2.000	8.350	3.490	0.092	4.801	2.461	3.095	0.931	3.030
March	10.000	1.880	6.400	7.000	3.250	4.300	1.463	0.675	2.876	1.037	5.110	3.220
April	4.250	1.140	0.300	1.450	1.500	4.299	2.138	Sprinkle.	1.274	0.981	2.491	0.475
May	0.000	0.090	0.100	0.001	0.210	1.150	1.841	Sprinkle.	0.203	1.037	2.401	0.590
June	0.000	0.000	0.000	0.001	0.310	0.010	0.853	0.000	0.098	0.000	0.549	0.135
July	0.000	0.000	0.000	0.000	0.000	0.000	0.000	0.000	0.650	0.000	0.000	0.000
August	0.000	0.000	0.000	0.000	Sprinkle.	0.000	0.000	Sprinkle.	Sprinkle.	0.000	0.000	0.000
Total	38.000	4.710	17.980	36.385	20.085	16.630	13.770	10.443	16.091	16.041	22.026	15.548

Months	1861.	1862.	1863.	1864.	1865.	1866.	1867.	1868.	1869.	1870.	1871.	Mean.
September	0.000	0.900	0.003	0.004	0.089	0.036	0.030	0.006	0.000	0.000	0.001	0.002
October	Sprinkle.	0.355	0.000	0.130	0.480	2.010	0.001	0.000	0.000	0.020	0.210	0.469
November	2.170	0.005	1.490	6.718	2.427	1.010	2.426	3.896	3.630	0.690	1.220	1.092
December	8.657	2.327	1.815	7.867	0.364	0.470	9.311	12.b50	2.612	0.971	10.500	5.409
January	15.036	1.723	1.077	4.76	7.459	3.440	3.036	4.790	6.030	2.075		3.510
February	4.900	2.751	0.186	0.719	2.010	7.104	2.147	3.630	2.147	1.910		2.700
March	4.500	2.360	1.303	0.491	2.018	1.895	4.348	2.120	1.649	0.690		3.038
April	0.891	1.050	1.050	1.370	0.470	1.695	2.300	1.240	2.120	1.454		1.670
May	1.896	0.355	0.742	0.460	0.523	2.196	0.270	0.648	0.270	0.756		0.208
June	0.011	0.000	0.087	0.000	0.160	0.008	Sprinkle.	0.008	Sprinkle.	0.001		0.053
July	0.000	0.000	0.000	0.004	0.018	0.000	0.000	0.000	Sprinkle.	0.000		0.027
August	0.000	0.000	0.085	0.000	0.000	0.000	0.000	0.000	0.001	0.000		0.004
Total	34.549	11.579	7.868	22.512	17.924	25.305	32.709	16.644	13.572	8.470	12.021	19.767

The records of the rain-fall of the State differ greatly in various localities, depending on the topography of the country. In San Francisco and in the coast counties it appears to be about twenty per cent. more than in the great valleys of the Sacramento and San Joaquin. In Nevada City, at an elevation of about 2,500 feet above sea-level, the fall of rain has been more than double that of the valleys. Higher up in the Sierras deep snows fall, which do not melt until late in the summer, when they furnish in a wet year the supplies of water necessary to hydraulic operations and quartz mining.

It will be observed that during the past three years (1868, '69, '70,) the aggregate rain-fall has been only 38.68 inches, which is but little more than that of many single years, and much less than fell in any other three consecutive years since the settlement of California by the Americans. The consequence has been a material diminution of the gold product in the face of a greater amount of development and exploration. Hydraulic operations have been almost suspended, and even the quartz-mills have been obliged to shut down or run at half their capacity for want of water.

The present season, however, has opened auspiciously, and promises to equal our best years in the supply of water. The rain-fall for the first three months of the season, up to end of December, has reached 12.02 inches at Sacramento. It is now believed the fall for the season, October, 1871, to May, 1872, will reach thirty-six inches.

The following table prepared by Mr. F. B. Pilling, of the United States Signal Service Corps of San Francisco, will show the distribution of the rain-fall for December, 1871. It will be observed by comparison with Dr. Logan's table, that rain-fall at San Francisco was 3.77 inches in excess of that of Sacramento for same period.

Table showing daily and monthly mean of barometer and thermometer, and amount of rain-fall for the month of December, 1871.

Date.	Mean daily barometer.	Mean daily thermometer.	Rain-fall.
	Inches.	*Degrees.*	*Inches.*
December 1	30.36	50	....
December 2	30.27	51	0.10
December 3	30.30	51	....
December 4	30.38	54	....
December 5	31.31	50	....
December 6	30.17	49	....
December 7	30.10	49	....
December 8	30.19	48	....
December 9	30.39	47	....
December 10	30.32	45	....
December 11	30.31	45	....
December 12	30.38	46	....
December 13	30.32	52	....
December 14	30.33	56	....
December 15	30.22	53	....
December 16	29.87	50	....
December 17	29.80	51	0.25
December 18	29.72	50	2.83
December 19	30.12	61	3.13
December 20	30.65	61	0.30
December 21	30.69	52	0.91
December 22	29.91	53	....
December 23	29.70	58	2.48
December 24	30.02	57	0.10
December 25	30.17	57	....
December 26	30.03	57	0.04
December 27	29.80	58	1.05
December 28	29.75	58	0.40
December 29	29.70	56	0.82
December 30	30.13	54	0.17
December 31	29.89	55	1.70
Monthly mean	30.00	53	14.30

Mining inventions and improvements.—The past year has not been noted for the introduction of many new or important appliances useful to the mining industry, but several inventions have been brought to a greater degree of perfection, and are now coming into more general use, with the most favorable results to economy in mining.

Hydraulic machinery.—In the gravel and hydraulic mining regions of the State, the improved pipes and nozzles for projecting a large quantity of water under great head or pressure against the gravel banks, have been thoroughly tested and found to be valuable accessories in the extensive and economical working of our auriferous deposits, and may be said to have revolutionized this branch of mining. The pipes in general use are those of Messrs. Craig and Fisher, of Nevada City, and Mr. R. Hoskins, of Dutch Flat, which are described in the report for 1871, pages 63 to 65. I doubted last year whether 1,000 inches of water could be successfully thrown through one nozzle in hydraulic mining. That doubt can no longer be entertained. The Dutch Flat Blue Gravel Company (on the Taeff ground—see my last report, p. 84) have been throwing 1,200 inches through a Hoskins pipe, with 432 feet head, and are now constructing a pipe to throw 2,000 inches. In fact, hydraulic mining is assuming proportions heretofore scarcely dreamed of.

Quartz-crushing machinery.—In quartz-mills there has been one invention which possesses the merit of novelty in the application of crushing power. This is a trip-hammer quartz-crusher, invented by Mr. J. D. Crocker, of Virginia City, Nevada, and lately introduced in California. A small five-stamp crusher was exhibited at the mechanics' fair of San Francisco in September of this year, and attracted much attention and favorable comment from persons interested in mining.

The chief improvement in this invention consists in operating light stamps on the principle of the tilt or trip-hammer, whereby quick, sharp

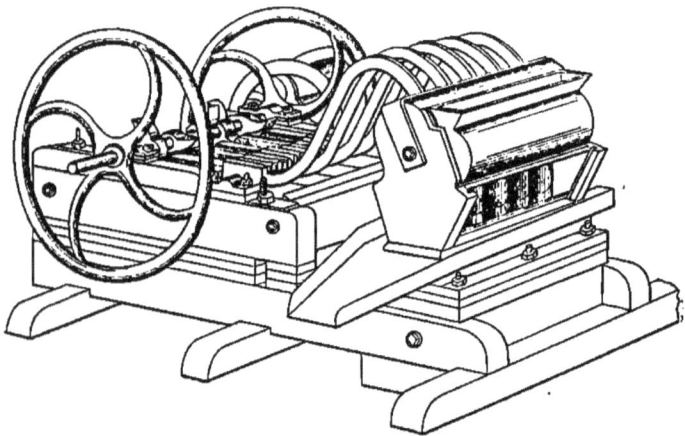

Crocker's Trip-Hammer Quartz Mill.

blows are made, with very light stamps, and with less proportionate power, according to weight of stamp and frequency of blows, than is required in operating heavy stamps with a direct lifting action. The advantage of sharp, quick blows with a light stamp, must be obvious to every one who has employed alternately a small light hammer, and a heavy one for breaking quartz before a battery. The inventor claims a decided advantage in his 400 light blows per minute over the 50 to 60 blows per minute delivered by the usual 600-pound stamp. A 5-stamp battery of this construction strikes 2,000 blows per minute; and while the first cost is greatly reduced in attaining the same crushing capacity, the power required to run it is also much reduced. The difference in freight to many places would also be very great. The machine is made of different sizes, and calculated to crush from 7 to 35 tons per twenty-four hours, according to size.

Whatever may be the results of its practical working on a large scale, there can be no doubt that it will prove an invaluable invention on account of its lightness, portability, and low price, to those who are engaged in the experimental development of quartz claims in localities where custom mills do not exist.

The capacity of the most powerful mills in California for crushing quartz is from one to one and three-quarter tons per day (24 hours) to each stamp—the usual weight of stamp being from 600 to 700 pounds, and the drops 60 to 90 per minute. While this method is acknowledged to be slow, it is not probable that it will be superseded for many years, if at all; but in the mean time attention is being attracted to all inventions which claim to work greater amounts of quartz in a given time and at less expense for cost of machinery. Of this character is an invention of Mr. T. R. Wilson, known as "Wilson's Steam Stamp." The principle on which this mill operates is that of the direct application of steam to the stamp, the stem acting as a piston. The best results have been obtained with 70 pounds of steam and 206 drops per minute for each stamp. The average amount of ore which it is claimed can be crushed in a day (of 24 hours) is 23 tons with a No. 6 slot screen; and the average consumption of fuel is about one cord of wood to 10 tons of ore. As instances of its capacity we are informed that some time since 10 tons and 800 pounds of rock were crushed in eight hours and forty-five minutes, using only one cord of oak wood. Again, in one run 51 tons were crushed in forty-seven hours, and 10,800 pounds in four hours fifty minutes with 68 pounds of steam. Several of Mr. Wilson's batteries are now in operation in San Diego County where they seem to meet with favor. A detailed description of this invention was given in the Mining Commissioner's Report for 1870, page 668, and it is briefly referred to here as one of the improvements apparently coming into more general use.

Not a year passes without the invention and introduction of numerous devices for the saving of fine gold and quicksilver in the tailings of our quartz-mills. These rarely achieve any reputation outside of the district in which they are invented, and many, after months of experiment and trial, are rejected as useless. One of these undercurrent sluices, invented by Mr. Evans, and lately improved by Dr. Frey of Sacramento, is made of cast iron, with transverse corrugations on the bottom, semicircular in shape, and three inches deep. At the bottom of each alternate corrugation is a narrow slit through which the heavier material falls down into another riffle below with larger corrugations. Both riffles are set on the same grade, which should be about one foot in twelve. The lower box is charged with quicksilver. These sluices do not cake, nor do they require any attention beyond cleaning up once a week. They have stood the test of use in quartz-mills below all the contrivances for saving gold, and have made large returns of gold,

H. Ex. 211——3

silver, concentrated sulphurets, and quicksilver, that would otherwise have been lost.

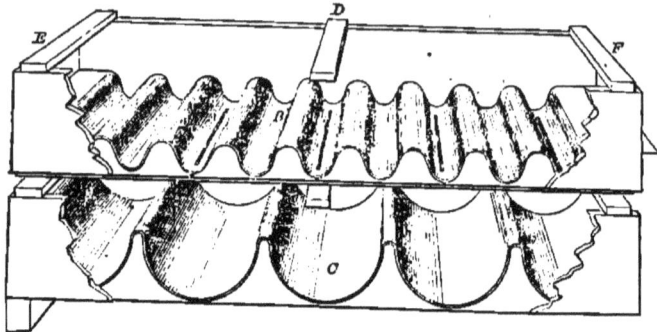

Evans' and Frey's Undercurrent Sluice.

The following is an assay of tailings saved at the Rhode Island Mills, Nevada, by these sluices during a run of one week:

Gold, $75.24; silver, $76.96; quicksilver, 250 pounds to the ton.

One of the most important appliances in the economical running of quartz-mills is Stanford's Self-Feeding Apparatus, which is rapidly gaining favor with miners and millmen. This consists of a hopper-shaped box placed above and in front of each battery. The feed-shovel at the bottom of the box is connected with a lever and a rod, and is shaken at each blow of the middle stamp of the battery by means of an upper tappet which strikes upon one arm of the lever. By this means motion is communicated to the forward end of the feed-box. One man can attend to twenty stamps, and the feeding is more regular than by hand, and materially increases the crushing capacity of the mill.[*]

Cement and gravel mining by machinery.—The process of treating hard cement and gravel by grinding and friction, instead of crushing under stamps, was introduced several years since by Captain J. B. Cox, who invented for this purpose the machine known as the "Cox Pan." This invention was intended to supersede the use of stamp-mills, by which method the bowlders, rarely containing gold, were necessarily crushed under the stamps at a great waste of time and power. This invention was not for a time received with favor, and in some cases was rejected after trial; but recent practical workings with the machine, since some improvements have been made, and after some changes in the manner of feeding, have demonstrated its utility and economy in the working of hard-cemented gravel. This pan is about five feet in diameter and two feet in height, and is intended to hold a charge of half a ton, or from 1,000 to 1,200 pounds. The rim of the pan is made of boiler-iron, and it has a perforated cast-iron bottom, through which the finer sand and auriferous material fall into the sluice-boxes or other gold saving appliances. The bowlders and large pebbles, which constitute so large a pro-

---

[*] On the other hand, some of the most experienced millmen in California prefer feeding the batteries by hand to any form of automatic apparatus. The self-feeders, they say, may be superior to careless or unfaithful workmen; but a skillful feeder can, if he chooses, by giving to each stamp at exactly the right instant exactly the amount of rock it requires, increase the capacity of the battery to an extent which more than compensates for the extra outlay in wages.—R. W. R.

portion of the cemented gravel, are discharged at intervals through a section of the bottom of the pan, which is opened like a trap-door, by means of a lever, when they have accumulated to such an extent as to retard the grinding and pulverizing of the cement in the pan. Four revolving-arms are attached to a shaft which passes perpendicularly through the center of the pan. On these arms are fastened the steel teeth (in appearance like plowshares) which, in the rapid revolution of the arms, break up the cemented gravel. An abundant supply of water is distributed while the pan is in motion, and materially aids in the disintegration of the gravel. Better results are claimed when the pan is fed continuously instead of by charges, which was the practice on its introduction. Details of the workings of this process will be found in the present report, in the matter descriptive of gravel operations near Jamestown, Tuolumne County, and at Dutch Flat, Placer County.

Device for transportation of ore in mountainous regions.—Mineral countries are nearly always mountainous, and the transportation of ores from the mine to an available mill-site or location for a furnace frequently requires a large outlay of money in the construction of a wagon-road or tramway, and a consequent expense in the hauling and transportation of ores, which prevent the profitable working of mines so situated. In order to obviate this difficulty, Mr. A. S. Hallidie, of San Francisco, has perfected an invention which he terms an "Endless Wire-Rope Way," a model of which, on a practical working scale, was exhibited in the fair of the Mechanics' Institute in San Francisco, in September, 1871.

This invention is one which promises to materially facilitate the working of quartz and other lodes located at inaccessible points, or where it is difficult to transport the ores from the ledge to the mill. It can be used at such locations for the delivery of material or ores in mountainous places or deep gorges; from the shore to vessels in the offing; for working up hill and down, and, where there is much descent, for economizing the power thus obtained by gravitation for any desired purpose. The invention consists in the use of endless iron or steel wire ropes, supported on peculiar sheaves, placed on posts, actuated by the gravity of the descending loads, or by an engine attached to a grip-pulley. The function of the grip-pulleys is to hold the rope so as to prevent its slipping in the groove. In general, the difference in altitude between the mill and the mine is sufficient to obtain by gravitation quite an amount of power, which is transmitted by the grip-pulleys for whatever purpose it is required, (and where there is no power obtained in this manner, it is given by a steam-engine.) The receptacles are small, self-dumping boxes which contain from 50 to 150 pounds of ore or other material. The rope travels two hundred feet per minute; the posts with the bearing pulleys on are usually two hundred feet apart, the hooks holding the ore-sacks being placed about fifty feet apart, and holding one hundred and fifty pounds of ore each. The rope is actuated by steam or water power whenever there is not sufficient descent to run it by the gravitation of the descending loaded carriers. In any event it is better to connect with the motive-power by gear, so as to secure uniformity of speed. The amount of ore delivered being four sacks of one hundred and fifty pounds, making six hundred pounds per minute, or 36,000 pounds per hour, or one hundred and eighty tons per day of ten hours. This can be increased or diminished at will. The cost of delivering the ore, including wear and tear of machinery, interest on outlay, and running expense, is about twenty-five cents per ton per mile. The cost of constructing such a line is from $4,500 to $7,000 per mile, according to the topography of the country. By its use material can be transported from a higher to a lower, or from a lower to a higher point. In the last case, power must be applied, which can be done directly from a stationary engine at one end by means of the grip-pulley; in the first case, often no extra power will be needed, the gravity of the descending loads being sufficient to keep the rope in motion. Similar inventions have been made before, and the merit of this, therefore, depends on the peculiar construction and adaptation to the wants of the localities. One can readily see the great many advantages that this method possesses, from the fact that it requires neither road to be built nor expensive machinery; that it can be run at all seasons of the year, even when there are five or eight feet of snow on the ground; that it can be rapidly and cheaply constructed in the worst possible country; and that when there is sufficient grade, not only does it run itself by gravitation, but produces a motive power at both ends of the line. Running at four miles per hour, the boxes are carried down at one side of the posts and up the other, 400 to the mile, each delivering fifty pounds of ore, or ten tons per hour, and they can be run twenty-four hours per day.[*]

Paul's process.—The "electric dry amalgamating process," introduced by Mr. Almarin B. Paul, of San Francisco, has been tested in a number of localities within the past two years; and though the reports of its value vary, it is but fair to say that the inventor has received many certificates of a very positive and favorable character.[†]

---

[*] This device is more fully described in a separate chapter of the present report, *q. v.*
[†] The latest in date of these is given herewith:

DUN GLEN, NEVADA, *March*, 1872.

Mr. ALMARIN B. PAUL—SIR: Yours asking for statement of tests made at Sprague & Co.'s new mill at this place is received, and we here cheerfully comply as follows: We reduced 22 tons of ore from the Auburn mine, owned by Messrs. Wright & Wentworth, containing besides gold and silver, magnetic iron, carbonate oxide, and sulphuret of lead, and has always been considered of a very refractory nature. The 22 tons worked by your process paid $107 per ton. Eleven tons of same ore worked in pans paid $53 per ton. The bullion from your process runs as high as 989-1,000 fine.

We also worked the tailings of a lot of 20 tons of ore from the Monroe mine, which ore in the first instance was reduced by battery and pan process, at Essex Mill, and paid $8.14 per ton. The tailings from this lot yielded by your process $293.27. The bullion per battery and pan working was 584 fine; by your process, as per certificates of San Francisco Assaying and Refining Works, was 960 fine. The loss of mercury was less than half a pound per ton.

We also made a test on a small lot of iron sulphurets from Monroe mine, very closely concentrated, and which had been previously worked by battery and pan amalgamation, and which had been salted, oxidized, and worked several times previously—each time yielding some low-grade bullion. This lot was treated by your process, and to our astonishment obtained more metal in value than we had gained by all the several previous working, though very carefully done. The bullion by last working, as per certificate of California Assay office, was 999-1,000 fine.

We have now commenced on the "Lang Syne" ore, from a mine belonging to the "Great Central Mining Company of San Francisco," the results of which speak more for your process, and the great revolution the system is likely to inaugurate, and of its inestimable value, than any tests we can give you. That you may fully understand this we must particularize.

In 1863-'4-'5, the Lang Syne mine was worked, and in 1864 a mill erected for reduction of its ores, and over $100,000 expended and lost. The ore could not be made to pay.

The operation was pronounced a failure, and the mill removed. Tests of ten and twenty tons were also worked at the Essex Mill, having all the then modern improvements, with like unsatisfactory results. The mill just erected for the working of your process is built upon the same spot where the old mill stood. We have now worked a number of tons of a class of ore of which there are hundreds of tons in sight, with the result of $30 per ton and bullion 966 fine, being more than double the result obtained in the other mills.

The same renewed life which this locality has received through the introduction of

The process and machinery are described in Mr. Paul's circular as follows:

I start out with the idea, speaking only of mill ores, that the precious metals, in bulk of value, are in the most simple but delicate conditions, and require thoroughness in reduction, care, and chemical affinities, to unlock. That in their metallic state they are incased in their matrix in atoms so minute that the word "infinitesimal" hardly expresses the fineness. To meet these conditions it requires thorough pulverization, then delicacy in preparation, amalgamation, and precipitation. You will understand by this that as far as *gold ores* go—excepting the advantages of calcination—I regard desulphurization, with all its appliances, as an unprofitable expenditure of time and money, and often complicating nature's simplicity. I would further add that chlorination is only an expensive mode of getting a high percentage of gold, which can be obtained by the *perfect* working of this system, at less than half the expense.

In all ores are gaseous and refractory substances, atmospheric or other films, which coat the metal and create repulsion between it and the mercury. An increase of these repulsive elements takes place where the grinding of ore, iron, and mercury are carried on together. There is sufficient evidence to establish this point—every intelligent miner has experienced it. I must, however, for the less experienced, quote from one in high authority, (Overman, p. 260:) "All metals appear to have a tendency to float in water, when in fine particles, some more than others. This is caused by a particle of gas, either air or water gas, adhering to the particles of metals, which *causes it to be light and float. Precious metals appear to possess more of this quality than others.*"

Again, the same author says: "Gold is by far heavier than silex, but we may observe, by means of a microscope, a multitude of fine particles of gold suspended in water, when we cannot detect the slightest particle of silicious matter." To sum up, as has been well expressed by another: "We have too long attempted to do by brute force what must be done with care and the gentle persuasion of affinities."

The practical working of the process is as follows: The ore is first heated, dried, then reduced dry by such machinery as best suits the views of parties and accomplishes the object of bringing the ore to the fineness of No. 14 wire cloth. The crushed ore is then conveyed to an iron pulverizing, preparing, and self-discharging barrel, where it is pulverized to flour in fineness and prepared for amalgamation, *under heat.* By heat, friction, and chemicals, it is put in what is termed an infinitesimal, electrical, live, and pure condition. Thus there is a combustion of all gases, destruction of rebellious films, and expulsion of atmospheric dampness, instead of which are created the greatest activity, attraction, and cohesion between the precious metals *only* and the mercury.

The ore thus prepared being so fine and light, and the metal to be operated upon infinitesimal, the question now comes how to produce effectual contact for amalgamation. To meet this point the ore is conveyed in its dry, heated, prepared, and electrical condition to an iron, wood, or earthen cylinder, to which is added from 20 to 25 per cent. in weight of mercury. The condition of the ore raises the temperature of the mercury, lessens its density, increases its volume, and the result is, the ore and mercury play together like water, and create the most thorough and complete intermingling. So perfect is the blending, that on examination with the naked eye hardly a particle of mercury is disclosed, notwithstanding the percentage to ore is so large. The mercury, in this finely-divided state, having been in continuous motion, rolling over and over on the surface and through the entire mass during the one hour given for amalgamation, it is reasonable to suppose has effectually done its work, and that the precious metals, no matter how fine, cannot escape a contact in this searching.

The harmony created between precious metals and mercury is finely illustrated by the fact that the baser metals are placed in antagonism, and consequently leave the mercury free from the fouling or sickening properties of the ore, no matter what it contains. It will be observed, too, that every infinitesimal particle has had its weight

your radically improved system, we are certain can be extended to others similarly situated by the introduction of your mills.

We will add that we have no difficulty in saving the mercury, and assert that the loss on all ores can be brought within that of pan amalgamation, and ordinarily very considerably less. Under all circumstances it is bright, active, and pure, and is used over and over again without any cleansing or retorting.

Your self-feeding and self-discharging pulverizing barrel is, without exception, one of the most complete reducers that we have ever seen. Its capacity is fully equal to your estimates of it, receiving ore from No. 14 wire cloth.

The pulverizing of ore by quartz, instead of iron, is not only cheaper but much better. If desirable, we will, at any time, give results of our working.

Yours,   SPRAGUE & CO.

CHARLES D. SMYTH,
Superintendent of Sprague & Co.'s Mill, also Great Central M. Co.'s Mines.

increased by contact with mercury, besides being placed in a state to receive, more actively, the precipitating element.

The operation, it will be admitted, is so far perfectly done, and the next question is the separating of this mingled mass. To accomplish this, the ore from the amalgamating barrel is discharged into a large wooden settler, of especial construction, and where water for the first time is introduced. The greater portion of mercury, carrying the precious metals, is soon precipitated and collected for drawing off into a receiver. The lighter portions of mercury and metal, in due time are drawn off into an electric settler, where they are precipitated by electricity upon the principle of electro-plating. excepting there is no adherence of metal or mercury, both of which are drawn off together. The residue is then allowed to flow off as wastage, or, if desired, for concentration of base metals. A one-thousand-pound charge is worked every hour, and yet every one-thousand-pound charge has over four hours of varied treatment.

It will be observed that though considerable mercury is used, it does not involve having so large an amount on hand as at first thought would seem, in consequence of its being in continuous use.

There are other essential points of equally practical character, but which are only given to those who engage in working the process.*

COST, WEIGHT, AND EXPENSE OF WORKING.—By way of giving definite data as to expense of machinery, I will contract to furnish in San Francisco, until further notice, all material (outside of power and timber) requisite for working according, and up to my idea, which embraces crushers, pulverizing, preparing, and amalgamating barrels, electric settlers, concentrators, shafting, gearing, pulleys, belts, boxes, bolts, copper, zinc, conveyers, and bolters, as follows:

For mill of 1½-ton capacity per 24 hours.. $1,000
For mill of 3-ton capacity per 24 hours.. 1,750
For mill of 6-ton capacity per 24 hours.. 3,000
For mill of 12-ton capacity per 24 hours.. 5,500
For mill of 24-ton capacity per 24 hours.. 10,000

Royalty is included in the above figures. Or parties can have the machinery built on my order and after my plans, at such founderies as best suit their business and local relations, by allowing my charges for patents, which are set at low figures, as the above estimates show.

By way of further insight, I give the following as a close though approximate estimate of weight, power, cost per ton of working. Of course these figures vary according to locality, wood, and labor:

	Weight.	Power.	Cost per ton of working.
1½-ton mill	2½ tons.	5-horse.	$7 50 to $10 00
3-ton mill	4½ tons.	8-horse.	5 00 to 7 50
6-ton mill	7 tons.	15-horse.	4 00 to 6 00
12-ton mill	12 tons.	25-horse.	3 50 to 4 50
24-ton mill	24 tons.	45-horse.	3 00 to 4 50

The putting up of machinery is not expensive, it being mainly framing timber.

I offer no objections to stamp batteries; on the contrary, for larger mills the additional cost has an equivalent for work executed for pulverizers.

Stamping machinery of mills already erected can at a small expense be changed to answer my purpose as reducers.

The principal objection heretofore urged against dry working is dust. This I have overcome by using incased batteries or crushing-machines, then pulverizing ore in incased barrels, which are both self-feeding and self-discharging. From this barrel the ore is conveyed mechanically and deposited in another—the amalgamating barrel— which is closed during the operation of amalgamating. From this it is delivered into a closed hopper, and from thence gradually into settlers, all being performed with an ease and cleanliness not found in any mode of working.

The system has a great advantage where water is scarce, as the capacity of machinery may be rated according to the amount of water for power. Mill-owners having their own ideas as to what machinery is best suited to their wants and circumstances, and as some may desire single portions which make up the process as a whole, to accommo-

---

* This description is not sufficiently definite to permit a discussion. The theoretical explanations are partly unintelligible and partly untenable. The production of very fine bullion is not of itself a proof of thorough extraction, but rather an indication to the contrary, particularly in silver ores. I am inclined to ascribe any practical success which the process may have achieved, to the fine dry crushing, the use of chemicals, and the large quantity of mercury continually employed. But all millmen know that these methods have their drawbacks, in cost, inconvenience, and other respects. I have never seen a thorough discussion of the economical results of this process; and though I am prepared to believe that it may be advantageous, under certain circumstances, the claims in its behalf, that it involves some novel electrical action, and that it works equally well for all ores, no matter what base metals they contain, are undoubtedly without foundation.—R. W. R.

date these views, I will sell separately electric settlers singly or in pairs, (as they should properly go.) These settlers precipitate flour mercury and metal, by the agency of electricity, and at any time within ten minutes can be *cleaned up quite thoroughly without stopping machinery.* The action of electricity has a remarkable cleansing effect upon mercury. Their efficiency in this respect is such that where ores contain sickening properties, they will soon pay for themselves. They are also admirably adapted for working slums from mills; and for gathering the flour mercury therein they are superior to anything ever introduced.

The self-feeding and self-discharging pulverizing barrel is another desirable article. It is well understood by quartz miners, that down to a certain size, stamps or crushers are the most efficient reducers; but for reduction to a powder at a practical expense, great difficulty has been experienced, on account of slowness and large power required.

In these barrels these defects are remedied in consequence of making it both self-feeding and self-discharging.

It takes but little power, comparatively, from the fact that but little ore is required at a time, it going in at the ratio at which it is discharged.

Again, on account of there being a less amount of ore at a time for reduction, a less amount of reducing iron is required. Again, more work is executed at less expense, from the fact that ore, as soon as reduced to its given fineness, is out of the way, leaving no idle work to be done. This barrel can be made of any size to work from one ton to twenty tons per day, and to reduce to any given fineness from No. 20 wire-cloth to 100. It is made for strength and work, and will reduce faster, cheaper, and with less expense for wear and tear, than any class of pulverizing-machinery. In fact I am convinced that where power is no expense, the cheapest way of reducing quartz below a given fineness is by attrition, and making quartz reduce quartz, in these self-feeding and self-discharging pulverizing barrels.

This system is secured by four letters-patent, issued June 29, 1869; October 19, 1869; April 5, 1870; May 10, 1870.

Ambler's blow-pipe furnace.—This apparatus was patented during the year by Mr. Stephen S. Ambler, of Monitor, Alpine County, California. It has not yet been successfully put in practice, though several furnaces of the kind are reported as in course of construction in California and Nevada. The following description is taken substantially from the Scientific Press of San Francisco.

The main principle in this furnace, which differs from those of ordinary construction, is that a stream of heated air is passed into the ore without any loss of oxygen by combustion, and that the wood is converted into charcoal before it reaches the fires, to be used as fuel. These are two important points. Generally when air is introduced it passes through the fire, thereby losing a large proportion of its oxygen; by the use of this furnace all of it is utilized, from the fact that it is merely heated, not burned. This excess of oxygen prevents concentration or melting of the sulphurets in the cylinder, another point of great importance. When the ore contains a large proportion of sulphurets, the heated air may be passed over it by means of the blower and pipe and they may become more completely oxidized, whereas if only a small proportion of sulphurets are present, the air can, by means of the dampers, be turned under the grate into the fire. These draught can be regulated at will to throw either all or a certain portion of the air in the direction required, either into the fire or the revolving cylinder containing the ore, according to the class of ore under treatment. The hot-air chamber is arranged so as to permit the introduction of a jet of steam, or of water which is instantly converted into steam, and aid in the decomposition of certain classes of ores.

In order to explain this more fully, reference is made to the accompanying cuts, of which Fig. 1 is a side elevation, and Fig. 2 a plan.

A, represents the furnace, constructed of brick in the usual manner which is provided with a grate over the lower door, and above which is the carbonizing chamber, E. This chamber is kept constantly filled with wood which is fed through the upper door, C. When the doors, C and D, are closed, the wood in the upper part will be carbonized or converted into charcoal, and, as the fire below burns out, the charred or carbonized wood will settle down and continually feed the fire. At the back of the chamber, E, is a hot-air chamber, F, into which the heat, gases, and flame from the chamber, E, and also steam,

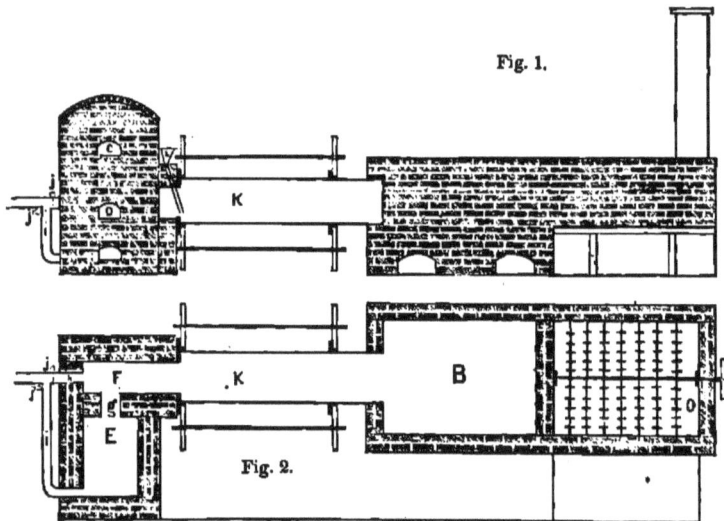

Fig. 1.

Fig. 2.

Ambler's Blow-Pipe Furnace.

when desirable, pass through the flue, g. A blast pipe, h, enters the chamber, F, at about an equal height with the flue, g, while a branch-pipe, i, passes to the front and enters the chamber, F, so as to deliver its blast directly under the fire in the grate. The blasts from these pipes are regulated by dampers, j, j'. When the damper of the pipe, h, is closed, and the damper, j', opened, the blast will be delivered upon the grate, and a reducing flame produced which will pass through the flue, g, into the chamber, F, and when the damper, j, is opened, oxygen will be supplied to the revolving cylinder, K, which contains the ore, through the chamber, F, and an oxidizing heat produced. In communication with the chamber, F, opposite the blast-pipe, h, is the revolving cylinder, K, into which the ore is fed through the hopper, L, so that as the ore meets the blast and heat from the chamber, F, it will be carried into the revolving cylinder, and there subjected to heat and roasted while passing through. In connection with the furnace, A, and revolving cylinder, K, is the dust-chamber, B. The heavy ore passes from the cylinder, K, into this chamber, and is taken away from the doors below. The light dust is carried by the current of air against the revolving perforated disks in O, one-half of which are submerged in water. These disks permit the passage of air, but the wet surfaces catch the dust, which, by the revolution, is carried under the water and washed off into the vat below, where it can be taken out when required.

The process of feeding the wood into the closed chamber, where it may be subjected to heat without air, is an important improvement, and, as can be seen, it will descend to the fire as it is needed.   The cylinder is worked by friction gearing.*

Rock-drilling machines.—This class of inventions was introduced in California in 1870, and is now extensively used in this and adjoining States and Territories.   The only machines of this description in use in California are the diamond drill (Leschot's patent, as improved by Severance & Holt) and the Blatchley drill, invented by Dr. Blatchley, of San Francisco, the Burleigh drill used in the eastern States and in Colorado never having been introduced here.   To the successful operation of these drills we are in a great measure indebted for the recent investment of capital in and consequent development of our great mineral resources.   By the use of these machines bed-rock tunnels can be run in from one-half to one-sixth of the time required by hand-drilling, so that one of the greatest objections to this kind of mining (the great length of time required to drive a long tunnel) is obviated.   All over the Pacific coast are innumerable mines that will almost pay for working by the ordinary method, which, by the use of these drills, can be made to yield a large profit.   A cheap, simple, durable, and efficient rock-drill, whereby the power of fifty or one hundred men can be concentrated in driving one drift or tunnel, has long been a desideratum, and has long exercised the ingenuity of our mechanics and miners.   The system of drilling by machinery used in the construction of large tunnels, such as the Hoosac and the Mount Cenis, was not adapted to our mining tunnels, which are rarely more than 6 by 4 feet in dimensions.   The great difficulty was in the application of power.   Steam was tried, but the pipes conveying steam to the drills at the face of the tunnel created an unbearable heat in the tunnels, and this plan was abandoned.   Compressed air was next tried, with better results, but the construction of compressors involved a great additional expense, which neutralized the utility of drilling by power.   We have reason to believe that all obstacles have now been overcome by the use of water under pressure as a motive power.

The diamond drill.—A. J. Severance, after two years of constant labor and experiments in building and running the diamond drill in tunnels and open-cut rock-work, has at last brought this drill to a high state of perfection.   He has been engaged in running some of our hardest bed-rock tunnels, and has proved by actual demonstration that he has run the same tunnel many hundred feet with one of his improved drills at a cost of $30 per foot, the same tunnel costing by hand-labor $46 per foot, besides running twice the number of feet per month as was run by hand.   By recent improvements his drills can be placed at any angle and adjusted so as to be able to bore holes in any desired direction as easily as by hand-drilling; and during the last 400 feet run in a tunnel 6½ by 9½, not a hole was drilled by hand-labor.   Heretofore these drills have been run by compressed air, or by steam power; but recent modifications have been made by him doing away with the great cost of an air-compressor or steam-power, and in its place the application of water-power has superseded steam or air.   This has been accomplished by attaching an ordinary hurdy-gurdy wheel or a small turbine upon the

---

* There is no feature in this furnace which can fairly be called new, taken by itself.   It is an ordinary gas-generator; the introduction of air to the carbonic oxide from the generator is necessarily involved in the use of the latter; and the use of the cylinder is equally familiar.   Novelty may, however, be claimed for this combination of well-known contrivances.   But it is open to another objection, which concerns me more than any question of novelty.   It is an arrangement for procuring a blow-pipe heat when no such heat is required, indeed, for a process (namely, that of roasting) to which such a heat is fatal.—R. W. R.

back end of the car, upon the front end of which the drills are firmly secured. The drill-car is made to suit the same track as the rock-car, and when eight to twelve holes are bored the drill is disconnected by detaching the feed-hose, (rubber or canvas, three inches in diameter,) which simply connects the water-pipe to the nozzle, which plays upon the water-wheel, and is geared to the drills upon the front end of the car. The car is then run back to a chamber in the tunnel, the holes loaded and filled, or exploded with a battery simultaneously, thus utilizing the whole force of the powder. As many as 663 cubic feet of loose rock have been obtained by a single shot of twelve holes in a tunnel 6½ by 9½ feet. This new method of using water-power in the tunnel has completely satisfied the miners that the year 1871 has been a year of progression in mining, and that the day is now close at hand when their tunnels can be run much cheaper and at a great saving of time, most of the mines having already hydraulic pressure by means of ditches, affording them plenty of water. The amount of water required depends upon the head obtained. For a two-drill machine, under 300 foot head, about ten to fifteen miners' inches is sufficient; but the more head the less water is required. The water may be taken from the ditch at any distance from the drill, and conveyed down to the mouth of and up the tunnel to the drill at the face of the tunnel, and it may be conducted through a 4 or 6-inch pipe, made of No. 10 to 24 iron, according to the amount of pressure at hand. Two machines just sold to the Union and American Companies in Nevada County are being run under 300-foot pressure. Where there is no natural head of water, one can easily be created by means of Knowles's patent steam-pump, which can be placed at the mouth of the tunnel and supplied with water by means of a small reservoir, thus pumping the water up the tunnel through a pipe against the wheel, the water thus running back into the reservoir, and being pumped over and over again without exhausting the supply. A trial was made recently at the Miners' Foundery in San Francisco for the benefit of mining and scientific men. The pump used was a small-sized Knowles pump, steam-cylinder 10 inches, water-cylinder 5 inches; pressure raised upon the pump 70 pounds—equal to about 150 feet of water-head; and two drills were run at the same time through hard granite, boring a 1½ inch hole at the rate of one inch per minute. Mr. Sutro, of the Sutro tunnel, with his engineer, witnessed the workings of the drill, and at once purchased one for his tunnel, which no doubt will save at least three-fourths of the time required over hand labor. This new application has overcome all doubts and difficulties in the way; tunnels are being run which for years have been abandoned, and new ones are started.

The Diamond drill is now in use in tunnel operations at Smartsville, Yuba County; the Union Gravel Company's ground, Nevada County; the American Company's ground, North San Juan; Nevada County, Oregon City, Butte County; the Taeff and Franklin ground, Dutch Flat, Placer County, and in many other localities in perpendicular boring for prospecting purposes, or in boring for water. Some details of its operations in running tunnels will be found in this report, where operations at Smartsville, Yuba County, are described.

During the publication of this report a novel and important application of the Diamond drill has been successfully made at Saint Clair, Pennsylvania, in the sinking of deep shafts. This will be fully described in my next report; in this place I can only say that the peculiarity of the method consists in boring a large number of holes from the surface to the full depth of the proposed shaft, unless this is too great. Three hundred feet is the depth of the holes in the shaft referred to. These are then filled up with sand. When the drilling is over the machines are removed, and blasting commences. Four feet of the sand is removed with a common sand-pump from the upper part of each hole, and one foot of clay-tamping is put in. This leaves an ordinary three-foot hole, which is blasted out, (with dualine or giant powder.) The interior holes are fired first, and afterwards those which have been bored in the corners and along the sides of the shaft. This process is repeated until the holes have been "used up," and the shaft is down to the bottom of the borings. If additional depth is desired the machines are set at work again, and a new set of deep borings is made. The sides and corners are found to be remarkably true and smooth. The drills here used are not annular, but have full convex heads in which the diamonds are set, and which are perforated to permit the passage of water. A stream, passing down through the tube used as a drill-rod, and up on the outside of this tube, keeps the hole clean and the drill-head cool. The expense and the time required for sinking a shaft are by this method both greatly reduced. The average rate of drilling has been 34 feet per day, the maximum thus far for a single machine being 67 feet in 8 hours. The shaft has been blasted out at the rate of over 25 feet per week. Mr. M. C. Bullock, engineer of the American Diamond Drill Company, 61 Liberty street, New York, is one of the patentees of the process, and Mr. Henry Pleasants is the engineer in charge of the work.

The Blatchley drill.—This drill, which was briefly noticed in my last report, is rapidly growing in favor. It has but recently been perfected, and is unlike any other rock-drill, both in principle and construction. It operates by percussion, and the blow is like that of the churn drill. It gives from three to six hundred blows per minute of as great a force as the drill-point will sustain. This is a greater degree of speed than has heretofore been obtained; consequently it drills more rapidly than it has been possible to drill before this machine was invented. It has an automatic feed, whereby the drill is fed forward just as fast as it cuts, and no faster; in hard rock, slowly; and in softer rock, more rapidly, precisely as it is required for its most efficient operations. At each blow the drill makes a part of one revolution so as to strike in a different place at each blow, as a miner turns his drill in hand-drilling. It contains only about one-fourth as many pieces as other power-drills, and does not depend for its action on any springs or pivots, liable to get out of order. It has no steam or air-engines attached to the drill with delicate parts and nice adjustments to be destroyed by the concussion and recoil of the blow. The construction is such that the connections between the drill and the driving machinery cease at the moment the blow is delivered, so that there is no recoil on the machine. On this account it can be set up in a mine on a plank, and does not require a car and heavy fastenings to hold it in place when in operation. It is estimated that a miner working at ordinary speed strikes twenty blows per minute; but this machine will strike five hundred in the same time, all of equal force, and all precisely square against the rock, thus doing the work of thirty men; and as four or five, and even more of these machines can be run in an ordinary-sized drift or tunnel, at the same time, the work of a hundred men can be done in the space where only four could work by the old method of hand-drilling. In form it is composed of two cylinders, the shorter and larger one being placed on the top of the other. In the small size, the large cylinder in which the drill moves is twenty-two inches in length and three inches in diameter;

it has flanges on the bottom by which it is secured when in operation. The other cylinder is seven inches long and five in diameter, and is secured on the top of the other at one end and parallel with it, the two being arranged somewhat like the barrels of an opera-glass. The upper cylinder revolves and communicates a reciprocating motion to the drill. The length of the whole machine, excepting the drill, is twenty-two inches, height eight, and the width five inches, excepting the flanges, which are ten inches; and the weight is seventy-six pounds. The drill is of the length required to reach the bottom of the hole, and may be of any length, from one foot to six feet or more. This small size enables it to be used in any tunnel, shaft, or place in a mine where a miner can enter, and is so arranged that it will bore a hole in any direction. It can be put in operation in a tunnel in a few minutes after a blast is exploded, and before the broken rock is removed, and while it is running the *débris* can be taken away. In a small space one man can operate one machine, but in a quarry where there is sufficient room he can manage several. It requires no more skill to run it than is required to operate a sewing-machine, and any miner of ordinary intelligence can learn to run it in a couple of days. It takes from one-half to one horse-power to run it. The motive-power can be steam, compressed air, water, or horse-power. In an ordinary-sized tunnel a tread-mill horse-power can be placed on one side of the track in the tunnel, and be moved in as the tunnel is driven forward. Where a steam-engine is at the surface, power can be taken from it to operate in the deepest and most extensive mine. Where a high fall of water can be obtained, a small wheel attached to the drill gives a very convenient power. This machine has not been completed for a sufficient length of time to have worn out any of them, but the first one made has drilled nearly two thousand feet, and is apparently as good as ever. All of the parts of the different machines are alike, so that one can be substituted for another, making it very simple to repair, in case a machine should get out of order.*

The Von Schmidt drill, noticed in my report of 1871, is being constructed for running a tunnel through the Sierra Nevadas for the Lake Tahoe Water Company, and though probably of great utility, its merits have not yet been tested by actual experiment.

The sale of mineral lands and quartz ledges.—The investment of large amounts of foreign capital in the purchase and development of our gravel mines and quartz ledges is to a great extent due to the operation of the various congressional acts throwing the mineral lands in the market for occupation and purchase, whereby title may be secured, instead of holding mining property, as formerly, subject to the insecure tenure of local mining laws and usage. The following table will show the extent to which our miners have availed themselves of these acts:

---

* This account is taken from statements made in behalf of the inventor. I do not wish to discredit it, but merely to say that I have not verified its claims by personal examination.—R. W. R.

List of mines in California surveyed from October, 1870, to January, 1872.

Name of mine.	Description.	County.	Name of mine.	Description.	County.
Little York claim	Placer	Nevada.	Bradley & Gardner	Placer	Placer.
Greenwood	Gold quartz.	El Dorado.	North Fork and Bear River claim.	...do	Placer and Nevada.
Davidson	...do	Do.			
Rock River	...do	Butte.	Pond & Constable	...do	Placer.
Nevada	...do	Nevada.	Briggs, Roberts & McGuire	...do	Nevada.
Pond & Co.'s claim	Placer	Placer.			
Henry Dorr's claim	...do	Do.	Franklin Miner	...do	Placer.
Spring Hill	Gold quartz.	Nevada.	Wildcat	...do	Amador.
Rough and Ready	...do	Plumas.	S. Bright	Gold quartz.	Do.
Eureka	...do	Do.	Oakville	Quicksilver.	Napa.
Pioneer Chief	Gold and silver quartz.	Calaveras.	Lincoln South	Gold quartz	Amador.
			Norambegna	...do	Nevada.
New York Hill	Gold quartz.	Nevada.	Luetje & Schwartz	Placer	Do.
Lincoln	...do	Amador.	Napa	Quicksilver.	Napa.
Zeile	...do	Do.	Red Hill	Placer	Tuolumne.
Oneida	...do	Do.	Crane's Gulch	...do	Nevada.
Cozzens's claim	Placer	Nevada.	Union Company	Gold quartz	Calaveras.
Slate Lodge	Gold quartz.	Do.	Hilton & McPherson	Placer	Tuolumne.
Wolcott claim	Placer	Do.	Baltic Gravel	...do	Nevada.
Hancock & Tibbits	Gold quartz.	Calaveras.	Picayune Gravel	...do	Do.
Kennedy	...do	Amador.	Saint Lawrence	Gold quartz.	El Dorado.
Maxwell	...do	Do.	Stanislaus	...do	Calaveras.
Amador	...do	Do.	Manhattan	Placer	Placer.
Blue Jacket	...do	Do.	S. H. Dikeman	...do	Nevada.
Socrates	Quicksilver	Sonoma.	Dutch Flat and Franklin	...do	Placer.
Railroad	Gold quartz.	Amador.	Sailor	...do	El Dorado.
Allison Ranch	...do	Nevada.	Olson & Donaldson	...do	Do.
Red Hill	Quicksilver.	Napa.	Washington	Gold quartz.	Plumas.
Cinderbergh claim	Placer	El Dorado.	Moorhouse	...do	Nevada.
Sliger claim	...do	Do.	Enterprise	...do	Calaveras.
Wesko claim	...do	Placer.	Richards	Placer	Tuolumne.
Newton	Copper	Amador.	Eclipse	Gold quartz.	Amador.
Staples & Co	Placer	El Dorado.	Reserve	...do	Calaveras.
South Yuba Canal Company.	...do	Nevada.	Santa Cruz	...do	Do.
			Spagnoli	...do	Amador.
Sargent & Jacobs	...do	Do.	Deadhead claim	Placer	El Dorado.
Georgia Slide	...do	El Dorado.	Powell	...do	Placer.
Town Talk	...do	Nevada.			

Total claims surveyed, 72. Placer, 34; quartz, 33; quicksilver, 4; copper, 1. Many of these have been patented, and on the others patents are pending.

Statistics of quartz-mills and mining ditches.—By the laws of California it is made the duty of the county assessors to return each year to the surveyor general of the State, among other statistics, a list of the quartz-mills and mining ditches in their respective counties. This duty is very imperfectly and carelessly performed, as will appear from an examination of the returns for the years 1867-'68 and 1869-'70, on page 15 of the Mining Commissioner's Report for 1870, where the errors in these tables are pointed out and commented on. I believe the following table, compiled from the surveyor general's report for 1871, though evidently more accurate than the two others referred to, is open to much criticism, and with this explanation I give it for what it is worth. It will be observed that there are no returns from two of our leading quartz-mining counties, Tuolumne and Plumas. According to Langley's Pacific Coast Directory, the former of these counties has 41 mills, with an aggregate of 445 stamps, and the latter 19 mills, with an aggregate of 233 stamps. Returns from those counties would have materially run up the total under the heading of "tons crushed," and shown, by comparison with the table of 1867-'68, a great increase in this branch of mining. The table of 1869-'70, (page 15, report of 1870,) is so manifestly erroneous under this heading in the amount credited to Placer County, as to be useless for comparison.

Table of quartz-mills and mining ditches, as reported by surveyor general of California, for the years 1870 and 1871.

Counties.	QUARTZ-MILLS.		MINING DITCHES.		
	No.	Crushed.	No.	Length.	Amount of water used per day.
		Tons.		*Miles.*	*Inches.*
Alpine	4	100			
Amador	27	70,360	32	405	5,450
Butte	18	8,000	32	180	41,000
Calaveras	29	*130,000	10	515	5,800
Del Norte			30	112	6,912
El Dorado	44	21,645	58	966	9,450
Fresno	2				
Inyo	4	600	1	3	40
Kern	11	500	2	5	1,728
Klamath	4	6,000	80	90	16,000
Los Angeles	2	5,560	3	25	800
Mariposa	34	24,000	66	66	Unknown.
Mono	5	390	2	13	720
Nevada	60	190,000	70	946	42,400
Placer	14	5,000	10	302	14,000
Plumas		No returns.			
Sacramento			2	45	4,000
San Bernardino	1		2	5	500
San Diego	2	900			
Sierra	21	31,000	49	208	21,000
Siskiyou	3	270	24	290	5,010
Stanislaus			3	12	1,000
Tehama			1	7	100
Trinity			110	340	45,773
Tulare	2				
Tuolumne		No returns.			
Yuba	4	2,000	18	69	4,500
Total	290	496,265	564	4,614	222,243

* Probably an error; placed too high.

Great discrepancies exist between these returns and the list prepared by Mr. Langley for his directory, published in last year's report, (pages 403–409.) For instance, Langley gives Amador County 36 mills, or 9 more than the surveyor general's report; Mariposa County 29 mills, being 5 less than the surveyor generals report. He gives Nevada 74 mills, to 60 in this table, and Placer County 32 mills, while the assessor's returns show only 14. So in Sierra County, Langley 36 mills, and the assessor returns only 21. I cannot reconcile this difference except on the hypothesis that the assessors have not returned some mills which have not run for several years, while Langley has kept them on his list.

THE SOUTHERN MINES.

The term "southern mines" is an indefinite one, but is generally understood to embrace the country between the Cosumnes River on the north and the Chowchilla River on the south, a distance of one hundred miles, and include the counties of Mariposa, Tuolumne, Calaveras, and Amador. In width the mineral belt extends from the eastern edge of the San Joaquin Valley to an average altitude of 2,500 on the Sierras, a distance of forty miles east and west, thus embracing an area of 4,000 square miles.

This region of country was the scene of the earliest mining operations in California, as the surface placers were here more accessible and productive than further north; and within its limits are found the once populous and thriving mining towns of Mokelumne Hill, Columbia, Sonora, and Mariposa. Its population, as estimated in 1851, by Abbé

Alric, then parish priest of Sonora, was not less than 50.000, nearly all of whom were engaged in mining.

This extensive territory is cut and eroded to great depths by four principal streams, running from east to west, and crossing the course of the ancient streams, viz: the Merced, Tuolumne, Stanislaus, and Mokelumne rivers, which, with their tributaries, have acted as distributors of the auriferous deposits, and carried the gold from its original place of deposit to the banks and bars which yielded such enormous sums during the early days of mining. The waters of these rivers have since been diverted into ditches and flumes for mining purposes, and their principal tributaries run dry in the summer, giving the country a parched and desolate appearance during the greater part of the year.

Within the area above described, scarcely a square mile can be found in which, even at this late day, a "prospect" cannot be obtained, although placer mining as a business has ceased to yield large returns, except in the opening of new ground at points where water has been lately brought in, or in the development of the ancient channels. Superficial placer, as well as river and bar-mining may be considered as practically exhausted, although operations are still prosecuted, on a small scale, on the limestone belt, and on the rivers during the short season of abundant water.

The principal mineral resources of the southern mines, at the present time, are vein-mining in the gold-bearing quartz belts and gravel-mining on the ancient channels. Within the limits of this region are found the rich and extensive copper belt, of which the town of Copperopolis is the central point; the Mother lode of California, one of the most thoroughly developed veins of gold-bearing quartz in the world; and a system of veins running in the granites, high up in the Sierras, which are comparatively undeveloped. Other systems of veins, unconnected with either of the above, and on which extensive and profitable mining operations have been carried on for several years, occur near the contact line of the slates and granites. All these lodes and systems of veins have a general course of northwest and southeast, and are evidently true fissures. At many localities, among which we may instance Bald Mountain, east of Columbia, in Tuolumne County, smaller veins occur, with an east and west course, generally in the slates. This latter class have the peculiarities of gash-veins, most of them "pinching out" at a depth of from fifty to one hundred feet, and being generally richest above the water-line and near the surface. This class is known as "pocket-veins." Although many so-called pocket-veins exist near to and parallel with the Mother lode, as at Angel's Camp, Calaveras County, where this class of mining is extensively carried on, it is generally believed they are outlying "stringers" of the main lode.

Copper mining.—An extensive belt of copper veins exists in the slates in the western part of Calaveras and adjoining counties, at an elevation above sea-level of about 1,000 feet. These veins were discovered in 1861, and subsequently developed to a depth of 500 feet, at Copperopolis, Calaveras County, maintaining their character for width of vein and grade of ore at lowest levels opened; but in 1867 operations ceased in consequence of the depreciation in the price of copper, and have not since been resumed. Up to the cessation of work the copper mines of California (the most productive of which are in the southern mining region) had exported 68,631 tons of ore and 847 tons of metal in bars. Up to that period nine smelting-furnaces had been erected in California at an aggregate cost of $236,000, all of which proved financial failures, and a majority of which were technical failures. When the copper mines of

Copperopolis were in operation, the price of freight to Stockton, the nearest shipping point, was $8 per ton, the distance being thirty-six miles. From that point it cost $2 per ton to San Francisco, whence it was sent to Swansea, Wales, at a further cost of $15 per ton. The total cost on each ton, mining, etc., included, when landed at Swansea, has been estimated at $50. Labor was then $4 per day. At this rate only first-grade ores paid a profit; and with increasing depth, second-rate ores predominated. The consequence was the suspension of operations and the decay of the once flourishing town of Copperopolis. The principal mines are the Union, owned by Glidden & Williams, of Boston, Massachusetts, and the Keystone, the title of which is in litigation. Both of these claims are opened to a depth of from 400 to 500 feet, and have expensive and powerful hoisting and pumping-works.

The reopening of these mines will depend on the price of copper. This has been steadily decreasing in England since 1864, but the prospects of the trade are now much better than they were one year ago. The production of Chile, in 1869, reached the enormous amount of 55,000 tons, but it has since diminished to 49,000 tons, with prospects of a continued diminution. On the other hand, the demand for this metal is increasing.

Should the Stockton and Copperopolis Railroad be completed to Copperopolis, it is believed that with present prices of labor ($2 to $3 per day) these mines could be re-opened, and the second-class ores, of which large quantities are in sight in the mines, could be extracted and shipped with profit, as it is believed that the price of freight to Stockton would not exceed $1.50 per ton, and the total cost of laying down ores in San Francisco would not exceed $20 per ton, instead of $32 per ton, as heretofore. The re-opening of the mines, however, would require a large outlay of capital, and perhaps involve the erection of new hoisting-works. The Stockton and Copperopolis Railroad was completed as far as Milton, at the base of the Sierras, when operations were suspended in consequence of the closing of the mines. Ten miles more would complete the road, but its completion will depend very much on the price of copper in the marts of the world.*

The Mother lode.—The Mother lode of California is a vein, or, more properly, a series of veins of quartz which has been traced on a longitudinal line, with occasional interruptions, for a length of about seventy-five miles, from Bear Valley, Mariposa County, to Amador City, Amador County. Throughout the entire distance it has a general northwest and southeast course, and an almost uniform dip to the northeast of eighty degrees.

Whether this singular formation is a "lode" or a mere accidental occurrence of a series of veins on a longitudinal line in the same belt of slates, is a question on which eminent mining engineers have differed. It has been discussed in the report of J. Ross Browne of 1868, and in my report of 1869. Mr. Skidmore considers it as a defined lode, believing that recent developments, at various points, have a tendency to confirm this theory.

The most southerly well-defined outcrop of this remarkable vein is at the Pine Tree and Josephine mines, on the Mariposa estate, at an elevation of about 2,500 feet above sea-level. From this point it takes a northwest direction, striking across the numerous spurs of the Sierras which form the divides between the Tuolumne, Stanislaus, and Moke-

* Since writing the above, I learn that operations have been resumed on the Union Mine, at Copperopolis.

lumne Rivers and their tributaries, and terminates in the foot-hills of Amador County, the most northerly deep-developed claims being the original Amador and Keystone at Amador City, although many locations between these mines and the Cosumnes River are supposed to be on the same lode. Beyond the Cosumnes the lode is not traceable.

Between its southern and northern extremities it is frequently broken up and lost, (invariably so at the intersection of the principal rivers,) making its appearance again at a distance of several miles, frequently in the form of a solid wall of quartz on the summits of isolated hills on the line of its strike, these croppings being visible for many miles. The most prominent of these hills are Piñon Blanco, Quartz Mountain, Whisky Hill, and Carson Hill. At these points the lode has widened with the "blossom" of the mountain, and presents the appearance of a system of parallel veins separated on the surface, and to an indefinite depth, by "horse" matter, composed of nearly equal parts of slate and broken quartz. Locations have been made in the majority of such instances on the croppings, parallel to each other, one on the hanging wall and another on the foot-wall of the lode, these walls being sometimes separated, as at Quartz Mountain, Tuolumne County, by a distance of from 200 to 250 feet; but recent deep developments at various points would seem to indicate a tendency of these walls to narrow, which, at depths of from 1,000 to 1,200 feet, and in some cases less, would shut out the "horse" and develop a permanent fissure of from fourteen to eighteen feet in width.

The principal points at which mining has been prosecuted on the Mother lode, are Bear Valley, Princton, and Mariposa, in Mariposa County; Quartz Mountain and vicinity, near Jamestown, Tuolumne County; Rawhide and Tuttletown, in same county; Carson Hill, Angel's Camp, and Paloma, in Calaveras County; and Jackson, Sutter Creek, and Amador City, in Amador County; the deepest development having been made at Sutter Creek, where, at the Amador mine, a shaft has been sunk, and levels opened to a depth of 1,300 feet.

The entire length of locations made on the Mother lode is estimated at 180,000 feet, equal to half the distance between its northern and southern extremity. Many of these locations, however, run parallel to each other, and the ground continuously located would not exceed 100,000 feet, while the ground now in process of development (including only those claims on which work has been done in 1871) will not exceed 40,000 feet, exclusive of the Mariposa estate, on which operations are temporarily suspended owing to financial difficulties and litigation.

The longest break of the lode is between Angel's and Jackson, a distance of twenty-three miles, on which only one mine, the Paloma, near the south bank of the Mokelumne River, is generally acknowledged to be on the lode, though recent discoveries tend to prove the continuity of the lode between these points. At various other points the lode "dives", for several miles, and at one point, between Whisky Hill and the Rawhide mine, it is covered by the lava flow which constitutes Table Mountain.

Throughout the entire course of the lode we find many instances of failure, manifested by abandoned works and idle mills, but these are to be attributed either to mismanagement or to the injudicious location of works at points where no pay-chimneys exist. The early locations were made at any point where croppings appeared, and deep shafts were sunk, and mills erected on barren ground, without any effort to find the chimneys of the vein. In this way thousands of dollars have been uselessly squandered by men utterly ignorant of mining, and these

H. Ex. 211——4

monuments of their folly have discouraged those who otherwise would have invested in quartz mining. Another cause for the stagnation of mining on this lode is the improvident manner in which work has been carried on, resulting in the exhaustion of the surface deposits, which were worked by open cuts at various places down to the water-line, when operations ceased until shafts could be sunk and hoisting-works and pumping-machinery erected, the owners of the ground having committed the common error of failing to keep their ground developed in advance of their milling capacity. Perhaps a third reason may be found in the superior attractiveness of the mines of the neighboring State of Nevada, where fortunes are made (and lost) with greater rapidity in mining operations.

The amount of ore crushed by mills on the Mother lode has been less for this year than for several years past. This is owing to a combination of circumstances not likely to occur again. The Mariposa estate, containing several of the most productive mines, has been involved in a litigation which has temporarily suspended operations. An unexampled drought prevailing over the State has closed many mills for want of water, and others are only running half their stamps. These claims are supplying ore far in excess of their present means of crushing. In addition to this a "strike" took place during the month of April, in Amador County, and the Oneida, Keystone, Amador, and other leading mines were closed for a period of nearly three months, resulting in a loss to the owners of nearly $2,000,000. This latter difficulty has now been adjusted, and the present season promises to give an ample supply of rain.

The prospects for the future are encouraging. At all points where great depth has been attained there has been a steady improvement in the quantity and quality of the rock. At Sutter Creek, the Amador mine has attained a depth of 1,300 feet, disclosing a vein of from 12 to 14 feet in width. The quality of the quartz here has improved with each successive level below a depth of 500 feet. At the Oneida, Keystone, and other deep mines, the same encouraging features are met with, while many claims heretofore partially opened are now erecting mills and powerful hoisting-works. The development of the Paloma mine, owned by ex-Senator Gwin, near Mokelumne Hill, and of the Angel's Quartz Mining Company's mine at Angel's Camp, both of which, at lowest levels—400 to 500 feet—have opened rich and extensive zones, is exerting a strong beneficial influence on quartz mining, which is already being felt in the investment of home capital in this heretofore neglected branch of business.

Quartz veins in the granite belt.—In addition to the Mother lode, this region possesses other quartz mines which are yielding large amounts of gold. At an elevation of 4,000 feet, in the Sierras, an extensive series of veins has been opened in the granites, the most noted of which is the Confidence, sixteen miles east of Sonora, Tuolumne County, a mine scarcely known outside of Tuolumne County, yet producing, with forty stamps, from $30,000 to $40,000 monthly. Other claims of like characteristics are being developed to the north and south of the Confidence, but whether they belong to the same system, of which there is a strong probability, remains to be demonstrated. Among these is a recent discovery in the upper part of Calaveras County. The veins here present many striking features of interest. Like many of the most noted veins of California, they occur in fissures which have been opened in the earth's crust by the rending asunder of the rock formation across its stratification. The country rock is slate. This contains numerous dikes of traps, porphyritic green-stone, etc., occupying fissures which run transversely across the slate. There are several parallel fissures, nearly vertical, having a course of N. 40° E., with very solid and smooth walls, and from 5 to 12 feet wide. In these occur the quartz veins, the quartz occupying only a portion of the space, (3 to 9 feet,) the balance being filled by an accompanying vein matter differing from the country rock, and not found outside of the fissure walls. The whole of this vein matter is full of base metals, particularly the sulphurets of iron, deposited in a way that indicates a previous state of solution or possibly vapor ; for besides being disseminated throughout the body of the rock, the faces of broken pieces, which had no seams visible to the eye, are often found coated with particles of metal, forming flakes which can be removed with a knife-blade. All of this material contains some gold, but the pay rock proper occurs in chutes of a peculiar kind of quartz, which is held by many of the miners to belong to the true chimneys of the precious metals. It is so thoroughly impregnated with the various base metals, especially the sulphurets of iron, lead, and zinc, that not an ounce of it can be found destitute of these. The gold is diffused in fine particles through the ore, as if an element of its composition. The ore of these lodes is of high grade.

Further west, in Calaveras County, we find the quartz mines of Railroad Flat and West Point, and in Amador County a group of promising mines, near Volcano. At these places hundreds of veins exist, yielding high-grade rock, but few of which are developed below the water-line. The future of these districts depends on the development of a few claims, considered as representative mines, on which work-is now being prosecuted with the best indications of success. The discovery of pay rock in a mine, such as the Petticoat at Railroad Flat, on which a deep shaft is being sunk, will have a tendency to open more than fifty claims in the same vicinity.

Gravel mining.—The gold-bearing gravel deposits of the southern mines are not as extensive in area and depth as those of the central mining region, (described in my report of 1871,) neither is the topography of the country so favorable for the opening and draining of ground; but, on the other hand, the gravel will yield a larger amount of gold per cubic yard. This opinion is based on personal observation, and on the carefully prepared tables furnished by Mr. J. Rathgeb, of San Andreas, which will be found in this report.

The modes of occurrence of gravel deposits here are various. They occur, first, in well-defined ancient river channels, under a capping of lava which has filled the rivers of past ages ; second, in isolated mounds or hillocks, evidently the remains of such channels, which, being unprotected by a covering of lava, have been broken up by the action of air and water; third, in basins or flats which have received and held the wash of these disintegrating river beds; and, fourth, in low, rolling hills, near the base of the Sierras, and beyond the reach of the lava flow. The richest deposits have been found on the flats on the east side of Table Mountain, at places where bars evidently existed in the ancient river, and the lava flow sought the deep channel, forming an elbow or curve on which the lava crust was very thin, or sometimes entirely denuded.

Table Mountain.—The most remarkable feature in the landscape of the southern counties is the great Table Mountain. This is a flow of basaltic lava covering an ancient river bed several hundred feet higher than the modern river—the Stanislaus—by which it is crossed through deep cañons at two points—Abby's Ferry and Byrne's Ferry. The basaltic matter has a width of from 1,700 to 2,000 feet, and a thickness of about 150 feet; its elevation at Jamestown is about 500 feet above the surrounding country. Beneath this capping of lava is a stratification of sandstone sixty to one hundred feet in thickness; and underneath this, on the bed-rock of the ancient river, is found the cemented gravel from one to five feet in thickness, and in some localities exceedingly rich in gold. The distinctive tabular appearance of the mountain can be traced from Murphy's, in Calaveras County, where it has an altitude above sea-level of about 3,000 feet, to Knight's Ferry, Stanislaus County, where it is lost in the foot-hills of the Sierras on the eastern border of the San Joaquin Valley, a distance of thirty-six miles. At its upper end it is much broken up, but in Tuolumne County, below Abby's Ferry, continuous stretches of many miles occur with a gentle grade. Below Byrne's Ferry, where it is again cut by the Stanislaus, it is much broken, and is finally lost in the foot-hills.

Prof. J. D. Whitney, in his General Geology of California, (vol. 1,) says of this mountain:

On approaching Table Mountain and examining the material of which it is composed, and the position which it occupies, it is seen at once that it is a vast lava flow, of which the upper surface remains very nearly at the level and with the form which it originally had at the time of its consolidation, while its edges and the surrounding country have been denuded and washed away, so that the topography of the region is entirely different from what it once was—in fact, it is almost the reverse of it. No one can deny that a stream of melted lava, running for forty miles down the slope of the Sierra, must have sought and found a depression or valley in which to flow, for it is impossible that it should have maintained, for any distance, a position on the crest of a ridge. Nor could the valley of the Stanislaus, now two thousand feet deep, have existed at that time, for this flow of lava is clearly seen to have crossed it at Abby's Ferry. The whole face of the country must, therefore, have undergone an entire change since the eruption took place, during which this mass of lava was poured out. The fact that the lava flow of Table Mountain took place in a pre-existing valley is not only capable of being demonstrated on general principles, but is confirmed by what has been shown, by numerous excavations beneath it, to be the character of the formation on which it rests.

Professor Whitney estimates the amount of denudation which has taken place during the period since this lava flow took its present position at not less than three or four thousand feet of vertical height. The excessive hardness of the basaltic capping of the mountain has protected it from any appreciable amount of denudation and erosion. This is manifested by the scarcity of vegetation on its summit, where there is no soil of sufficient depth to support more than a few stunted shrubs.

The discovery of the auriferous character of the bed of this ancient river was made accidentally by some placer-miners working in the vicinity of Shaw's Flat, in 1854, at a point near the rim rock of the channel where the lava capping had been denuded. In the excitement which followed almost the entire length of the mountain was located, and hundreds of tunnels run to strike the channel. From the best data at our disposal we estimate the total length of tunnels run in this mountain at forty thousand feet, at a cost of not less than $800,000. The number of feet located was between sixty thousand and eighty thousand, many of these locations being parallel on the supposition that there were two channels. This idea probably originated from the fact that the river frequently changed its channel, as do modern streams. At the present time the wild spirit of speculation, which induced the expenditure of such large sums in running tunnels, has subsided, and mining is only carried on at a few localities, but generally with handsome returns. Some details of present mining operations in this mountain will be found in the description of Hughes's claim, (formerly Maine Boy's tunnel,) and the ground of the Table Mountain Tunnel Company, under the heading of "Tuolumne County."

Gravel deposits in the foot-hills.—The La Grange Ditch Company.—Extensive auriferous gravel-beds, apparently of secondary deposition, are found in the low rolling foot-hills of all the southern counties, but generally in localities remote from water, and in consequence their development has been retarded. During the past year thousands of acres of ground of this character have been purchased and located by San Francisco capitalists, and vast projects are under way for bringing water to them by tapping the rivers high up in the mountains. Of this character is the enterprise known as the "La Grange Ditch Company," which is one of many having like objects. This company own between four and five thousand feet frontage on the Tuolumne River, near the town of La Grange, Stanislaus County, formerly known as French Bar, their ground running back from the river a distance varying from one-half mile to one mile. The bed-rock lies at a sufficient height above the Tuolumne River to make a tunnel necessary for opening their ground. The gravel is soft and but little cemented and will wash easily. The banks (or gravel-beds) average about one hundred feet in thickness and prospects largely from top to bottom, several pan tests having shown the extraordinary average of three to five cents to the pan. The company are now building a ditch, taking the water out of the Tuolumne River, at a place known as Indian Bar, seventeen miles above their claims. This ditch is six feet wide on the bottom, eight feet on the top, and four feet deep, with a fall or grade of from six to ten feet per mile, and will carry four or five thousand inches of water, delivering it at a height of over two hundred feet above the level of the Tuolumne River. The ditch was commenced about the middle of July, 1871, and one thousand men have been constantly employed in its construction; eight hundred of these being Chinamen at one dollar per day, the white men of course receiving higher wages. The total cost of this ditch is estimated at $200,000, and it is expected to be complete in time for the company to avail themselves of the rise of the river early in 1872. Mr. Edmund Green, the superintendent, has adopted a novel method of taking the water from the river by which he avoids the risk incidental to a high dam. This is by digging a cut along the side of the river large enough to contain a box flume which will be covered with rock and dirt, leaving the bank of the river in its natural state. But little fluming will be used. Around rocky points a stone wall is commenced far enough down the banks to get a perfectly solid foundation. This is continued up to the top of the ditch and is made four or five feet in thickness, then an inner wall two feet wide and two feet distant from the outer one is built four feet high from the bottom of the ditch; clay is then tamped between these two walls so as to make it water-tight, the water running in the ditch inside both walls. They have in addition a winter ditch with a capacity of three thousand inches of water, taken from Dry Creek, three miles distant from the mines. The company will commence washing early in 1872 with five four-feet flumes, using four eight-inch nozzle improved hydraulic pipes under a pressure of one hundred and fifty feet. Should this enterprise prove a success, of which there is no reasonable doubt, capital will seek investment in like projects, and many thousands of acres of equally valuable ground will be developed in the southern mines.

Other gravel deposits.—Turner's and Kincaid Flat.—Many auriferous gravel deposits of great richness exist in various parts of the southern mines, particularly near the limestone belt, in basins or flats where, from the nature of the surrounding country, drainage is impossible except by the construction of long and expensive tunnels. These gravel-beds, like the detrital matter in the foot-hills, are probably the result of secondary deposition in a recent geological epoch—the effects of the distribution by water of the ancient channels, and the denudation of the surrounding country. This theory is founded·on the fact that in these basins the remains of the mastodon and elephant are found in great abundance, while in the old river-beds they are rarely discovered. These basins were worked in early times as deep as was possible by means of shafts, whims, and pumps, but as the bed-rock was approached the water was found to be an insurmountable obstacle, and they were temporarily abandoned. Subsequently, drain-tunnels of great length were run at various places where these basins occur. Two of these enterprises, the first in contemplation, the second nearly completed, will be briefly described here as an illustration of this branch of mining.

The Turner's Flat and Table Mountain Mining Company early in 1870 acquired by purchase and location five hundred acres of ground at Turner's Flat, near Jamestown, Tuolumne County, this location giving them also 5,000 feet in length on the Table Mountain channel; the "flat" being on the eastern side of Table Mountain, and their location running down on the western side of the lava-covered mountain toward the Stanislaus River. They are now making application for a patent to this tract and negotiating for capital to open their ground. Turner's Flat is an extensive deposit of auriferous gravel, which has proved very rich, and is of unknown depth. Between the years 1853 and 1857, when in the possession of former owners, several shafts were sunk in the flat, near the line of the lava, to a depth of eighty feet, when a large quantity of water was encountered, of such volume as to resist the power of the best pumping machinery then in use. About the year 1855 or 1856, an English company acquired large interests on the flat, and endeavored to drain it by running a tunnel from Slate Gulch, a small stream tributary to Wood's Creek, lying far to the east of Table Mountain. This tunnel, if completed, would have been about three miles in length, but after spending $30,000 it was ascertained that an error had been made in the surveys, and the completion of the tunnel on the grade they were running would have brought them to the surface instead of the bottom of the basin, and the work was abandoned. No tunnel except one from Wood's Creek seemed feasible for the drainage of this ground, and this would have been nearly five miles in length with a light grade. This valuable property then remained·unoccupied and undeveloped for a period of nearly ten years, when the present company acquired the ground and decided on a new point of attack by opening the bottom of Turner's Flat from the west side of Table Mountain, where the formation is very precipitous, instead of on the eastern side, where it slopes gently to Wood's Creek. This tunnel will enter the west side of Table Mountain about three hundred feet below its summit and about one hundred and fifty feet below the lava line. It is estimated from surveys that its total length will not exceed three thousand feet, and at that distance, with a grade of ten inches to twelve feet, it will tap the gravel deposits of Turner's Flat at a depth of one hundred and twenty feet. A deep gorge, setting back on the western face of Table Mountain, opposite the flat, greatly facilitates the construction of this tunnel. The Turner's Flat gravel deposit is evidently an outbreak from. the Table Mountain channel, and had yielded about $1,000,000 before operations were suspended. The gravel on the bottom is said to have paid as high as $5 per bucket. The tunnel will be laid with flume, and the grade will be sufficient to effectually break up the gravel and release the gold.

The Kincaid Flat Mining Company is a San Francisco company, who have been engaged for several years in an enterprise similar to the above, and are now approaching its completion. Kincaid Flat is a basin-like depression, surrounded by low hills, situated a few miles from Jamestown, Tuolumne County, on the eastern verge of the great limestone belt, and contains about one hundred and fifty acres of gravel of unknown depth. Mining was commenced here shortly after the discovery of the placers on the limestone belt, and immense sums were taken from the flat. Rumor says that one claim fifty-five feet square yielded $100,000, and that the total product of the flat was not much less than $2,000,000. While any estimate of the yield of a tract of mining ground in early times must be accepted with due allowance for local exaggeration, it is undoubtedly true that this ground was notoriously and fabulously rich. The present company having acquired this ground several years since, commenced a tunnel from Sullivan's Creek, the nearest depression, and have been working on it ever since. After running an open cut two hundred and fifty feet, they commenced a tunnel, and had run one thousand three hundred and thirty-five feet up to November, 1871. They had four hundred feet more to run to strike the bed-rock of the basin. Their tunnel is about five by seven, and for one thousand three hundred feet runs through limestone, when they encountered a trap-dike of excessive hardness, which will materially delay its completion. The tunnel has already cost $60,000, and is in all respects a model piece of work.

The labor has so far been done by hand, but it is probable it will be completed by machinery. This tunnel will not drain the flat below a depth of eighty feet from surface. The construction of such an expensive work shows the confidence of our local capitalists in the undeveloped resources of our mines.

At Mokelumne Hill, Calaveras County, extensive operations are carried on in the ancient channels, which are not yet exhausted. They will be noticed at some length in this report. Near San Andreas, Calaveras County, and in the vicinity of Angel's, Murphy's, and Vallecito channels, mining will be prosecuted with vigor during the coming year. At Byrnes' Ferry a New York Company have tunneled Table Mountain and are said to be taking out good pay. Extensive gravel operations are also in progress at Garrote, in Tuolumne County, and in various parts of Mariposa County. There is much similarity in methods of mining in these various localities, and my time and space will not admit of more than general descriptions of some of the representative claims, selected on account of the magnitude of their operations. These will be found scattered throughout this article.

The formation of gravel deposits.—As frequent mention has been made of gravel operations in isolated patches which were formerly parts of ancient river-beds, I present the section of a shaft in a claim near Angel's Camp, Calaveras County, in which a human skull was found at a depth of one hundred and thirty-two feet, imbedded in a stratum of gravel five feet in thickness. The figures on the left-hand side represent feet, and show the thickness of the different strata. The skull was found in the five-foot gravel-bed above the "red lava."

40	Black lava.
3	Gravel.
30	Light lava.
5	Gravel.
15	Light lava.
25	Gravel.
9	Dark brown lava.
5	Gravel.
4	Red lava.
17	Gravel.
153	Slate.

When the State geologist shall have completed his labors and issued his reports through the press, we shall be in possession of many facts which will elucidate and perhaps reconcile many apparent contradictions now to be observed in the formation of these deposits. Mr. Goodyear, one of Professor Whitney's assistants, in a letter to the Mountain Democrat, thus foreshadows a new theory:

With reference to the general question of the origin and distribution of the auriferous gravel itself there have been no end of theories, and every agency that is capable of moving rocks, from salt-water oceans to enormous glaciers and floating icebergs, has been called in to account for the phenomenon. With reference to most of these theories, it will only be necessary here to state the fact that no well-informed man can study carefully for himself in the field over any considerable extent of this country the character and distribution of the gravel, and the detailed structure of the banks with the fossils which they contain, without being led to the irresistible conclusion that there is but one possible agency which is at all capable of satisfactorily accounting for the complex and intricate phenomena, and that this is to be found in the action of fresh and running water. This agency was involved in the old "blue-lead theory" which has been for so many years a favorite, not only among the best-informed practical men, but among leading geologists and mining engineers as well. The gist of this theory may be stated in a few words by saying that it involved the supposition of the former existence here of a great river with its branches, the main trunk of this river being supposed to hold for one or two hundred miles, if not more, a general southeasterly course, nearly parallel with the present main crest of the Sierra, before the mountains were uplifted. But the detailed and extensive explorations of the gravel mines which have been made during the past two years by the State geological survey, have developed among other things the fact that this theory, too, is not only inadequate to account for the complex facts, but that it is not unfrequently in direct conflict with them. The questions involved are extremely complex, and it is no wonder that in the absence of systematic and extensive investigation of the facts in the field, the theories at first propounded should have been wide of the mark. And our own work in this direction is by no means as yet complete. What the true theory is, therefore, it would be premature for me to attempt to develop here. But it is rapidly assuming shape in our minds, and the whole subject will be thoroughly discussed in the forthcoming volumes of the geological report of Professor J. D. Whitney, provided the coming legislature shall furnish the requisite means for their completion and publication.

The limestone belt.—Placer mining.—Another prominent feature in the geology of this part of the country is the limestone belt, on which are found the early placers, noted for their immense yield from 1850 to 1855. This belt runs through all the southern mining counties, and can be traced continuously for nearly one hundred miles. Its course is northeast and southwest, and its width varies from half a mile to three or four miles, in some places contracting, at others expanding. At Sonora it is narrow, while at Shaw's Flat and Columbia, a few miles farther north, it is several miles in width. Throughout its entire length it was noted for the richness of its placer deposits, which were, however, merely superficial, rarely exceeding in depth six or eight feet, except in places where the limestone formation contracts, and at these points it has been worked to a depth of from forty to one hundred feet by following and cleaning the crevices. Throughout its length the limestone bed-rock has been deeply worn by the action of swiftly running water carrying bowlders and *débris*, which have cut and carved it in the most singular and fantastic shapes to a depth of many feet. In many places remarkable underground caverns of unknown extent are found. One of these exists near Cave City, El Dorado County, of many acres in extent, which has never been thoroughly explored, although discovered as early as 1852. The rich flats near Columbia and Springfield, when discovered by the early prospectors, were covered with dense growths of pine, and the entire face of the country has been so changed by mining operations as to be unrecognizable to the miner of '49-'50. The richest portion of the lime-stone belt has been found on the east side of Table Mountain, in Tuo-lumne County, and it is very probable that the placers owed much of their wealth to the scattering and distribution by water of portions of the Table Mountain channel not protected by lava. There is strong evi-dence of the correctness of this opinion near Springfield, Columbia, and Shaw's Flat—all of these places being but slightly below the level of the ancient channel at its exposed points. These towns, and Sonora, Jamestown, Montezuma, and Chinese Camp, owe their existence to this class of placers. Near the head of Table Mountain, where the basins and crevices on the limestone belt are deep, we find the towns of Mur-phy's and Vallecito, in Calaveras County, where mining is still prosecuted on a small scale by means of whims and pumps, with a fair profit, but the ground remaining to be worked is limited. At these points the Table Mountain is much broken up and loses its identity as a continuous range. The "flats" between the mound-like elevations have proved exceedingly rich, but all efforts to drain them have proved pecuniary failures in con-sequence of the great length of tunnels required. The towns of Colum-bia, Springfield, and Sonora, once the most populous of the southern mines, were built on the best placer ground, and town lots are now more valuable for mining purposes than for business and residence. Placer mining in their vicinity has been virtually abandoned to the Chinese, who are satisfied to work ground which has been passed through the sluice-boxes two or three times. It often happens in these towns that a lot with a brick house on it is bought and the house torn down, merely for the purpose of taking the gold from the ground. As some of these towns are very much decayed, property of this kind can be bought for prices which leave an ample margin of profit after sluicing out the ground. In confirmation of this statement, the following item from the Sonora Democrat of April 15, 1871, is given:

The fine brick store occupied so many years by Condit has been taken down with the store-room next south of it, and now the brick building next north is being taken down for the purpose of mining the ground under it. Every day pieces of quartz are found that are very rich in gold. The store was built on ground that had not been mined; it is proving so rich now that a mining hole will soon take the place of the building. Pieces containing from one to three hundred dollars each have been taken out within a week. One week's washing has averaged $10 per day to the hand em-ployed, running one wheelbarrow, and only as yet washing top dirt. Several pieces were found ranging from one to three ounces; 12 wagon-loads (to test the claim before erecting sluices) paid him $150. In the rear of this same building, a few years since, one 25-pound chunk was found, and several of nearly that weight. A dog, digging for a gopher, at one time scratched out a piece of quartz for which Mr. C. obtained $70. Small pieces of float quartz are now daily found in this claim containing from $1 to $10 in free gold.

In fact, it would doubtless prove a paying investment to buy the land on which several of these towns are situated, tear down the buildings, and sluice off the ground. The town of Sonora, however, has other resources, and is now experiencing a return of its former prosperity through its great agricultural and horticultural interests and the numer-ous quartz-mining districts of which it is the basis of supply.

MARIPOSA COUNTY.

This is the most southerly of the counties included within the limits of the "southern mines." The placer interests of the county have been neglected for years, but this branch of mining will be revived with the introduction of water, for which purpose several ditches are being con-structed.

The quartz interests show marked improvement, manifested by the opening of many mines abandoned for years, the discovery of new ledges, and the investment of San Francisco capital in developed mines. The mines belonging to the Mariposa estate are so situated that they require water for milling purposes; the mines are proven to be abundantly rich and comparatively inexhaustible, but cannot be worked successfully the year round without taking the ores to the Merced River. They now propose bringing the river to the mines, and the long-talked-of Mariposa ditch, located in 1852, is no longer a myth but a fixed fact, as the work is being actively prosecuted by the company.

Of the celebrated Mariposa estate and its mines there is nothing this year to be said. A new set of legal complications has paralyzed operations, but will terminate, it is believed, in a complete reorganization free of incumbrances.

The county of Mariposa possesses many valuable claims outside of the boundaries of the Mariposa estate, some of which are among the most productive in the State. Among these we may instance the Ferguson mine, the Eclipse, and Hites' Cove mines.

The Ferguson mine has recently been sold to an English company for the sum of $100,000. It is situated on the main fork of the Merced River, not far distant from the Yosemite Valley. The company own 3,700 feet, and intend erecting powerful machinery, as there is water-power enough to drive an unlimited number of stamps. This mine has been in successful operation for over ten years without levying an assessment, and during the present year has paid dividends of $4,000 per month with an eight-stamp mill. The vein is from 1 to 8 feet thick, averaging $2\frac{1}{4}$ feet. The average pay of the rock is now about $44 per ton. Their tunnel is in 1,100 feet, on a level with the mill, to which the rock is easily taken by car. At the back of the tunnel they have sunk a shaft 100 feet deep; at this point it is 800 feet below the surface.

The Hites' Cove mine (described in report for 1871) is engaged in driving a cross-cut which is expected to reach the lode in May, 1872. This cross-cut will give them three hundred feet of backs. This mine has always paid large returns to its owners.

The Washington mine has a large vein and supplies a forty-stamp mill. It has never levied an assessment, but has paid for all improvements from its opening.

The Francis mine is a notable instance of a successful mining operation without capital. Mr. Francis, the recent owner, purchased the mine six years since on credit, and erected on it a five-stamp mill. The mine and mill soon yielded its owner a large advance on the original cost. The mine has recently been purchased by parties in San Francisco who propose erecting a sixty-stamp mill, as the mine has been developed sufficiently to warrant it, having three years' ore in sight and opened up. The lode is from 4 to 12 feet thick, and is traceable for miles through the country—is opened by an adit level driven on the mountain side on the lode with constantly increasing backs. The rim will pay in free gold milling process from $12 to $15 per ton, with from 10 to 60 per cent. of sulphurets, worth from $100 to $300 per ton. The pay chute is of unusual length, and is already traced and tested for 1,900 feet. The company have 3,000 feet of lode.

There are many other mines in the county in various stages of development, but their opening has been retarded by the unfortunate condition of the Mariposa estate, which has been erroneously supposed to contain the best mines in the county, and which, considered abroad as an unsuccessful example of mining on a large scale, has proved detrimental to all attempts to induce capitalists to invest in this county. The condition of this estate is to be attributed to the mismanagement incident to the control of all large and unwieldy corporations whose stock is elevated or depressed at the pleasure of operators who never saw the property of the company, and feel no interest in its development beyond the temporary enhancement or depression of the value of shares in the stock market for the purposes of speculation.

This state of affairs, however, will not be of long duration, as the resources of the county in gold-bearing quartz have attracted the attention of some of the leading mining operators of San Francisco, who have purchased several mines and are now engaged in the development of the long-neglected wealth of this county.

TUOLUMNE COUNTY.

This county, adjoining Mariposa on the north, has an area of 915,000 acres, of which but a small proportion is under cultivation, although the soil of the western portion of the county, from the foot-hills to an elevation of 2,500 feet in the Sierras, is unequaled in the productiveness of its orchards and vineyards, both as to the quantity and flavor of its fruits. The county has promising quartz interests which have for a few years past been dormant, but are now reviving under the impulse of successful operations in this branch of mining, while gravel-mining at various places on Table Mountain has proved, under good management, both safe and profitable as a business.

Chinese Camp.—The town of Chinese Camp is situated in a basin or flat, east of Table Mountain, at an elevation of 1,300 feet above sea-level, the Mother lode lying about one mile east of the town. The present population does not exceed five hundred persons, of whom three-fifths are Chinese. The main interest of the place has been placer-mining on the flat on which the town is situated. The ground has been worked over to bed-rock, a depth of three to five feet, and the best part exhausted, although the Chinese still carry on mining on a small scale when they can find dirt paying one dollar a day to the hand, of which enough remains to last for several years. The supply of water is limited, rarely more than one hundred inches being available, and this is sold at fifteen cents per inch for ten hours' use. Great difficulties were met with in obtaining an adequate supply of water for this place, as a deep ravine (Wood's Creek) ran between the town and the nearest ditch. These were finally surmounted by running a pipe, eleven inches in diameter and one mile in length, from the Phœnix Water Company's reservoir across the ravine of Wood's Creek to a hill above the town. The head of the water-supply (at the reservoir) is two hundred feet higher than the discharge-box on the hill above the town, and an intervening depression—the ravine of Wood's Creek—having a depth of seven hundred feet, measured from the discharge-box, has been overcome, not by the construction of an expensive trestle-work, as was formerly the practice in carrying water over depressions, but by laying the pipes on the surface and relying on the pressure. Two hundred inches have been safely run through this pipe. The gold found here is noted for its fineness, and is probably the result of the breaking up of old channels, of which remnants are found on the edges of the basin. The pay-dirt extends entirely across the flat or basin, a distance of nearly three miles. Some spots proved very rich, but were quickly exhausted, and the future of this class of mining is not promising.

Detached masses or patches of cemented gravel are found on the summits of a few mounds or spurs of hills in this basin. These patches, on account of their hardness, seem to have resisted the disintegrating influences of air and water, which have swept away and scattered the original deposit of which they formed a part, depositing the released gold in the adjacent gulches and streams and over the flat. They are but isolated monuments, indicating the existence, ages since, of an extensive belt of gravel deposited by the action of water, but whether in the channel of a running stream, or in a lake-like depression, or whether these deposits were formed from the "wash" of the ancient river now covered by the lava of Table Mountain, (which is probably the case,) cannot be ascertained without close observation and patient investigation. One of these patches, situated immediately to the east of the town, and about one hundred and eighty feet above the level of the basin, covers an area of about ten acres and proved exceedingly rich, most of the pay being found in a blue streak, varying from one to twenty inches in thickness and lying immediately above the bed-rock. The depth of the gravel in this tract did not exceed thirty feet, and the best surface-diggings of the vicinity were found on the slope to the east of this mound and in the bed of Wood's Creek. These detached masses of gravel are said to extend southerly into Mariposa County, where they exist in larger bodies, and will be worked immediately on the completion of several ditches now in process of construction. Toward the north we find the great surface deposits of the limestone belt which probably owe its origin to the same causes.

The quartz ledges in the vicinity of the town are all supposed to be on the Mother lode, which lies immediately to the east. They are, beginning with the most southerly and proceeding north: The Clio, 10 stamps—not running for two years; Orcutt's mine (supposed to be on a spur of the Mother lode)—a very rich vein of decomposed quartz; Eagle, 10 stamps—operations suspended in mill pending completion of drain tunnel; the Shawmut—idle for two years; belongs to a Boston company and is closed on account of mismanagement and defective machinery.

Two miles northerly from Chinese Camp, and on the edge of the same basin, is situated the decayed and deserted town of Montezuma, which was once noted for its rich placers. Wood's Crossing, on Wood's Creek, one mile south of Jamestown, is noted as being the first ground worked in the southern mines, early in 1848. The creek here was very rich, and was worked by the Indians, of whom there were many here at that time. It is said that the first traders who came into this part of the country bought gold-dust from the Indians in exchange for its weight in beads, raisins, &c.

Jamestown.—Between Montezuma and Jamestown the country is broken up by rolling hills of no great elevation. The Mother lode is intersected and cut half a mile south of Jamestown by Wood's Creek, a stream carrying but little water in the summer, but very turbulent in the rainy season. At this point the croppings of the lode have a width of from twelve to sixteen feet, compact and boldly defined, with numerous spurs and parallel veins on the east side, but the quartz is barren at the surface. These parallel veins have proved very rich but not continuous. Abandoned excavations and tunnels show that they have been followed till they "pinched out" or became merged in the main lode. Near here, on the line of the Mother lode, are situated Quartz Mountain and Whisky Hill, famous for both rich quartz mines and great failures. The most noted mines are the Golden Rule, the App mine, the Heslep mine, and the property of Rosencrans, Preston & Co. on Whisky Hill.

About one mile and a half west of Jamestown the Mother lode has been crossed by the lava flow of Table Mountain, and makes its appearance again in prominent croppings on the west side of the mountain at the Rawhide mine, one of the oldest locations in the county.

The past history of Jamestown is that of all mining towns relying mainly on placer diggings—a hasty growth and a slow but sure decay. The placers in this vicinity and the bed of Wood's Creek were of great richness, and for a few years the sales of gold-dust averaged nearly $1,000 per day. The placers were of the same character as those of Chinese Camp, rarely exceeding two feet in depth,and have been worked over several times. The present population does not exceed three hundred. The future of the town depends on the quartz-mining on the Mother lode and the opening of the old channel underlying Table Mountain. Both of these classes of mining require large expenditures, but promise large and profitable returns.

Quartz Mountain is a cone-shaped hill rising abruptly to a height of four or five hundred feet from a rolling country. Its isolated position makes it a prominent landmark for miles around, this prominence being greatly enhanced by an immense cropping of quartz which crosses the summit of the mountain on the line of the Mother lode, and at a distance presents the appearance of a great wall. Most of the mines are opened about half way between the base and the summit. The vein-matter here seems to have widened with the expansion of the mountain, and veins are found both on the foot-wall and the hanging-wall, separated by masses of broken slate and quartz, known as "horse-matter." These veins have a pitch to the east of eighty degrees.

The Heslep mine, on Quartz Mountain, probably the most thoroughly developed mine on Quartz Mountain, is a location of 1,555 feet on the hanging wall of the lode. The mine was opened in 1852, and was then worked by an open cut on the face of the ledge. This cut has been continued for nearly 100 feet, and a tunnel run from its face, on the ledge, a distance of 600 feet. Winzes have been sunk from the floor of this tunnel to a depth of 65 feet and a new level opened. On this level the vein is 14 feet wide; 8 feet of this vein-matter is hard white quartz, and 6 feet are a mixture of soft slate and broken quartz, highly sulphureted. The foot-wall of the vein is greenstone and the hanging-wall slate. The average pay of this mine, which may be taken as a fair representative of the mines on Quartz Mountain, is about ten dollars per ton for rock taken from the tunnel, and a slight advance for rock from the lower level. This is exclusive of specimen rock, which is now often encountered as depth in the mountain is attained. The pay runs in chutes or chimneys which rarely exceed 100 feet in length, but several are found on the line of the location. The softer material can be mined and milled at an expense of $3 per ton. Of this class of rock large reserves are in sight on the surface which can be cut down on an open face and run through the mill in large quantities, with a profit of $3 to $5 per ton without interfering with the development of the mine. The property is owned and managed by Mr. B. F. Heslep, and is an example of prudent and skillful management. He has commenced sinking a shaft from the surface to connect with the winze from the tunnel, and on its completion will open a new level 160 feet below the tunnel, or 300 feet in depth from the surface. The mill has fifteen stamps of 650 pounds, with a crushing capacity of eighteen or twenty tons per day, and is run by an overshot wheel forty feet in diameter, supplied with water all the year, at an expense of $140 per week, night and day. The mill is built for thirty stamps, and the other fifteen will be put in as soon as the shaft is sunk and the lower level opened. The mill, situated a short distance below the mouth of the tunnel, is provided with Stanford's self-feeding apparatus, requiring only one man to attend to the fifteen stamps. No pans are used, as the main reliance is on the free gold. The sulphurets are preserved for future treatment.

The App mine next adjoining and parallel on the east is owned by Griffin & Co. The ledge is 1,000 feet long, and from 15 to 30 feet wide. They have a 25-stamp mill, but it has only been run at intervals during past year. Their rock will run from $15 to $20 per ton. They have a shaft down 580 feet, on an incline of 60°. The hoisting-works are very complete, and run by an engine of 25 horse-power.

The Knox mine, one-half mile from the Heslep, is 1,200 feet in length, with a vein 10 to 20 inches in width. It is owned by Preston & Co. They have hoisting-works run by a 30 horse-power engine, and a mill of ten stamps and two arrastras, now crushing ten tons per day. The quartz yields an average of $30 per ton exclusive of sulphurets.

The Golden Rule, near Poverty Hill, is owned by a San Francisco company. Their mill of 15 stamps is run by water-power, (50-foot wheel,) and will crush 15 tons per day (24 hours,) which average $10 per ton. A tunnel 500 feet in length is run in on a grade with the mill, to which the rock is brought by car. This tunnel runs nearly east. At this point, the tunnel runs south 75 feet, where their hoisting-works are situated, 87 feet underground. The hoisting-works are run by an engine of 12 horse-power. At the hoisting-works there is a vertical shaft 225 feet deep from the surface. The hanging-wall is of serpentine, and the foot-wall of slate formation. The vein is from 7 to 9 feet wide. The Golden Rule has not been worked regularly during the year 1871, and no dividends were paid.

Table Mountain.—This remarkable remnant of the basaltic overflow overlies, as is well known, the ancient channel of auriferous gravel. The accompanying diagram is made from observations with a Green's barometer, and shows the interesting fact that the rim-rock of the ancient river is higher on the west side than on the east. The difference is 33 feet.

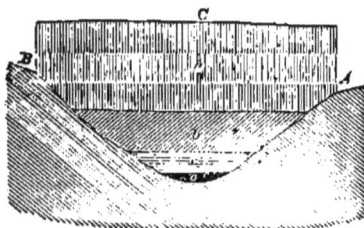

Section of Table Mountain, California.

The observation taken at A was made at the hoisting-works of the Table Mountain Co.'s claim, on the eastern side of Table Mountain, (formerly known as the "Humbug" claim,) 87 feet above the rim-rock of the ancient river, covered by the lava-flow known as Table Mountain; elevation, 1,680 feet.

Observation B was made at the hoisting-works of Hughes' mine, on the west side of the mountain (formerly known as the Maine Boys' tunnel) 110 feet above the rim rock of the ancient channel; elevation 1,736 feet. This claim is next adjoining and south of the Table Mountain Tunnel Co.'s ground.

Observation C was taken on the summit or plane of Table Mountain and at a point equidistant from observations A and B; elevation 1,882 feet.

The diagram, a cross-section of Table Mountain, is a representation of the stratification at this point: *a* is the basaltic lava rising in precipitous cliffs from the surrounding country; *b* a stratum of compact sand nearly hard enough to be designated as sandstone; underneath this is a layer of gravel, represented by *c ;* and *o* shows the position of the auriferous gravel.

The auriferous gravel is extracted by means of inclines running respectively from A and B to the bottom of the channel.

Mining in Table Mountain.—The ground of the "Table Mountain Tunnel Company" is situated on Humbug Flat, about one mile north of Jamestown on the east slope of Table Mountain. The company own 4,500 feet of ground, north and south on channel, and claim from base to base of the mountain. Their gravel deposit is partly under the lava capping, but a large portion of the best ground lay to the east of the lava flow and seemed to owe its origin to a deflection of the ancient river channel, which here took a slight bend, followed by the lava stream, leaving exposed a large portion of the old river bed. The ground was first worked in 1852 by a tunnel running from Peppermint Creek, an insignificant stream on the east side of Table Mountain. This tunnel was run 1,000 feet in slate bed-rock, and on rising twenty-five feet pay gravel was struck near the rim of the channel. This tunnel cost $60,000, and ruined its projectors, only four of them holding a small interest when pay was struck. Subsequently the ground was opened by inclines higher up the slope and near the edge of the lava capping of the mountain, and is still worked by this method, the. Peppermint Creek tunnel being used for drainage. Drifts are run from these inclines by means of which the ground has been breasted out for a length of 2,500 feet, and 2,000 feet yet remain to be opened. The channel of the old river proved to be from 100 to 150 feet in width, and the breasts were opened to a width of 30 to 80 feet according to the yield of the gravel. The thickness of the pay gravel varies from four to six feet—the richest being on the bed-rock, which is "picked" down to a depth of ten or twelve inches. The main incline, through which the mine is now worked, is 320 feet in length, the perpendicular depth from the bottom of the incline to the surface being eighty-seven feet. The strata passed through in running this incline were compacted sand alternated with thin layers of "pipe-clay." Underneath this was found the pay gravel—four to six feet in thickness, then a soft slate bed-rock. The main north drift is now 1,600 feet in length on the channel, but "breasting" has not been made on more than half this distance. The gravel is brought from the "head" or "breasts" of the drifts by a car drawn by a mule to the foot of the incline, whence it is raised to the surface and emptied in a large dump-box.

The method of treatment here is by the "Cox pan," a process which is fully described in another part of this report. The hoisting-works and mill are run by eighty inches of water, by means of an overshot wheel thirty feet in diameter. A test of several hundred car-loads of gravel at this mine showed a yield of five to six dollars per car-load, equivalent to $10 per cubic yard; but this "run" was on very rich ground. The average yield for a run of several successive months would probably not exceed half the amount obtained when the test was made, and with this yield a large profit accrues to the owners. The gravel of this part of the channel of Table Mountain is soft and easily extracted with the pick. In using the "pan" the gravel is fed continuously from a "hopper," the feeding being interrupted only long enough to discharge the bowlders from the mill or pan. The dirt released by the action of the pan passes through the apertures in the bottom, whence it runs through three hundred feet of sluice-boxes, and the proprietors are confident from experimental tests that they save as much of the gold as by any process in use. They speak highly of the "Cox pan" after two years of constant use.

The Hughes mine, comprising the ground formerly known as the Maine Boys' tunnel, one of the earliest locations on Table Mountain, is situated north of, and next adjoining the claim above described, but is opened from the western instead of the eastern face of the mountain, which is here nearly 2,000 feet in width. The main drifts of the two companies are now approaching each other, one running north and the other south on the channel, and before the close of the present year the communication will be opened through the mountain under the lava crust which filled the ancient channel of Table Mountain. This will afford complete ventilation for both mines, and greatly facilitate future explorations. Barometrical observations taken on a line across Table Mountain at this point give the following results:

Altitude above sea level.

Mouth of Table Mountain Company's incline, east side Table Mountain.....	1,680 feet.
Mouth of Hughes's incline, west side Table Mountain......................	1,736 feet.
Summit, or plane Table Mountain between these points...................	1,882 feet.

It will be observed that the rim rock of the old river bed is thirty-three feet higher on the western than on the eastern side, showing that the river wore its deeper channel on the western side. The bottom of the river (bed-rock) seems to be about 300 feet below the plane of the mountain. It is an interesting question which we submit to the consideration of the State geological survey, whether these facts have any bearing on the great distribution of gold on the eastern side of the mountain as manifested on the limestone belt from Columbia at Kincaid Flat, and in the bed of Wood's Creek. The Maine Boys' ground, after many vicissitudes of good and bad fortune for nearly twenty years, passed into the hands of its present energetic owner, Mr. David Hughes, who has demonstrated that this class of mining can be carried on with uniform profit when managed with economy and skill. Mr. Hughes owns 1,900 feet on the channel, and his ground is opened by an incline four hundred and ninety-six feet in length, and 110 feet in perpendicular depth. The width of the channel varies from 150 to 300 feet, while the pay-dirt varies from thirty feet to the entire width of the channel. The gravel is about twenty feet in thickness, but the paying portion seldom exceeds four feet on the bottom. The stratification is the same as on the other side of the mountain. Several "channels" are supposed to exist in the claim, but they are probably the results of the changing of the river bed from east to west as demonstrated by the barometrical observations above noted. The claim is worked through two tunnels, the drifts in one (the old Maine Boys') being 2,600 feet in length, in the other 1,500 feet. The gravel is taken out and hoisted in the same manner as in the claim last described. Mr. Hughes uses two "Cox pans," run with fifty inches of water by an overshot wheel thirty feet in diameter. Each pan will treat forty car-loads of gravel per day—one ton to the car-load. The gravel is fed continuously as at the Table Mountain Co.'s mill, about ten inches of gravel being kept constantly in the bottom of the pan, but it is here charged with a small quantity of quicksilver. From one hundred to two hundred feet of sluices are used below the pans. The yield is from $2 50 to $3 per car-load, and is considered good pay. The gravel sometimes yields much higher, rarely lower. This is a little less than $5 per cubic yard. Mr. Hughes has used these pans for several years in succession, and has made several improvements in their construction and manipulation, particularly in the distribution of the water in the pulp while in motion. He says they will successfully work any gravel soft enough to yield to the pick, and work it closer than any other process, though he admits the loss of a considerable percentage of the finer gold.

Sonora.—The country between Jamestown and Sonora, a distance of four miles, is more hilly, and mining has been confined to the bed of Wood's Creek. The town of Sonora, the largest town in the southern mines, was settled in the summer of 1848 by a party of Mexican miners from the province of Sonora, and was originally known as the "Sonoranian Camp," to distinguish it from Jamestown, which was settled at about the same time and known as the "American Camp." In the fall of 1849, with the discovery of the great wealth of the placers on the limestone belt, the population of Sonora increased with great rapidity and was estimated to be at least five thousand persons. This population, however, was much diminished in the succeeding year by the enforcement of the foreign miners' tax, which was then thirty dollars per month, but revived in 1852, and stood at about three thousand for many years, until the gradual decay of the placers, when it settled down to its present population, about twelve hundred souls. With the exhaustion of the placers the town underwent a period of decay nearly as rapid as its growth, but is now recovering its prosperity on a more stable basis. This town, like all mining towns in California, has suffered severely from repeated and destructive conflagrations. The losses by fire from 1849 to 1853, during which period four great fires devastated the town, were estimated at one million of dollars. The first newspaper published in the mines of California was the Sonora Herald, the first number being issued on July 4, 1850. Nos. 1 to 7 of the Herald were printed on foolscap; single copies sold at fifty cents. The state of society during the flush times—from 1849 to 1853—was of the worst description, and crime and lawlessness were much increased by the ill-feeling engendered between the Mexican and American population. Murder and robbery were crimes of frequent occurrence, and these were punished by the stern discipline of lynch law. In 1855 a vigilance committee was organized, in consequence of the murder of some Americans by Mexican desperadoes, and after several summary trials and executions the entire Mexican population was banished, and the guilty and innocent suffered alike. The winter of 1852 was remarkable for its severity, the rain-fall exceeding that of any year since the occupation of the country by the white races. The streams rose and swept away the few ferries, the roads were destroyed, and great sickness, destitution, and suffering resulted. Some idea of the price of living in Sonora at that time may be formed from the following extract translated from a work on California, published in the city of Mexico in 1866, by l'Abbé Alric, who was the parish priest of Sonora from 1851 to 1855.

"During the months of November and December, 1852," says the abbé, "rain and snow fell in great abundance, and the roads were rendered impassable for several months. The consequence was that the price of living advanced to exorbitant rates. Eggs were worth $1 each; bear's meat, $1 per pound; bread, $2 per pound; chickens, $10 to $12 each; and a turkey sold for $20. Everything else was sold in proportion."

I quote a few more extracts from this work to show the social condition of Sonora at this period:

"On the 8th of June, 1851," says the abbé, "I was compelled to witness a terrible spectacle at a neighboring camp where I was called to administer the last consolations of religion to two Mexicans who had been condemned to death by the Yankee miners on the bare suspicion of having been implicated in the murder of two Americans whose dead bodies had been discovered near the tents of the Mexicans. After these miserable men had received the rites of the church, a noose was placed around their necks, the end of the rope was thrown over a limb, and they were placed on the back of a mule which was driven from under them, the men remaining suspended until life was extinct.

"On the same day several Americans discharged their revolvers into a gambling-house where seven Mexicans were seated at play, on the pretense that the Mexicans had cheated at cards. On arriving, I found three dead and the others wounded. These died the next day, and a common grave received the remains of all. The murderers escaped unpunished.

"These scenes were repeated daily. In fact, on the next day I was sent for from Melones (now Carson Hill) to confess two miners accused of robbing a sluice-box. It was necessary to cross the Stanislaus River, where I was detained for some time, and on my arrival found I was too late; the only thing to be seen was two corpses swinging from a limb projecting over a precipice.

"One night I was called to visit two sick men, one a Frenchman, the other a Mexican, who lived in a tent some distance from the town. They did not seem in immediate danger of death, and I left them at 11 p. m., promising to return early in the morning. Soon after my departure they were robbed and murdered for the gold dust in their possession, and their bodies disposed of, and on my return in the morning I found neither tent, sick men, nor corpses.

"Such," remarks the abbé, "is life in California." This journal, extending over a period of four years, is a constant record of such scenes as the above. Finally a vigilance committee was organized, whose first proceeding was the banishment of all the Mexican population who refused to surrender their arms. The flock being dispersed, the abbé sought new fields of labor.

The Sonora of to-day, with its churches, schools, stores, and pleasant residences surrounded by gardens and orchards, presents a striking contrast to the rude collection of tents and brush houses described by l'Abbé Alric, and the rude and lawless population of 1849-'50 has been succeeded by a thrifty and energetic population engaged principally in the development of the wonderful resources of this locality, in fruit-raising and vine-culture, the orchards and vineyards here producing fruit in a profusion and with size and flavor unequaled in any other part of the State.

Quartz-mining near Sonora.—To the east of Sonora, high up in the Sierras, is found a belt of quartz mines, inclosed in the granite formation, which are rapidly assuming an importance second to none in the State in point of extent and productiveness, and but little known outside of the boundaries of the county. We will briefly notice a few of these claims.

The Confidence mine has been opened for more than ten years, during which period work had not been prosecuted with any degree of regularity until it passed into the hands of the present owners, Ben Holladay & Co., who purchased it for the sum of $15,000 on the recommendation of Mr. L. Gilson, a miner of experience and skill. The claim is 1,050 feet in length; its course northwest and southeast, with a dip to the west of twenty-eight to thirty degrees. Both walls are granite, and the fissure varies in width. At the north end of the drifts the vein matter is about eight feet wide and yields an average of nearly $75 per ton. The south drifts are wider, sometimes attaining a width of sixteen feet, and the yield from this part of the mine is from $12 to $15 per ton. The pay-chute, or chimney, as far as developed, has a length of 200 feet. The rock is a hard, white, sulphureted quartz, presenting no indication of its richness, and "specimen rock" is rarely found. The mine is opened by an incline, 500 feet in length, running on the vein—the perpendicular depth from surface to bottom of incline being about 240 feet. Three levels have been opened, each running 100 feet north and 100 feet south from the incline. The present owners took the mine with a ten-stamp mill. They now have a first-class forty-stamp mill and three arrastras, capable of reducing fifty tons per day. The mine is kept developed in advance of the capacity of the mill, and it is estimated that the reserves in sight will run the mill six months, without opening another level, and that these reserves will yield $200,000. About sixty men are employed. This company own 2,100 feet of a similar ledge, and as far as opened the ledge has gold visible in every part. Under its present management it looks as if it would last for years to come. The extraordinary fineness of the gold in this mine makes it the more valuable, the gold running from 900 to 920 fine. For the past two years the mine has been yielding at the rate of from $25,000 to $30,000 per month, sometimes as high as $40,000. The mill and hoisting-works are run by powerful steam machinery.

The Excelsior mine is situated a short distance to the northeast of the Confidence, and the geological and mineralogical features are similar. This mine produced $525,000 in twenty months with a ten-stamp mill. The mine is now closed and has been for the past five years, on account of a personal difficulty among its owners. Miners in the vicinity, who formerly worked in the mine, think it among the richest in the State. Explorations have reached a depth of 175 feet, the mine being opened laterally about 400 feet. Mr. G. F. Wright, who has been absent from the State for several years past, has returned from the East, and it is thought the mine will be re-opened early in 1872. During the working of this mine, the company's little ten-stamp mill made several runs ranging from $14,000 to $25,000 per week. One lot resulted in a clean-up of $25,800 from fifty tons of rock.

At Big Basin several ledges are worked, by means of arrastras, by Lewis & Brother. These ledges are found in talcose slate. Besides these claims, numerous veins are being opened on the same belt, which is believed to extend northerly into Calaveras County, where it outcrops at the Sheep Ranch claims, Washington district. Near Sonora and Columbia, on the slopes of Bald Mountain, numerous "pocket" veins of great richness, but limited in extent, occurring in slate, are found and worked by the adventurous class of prospectors who depend on this precarious method of mining for a subsistence. On the west side of Table Mountain we find the Mother lode, with the small mining villages of Rawhide Ranch and Tuttletown. Numerous claims at these points on which work has been suspended for several years will soon be re-opened. Among this class we may mention the Rawhide, possessing one of the finest mills in the State, closed for several years in consequence of mismanagement and litigation, and the Waters mine, near Tuttletown, which is said to contain a large amount of low-grade ore in sight, but operations have been suspended for want of machinery.

Columbia and vicinity.—Four miles north of Sonora we find the towns of Columbia, Shaw's Flat, and Springfield. The first-named was once an important place, second only to Sonora in population and wealth. All these towns are situated on the limestone belt, which is here nearly three miles in width, immediately to the east of Table Mountain, and have an elevation above sea-level of about 2,100 or 2,200 feet, being but little less than that of the plane of the mountain. The surface diggings here were from three to thirty feet in depth. The limestone bed-rock, everywhere exposed by the washings of the miners, is deeply eroded and cut in the most fantastic shapes by the action of running water at some remote period. The placers here were formerly extensive and rich, but at the present time they do not yield more than two dollars a day, and the unworked area is very limited. The best ground is covered with buildings in the various towns. These buildings are being removed for the purpose of mining the ground. The auriferous character of the ancient channel under Table Mountain was first discovered in 1851, near Shaw's Flat, where the denudation of the lava crust exposed the richest ground worked in this part of the State. As late as 1855–'56 it is said that some claims at Caldwell's ranch yielded ten or twelve pounds of gold per day, for many consecutive days. At that time the town of Shaw's Flat contained two hotels and a dozen stores. Now there are not a dozen houses occupied. Indian antiquities, such as pestles and mortars of stone, are here found in abundance, and the remains of the mastodon and elephant are frequently unearthed by mining operations. These remains are found at depths varying from ten to thirty feet beneath the surface. At Gold Springs, near Columbia, an elephant's tusk nine feet in length, and of proportionate thickness, was found in a good state of preservation, but did not long resist exposure to the air. Brown's Flat, situated on Wood's Creek, about half way between Sonora and Columbia, has a population of one or two hundred persons, who are engaged in mining on the creek. The limestone belt is here pinched down to very narrow limits, and the depth of pay-dirt in the creek varies from twenty to sixty feet, the crevices in the rock being still deeper. The dirt is hoisted to the surface by derricks, emptied in dump-boxes, whence it is run through sluice-boxes. Several claims are yielding from $3 to $5 per day to the hand, but the available ground is now very limited. Yankee Hill, near the head of Wood's Creek, and at the base of Bald Mountain, has been noted for the existence of coarse gold. Among the nuggets found here was one of twenty-three pounds, one of seventeen pounds, and many pieces varying in weight from one to four pounds. Mining is still carried on in the bed of the creek, but on a small scale. The basin in which these towns are situated has on one side Table Mountain, with its channel of auriferous gravel, and on the other Bald Mountain, which is noted for the richness of its "pocket" claims. The gulches and ravines of this mountain still contain much gold released by the decomposition of the quartz veins.

The following is an extract from the early mining laws of Columbia district. Those of other localities described did not materially differ.

"A full claim for mining purposes on the flats or hills of the district shall consist of an area equal to that of one hundred feet square.

"A full claim on ravines shall consist of one hundred feet running on the ravine, and of a width at the discretion of the claimant, provided it does not exceed one hundred feet.

"No person shall hold more than one full claim within the boundaries of this district, nor shall it consist of more than two parcels of ground, the sum of the area of which shall not exceed the area of one full claim; provided, nothing in this article shall be so construed as to prevent miners from associating in companies to carry on mining operations, such companies holding no more than one full claim to each member."

Another article provided that no claim should be sold to a Chinaman under penalty of forfeiture. Those owning ground now are very willing to sell to "John."

List of quartz-mills in Tuolumne County running during 1870 and 1871.

Name.	Stamps.	Owner.	Remarks.
Blue Gulch Eagle Mill.	10	Eagle Mill and Mining Company.	Work has been recommenced during present year with favorable prospects; tunnel being run to drain mine at low level.
Lombardo Mine and Mill.	10	Lombardo Company ..	Running at intervals.
Confidence Mine and Mill.	40	Ben. Holladay........	One of the best mines in the State; said to be yielding from $1,000 to $1,200 per day. The present owner took possession in 1867, since which time the mine has paid largely, and has been developed to a depth of 500 feet by an incline. Situated in the granite belt.
Deer Creek Mill	5	H. P. Gould	Running at intervals for two years past; ledge much broken up and pay spotted.
Buchanan Mine and Mill.	10	Buchanan Mining Company.	Suspended for many years; prospecting resumed this year; mine valuable, but will require a large expenditure to put it in paying condition.
American Camp Mine and Mill.	10	Jones & Woodman. ..	Running at intervals.
Golden Rule Mine and Mill.	10	Bosworth & Co	Work on mine and mill temporarily suspended during past year; will be resumed in 1872.
Heslep Mine and Mill..	15	B. F. Heslep...........	Running constantly, and paying well; vein wide, but of low grade; situated on Mother lode, Quartz Mountain.
Preston's Mine and Mill.	10	Preston & Co	Formerly known as Knox Mine; vein small, but rich; running steadily.
App Mine and Mill	25	Griffin & Co	An old location; work temporarily suspended.
Rawhide Mine and Mill.	20	Preston & Co	Operations suspended on account of change of management; work will be resumed. This mine has been very productive.
Spring Gulch Mine and Mine.	10	Sharwood & Co	First-class 10-stamp mill just completed.
Whitman's Pass Mill ..	5	Colby & Co	Vein small, but rich; in granite belt.
Starr King Mine and Mill.	5	Inch & Co	Working at intervals.
Clio................:...	10	Clio Mining Company.	Will resume work as soon as supply of water will admit.
Soulsby Mine and Mill.	20	Davidson & Co	A productive mine, but now in pecuniary difficulties and litigation.
Lewis Brothers' Mine..	2	Lewis Bros	Uses two stamps and arrastras; paying well.
Excelsior Mine and Mill	10	Wright & Co	Mill burned down; will be rebuilt.
Tullock & Co.'s Mine and Mill.	5	Tullock & Co	A vein on west side Mother lode, parallel with it.
Patterson's Mine and Mill.	10	Patterson & Co	Running arrastras constantly; will have 10-stamp mill in operation next year.

It appears from the above table that twenty mills, with an aggregate of two hundred and sixty-two stamps, have been running during the past two years, and but few of these with regularity. Langley's Pacific Coast Directory for 1870–'71 gives a list of forty mills with four hundred and forty-four stamps in this county. The county assessors for that year made no returns, so that we cannot compare the number of tons crushed with that of previous years.

CALAVERAS COUNTY.

This county adjoins Tuolumne on the north. It possesses gravel and placer deposits of considerable extent, but not of great depth, and labors under the disadvantage of scarcity of water. It contains, also, near the foot-hills, rich mines of copper, now unworked, and numerous veins of quartz of great richness. Many of the richest claims on the Mother lode are found in this county. The county assessor, in his report to the surveyor general for 1871, says: "The placer mines of this county are 'things of the past.' True, now and then, in one's journeyings, you will startle some old 'forty-niner,' in his secluded ravine, with pick, pan, and shovel, mayhap a rocker; now and then a squad of Celestials working, for the twentieth time, old tailings. But if the bright yield of placer mines has paled, we are content with the more resplendent glories of cement and quartz; of the first, we are but in our infancy." Throughout various parts of the county gravel diggings are found, and in several basins near the head of Table Mountain the gravel has a depth of thirty or forty feet, but drainage is difficult and expensive. Mr. J. Rathgeb, of San Andreas, has made close investigation into the character and yield of many of the gravel claims of the central part of the county, and furnishes the following valuable description and data:

The mining districts of San Andreas, Lower Calaveritas, and Fourth Crossing, in Calaveras County, are situated on the range of the Mother quartz lode of California, and are about eight miles in length by three miles in width. Since last year these districts have made progress in the mining industry and have advanced by their self-sustaining capacity to profitable operations.

The principal mines of these districts are the Davis, Sceiffart, Thorn, Rhead, Union, Anton, Miner's Dream, Bachman, and Thorpe.

The principal tunnel and hydraulic claims are the Garnet, Clarks & Co., Wilson & Co., Hedrick, Johnston, Bennet, Worthmann & Co., Pfeffer & Co., Cloyd, Raggio & Co., Rivera & Co., Oneto & Co., Waters, Gay, Peregrini and Driscol.

For the year 1871, the estimated product of bullion for these districts is fifty thousand dollars.

The ruling rates of wages are, for first-class miners, $3 per day, also $2.50 and $2; common laborers, $1.50 to $2 per day; Chinamen, $1.25 to $1.50.

List of stamp mills:

Demarest, 10 stamps, overshot water-wheel.
Union, 10 stamps, overshot water-wheel, 1 pan.
Thorpe, 5 stamps, overshot water-wheel, 3 arrastras.
Irvine, 10 stamps, overshot water-wheel.
Garnet, 5 stamps, overshot water-wheel, 1 arrastra.

A 10-stamp quartz-mill at the Thorn mine, and a 10-stamp quartz-mill at the Union mine, are in contemplation.

The Thorn mine has been sunk since last year to the depth of over 200 feet. There is in the shaft a well-defined gold-bearing quartz-vein of five feet in width. The total width of vein-matter from the foot-wall to the hanging-wall is twelve feet. The mill-rock is a blue-ribbon rock, similar to that of the celebrated Eureka mine, Amador County. The hanging-wall is a blue slate, the foot-wall a variety of granite. Substantial hoisting-works and other improvements are erected on this mine.

The Union mine is yielding well, and is being sunk to another level, having a good supply of timber on hand.

The Rhead mine has two shafts, thirty-three feet each. The vein is two feet wide at the bottom. Gold is impregnated in the foot-wall, and forming a thin coating on the talcose slate or rock of the foot-wall.

The Anton is a new mine, paying well.

The Thorpe mine has been worked till June; had to stop for want of water to drive a five-stamp mill and the pump attached to it to free the mine of water.

The Bachman mine is yielding well.

Auriferous deposits of gravel occur in a nearly parallel line with these quartz-lodes, to the east of them, and arising probably from the disorganization of the net-work of quartz-veins and rocks bearing gold. Tunneling and hydraulic washing in these deposits have been lucrative. Water may be obtained in these districts from four to six months of the year.

The Garnet claim has been drifted upon 400 feet in length by 80 feet in width, has a five-stamp mill with hoisting-works, a heavy Chile mill (arrastra) attached, with 32-feet overshot water-wheel. The yield has been $18,500.

Clark & Co. have extracted rich pay-dirt from a channel on the northern bank of San Andreas Gulch.

Wilson & Co. are getting out cemented gravel, hard to wash, because it will not slake for a long time.

The Harvey Hedrick claim is worked by hydraulic washing, and drifting is done when no water can be obtained for washing.

The Wade Johnston claim is a gravel formation, a channel through a hill, worked by drifting and hydraulic washing. The Bennot is on the same channel.

Raggio & Co. are excavating deep crevices in a limestone range in the river by means of a derrick and horse-power, getting good pay.

Worthman claim is a tunnel claim now; formerly a hydraulic claim; still yielding well.

Driscol & Co. are cutting a tail-race and tunnel into good diggings.

Adjoining, southeast of Fourth Crossing district, are the cement-gravel tunnel claims of Dogtown.

The Hammerschmidt & Hensel tunnel has been yielding very well; they intend to build a mill for the extraction of the gold held by the cement, which slakes hard.

The Barney Hurle tunnel is going further into good ground.

The Bully Tunnel Company have struck rich pay—$200 on 75 superficial feet—but water interferes badly.

Hydraulic claims, Calaveritas district, Calaveras County.

	Bonnet's claim.
Inches of water used in hydraulic washing	40.
Cost of water	$20 per week.
Fall of the water, (head)	30 feet.
Supply of water lasting in a year	4 months.
Number of cubic yards washed in ten hours	3 to 4.
Yield of gold per day	$10 to $12.
Height of gravel washed this year	12 to 14 feet.
Ground worked in the claim, total	1 acre.
Ground unworked	5 acres.
Ground worked this year	40 by 50 feet.
Number of men working	5.
Wages paid to hired men per day, per man	$1.25.
Composition of auriferous deposit, from top	Red soil, round, heavy quartz boulders, lava, sand, silicious slate fragments, white rounded quartz, granite, syenite boulders, gneiss, fine gold, iron-sand, and rubies.
Direction of channel	N. W., S. E.
Bed-rock	Blue and gray slate, and strata of quartz.
Curiosities found	

Hydraulic claims, Calaveritas district, Calaveras County.

	Johnston's claim.
Inches of water used in hydraulic washing	40.
Cost of water	$20 per week.
Fall of the water, (head)	80 feet.
Supply of water lasting in a year	4 months.
Number of cubic yards washed in ten hours	3 to 4.
Yield of gold per day	$12 to $14.
Height of gravel washed this year	10 to 15 feet.
Ground worked in the claim, total	5 acres.
Ground unworked	15 acres.
Ground worked this year	70 by 70 feet.
Number of men working	3.
Wages paid to hired men per day, per man	$1.25.
Composition of auriferous deposit from top	Red soil, round, heavy quartz boulders of all colors, lava, sand, fragments of slate, gneiss, granite, syenite, fine gold, magnetic iron sand.
Direction of channel	N. W., S. E.
Bed-rock	Blue and gray slate.
Curiosities found	Petrified turtles.

Hydraulic claims, Calaveritas district, Calaveras County.

	Hedrick's claim.
Inches of water used in hydraulic washing	40.
Cost of water	$20 per week.
Fall of the water, (head)	60 feet.
Supply of water, lasting us a year	3 months.
Number of cubic yards hydraulic washed in ten hours	3 to 4.
Yield of gold per day	$15.
Height of gravel washed this year	20 feet.
Ground worked in the claim, total	2 acres.
Ground unworked	60 acres.
Ground worked this year	40 by 100 feet.
Number of men working	5 to 7.
Wages paid to hired men per day, per man	$1.25.
Composition of auriferous deposit, from top	Red gravelly soil,sand,quartz boulders, cemented fragments of slate and other rocks, gold, fine black sand, &c.
Direction of channel	N. W., S. E.
Bed-rock	Blue and gray slate.
Curiosities found	Petrified wood and foliage.

Shaft claims at San Andreas, Calaveras County.

	Garnet Company.
Inches of water used in sluice-washing	20.
Cost of water	$10.50 per week.
Supply of water lasting	4 months.
Depth of shaft	42 feet.
Height of drift on an average	4½ feet.
Area of ground drifted upon	400 by 80 feet.
Length of ground unworked on channel	1,200 feet.
Average of cement-gravel extracted per day	2,500 pounds per man.
Quantity of refuse (boulders) left in the drift	One-half.
Quantity of gold-bearing cement-gravel crushed in twenty-four hours' run of 5 stamps of 500 pounds	18 tons, or about 7 cubic yards.
Crushed with heavy Chile arrastra	25 tons, or 10 cubic yards.
Expense of milling, per cubic yard	70 cents, (wages.)
Expense of extracting, per cubic yard	$2.50.
Yield of one cubic yard of ground	$8.
Total yield of claim in one year	$18,500.
Wages paid to underground-drifters	$2.50 per day.
Number of men working	12.
Composition of deposit from surface downward	Lava or sedimentary formation; hard cement above ledge or bed-rock ; auriferous gravel cement, hard quartz, black and dark colored.
Direction of channel	N. W., S. E.
Bed-rock	Very hard, compact, gray quartz slate.

Shaft claims at San Andreas, Calaveras County.

	Wilson & Co.
Inches of water used in sluice washing	15.
Cost of water	$1.50 per day.
Supply of water lasting	5 months.
Depth of shaft	33 feet.
Height of drift on an average	6 to 10 feet.
Area of ground drifted upon	300 by 70 feet.
Length of ground unworked on channel	900 feet.
Average of cement gravel extracted per day	2,500 pounds per man.
Quantity of refuse (boulders) left in the drift	One-half.
Quantity washed in sluices per day, ten hours	3 to 5 cubic yards.
Expense of milling, per cubic yard	
Expense of extracting, per cubic yard	1 day's labor.
Yield of one cubic yard of ground	$5.
Total yield of claim in one year	$3,000.
Wages paid to underground-drifters	
Number of men working	3.
Composition of deposit from surface downward	Lava, gravel cement, fragments of slate, granite, and gneiss, rounded, loose sand.
Direction of channel	N. W., S. E.
Bed-rock	Slate, blue.

Tunnel claims near Dogtown, Calaveras County, California.

	Hammerschmidt, Hensel & Co.
Inches of water used in sluice washing	25 to 30.
Inches of water used in hydraulic washing	20.
Height, or fall	100 feet.
Supply of water lasting in the year	6 to 7 months.
Cost of water, sluicing, per day	$4.
Cost of water, hydraulic, per week	$20.
Air-shafts	One of 70 feet, one of 148 feet.
Length of tunnel at present	1,500 feet.
Height of drift, (all pay-dirt)	6 feet.
When this claim commenced	In 1865.
Area of ground drifted out and worked	9,166 superficial square yards.
Number of cubic yards drifted out	18,332.
Length of ground unworked	400 feet.
Average of cement gravel extracted per day	20 to 25 car-loads, about 25 square yards.
Quantity of refuse left in the stope	Over one-half.
Quantity of gravel cement washed during last year, and extracted	2,200 square yards.
Yield of cement gravel washed	$4 to $5 per square yard.
Total yield last year	Over $8,000.
Total yield since commencement of this claim	$70,000.
Gold, description	Small flat particles.
Wages paid to underground-drifters	$2.50.
Number of men hired	4.
Number of men working	8.
Composition of deposit	Red soil, small and large white and blue quartz boulders, pieces of dark-blue slate, quartz and mica slate, cemented granite gravel.
Bed-rock	Dark blue slate full of cubical pyrites.
Direction of tunnel and course of deposit-channel	E. N. E.

Tunnel claims near Dogtown, Calaveras County, California.

	Barney, Hurle & Co.
Inches of water used in sluice washing.................	30.
Supply of water lasting in the year....................	6 to 8 months.
Cost of water, sluicing, per day......................	$4.
Air-shafts...	One of 105 feet.
Length of tunnel at present.........................	One of 200 feet, one of 400 feet, one of 300 feet.
Height of drift, (all pay-dirt)......................	4½ to 5½ feet.
When this claim commenced	In 1862.
Area of ground drifted out and worked	5,600 superficial square yards.
Number of cubic yards drifted out....................	7,500 square yards.
Length of ground unworked	600 feet.
Average of cement gravel extracted per day...........	15 to 20 square yards.
Quantity of refuse left in the stope	Over one-half.
Quantity of gravel cement washed during last year, and extracted	Not washed.
Yield of cement gravel washed	$4.
Total yield last year...............................	Not washed out.
Total yield since commencement of this claim........	$30,000.
Gold, description	Both coarse and fine.
Wages paid to underground-drifters	$2.50.
Number of men hired	
Number of men working	4.
Composition of deposit.............................	Red soil, quartz, gravel, slate, granite, sand, all cemented.
Bed-rock	Dark blue slate, veins of quartz.
Direction of tunnel and course of deposit-channel	East.

Tunnel claims near Dogtown, Calaveras County, California.

	Bully Company.
Inches of water used in sluice washing	30.
Supply of water lasting in the year	6 to 8 months.
Cost of water, sluicing, per day.....................	$4.
Air-shafts	4.
Length of tunnel at present.........................	700 feet.
Height of drift, (all pay-dirt)......................	4 feet.
When this claim commenced	1870.
Number of cubic yards drifted out....................	3,000.
Length of ground unworked	1,000 feet.
Quantity of refuse left in the stope	Over one-half.
Quantity of gravel cement washed during last year, and extracted...	2,000 square yards.
Total yield last year...............................	$700.
Total yield since commencement of this claim..........	$700.
Wages paid to underground-drifters..................	$2.50.
Number of men working.............................	4.
Composition of deposit.............................	Gravel, cemented.
Bed-rock	Blue slate.
Direction of tunnel and course of deposit-channel......	S. E.

Tunnel claims near Dogtown, Calaveras County, California.

	Buckeye Company.
Inches of water used in sluice washing	30.
Supply of water lasting in the year	6 to 8 months.
Cost of water, sluicing, per day	$4.
Air-shafts	4.
Length of tunnel at present	1,000 feet.
When this claim commenced	1870.
Number of men working	4.
Composition of deposit	Gravel, cemented.
Bed-rock	Blue slate.
Direction of tunnel and course of deposit-channel	S. E.

Tunnel claims near Dogtown, Calaveras County, California.

	Dietrich & Co.
Inches of water used in sluice washing	30.
Inches of water used in hydraulic washing	
Height, or fall	
Supply of water lasting in the year	6 to 8 months.
Cost of water, sluicing, per day	$4.
Cost of water, hydraulic, per week	
Air-shafts	10.
Length of tunnel at present	1,200 feet.
When this claim commenced	1870.
Number of men working	4.
Composition of deposit	Gravel, cemented.
Bed rock	Blue slate.
Direction of tunnel and course of deposit-channel	S. E.

Angel's Camp and Carson Hill.—These two places have been the scene of extensive and profitable quartz-mining on the great Mother lode, and the first-named town now possesses several of the best mines found on the lode. In 1851, Carson Hill was the scene of one of the great "rushes" which then periodically occurred throughout the State. Rumor said that a mountain of quartz had been discovered, with gold enough visible for the coinage of a nation. Companies were formed, claims taken up, and hundreds of men were soon at work. Disputes as to boundaries arose, and much bloodshed followed. Most of the claims were thrown in litigation, and the rich yield of the surface was wasted in the expenses incident to protracted lawsuits. Many years elapsed before decisions were had, and in the mean time the character of the quartz changed, as the water-level was reached, and the owners, impoverished by litigation, were not able to open their mines systematically. At the present time (December, 1871) no work is being done on Carson Hill, though there are several fine mines which will be opened next year. The principal mines here are the Finnegan, or New York, the Reserve, or Stevenot, the Morgan Ground, which is said to have yielded nearly $3,000,000, the Union, and the Kentucky—all closed. Proceeding in a northwesterly direction from Carson Hill, and following the outcrop of the Mother lode, which was here very prominent, we find the following claims: Two claims at Albany Flat, name unknown, Cogswell's, Cameron's, and the Raspberry. We are now at Angel's, where we find the Union, or Stickle's, the Utica, Lightner's, the Angel's Quartz Mining Company, Hill's Ground, and Bovee's Ground—the two latter parallel veins. None of these claims, except the Angel's Quartz Mining Company and the Union, are being developed at present. East of the above, at Angel's, we find a series of narrow pocket-veins, known as the "Dead-horse" claims. On these claims a great deal of work is being done, the rock being principally treated by arrastras, and yielding a fair profit, while occasionally a rich pocket of free gold is struck. About twenty of these arrastras are in constant operation at Angel's.

The "Angel's Quartz Mining Company" is a San Francisco corporation. They own 900 feet on the ledge, and purchased the property five years since. The ground was formerly known as the "Big mine" and the "Potter mine." The surface width of the vein here was about 50 feet, including "horse-matter," but the horse ran out at a depth of 300 feet, leaving a compact vein of 12 to 15 feet in width, which has been followed in the main shaft to a depth of 600 feet, though no level has been opened below 500 feet. They have two shafts 100 feet apart; both are sunk on the vein, which pitches east at an angle of 76°. Their mill cost $30,000, has thirty stamps, and is in all respects one of the most complete in the State. They use a Blake crusher, or rock-breaker, and Stanford's self-feeding apparatus. Mr. Potter, the superintendent, estimates that the crushing capacity of the mill has been increased six tons a day by the use of this contrivance, besides saving much expense for labor. The mill now runs about fifty-five tons per day, and the yield of the quartz is from $6 to $20 per ton, with a constantly increasing yield as depth is attained. This latter fact seems to be characteristic of the Mother lode. The sulphurets here are treated in pans and settlers, and yield about $70 per ton. Two and a half tons are treated daily. The mill is run by an 80 horse-power engine, and the hoisting-works by an engine of 25 horse-power. The water in the mine is raised by two of Blake's steam-pumps, one situated on the 200-foot, and the other on the 400-foot level. They regularly employ sixty men. The levels of the mine have been opened for 200 feet on either side of the shaft.

The entire surface of the Mother lode, from Angel's to Carson Hill, has been worked in early times as an open cut or trench, as far down as the water-level. It would seem, from appearances, that all the rock between the walls was sent to the mills or arrastras. At present operations are confined to two claims, the Angel's, above referred to, and the Union.

The Union Quartz mine is better known as the Stickle's Ground, and consists of 400 feet of ground. This ground was sold, in the fall of 1871, to a San Francisco company for $20,000. They have a ledge varying in width from 30 feet on the surface to 15 feet at lower levels, where the vein becomes compact. The mine is worked through a vertical shaft, 300 feet deep. They are running a 10-stamp mill by water-power, crushing daily about fifteen tons. Their rock will average about $10, exclusive of sulphurets, per ton; it is highly sulphureted, and the sulphurets are quite rich. The men are regularly employed on this claim. Wages here are $2.50 to $3 per day; wood costs $5 per cord. Giant powder and single hand-drills are used. Extensive improvements are in contemplation by the present owners, who will make this one of the leading mines of the country.

The Bovee mine is situated a short distance east of the Angel's Quartz Company's mine. It was opened to a depth of 300 feet, and had a first class mill, which was burned down by an incendiary. Operations were then suspended, and the mine filled with water, and caved in the upper levels. It is probable that during the present year the mine will be opened by a new vertical shaft, which will be sunk at a considerable distance to the east of the old shaft, and will strike the ledge at a depth of 500 feet, where the first level will be opened. This course, although involving a large preliminary outlay, seems best adapted to the development of much of the ground on the Mother lode.

Two miles northeast of Angel's we find several gravel-claims. The most prominent of these are known as the North Star, and McElroy's claim. The former has been prospected and will be opened during 1872, while the latter has been opened and yielding a fair profit for several years past.

Murphy's and vicinity.—The towns of Murphy's, Vallecito, and Douglas Flat are situated at the head of Table Mountain, and were noted for the great richness of the basins and flats situated in the elbows of the mountain and between the summits of the basaltic hills, which once formed a continuous lava-stream. Mining is still carried on by a few companies. All these towns are on the great limestone belt, and the characteristics of the placers do not materially differ from those of Columbia and vicinity, already described, except that here long bed-rock cuts and tunnels have been run to open and drain the flats and basins. Between Columbia and Murphy's, both Table Mountain and the limestone belt have been cut by the Stanislaus River, which runs through a tremendous gorge, exposing at one point nearly 1,500 feet in depth of the limestone strata. The drain-tunnels at these places are approached by an open cut or trench of 1,000 feet in length. These enterprises have proved failures in a pecuniary point of view, on account of their immense cost. The Union Water Company, at Murphy's, derive their water-supply from the North and Middle Forks of the Stanislaus River, about 50 miles above Murphy's. The aggregate length of the two ditches is 60 miles. They originally cost over $250,000; their receipts for the sale of water for mining and agricultural purposes average from $30,000 to $40,000 per annum.

The Washington quartz district is situated about four miles north of Murphy's, and was discovered in 1866 by Mr. Jaquith, who opened a series of claims, which he subsequently sold to the "Calaveras Quartz Mining Company," a San Francisco company, for $40,000. They immediately erected a 20-stamp mill, and are now obtaining large returns, having, it is said, already taken out more than the purchase-money. The principal ledge is opened to a depth of 150 feet. Two of the veins belonging to the company have a width of four feet each; another is ten feet wide—all in granite. The average pay is said to be $30 per ton at the mill of the company, though several tons worked by Mr. Potter, at Angel's, yielded a higher return. These claims have a northwest and southeast course and a dip of 33°. The rock is a dark-colored, sulphureted quartz, occasionally showing streaks of free gold.

Parallel to the above, but higher up in the mountains, are found another series of veins, known as the Sheep Ranch claims, which, as far as prospected, promise to yield largely. These are supposed to be a continuation of the famous claims in Tuolumne County, before referred to.

A correspondent of the Mining and Scientific Press, who visited this locality in the summer of 1871, thus describes operations in these districts:

The Ferguson & Wallace claim, six and a half miles north from Murphy's, is owned by the two gentlemen after whom it is named. They own 1,400 feet of a ledge, from eight to twenty inches wide. They are running two arrastras by water-power, working five men. Their rock thus far has averaged from $34 to $44 per ton. They have just completed a drainage-tunnel, 236 feet in length, at a cost of from $10 to $40 per foot. The rock is hoisted by a whim, run by horse-power, through a shaft 94 feet deep. The Mountain Quail claim, two and a half miles from the above, is owned by Mr. Samuel Woods. He has 1,100 feet of a 20-inch vein, which is much broken. It has never paid less than $75 per ton. He is running two arrastras by water-power, employing three men, working through a shaft by a windlass.

One mile from the above, a San Francisco company are now engaged working three different ledges, known as the South Bank, Enchantress, and Oro Minta. On the latter a shaft has been sunk 130 feet. One 110 feet deep has been sunk on the Enchantress. On the South Bank, a shaft 12 feet deep reveals a well-defined ledge, 8 feet thick. The few tons of rock crushed from the above-named ledges have averaged from $20 per ton up into the thousands, for small lots. Neiswander, Jaquith, Parsons, and others are the proprietors.

This is the claim above referred to as belonging to the Calaveras Mining Company. The spring of 1872 will probably witness some extensive exploration and development in these districts. Mining here may truly be said to be only in its infancy.

A few miles north of Murphy's is found one of the remarkable natural wonders of the State—the great limestone cave. This cave was discovered by accident, in 1850; but on account of its nearly inaccessible situation it has been rarely visited and never thoroughly explored. The length of the explored passages is about 1,500 feet north and south, by 1,000 feet east and west. The height of the chambers varies from fifty to one hundred feet, and they present the most wonderful display of stalactites and stalagmites, grouped in the most fantastic shapes.

Railroad Flat and vicinity.—The town of Railroad Flat is situated at an elevation of about 4,000 feet, and lies to the north of the districts above described. A few years since the town was scarcely known, but now it is becoming a place of much importance on account of the development of numerous quartz-veins in the immediate neighborhood. These claims were located many years since, but operations were not carried on below the water-level on account of a want of capital and limited milling facilities. The "flat" comprises about 160 acres of ground, which was found to be auriferous, and was worked on a small scale by the hydraulic process. While engaged in washing, the croppings of the "Petticoat," the leading mine of the district, were discovered. The detrital matter here is not over twenty feet in thickness, and its auriferous character is the result of decomposition of quartz; but the principal interest centers in the numerous quartz-veins. These veins are generally narrow, and the pay zones or chimneys of very limited extent in the majority of claims, but the quartz is of high grade, and present workings indicate an increase in the length of the chimneys with increasing depth. The country-rock is generally slate, and the quartz yields from $20 to $60 per ton. There are two custom mills here, Randolph's and Clark's, both of which are run almost constantly. The price of milling is from $3.50 to $4 per ton. As wood is cheap—$2.50 per cord—and water is dear—20 cents per inch—steam-power is used in preference to water. Randolph's mill has ten stamps of 680 pounds each, eighty blows to the minute, and crushes fifteen tons per twenty-four hours. He amalgamates in battery and uses very fine screens, having three hundred apertures to the square inch. Each battery has 25 feet of amalgamated copper plates. The Clark mill has eight stamps. No sulphurets are saved except by special agreement, and then an extra price is charged.

The "Petticoat," so named from having been located in the names of the wives of the discoverers, was accidentally discovered by the uncovering in a placer claim of its outcrop. At this point it was very rich, and is said to have yielded, down to the water-line, an average of nearly

$100 per ton. In consequence of an error in the location of the hoisting-works, and general improvidence and mismanagement, the mine ran in debt after the exhaustion of the surface quartz, and operations were suspended for about two years. Work was resumed in the summer of 1871, and a contract let for the sinking of a shaft to a depth of 350 feet. Previous to this work the lowest level was at 200 feet. Pay-ore has been struck in the shaft now being sunk. The lead shows a broken body of quartz from three to four feet in width, sufficiently rich in gold to pay the expenses of sinking. A drift will immediately be run toward the north, so as to intersect the rich chimney which cropped out on the surface. The company have powerful hoisting-works, and exploration will be vigorously prosecuted. The future of the district depends on the success of this enterprise.

The Prussian Hill mine, in this vicinity, is being rapidly developed by a San Francisco company, who are negotiating for its purchase. It is located south and east of Glencoe, on the narrow divide between the North and Middle Forks of the Calaveras, and west of the great soap-stone belt passing through the county. Some very rich rock has been taken from this mine, and it is believed, if properly developed, it will rank as one of the No. 1 mines in this county. Thirty tons of Prussian Hill ore, crushed, yielded about $1,100. The rock was taken from a shaft 70 feet in depth, and the lead improves as they go down upon it. The company have purchased the "French" mill at Rich Gulch Flat, which they intend removing to their mine as soon as practicable. The battery consists of 15 stamps, a 35 horse-power engine furnishing the motive power. A crushing in their mill has yielded $2,263 for 60 tons, or nearly $40 per ton.

West Point, another quartz-mining town, is situated six miles north-east of Railroad Flat. Its general features, both as to placers and quartz-veins, are similar to those of Railroad Flat, except that the country-rock is granite. None of the mines have been opened to a greater depth than 150 feet. The veins are narrow, rarely exceeding 15 to 20 inches, but the ore is of high grade, the yield varying from $40 to $50 per ton. There are five mills here, but only two have run constantly this season. Gravel-beds are rarely found so high in the Sierras. In plain sight of Railroad Flat and West Point there is a remarkable landmark of tabular shape called Fort Hill, which seems to be a portion of an ancient channel covered with lava. It has not been prospected.

El Dorado is a decayed mining town situated on the eastern edge of the limestone belt, a few miles south of Railroad Flat. There are numerous promising quartz-ledges near this town which are but little developed. George Rodocino is engaged in running a tunnel on a ledge 10 to 12 feet wide. The rock yields $10 to $12 per ton, and the vein is in slate. The tunnel will be 450 feet in length, and will strike the ledge at a depth of 150 feet. A 10-stamp mill run by water has been put on the claim.

Rich Gulch district, situated in the foot-hills of the county, about seven miles southwest of Mokelumne Hill, contains many valuable quartz-claims in a partial state of development. A correspondent of the Mining and Scientific Press, of San Francisco, whose statements we have found invariably reliable, notices several of these claims as follows.:

"'The Quartz Glen mine has been opened to a depth of 200 feet, and a tunnel run in on the lead 800 feet has just been completed, developing a ledge at this point (130 feet below the old works) of twelve feet average thickness. This rock is highly sulphureted. In a two years' run, with the use of an ordinary 10-stamp mill to crush the rock, some $200,000 were taken out in bullion. Mr. H. Atwood is the present proprietor.

"Several other apparently good mines are only partially developed for want of capital. The Oak Ridge claim has been opened to a depth of forty feet, shows a vein five feet thick, abounding in sulphurets, and prospects well. Work is being done on this claim by Messrs. Hoey & Co., its proprietors.

"The Poor Man mine, situated twelve miles southwest from Mokelumne Hill, is owned by the Lewis Brothers & Co. This company own 1,200 feet of one of the finest ledges in this county. Their hoisting-works, which are very complete, are run by an engine of 35 horse-power, now working through a shaft (nearly vertical) 260 feet deep. At this point the ledge is 20 inches thick, and growing larger. They have opened, lately, 30 feet at their 160-foot level—the ledge was well defined at three feet—in the cross-cut, 300 feet from their main shaft. The same machinery that runs their hoisting-works also runs their 5-stamp mill, crushing five tons daily, (24 hours.) The machinery is arranged for 10 stamps."

The Poor Man Company own 1,200 feet on the ledge, and have applied for a United States patent. It was discovered in 1868, by Wesley & Lewis, who have expended $55,000 in labor and improvements. The yield of the quartz has not been uniform. The best grade of rock has yielded $250 per ton, and the lowest grade from $20 to $40 per ton. The pay chute or chimney has steadily increased in length as depth was attained, and on lowest level has a length of 150 feet.

The Wolverine claim runs parallel with the above-mentioned, and only 100 rods distant. It is 1,000 feet long, has a shaft down 140 feet, and a tunnel in 300 feet. At this point they have a ledge four feet thick that prospects $30 to the ton; they have crushed 60 tons. San Francisco parties have thoroughly prospected this mine, and are about purchasing at $35,000. A recent number of the Calaveras Chronicle thus speaks of the development of this mine:

The work of prospecting the lead has proved the existence of three auriferous "chimneys," known as the north, middle, and south chimneys. The latter has been "stoped" out to the depth of 126 feet, the quartz extracted, good, bad, and indifferent, averaging something over $25 per ton. The new engine-shaft passed through a part of the middle chimney 221 feet from the surface. At that point some excellent quartz was taken out containing gold in visible quantities, associated with galena and iron sulphurets. The north chimney was intersected by the shaft at a depth of 182 feet. To the depth now reached, 220 feet, the quartz composing the chimney has rapidly improved in productiveness. The fissure, at present, is two and a half feet wide between walls, and is gradually widening as the shaft increases in depth. The rich stratum recently struck is now about six inches wide, and consists of a beautiful dark blue quartz, rich in gold, and the sulphurets of iron and galena. A level is to be run under the chimneys as soon as the shaft is down 230 feet, which will give "backs" for "stoping" as follows: North chimney, 130 feet; middle, 105 feet; south, 104 feet. Competent judges estimate that the best ore taken from the rich stratum recently discovered will pay as high as $150 per ton, while none of it will fall under $20.

Wet Gulch mine, three-fourths of a mile southwest of the Poor Man mine, is owned by Messrs. Bandman, Nielson & Co., (of the Giant Powder Company, San Francisco.) This mine has been thoroughly prospected by a shaft, 200 feet deep, as deep as could be done without machinery, which will likely be put in operation early in 1872.

Thorp's mine, near San Andreas, is looked upon as a promising mine. It is owned by Captain M. Thorp & Sons, who claim 1,000 feet of a ledge that averages four feet in width, (i. e., the principal vein.) It has three spurs, possibly three different veins. They have an incline shaft down on the ledge 65 feet deep, on an angle of 45°. With three arrastras and two stamps they have been crushing about five or six tons per day. They have just completed a five-stamp mill, and now expect to crush eight tons daily. Their rock averages about $6 per ton, running as high as $10.

The Thorn mine, in the same vicinity, is developed to a depth of nearly 300 feet, and is yielding very high-grade rock. The ledge is 12 feet wide, of which nearly one-third is high-grade rock.

Mokelumne Hill and vicinity.—The town of Mokelumne Hill was formerly one of the most populous in the southern mines, and its past history does not materially differ from that of Sonora. It was settled at about the same time, and like Sonora had a rapid growth and a slow decay. The principal mining interest here was in the old channels, found in the adjacent hills and ranges, which are now nearly exhausted, the future prosperity of the town depending on quartz mines in the neighborhood. The geological features of this region possess peculiar interest, and are thus described by Professor Whitney, in his volume on General Geology:

The sedimentary and volcanic deposits near the town are about 250 feet thick, and rest on a bed-rock which has an elevation of about 1,800 feet above the bed of the Mokelumne River, which is only a mile distant to the north. The upper part of the volcanic ridge is a mass of boulders or fragments of trachytic lava, not polished or smooth, but roughly rounded, as if by friction with each other, unaided by water. There are no other kinds of rock than volcanic represented in this bed, and no stray pebbles, even of quartz or slate—a fact that has been observed in many other places in this region. Beneath this bed of lava-boulders is a mass of strata, in some places nearly 200 feet thick, consisting chiefly of volcanic ashes, stratified and consolidated by water. These sedimentary volcanic strata are often fine-grained and homogeneous, having a light pinkish-red color, and breaking with a conchoidal fracture. The upper beds have much pumice mixed with them. Immediately on the bed-rock lies the stratum of pay-gravel, which in some places has been of the most astonishing richness.

The early mining laws of Mokelumne Hill limited each location to 15 feet square on account of the great richness of the ground. The channel was reached by deep shafts, sunk through the overlying volcanic matters or by tunnels run into the mountain-side, and success was a matter of chance. Sometimes a fortune would be taken from one claim, while adjoining claims would prove entirely unproductive. Some of the shafts were sunk to a depth of 300 feet, and many tunnels were run more than 1,000 feet. The thickness of the pay-gravel rarely exceeded 2½ feet. One of the layers of the channel on Stockton Hill, immediately southwest of the town, contains opals in great abundance. These opals were sent to France in large numbers, and there cut, but we cannot learn that the enterprise proved remunerative, and the casual visitor is welcome to dig out from the claim all the crude opals he may desire.

Most of the claims in the mountain have been practically exhausted, and such claims are now engaged in working over their "tailings," which have accumulated to great depth. Some idea of the great richness of the gravel may be formed from the results of the washings of their tailings. At the claim of Paul & Co. in cleaning up around the old dump-box—less than a week's labor—300 ounces of gold were obtained. At that rate the tailings will prove as valuable as the original mine, making it one of the very best gravel claims in the county.

Below the town, and to the south, is a deep depression between the mountains known as Chili Gulch. Near the head of this gulch hydraulic mining has been carried on by Shaw & Co. with profitable results. As the price of water here was 20 cents per inch, the ground must have been remarkably rich to afford a profit. The hydraulic ground was neither deep nor extensive.

The bottom of the ravine or gulch was exceedingly rich in gold, and in early times supported a large mining population. There are now but

H. Ex. 211——6

few companies engaged in mining. The principal claim of the neighborhood is the Indian Ravine Tunnel claim, situated one mile southwest of Chili Gulch, and owned in San Francisco. This is a blue-gravel, deep-channel claim, 1,200 feet in length, worked through a tunnel 2,800 feet long. The pay-gravel is about 60 feet wide and 6 feet deep. It is brought by car to the dump, at the mouth of the tunnel, and washed by hydraulic process. The pay averages about $1 per car-load, and 150 car-loads are taken out daily, (24 hours.) The tunnel cost $28,000. Air is forced into this tunnel by a water-blast. This is a fall of water arranged at the mouth of the same, 70 feet high, falling through a pipe into a tub, with another tub (inverted) of less diameter over the same, so arranged as to let the water escape but not the air. From this upper tube the air is conducted into the tunnel by a pipe.

The What Cheer Mining Company have recently opened some ground on an ancient channel about two miles south of Mokelumne Hill. They own 3,000 feet of ground, which was prospected in 1862 by a shaft 190 feet in depth, which struck the channel, but the claim was abandoned in consequence of the difficulty in pumping out the water. The present company have opened the ground by means of an incline 500 feet in length. The rim of the channel was struck at a perpendicular depth from surface of nearly 200 feet. The rim-rock was followed down until they reached the bottom of the channel, when drifting was commenced. This is known as the Corral Flat channel. The company have erected hoisting-works, run by water-power, using only twelve inches of water under a fall or head of nearly 300 feet, with a six-foot "hurdy-gurdy" wheel. Their water-power costs but $3 per twenty-four hours. The machinery for running the claim cost but $1,500, and the incline about the same amount. Two large dump-boxes have been erected, with a capacity of 1,000 tons each. The dirt will be run through 200 feet of sluice-boxes, by water raised from the mine by their pump. The total running-expenses of the claim, including labor, are said not to exceed $75 per day, which leaves a large margin of profit on low-grade dirt. The claim is owned by practical men, and is a notable instance of economical management.

Bates & Co. are working the old channel near Mokelumne Hill, through a tunnel 900 feet in length, with large returns. The Paul claim was opened by a tunnel 1,100 feet in length, which was found too high to reach the bed of the channel, and was cut down and run 1,000 feet further. Notwithstanding this great outlay it is said to have paid well.

Quartz mines near Mokelumne Hill.—The Paloma or Gwin mine, situated six miles west of Mokelumne Hill, at Lower Rich Gulch, is unquestionably on the Mother lode. Between Angel's and this point the croppings of the lode disappear for several miles, though recent developments are in favor of the continuance of the lode. The Paloma is the property of Messrs. Gwin & Coleman. It is said they spent nearly $100,000 on the mine and machinery before it became productive, and were once on the point of suspending operations. It is now one of the leading mines of the State. This claim is 2,800 feet long, and the vein is on an average 10 feet thick. They are working through a shaft 500 feet deep on an incline of about 60°, employing 50 men, running two mills, one of 20 and the other of 16 stamps. Only 93 inches of water are required to run both mills. It is accomplished by conducting the water to a 7-foot hurdy-gurdy wheel at each mill, through a hydraulic pipe under a 280-foot pressure. They crush daily (24 hours) 65 tons of rock, that averages them $10 per ton. Work will be commenced upon the 500-foot level as soon as the "sump" is sunk 10 feet deeper. The lead is looking first-rate in the bottom of the shaft, better than at any other point. When the whole battery is put in motion the mine will pay $1,000 per day. The owners expect to have 100 stamps in operation early in 1872. A late number of the Calaveras Chronicle says of a recent "strike" in this mine: "A stratum of rock of remarkable richness has been discovered at a depth of 400 feet. The rock closely resembles the slate of which the walls of the lead are composed. The stratum mentioned lies next to the foot-wall, and is about a foot in width and thickness. The inclination of the 'streak' differs slightly from that of the shaft. Commencing at a point near the surface it crosses the latter diagonally, leaving the shaft at the 400-foot level. By 'drifting' a short distance on the lead, however, from the 500-foot level, the stratum will undoubtedly be struck again. By the merest accident it was discovered that this singular deposit, instead of being slate, was simply a mass of sulphurets, carrying free gold in abundance; 2½ ounces yielded 166 grains of gold, or at the rate of $53,140 per ton. The rock assayed was broken from a chunk at least a foot square, and is considered a fair test of the richness of the whole stratum."

The Whisky Slide mine, six miles southeast of Mokelumne Hill, was located ten years since, but has been worked only at intervals until recently, when a 10-stamp mill was erected and the mine opened by a tunnel 350 feet in length, which cuts the lead at a depth of 100 feet. The ledge is in slate and has a width of from four to sixteen feet, the width gradually increasing with depth, but, as in all slate formations, the vein pinches and swells as it is opened on a horizontal line. The pay has been of low grade, but in consequence of the great facilities for extraction of quartz and the width of the vein, it has been milled and mined for $6 per ton. This includes the expense of hauling the rock one mile from the mine to the mill. The treatment is by amalgamation in battery, copper plates, and blankets.

This county possesses numerous other promising veins of quartz in various stages of development, and is destined to rank as one of our leading quartz counties. Its copper veins are also extensive and rich. They have been noticed elsewhere in this article.

The county assessor's report for 1870–'71 returns the number of quartz-mills as twenty-eight, which is probably correct, as it does not materially differ from Langley's list. The aggregate of quartz crushed is reported by the same authority as 130,000 tons, but this is probably a clerical or typographical error. The total number of stamps will not exceed 300, of which about 200 have run during the past year, and these not regularly.

AMADOR COUNTY.

This county is the smallest of the group we have included under the general designation of the "southern mines." It has a width, from north to south, of only twelve miles. The principal interest of the county is quartz-mining, some of the best mines being on the Mother lode. In this respect the county is second only to Nevada, both as to the number of tons crushed and the yield of gold from this source. The placer diggings, river mining, and other branches of mining of this character, are very limited and not worthy of special mention. The limestone belt noticed in the description of Tuolumne and Calaveras Counties extends into this county, where it is found at Volcano, but to the northward of this point it loses its continuity. A few remnants of volcanic action are found in the southern part of the county, near the Mokelumne River. The Mother lode enters the county near the Mokelumne River, the southern boundary, and has been extensively and almost continuously developed to the Cosumnes River, the northern boundary of the county, where it loses its characteristic features, and is known no further by that name.

During the past year a remarkable stagnation has existed in quartz-mining. Many of the principal mines were closed for several months in the spring and summer on account of a strike of the miners. This strike occurred during the very brief season of water-supply, so that when the miners resumed work water was becoming very scarce. During the latter part of the summer, owing to the drought, all the quartz-mills were idle with the exception of the Oneida, which, fortunately, obtained enough water to run forty stamps during most of the season. The fact of the quartz-mines not being worked has been a great calamity to the entire county. From seventy-five to eighty thousand dollars, which would otherwise have been dispensed to the laborers monthly, has remained in the bowels of the earth, consequently working a great hardship to all branches of industry. The leading mine, the Amador, which paid in dividends during the year 1870 the sum of $155,400, only disbursed $24,000 the present year, (1871.) Many other mines were forced to suspend operations during the most favorable season of the year, and great loss was entailed on the owners. Notwithstanding this, active operations were conducted in a number of claims which have for several years remained idle. Deep shafts have been sunk and new and powerful mills erected, and it is certain that the year 1872 will witness an unparalleled yield from this section.

The town of Sutter Creek contains a population of about 2,000, and is the largest and most prosperous of the southern mines. Here are found the principal mines. These have been fully described and illustrated by cuts in former reports, and we will simply note the operations of the past year.

The Amador mine is now in complete working order for the first time in nearly eighteen months. In April, 1870, the main shaft was burned out, and the time until the present has been devoted to repairing the shaft and putting up new hoisting-works, and it is now the best and most complete shaft and hoisting-works on this coast. In place of the old bucket plan of hoisting rock, they have a substantial "cage," on which iron cars, capable of holding eighteen hundred pounds of quartz, are run and speedily hoisted to the surface, and many other modern improvements have been introduced. The company claim 1,800 feet on the ledge. They have two sets of hoisting-works, each of which is run by an engine of 80 horse-power. The upper (or old) shaft is down 1,250 feet, on an incline of 70°; at that depth the ledge is 10 feet thick. Their new shaft is down 1,300 feet, on the same incline as the old shaft, and built in three compartments.

The company own three quartz-mills: one of 40 stamps, run by steam-engine of 75 horse-power; the other two, of 16 stamps each, run by water-power. The 40-stamp mill is run by the company, and crushes daily (24 hours) 80 tons of rock. One hundred men are regularly employed.

The mine was visited in July by an intelligent correspondent of the San Francisco Bulletin, who gave the following account of its appearance at that time:

Through the kindness of General Colton, vice president of the board of directors, and Mr. Steinberger, superintendent of the mine, we were permitted to descend into the Amador mine on Monday last, for the purpose of inspecting the immense underground excavations. This mine is situated on the great Mother lode, which extends through Amador to Calaveras and Mariposa Counties, and has been in active operation for nearly twenty years. The surface-workings overlook the quiet little city of Sutter Creek, which is located in a basin beneath, surrounded on all sides by an unbroken circle of hills. It is upon the Amador mine that this place is most dependent for its support.

Entering the dressing-room we were furnished with a suit of miners' working-clothes from among the many that were to be seen hanging on the walls all around, and at once proceeded to descend the north shaft. This shaft, it will be remembered by those interested in mining matters, was destroyed by fire, in May last, the same lasting thirteen days. The repairs are now nearly completed. The shaft, as it now is, is much larger than it formerly used to be. It is now twelve feet long, lying in the same manner as the vein runs, north and south, and embracing the entire width of the lode. The timbers employed for lining it are all square-sawn, measuring twenty inches in diameter. These timbers are placed at a distance of five feet apart, from center to center. Stout planking is used to support the foot-wall, which is a soft, rotten slate. The hanging-wall, being granite, needs no such support. This shaft is again subdivided into three compartments, in two of which patent cages will be employed in hoisting ore, instead of the iron buckets now in use. Entering a large iron bucket, three of us descended in charge of W. Jones, the foreman. We soon discovered that the shaft is nothing more or less than an immense underground incline, as it follows the dip of the vein from west to east, at an angle of about 75°.

For several hundred feet down the lode has been completely exhausted. The present working-levels are named after the principal stations on the way to New York, the first being named the "Latrobe;" the second, "Folsom;" the third, the "Sacramento;" the fourth, "San Francisco;" and the fifth, or lower level, "Panama." By these various names they are all known and readily distinguished. Our transit from the surface to "Panama" level, a distance of 1,250 feet, occupied about fifteen minutes' time. To raise ore from this level only occupies about one minute. This is the prospecting-level in the mine. At present, the vein measures twelve feet in width, and yields an average of about $10 per ton. Although "Panama" level is some 150 feet below the level of the sea, it is proposed to sink 200 deeper in the course of a few weeks. It is now the deepest excavation in search of gold existing throughout the world.

The Badger or south shaft is used for draining the mine. Large iron cylinders, of about twelve feet in width, having a carrying capacity of 300 gallons, are used for that purpose. These are hoisted by means of steam-power, at the rate of one per minute. During the summer season the mine is kept quite dry by running eight hours only of the twenty-four. In the winter or wet season the increased amount of water forcing its way into the mine nagally compels them to abandon the lower levels.

Every possible precaution is taken to insure safety and to prevent accidents occurring; nevertheless, I am informed that as many as thirty-eight men have been killed since the opening of the mine, a proportionately large number having been injured also.

The ore, as extracted from the mine, is at once conveyed to the mill, situated at a distance of about one-quarter of a mile from the shaft. This mill contains seventy-five stamps, forty of which are usually employed when the mine is in full blast. At present, twenty only are at work, inasmuch as there are only forty miners engaged in the underground excavations. It is here that an average of about two thousand tons of ore per month undergoes a thorough reduction, yielding an average of about $22 per ton, or an aggregate of about $45,000 per month. Previous to the present proprietorship, the reduction of the ore was very much neglected. The concentration of sulphurets was altogether disregarded, and they were allowed to escape. Hundreds of thousands of dollars have thus escaped into the bed of Sutter Creek, and are now probably irrecoverably lost. The concentration of sulphurets is now carefully attended to, and the gold extracted from them brings in a large revenue, yielding at the rate of $100 and upwards per ton. All the sulphurets saved at the various mills in the neighborhood are reduced at the Chlorination Works of Messrs. Jones & Belding. The yield of the sulphurets of the various mines ranges from $90 to $130 per ton.

Unfortunately, this promising state of affairs was again disturbed by trouble with the Miners' League, resulting in destruction of property and life. Consequently the annual report of the Amador mine shows only four months' running, 17,790 tons of ore raised, 16,490 tons crushed, giving $12.21 per ton, or $201,357, of which $24,000 was paid to stockholders.

The following extracts comprise the substance of the reports of the president and superintendent, January 1, 1872:

PRESIDENT'S REPORT.

It is a matter of no little regret to me that events, over which the officers of this company could have no control, have prevented my making as favorable a report, as to the product of the mine for the past year, as we all hoped. It is quite satisfactory, however, to know that the small yield has not been the fault of the mine, or its management. The large amount of dead work, (the result of the late fire,) the unusual scarcity of water, which compelled the mills to remain idle for several months, have been the main hinderances of a satisfactory return to all parties directly interested.

I would recommend the erection of a first-class 60-stamp mill at an early day, in order that we may be able to crush the large amount of low-grade ore which has accumulated in all parts of the mine, especially north of the new shaft. My predecessor, Colonel Fry, wisely made the same recommendation at the close of last year; but misfortunes, which I have before noted, have prevented this most important improvement. In addition, the Canal Company, on whom we would be dependent for water to run so large a mill, have been unable to complete the large ditch which we had hoped would furnish us an abundant supply of water. Although disappointed in its completion, it is but justice to say that it is one of the best-constructed ditches, to the point now reached, I have ever had the pleasure of seeing on the Pacific coast. And the officers of that company assure me, in the most earnest manner, that they will be able to furnish us, from this ditch, all the water we may require for our new mill, by the time it is erected.

In all the extensive improvements consummated during the past year, such as the sinking of the north shaft and erection of the new hoisting-works, the result has been most satisfactory.

The unceasing care and attention of Superintendent Steinberger to all the varied interests of the company under his management, during the past year, warrant me in saying that no mining property west of the Rocky Mountains has been more faithfully, honestly, and economically conducted. I say this from personal observation, having visited the mine during the past year two or three times each month. In conclusion, I would most earnestly recommend that the sinking of the new or north shaft be continued, as the shaft is constructed with a view of continuously going down on the vein, without interfering with work in the other parts of the mine. And should the vein continue, as it now shows at the bottom of this shaft, we hope to develop, in the next year, a large amount of valuable ore.

Hoping that my successor may have less care and anxiety in the management of this extensive and valuable property for the coming year than I have had in the past, with more satisfactory dividends,

I remain, very respectfully, yours,

DAVID D. COLTON,
President.

SUPERINTENDENT'S REPORT.

Herewith I send you statement of operations for the past year at the company's mine.

During the year 1871, 17,790 tons of quartz were extracted from the mine, and 16,490 tons were crushed at the Eureka (40-stamp) mill; 1,300 tons now on the dump at Rose mill.

A large amount of dead work has been done in various parts of the mine during the past year. The cleaning out and partial retimbering of the Panama drift, from north shaft to its face, (175 feet,) which was filled up with *débris,* retimbering old levels, opening drifts under pillars, cleaning out and retimbering the Latrobe level, &c., besides the sluking of the north and middle shafts.

The Badger shaft (which is the water-shaft) has been kept in good repair, and is now in condition, as is also the machinery, to contend against any reasonable increase of water. A prospecting-drift has been driven south from Badger shaft on the Panama level 220 feet, but no rock in paying quantities was found.

The Middle shaft has required considerable repairs, but is now in fair condition. This shaft has been sunk below the Panama level 155 feet, 105 of which was sunk the past year. Sixty feet below the Panama level the New York level was started, and 60 feet below the New York was started the Green level, leaving a sump of 35 feet under the Green level. The cost of sinking this shaft, including the turning off of Green and New York levels, was $4,853 50, or 46.22\frac{2}{7}$ per foot. The vein in the bottom of shaft is four feet in thickness, and the quality of the rock good.

The North or new shaft is now below the Panama level 95 feet, and cost $5,056.06, or 53.22\frac{3}{4}$ per foot.

The turning off of the New York level is included in the above figures. The average thickness of the vein in the bottom of shaft is seven feet. On the 1st of September last the Panama level was reached with this shaft, being 1,165 feet from surface, 460 feet of which was sunk the past year.

The entire cost of sinking this shaft from within 180 feet of surface to Panama level, a distance of 985 feet, has been $33,299.88, or 33.06\frac{88}{98}$ per foot. These figures embrace lumber, timber, lights, smith-work, labor &c., also the turning off of levels at Latrobe, San Francisco, and Panama.

The time occupied in the sinking and turning off of these levels was eleven months, two months having been lost in consequence of the strike. The timbers used for sills and legs are sawed red spruce, twenty inches square. The center-pieces and caps are 12 by 20 inches. The timbers are five feet apart from center to center and logged between with three-inch red spruce plank. The shaft is divided into three compartments, two cage-ways and one bucket-way. The size of shaft is in length in the clear thirteen feet, and in width is from six to twenty feet. Over the shaft, on the surface, have been erected first-class hoisting-works. Everything in connection with these works is in splendid condition and works admirably.

The north drift on Panama level is now north from North shaft 395 feet, 230 feet having been driven the past year. The vein in the face of drift is over sixteen feet in thickness, with a good regular hanging-wall. The quality of the quartz has improved very much in the last few feet that have been driven. From the general character and regularity of vein and wall we have every confidence in the permanence of the vein, and have reasonable assurance that, from the developments thus far made in this drift, there exists a large quantity of quartz above this level, as none of the levels above the Panama level have been driven over 180 feet north of North shaft. A level has been driven north under the old San Francisco level, and is north from North shaft 180 feet. From this level considerable bowlder-rock has been taken out, and large quantities still remain in place.

The Latrobe level has been cleaned out and retimbered to face of drift, being 80 feet north of North shaft. Large quantities of bowlder-rock are developed on this level.

The only stoping that has been done in the mine the past year has been above the Panama level, north of North shaft, on the bowlder-vein.

The only ledge-rock that has been taken out (excepting that from the sinking of the shafts) has been taken from pillars under the fourth, fifth, and sixth-hundred feet levels.

The great bulk of the rock mined and milled, the past year, was from the bowlder-vein.

The only quartz taken out below the Panama level has come from the sinking of the North and Middle shafts. But little has come from the Green and New York drifts, as they are as yet but a few feet from shafts.

The quartz is now in place from Panama to Green level, excepting what has been taken out from the sinking, giving us 120 feet of backs.

The 40-stamp mill is in good running condition. The water-wheel, which is old, has been repaired, and will, with occasional slight repairs, run this winter. The Rose mill has been repaired, and is running well. The Badger has also been repaired, and will commence crushing in a few days.

The working of the mine the past year has been attended with many difficulties and heavy expenditures, the strike and scarcity of water forming a part.

The vein in the lower part of the mine (Green and New York levels) looks most promising, and, so soon as the drifts are driven ahead far enough to open stopes on these levels, we will be able to get out sufficient ledge-rock to keep the mills running; until then we will be obliged to take out considerable bowlder-rock to keep up the supply.

The entire mine and machinery (as is everything else in connection) is in good working condition, and, unless some unforeseen accident takes place, no fears need be apprehended of the future profit of the mines.

Respectfully, yours,

JOHN A. STEINBERGER,
Superintendent.

The Oneida mine, south of the above, is one of the best developed in the county, and has been described in the Mining Commissioner's report for 1869, since which time no material change has occurred in the character of the rock. This mine likewise suffered severely from the strike of the miners and the drought. It is situated about half way between the towns of Jackson and Sutter Creek, and is owned principally in those towns. The length of the claim is 3,000 feet. They have three different incline shafts down on an angle of 65°. The first is down 300 feet, the second 700 feet, and the third 800 feet. They are working only through the two latter. The ledge will average 12 feet thick. They are working 100 men. Their 60-stamp mill is run by an engine of 60 horse power, and crushes daily 90 tons of rock, (24 hours,) which averages $10 per ton. The hoisting-works are run by two engines, one of 30 and one of 15 horse-power. This mill is complete with all the accompanying pans, machinery, &c., to successfully run a mill of its caliber. Drifting was commenced on the 800 feet level in the latter part of 1871. On the 700 feet level the drifts have been run 400 feet on either side of the shaft. Two pay chutes or chimneys run through this location. One has a length of 700 feet, the other of 400 feet, and both dip to the north. The mine has immense reserves in sight, and is a very valuable property.

The Kennedy mine, situated next adjoining the Oneida, on the south, will doubtless soon become one of the leading mines of the country. They own 2,300 feet of ground. Up to the present year hoisting was done by means of a whim, but the claim passing in the hands of energetic owners in the early part of 1871, powerful hoisting-works were erected and the shaft sunk to a depth of 500 feet. In sinking, the vein was followed at an angle of 60°. The first level was opened at 350 feet, and drifts were carried 200 feet south and 180 feet north. The next level was opened by the present owners at 500 feet. The vein is here four feet wide, the quartz averaging $20 per ton, though rock of a much higher grade has been struck since our visit. This claim has a foot-wall of granite and a hanging-wall of slate. An old shaft, sunk many years since at the north end of the ground, near the Oneida line, developed the existence of a rich chimney in that part of the mine. It is in contemplation to clean out this shaft or sink a new one at this point. Three distinct pay chutes have been developed by the drifts. The company have erected a first-class 20-stamp mill, which is constantly supplied with rock.

The Keystone mine at Amador City, one of the extreme northerly claims of the Mother lode, and considered by many second to none in Amador County, has also been described in the report for 1869. The mine is owned in San Francisco. It consists of 3,000 feet of ground, and their improvements consist of two sets of hoisting-works and a mill. The latter, containing forty stamps, is run by an engine of 125 horsepower. The hoisting-works are also run by steam—one by an engine of 20 and the other by one of 80 horse-power. They have two principal shafts sunk upon the ledge. The north shaft is down 346 feet, on an angle of 30°, and the south shaft is down 500 feet on the same angle. Four hundred feet down the vein runs from 10 to 30 feet thick. They regularly employ 100 men, and crush daily (24 hours) 80 tons of rock, which has averaged $16 per ton.

The following is condensed from the correspondence of L. P. McCarty to the Mining and Scientific Press of San Francisco:

The Summit mine, (south extension of the Amador,) near Sutter Creek, is owned in San Francisco. This claim is 1,400 feet in length, and has fine hoisting-works, run by an engine of 45 horse-power. Two shafts are sunk within 110 feet of each other. One is down 300 and the other 500 feet, on an incline of 45°. This mine was first struck in 1869, and from a chimney down 165 feet some $10,000 were taken out. At that time the ore ran from $16 to $32 per ton. Work is now suspended on this mine.

The Maxwell mine, situated close by the above-mentioned, is owned by an incorporated company in San Francisco. This claim runs north and south, and is 2,000 feet long. The hoisting-works are run by an engine of 20 horse-power, over a shaft down 750 feet, on an incline of 75°. At that point a cross-cut of 70 feet reveals a ledge 9 feet thick. They get out from 45 to 50 tons per day, and are working 25 men. The rock is crushed at the Badger and Rose mills, the two 16-stamp mills which they rent of the Amador Company, and are situated one mile distant. This rock is low-grade ore, averaging from $5 to $6 per ton, but is easily mined and crushed.

The original Amador mine is situated in the vicinity of Amador City, and is owned by J. A. Faull & Co., (a joint-stock association.) They claim 1,200 feet of ledge; have a fine set of hoisting-works, run by a steam-engine of 25 horse-power. A shaft is sunk 360 feet; at that point a ledge 3½ feet thick is found. Twenty-one men are regularly employed, getting out from 10 to 15 tons of rock daily. They are now erecting

A splendid 40-stamp mill, which will be completed in a short time. They have also opened a new shaft from the 200-foot level, which is intended for the main rock-shaft. Their mine is doing finely, with an abundance of excellent quartz in sight.

The Medeon mine, situated midway between Sutter Creek and Amador City, is owned and superintended by L. R. Poundstone. Hoisting-works are erected upon the same, and run by horse-power. This mine was worked as early as 1857, and paid, as far as worked at that time, about $30 per ton. Since that time it has been idle until now. Six men are at present employed sinking a new shaft, which is now down 75 feet. At this point they have found a ledge four feet thick, and still increasing, the prospects of which are quite favorable.

The Little Amador mine is being rapidly developed. The main shaft is down nearly 400 feet, and still progressing. The ledge is well defined to the lowest depth reached, and the rock taken out in sinking is rich. The mill connected with the mine will be completed and ready for crushing early in 1872.

The Maxon mine is located a short distance south of Amador City. The shaft has been sunk to a depth of 200 feet, with a drift of 80 feet, showing a good ledge of rock. Some of the rock crushed pays very satisfactorily. They have several hundred tons now on the dump.

The Zeile mine (formerly known as the Coney) is situated one-half mile south of Jackson, and is owned by San Francisco capitalists. This company own 800 feet of a ledge, and have developed it to a depth of 515 feet. Their hoisting-works are run by a 20 horse-power engine; the shaft is down on an angle of about 50°; their ledge is about 11 feet thick.

The town of Volcano, situated about twelve miles east of Jackson, contains 800 inhabitants. It is situated on what is known as the great limestone range. Its mines have been worked continuously since 1849.

The Amador County Canal and Volcano ditch, both of which are owned by San Francisco capitalists, supply the district with water. They take their supply of water from the North Fork of the Mokelumne River. This canal was originally a flume, thirty-one miles in length, with a carrying capacity of 900 inches, and cost $450,000. It is now being replaced by a substantial ditch. The latter (the Volcano ditch) cost in the neighborhood of $200,000, and has a carrying capacity of 700 inches, the income of which satisfies its owners.

The Markley mine is four or five miles from Volcano. About two years ago operations were commenced, and a depth of 280 feet has been reached. At the surface the main chimney is 2 feet thick and 60 feet long; at the 200-foot level, 4 feet thick and 280 feet long. From rock taken out of the 200-foot level they obtained $14,000, after which, in four weeks, they "stoped out" rock from which they obtained $8,000. Last fall they sank 60 feet, and put up steam hoisting-works of 15 horse-power. The company recently started up their twelve stamps, 350 pounds each, with which they pounded out $1,500 per week, and ran ten hours per day. This mine has recently been sold to an English company for $35,000.

Volcano and vicinity contains several other quartz mines of great promise, in various stages of development.

The assessor's returns report 70,360 tons of quartz crushed for the year 1870, by 27 mills. These returns are probably more correct than those of any other county in the State, as the county is small and the principal mines situated near the county-seat. The returns for 1871 will probably show a material decrease for that year, on account of the strike and the drought, but the year 1872 will undoubtedly be one of unparalleled prosperity. Three or four new first-class mills, of forty stamps each, will be ready for crushing early in 1872. This county possesses, according to the Pacific Coast Directory, 35 mills, with an aggregate of 579 stamps. Of this number, 384 stamps have been in operation during 1871, when water was available—a larger proportion than is shown by any other county in the southern mines.

SAN DIEGO COUNTY.

The promises held out last year by the new mines of this county have been realized to a considerable degree. Not alone have the mines then discovered and worked continued to yield encouraging amounts of the precious metals, holding out well at greater depth, both as to width of veins and contents of gold in the ore, but a large number of new ones have been discovered, some of which have turned out very satisfactorily.

Julian district.—The control of the California claim was bought in the early part of the summer by Messrs. Snyder, Morris & Co., of San Francisco. They straightened and timbered the shaft, and in sinking it deeper the vein was found to vary from two to four feet in width. The ore taken out during this work is reported to have yielded $57 per ton in the mill. The company continued sinking during the rest of the year, intending to put the shaft down 400 feet. I do not know whether they reached this depth. In December, they were reported to be ready to commence drifting and stoping.

The Owens mine, the original discovery on the same lode with the California, had a shaft 130 feet deep in August, in which the vein was four feet wide and showed some fine gold. The ore is raised by a whim, the first one built in the district. The shaft was sunk to 180 feet by October, and the mine is said to have paid very well. In October the air in the shaft became bad, and it was necessary to run a drift to connect the mine with adjoining works, in order to improve the ventilation. The vein exposed at that time in the lower part of the shaft was larger than above and contained excellent ore.

The Helvetia, which by many is considered the best mine in the district, had its shaft down 140 feet in August and raised good ore steadily, though the quantity was comparatively small, as from most mines in the district. The mines of the Lone Star Company and the Big Blue ledge are also considered very good property. The Stonewall had a shaft 100 feet deep, and at a depth of 60 feet a level 180 feet long to the north, and one 100 feet long to the south, in October. Stoping had not been commenced. The ore found was all decomposed, and there was no water in the mine. The mill of the company could only run five hours per day for want of water.

Banner district.—A number of mines appear to have been worked in this district. A correspondent of the Scientific Press, writing July 20, says in regard to them:

The Kentuck Company, under the superintendence of M. A. Lewis, controlled by McDonald & Whitney, of San Francisco, who have recently bought the mine, started up the work of sinking, this week, with full force, and on yesterday they struck some more of the rich ore for which that claim has been celebrated, being literally filled with gold. The Madden Company, on same ledge, have their usual amount of rich ore, and the ledge is improving every foot they sink. The last ore worked paid over $40 per ton. The Antelope Company, on the same ledge, are still running their new five-stamp mill on their own ore, paying $70 per ton. The Rodman mine, under the supervision of J. N. Tiernan, is looking well. In sinking the shaft the ledge pitched to the west into the hill, where it was about four feet wide. Last week a new and small ledge made its appearance in the shaft, showing free gold and silver sulphurets. It has increased until it now covers the bottom of the shaft, with slate intermixed. The prospects are very flattering, indeed, for a large and rich ledge, when the two ledges come together. The additional machinery brought here and added to McMechan's mill, (the Wilson Steam Stamp mill,) will probably be started up for regular work to-day. It consists of a 16 horse-power engine, a Varney pan, (latest improved,) a large wooden concentrator, and copper shaking table, of Tiernan's own invention. If they do not do first-class work, there is no use in having good machinery. Tiernan has shown himself a man of energy, and competent to carry on such work. The Bayley Company, on the south end of the Redman ledge, are taking out ore to work by the new process in McMechan's mill. They have a large ledge and rich ore. The King ledge, Golden Chariot Mine and Company, are still sinking in rich ore on their big ledge, which, at the depth of 50 feet, is four to five feet wide. The Little Joker, which was three or four inches wide on top, of soft, decomposed, chalky-looking ore, rich from top to bottom, is now 48 feet deep and two feet wide, with some ore that, report says, is worth from $2 to $5 per pound. There are many other mines that I would like to mention, but it will require too much of your valuable space. I will close by simply saying, of all the mining districts on the coast, I have not heard of one that shows as many paying ledges and rich prospects, or as much work done and paid for out of the mines themselves, or from a like quantity of ore worked—that can show as good returns, with as good a climate, timber, water, and everything to make the locality desirable for health or a home, as Julian and Banner districts, San Diego County, California.

The high opinion of this correspondent in regard to the districts named seems to be well founded, if the returns of the mills from ore of some of these mines can be accepted as a criterion. Twelve tons from the Golden Chariot, for instance, yielded $181 per ton; 40 tons of ore from McMechan's and the Redman mines yielded $60 per ton; and 15 tons of ore from the Antelope, and 25 from the Madden Company's mine gave $50 per ton; while a lot of ore from the Kentuck is reported to have yielded $76 per ton. On the whole, however, the mines in Banner, as well as in Julian district, are too little developed yet to give a regular yield. Curiously enough a number of new mills have been erected in addition to the older ones, only, however, to be forced to lie idle a great part of the time for the want of ore. For this reason the want of water, which has been experienced throughout the greater part of the year, has in reality not injured the mining industry as much as might be supposed. My inquiries in regard to the total yield of the county have so far not been answered, but it must have been many times larger than last year. As some San Francisco capital has during the year been invested in the San Diego County mines, it is probable that another year will witness a better development of the mines, and a more regular yield.

SAN BERNARDINO COUNTY.

The mines in the Clark district, which were mentioned in my last report as new discoveries of some importance, have been further explored and partly developed during the year. A considerable amount of high grade ore has been forwarded in small lots to Los Angeles, whence it was shipped to San Francisco. All this ore was of high-grade, but as the richest must, of course, be selected to pay for the enormous land transportation, this gives no criterion as to the general value of the ores or the mines. A number of new locations are mentioned in my advices from that region, but the whole district appears to be yet so much in its infancy that I reserve a detailed description of it for a future occasion.

Of the gold-mining enterprise at Belleville, in this county, I have no advices this year.

INYO COUNTY.

The base-metal mines of Cerro Gordo and vicinity have, according to all accounts, maintained their yield, giving ample employment to the various smelting-works. There has, however, nothing new of importance been developed during the year, and my description of the mines and furnaces in my last report exhausts the subject. My letters to prominent men in Cerro Gordo, asking for information in regard to the correct shipments of base bullion and bullion, have, I am sorry to say, not been answered satisfactorily, and I am therefore not able to give them. From other information I am inclined to believe that they have not been materially increased. The amalgamating ores have received more attention than the year before, and an English company, the Eclipse, has erected a new mill. A series of mines called the Silver Sprout has attracted some attention. An article in the Inyo County Independent, which appeared in October, says, in regard to these mines:

The series comprises some thirty different ledges, together forming a complete labyrinth or net-work of quartz-veins, such as is seldom met with—perhaps nowhere else in the State. Some twenty shafts and tunnels have been opened on this property, within an area of 500 by 2,000 feet, and from the developments already made, the evidence is very conclusive that the property must be a very valuable and extensive one. The two principal ledges, called the "Lamb" and "Silver Sprout," are about 300 feet apart, parallel to each other and to the crest of the mountain. They cut through and across three ridges which extend down the side of the mountain. This gives the best possible position for prospecting at varying heights. Both of these ledges have been opened at various places, from 50 to 300 feet apart, covering, in all, some 1,500 feet. Several of the minor ledges have been prospected, each presenting the same general characteristics of the main, or Silver Sprout.

In order to determine the future course in regard to the management of the mine, Mr. Wingard, the manager, has selected for shipment and reduction in San Francisco about 3,000 pounds of fair, average ore, a portion from the several different openings of the mine. The location, considered in relation to the facilities for getting out and reducing the ore, may be considered as good as could be desired. The altitude is very great, probably not less than 10,000 feet above the sea. It is about three miles distant from, but fully 3,000 feet higher than the famous Kearsage mine, and on the same mountain. It is, however, an entirely different series of mines, though there is a great similarity in the nature and appearance of the ores of the two mines. Both carry a small percentage of base metals, but scarcely enough to be considered "rebellious." The mine is situated upon the southern face of the mountain in such a manner that the snows or storms cannot materially affect the transportation of the ore by tramway, which is the only cheap method practicable.

The mill now owned by the company is down the cañon nearly five miles. The proper place for the mill, and to which it will soon be removed, is immediately below the mine, and about one mile distant. Here there is an abundance of timber and water, a wide valley free from all dangers of snow-slides, and where, by means of a tramway, ore can at all seasons of the year be shipped directly from the mine to the batteries. The whole cost of mining, transportation, and reduction of the ore from the mine, with the proper and practicable facilities, need not exceed $10 per ton. The Silver Sprout is unquestionably a very valuable mine, and with proper management cannot fail to remunerate its owners and at the same time prove of great benefit to this section generally.

The ore is of the class named copper ores in my last report, carries much silver, copper, and a little gold, and a very small percentage of galena. The principal shaft on the mine was 56 feet deep in November, and the ledge was 7 feet wide at the bottom, 5 feet of which was ore.

ALPINE COUNTY.

My efforts to obtain direct communications in regard to the mining industry of this county have been unsuccessful. But the delay in the Public Printing Office in printing my present report enables me to introduce here an excellent letter of Mr. J. Winchester, of the Globe Company, which appeared in the New York Tribune in the early part of 1872, and which treats very exhaustively on the mining resources of this county:

Alpine County was set off from Amador, and lies upon the summit and both slopes of the Sierra Nevada range. Geographically the larger portion of the county belongs to Nevada. Through the central portion, on the east of the main divide, runs the east or main branch of Carson River. Silver Mountain, the county-seat, is about 8,000 feet above the sea. Silver Mountain Peak, a few miles further south, is over 11,000 feet in height. The mineral belt running through Alpine is geologically traceable from, and is believed to be a continuation of, that upon which the Comstock lode is situated, the courses of the veins being identical, and exactly in range from north to south. When the excitement following the discovery of the Comstock was at its height, Alpine County, then a part of Amador, was overrun with "prospectors," and in 1867 a large installment of the prevailing mining furore was transplanted to what was then known as the "Silver Mountain mines," that town being the center of attraction. In 1862–'63, two daily lines of stages, with frequent "extras," were required to accommodate the rush into Alpine County, and Silver Mountain became a mining camp of from 1,500 to 2,000 people. In 1863–'64, the speculative bubble burst, and a gradual decadence set in, which reduced the population in 1868 to less than 200. The gradual change of the mining center from Silver Mountain, on the edge of the granite formation, to Monitor, in the midst of the trap and porphyry belt, continued to depopulate the former town, till at this time the number hardly exceeds 30 persons of all ages, while Monitor has steadily increased from a hamlet of a few families to a lively village of some 300 inhabitants. This town is situated two miles east of Carson River, on the creek of the same name, which rises in the dividing range between Carson and Walker Rivers. It owes its name to the exploit of Captain Ericsson's "cheese-box" in the hand-to-hand contest with the rebel ram Merrimac, an event that so electrified the country, and which saved the nation from a great disaster. So the people named this new mining camp Monitor. The deep cañon of Monitor Creek, being almost due east and west across the rock "formation" of the country, exposed the mineral ledges in numerous places between the river and town, and every one, large or small, was "located" in 1862-'63, and most of them regularly organized under the general incorporation act of California. Without capital, except muscle, the miners set vigorously at work in developing their various properties by shaft and tunnel, levying small assessments every "now and then;" but gradually muscle gave up the contest, and, by 1865-'66, work on the mines had almost entirely ceased, after some of the tunnels had been run a distance of 1,200 feet or more.

The most prominent "location," in the pioneer Alpine days, as now, was the "Tarshish." Its outcrop of porphyry looms up in bold relief from near the summit of Monitor Mountain—an eminence of about 1,000 feet above the creek, on its northerly side—and runs southerly, crossing the creek, and again cropping out in massive proportions, when it receives the name of "Esmeralda," and once more shows its crests on the easterly slope of a ridge of Globe Mountain. The north side of the cañon, half a mile westerly (from Monitor to Red Mountain) is one grand "outburst" of mineral matter, in some places, as on that portion called the Hercules lode, rising several hundred feet above the surface in almost perpendicular cliffs. The "Tarshish" has been in course of development—lying idle sometimes a year or more at a time—for eight or nine years, showing a degree of faith and pertinacity on the part of its owners which is now on the eve of a munificent reward. The ore is a pure sulphuret of silver—silver glance—the "gangue" being often a putty-like kaolin, but more frequently a harder rock of the same character. When concentrated, the ore assays from $1,000 to $3,000 per ton; but, as it comes from the mine, ranges from $50 to $200. The Tarshish is a New York company, the "Schenectady Gold and Silver Mining Company" being the owners; and Judge Potter, of the supreme court of the State, one of its most active and efficient directors. Under the superintendence of M. Schwerin, who has resided here for three years, the mine has been steadily developed, and one of the best 20-stamp mills on the Pacific coast erected at a cost of from $60,000 to $75,000 in coin. It is, as I write, starting into operation, with good ore "in sight" sufficient to keep it running night and day for at least three years. The motive-power is a 120-horse engine.

The Monitor and Northwestern is a company organized under a special act of the legislature of Wisconsin, with headquarters at Milwaukee. Its property is located in several different places in the district. Their first and most largely developed claim is on the Tarshish lode, next adjoining the Schenectady Company's mine. It is opened by an upper and lower tunnel, connected by a "winze," and explored by drifts and tunnels of about 1,500 feet in length. The lower level is about 90 feet deeper than the present lowest workings of the Schenectady Company's on the lode; and, as the exploration progress to the north, on the course of the vein, further into the heart of the mountain, the quality and quantity of the ore steadily increase. The Monitor and Northwestern Company also own a series of claims on Red Mountain, which will be opened by a main tunnel from a point on Carson River, by which great depth on the lode will be gained. They have about 2,500 feet of tunnel yet to make before a sure thing is struck, though some very promising though smaller lodes will be crossed before. If a judgment may be hazarded from the "croppings" on Red Mountain, the body of ore which the main tunnel will open up will equal, if not exceed, in extent and value the famous bonanza in the Crown Point and other mines on the Comstock. A first-class ore-house, on the south side of the creek, opposite the mouth of the tunnel, and reached by a substantial trestle bridge, has just been completed at a cost of over $3,000. The mill of this company is a large and convenient one, situated on Carson River, half a mile below the junction of Monitor Creek, and completed a month ago. Its capacity is 30 tuns, and it is equipped with stamps, pans, &c., of well-tried kinds. Except the furnace for roasting or chloridizing the ores, which is rather experimental, everything is first class. The machinery is run by water, and a turbine wheel is used. The power is ample for the reduction of 75 tons per day, which will be utilized as soon as the mines are properly opened. The mill was lately started up to test the fitness of the furnace to do its work, and as soon as a decision is arrived at as to the best method of roasting, steady work will be begun, and bullion regularly shipped. Such delays seem unavoidable in almost every instance of commencing the work of reduction, and, though vexatious, militate nothing against eventual success. Very few companies but have had to pass through the same ordeal.

The Silver Glance is also a Milwaukee company. Its claim is next south of the Monitor and Northwestern, also on the Tarshish lode, and, though only recently explored with any force, is developing some splendid ore-bodies. Being deeper than either of the other claims on the same lode, the ore is of a better quality. The company have a working-fund stock sufficient to open up the mine fully, and erect first-class reduction-works, which latter they propose to do as soon as the Tarshish and Monitor and Northwestern mills have proved what kind of machinery and processes are best adapted to the treatment of the ores.

Crossing to the south side of Monitor Creek, still upon the same lode—here called the Esmeralda—are two companies whose works are in close proximity, and which virtually are aiming to develop the same vein of ore. The Marion, by a shaft on the margin of the creek, so long ago as 1863-'64, was the first to develop the rich ore of the Tarshish-Esmeralda lode. Ore was disclosed assaying over $3,000 per ton; but the want of pumping-machinery and the collapse of the mining excitement at about that time caused work to be suspended. The owners dispersed, and nothing has since been done upon the claim. The Chicago and Detroit—a San Francisco company—have a tunnel a little below the Marion, which was run some 400 feet, but failed to strike the vein at that distance. Operations are in abeyance, probably to await the developments now making by the Globe Company on the first extension south, known as the Worden property, which was acquired a year or two ago.

The original location of the Globe Company was 1,000 feet upon the Hercules lode, which outcrops in grand proportions on the north side of the creek, about 1,000 feet west of the Tarshish, and is plainly traceable up a spur of Globe Mountain, on the south side, into a still higher elevation, known as Mount America. Subsequently, 1,000 additional feet on the Hercules and 800 feet on the Tarshish-Esmeralda lode were obtained by purchase. The Globe's main tunnel was commenced in the summer of 1868, though the company was incorporated and some work done five years previously. The initial point is about 1,000 feet below the Chicago and Detroit and Silver Glance, and will intersect the lode not less than 90 feet deeper.

The course of the tunnel is east 25° south, and is intended to strike the lode, in a distance of 1,000 feet, at a somewhat obtuse angle, and not less than 400 feet below the surface, or 180 feet deeper than the lowest level of the Tarshish. The tunnel is a substantial, well-timbered, double-track one; is now in 840 feet, and progressing at the rate of over 50 feet per month. In driving the main tunnel a vein of copper ore was struck at about 500 feet from the mouth, last June, not more than a hand's thickness in the roof, but which increased to six inches in the floor of the tunnel. An inclined shaft below the level, 120 feet deep, developed the vein from twelve to eighteen inches wide, of fine gray argentiferous copper, assaying from 28 to 36 per cent., valued at the "dump" at from $60 to $75 per ton for the copper, and $20 per ton for the silver. The large outcrops on the north seam to be converging to a common center in Globe Mountain and Mount America, and there can be no reasonable doubt that the veins unite, forming a lode of great width and of high-grade ore, within the limits of the Globe Company's ground. The continuity of the Tarshish-Esmeralda lode, and the quality of its ore, have been proved to within 1,000 feet of the Globe claim, and the outcrop upon the latter indicates that the lode extends southerly a great distance. The mill of the Globe Company is a substantial building, and, when fully equipped, will have a capacity of forty tons per day. It is located one hundred yards below the mouth of the tunnel, and the ore is run direct to the crushers, saving the heavy expense of haulers, which is necessary at most of the mines. In addition to the main tunnel, some 700 feet of "drifts" for exploration have been made. The fact that there is a ready market for all the copper ore that can be raised, for the manufacture of sulphate of copper, (blue-stone,) gives great value to this development in the Globe property. This chemical is used largely in the amalgamation of silver ores, and the demand for it is rapidly increasing.

In Monitor district are many undeveloped lodes, that only require capital to make them as valuable as the best; while in Silver Mountain and Scandinavian districts, as well as in Alpine and Raymond, are numerous others, more or less opened. The I X L was one of the noted mines of Alpine County so long ago as 1865, when it produced a considerable amount of bullion. The Buckeye, now owned by an English company, and called the Exchequer, has developed very high-grade ore. This company own a mill on Silver Creek, near the town of Silver Mountain; but, having been built six or eight years ago, it is not up to the present condition of ore-reduction for profitable work. It is intended to erect a new mill this season, with all the modern improvements. The Sovereign, a new English company, reorganized from the Imperial, will soon recommence work. The former company ran a tunnel some 1,800 feet, starting from Carson River, and its course is across all the developed lodes of the district, including the Hercules and Tarshish, and must eventually become a productive property. Across Carson River, opposite the mouth of Monitor Creek, is the Mount Bullion Company, with a tunnel already 2,000 feet in length, and still pegging away in the hard rock, full of confidence in future results.

Good Hope mine, also on the west bank of the Carson, several miles below Bulliona, has been developed by a tunnel over 600 feet in length, disclosing a large body of low-grade ore, assaying from $12 to $17 per ton. It is capable of supplying fifty tons per day, but has not been deemed of much value hitherto, owing to the impossibility of reducing that quality of ore profitably. The improvements in metallurgy of the past two years make this a valuable property. Mr. Thornton, the present owner, has placed one-half the mine on the London market at $75,000 for a working capital, negotiations being carried on through an influential San Francisco house.

Though the mines of Alpine have been overshadowed by their near neighbor, the Comstock, upon which all the capital and influence of San Francisco that could be applied were profusely lavished, yet their developments thus far have proved them to contain ore of a better grade than ever was shown in the Comstock mines, even in those now worth tens of millions of dollars in the market. Nor have I any doubt but that Monitor contains silver lodes that will lead in extent and richness any now known in the mining regions of the Pacific. Where a thousand dollars have been expended in development here, a hundred times that amount have been invested in the Comstock. In no part of California, and certainly not in Nevada, can be found the facilities for cheap exploitation of ore as here, with wood and water without limit.

Parallel with the silver lodes of the mineral belt, and often in close proximity to them, as in the Globe mine, are found copper-veins of fine quality. The Morning Star is situated two miles to the north, and east of Monitor, in the Great Mogul district. It was developed six or seven years ago, and some hundreds of tons of good copper ore raised; but at that time there was no method by which the ore could be profitably secured, or a market found for it. The ore is highly argentiferous. A tunnel of 1,000 feet intersects the lode; and a shaft, with hoisting-works, was completed two years ago. It is owned by a San Francisco company, who have let the mine lie idle for some time, awaiting the developments elsewhere to open a method of profitable treatment of the ores.

The Leviathan mine is about six miles northeasterly from Monitor. It was opened in 1864, as a gold and silver mine, the croppings being metamorphic sandstone, impregnated with red oxide of iron, resembling quartz. The tunnel at 350 feet found no gold or silver, and the mine was abandoned, the copper found instead having no value at the time. In 1869, Mr. Dorsett, an English gentleman, purchased the property, and has since worked it regularly for copper, having taken out some $25,000 in value in the course of opening out the mine up to the present time. Two tunnels, 176 feet apart, have been run into the sandstone, which is over 200 feet in thickness, and a third is about to be started. There are from two thousand to three thousand tons in sight, growing better the further it is opened. The ore is a red oxide and carbonate, entirely free from sulphur; the latter mineral underlying the lode in a nearly pure crystalline form. The Leviathan is under the management of Professor W. T. Rickard, now the assayer and metallurgist of the Tarshish.

The copper ore of the Globe mine is yellow sulphuret in the upper portion of the lode, changing to peacock in depth, and the latter giving way to a fine quality of antimonial, argentiferous, gray ore. Experts who have visited Alpine say the Globe copper is the finest they have ever seen on this coast. Its percentage is from 28 to 36 for the gray variety, and with depth increases in silver-bearing. Steam pumping and hoisting works were erected on the lode last fall. The recent large advance in the Eastern and European prices for metallic copper gives additional value to this vein, as the facilities for reduction are of a superior character.

THE CENTRAL MINING REGION.

The central mining region of California contains the most productive and important districts in the State, both quartz and gravel, and embraces the counties of El Dorado, Placer, Nevada, Yuba, Sierra, and Butte. Within these counties are found the rich quartz mines of Grass Valley and Auburn, the deep placers of Yuba, Nevada, and Placer Counties, and the drift-diggings of Sierra County and of Forest Hill and vicinity.

The country differs materially in its topographical and geological features from that described under the heading of "southern mines." The streams are larger, the mountains higher, and the cañons deeper and more abrupt. The ancient channels are wider, and the detrital matter covers a larger area and is deeper. As in the southern mines, surface and river mining is no longer a source of profit for white men, and has been abandoned to the Chinese, who are yet found on some of the rivers mining with "long-tom" and rocker. Quartz mining has become the principal mining interest, while the deep placers have attracted the attention of capitalists, large tracts of valuable ground having passed into the hands of English companies. This class of mines was fully described in the report for 1871, and will not be alluded to here except to note the progress of some large operations.

EL DORADO COUNTY.

This county has an area of nearly 2,000 miles, a great portion of which is or has been auriferous. It was in this county that gold was discovered on the 19th of January, 1848. The auriferous gravel deposits are far from exhausted; in fact, in many portions of the county they are almost untouched, on account of the scarcity of water. The county has 966 miles of mining ditches, but the amount of water supplied is entirely disproportioned to the length of ditches.* This scarcity of water will soon be remedied by a great enterprise now being carried on by the California Water Company. This company will take their water-supply from the head of the South Fork of the Middle Fork of the American River, and a system of lakes situated at an elevation of about 5,600 feet in the Sierras, on the divides between the various tributaries of the American. It is estimated that the whole area of these lakes and the artificial reservoirs in process of construction will be 143,202,600 square feet, exclusive of the Rubicon River, which flows 5,000 inches of water in the dryest season. These streams and lakes will give an unfailing supply of from 30,000 to 45,000 inches daily, under a pressure of six inches, (miners' measurement.) The completion of this great work (which is also intended to furnish water for irrigation in the valleys, and probably supply some of the cities) will in a great measure revive the early prosperity of El Dorado County and make it one of the foremost in the yield of gold.

I am indebted to Mr. W. A. Goodyear, one of Professor Whitney's assistants, who has spent several months in El Dorado County in the prosecution of investigations for the State geological survey, for the following article on the gravel-deposits of this region:

The gravel-hills of Placerville.—Those who are well acquainted with the character of this section of the country, will find nothing new or strange in the remark that the ancient auriferous gravel-deposits, scattered over the hills in the vicinity of Placerville, form a very interesting field of study for the geologist, as well as a valuable and extensive field of work for the miner. Speaking generally, the hills in which these deposits occur may be said to sweep in a graceful and almost semicircular curve of three or four miles' radius around the head of Hangtown Creek, this curve being convex toward the east, while its two arms, or branches, extending westward to points nearly opposite the town of Placerville itself, form, one of them, on the north, the dividing ridge between Hangtown Creek and the South Fork of the American River, and the other, on the south, the ridge between Hangtown Creek and Weber Creek.

The structure of the banks which enter into the formation of this extensive gravel-range is varied and complex, and many of the questions which they present relative to their origin, the exact mode of their formation, and the distribution of the gold which they contain, are intricate and difficult to solve. A detailed description of them would be useless here, as it would require much space, and would involve little that is new to those who are already acquainted with the ground, while to those who are not it might prove uninteresting, if not unintelligible.

Three weeks of diligent labor expended by the writer in a careful and detailed investigation of these deposits for the State geological survey of California, have, however, developed some points which it is believed will be of general interest, and which it is the purpose of this article to communicate.

First. As to the quantity of gravel here exposed. No attempt has yet been made to form any definite estimate of the number of cubic yards of gravel which these hills contain, and which is so situated as to be capable of being worked by hydraulic process if water were available in sufficient quantity and under proper conditions. From the nature of the case, moreover, the hills being irregular in form and outline, and the

---

* See table of mining ditches from surveyor general's report, ante.

distribution of gravel in them being also irregular, it is impossible to make any close estimates of this nature without full topographical surveys of the ground where the gravel lies. And even if such surveys had been accurately made, and their results were already at hand, although they would give the correct volume of the ground so measured, yet there would be another large element of uncertainty in the results, due to the ever-varying ratio which exists in different hills, and even in different parts of the same hill, between the volume of the gravel itself and that of the volcanic matter which is sometimes to a certain extent intermingled with it, and which often overlies it in heavy masses. The quantity available here for hydraulic purposes has often, doubtless, been greatly overestimated. It is perfectly safe, however, to say that this quantity is great, amounting, in the aggregate, to many millions of cubic yards, and to a quantity much greater than is concentrated within any given area of equal extent at any point between here and the North Fork of the American River.

Much of this ground would require, however, in order to be worked with profit, that the water should be delivered at a higher level, and in much larger quantities, and probably, in a good many cases, at somewhat lower prices than can be done by the present South Fork Canal Company. There is no doubt that the construction of a new canal which should be capable of meeting the above demands, would do more than any other one thing can do towards renovating the somewhat waning prosperity of the town of Placerville. One peculiar feature in the character of the gravel formations in this vicinity is the presence of such immense quantities of a true volcanic gravel, so well known to the miners here under the local name of "mountain gravel." Ninety-nine per cent. of all the pebbles and bowlders which it contains, and probably more than ninety per cent. of its whole mass, consists of matter which is volcanic in its origin. Yet the material is a perfect gravel, and its pebbles are as smoothly and perfectly rounded by water as the pebbles of any bank of auriferous metamorphic gravel in the State. This material forms the whole of the long face of bank exposed along the northern side of Hangtown Hill to the west of Oregon Point, nearly all the upper portion and fully half the height of the present face of the bank in the Excelsior claim at Coon Hill, almost the total height of the central portion of the bank in the Webber claim a short distance east of the Excelsior, a large portion of Webber Hill between Chili Ravine and Webber Creek, and may be found at intervals for a considerable distance to the east along the southern slope of the hills facing Webber Creek. It also occurs in large quantity at Indian Hill, in Negro Hill, and in the spur running back from White Rock Point, and in smaller quantities at numerous other localities. It is rarely cemented together with any firmness, but generally works very easy under the pipe.

Another point in connection with this volcanic gravel, of no less interest to the geologist than to the miner, is the fact that it contains fine gold, distributed apparently through its whole mass, even where it is seventy-five feet or more in thickness; and this, too, not in inappreciable quantity, but in sufficient quantity, so that at various localities where large heads of water have been used, it has paid "wages and water-money" for piping it off, even at present high prices for water. This fact, taken in connection with the occasional pebbles of quartz and metamorphic rocks which are sparsely distributed through it, proves that the stream which brought it here, though flowing mainly at the time over volcanic matter, nevertheless at some point of its course flowed over the naked bed-rock, or else over or through beds of previously deposited auriferous gravel, from which it gathered this fine gold. Besides this volcanic gravel, and the metamorphic, or true auriferous gravel, there are two other well-marked substances which play a large part in the formation of these hills, both of which are volcanic in origin. They are locally known by the miners here under the names respectively of "black lava," and "white lava," or "white cement." It may be noted here, however, that the term "lava," without further modification or explanation, is not strictly applicable to either of these substances; for though almost exclusively composed of volcanic matter, they are not entirely so, and furthermore they have never reached here in the condition of "lava flows," i. e., of streams of volcanic matter rendered liquid by heat, and flowing in this condition over the country; but, on the contrary, they have both of them been brought and distributed here in the solid state in the condition of broken and more or less angular fragments and comminuted particles, and chiefly, if not entirely, by the agency of water. The "white lava" has furnished the stone which has been employed as a building material for many of the houses and stores of Placerville. It is a consolidated mass of moderately fine-grained and nearly white volcanic sand and ash, almost entirely free from pebbles or larger fragments of any kind, although it does occasionally contain very small and more or less water-worn pebbles of quartz and slate-rock. It generally contains more or less mica, distributed through it in black, shining scales. Its grit is very sharp, and its surface is generally quite harsh to the feel. Its localities are very numerous, and it often occurs in isolated patches of very irregular extent and outline, scattered all through this range of gravel-hills. At some points it occurs in very heavy masses, as in the south side of Webber Hill, and near the

II. Ex. 211——7

flume at the head of Cedar Ravine, also at Prospect Flat, and especially near the toll-house, just above Smith's Flat, and in the high bluff overlooking the road just above the toll-house. At this latter point its vertical thickness is probably more than 200 feet. It varies greatly in hardness at different localities, being sometimes so hard as to ring like steel when struck with the hammer, and again so soft as to resemble an ordinary sand-bed. The quality used as a building material dresses easily and well, both with the hammer and chisel, and makes a very fair quality of building-stone for this climate. It is not, however, particularly handsome, owing to the dinginess of its color, which is also in general more or less irregularly stained with yellow, due to the presence of a little free oxide of iron. Its internal character and texture exhibit plenty of evidence that it was originally a well-defined breccia, consisting chiefly of angular fragments of all sizes and shapes, from the almost impalpably fine particles which make up the matrix in which the larger fragments are volcanic sand, which has since become more or less perfectly consolidated into a rock.

The physical characteristics and appearance of the "black lava" are entirely different from those of the "white lava." Instead of being white, like the latter, its general color (when not stained red by decomposition and higher oxidation of the protoxide and the magnetic oxide of iron which it contains) ranges from an ashy or leaden hue to a dark iron gray. In texture it is generally a well-defined breccia, consisting chiefly of angular fragments of all sizes and shapes, from the almost impalpably fine particles which make up the matrix in which the larger fragments are imbedded, up to blocks of several tons in weight, the whole mass being oftener than otherwise, in this vicinity, firmly cemented together into a hard and stubborn rock. A certain proportion of the bowlders which it contains are more or less rounded by the action of water. But its general condition might be very aptly illustrated by taking a mass of mortar and stirring into it as large a quantity of brickbats and irregular broken bits of stone as it could hold, and then allowing it to set and harden; the chief difference being that in the case of the "black lava," the mortar, as well as the stones which fill it, are composed almost entirely of broken and comminuted volcanic rocks. When this "black lava" is hard, and overlies in any considerable thickness the auriferous gravel, it forms, of course, a complete obstacle to the profitable working of the bank by hydraulic means. Twenty or twenty-five feet of this stuff on the top of a bed of gravel, if as hard as it is at numerous points in this vicinity, are enough to prevent all profitable working of the bank by hydraulic, unless the gravel beneath is uncommonly rich, or else, being only moderately rich, the bank is very deep and easily worked when once uncovered.

One of the best opportunities in this vicinity for studying the character of the "black lava," as well as one of the best illustrations of the working of the bank under difficulties and disadvantages, is in the Oldfield and Rocky Mountain claims, at Negro Hill. In the Oldfield claim, now owned and worked by the Chinese, the "black lava" forms a capping from 20 to 30 feet thick over the gravel, and so hard that every pound of it has to be blasted, and most of it broken up into fragments small enough to be handled, either by hand or with the derrick, and thus removed before the gravel under it can be worked. Sometimes, however, if undermined, a crack will open from top to bottom through this capping, at some little distance back from the face of the bank, and the huge block so formed will topple over and roll down into the claim. Enormous blocks of this kind, which have happened to lodge in such positions that they could be worked around, instead of broken up and removed, cover a large portion of the area already worked off in this claim. Some of these blocks are many hundreds of tons in weight. One of the largest of them was measured, and found to be 55 feet in length, while in no direction through the center was its diameter less than 25 feet. Its volume must be very nearly equivalent to, if it does not exceed that of a solid cube, and its weight somewhere in the vicinity of two thousand tons. They do no heavy blasting here, i. e., they move but little rock at a time, and all the work is done by hand with the assistance of a single derrick and a small hydraulic pipe to wash the gravel after the capping is removed. It is not difficult to understand that the gravel must be rich which will pay even the Chinese for working it under such disadvantages, and in the face of such an obstacle.

With reference to gravel-channels, I have no doubt that the opinions of well-informed practical miners here, so far as they go, with respect to the existence of anything like distinct or well-defined channels, are very near the truth. It is extremely probable, for example, that the yellow gravel which forms the lower portion of the bank at Coon Hill, the banks a short distance east of Dickerhoff's mill, on the south side of Cedar Ravine, and the deep yellow gravel in Big and Little Spanish Hills, are all portions of what was once a continuous channel, and the work of the same stream. And it is more than probable that a deep, continuous channel, known here as the "blue lead," extends from White Rock in a general southerly direction, beneath Dirty Flat and the two intervening ridges, to the extreme south end of Smith's Flat. But whether this channel from there on continues its general southerly course, coming out on the Webber Creek side at or in the vicinity of old Try Again tunnel, or whether it makes a sharp bend to the southward and passes under Prospect Flat, is a question impossible to answer with certainty until developments shall have been pushed further underground.

There is a remarkable similarity, not only in the character and appearance of the gravel, and the form of the channel so far as yet worked at Smith's Flat, and at Prospect Flat, but also in the general character and structure of the two flats themselves. Moreover, the apparent direction of the channel at Prospect Flat is southwesterly, which is its proper direction in case it should connect above with the Smith's Flat channel. But there is plenty of fall between Smith's Flat and the Try Again tunnel to allow of a good grade in this direction, as well as in the direction of Prospect Flat. And as all work at the Try Again tunnel was long since abandoned, I do not know the precise character of the gravel, or the appearance of the channel there, or what farther evidence there may have been in favor of its being the same as the Smith's Flat channel.

The question may very naturally be asked, whence came the enormous quantities of volcanic matter which are spread so far and wide over the country, and cover so great a portion of the western slope of the Sierra Nevada? The answer is that they came from higher up in the mountains, having had their origin in the great belt of volcanic action, whose tremendous throes once shook the mountains for hundreds of miles along the line, and in the near vicinity of the present summit of the range, but which did not, as a general thing, extend very far down its western flank. There is not a particle of evidence that there has ever been any volcanic action within many miles of Placerville. The volcanic matter here has traveled far, and is, generally speaking, more recent in its origin and deposition than the auriferous gravel.

The following contribution from Mr. E. N. Strout, of Placerville, is a description, in detail, of the section of country referred to by Mr. W. A. Goodyear in his paper on the gravel-deposits of Placerville:

The district described below is near the center of El Dorado County, and comprises about sixty square miles, in all of which more or less gold can be extracted from the earth. This tract is bounded on the north by the South Fork of the American River, on the east by land covered with timber, on the south and west by Weber Creek, which forms a junction with the South Fork of the American River. Within these boundaries, at Coloma, gold was discovered by J. W. Marshal and others in the winter of 1848.

Commencing at the junction of Weber Creek and the American River, running east seven or eight miles, including Gold and Bald Hills, there are several hundred acres of placer or surface-ground that would pay well to work if water could be obtained for that purpose. Bald Hill is a gravel-basin, rich in gold, as gravel taken from a shaft which has been sunk as deep as the water would allow, has fully demonstrated. This hill will eventually be worked by machinery. Bank-diggings are worked along the ravines on the sides of Bald Hill with success.

The Falls Mining Company, situated on Weber Creek, near Cold Springs, has for several years past averaged from $13,000 to $18,000 per annum. This company cut through granite rock for a long distance to drain their claim and give fall to their flume, which is nearly a mile in length. The first cost of construction was over thirty thousand dollars. The company have yet several years of work on this claim. There are bank and bar diggings along the creeks and rivers, which frequently pay large sums to the owners, but they are mostly worked by Chinamen, and therefore it is impossible to obtain a correct account of the amount taken out.

The most important mining ground within this district is a gravel-hill or ridge, and its spurs and offshoots, commencing at the lower or west end of the city of Placerville, on Hangtown Creek, rising to the height of some three hundred feet, running a south-southeasterly course to near Weber Creek, whence it swings to the eastward, thence north-northwest, finally veering to the west, and terminating at Reservoir Hill, on the bank of the American River, more than 1,200 feet above its bed.

This ridge is more than ten miles in length, and with its spurs, offshoots, and flats, will average a mile in width. It is composed of blue and reddish gravel, from six inches to twenty feet thick, which is found at a depth of about 100 feet below the surface. A small portion of this ridge is covered with a thin stratum of lava, or volcanic rock, under which, in most places, there is a layer of rich pay-gravel. More or less gold can be obtained from all the earth, but as a general thing not in sufficient quantities to pay for working, but it is generally washed in connection with the gravel found on the bed-rock.

Coonhollow Hill is a spur that puts out towards the west from the main ridge, and is one of the richest claims in the State. Since 1852 some forty acres of the point of this hill have been mined, and the lowest estimate given by those who have been on the ground during the time is that there has been mined out more than five million dollars, averaging about three dollars per cubic yard. It was first drifted and then worked by hydraulic pipe.

The ground not worked, and next adjoining that which has been worked, to the extent of forty acres, is now owned by Hunt, Alderson & Co., and is known as the Excelsior mining claim, but little of which has been drifted, most of it having been prospected by shafts, and is known to be as rich as that already worked. The Excelsior claim has yielded a large amount in the last year, working only a portion of the time, when water could be obtained, the owners being reticent as to the amount taken out. A large amount still remains on the bed-rock not cleaned up.

Weber & Co. claim about twenty acres adjoining the Excelsior, going toward the east; worked by hydraulic pipe, but, for the want of water, little has been done on it for several months.

Next adjoining is the claim of Antone & Co.; hydraulic ground. They own ten acres of good mineral ground but little worked.

Then comes the Green Mountain Company, which claims 100 acres. This company have run a tunnel 1,800 feet, and have lately found pay-gravel. Ames & Co. are the owners.

J. Jeffers & Co. are next in line, and own about twenty acres, drift claim, which pays about five dollars per day to the hand. With water it could be made a fine hydraulic claim.

T. Hardey has ten acres, worked by hydraulic; pays about the same per man as Jeffers.

J. Williams, ten acres, about the same as the two last mentioned.

Hall & Stewart's claim, thirty acres, have a 10-stamp mill; water-power. An excellent claim, but, for the want of water, has not been worked for the last year.

Many other claims along the south and west side of the ridge are located, but as yet only prospected; yet sufficient is known to warrant us in saying that they will be worked as soon as water can be furnished.

On the north side of this ridge are located the following claims, which have been worked more or less during the past year. There are also many claims surveyed and held, which have been worked only enough to hold them.

McDonald & Co., 1,060 feet front and extending to the center of the hill. Seven hundred feet of tunnel run. Taken out enough to pay expenses in running the tunnel. This claim can be worked by hydraulic, or by milling the gravel.

Kunckey & Co. have five acres; have taken out $5,000 this year. Worked by hydraulic and drift.

Christian & Ely have twenty acres; hydraulic and drift. Have taken out several hundred dollars the last year.

Allen, Steely & Co. Hydraulic and drift. Much the same as Christian & Ely, above referred to.

Dichooff & Goen have about thirty acres of gravel, which can be worked by hydraulic. They have a 10-stamp mill, water-power, 500-pound stamps. This mill has run most of the time for the last year, and, when on full time will crush about sixty tons, or some thirty cubic yards, per twenty-four hours. Water to run the mill and use in batteries cost the owners about thirty-one cents per hour. Water is furnished from the South Fork canal. This claim has paid the two owners about $1,000 per month each, clear profit, since their mill went into operation one year and a half ago. Water-power costs less than one-half that of steam.

Altar & Co. have about forty acres on a flat about one-half mile above Dickeroff & Goen, in the bend of the ridge. A gravel-drift claim. They have a 10-stamp steam-power mill, and hoisting and pumping machinery. Depth of shaft 100 feet. They have taken out about $10,000 this year.

Macumber & Co. have six acres; 200 feet tunnel, drift and hydraulic. Have taken out $4,000 in four months' work during the last year.

Blacklock & Co. claim twelve acres. Hydraulic. From 50 to 75 feet bank. Three months' work yielded $11,000 within the last year.

Hook and Ladder Company, owned by J. & J. Blair & Co. Claim forty acres hydraulic: bank 170 feet. Have taken out $11,000 in five months this year, and paid $2,500 for water.

Robinson & Co., 10-stamp steam-mill and hoisting-works. This claim is on a flat, 330 feet below the top of the ridge, and about on a level with Smith's Flat, one and a half mile east. The shaft is 100 feet deep. At the bottom of the shaft there is a steam-engine which receives the steam from the boiler above. This engine works the pump, and it is claimed to be a successful operation. Since the mill started, about one year ago, it is estimated they have taken out from $20,000 to $25,000. The owners decline to give results. They have 160 acres, most of which is claimed to contain blue gravel that will pay to work.

Manyard & Bro. have a claim of ten acres, drift and hydraulic, which have yielded about $5,000 the last year.

Corpendear & Co. have a large and valuable claim at Smith's Flat, called the Deep Channel claim.

Creighton & Co., Smith's Flat, have twenty acres, shaft and tunnel; 10-stamp mill; water-power. Have taken out $11,000 during the few months they worked the last year.

Cruson & Co. Gravel and hydraulic. A good claim, and worked when water can be obtained. A small, five-stamp steam-mill is used in connection with this claim.

Hancock & Salter, at the extreme west end of this ridge, have a most excellent hydraulic claim; bank about 90 feet. They have taken out several thousand dollars the last year. There is also a large amount of gold on their bed-rock not cleaned up. There are many claims located in this vicinity which are not worked, but are known to be valuable, and will be worked as soon as water can be had for that purpose.

What we most need here is a ditch of sufficient size to give us plenty of water for mining eight months in the year. It is to be hoped that Government will donate land for that purpose, and that we may have a ditch that will throw 10,000 inches of water into this mining locality ere long. Such a ditch would be of incalculable benefit to the community.

Quartz is not worked to a very great extent in this district. Shephard & Co. have a 10-stamp mill within the city limits, water-power, and a ledge near the mill. They constructed the mill about two years ago. The first fifteen months the rock paid $18 per ton. The last nine months they have been sinking on the ledge and getting out rock, the mill being idle.

The Pacific mill and mine near Shephard's is at present idle. A 10-stamp steam-power mill is upon the ground of the company. This mine has yielded more than $480,00, and was very profitable to the original owners, but for a number of years it has not been worked. One shaft is down 320 feet. Water is troublesome. This property has lately changed hands, having been sold to a company who are preparing to work it at an early day.

• On the north side of the city, and on the west of Quartz Hill, is situated the Harmon mine and mill; 10-stamps; steam-power. The Harmon is considered a good mine, but has been badly managed, and no work is being done at present on it.

Quartz Hill is about one mile in length, commencing on the north side of Hangtown Creek, and terminating at Big Cañon, and is about one third of a mile in width, with an elevation of about 500 feet above the cañon. Much of this hill is composed of barren quartz, although small veins are found intermixed with porphyry which are exceedingly rich in gold. About eighteen months ago a vein of this description was found on the east side, near Big Cañon, which has since been worked with much success, more than $50,000 having been taken out within the last year by Lemon & Hodge, Alsbargo & Lewis, and Fisk, the last named being the discoverer of the lead. The gold is found in narrow seams of quartz, and usually the richest where other veins intersect the main vein, running north and south. The rock taken out is laid aside for milling, except the richest specimens, which are pounded in a hand-mortar, and often produce thousands of dollars from a day's pounding. Frequently a quarter or third of the weight is gold.

Poverty Point, lying north of Big Cañon, has produced considerable gold from broken and irregular veins of quartz, similar, although not as rich as that found on Quartz Hill. At one time—about nine years ago—five steam quartz-mills were in operation on this point, which is about one and a half miles square, but at present there are none, the owners having removed them to other localities.

The U. S. Grant mine (with mill, 10 stamps, steam-power,) about nine miles east of Placerville, is said to be a good ledge, but is not worked at the present time. Miners' wages average about $2.50 per day, boarding themselves.

The amount of gold taken out in the district of country above named, for 1871, will be at least $50,000 per month, or $600,000 for the year.

During the year 1870, Wells, Fargo & Co. shipped from this office (Placerville) $473,015. Within this district is the Coloma office, which ships considerable dust; besides which many persons carry large amounts of dust to the mint or assay offices. It is well known that Chinamen lay aside all specimens that come into their hands, to take with them when they return to China, preferring the native gold to United States coinage. A number of lots in this city are now being worked over by Chinamen, the buildings having been taken down, or the ground under them being worked out, and the buildings sustained by props and posts until again filled in. Large sums are frequently mined out by the Chinese from these lots.

DEEP PLACERS OF EL DORADO COUNTY.

The following able report on the deep placers of El Dorado County was prepared for the California Water Company by Mr. M. D. Fairchild. I am indebted to Messrs. Cronise and Crossman for a copy of the report.

Tunnel Hill.—Tunnel Hill, so called because pierced by the Pilot Creek ditch, is a high ridge, extending from Bald Mountain to the junction of Pilot Creek with the South Fork of the Middle Fork of the American River, the former-named stream flanking its northeastern base for sev-

eral miles. Higher by several hundred feet than the lateral ridges of
the main Sierra which lie east of it, and which it cuts at right angles,
it is a marked feature of the region. Its western declension is quite
abrupt, and its side is seamed by depressions, small ravines constituting
the sources of Otter Cañon, and Rock Creek, the latter into the South
Fork of the American. These streams, as their volume has increased in
their flow toward the west, have formed deep gorges, and but a mile or
two from the western base of Tunnel Hill we again find lateral ridges
shooting out from it and running with an easterly and westerly trend
as distinctive as in the higher region.

"*Ancient river*"-*beds*.—Immediately at the foot of Tunnel Hill, upon its
western side, occur immense beds of auriferous gravel. These have a
general course of north and south, lie in nearly parallel deposits with
each other, at short intervals, and extend many miles to the westward.
Their characteristics and general features leave no doubt but they are
identical with and a continuation of that "ancient river"-bed system
which traverses the counties of Placer, Nevada, and Sierra on the north,
and the results of which, as shown by the workings at Todd's Valley,
Forest Hill, Yankee Jim's, Sarahsville, Michigan Bluffs, Dutch Flat,
Red Dog, Nevada, North San Juan, Forest City, Camptonville, Minne-
sota, and other places, where the more modern hydraulic appliances
have been brought into requisition, have added such vast quantities of
the precious metal to the commerce of the world within the past twenty
years.

Though these gravel-beds have been pretty thoroughly tested, and are
known to be rich in gold, for the past ten years they have been but little
worked, solely on account of a lack of water; and the introduction of that
element in sufficient quantities to justify the fitting up of proper hydraulic
works along the lines of these detrital channels, will be the dawn of an
era of unexampled prosperity, both to those who introduce it, and those
who apply it to the gathering of gold, and a mining *furore* will be created
throughout the State, unequaled for many years.

The underlying rock throughout this entire locality is that which is
known by the common appellation of "auriferous" or "metamorphic"
slates, and is intricately seamed with veins of quartz, large and small,
to which, no doubt, the detrital deposits are in a large measure indebted
for the gold they contain; but as these seams of quartz in the slate have
in this section formed a new feature in mining, that branch will be noted
hereafter under an appropriate head. Here will be given a brief description of a number of the gravel-beds lying between the head branches of
Otter Creek and the Middle Fork of the American, which the waters of
the California Water Company's canal will command, and which the
anxious miner is waiting to develop.

Kelly's Diggings.—The locality of these mines is thus designated
because discovered and superficially worked by a man named Kelly, a
number of years ago. The openings made are not extensive, and the
gravel, when washed off, shows a depth, from surface to bed-rock, of not
more than six feet, and appears to be the extreme eastern edge of the
auriferous zone found below Tunnel Hill. Its extent south, so far as
known, is the bank of the north branch of Otter Creek, whence it runs
northerly with a westerly inclination until it mingles with a larger deposit finding its way to the south branch of the Middle Fork of the
American—as far as we have need of following it at present. The area
washed off is, perhaps, four to five acres in extent. The gold was pretty
generally diffused through the gravel, top to bottom, and it paid (according to Kelly's statement) from $15 to $20 per day to the hand employed,

while washing. As before stated, it seems to be on the edge of the pay-ing belt, for shafts sunk close by disclose the existence of a channel ranging in depth from 30 to 80 feet, one-fourth of a mile wide, and two miles in length, south to north. Upon the introduction of water, sev-eral large hydraulics will be employed.

Bell's Diggings.—Next west of Kelly's is a gravel-range, upon which an opening has been made, known in former days as Bell's Diggings. It lies at the extreme head of Missouri Cañon, a large branch of Otter Creek. It has been prospected by tunnel and many shafts; found to be rich, but not worked to any extent, as there was no water to be had. This gravel-bed is supposed to be wide and have some connection with the Kelly deposit at its outlet toward the river at the north, and also to blend with the Kentucky Flat channel at its southern extremity, that is, running diagonally from one to the other of these two parallel chan-nels. When opportunity offers, hydraulics will be brought to bear against it.

Kentucky Flat.—Next we arrive at an extensive drift-channel, running almost continuously in an unbroken course from south to north for a distance of about five miles, having an average width of three-quarters of a mile, mingling with the Mount Gregory gravel-ridge lying upon the south hill-side of the Middle Fork of the American River, the latter ridge appearing to be distinct from the others, and running from east to west.

The gravel-range upon which is located Kentucky Flat includes many other diggings, known as Bowlder and Tipton Hills, &c., and as it is proper that they should be mentioned separately, this paragraph will refer particularly to the diggings lying between the north branch of Otter Creek and Missouri Cañon. The first of these are those belonging to A. J. Wilton & Co., which have been prospected to considerable extent, and are known to be rich. The extreme southern end has been washed off, and large quantities of gold extracted. At this point the deposit was not deep, averaging, perhaps, 10 feet; but further north the bed-rock declines, while the deposit thickens until it cannot be less than 100 feet, average depth, throughout the greater portion of the claim. One tunnel pierces the gravel for a length of 1,000 feet, and an opening has been made near the mouth of this for hydraulic washings, but, from lack of water, nothing but the merest superficial workings have been carried on here for two years. The width of this deposit is about one-quarter of a mile at this point, but widens thence both north and south. One thousand inches of water will be required at this claim alone. North, a little more than a mile, upon the south bank of Missouri Cañon, this channel has been opened by Messrs. Knight & Jones, thoroughly pros-pected by them, is of great depth, and now waits the introduction of the much-needed water for its successful and remunerative working. Here, where these drift-channels have crossed the gorge of Missouri Cañon, and intersected the Mount Gregory Ridge, they seem to trend toward the west, along the southern base of that mountain, and form a distinct parallel deposit with the main one in that ridge, and it is per-haps better to note such under a separate head, which will follow under the designation of "Mount Gregory." By survey, the rim-rock at Knight & Jones's claim has been found to be 100 feet lower than at the open-ings of Wilton & Co., at Kentucky Flat. From this deposit, west to the junction of Missouri Cañon with Otter Creek, there are many small cañons, with streams flowing toward each of the larger ones, nearly all of which have been rich; and though there are no more heavy gravel-deposits above the junction, there are undoubtedly extensive "seam"

diggings that will require much water in opening and working, as all
the gold taken from the collateral branches has been of that peculiar
character indicative of recent freedom from the parent rock.

Mount Gregory mines.—Here we find a very heavy gravel-deposit,
fully two miles long, with an average width of more than half a mile,
and varying in depth from 10 to 250 feet. This ridge is flanked upon
the north by the Middle Fork of the American River, and on the south
by Missouri Cañon, above its junction with Otter Creek, both very deep
and precipitous gorges, affording splendid facilities for hydraulic operations, the arrangement of proper sluices, and the disposal of the vast
amount of tailings which must be run off. At one time many miners
had delved for gold, a "city" sprang into existence upon the ridge, and
an immense mercantile and express business was done, but, in the midst
of this scene of bustle and activity, an event transpired which drove
industrious people to other parts. The small ditches which furnished
the water for mining purposes were monopolized, that important con-
comitant of mining was interdicted, and that unfortunate locality par-
celed out as a "reserve" for the posterity of an illiberal ditch-owner.
Stagnation followed, and, with untold thousands in the earth beneath,
the miners sorrowfully took their departure, having merely prospected
the section. That was ten years ago, and, as at the other diggings of
which we have spoken, water in abundance is all that will be required
to renew its life again. Then, large sums were made by the most common
method of sluicing; now, by the aid of the improved apparatus in use,
the result would be astonishing.

Upon the southern slope of the ridge the surface-earth was washed
until the heavy deposit was reached, and thereafter drifting was resorted
to. Upon some of the lower branches, where water is obtained in some
seasons from the adjacent cañons, a few miners have lingered and endeavored to hydraulic, but have done but little from the irregularity of
their supply. Others drift, and by husbanding the waters of different
springs manage to subsist and even to make money by washing the
gravel they unearth a half day in the week. The great demand is for
an abundance of water for hydraulic purposes, in order to attain to that
wealth which their neighbors have upon this ancient river-bed further
north, where facilities have been at hand for the rapid removal of the
mass of auriferous gravel. But little washing has been done upon the
northern side toward the river—only enough to determine that it will
pay. The hill has been pierced on both sides by tunnels, and found
rich. From the gravel thrown out in sinking a well upon the ridge,
many pieces of gold have been taken, one worth $2.50. This is only
60 feet deep, about 100 feet above the bed-rock.

The different diggings upon this ridge now paralyzed by lack of water,
and which are worked when the least quantity is available, may be
enumerated as follows: Gravel Point, Captain Gardner's Point, Bitter's
Point, Nameless Point, Carter's Point, Drummond's Diggings, Red
Point, Ross's Diggings, Mackey's Diggings, Lloyd's Diggings, Webster's
Diggings, Cooley's claim, Worthingham & Bowman mine, Garner's
claim, the Hercules mine of the California Water Company, with a
front of over half a mile, Drummond's Diggings, Cooley & Murzuer's
and others, all requiring hydraulic.

Volcanoville.—Volcanoville is situated upon the same ridge as Mount
Gregory, is a mining locality of considerable importance, and its mines
are of noted richness. A great deal of water will be used at this place
when once introduced. Still further west are found gravel-beds of considerable extent, known as Miller's and the French Diggings, Buckeye

Hill, &c., the deposit to be washed varying in depth from fifteen to twenty-five feet.

Jackass Flat.—Leaving the divide between Otter Creek and the Middle Fork we will go to that which separates the waters of Otter and Cañon Creeks. First across the branch of Otter at Kentucky Flat we follow the immense gravel-bed southward, commencing at Chris ranch or Jackass Flat, where we find a large area of drift awaiting water to be washed off. This channel seems to have a trend from east to west, and connects with the Bowlder Hill deposit.

Boulder Hill.—Boulder Hill is an extensive and deep deposit, and has been pretty thoroughly prospected by shafts and tunnels. Its situation is extremely favorable for rapid working with hydraulic power, being located upon the ridge between two branches of the creek, which here forms large gorges with a sufficient fall for trailings. The California Water Company has an excellent location upon the northern end of this hill, which has been pierced by a tunnel a distance of several hundred feet.

Darling's Ranch.—West from Bowlder Hill occurs a large gravel-ridge near Darling's ranch. It has been well explored by various shafts and tunnels, shows gold in paying quantities, and only awaits water to be placed in the list of paying mines. It is favorably located for opening either upon Cañon Creek or Otter Creek.

Bald Hill.—Bald Hill shows a reef of talcose slate, which cuts the drift-channel at right angles, and its apex is considerably higher than the surrounding hills. The surface-earth is auriferous, but there is little of it, and it is principally noted for its "seam" diggings, which will eventually cause a demand for a large amount of water.

Harrison Hill.—This hill is a continuous gravel-ridge, very deep, extending east and west. Like other of its fellows it is but little worked from similar cause, lack of water. It will require at least six powerful hydraulic streams in its working.

Cement Hill.—The extent of Cement Hill is nearly three-quarters of a mile long by half a mile wide. Years ago it was pierced by several tunnels, much of the bottom stratum of gravel was extracted and washed, and immense sums of gold taken therefrom. It is deep, and will all be washed off when water can be obtained.

Nevada Flat is a "branch" or lateral ridge shooting from the southwestern side of Cement Hill. Considerable water must necessarily be used in washing its gravel-deposits.

Bottle Hill.—The diggings of Bottle Hill are perhaps half a mile square in extent, and have been celebrated for their richness. The North Star, Saint Louis, Cuyahoga, Gravoy, and Hopewell tunnels, each extensive works, have pierced it from both sides, and the great portion of the bottom stratum of gravel has been extracted. But as it is very deep, and as the different strata of earth composing the bulk of the hill still remain, and contain more or less gold, the application of hydraulics will render its more perfect working remunerative, and it will eventually all disappear before the attacking miner.

Mount Calvary.—These mines are owned principally by C. H. Calmes, who has held them for many years, unable to work them on account of having no water, satisfied that they would ultimately remunerate him for his untiring patience. A large hydraulic stream will be necessary to their successful working.

Gravel Hill.—The location of Gravel Hill is west from Mount Calvary. The paying gravel-deposit is deep, nearly one mile square, and will all be washed off upon the introduction of water.

Jones's Hill.—Jones's Hill is divided by a gulch called Jones's Cañon, that portion of the hill upon the northern side consisting of a heavy gravel-bed, while that upon the southern side is strictly " seam" diggings. The gravel is deep, has been drifted to great extent, but will be worked as soon as hydraulic appliances can be directed against it. The area covered by this deposit is three-fourths of a mile square. Below it, or further west, is a smaller deposit, with similar characteristics, known as Mitchell's Flat.

Gopher Hill.—Gopher Hill is situated upon a divide between two branches of Cañon Creek, and is favorably located for hydraulic mining upon the northern side, where the cañon is precipitous, and presents the most favorable features to open at great depth, having abundance of room for the *débris* carried away by washing. It extends north and south a distance of about one mile, while its eastern boundary is supposed to intersect with the great channel running from Tipton Hill to Mount Gregory, upon which we have placed Kentucky Flat. An unfinished tunnel, driven into the hill many years ago, is found at the northern end, where also several shafts are sunk, which yielded considerable gold. Upon the southern end Current & Cashman have made a small opening with the limited amount of water they were able to obtain, and the results were of an exceedingly encouraging character.

Tipton Hill.—The most extensive workings in the whole section of country, near the base of Tunnel Hill, are upon the southern end of the principal gravel-ridge of which Tipton Hill is the southern terminus, and the claim of Knight & Jones the most southerly, in the diggings of the Messrs. Schlein. With a small head of water, with a pressure of only 65 feet, a sluice-grade of 6 to 8 inches in 12 feet, and boxes but 16 inches wide, without the aid of quicksilver, the average yield a day to the hand employed has been $6. Water for working these mines has been brought in small ditches from the head of Rock Cañon, upon the northern bank of which the tailings flow. Its northern boundary constitutes the northern boundary of the Schlein Brothers' claim. From there north, upon the channel, the California Water Company has a claim one mile in length, upon which there is a shaft, not yet to the bottom, 120 feet deep, and upon the eastern side, debouching into one of the branches of Rock Creek, a tunnel pierces the ground to the length of 1,100 feet. With proper hydraulic appliances, the yield from this magnificent gravel-bed will be enormous.

Fort Hill.—Further west is Fort Hill, on which are many claims where drifting is carried on. In extent, this deposit must be at least one-eighth of a mile wide by two miles in length, from north to south.

Other mines.—Upon the limits which this report so briefly touches, are other and noted mines, which require a great quantity of water in working, as Georgia Slide, Mameluke Hill, Buffalo Hill, Georgetown, &c., all requiring large hydraulic streams, beside the innumerable small cañon and many surface-diggings demanding smaller sluice-heads. And yet the ground we have thus far traversed all lies in an area of ten miles east and west by six miles north and south. With the completion of the canal, and the assurance that the miners could rely with certainty upon what water was needed for constant work, within one year from its advent at least one hundred extensive hydraulic mines would be ready for operations, requiring each from five hundred to fifteen hundred inches of water. These immense placers cannot be exhausted in a period of twenty years, and then the demand for water would not diminish, for the denudation of the bed-rock by the removal of the gravel will expose countless seams of gold-bearing quartz, which from time to time would be developed into extensive mines and worked to great depth by the application of water, aided only by the occasional blast of powder.

Seam-Diggings.—As occasional reference has been made in the above to "seam-diggings," the following explanation is deemed proper. In nearly the entire region of country which is traversed by the canal of the California Water Company, in El Dorado County, the slates of which the bed-rock is composed are permeated by innumerable seams of quartz, thousands of them being exceedingly small, while some show large nodules, and assume such proportions as to be frequently mistaken for true veins, many of which carry gold. To this fact the placers undoubtedly owe much of their richness, particularly the bottom stratum of the gravel-deposits, which is only a detrital mass caused by erosion and attrition of the bed-rock. Partially denuded, and subjected to atmospheric influences, in many places these seams are found in a state of decomposition, and the gold is frequently freed from its matrix, while the friable condition of the slate renders it susceptible to the attacks of the miner, who not unfrequently can wash away whole mountains of soft bed-rock with as much celerity as the gravel-deposits are disposed of. A number of mines of this character are now working in the locality spoken of, which pay well, and a hundred more will probably be opened when a time arrives at which the miner is not restricted in his operations by a lack of water.

QUARTZ-MINING IN EL DORADO COUNTY.

This class of mining has not heretofore been conducted with any great degree of success considering the amount of capital invested. The county possesses 30 quartz-mills, with an aggregate of 390 stamps, less than half of which are in active operation, and by the county assessor 21,645 tons are reported crushed. This, I presume, includes cement, as there are several mills engaged in this business. The past year, however, has witnessed a revival of the quartz interest, and many promising claims are being developed near Georgetown and Placerville. The characteristics of most of the veins in the county have shown them to be "pockety," though the yield has been enormous. The future of this interest will be determined by the operations now in progress. Should they prove profitable to the companies who have recently purchased, it is safe to say that work will be resumed on nearly all the abandoned mines of the county. A gentleman who has been engaged for several years in mining operations in this county, furnishes me with the following notes relating to the quartz interests of the county. The class of mining called "seam-diggings" is peculiar to this county.

The Georgetown divide has been noted for its rich placer-diggings, auriferous gravel, and a class of diggings found in no other part of the State, known as "seam-diggings." The formation is a talcose slate interstratified with small quartz-seams coursing in every direction. The quartz-seams are invariably rich in gold, while the formation has been decomposed to that extent (friable) that it can be worked with the hydraulic pipe so far as has been explored in depth, say from 30 to 100 feet. A scarcity of water on the divide has hitherto prevented working this class of mines to any extent. A ditch project, however, is in contemplation that will furnish an abundance of water for the divide. Several ranges of hills are composed of rotten bed-rock of slate, seamed with numerous layers of quartz of various thickness, ranging from an inch to twenty feet, all of which is so soft that it can be piped down, with the aid of an occasional blast, and washed through a sluice. The gold seems to have been freed from its original matrix by decomposition, and is easily saved. The Whiteside claim, near Georgetown, is one of this character. From this, with seventy inches of water, $4,000 in a week has been obtained, and the general average is good.

Aside from this industry, the mineral wealth of the county contained in the numerous quartz-lodes coursing through it still remains intact.

The Saint Lawrence is the only mine that has been developed to any extent, some 200 feet in depth. So far, the mine shows great value, and a new 20-stamp mill is about ready to start, with good machinery for hoisting and pumping.

On the Placerville divide the gravel-range is extensive and rich in gold, but a scant supply of water prevents the mines being worked to any extent.

The principal quartz mine of note is the Pacific, near Placerville, which has recently been bought and is being worked by an English company. The Hanlah mine, near Shingle Springs, is being worked successfully with a mill of 40 stamps run by a turbine-wheel. Ore low grade, but ledge large. The Pocahontas is a mine of value. The Davidson mine is erecting a 20-stamp mill. The Woodside quartz mine, which created such an excitement a few years ago, when pockets were found showing about equal parts of gold and quartz, is now filled with water, and lies neglected.

Chromic iron of a high grade, (60 per cent.,) is abundant in this county and is being profitably worked and shipped to England and the Atlantic States.

THE DISCOVERY OF GOLD.

The credit of the discovery of gold in California has, until of late years, been universally and properly conceded to James W. Marshall; but as years elapsed and many of the actors in the stirring scenes of the early settlement of the country are passing away, new claimants arise to dispute the honor of the discovery. It is undoubtedly true that gold was known to exist in California prior to Marshall's discovery at Coloma, but it had never been obtained in sufficient quantities to influence the destiny of the country. Placers had been worked at or near the mission of San Fernando, in what is now known as Los Angeles County, but the *padres* in charge of the missions discouraged the digging of gold as having a demoralizing tendency on their flocks. Rumors of the existence of gold were from time to time heard on our then western frontier, which were traced to the hunters and trappers who had penetrated these distant regions, but it remained for Marshall to make the discovery which settled and populated the State. With a view of preserving a record of this memorable discovery, with all its details, we here reproduce the narrative of Marshall as it fell from his own lips. The narrative is taken from a biography of Marshall edited by Mr. John Frederick Parsons, of Sacramento :

James Wilson Marshall, the discoverer of gold in California, was born in Hope Township, Hunterdon County, New Jersey, in 1812. His father was a coach and wagon-builder, and he was brought up to the same trade. His early life presents no features of special interest; and he had arrived at man's estate, being just twenty-one, when he began to turn his eyes westward, and to experience the yearning which makes the pioneer. Presently his mind was made up, and with such leave-takings as poor men make when they start out into the world and turn their backs, perhaps finally, upon the place of their birth, he set forth and journeyed until he came to Crawfordsville, Indiana. Here he worked as a carpenter for some months; but the leaven of restlessness was at work within him, and he set out again shortly, this time reaching Warsaw, Illinois. After a brief stay here, he once more packed his few possessions and wandered off to the Platt Purchase, near Fort Leavenworth, in Missouri. Here, for the first time since leaving home, he appears to have had some idea of settling permanently, for he located a homestead, worked steadily at farming and trading, and was in a fair way to prosper, when he was attacked with fever and ague, from which he suffered so much that after struggling against the disease for six years he was compelled to prepare for another exodus, or make up his mind to die where he was, for the physician said he could not expect more than a two years' lease of life. Just at this time people were beginning to talk a good deal about a strange, new country, far away in the West, called California. It was said to be a desirable place to emigrate to. The valleys were broad and fertile; the rivers were numerous; timber was plenty; and game abounded; and there was a charm about the name and the uncertain legends told regarding the new region that whetted the curiosity of the border men. Marshall heard of California. If he stold in the low bottom-lands he must die. He could only be killed by the Indians if he went. He decided to go. A party was being made up in the neighborhood, and gathering together his stock he joined it and set out. They started about the 1st of May, 1844, with a train of a hundred wagons, but owing to the heavy rains, which had flooded the bottom-lands of the Missouri and its tributaries that spring, they were delayed considerably. At length they arrived at Fort Hall, and here a consultation was held, and it was decided that the safest way to enter California would be by way of Oregon. All did not agree to this, however, and the difference of opinion finally led to a disruption of the party. Some went one way, some another; but Marshall joined a band of about forty souls, and the company started (on horseback, and packing their provisions) about the spring of 1845. There was then, and had been for some time, much trouble with the Indians; but this party was not molested in any way; and this fact is worthy of remark, for the reason that it was the first case of perfect immunity from attack recorded up to that time.

The journey was unaccompanied by any special excitement, and after wintering in Oregon they reached California safely, via Shasta, in the month of June, and coming down the Sacramento Valley, camped at Cache Creek, about forty miles from the present site of the city of Sacramento. Here they separated. Some went below, to San Francisco, (then YerbaBuena;) some wandered off up the valley; some proceeded to Sacramento, where already Sutter's Fort was established, and regarded with envy by the Mexicans, awe by the Indians, and admiration by the foreigners, (as all Americans and Europeans then were.) Among those who proceeded to the fort was Marshall, and here, in July, 1845, he engaged to work for Sutter.

There were then very few white settlers in the northern portion of California. The missions were still the principal centers of business and population, but the whole country was inert, stagnant, undeveloped, barren, and almost desolate. The power of the mission fathers had been broken, and the good work they had done had been negatived by the rapacity, ignorance, and obstinacy of Mexican officials and legislators. The patient labors of a hundred years had been overthrown in a twelvemonth, and the Christianized Indians had been relegated to barbarism. At the missions, where the old fathers had exercised a mild despotism, and where, for generations, their every word had been law, they were cast down and despised. New rulers, secular by denomination, too often coarse and brutal by nature, tyrannical and cruel by disposition, occupied the places of authority, and ground the faces of the poor. Brigandage and lawlessness had become established in some parts of the State, and progress there was none, save here and there where some enterprising American or other foreigner had procured a grant of land, and was cultivating a portion of it, or raising stock. The republic of Mexico, impotent as it was to govern the country properly, had, nevertheless, inflicted real injuries upon it which nothing but the subsequent annexation to the United States could have repaired.

Sutter had built the fort on the Sacramento River, and was engaged in raising grain and stock, and doing a small trading business. He also made blankets, having secured the services of a number of Indians who had been taught to spin by the mission fathers of San José, and one of the first tasks in which Marshall was engaged was the construction of a number of spinning-wheels for these blanket-weavers. The life at the fort was a rude one, destitute of comfort, and ill-supplied even with necessaries. The men soon wore out what clothing they had brought with them over the mountains, and thenceforward were compelled to trust to their rifles for their garments. Antelope were plentiful at that time, and from the skins of these animals most of the clothing was made. Sutter employed a band of hunters and trappers, mostly Indians, and these supplied the fort with meat, taking their pay generally in ammunition. Everything was conducted in the most primitive style. Tea, sugar, coffee, &c., were luxuries wholly unknown. Flour there was, of a kind; but rudely as it was prepared, the fort had the honor of introducing the first improvement in grinding wheat. The custom of the country was sufficiently barbarous. The grain was placed on a flat stone and pounded with another stone, the operators being generally women. Sutter, with the assistance of his men, constructed a rude mill, which was worked by a mule, which walked round and round, causing the upper stone to revolve. The flour thus produced was coarse, but the men thought themselves lucky when it contained no lumps larger than a nutmeg. There were no candles, and consequently all hands retired as soon as it was dark, save when some enterprising individual hunted up a pitch-pine knot, and thus secured an hour or so of smoky illumination.

* * * * * *

Up to this time the class of emigrants that had settled in California had consisted mainly of that restless vanguard of advancing civilization which always hovers on the frontiers, and whose mission seems to be to keep moving from place to place, from Territory to Territory, never staying anywhere long enough to reap the full fruit of their energy and toil, until the great settler, death, appears and ends their uneasy career by a final remove into another world. Some few had secured large tracts of land under Spanish grants, and had affiliated with the native Californians, by marriage or otherwise, but the majority were as ready as ever to "pull up stakes" again and journey on to some newer country, if such could have been found. The California of that time— 1847—was altogether unlike the California of a year after, or of any subsequent period. The influence of the old *padres* has been broken, and the clash of arms had rudely interrupted the sleepy placidity of their lives. The American, whose restless energy and unquenchable ambition rendered him an object of terror and perplexity to these staid old souls, had, it is true, conquered the country, but he was scarcely yet prepared to possess it. There seemed, indeed, to be a lull in the stirring life of the previous years.

The people were waiting, unconsciously to themselves, for something which was to change the aspect of affairs, and was to draw the eyes of the whole world upon this little-known region.

* * * * * *

The disturbances known as the Bear Flag War now broke out, and in these Marshall took an active part, rendering material assistance to the American forces through his knowledge of the country and the natives. On the cessation of hostilities Marshall returned to Sutter's Fort, (the present site of the city of Sacramento,) and determined on engaging in the lumbering business. He asked Sutter to furnish him with an Indian interpreter, purposing to explore the foot-hills for a suitable location for a saw-mill, and foreseeing the necessity of being able to converse with the mountain tribes of Indians. Sutter was at first reluctant to comply with this request, having need of Marshall's services, but after the latter had agreed to perform certain mechanical work for him, he consented, though it afterward turned out that the Indian who accompanied him knew more of the country than he did himself. Marshall set out on his quest, and followed up the banks of the American River for several days, examining the country all around, but not finding what he considered a suitable site for his mill. The country through which he passed became more diversified as he traveled upwards. Steep cañons and considerable ranges of hills broke up the landscape, and while contributing nothing to the ease of travel, added much to the picturesqueness of the route. Presently he branched off on the South Fork of the American River, and at length reached a place which he found was called Culloomah by the Indians, and which was afterwards known as Coloma. The river here flowed through the center of a narrow valley, hemmed in on both sides by steep, and, in some parts, almost precipitous hills. On the south side the declivity was the gentlest, and here a tolerably level stretch of land invited the erection of the town which sprung up there after the discovery of gold, while the slopes beyond afforded opportunities for cultivation, which in later years were fully availed of. The river makes several bends in its course through this valley, and on the south side a point of land formed by one of these curves in the stream presented the explorer with the mill-site he was in search of. The water-power was abundant, and the surrounding hills furnished timber in apparently inexhaustible quantities. Previous to this it had been supposed that the difficulty of bringing lumber from any point in the foot-hills was insurmountable, and Sutter's hunters had so impressed him with this idea that he considered Marshall's expedition little better than a waste of time. A careful examination of the locality, however, satisfied our hero that there would be no difficulty in transporting the products of the mill to the lower country, and having marked out a favorable site, he returned to the fort and acquainted Sutter with the successful result of the journey. At the same time he stated that he was in search of a partner with capital to assist him in building and running the mill, and Sutter at once offered to join him in the undertaking. This was about the 1st of June, 1847, and after many delays, caused principally by the attempts of others to interfere in the business, a partnership agreement was entered into between the two on or about the 19th of August. The terms of this agreement were to the effect that Sutter should furnish the capital to build the mill, on a site selected by Marshall, who was to be the active partner, and to run the mill, receiving certain compensation for so doing. A verbal agreement was also entered into between the parties, to the effect that if, at the close of the Mexican war, (then pending,) California should belong to Mexico, Sutter, as a citizen of that republic, should possess the mill-site, Marshall retaining his rights to mill-privileges, and to cut timber, &c.; while, if the country was ceded to the United States, Marshall, as an American citizen should own the property. The formal articles of partnership were drawn by General John Bidwell, who was then acting as clerk in Sutter's store, and were witnessed by him and Samuel Kyburg, Sutter's business manager. Shortly after these arrangements had been made, Marshall hired a man named Peter L. Wemer, with his family, and six or seven mill-hands, and with several wagons containing material, provisions, tools, &c., started for Coloma. Work on the mill was at once commenced, and prosecuted with energy and rapidity.

The names of the men who were then working at the mill, and who, if living, can substantiate the accuracy of this narrative, are as follows: Peter L. Wemer, William Scott, James Bargee, Alexander Stephens, James Brown, William Johnson, and Henry Bigler. Wemer was in charge of some eight or ten Indians, whose business it was to throw out the larger-sized rocks excavated while constructing the mill-race, in the day-time, and at night, by raising the gate of the fore-bay, the water entered and carried away the lighter stones, gravel, and sand. This was the work that was going on at the mill on the 19th of January, 1848.

On the morning of that memorable day Marshall went out as usual to superintend the men, and after closing the fore-bay gate, and thus shutting off the water, walked down the tail-race to see what sand and gravel had been removed during the night. This had been customary with him for some time, for he had previously entertained the idea that there might be minerals in the mountains, and had expressed it to Sutter, who, however, only laughed at him. On this occasion, having strolled to the lower end of the race, he stood for a moment examining the mass of *debris* that had been washed down, and at this juncture his eye caught the glitter of something that lay, lodged in a crevice, on a riffle of soft granite, some six inches under the water. His first act was to stoop and pick up the substance. It was heavy, of a peculiar color, and unlike anything he had seen in the stream before. For a few minutes he stood with it in his hand, reflecting, and endeavoring to recall all that he had heard or read concerning the various minerals. After a close examination he became satisfied that what he held in his hand must be one of three substances—mica, sulphurets of copper, or *gold*. The weight assured him that it was not mica. Could it be sulphurets of copper? He remembered that that mineral is brittle, and that gold is malleable, and as this thought passed through his mind, he turned about, placed the specimen upon a flat stone, and proceeded to test it by striking it with another. The substance did not crack or flake off; it simply bent under the blows. This, then, was gold, and in this manner was the first gold found in California.

The discoverer was not one of the spasmodic and excitable kind, but a plain, shrewd, practical fellow, who realized the importance of the discovery, (though doubtless not to its full extent, since no one did that then,) and proceeded with his work as usual, after showing the nugget to his men, and indulging in a few conjectures concerning the probable extent of the gold-fields. As a matter of course, he watched closely, from time to time, for further developments, and in the course of a few days had collected several ounces of the precious metal. Although, however, he was satisfied in his own mind that it *was* gold, there were some who were skeptical, and, as he had no means of testing it chemically, he determined to take some down to his partner at the fort, and have the question finally decided. Some four days after the discovery it° became necessary for him to go below, for Sutter had failed to send a supply of provisions to the mill, and the men were on short commons. So, mounting his horse, and taking some three ounces of gold-dust with him, he started. Having always an eye to business, he availed himself of this opportunity to examine the river for a site for a lumber-yard, whence the timber cut at the mill could be floated down; and while exploring for this purpose he discovered gold in a ravine in the foot-hills, and also at the place afterwards known as Mormon Island. That night he slept under an oak tree, some eight or ten miles east of the fort, where he arrived about 9 o'clock the next morning. Dismounting from his horse, he entered Sutter's private office, and proceeded to inquire into the cause of the delay in sending up the provisions. This matter having been explained, and the teams being in a fair way to load, he asked for a few minutes' private conversation with Colonel Sutter, and the two entered a little room at the back of the store, reserved as a private office. Then Marshall showed him the gold. He looked at it in astonishment, and, still doubting, asked what it was. His visitor replied that it was gold. "Impossible!" was the incredulous ejaculation of Sutter. Upon this Marshall asked for some nitric acid to test it, and a *vaquero* having been dispatched to the gunsmith's for that purpose, Sutter inquired whether there was no other way in which it could be tested. He was told that its character might be ascertained by weighing it, and accordingly some silver coin ($3.25 was all the fort could furnish) and a pair of small scales or balances having been obtained, Marshall proceeded to weigh the dust, first in the air and then in two bowls of water. The experiment resulted as he had foreseen. The dust went down; the coin rose lightly up. Sutter gazed, and his doubts faded, and a subsequent test with the nitric acid, which by this time had arrived, settled the question finally. Then the excitement began to spread.

*　　　　*　　　　*　　　　*　　　　*

Statements have been published in newspapers and allusions have occasionally been made to the almost fabulous cost of living at that time, but the following extracts from one of the books kept at Sutter's Fort will, perhaps, convey a better idea of the actual state of things. We append a few items at random :

PRICES IN 1849.

1 canister of tea	$13 00
2 white shirts	40 00
2 kits of mackerel	60 00
1 fine-tooth comb	6 00
1 hickory shirt	5 00
3 pounds of crackers	3 00
1 barrel of mess pork	210 00
2 pounds of mackerel	5 00
1 bottle of lemon-sirup	6 00
4 pounds of nails	3 00
1 paper of tacks	3 00
1 dozen sardines	35 00
1 dozen Sedlitz powders	17 00
1 pair of socks	3 00
1 pound of powder	10 00
1 bottle of ale	5 00
1 bottle of cider	6 00
1 hat	10 00
1 pair of shoes	14 00
1 bottle of pickles	7 00
1 can of herrings	30 00
13 pounds of ham	27 00
1 bottle of mustard	6 00
2 pounds of sauerkraut	4 00
55 pounds of tarred rope	75 00
1 tin of crackers	24 00
1 candle	3 00
30 pounds of sugar	18 00
1 Colt's revolver	75 00
1 pound of onions	1 50
1 tin pan	9 00
1 keg of lard	70 50
1 pair of blankets	24 00
1 dozen of champagne	40 00
1 pound of butter	2 50
50 pounds of beans	25 00
200 pounds of flour	150 00
13 pounds of salmon	13 00

Such is a sample of the prices of necessaries and luxuries at the time of the great rush. It will be seen that though the rates of labor were enormously high, the opportunities for saving were not much above the average.

 * * * * * *

Two-and-twenty years have passed over Coloma since the day when James Marshall stood at the end of the tail-race and pondered over that bit of yellow metal. That bit of yellow metal has been multiplied by millions upon millions. The trifling acceleration of the pulse that marked the first emotion of the discoverer has swelled into a wave of maddening excitement whose roar has re-echoed round the world. The spring struck in that little mountain valley has flowed and spread until mighty cities have been built upon its banks and communities have been refreshed by its waters. From out that wonderful vale has risen all of good and evil that can affect humanity. At first the center of the swarming adventurers, leaping, as it were, in a moment from the quiet humdrum of its early settlement into the full glare and crash of a mighty mining excitement, it has passed through the prosperity, the fever, the noise, the hurly-burly, and the slow decline, and has settled at last into the peaceful semblance of some New England village.

Picture it to-day as a pretty hamlet of some two hundred inhabitants, its broad single street so overshadowed with great, heavy-foliaged trees, that the sidewalks are scarcely visible; its modest, low-roofed houses, gracefully bedecked with bright flowers and fresh green creepers; its main thoroughfare silent throughout the day, save when the daily stage dashes gallantly in, and draws up with a rattle and a crash at the door of Wells Fargo's office, where the courteous agent sometimes might find time lie heavily upon his hands did he not also undertake the duty of telegraph operator, besides doing a little something in trading. Upon the hill-side the vineyards flourish, and the orchards. In the warm summer air the peaches mellow and grow golden and ruddy, and the great bunches of grapes swell out from behind their leafy screens, and give promise of that "wine that maketh glad the heart of man." Around the modest houses of those few who are content to pass their days in this celebrated yet little known spot the roses and honeysuckles clamber, and the air at evening is heavy with perfume. Up among the bends of the river some mining is still going on, but there are few claims which now yield high wages, and the Chinaman, patient and content with little, has set himself to pick up the crumbs that have fallen from the rich (white) man's table. One striking evidence of what the town has been is visible in the rear of the houses nearest the river. Close up to the back doors the bowlders are piled. It is a Titanic beach—the *débris* of the mining of twelve or fifteen years. Gazing upon these stones, so completely divested of earth, so white and bare and ugly, one is tempted to imagine them the bones of the skeleton of gold, which has here been picked clean by the active fingers of ambitious man.

And across the river we look in vain for the site of Sutter's mill. Years have passed since the last vestige of that structure was removed by some miner, careless of tradition, but needing timber. Even the man who first found the gold there has to scrutinize the place carefully before he can put his foot down and say, "Here is the spot. It was within a yard of where I stand that the first *chispa* was picked up." So mangled and torn and mined away has the face of nature been in this historical locality that those who knew her best would fail to recognize the scarred and disfigured lineaments. Yet it is Coloma; and yet the site of the gold discovery can be pointed out. In a few years more, however, the oldest inhabitant will have lost all trace of the spot, and the visitor will be only able to discover that the gold was found "somewhere hereabout." There is need of a monument at Coloma, and the site of Sutter's mill should be marked in an enduring manner. California has been far too careless in such matters heretofore, and she will regret in the future the vandalism that has left her no relics of a time which grows in interest and in value as it recedes into the past.

PLACER AND NEVADA COUNTIES.

These two counties are the seat of the most extensive mining operations in the State, both in quartz and gravel. The operations on the deep placers, which are now attracting the attention of capitalists at home and abroad, are fully described in last year's report, (pages 55 to 90,) and require no further notice here. A large area of auriferous ground between the Middle and North Forks of the American River, in Placer County, has for years remained undeveloped for want of water. This want is about to be supplied by taking the water from Lake Tahoe, situated near the summits of the Sierra Nevada, at an elevation of 6,000 feet above sea-level. The surveys have been completed and dams built at the outlet of the lake into the Truckee River. In order to bring the water from the lake in the most convenient and desirable manner, it was found that a tunnel would have to be cut through the western summit of the Sierras. This tunnel would be of about five miles in length, and even if of small dimensions would have involved a very heavy outlay. A community of interests has led to a contract between the Central Pacific Railroad Company and Colonel Von Schmidt, by which it is arranged that the latter shall construct a tunnel of such dimensions as shall admit of the passage of trains, thus enabling the Central Pacific to shorten its road seven miles, lower the line of the railroad upwards of 1,000 feet, and dispense with twenty miles of snow-sheds, which last are, from their expense and danger, the most objectionable feature of the line. The precise points at which the tunnel will enter the mountains have not yet been precisely located. It will, however, enter the mountain on Cold Stream, close to Truckee, on the eastern slope, and on the North Fork of the American River on the western side. The entire length of the tunnel will be about, or a little less than, five miles. For one-third of this distance it will be ventilated by shafts sunk from the slope of the mountain. The stipulated size of the tunnel is 19 feet high by 21 feet in width, and it is to be completed in five years; but Colonel Von Schmidt fully believes he will complete it in three years. The cost has been estimated at $1,500,000. Carefully made examination shows that less than a mile of the boring will be through granite, which is very much less than was expected. The remainder is, for the most part, cement, easily removable by the pick, without resort to blasting. The boring will be performed by an instrument of Colonel Von Schmidt's invention. This machine is constructed upon the Severance diamond-drill principle, but in the mode of application the machine differs materially from all others at present in use. It consists of a circular wheel, eight feet in diameter. Imbedded in the rim of the wheel, each revolving on its own account, will be twenty-four diamond drills, one foot apart. In the center of the wheel is a single drill, and this is kept one foot in advance of the other drills. The wheel is calculated to make 800 revolutions per minute, the drills revolving at a higher rate of speed. The periphery of the tunnel will be on the scale of eight feet; the groove cut by the drills will be two inches wide and three feet deep. It is intended to. load the center hole alone, then run the machine back on the track, and raise the lower half of the wheel on hinges. The blast is fired, and the great cheese of rock crumbles to pieces. The machine is so constructed as to admit of three feet space inside of the wheels, between its framework and the bed of the tunnel, and facilities for removing the *débris* are afforded by an inner car track. The machine will be run by compressed air. Two pipes, each six feet in diameter, will be laid between and under the railroad track. The supply of water is estimated at 200,000 gallons per day.

Gravel and hydraulic ground.—Some idea of the enormous richness of the gravel-deposits between the North and Middle Forks of the American River may be formed from the following extracts from the local newspapers. I have taken the trouble to verify these statements, so far as the yield is concerned, by personal inquiry from reliable sources.

The Auburn Stars and Stripes of June 15 says:

From Michigan Bluff, Turkey Hill, and Last Chance the reports are encouraging in the extreme. From the Weske claim—twenty men working six days—the yield left to the owner a dividend of $4,030 for the week. Weske now has about four feet in depth of pay-dirt, and there is every indication that he is on the eve of striking the main channel, when he is sanguine of a deposit far surpassing anything he has yet worked. Last Saturday John Yule brought over from his claims, near Last Chance, to Michigan Bluff, $1,740, the product of 138 days' work on the time-table. This gives $36.75 per man per day as the yield of the Weske, and $13.60 per man per day as the yield of the Yule claim, making no allowance for considerable dead work in both claims. In addition to the above, the Van Emmon Brothers last week cleaned up 106 ounces in the Big Gun claims, Michigan Bluff.

The Weske claim paid for the first seven weeks of the year 1871, over and above all expenses, as follows: January 1, $752.50; January 8, $510; January 15, $711; January 22, $500; January 29, $904.50; February 5, $900; February 12, $900. Aggregate net yield for seven weeks, $5,178, or at the rate of $38,465 per annum. The claim embraces about 1,600 feet of the ridge between El Dorado and Volcano Cañon. A tunnel from the El Dorado Cañon side has been driven 1,700 feet; straight for the first 500 feet, and since that following a rich channel parallel for some distance with El Dorado Cañon, then diverging into the ridge at right angles, and apparently leading towards a main channel supposed to be perhaps 1,000 feet further in, and doubtless immensely rich.

Gold-dust to the amount of 84 ounces, valued at $1,575, was shipped from Michigan Bluff to W. H. Watson, secretary of the Yule Gravel Company, San Francisco, said sum being the yield of the above-named claim for the week ending September 30, with eight drifters and two car-men at work, and a drawback of a considerable percentage of what is known among miners as "dead work," i. e., in "squaring up," "straightening track," and the like. Owing to these drawbacks, the above yield represents but 52 days' work in the actual operations of taking out and washing pay-dirt, which is over $30 per day to the hand.

The Yule claim, two miles from Last Chance, at Startown, is the best claim in that section. The scene of present operations is 1,300 feet from the mouth of the tunnel, where a "breast," measuring 130 feet parallel with the tunnel, is being driven toward the west line of the Yule, which is the east line of the Morning Star. Thirteen men, beside the superintendent, are employed. This claim yielded a fraction over 102 ounces for the week ending July 1, and dividends exceeding $6,000 for the month of June. The Weske dividends exceeded $20,000 for the month.

The Morning Star Company have realized returns equal to those obtained in the Yule claim. A clean-up (July 1) limited to the three upper sluices, resulted in a yield of 12 ounces 5 pennyweights of coarse gold, salable for $18.12½ per ounce at Michigan Bluff. Including the above, the yield for five days amounted to 64 ounces, netting, over and above all expenses, more than $900. Considering the uniform results obtained, we may calculate on 700 ounces as the yield from the block of ground, 90 by 130 feet, between the present line of operations and the west line of the claim.

The Weske ground is two miles above Michigan Bluff. It is a gravel claim, and contains 210 acres, the property of Adolph Weske. In June, 1871, a clean-up of six days' work of twenty men was 261 ounces, worth $17.50 per ounce, or $4,582.50. This shows over $38 per day to the man, and if we deduct $360 wages for the men, at $3 per day, Mr. Weske has cleared in one week $4,222.50.

The same paper gives an extract from a private letter from Michigan Bluff, (June 8:) "All the talk here is about big pay in the Weske claim, Turkey Hill. They cleaned up last week 264 ounces, and picked up 60 ounces yesterday before dinner. The dirt is a sort of blue cement, and is the richest ever discovered here. The tunnel is in 1,800 feet; the paying gravel is about 2½ feet thick. This claim yielded $4,033.50 for the week ending June 26, and $4,404.25 for the week ending July 1, giving a fraction over $23,000 for the four weeks ending with the latter date."

Near Forest Hill and Bath are several claims worked by drifting on ancient channels, crushing the cement by mill process. There are six mills, with an aggregate of ninety stamps, erected for this purpose, but none of them have been run regularly for several years past. The owners of ground are awaiting the introduction of water, which will supply a more economical method of treatment. Quartz-mining in this vicinity has not been prosecuted with much vigor. Todd's Valley and Iowa Hill have immense tracts of hydraulic ground and gravel, which cannot be worked to advantage with the present limited supply of water. This part of the country is cut up into immense cañons by the erosion of the two forks of the American River and their tributaries. The Von Schmidt enterprise, above noticed, seems to be the only feasible one for procuring a large supply of water. Mr. Charles Fett, of Forest Hill, writes as follows:

There is nothing new to report for this year. Many of our mines do not produce as much as in the previous year, and no new mines have been opened. In my last year's report I estimated the total product of our district at $200,000, but I found afterwards that my figures were somewhat too low. The above-named amount will be about correct for this year's product.

A *large* supply of water the year round would make our place one of the liveliest camps in the State, and we have some hopes of getting a large ditch in here before long. We have a belt of land between this place and "Shirt-Tail Cañon," from two to five miles wide and eight miles long, the largest portion of which will pay well for hydraulic mining with a *large* supply of water. I base my judgment on the hydraulic claims of Nevada County, and Gold Run in Placer, where much poorer ground pays a handsome profit. An estimate of the yield per cubic yard I am not prepared to give, because heretofore our supply of water has been small, and only for two or three months in the year, and consequently only selected ground could be worked under such circumstances.

We have also deep deposits of gravel to a large extent, which, however, require heavy capital for their development.

Of crushing-mills we have but one in operation, the Paragon mill at Bath. All the others are idle, and offered for sale, except the Rough Gold Company's mill at Bath, for which the company still hope to have use.

At Dutch Flat the celebrated Taeff and Franklin ground, comprising forty acres of gravel with a depth of 240 feet, which has been noticed in previous reports, was sold during 1871 to a San Francisco company for $100,000, half the amount being paid down. The company will run a bed-rock tunnel from Bear River, for which purpose they will use a

Severance diamond drill. Operations in this vicinity are netting as good returns as in any previous year, and better than for several last past. Rablin, Taeff, and others in the Summit claims, Plug Ugly Hill, recently patented, realized fifty-five and forty-eight pounds respectively, from two clean-ups, after two weeks' washing. This company owns the fee to over one hundred acres of ground supposed to be as good.

Andrew Larson has opened a set of claims known as the "Central." He has put in 2,200 feet of flume 40 inches wide, and 400 feet of flume 44 inches wide, the latter in a tunnel which has been just completed through soft clay, with massive trap-bowlders, and so moist as to require the use of false timbers and a boarded breast to work it at all. The tunnel is timbered with giant posts, standing on solid soils, the whole lagged with heavy lagging. This tunnel gives an additional fall of 60 feet, and his shaft at its head is cribbed with strong timbers, then lined with planks, and then again with sheet iron upon a portion. He will have a bank 200 feet high, which he will attack with 500 inches of water from Hoskin's Dictator and Little Giant.

Many other claims at Dutch Flat and vicinity are being prepared for extensive operations in the spring of 1872, and a larger yield may be expected from this locality than for years past. The Cox pan, noticed in my description of the southern mines, has lately been introduced here. One of the owners of the Baker ground furnishes the following account of its operations:

On the Baker claim at Dutch Flat is one of Cox's cement-mills for working cemented gravel. The cement worked by this mill is the blue cement, and probably as hard as any in the State.

The cement and gravel are loaded in a car in the claim and run to the mill on a track, where it is dumped into a hopper with an inclined bottom, from which it is loaded into the mill by means of a gate operated by the man in charge of the mill. It thus requires but once handling—a great saving of cost.

There are usually put into the mill at one time from ten to twelve hundred pounds, which is done while the mill is in motion, as will be described below.

In the top of the rim of the mill there is constantly a stream of water of four or five inches, which carries the pulverized cement down through the small openings of the bottom of the mill into the sluice-boxes provided for saving the gold.

The mill is set in motion and the gate of the hopper raised to gradually let the cement enter the mill, which generally occupies about two minutes, when it is worked about four minutes longer, when the mill is stopped and a trap-door opened by a lever, the mill set in motion, and all the rock is driven through the trap-door into the rock-sluice, (the cement which contains the gold having been thoroughly disintegrated and pulverized by the friction, &c., and passed into the sluice-boxes.) The operation is then repeated. Softer material requires less time to work each charge.

The mill will readily work from 100 to 125 tons (of 20 cubic feet to the ton) in 24 hours, and at an expense, including water-power, labor, &c., of about 10 cents per ton. The hardest cement requires the mill to be worked at a speed of 65 revolutions per minute, which requires about 8 horse-power. This mill is run by a hurdy-gurdy wheel 10 feet in diameter.

The mills working in Tuolumne County are also worked by hurdy-gurdy wheels, but it only costs from 7 to 9 cents per ton, the cement not being so hard as that at Dutch Flat.

This machine certainly does its work very thoroughly and cheaply; every stone is thoroughly cleaned and washed, and the cement so pulverized that it is very difficult to find a color in the tailings.

The inventor claims that one mill will do as much work as a 25-stamp mill.

The cost of this mill is very small compared with the stamp-mill, the price being $1,200. The cement can be worked for about one-tenth the expense of the stamping process, as it costs from $1 to $1.75 by stamp; besides, it does its work more thoroughly. The wear and tear is estimated at 10 cents per day.

The yield for Gold Run district for the past year has been unusually light, on account of the drought. A tunnel to open and drain this ground has been commenced by the Gold Run Ditch and Mining Company. It will be run from Cañon Creek, and will tap the mines 243 feet lower than the present benches. It will be 2,200 feet in length, 9 feet wide, and 8 feet high.

Nearly all the extensive and valuable hydraulic and gravel ground between Dutch Flat and Nevada City, lying on Bear River, Steep Hollow, and Greenhorn Creeks, (see Report for 1871, pages 81–84,) has passed into the hands of English companies at prices which have yielded fortunes to their former owners, but which will prove highly remunerative to the purchasers. The "You Bet" ground, formerly belonging to Edward Williams, is now incorporated in London as the Birdseye Creek Mining Company. The Little York ground is about to pass under the control of a London company, and it is rumored that the vast and valuable interests of Messrs. Sargent & Jacobs, near Quaker Hill, will likewise soon change hands. No facts or figures could be obtained from the new owners this year, which is much to be regretted, as their systematic and careful management (as manifested in the conduct of some of their early purchases) would throw some light on many questions of economical mining. The earliest purchase of this kind of ground was made at Buckeye Hill, near Sweetland, in Nevada County, and the profits realized have produced the natural results of turning the attention of English capitalists to these enormous and comparatively undeveloped resources of our State. The result has been the investment during the past year of over $1,000,000 in our gravel mines alone.

The following report from French Corral mining district, Nevada County, was furnished by Mr. W. M. Eddy, of French Corral:

Estimated area of mining ground to be worked, 275 acres.
Estimated average yield per cubic yard of hydraulic ground, 15 cents.
Average yield per cubic yard of cement ground, $3.50.
Cost per cubic yard of mining and working cement ground, $1.10.
Net yield of cement ground, $2.40.
Principal mining companies in this district as follows:
The French Corral Mining Company having about 75 acres of ground, both hydraulic and cement, own and are running steadily one 15 and one 10-stamp mill, crushing cement. Bed-rock tunnel in contemplation. Present tunnel low enough to work hydraulic ground several years. Own valuable water-right in connection with claims.
Kansas Company own valuable cement claims. Have one 10-stamp mill running steadily. Have ground for about one and a half or two years' work.
Nebraska Company—Cement ground. Have one 10-stamp mill running steadily, and ground to last about six months.
Kate Hayes and Troy Company own about thirty-five acres of ground, (hydraulic and cement.) Not worked at present for want of water. Deeper bed-rock tunnel contemplated.
Trust and Hoper Company have about twenty-five acres of hydraulic and cement ground. Bed-rock tunnel contemplated and necessary to work it advantageously.
Bell, Alexander & Co. have about forty acres of ground, both hydraulic and cement. Bed-rock tunnel necessary to work it.
Allison and Co. have about twenty-five acres, both hydraulic and cement ground. Tunnel necessary to work it.
Monte Cinto and Railroad have about twenty-five acres cement and hydraulic ground. Tunnel necessary.
Bed-Rock Tunnel Company have some twenty-five acres ground, both hydraulic and cement, completed, and opened up a bed-rock tunnel this year 2,700 feet in length. The Allison and Monte Cinto Companies' claims join this company's ground, and can be worked through their tunnel.
Estimated gross yield of mines in this district this year $250,000; ruling wages for skilled labor, $4; for unskilled, $3.
Estimated average depth of mining ground, 100 feet.

The above-described district is situated between the South and Middle Yuba Rivers, a section of country described in Report for 1871, pages 72 to 78. The North Bloomfield Gravel Company, further up the same ridge, having demonstrated the great richness of their ground by prospecting-shafts, have commenced a tunnel to open their claims to the bed-rock. This tunnel will be run from the Yuba River, and will reach the claims at a point estimated 200 feet below the bed-rock, giving ample fall for working. In shaft No. 1, they have run 500 feet each way, demonstrating the channel to be 1,000 feet wide. Here they have taken out as high as $1,000 a day, the pay-gravel being equally distributed. In No. 2 they have struck the mine, also in No. 3, and find prospects sufficient for hydraulic mining all the way down. The distance between the two shafts is three-quarters of a mile, and the striking of the last prospect is important as demonstrating a continuous channel from one to the other. The dam at Bowman's is heavy enough to keep the ditch full, and 8,000 inches now run to waste. The ditch is running 2,800 inches. They are using three or four pipes with six-inch nozzles day and night.

A Grass Valley paper of recent date, in alluding to the prospects of this ridge, says:

The Union Hill Gravel Mining Company, at Columbia Hill, is making extensive preparations for hydraulic mining. Heretofore they have used about 400 inches of water day and night. As night-work is not as profitable as day-work, the company concluded to construct a large reservoir, and use 1,000 inches ten hours daily. The reservoir will be completed in about three weeks, at a cost of between $3,000 and $4,000. Sixty men are now employed in its construction. The company has run a bed-rock tunnel 1,050 feet, and has 600 feet yet to finish. The rock is hard syenite, and only 12 inches' advance can be made in twenty-four hours, with three shifts. To expedite the completion of the tunnel, the company is making arrangements to put in a diamond drill. This will be driven by water-power introduced into the tunnel by means of an iron pipe. The pressure will be something over 300 feet. The power is to be applied to the drill by means of two hurdy-gurdy wheels. The tunnel has cost thus far $13 per foot. It is estimated that the diamond drill will facilitate progress in the tunnel 66 per cent. over the hand-drill. This winter the company will use in their claims two nozzles, one five-inch and one six-inch. The smallest will discharge 400 inches, and the largest 600 inches of water, under a 300-foot pressure. When their bed-rock tunnel is finished it will be 297 feet from the surface, or from the first bench that has been washed off. This bench is 142 feet from the original surface. This will give the reader an idea of the original depth of the claims. The Union Gravel Mining Company have a vast amount of mining ground in one body, their claims extending two and one-fourth miles on the gravel-channel, which latter has a width of from 1,000 to 2,000 feet. They use for hydraulic purposes in their claims 1,200 feet of 18-inch pipe and 3,800 feet of smaller pipe for their diamond drill.

The claims at Relief Hill are yielding excellent returns to their owners. The channel proper of the Great Blue Lead at this point is over 2,000 feet in width, and recent developments prove that the pay-gravel continues from rim-rock to rim-rock. The owners are just getting fairly into their mines, after years of persevering labor. The recent "clean-ups" of the What Cheer and Walkinshaw Consolidated, Eagle, North Star, and Union Companies, showed a yield of from $16 to $30 per day to the hand. The above mines cleaned up, September 27, over $20,000, middle of October, $5,000, and November 13, $14,673.34.

The gold in this ancient river-bed is very heavy, especially in the center of the channel, where pieces are frequently found weighing from one to five ounces, and sometimes as high as 15 ounces. The above claims have from 300 to 1,500 feet frontage, and one mile in length. The banks vary from 75 to 200 feet in height. The hill known as Relief Hill is situated three miles from North Bloomfield, and 17 miles from Nevada City, California. The net profits of these claims for less than two months exceed $30,000.

At Omega, during the past summer, unusual preparations have been made in placing the best gravel-claims in a complete state of readiness for hydraulicking. A large amount of drifting and blasting has been done. Tully & Co. have two sets of claims, and will be able to use from 1,500 to 2,000 inches of water; Burwell & Fuller will use from 500 to 700 inches; Sale & McSorley also 500 inches. Fuller, Pease & Co. have about 1,000 feet of pipe down, and will run 750 inches of water. Evans & Co. have about 600 feet of pipe down, and will use about 600 inches of water. S. Kyle has about 500 feet of pipe laid, and will use about 500 inches of water.

The prediction in my last report of the increasing yield of the gravel-channels near Grass Valley has been fully realized. The Hope Company, which was then the leading company in this branch of mining, have been constantly at work, and are now using an eight-stamp mill. The yield from this claim has not, however, been uniform, and has disappointed the expectations formed from the condition of the mine at the period of our visit in July, 1870. The mine is now looking better. This company took out $10,000 in November last, at an expense of $2,500. Many claims there engaged in prospecting have been successful in striking rich spots in the channel. The Webster Company are reported to have taken $27,000 from a piece of gravel in this claim 25 by 100 feet in dimensions. The Picayune Company also have rich ground opened by a tunnel 725 feet in length. The Town Talk, lying to the east of the town, has made some extraordinary runs. A month's run in September last yielded 279¼ ounces of retorted gold, worth $5,000. The channels here seem to be "spotted" in their character, and difficult to trace, as the topography of the country gives no indication of their subterranean course. The surface is gently undulating hills and drainage is difficult. Most of the claims are opened by shafts or inclines, and the water and gravel raised by machinery.

The yield of the deep placers.—The question of the yield of hydraulic and gravel claims is one which has lately attracted the observation of some of our practical miners. Some interesting data relative to the yield of Gold Run district, Placer County, were furnished the writer last year by Professor W. H. Pettee, then connected with the State geological survey. The facts were known too late for insertion in the last report, and were first published as a communication to the Engineering and Mining Journal, of New York. This communication called forth the following editorial in the Mining and Scientific Press, of San Francisco:

It would appear that heretofore the yield of the placer-dirt, at least in several localities in our State, has been generally overestimated. An example of this is with regard to the placers of Gold Run district. The Engineering and Mining Journal lately had a communication with regard to an interesting calculation of the average yield per cubic yard of the dirt washed in this district, made by W. H. Pettee, of the California State geological survey. The superficial area here, from the Central Pacific Railroad southerly to the place where the deposit has been broken off by the cañon of the North Fork, is estimated at 860 acres, of which about one-half has been worked over, not worked out, as the bed-rock has been reached only at the southern extremity, in the ground of the Indiana Hill Cement Mining Company. It is estimated that 43,000,000 cubic yards of dirt have been removed by hydraulic process, and the gross product of the district, calculated from statistics furnished principally by Messrs. Moore & Miner, is given as about $2,000,000. The average yield, therefore, has not been over 4½ cents, and yet hydraulic mining has been carried on with large profit. This calculation, however, embraces only the product of the surface-dirt, as there are still from one hundred to two hundred feet of gravel and cement underlying the excavation. As the richest dirt is generally found near the bed-rock, future yields will probably bring up this average considerably higher. Several estimates have been made of the average yield of the claims between the Middle and South Yuba. Our readers will remember Laur's estimate of about 16 cents, and Silliman's of about 30 cents per cubic yard (in Ross Browne's Report, 1868) for this last region. We may say in addition that we believe Mr. Pettee's calculations to have been as carefully made as any others, probably more carefully than any before.

The editor is correct in assuming that the calculations of Professor Pettee were made with care. They were made after a careful and detailed measurement of the banks of the basin, and an estimate from such measurement of the quantity of dirt removed, taking into consideration the topography of the surface of the country before hydraulic washings commenced. For this latter purpose he had to rely, of course, on the information of the miners of the district. The gross yield of the district was obtained from the books of Messrs. Moore & Miner, bankers, of Gold Run, and will be found on page 85 of the Report for 1871. The method of calculation adopted, while it cannot be claimed to give accurate results, will at least afford an approximation to the yield of the district per cubic yard of the amount of gold extracted from the *upper strata of hydraulic dirt*, but cannot be accepted as the average yield of the district, as none but hydraulic dirt was embraced in the calculation. At that time the harder gravel and cement had not been reached, and subsequent runs indicate an increasing yield in the lower strata, while the bottom, as developed in the mine of the Indiana Hill Cement Company, is proving of great richness. In view of these facts we feel justified in believing that the average yield of Gold Run will equal, when bed-rock is reached, that of the placers described by Professor Lauer, between the Middle and South Yuba. The successful mining of low-grade dirt at this place is owing to the softness of the dirt, its great depth, (average of 200 feet,) the abundance and cheapness of water, and the facilities for running off with plenty of fall. The outlets are now becoming filled, and a bed-rock tunnel is necessary.

Mr. W. M. Eddy, of French Corral, Nevada County, has made some estimates, based on experiments undertaken at our request, of the yield of hydraulic ground near French Corral, and gives the following as the results:

Average yield per cubic yard of hydraulic ground............ $0 15
Average yield per cubic yard of cement ground............... 3 50
Cost per cubic yard mining and working cement ground, (mill process).. 1 10

This gives a net yield for the cement ground of $2.40 per cubic yard. The mills are run by water-power. The hydraulic dirt, it will be observed, is much richer than at Gold Run, but it is not as deep, and is much more compact.

A correspondent of Stars and Stripes (Auburn, Placer County) estimates the average past yield of ground per cubic yard, for districts in that county, as follows: Iowa Hill, 71 cents; Independence Hill, 25 cents; Roach Hill, 60 cents; Richardson Hill, 15 cents; and Wisconsin Hill, 12¼ cents; and says parties in Gold Run district estimate the cost of working there at 2 to 3½ cents per cubic yard. It is estimated that the cost of working at Iowa Hill will be 2¼ cents; Independence Hill, 2 cents; Roach Hill, the gravel on which is much harder than in the other places named, 6 cents; Richardson Hill, 3 cents; and Wisconsin Hill, 2 cents.

In this connection attention is called to the tables prepared by Mr. J. Rathgeb, of Calaveras County, relative to yield of mines near San Andreas, also to description of claims in Table Mountain, under the heading of "Southern Mines."

Quartz mining in Placer and Nevada.—The gold-bearing quartz-ledges of these counties are so numerous as to render a detailed description, or even mention, impracticable within the limits of this article. Most of the leading mines, particularly those of Grass Valley and vicinity, have a world-wide reputation, and have been repeatedly described in the reports of the Mining Commissioner, so that their characteristics are familiar to all persons interested in mining. Within the limits of these counties are three important districts, containing groups of valuable mines, viz., Ophir district, in Placer County, and Grass Valley and Eureka, in Nevada County.

Ophir district lies near the town of Auburn, on the Central Pacific Railroad, thirty miles from Sacramento City, at an elevation of only two or three hundred feet above the Sacramento Valley. This group of mines lies in a southerly direction from the celebrated Grass Valley mines, and, it is claimed, are on the same belt of formation, though we doubt whether this is capable of demonstration. This locality was famous in early times for the yield of its surface placers. These placers were of great richness, but very shallow, and the gold was undoubtedly the result of the decomposition of the numerous scams of quartz which cropped out of their slate casings. As the placer-ground was exhausted by the improvident miner of those days, he turned his attention to the quartz-veins, and explored them to a depth of from thirty to fifty feet, as far as he could work without pumping and hoisting-machinery. These veins run parallel to each other, and can be traced by these workings for thousands of feet in length. Most of them being narrow, and the gold lying in "pockets," they were only explored to the water-line, where they were abandoned and lay dormant for years. The towns of Ophir and Auburn were, for several years, nearly deserted, the latter, owing to the location of the county-seat, preserving something of its former prosperity. It is only within the past two years that some of these mines have been opened systematically, and the results have been, in every instance, satisfactory.

This metalliferous belt, varying in width from three to five miles, courses through the county, and consists of slates highly metamorphic, trap, (diorite,) porphyry, and granite.

The ore-bearing veins vary in thickness; when in slate, from two to five feet, but when occurring in trap or granite their strike is at right' angles with the formation, and rarely exceed twenty inches in thickness. The ores contain gold and silver, pyrites of iron, and copper blende occasionally in small quantities, and tellurets. Pockets and nests of nuggets of gold are frequent, often containing from $10 to $100,000 in a single nest; but the mass, aside from the pockets, rarely exceed $10 to $70 per ton of 2,000 pounds of ore.

The St. Patrick Mining Company own three parallel veins in trap. The St. Patrick proper, near the surface, contains a pocket from which $75,000 was taken. The present depth is 220 feet, with a persistent yield of milling ore. An Auburn paper of March, 1871, thus refers to a crushing made by this company: "One hundred and ninety-four tons of rock were crushed, yielding within a small fraction of 60 pounds of retorted gold, valued at about $11,000. In addition to the above, the company sent below a lot of specimen rock—about half a ton—the value of which is variously estimated at from $1,000 to $2,500. Thus we have a test by the crushing of nearly 200 tons of rock just as it came out, good, bad, and indifferent, that shows an average of about $67.50 per ton. Considering that the bottom of the shaft is 80 feet below the water-line, and that the rock at that depth is the best they have struck, the above must establish the reputation of the St. Patrick."

The Big Doig, a parallel vein, and only 120 feet south, also contains rich and remunerative milling ore.

The Peachy, traceable on the surface for 1,500 feet, will average 18 inches thick, and pay from $30 to $40 per ton.

The Bellvue has eight parallel veins in trap, coursing east and west, and within a distance of from 30 to 100 feet of each other. The ores from this mine have milled from $33.50 to $40 per ton, without regard to the sulphurets, which are of a very high grade, and assaying from $50 to $6,000 per ton. The ore contains about 2½ per cent. of sulphurets.

The Greene mine is about two feet in width, and has paid dividends from ore taken from the shaft without stoping. A local newspaper of May, 1871, gives the following particulars of the yield of this remarkable mine: "The owners have placed on the mine a small crusher, pans, and other reduction-works, and have made two small runs of the rock. They first reduced ten tons, which yielded $500, or $50 to the ton. Another small run of 6 tons and 1,600 pounds from a different class of ore was then put through, which turned out $1,255 in gold, or near $200 per ton. Subsequently they ran through 12 tons of ore, which yielded some $15,000 in melted gold, over $1,100 to the ton. This is the most astounding yield we have ever chronicled, and we doubt if it has ever had its equal from the same amount of gold-bearing quartz. There are now on the dump at the shaft 100 or 200 tons of ore, fully half of which is rich. This mine has been extensively prospected by shafts and drifts, and all the expenses, including the mill, pans, and machinery, have been met by running small portions of the quartz through an arrastra occasionally, or pounding out gold in a hand-mortar." Another paper of later date says: "At Greene's (the old 'Mallet') ledge, supposed by some to be on the Good Friday ledge of the Ophir Company, they have within the past week taken out a large quantity of surpassingly rich rock. Many specimens seem to be almost half gold. The amount of gold in the quartz extracted within the week is variously estimated at from $12,000 to $20,000. Two and a half days' crushing last week, with four stamps and one Hepburn pan, realized 185½ ounces of retorted gold. This is at the rate of $1,987 per day, and gives a total of about $28,000 within the past month. A two-weeks run of the mill—four stamps—has yielded within a fraction of $50,000, which is reckoned the heaviest yield on record." These yields, astounding as they may seem, are well authenticated.

The Peter Walter, an extension of the Greene easterly, has been noted for its rich pockets. The present owners are working the mine for its value for milling ores with good results. The company have complete steam-works on the mine, propelled by a 25 horse-power engine, with friction gear and everything working in most perfect order. The main shaft is down about 120 feet, and work is being prosecuted night and day with 3 shifts, 12 men in all working under ground. Levels are being run both east and west from the shaft at a point 116 feet deep. These levels are in some 10 or 12 feet, and good ore, showing plenty of free gold, is being taken from them and hoisted to the surface. The value of the quartz taken from these drifts is placed at from $40 to $50 per ton. It is in contemplation to erect a 10-stamp mill.

The St. Louis. This mine is inclosed in granite. Vein from 2 to 6 feet in thickness, and ores varying in value from $20 to $60 per ton.

The Crandall. This vein is in a highly metamorphic slate, closely resembling trap, and is an exception in its strike to most other veins in this formation, as it runs easterly and westerly, averaging in strength 3½ feet. Ore valued at $20 per ton. Recent developments in depth denote a valuable and extensive mine.

The Shipley mine has a powerful vein carrying low-grade ores.

Mr. J. H. Crossman, to whom I am indebted for notes on this district, furnishes the following respecting two well-known mines:

The Buckeye, with a parallel lode 30 feet distant called the "Big Vein" or the "Elizabeth," used to be worked by shafts from 30 to 50 feet deep, or until water or hard rock was reached; then the shaft would be abandoned and another sunk. In that way it was prospected or worked for about 1,600 feet in the two veins to the depth stated; the ore taken out paying from 30 to 100 ounces per ton by arrastra reduction. The Fraser River excitement broke out and the owners all sold out or abandoned their claims. Some San Francisco capitalists recently bought the entire hill containing eight parallel lodes within a compass of 300 feet at right angles, and extending on their line of strike 2,500 feet easterly and westerly, dipping northerly; all true fissure-veins cutting the formation, which is greenstone (diorite) and metamorphic slate. The veins are small, say from 12 to 30 inches thick, but carry uniform ore of a very high grade, working by mill-process from 35 to 50 ounces per ton. The concentrated sulphurets assay from $434, the lowest, to $17,000 per ton. The superintendent has inaugurated systematic workings by shafts and levels, and erected good hoisting-works. He has taken out and milled 200 tons of ore, which was extracted at an expense of $8 per ton and paid $35 per ton. He has nearly 400 tons on the dump, ready for milling, that he estimates worth $40 per ton. He is driving an adit level from Doty's ravine (now in 123 feet with 50 feet of backs) on the "Elizabeth lode" 30 feet from the Bellevue mine north, parallel with it and 1,500 feet from the main shaft of the Bellovue on the line of strike of the lode. This level when opposite the Bellevue shaft will drain the eight mines to a vertical depth of 175 feet. A cross-cut driven north and south will tap all the lodes and furnish many thousand tons of high-grade ores. The vein, which was from six to eight inches thick, and what is known as ribbon-rock, exhibited rich specimens containing more gold than quartz; the parallel ribbon adjoining showed bunches of wire and horn silver; then came a ribbon of massive sulphurets, containing galena, blue sulphuret, silver, and sulphuret of iron. The formation is easily worked, drifts being driven for $6 per foot, and shafts sunk for $12 per foot.

Mills in the district.—The Green Emigrant, 20 stamps; the Ophir, 10 stamps; the Shipley, 10 stamps; the St. Patrick, 15 stamps; and several others contemplated.

Among other mines of note in Placer County is the Rising Sun, near Colfax, owned principally in Grass Valley. This mine has been managed with great economy and prudence. The ledge is narrow, but well defined and rich. A remarkable strike was made this summer in the east drift of the 300-foot level (the lowest level.) The rock is deep blue in color, and heavily seamed with gold. This level has proved uniformly rich. The Auburn Herald says of the recent strike:

The rock is seamed with gold, which, if all in one, would make a slab of gold about half an inch thick, running longitudinally through the ledge, which is about a foot thick at this point. The gold is very pure and worth $18.50 per ounce. Such rock as we saw should yield $15,000 or $20,000 to the ton. Some distance from this point the ledge widens to four feet, and is an almost solid mass of rich sulphurets. This very rich body of ore was opened at a depth of 310 feet from the surface, a point far below any atmospheric influence, and is for this reason more reliable as a permanent and extensive deposit. This mine has been worked for several years, and within the past year the company have sunk the shaft from the depth of 230 feet to the present depth of 310 feet, aside from the drifting and prospecting the ledge as they went down. The company have a five-stamp mill, steam-power, 800-pound stamps, nine-inch dies, eleven-inch drop and sixty-five drops per minute, with a capacity of ten tons per day. There are 3,000 feet in the claim, and within the past three months dividends to the amount of seven dollars per foot have been declared and paid, being a net profit on the working of the mine in ninety days of $21,000. This is the deepest working mine in the county, and although the owners had at one time expended over $60,000 over receipts, they kept on with undaunted courage, knowing no such word as "fail," and now richly deserve their success and the fine prospect in sight.

At Grass Valley and Nevada City, the past year has been one of unusual prosperity, and prospecting has been vigorously carried on between the two towns on the ridge, where several "blind" ledges of great promise have been discovered this season.

Report of the North Star Company for the year ending September 5, 1871.

The cash balance on hand at the close of the previous fiscal year was $1,389.33. The bullion product from ore and skimmings was $205,101.50; and from sulphurets, $13,797.55; receipts from tributors, mill-tolls, &c., $5,518.12. The disbursements embraced $76,500 in dividends to stockholders, $150,884.70 for general expenses, including mining and milling labor, supplies, salaries of officers, &c.; $5,988.08 for mine and mill improvements and wood ranch; $3,567.25 for unpaid debt of the previous year; and $2,745.55 for gain in stock account and supplies. The receipts from all sources exceeded the disbursements for current expenses, mine and mill improvements, and wood ranch, by $67,544.39. As the company paid their stockholders $76,500 in dividends, and liquidated an unpaid debt of the previous year of nearly $3,600, this balance was of course more than absorbed. In other words, the result for the year may be briefly stated as follows:

Disbursements for all objects		$239,686
Cash September 5, 1870	$1,390	
Receipts for the year	224,417	
		225,807
Liabilities		13,879

Against the liabilities the company report assets in mine and mill supplies amounting to $21,041, thus showing a surplus of assets, in supplies, of $7,162. The receipts since the commencement of the new year, September 5, 1871, aggregate $3,662. It is expected that dividends will be resumed in the course of a month or two. The total amount of ore raised was 9,212 tons; amount worked, 9,172 tons; leaving 40 tons on hand. The total yield of ore worked was $205,102, showing an average yield of $22.25 per ton, exclusive of sulphurets. The aggregate net profit of the ore worked was $48,129, or an average of $5.25 per ton. The gross yield of 341 tons sulphurets worked was $21,491; cost of working the same, $7,694; net profit on same, $13,798.

Report of the Eureka Company for the year ending September 30, 1871.

The claim is at Grass Valley, and has long been known as one of the best gold-quartz mines in California, though recently the grade of ore has unexpectedly run quite low. The report of Superintendent Watt, for the year ending September 30, shows 17,447 tons of quartz raised. The amount of ore crushed was 18,560 tons, in 305 running days, with a 30-stamp mill. On account of a scarcity of water only fifteen stamps are now used. The amount of drifting and cross-cutting made was 950 feet. The main shaft is sunk 849 feet on the ledge, or 786 feet vertically, of which 120 feet was sunk and timbered during the past year, besides retimbering 120 feet and sinking 200 feet of winzes. The report closes with the remark: "Knowing that the future prosperity of the Eureka mine depends upon developments, I am sinking the main shaft and driving the drifts and cross-cuts as rapidly as possible."

RECEIPTS.

Cash on hand October 1, 1870	$84,350
Bullion	556,951
Sulphurets	6,104
Premium and discount	2,916
Miscellaneous	1,378
Total	651,708

DISBURSEMENTS.

Dividends to stockholders	$360,000
Mining account	154,551
Milling account	36,702

```
Construction...............................................................   $22,563
Prospecting................................................................    15,469
Mine purchase..............................................................     8,747
General expenses...........................................................    16,349
Miscellaneous..............................................................    17,459

     Total expenditures...................................................   625,840
Cash on hand October 1, 1871.............................................    25,868

     Total................................................................   651,708
```

The company have no liabilities. Their assets aggregate $127,477, as follows :

```
Cash on hand............................................................... $25,868
Ore and supplies..........................................................   23,109
Mill, estimated value.....................................................   40,000
Mine improvements, &c.....................................................   30,000
Sulphuret works...........................................................    3,000
McDougal works............................................................    4,000
Wood ranch, 160 acres.....................................................    1,500

     Total assets.........................................................  127,477
```

The average yield of the ore for the year was $30 per ton, and of the sulphurets, $158.23. There were 275 tons of sulphurets worked during the year. The average cost of mining the ore was $8.82, and the average cost of milling $2.02, or a total of $10.84, leaving a profit of $19.16 per ton. The average cost of concentrating sulphurets was $15.88, and of reducing the same $22.16, or a total of $38.04, leaving a profit of $120.19 per ton. The net profits on the operations of the mine for the year were $330,763, or nearly $30,000 less than was paid in dividends, showing a draft to that extent on the surplus carried over from the previous year. The mine went into operation on the 1st October, 1865, since when the receipts have been $3,382,343, of which $3,362,234 was from bullion taken out. The disbursements for the same period were $3,342,495, of which $1,694,000 was in dividends to stockholders, $133,105 for construction, $1,219,492 for mining, milling, and other current expenses, and $295,808 for mines. We annex the monthly dividends of the company for the last four fiscal years :

Month.	1867-'68.	1868-'69.	1869-'70.	1870-'71.
October..	$20,000	$20,000	$30,000	$40,000
November...	20,000	20,000	30,000	40,000
December...	40,000	30,000	30,000	40,000
January..	20,000	20,000	30,000	40,000
February...	20,000	....	30,000	40,000
March..	20,000	20,000	30,000	40,000
April..	20,000	30,000	30,000	40,000
May..	20,000	30,000	30,000	40,000
June...	30,000	30,000	60,000	20,000
July...	30,000	20,000	30,000	20,000
August...	30,000	20,000	30,000	....
September..	20,000	24,000	40,000	....
Total...	290,000	264,000	400,000	360,000

No dividend has been paid since July, and there is no immediate prospect of a resumption of such disbursements. On the 1st instant there were 85 tons of quartz on the surface, and 950 tons broken in the mine, ready for hoisting, the value of which, as put down in the assets, is $9,129. The company hope to crush sufficient ore to meet current expenses until new bodies can be discovered and opened.

Report of the Idaho Company for the year ending December, 1871.

Work during the year, 1,057 feet of drifts, 79 feet of shaft sunk, 145 feet of winze raised. The 400-foot level was thought to be worked out a year ago, but good ore has since been struck east of the shafts, and 122 feet of length exposed on that level, which promises as good an average as that of last year.

The 600 east level is in 460 feet from the shaft, and still in pay-rock; the 600 west drift is in 266 feet from the shaft; and the 700 west drift is in 25 feet from the shaft.

The new shaft is completed 48 feet up from the 200 level, and a small working shaft is through to the surface.

The amount of rock crushed during the year is 11,133 tons, of which 729 tons came from the 400-foot level, 1,478 from the 500 level, 8,382 from the 600 level, 69 from the 700 level, and 475 from the shaft, giving a total yield of—

Bullion	$385,017 90
Ninety-two tons of sulphurets	10,041 23
Specimens and tailings	296 50
	395,355 63

Or an average of $35.50 per ton.

The average cost of mining and milling (not separated) was $10.20 per ton.

EXPENDITURES.

Mill and mining account	$113,630 47
Sulphurets	4,229 00
Construction—new shaft	11,466 87
Drain-tunnel	391 25
Drifting 350 feet in 200 level and sinking 45-feet winze	5,241 00
Repairing account	5,745 02
Law expenses in defending title to 300 shares of stock	14,919 45
General account	13,564 10
Dividends—$75 per share on 310 shares	232,500 00
	401,687 16

RECEIPTS.

Cash from last year	$11,358 40
From bullion on hand last year	7,070 18
Fourteen tons sulphurets on hand	1,446 76
Proceeds from 11,133 tons of quartz crushed during the year	385,017 90
Proceeds from 92 tons of sulphurets	10,041 23
Specimens and tailings	296 50
Water	650 00
Scrap iron	13 84
Charcoal	75
Pump column	64 00
Lease of gravel mine	200 00
J. W. Gashwiler	2,500 00
Total receipts	418,659 56
Expenditures	401,687 16
Cash on hand	16,972 40

The Empire Company has been busy for a great part of the year in extending underground workings, and the new mill has not been steadily run at full capacity. No detailed report has been received. The company has obtained a United States patent.

The New York Hill, one of the celebrated old mines of Grass Valley, has been re-opened, after a long period of idleness. This mine is reported to have produced in earlier years nearly $1,000,000.

The Daisy Hill, Seven-Thirty, Perrin, Grant, and South Star mines are reported as actively operating, with favorable prospects.

The Banner and Pittsburg mines at Nevada City have also been worked with success. The latter mine was offered for sale to an English company, who obtained a report, with reference to this sale, by J. D. Hague, esq., which is a model of comprehensive clearness and judgment. I extract the following passages, which give a good description of this important property:

The vein or lode on which the Pittsburg mine is worked crops out upon the slope of a hill-side that rises just to the right of the road leading from Grass Valley to Nevada. The course of the vein corresponds generally with the trend of the hill, being about northwest and southeast; and its dip, being generally southeast, at an average inclination of 43° from the horizon, takes it in under the higher portion of the hill. The slope of the ground is gentle, or only moderately steep, and is inclined to the northwest. That portion of the surface which covers the Pittsburg mining claim is cut by two or three shallow ravines or water-courses, which give the profile of the ground along the line of the vein an uneven character. The works of the mine are perhaps two or three hundred feet, more or less, above the road in the valley below; and the hill rises above the mine to a considerably greater altitude. The country beyond rises to the eastward, towards the summit of the Sierra Nevada Mountains.

* * * * * * *

The works of the mine are at the north end of the claim. The newer, or north shaft, which is sunk on the inclination of the vein and coincides with it in dip, is about 300 feet from the north boundary. The ground north of this shaft, and between it and the northern limit, has been developed to a considerable depth, and in places worked out close up to the line. The older, or south shaft is located at 660 feet from the north boundary, measured on the surface; but it is so inclined, with reference to the dip of the vein, as to extend further and further to the south, increasing the distance between it and the north boundary, or the north shaft, by about 100 feet horizontally for every 300 feet sunk. The ground between these shafts, to the depth of the 537-foot level, is worked out. South of the south shaft the vein has not been much developed. Several levels have been driven a little distance, from 50 to perhaps 200 feet, encountering a series of breaks or faults, showing the ground to be considerably disturbed in the neighborhood of the ravine, shown on the surface. The vein has thus far been clearly traced through three of these faults, and appears to be somewhat enriched by them, as streaks of free gold occur with greater frequency in their neighborhood than elsewhere; but, owing to various reasons, the developments have been chiefly made on the north end. The length of vein developed by the work thus described, measured from the north boundary to the south end of the mine, is about 800 or 900 feet; and the depth already reached in the north shaft is 783 feet. The vein has been opened slightly both on the north and south extensions of the Pittsburg mine; but either for want of capital on the part of the owners, or for other reasons, has not been much developed. Our knowledge of its characteristic features is therefore based on observations made within the limits above mentioned. The vein is inclosed in a hard, greenstone. This rock is usually fine-grained, compact, and of homogeneous character, but slaty enough in structure to cleave or break well on blasting. The walls of the vein are pretty well defined; the foot-wall especially so. They are readily distinguished and followed, even where the vein is pinched to a mere seam in thickness. Their course is about northeast and southwest, and is tolerably regular, though curving here and there in one direction or the other. The dip varies also in different places, but has an average inclination to the southeast of 43° from the horizon. The vein itself is a clear, hard, compact seam of white, or bluish-white quartz, carrying some free gold and considerable quantities of gold-bearing iron pyrites. The quartz is usually separated from the walls by a thin layer of clay or "gouge." The thickness of the vein is variable. The walls sometimes approach each other closely, pinching the vein to a mere seam, and again expand to a width of three or four feet. In some places they are five or six feet from each other, but in such cases there is usually a mass of country-rock included between two separate seams of quartz. The average thickness of the vein is estimated by the mining captain, who has been working the mine during several years, at 18 inches. I do not think this unlikely, although, for safety, I prefer to estimate it at 12 or 15 inches in thickness. The distribution of the gold in the quartz, if not entirely uniform, is somewhat remarkable, in so far that all the quartz of the vein is considered as milling rock. None of it is rejected, but every pound, apparently, is taken out and sent to the mill for crushing. In fact, the old dumps are now being overhauled for the purpose of selecting such pieces of vein-rock as have escaped the ore-sorters in past time. It also appears from the old work-ings in the mine, that all or nearly all the ground in that portion of the mine that has been worked has been stoped out, indicating that the pinches in the vein, or courses of poor rock, are of comparatively small extent. This accords with the state-ments of the mining captain, who says that the pinches or contractions in the vein are seldom more than a few feet in horizontal measurement, while the expanded por-tions of continuous quartz-vein are very much greater, in one case, in the south end of the mine, over 200 feet. It has therefore been the custom, even where the vein is pinched, to carry the stopes through, as there are sometimes small bunches of quartz found within those areas rich enough to pay for the work. The free gold in the quartz is sometimes though not everywhere visible. The sulphuret or pyrites, with which much of the gold is associated, is generally present in greater or less quantities throughout the vein. Sometimes it is sparsely distributed in bunches or specks, and in other places it forms solid seams of several inches in thickness. The vein and its inclosing rock are both hard. This condition increases the cost of mining somewhat, though there is some compensation in the diminished cost of timbering. Work is generally done by the day, but when contracted for it costs as follows: Drifting, $4.50 to $10 per foot; sinking shafts, $28 per foot; sinking winzes, $7 to $9 per foot; stoping, from $12 to $20 per fathom.

After a description of the underground workings, the condition of the different stopes, and the several blocks of ground available as reserves, Mr. Hague continues:

Taking 13 feet of solid vein as equal to one ton of 2,000 pounds, and assuming the thickness of the vein over all this area will average 12 inches, or one foot, we have as the number of tons available from this ground, 164,000 cubic feet, divided by 13, equal to 12,715 tons. This amount is increased to nearly 16,000 tons, if we assume an average thickness of 15 inches; or to over 19,000 tons at an average thickness of 18 inches. The probable yield or value of the reserves will be discussed farther on. Of the ground beyond these reserves but little can be affirmed in advance of development, except that all the indications of its value are very favorable. The vein at the bottom of the mine appears as well as ever, and affords satisfactory evidence of continuance in depth. In the ground south of the slides but little work has so far been done, but the quartz has been rather richer in their vicinity than on the average elsewhere in the mine. Free gold is said to occur frequently. On the surface, also, it is said that the ravines furnished rich dirt for washing up as high as the crossing of the vein, indicating that the source of the gold so obtained was in the vein near and south of the slides. More-over, a tunnel was driven years ago into the vein from the surface, and is reported to have found a good lode; it is now filled up and inaccessible. These conditions all indicate that the south end of the mine will be found productive; and it is accepting nothing more than an ordinary mining risk to assume that the vein will be found there in place and valuable.

The equipment of the mine is excellent, and amply sufficient for greater duty than hitherto required. The drifts are sufficiently timbered, and are furnished with tram-ways throughout. Both the shafts are provided with double tramways, (wooden stringers with iron band 1¼ inch wide by ¾ inch thick;) each shaft is 14 feet long by 5½ or 6 feet high, furnished with substantial footways and with abundant space for the pumps, which are well set. In the north shaft the water is raised from the bottom of the mine by a 6-inch drawlift to the 537-foot level, and thence by a 6-inch plunger-pump to the 345-foot level, where it is allowed to run across to the south shaft. In that shaft the water is raised from the bottom by a 6-inch drawlift to the 345-foot level, and thence with the water coming from the north shaft by a 10-inch drawlift to the surface. The mine is comparatively dry, and the pumps only run a portion of the time. The 10-inch drawlift in south shaft, which raises all the water from the mine, runs on a 3-foot stroke, 4½ strokes per minute. The pump-rods are well put together, and the bobs and other appurtenances are provided in a workmanlike manner. The North-Hoisting Works comprise two engines, one for pumping and the other for winding. The former is 10 inches by 24 inches; it drives the pumps by means of a wheel and pinion and oscillating bob; it also drives a fan-blower for ventilation. The hoisting-engine is 12 inches by 24 inches, and drives two reels, about 5 feet diameter, by means of friction-gearing. The reels, or drums, are fur-nished, one with ⅞ inch iron-wire rope; the other with ⅞ inch steel-wire rope, sufficient for but little more than present depth of shaft. The two engines take steam from one 16-foot boiler, (tubular,) 48 inches in diameter. Connected with the works of this shaft is the blacksmith's shop. The north shaft-house is 300 or 400 feet from the south shaft-house and mill. A level trestle-work and tramway leads from the mouth of the shaft to the mill, by means of which the cars raised from the north shaft are run over to the mill, delivering the ore at the rock-breaker. The South Shaft Hoisting-Works are fur-

nished with an engine 12 inches by 24 inches, which drives the pumps in that shaft, two
friction-wheels winding Manila ropes, and a few for ventilating the south drifts. This
engine takes steam from two tubular boilers 16 feet long by 48 inches in diameter, which
also supply steam to the engine driving the stamp-mill. This last-named engine is
12 inches by 24 inches, and is placed in the same room with the hoisting-engine, and
under the care of the same driver. The houses covering these works are substantial
wooden buildings, spacious and well adapted to their purpose. The mill is connected
with the south shaft, and is conveniently located with reference to the supply of ore.
It contains a Blake's rock-breaker, 10 stamps, in 2 batteries of 5 each, weighing 650
pounds per stamp, dropping 10 inches 80 times per minute. The crushed rock is discharged through punched sheet-iron screens, and the capacity of the stamps, with the
screen now in use, is estimated at little over 1½ tons per day per stamp, or, say, 16 tons
per day for 10 stamps. Amalgamation is carried on in the battery, where about one-
third of the product is obtained. Beyond that are amalgamated copper-plated aprons
6 feet long; and further, tables or sluices 16 feet long, likewise provided with copper
plates, over which the pulp passes to the "Eureka amalgamators," of which there are
two. The bulk of the amalgam produced comes from the copper plates of the aprons
and tables, only a small portion being found in the amalgamators. Leaving these,
the pulp passes on to concentrating-tics or sluices, in which the sulphurets are saved
or cleaned; thence, beyond the mill, to a shed, where this operation is repeated; and
finally the tailings are worked in a round buddle, for the purpose of saving the sulphurets. These are collected and sold to the Chlorination Works. The product in
sulphurets averages 20 tons per month, worth about $95 per ton, less $25 for cost of
treatment, (chlorination.) The cost of milling is placed at $3.25 per ton. The process
employed in milling is the same in its general features as that usually employed in
gold-quartz mills in California, although the details of the methods in use vary in different mills. It would be impossible to say, without some experiments or closer or
longer observation of the results obtained, whether its efficiency can be increased or
not; but it is not improbable that careful study might improve the operations of the
mill, either in the amalgamation of the free gold or in the concentration of the sulphurets. The mill is well built, compact, and substantial; the stamps are in good
order, and well put up. The power provided is ample for twenty stamps, and might
possibly be made sufficient for thirty; though this would be doubtful, especially if any
auxiliary machines requiring power, such as concentrating or grinding appliances,
were added to the mill. The building is spacious, light, and convenient, and furnished with retorting and melting furnaces. The additional stamps can be very conveniently added, without large cost for preparing foundations or much increasing
the size of the building. The cost of supplying new stamps and putting them
in running order would probably be from $1,200 to $1,500 per stamp, assuming
no extraordinary expenses for increase of power or building. Two thousand dollars
per stamp is a fair estimate for a new quartz-mill in that part of the State, including
building, power, and all ordinary appurtenances. I think $1,500 per stamp would prob-
ably cover the cost of adding twenty stamps to the Pittsburg mill.   *   *   *   *
The net proceeds actually received during this period by the present owners are as
follows: Product, $288,197.59; expenditures, $221,019.19; net receipts, $67,178.40; in
addition to which the amount of real profits that may be regarded as applied to other
purposes, as before noted, is $15,000; making in all, say, $82,000. The foregoing accounts cover 23 months. During the last 12 months the profits appear to have been
somewhat larger than during the preceding eleven months, although the accounts for
each month are not so closely separated as to show exactly the work of any given
period. As nearly as can be ascertained from the books, the bullion receipts, from August 1, 1870, to August 1, 1871, amount to $160,269.06; and the expenditures for same
time to $119,408.41; leaving, as net receipts, $40,860.65. And as the larger portion of
the expenditure for opening ground in advance, and for accumulating supplies, has
been made during this past year, it follows that the real profits have been between
$45,000 and $50,000. It therefore appears that the profits of ten stamps during the
past year have been between £9,000 and £10,000; and, assuming the same relation of
cost and yield for thirty stamps, the expected profits might be from £25,000 to £30,000.
    *   *   *   *   I fully concur in considering the property one of established value,
and one which promises to sustain a high reputation in the future. Its locality and surroundings are all favorable. Its equipment both for mining and milling is excellent. The
lode is well defined and productive; its average thickness is from 12 to 16 inches, and there
are no present indications of any diminution either in size of vein or yield of ore. The
yield of the ore I place at $36 per ton. The cost per ton I find to have been, during two
years past, $25.75; and basing an estimate on these results, I allow a future profit of
$10 per ton. The developed reserves are sufficient to supply a 30-stamp mill at least
one year and perhaps more. It has been already shown that, taking the vein at 12
inches thick, the ore in sight may be estimated at 12,700 tons, and the net value of
this at $10 per ton would be $127,000. A greater thickness, say of 15 or 18 inches,
would increase this value to $160,000 or $190,000. The undeveloped reserves in the

south ground and the proposed north extension give promise of large supplies of ore for the future. * * * The principal reason for so high costs of working is that the rock is hard and the vein is small. I do not place the average thickness of the vein at over 18 inches, and base my estimate of reserves on a thickness of 12 inches, and as about 4 feet thickness of ground must be broken in mining, it is obvious that it costs two or three times as much per ton of quartz obtained to work a 12-inch or 18-inch vein as it would to work a 3 or 4-foot vein.

The report concludes with a frank recommendation of the purchase of the property; but its statements were not sufficiently extravagant to suit the speculative taste of the "promoters," and the sale fell through.

Eureka district, Nevada County, is situated twenty-two miles east of Nevada City, at an elevation of 6,000 feet above sea-level. The veins here are numerous and well defined. Most of them are in granite. The district is now yielding large amounts of gold, and is of sufficient importance to deserve a personal visit and detailed description, but this was impracticable this year.

The owners of the Lindsey ledge, between Fall and Diamond Creeks, are running their ten stamps on $20 rock. They have a ledge 12 feet wide, and all the rock pays. The company will enlarge their mill next spring, and add ten more stamps. An offer of $100,000 has been refused for the mine.

The Jim mine, about three miles south from Eureka, is being systematically developed. A tunnel was commenced last spring 1,000 feet distant, and below the old incline shaft. This tunnel has been run into the hill 600 feet, and recently struck the ledge, which is twenty inches thick, and prospects well in free gold and good sulphurets. This winter a drift will be run on the ledge from the tunnel to the old incline, a distance of 500 feet. This will develop the mine thoroughly, and open an immense quantity of ore, as the "backs" from where the tunnel strikes the ledge are 225 feet from the surface, and they increase in depth toward the bottom of the old incline, to which point the drift is to be run. At the latter point the ledge will be 275 feet from the surface. For a period of nine months, rock near the surface averaged $20 a ton, and $1,000 a stamp for each month. On account of the severity of the winter at the mine, the mill will not commence crushing ore before next May or June.

The Erie, another mine of prominence, has paid dividends during the greater part of the year.

The county assessor, in his report to the surveyor general, thus sums up the situation in respect to quartz mining: "In quartz mining a general prosperity is shown. Some mines of that kind fail, or seem to fail, but on the whole the business is remarkably prosperous. Especially is this the case in Grass Valley district. In that district many quartz mines which have been idle for years are now being worked, with prospects of good results, while only one noted mine has failed to continue paying large dividends. In the case of the one large mine which has ceased to yield gold in paying quantities, work is being prosecuted with vigor, and with every indication that its old prosperity will be restored."

YUBA COUNTY.

This county is celebrated for its "dead rivers," and the extensive mining operations which have of late years been undertaken for the purpose of their development, and the great yield of gold which followed these operations and will continue for many years. Those who feel an interest in this class of mining and the phenomena of the dead rivers in this county are referred to pages 64 to 67 of the Mining Commissioner's Report for 1870, and pages 68 to 72 of the Report for 1871, where the gigantic enterprises for the development of the old channels at Smartsville are fully described. A correspondent says of this locality:

The gravel mines of this section are regarded as the richest in the State, and, in fact, in the world, which idea is very naturally obtained from the success attending their operations. But this idea is to a great degree erroneous. The extent of these mines is exceedingly limited, and the yield per cubic yard is not extraordinary. Indeed, there are probably thousands of acres of gravel-beds in California equally as rich, and no doubt many that would yield more. But the great secret underlying the profit of these operations is in the bountiful supply of water at their command. The Excelsior Canal Company, that supplies nearly all the demands, is a consolidation of five companies, whose works have been projected from time to time since 1851. The company now has three ditches running from the South Yuba and Deer Creek, which furnish all the water for which they are called upon, while the Nevada Reservoir Ditch Company, having its source at Wolf Creek, brings in large volumes in the winter and spring months, most of which is used in the Blue Point and Smartsville Consolidated claims. The works now owned by the Excelsior Company cost, originally, close upon a million dollars. Of course they could be constructed cheaper now—perhaps for half the amount—yet this is the amount of capital that has been expended in the works of this company alone, and upon which a fair interest is sought.

Within the past year work has been vigorously prosecuted at Smartsville and Timbuctoo, and further up in the county, at Camptonville, on the hydraulic and cement ground. There is but little quartz in the county, most of the mines of this character having proved failures.

The Union Hydraulic Mining Company of San Francisco are engaged in opening valuable ground at Park's Bar. It is believed to be the same channel as that found at Smartsville and vicinity, on the opposite side of the river, the latter having cut through it.

The different claims have been worked in a small way for fifteen years. Water has been furnished by two ditches taking the water from Dry Creek, which is quite a large stream in the winter time.

The present company have purchased all the claims, giving them fully 160 acres of gravel-bank, averaging over 100 feet deep, all of which prospects well, portions giving as high as 50 cents and $1 to the pan. The company have just completed a cut, from 10 to 20 feet deep, fully one mile in length, and will soon have a four-foot "flume." They will commence washing with two eight-inch pipes as soon as the rains commence, and expect fully as good returns as are obtained on the opposite side of the river.

The completion of the tunnel of the Blue Point Gravel Company, which developed the great value of that portion of our mining territory lying on the Old River channel, (nearly three miles of which is well defined,) is one of the most important additions to the mining interest of this section and to the State. This tunnel is 2,270 feet in length, and was three and a half years building, at a cost of $146,000. Two thousand kegs of powder were exploded in one blast in opening the mine. The depth of the mine from surface to bed-rock is 73 feet. The tunnel is 35 feet below bed-rock, coming up into the mine on an incline. After firing the blast and opening a side incline, 1,000 inches of water, at a cost of $100 per day, were turned on the pulverized mass of gravel, and after 49 days' washing, 100 boxes of the head of the flume were cleaned up, realizing $43,000. Then, after working 41½ days, the flume, where cleaned up the entire length, produced $73,000, at a cost of $12,190. From the fairest estimate, this company has about 1,400 days' washing, that will average $1,000 per day.

The Smartsville Hydraulic Mining Company, whose mining ground adjoins that of the Blue Point Company on the east, have, for three years past, been washing off the upper strata, so as to be certain of the course of the Old River, preparatory to opening at another point. The last year's work has developed some 2,000 feet of the channel as being on the Smartsville Company's claims, and shows equally as good as any that has been worked in the Blue Point Company, adjoining.

The Blue Gravel Company, lying west of the Blue Point Company, are running a lower tunnel, their first one not being deep enough to work the old channel to the bottom., This tunnel is 1,800 feet in length. About 180 feet remain to be finished, which will require some eight months to complete it.

West from this is the Pittsburg and Yuba River Mining Company. They have suspended work upon the upper strata preparatory to running a tunnel to work the Old River channel.

Next west from this is the Rosebar Mining Company. During the past year they have been washing off the upper strata. They have commenced a bed-rock tunnel, which will be some 1,800 feet in length before reaching the old channel.

Next west from the Rosebar comes the Pactolus Gold Mining and Water Company. This company have their tunnel nearly completed. They have put down their shaft in the Old River channel preparatory to opening their mine. The prospects from shaft are fully equal to the gravel being washed in the Blue Point. Their tunnel brings them into the channel 160 feet below the present working level, and the extent of their mining ground, their receipts from working the upper strata and the prospects from the shaft, to be connected with the tunnel. point this out as a mine of great importance to this vicinity, the county and State.

Next west is the Babb Company mine. The entire length of this mine is on the Old River channel, some 1,400 feet, and it is the only mine which has been worked to the bed-rock. The upper stratum is worked off in two benches, when a stratum of very hard cement and gravel is found on the bed-rock, varying from three to eight feet in thickness. This stratum has usually been worked off with the upper strata, but during the past year has been left and worked through a cement-mill. paying from $2.50 to $6 per ton, from fifty to seventy tons per day being crushed at an expense of about $70. This company is owned by the same parties as the Pactolus mine, but in different proportions. These comprise the only mines in this vicinity now operating.

There are two tunnels which have been completed during the past year—the Blue Point and Pactolus; in progress two—Blue Gravel and Rosebar Company; in contemplation, Pittsburg and Yuba River. The condition and prospects of the mines are very encouraging. The product of bullion cannot be ascertained, each company keeping their affairs very close on account of heavy taxation. The amounts named are, however, correctly given. There is not a quartz-mill running in the county. Wages in the neighborhood above spoken of are $3 to $3.50 per day; three miles from there, $2.50.

The district is all gravel and hydraulic mining, (except the lower stratum spoken of in the Babb company.) Area of ground being worked, 100 acres; extent of gravel-deposit, 1,000 acres, varying in depth from 50 to 500 feet. The yield per cubic yard of hydraulic ground varies from a few cents to as much as $3. Very often the surface part of the bank will contain no gold; then a pay stratum is met with, and the value of the yard depends upon the hardness of the gravel and the ease with which it will wash. Gravel has been washed worth less than 30 cents per yard, still it has paid $400 per day.

The diamond drill at Smartsville.—A correspondent of the Mining and

Scientific Press furnishes the following particulars of the work accomplished on the tunnel of the Blue Gravel Company:

The tunnel in question is to be 1,563 feet long, 8 feet high, and 6 feet wide. Already (July, 1871) 1,285 feet have been cut by hand in three years, at a cost of $40,000. Eight men were employed in the work daily, making about one foot per day, at a cost of about $40 per foot, blasting with black powder. The diamond-drill machine has now cut 50 feet, with the aid of two men in the tunnel to operate it. The motive-power is compressed air, supplied from a 15 horse-power engine at the mouth of the tunnel. The bore-holes are 1¼ inches in diameter and 30 to 50 inches deep. Giant powder is used, being fired by an electric fuse and battery. During a space of 4½ days, accurate observations and estimates of cost, work done, &c., were made. During this time, 11 feet 6 inches of tunnel were cut through the hard syenite bed-rock, running out the *débris* 1,520 feet from the face of the tunnel. In this work 7 runs were made with the drill, (and 7 blasts,) 8 to 13 holes being bored at each run, and 72 holes being driven in all. These holes varied in depth from 2½ to 4½ feet. To make one foot of tunnel 6½ by 9½ feet, by hand-drilling as before done, required about 30 holes, or from 300 to 330 holes for the lineal distance of 11½ feet, while with the diamond drill and giant powder only 72 holes were required. During the last 40 feet not a hole was drilled by hand, nor a steel drill used. During the 108 hours (4½ days) named, 50 hours were consumed in running out the rock, leaving 58 hours for boring, traveling to and from meals, &c. The average rate of boring, therefore, was nearly 2.4 inches per hour; or, of boring, running out rock, &c., nearly 1.3 inches per hour. Or, the tunnel was run in at the rate of 2.6 feet per day, while before only 1 foot was made. The electric fuses and battery were furnished by the Electrical Construction and Maintenance Company, of San Francisco. The battery was placed about 800 feet from the face of the tunnel, in a convenient place, and the charges were exploded without a single failure. One great advantage found by blasting by electricity was that, while 10 to 20 holes of giant powder were fired simultaneously, and while the effect was greater than where a powder-fuse was used, the offensive smell was much less. This last is agreed to by all the workmen. With the electric fuse but little or no difficulty was experienced in this respect. It is thought that the offensive odor is caused by the combinations formed by the combustion of the black powder and nitro-glycerine.

The cost of running the 11½ feet of tunnel, including oil, wood, and all other expenses, excepting the interest on the cost of the machine, amounted to $24.01 per foot, being about three-quarters of the cost of hand-labor, while the work progresses nearly three times as fast. It is thought that these expenses can be reduced still further. The saving of time by the speedier completion of the tunnel, effects an important item in the economy of the enterprise.

Since the above was written the use of compressed air has been superseded as a motive-power, and a new appliance of water-power adopted. We refer the reader, for a description of this improvement, to the matter under the head of "Rock-drilling Machines" in this report.

BUTTE COUNTY.

This county has valuable mining interests both in quartz and gravel. The foot-hills near the Sacramento Valley were formerly very extensively worked as placers, but are now considered exhausted. From a late number of the Oroville Record we learn that some very promising ledges containing gold and silver ore have been discovered on the upper waters of the Feather River, near the dividing line between Butte and Plumas Counties.

"These ledges," says the Record, "are of recent discovery, and the silver ore we have seen promises as well as the famous silver mines at Nevada. They are near the head of Big Kimshew and Philbrook Valley. A new mining district has been formed, under the name of Golden Summit district, and rules and regulations adopted. The silver ledge is fifteen feet in width. A shaft has been sunk to the depth of only some eight or ten feet. Some of the rock thrown out by the blasts has been sent below, and assays $450 to the ton. Parallel to this silver ledge is a vein of gold-bearing quartz, which abounds in pure gold, and produces some very fine specimens. Some four miles in a northerly direction is another very rich quartz-ledge, from which $500 were taken recently in one day with a hand-mortar. These new discoveries are in the vicinity and partially surrounded by the bank-diggings that have long been known as the Gravel Range. Being near the summit of the mountains, between Inskip and Humbug Valley, Plumas County, the winters are very severe, and but little progress has been made with the bank-diggings. There can be no longer a doubt of the mineral wealth of that section. As the snows of winter will soon envelop it, it is probable that but little will be effected the present season; but another summer will witness developments that will show Butte County to be one of the richest mineral sections on the Pacific coast. These claims are on the headwaters of Feather River, but so high that it is impossible to carry water to them, except from the melting snows. These ledges and bank claims, being found at the head of what is known as the Dogtown ridge, or the divide between the waters of the North Fork and those that flow westerly into the valley of the Sacramento, seem almost to confirm the belief formerly entertained by miners that there was some original place of deposit, from whence the gold had washed into placers and gulches where it was first discovered. It requires but a small stretch of the imagination on the part of those familiar with the mineral wealth of the ridge, as heretofore developed, to locate the original deposit on these bleak, bold, and barren hills, and then follow the wash down to Inskip, Kimshew, the rich placer mines that have been worked in the gulches leading into West Branch; to Cherokee Flat, Morris Ravine, at the mouth of which is Feather River, and where was found the richest part of the celebrated Cape claim in 1857."

Mr. F. A. Herring, of Forbestown, writes the following interesting letter, which is inserted without comment on its geological theories:

FORBESTOWN, CALIFORNIA, *October* 25, 1871.

Forbestown is situated about midway between two lines of ancient gravel-deposits: the one at La Porte and vicinity, near the summit, the other at the foot-hills, upon which the mines at Oroville and Cherokee are located. The gold is found in this district in what are termed surface-diggings, and is no doubt derived from the quartz-veins, which are quite numerous. The ravines and water-courses are mostly worked out, save such as are not easily drained. The gold is now mostly found in the vicinity of quartz-veins, on the ridges between the ravines. New York and Ohio Flats, a continuation of the same valley, were once worked by drifting, the pay-gravel being brought to the surface to be washed. It was in no place more than 25 feet in depth. The pay-gravel is no more than from 2 to 4 feet in thickness, and is covered from 8 to about 20 feet in thickness with alluvium, destitute of boulders, or even pebbles. The bed-rock, where it has been stripped, has a fall of but about one inch to the rod. The valley is about two and a half miles in length, and its general direction W. by N. W. and E. by S. E., the water running, in an easterly direction, diagonally across the line of the Sierra Nevada, and toward the summit. Before the open cut was made the lower end of the flat was a basin; and, as the auriferous stratum is made up of gravel and boulders, such as are found in the beds of swift-running streams, there has evidently been a change of level, which dates back to the upheaval of the Sierra Nevada. The miners discover indications that this valley was inhabited by man before the change of level occurred. Drinking-cups of stone, and stones used by the Indians for grinding food, have been found upon the bed-rock and upon the gravel, covered to a depth of 20 feet with alluvium.

There has been, within the past year, about $120,000 worth of gold dust taken out of the scope of country about ten miles square, with Forbestown for its center. The yield is about $30,000 less than usual this year, on account of the extreme drought that has prevailed. The Gaskill & Bowers claim on Ohio Flat yielded about $36,000 the past two years on 54 square rods of ground. The pay-gravel was about four feet in thickness, and had formerly been worked by drifting, at which time it yielded about the same amount.

It is believed by the miners that there are some rich quartz-veins in this district. Some very rich pockets have been found. There are three quartz-mills in this district, two of five stamps each, and one of four stamps. One of them, called the Slater mill, has been running the past year, but with what result I am unable to state. The other mills have been idle.

It is a matter to be regretted that our learned men who have visited this coast should be so blinded by their theories as to be unable or unwilling to conduct their observations in such a manner as would enable them to arrive at the truth in regard to the age and origin of our auriferous gravel-deposits. When a certain learned professor, well known to the scientific world, was at Forbestown, some five or six years since, he had some conversation with one of our citizens, Mr. Gaskill, in regard to the age of gravel-deposits, who informed him (the professor) that the Bangor blue lead in Butte County runs in a north-of-northwesterly direction, and at nearly right angles to the present water-courses. The Bangor deposit is in the lowest range of foot-hills. The old river-bed is one hundred feet and upward below the beds of the present ravines and creeks, which cut their way to varying depths through the ancient gravel-deposit and the schistose rocks of the county. It has been found impracticable to drain this ancient deposit so as to work it by the hydraulic process. The professor said if such was the fact it implied such a change in the topography of the country as to upset the theories of geologists, and it was much easier for him to believe that Mr. Gaskill and a few miners were mistaken than that so many learned men should uphold a false theory! He even declined to visit the site of this ancient deposit, when Mr. Gaskill offered to take him there that his statement might be verified! Thus the professor lost the privilege of making a valuable contribution to science.

It seems to be fashionable with late geological writers, and, so far as I know, they all run in the same groove, to hold that our auriferous deposits belong to what they term the post-tertiary age, when the independent observer finds the rocks teeming with facts showing them to be much older than the tertiary age.

Mr. John Nisbet, of Oregon City, sends me the following, in answer to a circular:

The entire status of mining in this district has changed for the better since your last report.

The Oroville Gold and Silver Mining Company, near Long Bar, have erected a 12-stamp mill, and started to work early in the summer with favorable prospects, but were compelled to shut down for want of water. Since then they have been engaged taking out quartz from two distinct veins, both rich in gold. They will resume work as soon as the rains afford water for the battery.

The Cambria mill, in Oregon City, has run steadily for over a year, and paid from twelve to twenty dollars per ton, exclusive of sulphurets that are abundant after passing the water-level.

The Cambria and Nisbet Mining Companies have consolidated and are putting up a 16-stamp mill on the site of the Nisbet mill which was burned down about three months ago. The design is to work both mines from the Nisbet shaft, and economize by a different system of drainage and increased crushing-power.

The above mills are all that have worked the past year. The Oroville Company, from the lack of water, and the Nisbet Company, from the loss of mill, have done very little, while the water-power arrastras have done nothing at all. The water-ditches being constructed for this neighborhood will in future obviate this difficulty and afford an abundant supply for Cherokee and Morris Ravine and the gravel-deposits underlying the table-land. The ditches completed this fall are those of the Spring Valley and Indiana Companies. The Cherokee Company will not get theirs completed till next summer, although the greater part of the work is done. The Blue Gravel Company will start in on the same scale next summer also, so that the joint supply for the district will amount to at least twelve thousand inches, amply sufficient for hydraulic and milling purposes, not only to give stability, but to thoroughly develop the placer and quartz mines in the district.

The Table Mountain of Butte County.—This remarkable formation, celebrated for its past yield of gold and the extensive workings now in progress, seems to be of the same formation and nature as its namesake in Tuolumne County, described in our article on the southern mines. The following paragraphs are extracted from an interesting report made by Mr. Charles Waldeyer, general superintendent of the Cherokee Flat Blue Gravel Company, to the president and trustees of the company:

Amongst the great gravel-deposits of California the "blue lead" has attracted the greatest attention. Not only its position, which is immediately upon the bed-rock, but also its color, varying from a dark green to a bright indigo color, proves it to be the oldest gravel-deposit on the Pacific coast; this color evidently resulting from the substratum of bed-rock, the grinding away of which afforded the pigment to dye the quartz and other *debris*; and the sulphurets of iron, by a chemical change, assisted in making this color fast.

It is hardly reasonable to ascribe the creation of the blue lead and the gravel-deposits overlying it merely to the action of water, and we must call to assistance the glacial theory.

Before entering upon the above-mentioned task it will be necessary to inquire, how was the quartz, the matrix of almost all the gold found in California, supplied? Can we really believe that enough quartz-ledges were destroyed to cover from fifty to a hundred square miles with nothing but quartz-gravel and quartz-sand from two to three hundred feet deep?* Would all the known quartz-ledges of California, rising from unknown depths, even ten times higher than Mont Blanc, have been sufficient to furnish one ten-thousandth part of the quartz-gravel deposits now found in this country?

After considerations like the foregoing, it must become evident that the quartz, before being ground into pebbles and sand, existed in huge masses, and in such a position that any great grinding force, or power of attrition, could be applied more or less directly to it.

Any other theory about the creation of quartz-ledges will do just as well as the igneous, for it must be presumed that even if the fissures, now containing the quartz, were filled from the surface, then there is no reason to doubt that immensely more siliceous matter was at hand than to fill merely the comparatively few fissures, and we have again enormous deposits of quartz on the surface, waiting for nature's great crushing process.

The glacial-motion theory accounts at once for the existence of the great deposits of drift, not only in and near the valleys of our State, but also on the highest mountain-tops of the Sierra Nevada.

According to the glacial theory, immense ice-fields, thousands of feet deep, covered the continents, and commenced at a certain period their slow motion from north to south, crushing and grinding under their immense weight whatever might be subjected to their pressure.

In support of this theory, formed upon a broader basis than the evidences of the physical geography of California, it must be stated that these ancient channels or dead rivers of California, now the depositories of the different blue leads and their overlying quartz-gravel strata, run all in the same direction, to wit, from north to south, when all the modern rivers of California debouch into the Pacific Ocean to the west.

It must be evident that this glacial motion was the beginning of the latest thermal change, and that when the ice had disappeared, immense floods succeeded, washing the crushed pulp into channels and stream-beds, and filling them to the depth of hundreds of feet, as evidenced by existing deposits.

Whether the views advanced in the foregoing lines are correct or not, will be immaterial. However, they may illustrate that—1. The great gravel-deposits of California are only an accumulation of the same gold-bearing material which is now extracted from the quartz-veins. 2. That nature, though in a crude manner, performed the operation of crushing, and facilitated thus the extraction of gold. 3. That gravel-mining, under the modern improvements, will be the most extensive, the most lasting, and the most remunerative mining operation carried on in the State.

The great blue lead and gravel-deposit in Butte County, claiming our particular attention, will be made the object of a careful description.

This deposit occurs under a "plateau" known as the Tuscow or Butte County Table Mountain, extending for a distance of about eight miles from Cherokee Flat, in the north, to Thompson's Flat, in the south, having a width of from two to four miles. It is bounded on its western side by the Sacramento Valley, and on the south, east, and north by Feather River and its tributaries, the North Fork and West Branch, so that from all sides an ascent of nearly one thousand feet has to be overcome.

Ever since the earliest days of gold-mining in California the surroundings of the above-mentioned "plateau" or table-land offered the richest harvest to the miner. Every ravine descending from this plateau or near it became celebrated for its richness. Cherokee Flat, at its northern point, and Morris's Ravine, near its southern point, yielded their millions to the most primitive applications. The miner in those days was satisfied to find the gold, and did not inquire whence it came; but when these

---

* Mr. Waldeyer's theory that vast masses of quartz overlaid unconformably the upturned edges of the schists (for that is what his remarks must suggest, if anything) is neither tenable nor necessary. These schists abound in quartz seams and segregations too small to be worked as "ledges" by the miner. As to the glacial theory, this application of it can scarcely be admitted. The leading phenomena of the "blue lead" channels are not glacial, but fluviatile, though the operation of glaciers in the denudation of the sierra is not improbable. But it is water, and water only, which arranges gravel according to size and specific gravity, and concentrates in such deposits the heavier and less oxydizable metals.—R. W. R.

ravines and gullies, which were actually nothing but nature's "sluice-boxes," had been worked out, then the question arose, where is the storehouse from which all this gold has been drawn? and every indication pointed toward the great Taucow or Table Mountain.

As the mode of mining improved, and particularly when hydrostatic pressure was employed, known in California as the *hydraulic process*, then the peculiar situation of Cherokee Flat, which justly can claim the richest and most extensive gravel-deposit in the State, was seriously felt. It was considered next to impossible to bring water in sufficient quantities to that place. The only way to bring water was either from the West Branch or North Fork, both tributaries of Feather River, and in either way across the chasms formed by those streams, which flow nearly a thousand feet below the level of Cherokee Flat.

Reservoirs and ditches were constructed, at the expense of hundreds of thousands of dollars, merely to catch and store away the rain-water during the rainy season, and thus the means for two or three months' mining were obtained. The splendid results of these short mining seasons challenged the energies of enterprising men, and several unsuccessful attempts were made to enlist capital in the introduction of living water to these mines. The idea of carrying in water by means of pipes across chasms of 900 or 1,000 feet vertical depths startled mining engineers of note in the United States and England, and their estimates were either so enormous as to cost, or so doubtful as to success, that even the most daring abandoned the scheme. And thus, up to the summer of 1870, the mines depended altogether on rain-water, but yielded so largely that a party of San Francisco capitalists, in connection with the Spring Valley Mining Company of Cherokee Flat, took the matter in hand, and, after six months' labor, a sheet-iron pipe, 30 inches in diameter, discharged in a steady and even flow a thousand inches of water. This water was taken from some small tributaries of the North Fork, known as "Flea Valley Creek" and "Camp Creek," and the unusual dry season of California lessened the supply of water to about a regular flow of 600 inches for the whole 24 hours. Besides this rather small supply of water, the Spring Valley Canal and Mining Company had not only difficulties on account of the insufficient depth of its tunnel—a lower tunnel not being finished—but also on account of those constant improvements which an altogether new *modus operandi* generally necessitates; but still this company produced a fraction over one thousand dollars for every day's washing, which yield will probably be doubled in the future by the greater supply of water and other improvements now under hand.

The enterprise of the Spring Valley Canal and Mining Company, being the most important of the kind in this State—in fact, being the first which has overcome a depression of between 800 and 1,000 feet, (such a one being formed by the chasm through which the West Branch flows)—deserves description.

The water is carried in a ditch past Yankee Hill to a point on the north side of West Branch, said point being 980 vertical feet higher than the point where the West Branch is crossed, when a sheet-iron pipe receives the water, carrying it down the slope of the mountain to the crossing of the West Branch, and thence up Cherokee Hill to a vertical height of 830 feet, where the water is discharged again into a ditch and carried to the mines for use. The pipe has a diameter of 30 inches, is over two miles in length, and constructed of the best boiler-iron—the heaviest being three-eighths of an inch. The pipe on the Yankee Hill side is 150 vertical feet higher than the pipe on the Cherokee Hill side, where the water is discharged, so that there is a head of 150 feet. The water, however, during its greatest supply and heaviest discharge, never rose more than 50 feet on the Yankee Hill side over the point of its discharge on the Cherokee Flat Hill side, and was then estimated equal to 1,500 inches, (miners' measurement, (i. e., water discharged under a head or pressure of six inches.)

The Spring Valley Canal and Mining Company estimates its expenditure for pipe, ditches, reservoirs, &c., at $300,000, independent of the old works, as tunnels, &c. The pipe alone is an item of between $55,000 and $60,000, delivered at the foundry.

The experience so far gained shows that, under ordinary circumstances, and provided that even a moderate supply of water, say one thousand inches, is received for the whole year, one year would be sufficient to clear the whole expenditure of $300,000. However, the expectations of this company go far beyond the results of its practical experience, for the reason that the gravel-deposits increase in richness as the deeper parts of the channel (or those parts further removed from the rim-rock, or river-bank) are reached, when particularly the "blue lead" develops finely, yielding from five to fifteen dollars per ton.

Since the Spring Valley Company led the way in introducing water to Cherokee and the rich mines of Table Mountain, other companies have commenced similar operations, among them Hendricks & Co., taking the water from the upper part of West Branch, and carrying the same by ditch and pipe to Morris's Ravine, a point where the same gravel-deposit is entered from the southeast; proving equally rich and extensive, and proving thus, to the most skeptical, the unbroken continuation of the great "gravel lead" under Table Mountain.

Morris's Ravine is about five miles below Cherokee Flat, and between four and five hundred feet lower, counting from the bed-rock exposed in the Spring Valley Company's claims to the bed-rock in Morris's Ravine, thus showing a fall of 100 feet per mile to the channel under Table Mountain. It may here just as well be mentioned that the "rim-rock" on the eastern side of Table Mountain is a great deal higher than the "rim-rock" on the western or valley side, and that the plateau of Table Mountain itself has a double inclination—one from north to south, evidently due to the natural fall of the old channel, and the other from east to west, very likely due to the fact that after the Table Mountain was formed the mountain-chain of the Sierra Nevada rose slowly under the many volcanic disturbances to which this country has evidently been subjected, and tilted the Table Mountain from east to west. The gold found in the deposits under and near Table Mountain is of unusual fineness, averaging from $\frac{9.60}{1000}$ to $\frac{9.90}{1000}$ fine, and differs in size from the finest flour-gold (generally contained in the upper gravel-deposits) to nuggets weighing many ounces, found on and near the bed-rock.

In opening the great "lead" under Table Mountain by expensive tunnels, without actually knowing the depth of the channel or basin, in which the richest part of the deposit rests, great risks had to be taken, the more so as only very few places for such tunnels combined the desirable advantages, to wit: 1. Sufficient depth to drain and work the basin under all circumstances. 2. Proximity to the mine, so that a tunnel of moderate length would suffice to operate. 3. Ample room for all tailings, or washings, after the gold was extracted.

The overlooking of these three requisites, or any of them, has been a source of much unprofitable labor and great expenditure in mining operations on the Pacific coast; nevertheless the desire to take the shortest road to fortune proved stronger than all dearly-bought lessons, and so we have here, after the expenditure of hundreds of thousands of dollars for tunnels and other "outlets," only one tunnel which combines all the requisites mentioned before. This tunnel belongs to the Cherokee Flat Blue Gravel Company. This company, originally called the Butte Table Mountain Consolidated Mining Company, has been fifteen years in possession of its mining ground, consisting of about one thousand acres, and has expended about $160,000 on improvements, such as tunnels, inclines, shafts, reservoirs, and machinery.

The mining ground of the company stretches in a southeasterly direction 10,000 feet across the channel or basin, from "rim to rim," (or shore to shore) and for 4,500 feet down the channel.

The Spring Valley Canal and Mining Company, owning the adjoining ground along the upper line, has worked within 100 feet of the boundary line of the Cherokee Flat Blue Gravel Company, and developed in the progress of work richer gravel deposits than ever seen before at this place. The same company bought a piece of ground being about 200 by 400 feet, paying for this ground $31,800 in United States gold coin, at which price an acre would be worth a fraction over $17,000, when the yield per acre, according to the statistics of more than fifteen years, would be from $60,000 to $80,000. But, considering that all the mining operations at Cherokee Flat have been merely confined to the outskirts of the "lead," or what, in miners' parlance, is termed the "outside wash," and that, therefore, only the lighter particles of gold were reached; that, furthermore, heavier and larger gold is reached as the channel deepens, and that, therefore, the yield increases and will increase for thousands of feet ahead, we must come to the conclusion that all the riches so far developed are but a slight indication of what the future will disclose.

The upper quartz-gravel, of which the main body of Table Mountain is composed, in a body from 250 to 300 feet deep, consists altogether of quartz pebbles and quartz sand, and carries very fine particles of gold. The two lower strata, however, the "rotten bowlders" and blue lead, or gravel, are mixed, more or less, with different bowlders, the former of clay-slate, the latter of talc-slate. Both strata are very rich in gold, and average each from five to twenty-five feet in thickness. The "rotten bowlders," as they are called, on account of the softness of the bowlders, have so far been the chief resources for gold at this place, as they were easily reached, lying above the blue lead. This stratum seems to be peculiar to the Table Mountain deposit, and is not found in any other mining locality. The Cherokee Mining Company, one of the oldest and most successful mining companies of this place, working, for want of a lower outlet, altogether the "rotten bowlders" and quartz-gravel above them, and depending, so far, altogether on rain-water collected in large reservoirs, produces from $80,000 to $100,000 every year, during the few rainy months. This company is now engaged in constructing a large ditch from Butte Creek to the northwest of this place, carrying it by pipe over a wide and deep depression to Cherokee.

SIERRA COUNTY.

The Sierra Buttes mine has been for some two years in possession of an English company. From the reports submitted at their last two semi-annual meetings in London, the following particulars are gathered: Dividends amounting to 20 per cent. annually are paid by this company. The property was purchased on the careful and elaborate opinion of Mr. Henry Janin, (published in part in my report of 1870,) which has been on the whole fully justified by the subsequent developments, although unforeseen delays and expenses have somewhat diminished the profit on the ore extracted, cutting it down, for the first year, from $8.50 to $6.50 per ton. At the meeting in September, 1871, the chairman expressed the expectation of an annual profit of £34,000. In the first three months of the year 5,500 tons were reduced in the mills, and in the second quarter about double that quantity, the average product being about £2 per ton, and the average cost of milling about 15s. Meanwhile about £3,200 had been expended in opening new ground, and in accessory works, putting the mine in a much better condition. The half-yearly statement for December 31, 1871, showed a balance available for dividend of £20,604. As there are 112,500 shares in the capital stock, paid up at £2 each, the above sum represents two dividends (January and April, 1872,) of 2s. each per share, or ten per cent. for the six months. The stock sells at £4 10s. to £5 per share in the London market. Experiments having been made with the view of saving gold from the tailings, it was finally determined to erect three pans for that purpose. The cost of the plant was $7,336; and the pans yielded from concentrated tailings, during the six months, $3,446, at a running cost of $534. The company has determined to put up additional pans for this purpose, and to postpone the erection of a new stamp-mill, which would be highly desirable, both for increased capacity of reduction and on account of the age and rickety condition of one of the present mills, (the Hanks.)

The product for the whole year 1871 was:

For the first six months, 15,700 tons, yielding		$162,000
In the second six months, 17,500 tons, yielding		227,000
Total	33,200 tons, yielding	389,000

Average per ton : for the first six months, $10.12; for the second six months, $12.64; for the year, $11.71.

The agent's report contains the following account of operations:

The working capacity of the mine having been so largely increased since the property passed into the hands of the present owners, it was found necessary last fall to erect additional boarding-houses, store-house, carpenters' and blacksmiths' shops, &c.; and, in addition to these, three pans and one settler, and a turbine water-wheel to drive the same, have been provided for working the tailings. The latter having only recently been completed, little benefit has yet been derived from them, but they promise to yield very satisfactory results, and it is a subject worthy of the consideration of the directors, whether additional machinery should not be provided early in spring (which is the best and only season of the year when surface-works should be carried on) for working the vast amount of tailings now being produced. There are about 30 arrastras at present at work on the tailings, and some more are to be erected as soon as the weather will permit. The above extraordinary outlays have, including the damming of the lakes and repairs to the flume and ditch, amounted to the large sum of $21,666.60, as per statement attached. During the period under review there were 17,356¼ tons of ore mined, and 17,501 tons crushed, which produced $221,300.78 in bullion by the ordinary mill process, being an average of $12.64¼ per ton, at a cost of $3.77½ for mining and 95¼ cents for milling, which leaves a profit per ton of $7.91¼ against $7.76¼ for the corresponding term of last year, and $6.42 for the six months ending June 30 of the current year, (1871.) In addition to the above bullion product, we have received from the arrastras and pans $5,609.60 net, which, added to the yield in the mill, would make the net profit per ton $8.23¼. The yield per ton has fluctuated, as usual, several dollars per ton, but it will be noticed that during the months when the water-supply was low the yield is higher, and vice versa. This is explained by the fact that since the last three or four months of the year are always the most expensive, ... necessitating heavy outlays to provide for the requirements of the winter, and the water-supply is usually slack at this time, it is found necessary to work less ore from the low-grade chimneys, and more from the richer deposits, in order to meet the extra expenses and at the same time leave a fair surplus. The average yield per ton was, in July, $9.25½; August, $10.30; September, $13.86; October, $16.80; November, $12.18; December, $14.91¼. In November the yield would have been higher had it not been for a cave in one of the rich stopes, and as soon as it was cleared an accident happened to the flume, which cut off the water just as the rich rock was being milled. The average yield per ton for the six months, however, compares favorably with that of previous ones, being $12.64½ against $12.41½ for the corresponding period of 1871, and $10.12 for the first six months of this year. The average cost per ton for mining expenses is heavy, owing to a large amount of prospecting having been done in the mine, amounting to $12,328.14, or, say, 71 cents per ton, and, in addition to this, many items were charged to the mine which more properly belong to surface improvements, &c., amounting in all to nearly $1 per ton. The sum spent in prospecting, &c., has been amply repaid by the large amount of ore-ground laid open, and, though the expense becomes heavy at present, it will tell very favorably in the future of the mine.

CHAPTER II.

NEVADA.

The State of Nevada has last year taken the lead, for the first time, in the production of the precious metals, outstripping even California. This result is due partly to the increased productiveness of the Comstock mines, and partly to the great prosperity of the Eureka and Meadow Valley districts. The figures of Wells, Fargo & Co. probably cover the whole production, since in this State practically all the movements of bullion are made by express or are easily ascertainable by the express-agents.

Mr. John J. Valentine, general superintendent of the bullion department of Wells, Fargo & Co.'s Express, has furnished the following important statement of the bullion product in the State of Nevada:

Aurora	$45,761	Piocho	$3,962,224
Austin	965,536	Pine Grove	137,672
Belmont	268,903	Palisade	27,130
Battle Mountain Station	129,441	Rye Patch	41,259
Carson	119,636	Reno	192,977
Carlin	27,811	Silver City	200,800
Eureka	2,173,106	Toano	40,031
Galena	206,357	Unionville	313,696
Hamilton	1,339,420	Virginia and Gold Hill	11,053,325
Mineral Hill	701,014	Wadsworth	20,276
Mountain City	149,273		
Mill City	4,485	Total	22,177,046
Oreana	6,900		

This agrees well, on the whole, with the returns made to me from the different districts; but it is impossible to make comparison in details, since the above table contains many items referred to the point of shipment, rather than the place of production. In making up the general statement quoted in the introductory letter to this report, Mr. Valentine has added $300,000, (I presume for amounts carried by private hands,) making the whole product of the State $22,477,046. The yield for 1872 will be still greater, as the most recent developments on the Comstock have largely increased its productiveness.

THE COMSTOCK MINES.

The history of these mines during the year has been one of unexpected improvement and unexampled speculation. At the beginning of 1871 the Crown Point had opened a body of ore on the 1,100-foot level, six feet in width. It is said that the section of the mine where this discovery took place had never been prospected from the 300-foot level down. The 1,100-foot level in this mine is about 1,700 feet below the highest point of the Comstock outcrop at the Gould and Curry shaft. The stock of the company had been, during the autumn of 1870, as low as $3 per share; but it advanced in January to $41, in February to $55, in March to $160, in April to $195, in May to $310, and in June to $340. At the beginning of June Mr. A. Hayward, already the representative of 6,300 out of the 12,000 shares, bought 4,100 shares at $300. At this time the ore-body was known to be 80 feet wide, 200 feet long, and 100 feet deep, as far as followed. By October it was known that this body was 270 feet long within the Crown Point ground, and that on the 1,200-

foot level it extended 70 feet farther north than in the level above. A 1,300-foot level was also opened in it, and the incline was vigorously pushed ahead. The assays of the ore were very high, ranging up to $150 for large quantities. By the early part of this year the mine was able to keep six mills in operation, reducing daily 375 tons of ore. The gross product, as will be seen from the tables given below, rose from $125,574 in the first, to $719,121 in the second, $526,565 in the third, and $599,623 in the fourth quarter of 1871.

A late account (January 13, 1872) thus describes the appearance of the mine:

Cross-cut in the 1,200 level is in 82 feet, all the way in very fine ore; face of it shows ore that will mill $60 per ton. No sign of the east wall has yet made its appearance. The south drift, same level, is in from the cross-cut 60 feet; the entire distance in ore that will mill at least $80. It is yet 80 feet on the line of the vein from the face of this drift to the Belcher north line. The 1,100 level is yielding its usual quantity and quality of ore. There yet remain a little over two-fifths of the superficial area of the ore-body on this level to be worked. The breast on the 1,000 level shows marked improvement as the Belcher line is approached. The superintendent judges that the ore on this level would mill on an average $35 to $40 per ton. There has been but little work done above the track-floor on this level. On the 900 level no ore of value is found.

The position of this body led at once to the expectation of its southerly continuance into the Belcher mine, the stock of which began to advance in consequence. On the 26th of March, the water in the shaft was about 26 feet above the 850-foot level, and they commenced hoisting water. In July they were at work on that level, in promising quartz, supposed to be the extension of the Crown Point body. This body was soon completely developed and recognized, and in August 680 tons from it were worked, yielding $77 per ton. It was found "in force" on the 1,100-foot level, and followed on that level for more than 100 feet before the middle of September. The stock of the company had risen meanwhile from $8 in January to $285 September 15. Before the end of that month it had reached $405. The Crown Point and Belcher body, as explored up to the first of October, was 533 feet in length, 50 to 70 feet in width, and known to be 300 feet in depth; that is, beginning at the 900-foot level. The ore from the Belcher proved singularly rich in gold, the bars running from 88 to 96 thousandths of gold, while those of the Yellow Jacket were but .028 to .032 fine. A sample taken about November 1 from the south end of the drift (1,000-foot level) in this mine assayed, gold, $186.05; silver, $40.13. Assays of selected ore gave, gold, $9,266.52; silver, $1,394.33. The last advices from the mine during 1871 were as follows:

December 2: The shaft is retimbered within 290 feet of the surface. The incline is down 95 feet below the 850 level. The east drift, from 900 level is in 46 feet, the face showing low-grade ore, but improving as they go east. The sill-floor on 1,100 level is 135 feet in length; the face is 60 feet in width and in very fair ore, assaying $70 to $100. The south drift is in 250 feet and still in good ore; the cross-cut from this point is in 10 feet, the face of it hard porphyry. The pay streak at this point is 15 feet in width, and will assay $80 on an average. The east cross-cut from south winze, 30 feet down, is in 32 feet; the face shows good ore, showing thus far 13 feet of ore, which assays $106. The north winze is down 34 feet, and in good ore. The raise is up 44 feet. Have made cross-cut to the west from top of raise 36 feet, the face of it in good ore. During November 5,664 tons of ore were shipped to mills.

December 16: The main shaft is now retimbered to within 267 feet of the surface. The incline is down 114 feet below the 850-foot level; the ground still remains hard. The incline is timbered up 119 feet above the 850 level, leaving 31 feet to be timbered to connect with the main shaft at 700 level. The cross-cut from south drift on 900 level is in 61 feet; the face shows good milling-ore, assaying from $60 to $100. On the 1,110 level the south drift is in 275 feet from north line; the face still in good ore, which assays $100 to $150. Have commenced to sink a winze in south drift, 243 feet south of north line, or 95 feet ahead of stope. Since last report have continued east cross-cut No. 2 from south drift. Drove through four feet of hard, barren material and got into good ore again this morning, which bids fair to be extensive. This cross-cut is 250 feet south of north line. The sill-floor is 158 feet in length, and we are working 50 feet in width and good ore still east of us. The breast looks splendid. The north winze is down 62 feet and looking well. Average assays $75.

The product of the mine for the four quarters of the year 1871 was: First quarter, $1,034.50; second quarter, no returns; third quarter, $212,038.56; fourth quarter, $985,848.31. There is no doubt that the value of this and the Crown Point mine will go still higher. The Virginia City Enterprise, of January 11, thus speaks of the mine, and it is not surprising that, in view of such developments, the stock should at that time have stood at $500, against $8 in January of the previous year:

This splendid mine has never looked better than at present. Throughout a length of 320 feet, every stope, cross-cut and opening of any kind on the great pay-deposit, shows magnificent ore. The north winze, now being sunk on the 1,100-foot level, is now down 135 feet. It descends at an angle of 40 degrees, and is all the way in ore of exceeding richness—ore, the average assays of which are from $150 to $200. On the ninth level of the old mine, where the ore-body has been cut through, the pay-ground is found to be 30 feet in width and very rich. Thus the deposit is seen to have width and depth as well as length. As yet we are unable to give the yield for the past month in exact figures, but it will vary but little either way from $325,000. Of this amount $200,000 may be set down as net profit. Taking into consideration stoppage of the mills and consequent falling off in the quantity of ores reduced, the showing for the month ought to be satisfactory to the stockholders. The future of the mine cannot be otherwise than brilliant. The company now have on hand over $700,000 in coin and supplies of all kinds in immense quantities. Besides the stores at the mine, they now have (paid for) in Carson City, timbers and lumber to the value of $17,000. Add to this the fact that all the ore-breasts throughout the mine are looking splendid, and that an immense body of ore has been explored and is now in sight ready to be raised, and we can see that all that has yet been done is as nothing when compared with what can now be accomplished. A very considerable increase in the working force employed in the extraction of ore will be made to-day, and should the weather prove favorable for the transportation and reduction of ores for the remainder of the month, we may look for a big yield for January.

The increase of the proportion of gold in the ore from the deepest levels of the Belcher and Crown Point is attended by a return in the appearance of the ore to the type made familiar in earlier years by the rich black sulphuret bodies of upper levels in the Gould and Curry, Ophir, and Mexican mines. This tends to remove the apprehension expressed in my former reports that the character of ore might change in depth, assuming a predominance of refractory base-metal minerals. Baron Richthofen's expectation of more widely disseminated low-grade ore in depth is likewise partially contradicted by this discovery; though it may hereafter be found that such low-grade ores do exist at these deep levels. In fact, the study of the ore-bodies on the Comstock hitherto has been mainly a study of excavations; and the portions of the vein which were not extracted are not usually laid down on the maps as ore-bearing at all. This naturally leads us to overlook connections between different ore-bodies which may nevertheless exist.

South of the Belcher, the principal activity has been in the Overman and Caledonia mines. In the former, a new vertical shaft, on the east side of the vein, necessitated by increasing depth of workings, has been in progress. It was down 500 feet February 14, 1872.

The Caledonia is claimed to be the farthest south of the mines on the Comstock. Whether it is actually on that vein or not, it has been worked with considerable activity during the year, producing 18,836 tons of low-grade ore, of the gross value of $244,890. This was formerly called the American Mining Company.

The Succor is a mine in Gold Cañon, about one and a half miles from Crown Point. It is supposed by some to be upon a vein east of the

Comstock, and by others to occupy a branch, or even the main continuation of the Comstock, which, they claim, has been warped in that direction. It is reported as producing, during the first three quarters of 1871, 8,400 tons of ore, of the gross value of $105,940. The San Francisco Stock Report says of this mine, (March 10, 1871:)

We learn that the main tunnel is in 1,250 feet, and was in ore that paid a profit over expenses for the whole distance; an upper tunnel is in 570 feet, and was also in ore all the way. The lower tunnel is connected with this by an upraise. From the upper tunnel a raise has been made to the surface, thus connecting the lower tunnel with the surface. These raises have demonstrated the fact that ore exists the whole distance, from the surface to the lower tunnel, which is over 500 feet. An incline has been sunk from the surface (1,000 feet from the mouth of the tunnel) to a depth of 140 feet, and was in ore all the way. This incline will intersect the tunnel 1,000 feet from its mouth. Still further to the east of the incline are good croppings that extend to the eastern boundary of the mine. On the opposite (or west side) of Gold Cañon, a shaft has been sunk 180 feet, and a drift is being run which will cut the Succor ledge at a depth of 140 feet below the lower tunnel on the east side of the cañon. The ore that is being extracted from the mine mills about $15 per ton, while the cost of mining it is only $1 and milling $6 per ton, thus leaving a profit of $8. The first ore crushed in Nevada came from this mine, and was milled at the Pioneer mill in 1859. From the proceeds of this mine they have been enabled to build a mill of the capacity of crushing 35 tons daily, 4 Stevenson's pans, and pay for all prospecting, improvements, and mining that have been done. The past sixteen months' workings have returned $226,000. The dip of White ledge is east and runs northerly and southerly, being the same course as that of the Comstock, and is in a direct line with the new shafts of the Hale and Norcross, Savage, Chollar-Potosi, and others, and many mining experts have set forth the assertions, based upon thorough knowledge of the dip and course of the Comstock lode, that this White ledge is on the lode.

The Kentuck mine, north of Crown Point, has at last been able to extract ore from the section of the mine burned out by the fire of April, 1869. The rock in May, 1871, was still so hot in many places as to burn the naked hand. The timbers were found to be converted into charcoal.

The Yellow Jacket developed a body of ore in the north drifts of the 1,000 and 1,100-foot levels during the spring and summer of 1871, and maintained a steady production of ore. The price of the stock was comparatively free from fluctuations during the year. The main incline of this mine has reached the level of the proposed Sutro Tunnel.

The Chollar-Potosi produced during the year 59,535 tons of ore, of the gross value of $2,067,827. The year ending July 1, 1871, was the most profitable in the history of the mine; but the great falling off in product during the remainder of 1871 indicates the rapid exhaustion of reserves. The product for each quarter was as follows: First quarter, $820,295; second quarter, $770,347; third quarter, $287,175; fourth quarter, $189,909.

The operations of the Hale and Norcross are described at length in the extract from the superintendent's report, given below.

In the Savage, a new body of ore was discovered, about the beginning of the year, in the 335-foot level. It is said that this body, which contained a large amount of low-grade ($23) ore, was strangely overlooked in former workings. Mr. Bonner, when superintendent of this mine, "drifted square up to it and struck the clay inclosing it, then 'raised' for a few feet along the clay, and stopped work without ever cutting through." The old Potosi chimney was also worked to some extent; but the operations of the mine were not specially profitable, as will be seen by the report of the superintendent, who declared in July that, so far as could then be seen, the mine was exhausted of paying ore. But the discovery of rich ore in the 1,400-foot level, on the 2d of February, 1872, has put a new face on affairs. Full particulars of this discovery are not at hand, but its significance is certainly very great.

I quote the following remarks upon it from the San Francisco Stock Report:

That the strike should occur in this mine is of far greater importance and value to base the belief upon that other mines will strike it eventually, than if the discovery should have been made in some mine more contiguous to the Crown Point. Adjacent to it on the north is the Gould and Curry, and on the south the Hale and Norcross. Further south is the Chollar-Potosi, and most of the other mines that are speculated in, and called Washoe stocks. The development of ore in the Belcher, and its discovery in the Savage, is the finding of a paying vein almost at what has been considered the two extremities of the lode. Between these two mines are located the most of the mines that have ever paid dividends, and the importance of the development aids to substantiate the belief that this lode is more likely to be a continuous than an irregular, broken body. The Crown Point's rich body was met at a depth of about 1,200 feet, although ore of a low grade was discovered 900 feet down, and that of the Savage is met at 1,400. The declivity of the hill is so great that the Savage mine, or where their new shaft is located, is about 200 feet higher than the Crown Point, and it would be adduced from this fact, that if the country where these mines are located was a level plain, instead of a hilly formation, these two mines would have struck ore at about the same depth. The continuity of this body may continue up to the Sierra Nevada mine, as lately in this claim a vein of metalliferous quartz has been discovered, similar to that found in the mines located on the Comstock lode proper—the walls composed of porphyry and clay. This new deposit goes as high as $29 in silver, and considerable free gold is perceptible in the rock. This claim is situated about one mile above the Ophir, and is at the extreme end of where any mining operations have ever been carried on. It has been a question of doubt, ever since silver has been discovered in that country, as to where this silver lode begins and ends. For several years back its terminus has been placed at about the Ophir mine; but these calculations are now apt to be upset by this recent discovery. Between the Ophir and Sierra Nevada mines there is a very large section of country that has never been worked at any great depth. An impetus will now be given to prospect this virgin ground, and who can foretell what the result will be?

The operations in the Gould and Curry comprised the retimbering of between 400 and 500 feet of the shaft in which the timbers were more or less decayed, and the continuation of explorations, which resulted, up to the close of 1871, in no discoveries of importance. The product of the mine was 1,713 tons in the first quarter, gross value, $34,946, and 345 tons in the third quarter, gross value, $12,074. This product is from bodies of ore in upper levels.

From the foregoing brief outline of some of the more important operations, it will be seen that the prospects of the Comstock mines for the immediate future are most brilliant. According to the Alta California the ore-bodies in Crown Point and Belcher contain at least $90,000,000; and this enormous value, together with the circumstance of their discovery at the greatest depth attained, will undoubtedly lead to more extensive prospecting, as well as largely increased production.

In view of the large amounts of low-grade ore which remain, and other bodies of low grades which are continually discovered in the mines, the question of the price of labor becomes important, though the richer discoveries, by restoring to some of the companies the prosperity which has heretofore rendered them careless of this question, may postpone its reasonable solution. The president of the Hale and Norcross, in his annual report, gives the following statement, showing the reduction of all expenses, except that of labor, within the four years preceding 1871:

	1867.	1870.	Reduction.
Cost of timber per 1,000 feet	$31 32	$21 32	31.99 per cent.
Cost of wood per cord	15 05	11 33	24.71 per cent.
Cost of milling per ton	14 21	11 16½	21.42 per cent.
Total	60 58	43 81½	27.67 per cent.
Yield of ore per ton	34 14	25 13	26.10 per cent.
Labor per day	4 00	4 00	

H. Ex. 211——10

It thus appears that while the items of timber, fire-wood, and milling expense show an average reduction of 27.67 per cent., and the value of the ore reduced a decline of 26.1 per cent., the wages of labor still maintain a high standard. Upon this point the president remarks:

In all mining communities where employment depends on the development of hidden wealth, there must of necessity be large numbers who have nothing to do, and who can be only a charge and expense upon their more fortunate co-laborers; but when circumstances, such as now exist in the large bodies of low-grade ore already developed on the Comstock lode, have placed it in the power of all to be employed at prices from 15 to 20 per cent. less than present rates, how much better is it for labor to conform to the necessities of the case than to insist on prices of ten and eleven years ago, thereby paralyzing work, and, in order to maintain its position, compelled to contribute a large percentage of its earnings toward the support of the unemployed.

The San Francisco Stock Report of March 15 says on the same topic:

Increased activity is perceptible in the stock market, based upon confidence being restored to the existence of deposits of ore in the lower depths of the Comstock, and that of a quality which will leave a fair margin for profit over expenses. The great cost of mining is now becoming a subject of much concern to interested parties, there having been no reduction made in this department since the organization of the Miners' League. From well-informed persons we learn that employment is given to about three thousand miners, at $4 per day; this sum aggregates but $12,000 daily, being about $360,000 per month. The quantity of low-grade ore existing in some of the mines is very extensive, while many outside claims are compelled to lie idle on account of high-priced labor. To extract these vast deposits and to work idle mines, employment could be given to at least treble the number of men now employed, but a reduction is necessary; though the amount is but trifling to each man, yet the aggregate sum would enable mines to be worked without calling monthly upon stockholders to meet the enormous demands for labor that they are now subject to. The strength of these trade combinations consists more in numbers and in the wealth that they collectively can command than in the threats of the destruction of the property of those who are their employers, if the employés' demands are not acceded to. If the miners would consent to a reduction to $3 per day, (which is more in proportion to what they were receiving before the completion of the railroad, which reduced the freight tariffs so much that the necessaries of life are furnished to them now at nearly 50 per cent. lower than previously,) employment could be given to nine thousand men, and thus, instead of these men controlling among themselves but $12,000 per day, they would control $27,000, and a far larger force, which, to displace, would cause more trouble, and it would be an almost impossibility to replace such a large number without causing losses that a year's time would hardly meet. And this reduction would result greatly to their good, and in a manner to which they pay but little attention. Most of these miners speculate to a greater or lesser extent in the stock of the mines in that district. When the mines have good deposits of ore, which enable them to meet expenses and disburse dividends, the miner that has stock in such mines is in far better condition than when he is called upon for money to meet assessments, for which there would be no need if he would consent to a slight reduction in his demands. If he will thus take into consideration the sum he has thus paid out and the losses he has suffered by the depreciation of the stock he has held, he will find out that a greater call than 25 per cent. upon his $4 per day has been made, and that these deposits still remain, which, if he but consented to make the price of his labor proportionate to the prices of other commodities, would have been worked, and by the extracting of them the necessity of many assessments would have never existed, and thus indirectly his labor would bring to him more than $4 daily by the maintaining of the value of the stock which he held, and the doing away of assessments which he has been called upon to pay.

The proceedings with regard to the Sutro Tunnel scheme have brought out a good deal of interesting evidence concerning the Comstock mines, and the methods of working now employed there.

A law of Congress, approved April 4, 1871, authorized the President of the United States to appoint a board of three commissioners, two engineer officers of the Army and one mining or civil engineer, to examine and report upon the Sutro Tunnel, with special reference to the importance, feasibility, cost, and time required to construct the same; the value of the bullion extracted from the mines on the Comstock lode; their present and probable future production; also, the geological and practical value of said tunnel as an exploring work, and its general bearing upon our mining and other national interests in ascertaining the practicability of deep-mining.

The commission, consisting of Major Generals H. G. Wright and J. G. Foster, and Mr. Wesley Newcomb, transmitted its report, December 1, to the Chief of Engineers, who transmitted it, January 4, to the Secretary of War, who transmitted it, January 6, to Congress.* This report contains much interesting information, though it must be pronounced in many respects superficial and unsatisfactory.

The commission estimates the total yield of the lode, from 1859 to 1871, at $125,000,000; and the present annual product at $15,000,000. As regards the probable future yield, it is declared that no claim can be made to anything like accuracy, except in the few instances in which ore-bodies are now developed. The commission declares its belief that the lode is a true fissure-vein, extending downward indefinitely in the crust of the earth; but whether it will continue to be ore-bearing is a matter of opinion, to be based upon probabilities, and the actual results experienced in deep-mining in other parts of the world. These, in the judgment of the commission, favor the finding of ore down to the lowest depths that can be reached.

The report declares that, as an exploring work for deep-mining, the Sutro Tunnel merits favorable consideration; indeed, that its value for this purpose is so evident as scarcely to be called in question. With regard to drainage and ventilation, it is concluded that the tunnel would not be a necessity for these purposes. This part of the report is remarkably wanting in definiteness and force. In the matter of drainage, the commissioners seem to have made no original investigations whatever, but to have accepted without question the conclusions of several superintendents of the mines—that is to say, of persons in the employ of the parties who are opposing the construction of the tunnel. They were thus led to the startling conclusion, among others, that the cost of raising water from the mines is already very slight, and will decrease still further as the workings advance in depth. This opinion is open to challenge on four grounds: First, it ignores the indirect cost of drainage by means of pumps, *i. e.*, the extra time and labor, aside from the mere cost of pumping, involved in this method, and the frequent loss and delay arising, particularly upon the Comstock vein, from sudden influx of large bodies of water.† Secondly, it is based upon

* See Senate Documents, Ex. Doc. No. 15, Forty-second Congress, second session.

† In illustration of this point, the following article from the San Francisco Stock Report of January 26 is given:

"Most of the difficulties that beset mining in its incipient stage have been overcome by the many improvements made upon the machinery then in use. The horsewhim has been superseded by the steam-engine, and by which appliance hundreds of tons of ore are raised to the surface in the same space of time that formerly but ten were extracted, and at a greatly less cost. New inventions have succeeded the old form of pans, and now 70 (and frequently 90) per cent. is returned by the mills, to but 40 and 50 in the early days of mining. Rock-breakers worked by steam are used in place of the heavy sledge-hammer propelled by manual labor. In all the departments great improvements have been made, which have proven of incalculable value to mine-owners. The expenses, by the application of steam and new inventions, have been decreased almost 50 per cent., and lives that were formerly constantly in jeopardy, are now almost always comparatively safe by the use of the safety-cage. In every minutia pertaining to mining matters there has been an improvement upon the primitive mode that was in use ten to twelve years ago. All the obstacles that nature seemed to present to prevent the success of human skill have been overcome, with one exception, and that is the prevention of water interfering with work in the mines. When water flooded upper parts of a mine, no little trouble was experienced to drain it out, while the expense was proportionately large. This was at a time when 500 to 800 feet were considered great depths for a shaft. Superintendents frequently complained of vexatious delays to which they were forced to submit by this observations referring to the third of a series of unusually dry years, when the amount of water was exceptionally small. Thirdly, it takes into account but a small portion of the vein, and it ignores the fact that the fear of striking bodies of water has been one cause limiting the amount of exploration in the mines. It has often been pointed out that the water in the Comstock frequently occurs in portions of the vein-mass surrounded with clay-seams, and that this clay is so tenacious as to support a heavy column of water, so long as it is not broken through in mining. Now it has repeatedly happened that, fearing to cut through such a seam of clay, the miners have failed to find a body of ore which lay behind it. This is apparently the only explanation of the fact that the ore-body discovered in January, 1871, in the 300-foot level of the Savage mine, had been followed for a considerable distance along the outside of its clay sheathing years ago; until the clay was nowhere cut through, and the presence of ore was consequently not at that time ascertained. Deliverance from the water-risk would, in my opinion, permit further explorations of the upper levels of several mines, which, notwithstanding the large amount of work which has been done in drifting upon them, I believe are still not so thoroughly prospected as to justify abandonment. The careful study of the mining maps suggests this opinion, and the history of the repeated discoveries of overlooked bodies of ore confirms it. Fourthly, the opinion that the cost of raising water will be still less at greater depths, ignores the phenomenon presented by artesian borings, namely, that subterranean supplies of water from distant sources are struck at depths between 2,000 and 5,000 feet. The origin of such streams and springs may be a hundred miles away; and the local dryness of the place where the boring is made has nothing whatever to do with it.

On the subject of ventilation, the report of the commissioners is still less satisfactory. It is almost incredible that educated engineers should lend their names to such statements as the following:

Even with all the aid that the tunnel can be expected to afford, it is the opinion or the commission that mechanical ventilation by blowers, operated by steam or other power, would still be needed at the headings and in the stopes where the air from the tunnel would not penetrate.

According to natural laws as at present understood and received, the air entering the proposed tunnel would pass through it and up the shafts of the mines by the

---

annoyance, and often work was stopped for weeks. Men of ingenuity set themselves to thinking, and pumps of considerably more force were invented, which, for a time, were successful in obviating this difficulty; but now shafts have been sunk to such depths that more powerful machinery must be put in use; and in a few years another change will have to be made, for the power will then be too light for the depths that will be attained. Vast outlays of money will be required to put this machinery in place, and work in the mine, for the time being, will likely be stopped. The aqueous fluid seems to have turned out to be the greatest obstacle that nature has presented to human skill to conquer in the mines of the Comstock; and even with the powerful machinery now in use, work has to be stopped in those portions of the mines where water has gained upon pumps; as, for instance, the superintendent of the Crown Point, in a letter of the 7th instant, in speaking of the 1,300 level, says: 'On the 1,300 level nothing is being done, and nothing will be done until the water is drained off, which will probably take three weeks.' This mine has one of the most powerful pumping-works upon the Comstock, and yet so great a headway was gained by the water, that work was forced to be stopped in a certain portion of the mine. No appliance now in use is of sufficient power to prevent the repetition of such an occurrence, and the skill of the superintendent, who is regarded as one of the best miners in Nevada, is put to a severe test to drain it out in the time specified above. Considerable portion of the expense incurred in mining is caused by the hoisting and draining of water, and he who will devise means whereby this difficulty will be conquered and which will prove permanent, will not only be a great benefactor, but will meet with such a compensation as will place him in a position far above want."

Certainly nothing can be "devised" more simple and efficient for this purpose than a deep tunnel.

easiest and therefore by the most direct channels, thereby conferring little if any benefit upon the stopes and drifts not in the line of such direct transit. Hence the necessity which is assumed for a continuance of mechanical ventilation for certain portions of the mines after the completion of the tunnel.

And here it may be proper to allude to certain anomalies observed in the ventilation of the mines on the Comstock lode, as well as in mines upon lodes lying to the eastward. According to the received laws of ventilation it would have been assumed that, in the case of two shafts connected at bottom by drifts, the air-current would pass down the lower and through the drifts up the higher, and that this rule would be without exception where not influenced by circumstances of situation or artificial causes; that, in the case of a long adit or tunnel, the inner extremity of which was connected with the surface by a shaft, the outer being directly upon the side of the mountain, the current would be through the tunnel and up the shaft. In the former case the current was found to be sometimes in one direction and sometimes in the other, it having been permanently changed in one instance, after the occurrence of a fire in one of the mines thus connected; the down-draught having been through the shorter shaft before the fire, and through the longer over since. In the latter case, which applies to two tunnels visited by the commission, the down-draught was into and downward through the shafts and out of the tunnels in a very strongly perceptible current. In view, therefore, of these anomalies, it would seem uncertain whether the current of air would pass through the proposed tunnel into the mines and out through the shafts, or the reverse. So far as the ventilation is concerned, it will be of little importance which way the current should pass. Probably the mines would be the more benefited by its passing downward through them and out of the tunnel, than in the reverse direction.

The author of these paragraphs seems to be ignorant that the air-current of a strong natural ventilation can be carried, by simple contrivances, to all the headings and stopes at will; that it can be split into separate currents, and these conducted wherever they are needed; that this is not only theoretically possible, but actually practiced in large mines; and that the power of the current supplied by a connection between a horizontal tunnel and a shaft 2,000 feet deep would be much greater than that of the mechanical blowers now employed at the Comstock mines.

The talk about "anomalies" in the direction of the current is absurd, as a plain statement of the theory of natural ventilation will show. The air at each opening of a mine (the downcast shaft or tunnel-mouth, as the case may be, and the upcast shaft) is under the pressure of a column of air, extending upwards to the top of the atmosphere. Taking for illustration the simplest case, that of a shaft connected at the bottom with a tunnel, we have at the two ends of the tunnel two different pressures. At the outer end there is the column of exterior air already mentioned; at the other end there is a shorter column of exterior air (terminating at the top of the shaft) plus the column of interior air in the shaft. It is evident that if the density of the air in the shaft is exactly equal to that of the air outside, the pressure will be the same at both ends of the tunnel, and there will be no current at all. If the air in the shaft,.being warmer, is lighter than the air outside, the pressure at that end of the tunnel will be less, and there will be a current in through the tunnel and out through the shaft, with a power and speed determined by the difference in weight, with a due allowance for drag or friction of air. In most mines in temperate climates, not exceeding 500 or 600 feet in depth, the draught is one way in winter and the other way in summer, because the temperature in the mine is at one time in the year higher, and at another time lower, than the temperature outside. With a mine so deep as to give an excess of temperature inside, over the summer temperature outside, the current will always be one way. Even a slight difference of heat being multiplied by the length of a very deep shaft, makes a mighty difference in *weight of air-column*, which is the moving force.

The course of ventilation between two shafts, the mouths of which are

on different levels, is (friction &c., aside) what it would be for a tunnel and shaft, the depth of which is equal to the said difference of level. When two shafts with their mouths on the same level are connected by a drift below, there is at first no impulse to a current in either direction, unless for some reason the air in one shaft is warmer than that in the other. But if the air in both is equally warm, *and warmer than the air outside*, then upon any slight cause starting the current in either direction, it will continue for an indefinite period, though with no great strength, in that direction, because the place of the warm air passing from one shaft into the other will immediately be filled by cold air pouring into the top of the first, and thus a permanent inequality in the weight of the two columns will be introduced. But should the air outside become warmer than that inside, the current will cease; because, in that case, the outside air entering the top of the shaft from which the movement was taking place through the drift, would make that
• column lighter, not heavier, than the other, and so would tend to stop or reverse the current.

The formula and calculation of the theoretical effect of the difference in temperature inside and outside of a mine are not difficult. What is difficult is the calculation of the drag and other hinderances which diminish the power and speed of the current. Leaving these, for the present, out of the calculation, it may be shown mathematically as follows:

Let T be the temperature in Fahrenheit degrees above freezing-point of the outer air; t that of the air in the mine; a the co-efficent of expansion, (according to Dalton, about 0.0023 for each degree Fahrenheit;) H the depth of the shaft above the tunnel; and M the pressure of the atmosphere at the highest opening of the mine, and let it be assumed that T is less than t.

The pressure at the mouth of the tunnel will then be—

$$P = M + H \text{ of the temperature } T,$$

and the pressure in the shaft,

$$P^1 = M + H \left(\frac{1 + a\,T}{1 + a\,t} \right) \text{ of the temperature } T.$$

The excess of pressure at the tunnel-mouth will therefore be—

$$P - P^1 = H - H \left(\frac{1 + a\,T}{1 + a\,t} \right) = H \left(1 - \frac{1 + a\,T}{1 + a\,t} \right) = H\,a \left(\frac{t - T}{1 + a\,t} \right)$$

This may be designated h.

For H = 2,000 feet, T = 33° above freezing-point, and t = 53° above freezing-point, we have—

$$h = P - P^1 = 2,000 \times .0023 \left(\frac{53 - 33}{1.122} \right) = 82$$

or the weight of a column of air at the temperature T, 82 feet high.

The general formula for velocity under a pressure P being $v = \sqrt{2\,g\,P}$, we have here for the pressure $P - P^1$ as above obtained,

$$v = \sqrt{2\,g\,h} = \sqrt{2\,g\,H\,a \left(\frac{t - T}{1 + a\,t} \right)}$$

or, for the special case assumed—

$$v = \sqrt{164\,g} = 72 \text{ feet per second.}$$

The quantity admitted must be found by multiplying the area of the tunnel-opening by this velocity. Thus, 140 feet area would allow over 600,000 cubic feet per minute to pass.

The effect of drag or friction will be presently considered. Disregarding resistances, we find from the foregoing formula—

1. The quantity of the entering air is proportional to the velocity;
2. The velocity is proportional to the square root of the depth, and the square root of the difference in temperature inside and outside of the mine.

A part of the surplus pressure $P — P^1$ is consumed in overcoming the resistance of friction. This loss, measured in height of air-column, is, according to D'Aubisson—

$$\frac{M}{g} \times \frac{C}{A} \times L\, v^2$$

M being an empirical co-efficient (about 0.01 feet,) A the area, C the perimeter, and L the length, of the air-passage, and v the velocity of the current. Velocity, and therefore power, is moreover lost in overcoming extra resistance when the current passes from a large passage into a smaller one, the new velocity being nearly

$$v^1 = v \times \frac{a}{A}$$

or the two velocities being proportional to the two sectional areas of passage.

For a given case, the pressure must outweigh all resistances. Assuming the sectional area to be constant, we have—

$$h = \frac{M}{g} \times \frac{C}{A} \times L\, v^2,\ \text{whence}$$

$$v = \sqrt{2\, g\, h\, \frac{1}{2\,M} \times \frac{A}{C} \times \frac{1}{L}}$$

Since the theoretical velocity, apart from resistances, is $\sqrt{2\,g\,h}$, we infer from this formula that the actual velocity is proportional to the square root of the relation between perimeter and area of the section of the air-way, and inversely proportional to the square root of the length of the air-way. On the supposition of a tunnel 20,000 feet long and a shaft 2,000 feet deep, with sectional area of 140 square feet and perimeter of 50 feet, the theoretical velocity due to a difference of 20° in temperature would be reduced from 72 feet to 5.8 feet per second.

It appears from the formulas, as I have said, that the velocity is proportional to the square root of the depth, and to the square root of the difference in temperature. Hence, all other things being equal, the ventilation through a shaft 2,000 feet deep, connected with a tunnel, compares in effectiveness with the ventilation through two shafts reaching to the same depth, differing, say, 220 feet in level at the surface, as $\sqrt{2,000}$ to $\sqrt{220}$, or nearly 3 to 1. Again, other things being equal, a difference of temperature of 5° would be half as effective in natural ventilation as one of 20°. Such conclusions as these may safely be drawn from the foregoing demonstration. The formulas I have given are not intended for any other use than to show the definite relations and conditions of the problem. The student who desires to find a more thorough mathematical discussion of the question may consult Peclet's Treatise on Heat and its Applications, or, still better for this special purpose, Combes' Aérage des Mines, Vol. II, p. 335, and elsewhere.[*]

This is the nature of the examination which should have been made

---

[*] In this brief discussion I have followed Lottner and Serlo, Leitfaden zur Bergbaukunde, Vol. II, p. 145, et seq.

by the Sutro Tunnel commissioners. In the absence of it no one can say, from the imperfect data at hand, exactly what would be the ventilating effect of that tunnel; and I confess to a feeling of more than disappointment at finding in a report which, it might reasonably have been anticipated, would seriously consider a question of such importance, nothing on the subject but the unscientific and inaccurate talk about "anomalies" and "violations of the laws of ventilation," in which engineers have no right to indulge.

It is due to the eminent gentlemen who composed this commission, to say that they construed the law under which they were appointed as directing them to do no more than they have done. If this was the case it was an unfortunate defect in the law. The only benefit to be derived from the skill and experience of such men was, I think, in requiring them to study the question as engineers, not as tourists. Their estimate of the cost of the tunnel and all branches is apparently careful and trustworthy. I subjoin it in full:

Estimates of costs of the Sutro Tunnel.

Cost of sinking shaft No. 1, 109 square feet area and depth of 530 feet, including tools, labor, and materials of all kinds, at $40.24 per foot of depth, (being the average cost in the seven principal mines of the Comstock lode) $21,327 20
Same, shaft No. 2, 109 square feet, 1,025 feet deep, at $40.24 .... 41,246 00
Same, shaft No. 3, 109 square feet, 1,319 feet deep, at $40.24 .... 53,076 56
Same, shaft No. 4, 109 square feet, 1,499 feet deep, at $40.24 .... 60,319 76
Same, shaft No. 5, 109 square feet, 1,465 feet deep, at $40.24 .... 58,951 60
Same, shaft No. 6, 109 square feet, 1,465 feet deep, at $40.24 .... 58,951 60

Preliminary tunnels or drifts.

Cost of labor, tools, and materials of all kinds, for drift of main tunnel, 6 feet wide, 7 feet high, and 19,790 feet long, at $16.90 per running foot, (being the average cost of 2,185 feet completed July 1, 1871) $393,821 00
Deduct value of one-half of timber of drift, which may be used again as the enlargment progresses 14,644 60
                                                                                                379,176 40
Cost of labor, tools, and materials of all kinds, used in drifts of branch tunnel, 6 feet by 7 feet, by 12,000 feet long, at $19.90 per running foot 238,800 00
Deduct value of one-half timber of drifts used a second time 8,880 00
                                                                                                229,920 00
Cost of enlargement of drift to full size of tunnel, 13½ feet by 12 feet, by 19,720 feet long, 2,366,400 cubic feet, at 25 cents 591,600 00
Same, of branch tunnel, 12,000 feet long, 1,440,000 cubic feet, at 25 cents. 360,000 00
Cost of timbering main tunnel full size, 19,720 feet in length, at $17.34 per running foot 341,944 80
Same, of the branch tunnel, 12,000 feet, at $17.34 .... 208,080 00
Cost of general material and sundries, including surveying instruments, large transit building for the same, boarding and lodging houses, barns, horses, carts, magazines, blowers, air-pipes, &c., for four shafts of main tunnel 66,439 00
Same, for branch tunnel, two shafts 20,000 00
Cost of hoisting and pumping engines and machinery for four shafts of main tunnel 121,679 00
Same, for branch tunnel, two shafts 108,930 00
Cost of boilers and parts, four shafts, main tunnel 33,736 40
Cost of boilers and parts, two shafts, branch tunnel 25,256 00
Cost of labor and materials for the erection of machinery, and temporary buildings to cover the same, for the four shafts of main tunnel 32,265 00
Same, for the two shafts of branch tunnel 21,510 00

Cost of material and time employed in attending machinery during the sinking of four shafts, main tunnel.. $138,734 27
Same, for the two shafts of branch tunnel.. 82,489 50
Same, during the running of the preliminary tunnel of the main tunnel.. 324,784 90
Same of branch tunnel.. 216,523 26
Cost of appliances for hauling rock and ore out of main tunnel............. 50,000 00
Cost of appliances for hauling rock and ore out of branch tunnel 35,000 00
Add for office expenses, superintendence, engineering, and contingencies, 20 per cent.. 736,388 25

Total cost in gold... 4,418,329 50

RECAPITULATION.

	Main tunnel.	Branch tunnel.
Sinking shafts...	$175,969 52	$117,903 20
Running preliminary tunnels............................	379,176 40	229,920 00
Enlargement of drifts to size of tunnel.................	591,600 00	360,000 00
Timbering full-size tunnel..............................	341,944 80	208,080 00
General materials and sundries.........................	66,439 00	20,000 00
Engines and machinery.................................	121,679 00	108,930 00
Boilers and attachments................................	33,736 40	25,256 00
Erection of machinery and temporary buildings for same	32,265 00	21,510 00
Attending machinery in sinking shafts.................	138,734 27	82,489 50
Ditto in running preliminary tunnels	324,784 90	216,523 26
Endless wire-rope, &c...................................	50,000 00	35,000 00
	2,256,329 29	1,425,611 96
Office expenses, superintendence, engineering, contingencies, 20 per cent......................................	451,265 86	285,122 39
	2,707,595 15	1,710,734 35

Time required to complete tunnel.

Depth of shaft No. 4, (the deepest)... 1,499 feet.
Average daily progress in the shafts of the Comstock.................... 3 feet.
Number of days required to sink shaft No. 4, 1,499 feet................. 500 days.
Whole length of main tunnel... 19,790 feet.
Distance penetrated by preliminary tunnel, July 1, 1871............... 2,185 feet.
Average daily progress in preliminary tunnel............................. 4$\frac{1}{4}$ feet.
Distance penetrated when shaft No. 4 reaches tunnel-level.............. 4,260 feet.
Distance remaining to be penetrated at that time........................ 15,530 feet.
Number of available working headings.................................... 9
Greatest distance to be penetrated by any drift to meet the drift from the adjacent shaft... 2,432 feet.
Time required to run above distance at 4.15 per day..................... 586 days.
Total time required to sink shafts and run drifts........................ 1,086 days.
Additional time required to enlarge tunnel to full size .... 100 days.
Total time required to complete main tunnel.............................. 1,186 days.
Number of years required to complete main tunnel .... 3$\frac{1}{4}$ years.
Number of feet of branch tunnel run from four shafts at bottom of shafts Nos. 5 and 6, 1,465 feet deep, when main tunnel is completed... 414 feet.
Additional time required to extend branch tunnel to 12,000 feet, working two headings... 50 days.
Total time to complete main and branch tunnels, (manual labor)........ 3$\frac{1}{2}$ years.
Total time to complete main and branch tunnels, (by machinery)........ 2$\frac{3}{5}$ years.

I cannot here discuss the question raised by the commission as to the feasibility of concentrating ores or tailings at the mouth of the tunnel. Their conclusion is that they do not know, and their recommendation is that a commission be sent to Europe to find out!

An investigation is now in progress before the Committee of the House of Representatives on Mines and Mining, in which the commissioners themselves and many other experts have been witnesses, and the results will, no doubt, form an interesting addition to the literature of the Comstock lode, the Sutro-Tunnel chapter of which is already quite voluminous. Pending this inquiry and the proposed action of Congress in the matter, the company has begun operations with considerable vigor, as will appear by the following extracts from the report of the superintendent for the months of December, 1871, and January and February, 1872:

We have met with many difficulties during that time, caused mainly by an unusually severe winter, the extraordinary fall of rain and snow having made the roads almost impassable. As spring approaches we may look for the removal of these obstacles; in fact, the latest accounts give favorable news in regard to the weather and the condition of the roads.

You will perceive by the annexed statements that the expenditures were—

For the month of December, 1871	$28,821 04
For the month of January, 1872	43,517 40
For the month of February, 1872	50,490 41
Or a total for the three months of	122,828 85

This does not include any expenditures incurred by the San Francisco office.

In December last work was commenced on all four of our shafts, and the same has been prosecuted since with due energy by day and night. On the 24th of this month the progress at the different points was as follows:

Length of tunnel	2,801 feet.
Depth of shaft No. 1	120 feet.
Depth of shaft No. 2	282 feet.
Depth of shaft No. 3	147 feet.
Depth of shaft No. 4	120 feet.

The slow progress of shafts Nos. 1 and 4 is accounted for by the fact that a considerable quantity of water has been encountered, and that the pumping-machinery was delayed on the road. Shaft No. 2, in which the quantity of water was small, has been progressing steadily ever since its first commencement.

In December last a contract was made with the Diamond Drill Company for the use of diamond drills in all portions of the works. One of these drills has arrived at the tunnel, and experiments are being made for the purpose of ascertaining the best mode of employing it. With these drills it is confidently expected that the monthly advance in the tunnel will be 250 feet, and that of the shafts 150 feet. We may, therefore, look for a more rapid progress as soon as these are in full operation, which we hope will be the case by June next.

Temporary steam-engines and buildings have been erected on all the shafts; also, extra boilers and steam-pumps have been placed in operation, all of sufficient capacity to reach a depth of 500 to 800 feet. After that depth is reached machinery of much larger dimensions will be required, both for hoisting and pumping.

We have received estimates for the hoisting-machinery from four of the machine-works at San Francisco, the lowest bid amounting to $65,000. The cost of transportation and erection, including buildings, will probably amount to a similar sum.

No specifications for large pumping-machinery have as yet been submitted. They will be made out shortly, and bids, based upon them, invited from the founderies. A rough estimate of its cost, and placing the same in running order, may be given at $200,000. All this heavy machinery should be contracted for within the next sixty days, since it will require at least four months to construct and erect the same, it being highly desirable for the rapid prosecution of the work that no delay should occur on that account.

The necessary tools for a first-class machine-shop at the mouth of the tunnel—such as lathes, planing-machines, drills, &c.—have arrived, and a suitable building and steam-engine have been erected.

We have almost completed an excellent wagon-road, commencing at the mouth of the tunnel, leading over the first summit, at an elevation of 1,350 feet, to shaft No. 2, situated in a ravine just beyond. From that point an old road to Virginia City has been placed in repair.

The poles for a telegraph-line from Dayton to the mouth of the tunnel, and from thence to the four shafts and Virginia City, have been placed in position, and instruments at seven different stations will be in operation before long.

We have erected commodious boarding and lodging houses for the accommodation of the men at each of the four shafts, also a new one of much larger dimensions at the mouth of the tunnel.

The number of men employed was—

During December	159 men.
During January	231 men.
During February	326 men.

The work is being carried on with commendable care and economy, under the super-vision of our energetic foreman, Mr. John D. Bethel. All the different departments receive their proper attention. The mechanical engineering department is under the superintendence of Mr. John Anderson, that of surveying and civil engineering under Mr. G. H. Haist, and the office and accounts under the charge of Mr. L. Cheminant.

Most important developments have of late been made in the Comstock lode. The discovery of rich and extensive bodies of ore in the Crown Point and Belcher mines, at a depth of 1,300 feet, estimated to be worth, as far as developed, over $30,000,000, have finally established the continuance of the southern part of the Comstock lode in depth. Discoveries within the last month in the Savage mine, at a depth of 1,400 feet, of the most valuable character, have settled this question beyond cavil for the northern por-tion of the lode. The importance of these discoveries can hardly be overestimated, since they remove all doubts of the brilliant future of the tunnel enterprise.

It is highly desirable to push the tunnel to final completion at as early a moment as possible, and it appears that this may be accomplished within the next two and one-half years, provided the company secures all the requisite funds.

I annex the reports of the assessor and of the different mining com-panies :

Abstract statement from the quarterly assessment-roll of the proceeds of the mines of Storey County for the quarter ending March 31, 1871.[*]

Name.	Number of tons extracted.		Value per ton.	Gross yield or value.	Actual cost of extrac-tion.	Actual cost of transportation and reduction or sale.	Total cost.
	Tons.	Pounds.					
American, (now Caledonia,) Gold Hill district........	2,233	250	$13 48	$32,116 66	$5,382 81	$20,098 93	$25,680 93
Belcher, Gold Hill district.	279	....	3 70	1,034 50	....	....	1,774 00
Segregated Belcher, Gold Hill district...............	1,731	....	21 08	36,493 57	18,472 00	20,772 00	39,244 00
Crown Point, Gold Hill dis-trict....................	5,908	1,000	21 23	125,574 69	71,665 98	62,566 65	134,232 63
Chollar Potosi,Virginia dis-trict....................	20,897	....	39 25	820,295 70	106,298 37	250,764 00	357,062 37
Empire, Gold Hill district..	2,012	....	14 08	28,341 57	18,047 64	15,488 80	33,536 44
Gould and Curry, Virginia district	1,713	750	20 40	34,946 12	68,877 34	20,560 50	89,437 84
Hale and Norcross, Virginia district	17,187	1,920	20 76	356,850 12	100,803 50	107,945 67	208,839 17
Overman, Gold Hill district.	6,208	1,458	11 22	69,686 10	51,921 00	68,296 00	120,217 00
Savage, Virginia district...	13,928	1,380	22 10	307,916 73	131,710 24	129,496 18	261,210 42
Succor, Gold Hill district..	1,700	....	11 46	19,484 00	....	....	24,251 00
Imperial, Gold Hill district.	2,651	....	9 00	23,838 00	17,263 00	15,997 00	33,260 00
Sierra Nevada, Virginia dis-trict....................	5,865	....	7 84	46,121 01	19,020 73	28,670 40	47,691 13
Yellow Jacket, Gold Hill..	15,037	....	34 98	547,155 28	131,767 44	187,644 00	319,411 44
Total.................	97,952	818		2,449,854 05			

[*] *Assessment under new law. See Statutes of Nevada, 1871, p. 87.*

Abstract statement from the quarterly assessment-roll of the proceeds of the mines of Storey County for the quarter ending June 30, 1871.

Name.	Number of tons extracted.		Value per ton.	Gross yield or value.	Actual cost of extraction.	Actual cost of transportation and reduction or sale.	Total cost.
	Tons.	Pounds.					
Savage, Virginia district...	13,412	1,310	$21 60	$290,670 59	$132,831 74	$146,138 18	$278,096 92
Overman, Gold Hill district.	3,391	1,550	11 40	36,355 09	22,146 00	35,975 08	58,121 11
Sierra Nevada, Virginia district	2,773	....	7 17	19,889 49	....	....	22,085 51
Yellow Jacket, Gold Hill district	15,787	....	37 90	598,916 05	162,674 47	189,444 00	358,811 49
Crown Point, Gold Hill district....	18,972	1,000	37 90	719,191 80	132,763 82	227,767 00	360,530 82
Kentuck, Gold Hill district....	1,620	....	19 37	35,245 00	18,680 42	20,020 00	38,700 48
Empire, Gold Hill district..	1,993	....	14 15	28,200 00	13,721 00	14,047 50	28,668 91
Succor, Gold Hill district..	2,100	....	10 22	21,473 00	12,083 00	10,500 00	21,188 00
Caledonia, Gold Hill district.	7,383	....	13 75	101,585 23	24,285 02	69,400 20	93,685 22
Chollar-Potosi, Virginia district......	20,663	....	37 75	770,347 62	94,526 66	247,956 00	342,482 66
Hale and Norcross, Virginia district	15,844	1,390	17 09	270,795 11	111,880 15	159,626 36	271,506 50
Segregated Belcher, Gold Hill district	1,319	....	17 50	23,604 00	13,663 83	15,828 00	29,491 83
Total	105,477	6,250	....	2,018,402 98			

Abstract statement from the quarterly assessment-roll of the proceeds of the mines of Storey County for the quarter ending September 30, 1871.

Name.	Number of tons extracted.		Value per ton.	Gross yield or value.	Actual cost of extraction.	Actual cost of transportation and reduction or sale.	Total cost.
	Tons.	Pounds.					
Belcher, Gold Hill district.	2,689	....	$78 85	$212,038 56	$62,480 13	$32,268 00	$106,748 13
Segregated Belcher, Gold Hill district.............	649	....	17 02	11,009 24	4,415 83	7,788 60	12,204 43
Crown Point, Gold Hill district......	17,240	1,000	30 54	526,565 00	148,548 02	193,925 00	342,473 08
Chollar-Potosi, Virginia district......	9,816	....	29 26	287,175 73	74,358 60	117,792 00	192,150 60
Caledonia, Gold Hill district......	4,802	....	11 52	55,689 86	28,287 64	42,910 07	71,197 71
Empire, Gold Hill district.	2,200	....	12 61	27,759 35	23,611 34	16,450 00	40,061 35
Hale and Norcross, Virginia district	9,409	1,720	15 21	144,337 22	102,349 07	112,918 32	215,268 29
Kentuck, Gold Hill......	3,430	....	17 43	59,791 00	25,312 67	41,160 00	66,472 67
Savage, Virginia district...	11,523	410	20 62	237,661 61	108,725 47	128,815 77	237,541 24
Succor, Gold Hill district..	4,600	....	14 12	64,983 07	13,000 00	12,900 08	25,900 00
Sierra Nevada, Virginia district..	5,006	....	10 54	....	....	....	24,348 97
Yellow Jacket, Gold Hill..	9,137	1,000	27 46	250,844 40	141,335 61	109,650 00	250,985 61
Total...............	76,533	4,130	....	1,916,015 09			

Abstract statement from the quarterly assessment-roll of the proceeds of the mines of Storey County for the quarter ending December 31, 1871.

Name.	Number of tons extracted.	Value per ton.	Gross yield or value.	Actual cost of extraction.	Actual cost of reduction or sale.	Total cost.
Belcher	15,779	$62 47	$985,648 31	$217,149 65	$189,348 00	$406,497 65
Crown Point	17,826	33 63	599,622 99	179,616 74	213,936 00	393,552 74
Chollar-Potosi...........	8,159	23 27	189,909 78	73,159 22	97,908 00	171,067 22
Caledonia...............	4,396	12 61	55,497 92	31,124 49	39,151 45	70,275 94
Hale and Norcross	11,680	15 04	175,718 04	97,983 69	140,168 76	238,152 45
Kentuck	3,933	11 45	45,033 76	31,464 50	47,196 00	78,640 50
Savage	10,891	19 23	209,238 41	125,278 71	130,575 90	225,854 61
Sierra Nevada	4,651	6 09	28,358 07	....	....	29,794 97
Yellow Jacket	7,023	25 56	179,602 50	104,273 05	82,052 00	187,225 05
Gould and Curry	345	34 99	12,074 21	44,596 42	4,140 00	48,736 42
Total...........	84,677	....	2,480,903 99			

Summary of assessor's returns for 1872.

	Tons of ore.	Gross yield.
First quarter	97,952.41	$2,449,854 01
Second quarter	105,477.12	2,018,402 99
Third quarter	78,533.07	1,878,015 09
Fourth quarter	84,677.00	2,480,903 99
Total	366,639.60	9,727,176 11

The incompleteness of the assessor's returns is shown by the fact that the shipments of bullion by Wells, Fargo & Co., from Virginia and Gold Hill, during the year 1871, amounted to $11,053,328. The difference is probably due to the product of other mines than those mentioned in the table, and partly, perhaps, to the shipment of bullion obtained from tailings and slimes.

Report of the Gould and Curry for the year ending November 30, 1871.

The superintendent reports 2,635 tons of ore extracted during the year ending November 30, aggregating $91,645, and showing an average of $26.21 per ton. The time has been mostly spent in prospecting, with no favorable results. The mine is in good order, shafts well timbered, and everything in working condition. The president and superintendent both express the hope that further explorations of the deep levels will yet reveal as good indications as have been found in adjoining claims.

RECEIPTS.

From bullion product	$91,645
Assessment No. 9, $12.50 per share	60,000
Assessment No. 10, $15 per share	72,000
Assessment No. 11, $15 per share	72,000
Rent of mill and materials sold	22,650
Road franchise sold, and tolls	3,868
Miscellaneous	922
Total receipts	323,185
Cash on hand December 1, 1870	26,853
Total	350,038

DISBURSEMENTS.

Labor at mine	$104,236
Supplies at mine, &c	68,427
Working 3,500 tons of ore at custom mills	41,950
Day and night watchmen	2,070
Real estate in Virginia	1,186
Taxes—State, city, and county	4,262
General expenses	19,740
Exchange	1,254
Legal expenses	1,950
Miscellaneous	1,087
Total disbursements	306,771
Cash on hand December 1, 1871	43,267
Total	350,038

The assets of the company aggregate $129,935, consisting of $62,674 for real estate, &c., at mine; $21,476 for materials in storehouse and yard; $1,517 for bills receivable, and $43,267 cash. The liabilities consist of few uncalled-for dividends, amounting to $606. The average cost of extracting ore was $8.43 per ton.

Report of the Savage for the year ending July 11, 1871.

[Extract from the Superintendent's report.]

The Savage mine has yielded during the past year, ending June 30, 39,715 $\frac{1148}{2000}$ tons of ore; in the first six months, 12,503 $\frac{1388}{2000}$ tons, and 27,211 $\frac{1783}{2000}$ tons in the last six months of the year. This ore has been extracted from the following sections of the old and new mines:

NEW MINE.

	Tons.	Pounds.
First level	2,412	540
Second level	5,145	710
Fourth level	268	1,700
Seventh level	5	1,200
Eighth level	7,958	1,680
Ninth level	12,696	1,560
Tenth level	1,622	1,800
	30,110	1,190

OLD MINE.

	Tons.	Pounds.
First level	1,698	810
Third level	106	1,900
Fourth level	437	1,250
Fifth level	567	80
Sixth level	6,373	720
Seventh level	421	1,200
	9,604	1,960
Total	39,715	1,150

There have been reduced 38,147$\frac{700}{1000}$ tons, and sold 802$\frac{333}{1000}$ tons, (in the mouth of July, 1870,) leaving, from the year's production, 766$\frac{433}{1000}$ tons, the proceeds of which have not yet been received. The aggregate amount of bullion and cash received exceeds the total expense by the sum of $8,510.68, which is the profit so far realized. The yield of these 766$\frac{433}{1000}$ tons of ore, less the cost of reduction, added to the above sum of $8,510.68, will constitute the actual profit of the whole year's operations.

The average yield of the ore reduced is $21.43 per ton; the average cost of reduction, which includes labor, materials, and all incidental expenses, is $11.06, and the average cost of production is $9.95 per ton, leaving a profit of forty-two (42) cents per ton.

The first and second levels of the new mine were re-opened, and ore extracted from them, in the month of February, and they still continue to yield some ore. The fourth level, also re-opened, continues to yield a small quantity. The eighth level was exhausted in the month of December, and the ninth level in the month of May. The tenth, or lowest level, was opened in the mouth of May, and cross-cuts have been run across the vein, but up to this time no ore of value has been discovered on this level. The ore (1,622$\frac{900}{1000}$ tons) which is designated above, for the sake of convenience, as coming from the tenth level, was extracted from between this and the ninth level. Some ore yet remains in that locality, near the south line, but the quantity is very limited.

In the old mine, the sixth level has yielded the largest portion of the ore which has come from the old works. This level was re-opened in the mouth of December, and it continues to yield some ore.

The main shaft has been sunk 100 feet, to the tenth level, and that level has been opened to our southern boundary, with four cross-cuts made on that level at intervals of 100 feet, extending from the west wall of the lode across the vein, and penetrating the east wall from 50 to 100 feet. Over 4,000 feet of new drifts have been run and old drifts re-opened in the new and old mines, and considerable ore of moderate grade has been found and extracted. Two new pumps have been placed in the main shaft in addition to those which are already in use. Air-connections have been made with the Hale and Norcross mine on the various levels, rendering the levels cool and comfortable, so that the workmen are enabled to perform a greater amount of labor, and at the same time the mine-timbers are protected from the rapid decay which occurs where there is a want of ventilation. Much work has been done on the main shaft, which is now in a state of good repair, and will probably remain so for six months to come. The hoisting-machinery, the pumps, and the pump-engines, at both the old and new shafts, are in excellent order. The power of the machinery is sufficient to hoist from a depth 200 feet lower than our present lowest level.

In re-opening the various old levels, by timbering them in a solid manner, and by placing good car-tracks in them, we have been able to handle a large quantity of ore at comparatively a light cost, and have made up in quantity what the ore lacks in quality.

Considerable ore yet remains in the upper levels, which further exploration may disclose. At this time, however, we have no knowledge of any extensive body of ore in any part of the mine. On the whole, the mine may be said, so far as can be seen, to be exhausted of ore of a paying value. By sinking deeper new bodies of ore may be exposed, but until such bodies are found, profit from the mine cannot be looked for.

[Extract from the Secretary report.]

RECEIPTS.

Cash, balance on hand July 11, 1870	$9,445 25
Superintendent, balance on hand July 11, 1870	5,126 23
Bullion, bars received from mine	826,338 17
Bullion, grains and slag sold	338 15
Premium on bullion sold	5,012 48
Ores sold at mine	4,044 95
Real estate, rent of houses	94 00
Virginia and Truckee Railroad Company, return freights on account	13,616 40
	864,015 63

DISBURSEMENTS.

Taxes and stamps—Federal, State, county, and city	$7,049 06
Timber and lumber	48,768 71
Labor and salaries—foreman, miners, engineers, carpenters, laborers, company's officers	272,528 34
Exchange, on superintendent's drafts	2,863 30
Surveying in mine	590 00
Materials—mining supplies, hardware, candles, oils, &c., and insurance on hoisting-works	49,096 18
Fuel—wood and charcoal	45,472 42
Books and stationery for Virginia and San Francisco offices	478 41
Freight on mining materials to Virginia	2,565 70
Freight on bullion, per Wells, Fargo & Co.'s express	3,359 24
Assay office—net cost of company's office	5,206 56
Assaying, at outside offices	1,138 71
Interest on overdrafts at bank	883 42
Legal expense—attorney's fees, &c	6,195 39
Mill-materials and labor—total cost of reduction of ores at company's mills, including repairs and insurance	79,556 27
Reduction of ores—paid custom mill	304,976 33
Incidental—sundry extraordinary expenses	3,477 35
Horse-keeping—horses, vehicles, and food	1,740 34
Water—Virginia and Gold Hill water-works	6,972 50
Expense—office rent, porter, &c	2,421 59
Real estate—city lots at Virginia, recording	376 00
Cash, balance on hand this day	18,299 81
	864,015 63

ASSETS, JULY 10, 1871.

Cost value of stores on hand at mine	$16,875 43
Cost value of stores on hand at assay office	447 40
Cost value of stores on hand at Savage mill	3,039 65
Furniture of office building at Virginia and San Francisco office	3,500 00
Net value of 1,182 tons of ore on hand at mills and mine, (cost of reduction allowed)	6,063 63
Bills receivable for assaying	239 50
Cash on hand	18,299 81
Total cash assets	48,494 42

Report of the Hale and Norcross for the year ending March 1, 1871.

During the past year $67 per share, aggregating $536,000, has been paid in dividends to stockholders, being $46,000 in excess of the amount disbursed in 1866, heretofore the most prosperous year of this company. The president, in his report, sets forth a very strong argument for a reduction in the price of labor, and to which remarks the laboring men should pay particular attention, as it concerns as much their future prosperity as that of the companies. The following statement extracted will show in what departments a reduction has been made within the past four years :

	1867.	1870.	Reduction.
Cost of timber per thousand feet	$31 32	$21 32	31.92 per cent.
Cost of wood per cord	15 05	11 33	24.71 "
Cost of milling per ton	14 21	11 16½	21.42 "
Total	60 58	43 81½	27.67 "
Yield of ore per ton	$34 14	$25 13	26.10 "
Labor per day	4 00	4 00	

Here we have the cost of the principal items in mining for the years 1867 and 1870. While the items of timber, wood, and milling show an average reduction of 27.67 per cent., and the value of ore shows a decline of 26 $\frac{1}{10}$ per cent., labor still maintains its high standard of valuation.

In his closing remarks the president says:

"In all mining communities where employment depends on the development of hidden wealth, there must of necessity be large numbers who have nothing to do, and who can be only a charge and expense upon their more fortunate co-laborers; but when circumstances such as now exist in the large bodies of low-grade ore already developed on the Comstock lode have placed it in the power of all to be employed at prices from 15 to 20 per cent. less than present rates, how much better is it for labor to conform to the necessities of the case, than to insist on prices of ten and eleven years ago, thereby paralyzing work, and, in order to maintain its position, compelled to contribute a large percentage of its earnings toward the support of the unemployed."

[Extract from the Secretary's report.]

RECEIPTS.

Cash on hand February 28, 1870	$118,051 57
Return freights Virginia and Truckee Railroad	11,326 01
Premiums on bullion	15,777 38
Proceeds of 64,974 tons ore	1,627,961 13
Other sources	30,777 76
Total	1,803,983 86

DISBURSEMENTS.

Labor on contracts	$278,063 59
Wood, 4,827¼ cords	54,535 12
Timber and lumber	49,006 86
Hardware, candles, powder, &c	45,076 07
Working ore	725,787 21
Salary assayer and assistant, &c	7,224 13
Dividends to stockholders	536,000 00
Sundry accounts	50,546 77
Cash on hand	56,654 11
Total	1,803,893 86

During the past year 64,974 tons of ore have been reduced, yielding $1,632,844.38, and there remain on hand 4,208¼ tons, valued per assay at $145,124.20, the cost of mining which is paid. During this period the shaft has been sunk from the sixth to the eighth (1,300-foot) level, and retimbered for a distance of 309 feet. The old mine produced 18,386 tons of ore; 5,633¼ tons of ore have been extracted from the eighth or lowest level, and there yet remains standing, in the stopes of this level, a quantity sufficiently extensive to require fully eight months for its extraction. The ore-body on this level is now opened 313 feet in length, and has for this distance an average width of 24 feet. The southernmost workings in this ore-body have not as yet reached the terminus of the ore. Although this level is but partially opened, the developments thus far made expose an ore-body greater in length and width than any before shown in the mine. The hard character of the ore—all of it requiring to be blasted—gives every reason to believe that the present good ore will continue to a much greater depth than is yet reached. The prospects of the mine are now more brilliant than at any previous date.

Report of the Hale and Norcross for the year ending March 1, 1872.

[Extract from the Secretary's report.]

RECEIPTS.

From bullion	$862,702
From assessments	200,000
February receipts	37,360
Miscellaneous	24,612
Total receipts	1,124,674
Cash, March 1, 1871	56,654
	1,181,328

DISBURSEMENTS.

Milling 49,625 tons of ore	$539,905
Mine account, labor and supplies	363,148
Bills payable	100,274
Dividends to stockholders	80,000
General expenses	35,253
Machinery account	22,207
Freight account	13,916
Assaying	6,132
Taxes	5,947
Miscellaneous	8,018
	1,174,800
Cash, March 1, 1872	6,528
	1,181,328

The superintendent's report embraces all the operations in the mine during the past year:

During the past year 48,571½ tons of ore have been extracted from the various levels of the mine, and 49,625 tons and 495 pounds have been reduced. The entire yield of bullion therefrom was $862,701.36, and there remain on hand, in the several ore-houses, 3,154 tons and 1,860 pounds of ore, of an assay valuation of $84,072.09. Within this period the main shaft has been retimbered in the most substantial manner from the eighth (or 1,300-foot) level to the surface.

This work has been attended with great expense, and has also retarded the exploration of the mine below the eighth level for at least one-third of the year.

The shaft is now in good repair, as are also the various drifts, adits, and winzes throughout the mine, and the chief air-passage and ladder-way connecting the upper and lower mines is in perfect condition, and of sufficient capacity to render all the levels cool and pleasant.

At a perpendicular depth of 1,234 feet, or 66 feet above the eighth level, the main shaft intersects the hard west wall of the ore-vein.

We have, therefore, been compelled to continue our operations at greater depths by means of an incline.

This incline is now sunk to a distance of 290 feet below the eighth (or lowest) level. It descends at an angle of 39 degrees from the horizon, and is sunk in the hard west wall of the vein.

It has evidently passed below the water-line, as the rock is extremely hard, requiring continuous blasting, and is entirely devoid of moisture.

It is strongly timbered with 14 by 14-inch timbers, three feet apart, and is divided into two compartments, each 7 feet in height and 6 feet in width in the clear. One of these compartments is used for the pumps and column and air-pipes, and for a ladder-way. The other is reserved solely for the passage of the large incline-car, which is designed to do all the work of hoisting from the openings below the eighth level up to the bins and chutes at that level. These receptacles are prepared for and receiving and transferring the contents of the incline-car into the smaller-cars provided for conveying the rock up to the surface through the vertical shaft. This incline-car is so constructed as to be self-adjusting when dumped, and is quadruple the capacity of the cars in use in the vertical shaft. It is worked by the powerful engine on the surface recently erected, the rope operating it passing through the pump-shaft, thus retaining the full use of the other two compartments of the shaft for hoisting from the ninth level, and from the various levels above the point of commencement of the incline.

At a perpendicular depth of 115 feet below the eighth level we are at present engaged in excavating for our ninth-level station; but we have not as yet reached the ore-veins.

Within the past six months we have expended a large amount of money in the purchase and erection of machinery suited to the incline method of working the mine.

We have erected for this purpose one engine of 22-inch diameter of cylinder, and 30-inch stroke, with reels and connections complete, and two boilers, 56 inches in diameter and 16 feet long, all of which are of the latest improved patterns.

With these additions our hoisting power is adequate to work the mine to a depth of 2,500 feet, and our hoisting capacity will not be diminished until that depth shall have been attained, as all of our regular machinery will be employed in raising the ore and waste rock from the several levels opening from the shaft, and the new and powerful improvements will be reserved for the especial service of the incline and contiguous workings below the termination of the vertical shaft.

In the workshops appertaining to the mine our mechanical facilities have been greatly augmented.

We now have in operation a lathe 21 feet by 30 inches swing, driven by an engine of 13-inch diameter cylinder, and 36-inch stroke. This engine was especially secured for the actuation of this machine, and for working the cross-cut, edging, and wedge saws, which we have also just erected.

The acquisition of these important tools will enable us to materially reduce the accessory expense of the mine.

The ores extracted during the latter half of the year have been unavoidably of a low grade, and, owing to the interruption of operation at the bottom of the mine, (pending the retimbering of the shaft,) were produced from the upper levels and from the section known as the old or upper mine. Much of this quality of ore yet remains in these localities, but as it yields but little profit our attention will be directed to new deposits at greater depths, so soon as stations can be opened from the incline now advancing downwards.

The machinery of both mines is in perfect order.

Report of the Chollar-Potosi for the year ending May 31, 1871.

[Extract from the Superintendent's report.]

The amount of ore raised during the year was 84,681 tons, from the following sections of the mine:

	Tons.
Belvidere Station	68,856
Blue Wing	10,971
New Tunnel	4,854
Total	84,681

The quantity raised for the previous year was 56,636 tons, of which 26,000 tons were from the Blue Wing, 12,000 tons from the Tunnel, 8,000 from the Belvidere, and 10,000 from Grass Valley and croppings.

The amount of ore milled during the past year was 83,775 tons, distributed throughout the year as follows:

1870-'71.	Tons worked.	Bullion.
June	4,470	$215,548
July	5,954	261,277
August	5,535	249,553
September	6,200	264,557
October	6,467	275,715
November	6,309	258,297
December	11,880	565,654
January	6,779	276,606
February	6,048	241,967
March	8,070	301,722
April	8,034	273,160
May	8,029	319,967
Totals	83,775	3,444,023

The average yield of the ore, and the cost of milling and mining, compare as follows with the previous fiscal year:

	1869-'70.	1870-'71.
Average yield of ore per ton	$24 86	$41 30
Cost of milling	12 81	12 00
Cost of mining	3 99	4 69
Profit	8 06	24 61

The receipts for the year amounted to nearly $3,476,000, the principal items being the following:

From bullion	$3,444,023
Reclamation from mills, (less $117 paid)	16,283
Premium on bullion	4,619
Return freight, Virginia and Truckee Railroad	4,715
Miscellaneous	6,295
Total receipts	3,475,935
Cash on hand July 1, 1870	128,253
Total	3,604,188

During the previous fiscal year the receipts amounted to $1,522,277.

The disbursements for the year aggregated $3,403,467, against $1,563,015 for the previous year. The leading items in the disbursement account for the year just closed are the following:

Dividends paid to stockholders	$1,946,637
Working ores	1,005,300
Labor account	234,496
Timber and lumber	75,303
General expenses	29,823
Taxes	21,971
Freight	17,162
Hardware	11,718
Miscellaneous	61,057
Total expenditure	3,403,467
Cash on hand July 1, 1871	200,721
Total	3,604,188

The business for the past year shows an increase of over 100 per cent. as against the previous year. The dividends paid to stockholders were unusually remunerative.

LIST OF MONTHLY DISBURSEMENTS.

Paid.	Per share.	Amount.
July 21, 1870	$1 00	$28,000
August 10	2 00	56,000
September 10	2 50	70,000
October 10	3 00	84,000
November 10	4 00	112,000
December 10	5 00	140,000
January 10, 1871	5 00	140,000
January 16	5 00	140,000
February 10	5 00	140,000
February 15	5 00	140,000
March 10	5 00	140,000
March 15	5 00	140,000
April 7	5 00	140,000
April 14	5 00	140,000
May 10	5 00	140,000
May 20	5 00	140,000
June 10	2 00	55,000
Totals	68 50	1,946,000
Balance on previous dividend		637
Total		1,946,637

The above is the most flattering exhibit of dividends for a single year in the history of the company. In July, 1870, the market value of the above stock was from $30 to $37 per share. Parties who then bought and who have retained their stock, have not only received double what they paid in dividends, but now can realize a very handsome advance on the original price.

Report of the Yellow Jacket for the year ending July 1, 1871.

[Extract from the President's report.]

The ore mined and milled during the past year was 59,875 tons, and came principally from the 900 and 1,000-foot levels; and there still remain in place, on the same levels, many tons of ore of a similar quality—enough, I am in hopes, at the present rate of extraction, to last until we develop some other body on the new levels we are now employed in opening.

During the year there have been upward of 3,800 feet of working and prospecting drift run, and 1,080 feet of slope and perpendicular winzes raised and sunk. The main shaft has been sunk, full size, from the 1,000 to the 1,100-foot level, and the pump-apartment opened between the 900 and 1,000-foot levels. We have also started an incline or continued main shaft, on an angle of 45 degrees, to the east, from a point 30 feet above the 1,100-foot level, which has been sunk to a depth of 30 feet below the 1,100-foot level. We were obliged to start the shaft on this slope, as it had penetrated the west country-rock at the 1,100 foot level.

The main working drift, 1,100-foot level, has reached a point 400 feet north of the shaft. There has been but one cross-cut run on this level, one east of the shaft, at which point we find flattering indications of the existence of an ore-body below.

By the middle of September, or before, I am in hopes that we will have the 1,200-foot level opened.

During the year there have been extensive alterations and improvements made in the hoisting and pumping machinery, substituting steel-wire ropes and gearing for the manilla ropes and friction-reels, putting in place, in running order, a new 18 by 27 inch hoisting-engine for working the new incline-car below the 1,100-foot level. We have also changed the three 12-inch pumps in the shaft to 8-inch ones, and also put in new pump-rods and column above the 1,100-foot level.

The hoisting and pumping machinery is now in first-class order, and capable of working the mine to a greater depth, with very little expense outside of necessary wear and tear.

[Extract from the Secretary's report.]

RECEIPTS.

Bullion account :

Proceeds of 59,875 tons of ore	$1,863,126 06
Proceeds of 1,955 tons tailings	18,197 75
Total	1,881,323 81

Virginia and Truckee Railroad :

Amount returned in freight	7,616 23

Sundries :

Sale mill-machinery, rope, &c	51,041 95
Amount for advertising	35 40
Balance on hand July 1, 1870	156,043 36
Total	2,096,062 45

DISBURSEMENTS.

For labor	$325,501 00
Mine supplies	74,545 42
Improvements	35,534 97
Tramway	3,905 15
Legal expenses	4,000 00
Expenses	5,391 74
Candles and oil	11,037 46
Powder and fuse	3,056 46
Wood	32,013 65

Iron and steel..	$5,097 63
Timber	69,466 73
Crushing	726,680 17
Assay, and discount on bars	28,303 80
Taxes on real estate...	3,187 87
Interest	2,354 92
Salaries	13,600 00
Assay office	3,926 39
Working tailings...	8,972 09
Total..	1,359,575 29

DIVIDENDS.

No. 18 to 23, inclusive	324,000 00
Amount paid on back dividends..	160 39
Cash on hand July 1, 1871..	379,363 91
Stock on hand, such as lumber and mining supplies	35,962 87
Total	2,096,062 45

(Total expenses for each ton extracted from the mine, $22.65.)

RECAPITULATION.

59,875 tons ore worked, yield $31.11 per ton................	$1,863,126 06	
1,955 tons tailings worked, yield $9.30 per ton	18,197 75	
		$1,881,323 81
Received from sale of mill-machinery, &c	51,077 05	
Received from Virginia and Truckee Railroad, advanced freight...	7,616 23	
		58,793 28
Total receipts..		1,940,017 09
Expenses of company..	1,356,575 29	
Profit for year ending July 1, 1871	583,441 80	
		1,940,017 09

Report of the Kentuck for the year ending November 1, 1871.

[Extract from Secretary's report.]

RECEIPTS.

Cash on hand November 1, 1870...	$6,144 80
Assessments Nos. 4 and 5	40,000 00
Bullion, premiums, &c..	119,683 35
Total	165,828 15

DISBURSEMENTS.

Real estate..	$271 19
Freight on treasure...	575 50
Interest...	7 92
Mine account, supplies, labor, assaying, transportation, milling, exchange, &c..	140,673 51
Gold Hill office, including salary of superintendent and clerk, rent, &c...	3,938 28
San Francisco salaries, office expenses, &c..............................	4,891 64
Cash on hand November 1, 1871	15,470 11
	165,828 15

Report of the Crown Point for the year ending May 1, 1871.

[Extract from Secretary's report.]

RECEIPTS.

Amount of bullion produced	$472,121 48
Premium on the same	474 92
Assessment No. 20	36,000 00

Assessment No. 21.. $42,000 00
Rhode Island mill.. 11,034 63
Sundries 11,777 19

Total receipts 573,468 22
Cash in hands of superintendent, May 1, 1870.......................... 1,735 48
Cash in treasury, San Francisco, May 1, 1870 .... 37,573 60

Total ....•.............. 612,777 30

DISBURSEMENTS.

Crown Point mine, labor and supplies $255,191 62
Rhode Island mill, supplies... 31,252 16
Rhode Island mill, improvements 3,565 32
Crushing 18,904 tons ore..·... 191,149 93
Legal expenses.. 1,450 00
General expenses.. 13,905 68
San Francisco expenses.. 9,802 48
Taxes 4,008 27
Assaying.. 3,664 00
Freight on bullion.. 1,913 02
Miscellaneous items 2,362 34

Total disbursements... 518,174 83
Cash on hand May 1, 1871 .... 94,602 47

Total 612,777 30

The quantity and average yield of ore worked are:

Mills.	Ore in tons.	Average.	Bullion.
Rhode Island	2,204	$34 34	$75,089 68
Brunswick.................................	13,773	17 41	230,802 91
Petaluma	650	9 90	6,438 87
Pioneer...................................	2,210	31 48	69,602 20
Pacific...................................	1,850	35 31	65,333 24
Kersey	420	37 75	15,855 28
	21,087	22 39	472,121 18

The average yield of the 5,392 tons of ore worked during the previous fiscal year was $14.12, and the cost of working the same $11.12. During the past year, 21,087 tons were worked, at a cost of $10.05, and showing an average yield of $22.39 per ton. Following is a statement of the assets of the company on May 1, 1871:

ASSETS.

Cash on hand May 1, 1871... $94,602 47
Rhode Island Mill.. 60,000 00
Mine and improvements... 80,000 00
Stock at mine... 13,053 05
Stock at mill... 6,961 40
Ore (622 tons) at mills... 19,195 00

Total assets.......,.. 273,811 82

The company have no liabilities. The mine is in a satisfactory condition. The new board declared a dividend of $10 per share, amounting to $120,000, carrying over a surplus equal to a dividend of the same amount. This is the largest dividend ever paid by the company, and the first since September 12, 1868.

Report of the Belcher for the eleven months from February 1, 1871, to January 1, 1872.

RECEIPTS.

From bullion	$1,199,135
Assessments Nos. 7 and 8	51,925
Virginia and Truckee Railroad	1,036
Total receipts	1,252,096
Cash, January 31, 1871	624
Total	1,252,720

DISBURSEMENTS.

Labor	$137,103
Crushing 18,468 tons ore	221,621
Hoisting 18,468 tons ore	19,203
Machinery account	11,000
Miscellaneous	150,847
Total disbursements	539,775
Cash, January 1, 1872	712,945
Total	1,252,720

The ore yield of this mine for the past year was all due to the last five months of the year. The quantities hoisted in each month, together with average and aggregate values, were as follows:

Month.	Tons.	Average.	Value.
August	680	$74 90	$50,936
September	2,009	80 19	161,103
October	4,200	74 52	313,041
November	5,717	60 40	345,419
December	5,853	55 83	328,036
Totals	18,468	64 26	1,109,135

The assay value of the bullion was 54 per cent. gold and 46 per cent. silver. From the balance of $71,945 held on the 1st of January, the company paid a dividend of $10 per share, aggregating $104,000.

For further ables, including dividends, assessments, rates of stock, &c., see the appendix to this report.

LANDER COUNTY.

The product o. this county has been notably increased over that of the preceding yer. This is principally due to the largely extended mining and smeltingoperations at Eureka, and to a resumption, to a large extent, of the acivity of several years ago in the Reese River district.

In *Reese River 'istrict* the refitting of two mills, the Mettacom and the Citizens', ad the introduction of roasting-furnaces and dry-crushing machiney, have exerted a favorable influence on a great number of mines whia had not been worked to any extent for years, although only one c these, the Citizens' Mill, does custom work. It was finished, I am iformed, in July. The roasting-furnaces used are White's rotary cyliders, a furnace very similar to the Brückner cylinder in operation in Corrado. The Mettacom Mill has been put in working order, and also fittd with a White's furnace, by the Pacific Mining Company, an English corporation. The operations of this company, which is only second in importance to the Manhattan Company, are best shown by the reports of Captain H. Prideaux, the superintendent, and Mr. J. Howell, in charge of the mill, rendered in January, 1872, to the board of directors of the company.

THE PACIFIC MINING COMPANY,
Austin, Nevada, January 20, 1872.

To the Board of Directors of the Pacific Mining Company :

GENTLEMEN : I beg to hand to you the following as my report: Since I have had charge of the Pacific Company's mines at Lander Hill our workings have been as follows : Levels driven on the course of the vein, 58 fathoms 2 feet; cross-cuts extended, 55 fathoms 3 feet ; rises on the course of the vein, 53 fathoms 2 feet ; rise not on the vein, 13 fathoms 2 feet ; winzes sunk, 36 fathoms ; stopes, 141 square fathoms on day work, 135 fathoms by tributers—making a total of 113 fathoms 5 feet of levels driven, 66 fathoms 4 feet of rises, 35 fathoms of winzes sunk, and 276 square fathoms of ground stoped. We have also shipped from the mine 356 tons 872 pounds of ore, 54 tons 375 pounds of which were milled by the Manhattan Company, and produced fifteen bars of silver bullion, value $9,653.19. The balance of the ore has been shipped to the Company's Mettacom Mill, and is now being treated. The silver bullion produced at the Mettacom Mill up to 4th instant was twenty-four bars, value $22,280.03. Besides this there is now on hand about $5,000 worth of silver, which will shortly be melted into bars. The estimated quantity of ore at the mine, on surface, is 100 tons. I arrived at the company's mines March 29, 1871 ; it took me until May to repair the machinery and shaft, so the actual work commenced in the mine in May, 1871.

Cross-cuts.—The 550-foot cross-cut is being driven north to intersect the Buel North Star ledge 150 feet below our present working on this ledge ; we are, however, shortly expecting to intersect it, after which this cross-cut will be suspended and levels extended on the ledge. Our last working on this ledge, which is 150 feet above the cross-cut, shows a well-defined vein of good ore from 1 foot to 2 feet in width. By cutting the ledge in this cross-cut we shall lay open a piece of new ground 150 feet high for the whole length of the mine. I have no doubt but that we shall cut a good ledge at this point. The north cross-cut at the 400-foot level is being driven with all speed. We expect to cut another ledge in this cross-cut in about six weeks. We have so far intersected two ledges in this cross-cut, one of which (the Datter's ledge) has and is producing good ore. The next ledge which we expect to cut here is considered one of the best ledges on Lander Hill. From the company's Lane and Fuller shaft we are driving a cross-cut north to cut a good edge, which is, I judge, within 50 feet of us. I am certain that this ledge, on cutting it, will reward us well for the outlay. After we have intersected the ledge, this cross-cut will be suspended. The south cross-cut at the 400-foot level: this cross-cut has been idle for some months, but we intend to again commence driving it in about two months' time. This cross-cut will intersect the ledges south of us, and it is necessary that it should be continued.

Stopes.—The ledge in the stopes east of sump winze will average 1 foot wide of good ore. These stopes will yield a larger quantity of ore when required. The stopes west of sump winze have scarcely been worked, owing to our having a supply of ore on hand, and the necessity of conveying air to the western part of the mine. We have here a piece of ore-ground, ready to be stoped, 100 feet by 130 feet.

We are extending our 500-foot level on the course of the vein, an sinking a winze from the 400-foot level west ; by so doing we shall soon have another piece of ore-ground ready for stoping, 100 feet square. The stopes above the 400-foot level are at present worked on a small scale ; the greater part of these stops will soon be let on tribute. The different stopes on tribute are producing ore more or less, some of which is very rich. The estimated reserves of ore in the mines at 500 tons ; as soon as we cut the different ledges in the cross-cuts this quantity will be greatly increased. Altogether, our prospects are exceedingly encouraging, and the mine is in a good working condition. In two of the cross-cuts we shall shortly intersect the ledges we are driving for, which will, I have no doubt, prove rich ones. We have a rich ledge at the 400-foot west level, and are making the necessary preparations to commence stoping it. There have been added to the company's property for ledges, which have yielded good ore ; we shall shortly add two more, which I think will exceed in value any of the others. In my former letters I have advised the removal of the Mettacom Mill. All I have to add is that the removal of the Mettacom Mills absolutely necessary. For more particulars on this head please notice Mr. Howe's report.

Waiting your further instructions, I remain, gentlemen, your obedient servant,

HENR PRIDEAUX,
Superintendent.

CONDITION OF MINING INDUSTRY—NEVADA.

169

AUSTIN, *January* 20, 1872.

To the Directors of the Pacific Mining Company, limited:

GENTLEMEN: In accordance with your request, I beg to hand you the following report on the Mettacom Mill. This mill was built in the year 1866, and was run about fifteen months in all, previous to its passing into the present owner's hands. The structure is principally of wood and stone, and cost originally about $7,500. It is divided into four distinct compartments, which are designated as engine-room, battery-room, amalgamating-room, and furnace-room. The motive power consists of two tubular boilers, 16 feet long, 44 inches in diameter, and 14-inch cylinder high-pressure engine, with 30-inch stroke, and nominally rated at 70 horse-power. The battery-room has two five-stamp batteries, each stamp weighing about 900 pounds. They are run at the rate of ninety drops per minute, and their crushing capacity is about 12 tons in twenty-four hours, passing the ore through a No. 40 screen, or wire cloth, with 1,600 holes or meshes to the square inch. From the battery-room the pulverized ore passes through elevators into the furnace-room, and is there desulphurized and chloridized in a "White's cylinder revolving furnace." This furnace is a cast-iron tube, 24 feet long, and 30 inches diameter, lined inside with fire-brick, and driven at the rate of eleven revolutions per minute. Its roasting or chloridizing capacity is about 15 tons in twenty-four hours. From the furnace-room the ore passes into the amalgamating-room, which is furnished with six amalgamating-barrels, of a capacity each of 1 ton in twenty-four hours. In addition to the mill there is on the premises a wood building, 14 feet by 40 feet, used as a boarding and lodging house for the workmen; also a blacksmith's shop, retort, and melting-room; also a brick building, used as mill-office and assaying-room, together with quantities of tools and various kinds of personal property, of which I sent you a complete inventory a short time ago. When the mill was built it was furnished with four reverberatory furnaces, which were at the time considered sufficient to chloridize ore from a ten-stamp battery, but after starting the mill it was soon found that the capacity of the battery was far beyond that of the furnaces, and while we could crush on an average 12 tons in twenty-four hours, we could not chloridize more than 7 tons, but since the mill has been supplied with this new furnace, it places the amalgamating capacity but one-half that of the balance of the mill—this, however, I explained fully to Mr. Sewell last fall, when I took charge of the mill—consequently we are able to work the mill at present to but half its capacity, and, at the same time, are compelled to work the engine day and night. This is, of course, working to a great disadvantage. To increase the amalgamating capacity of the Mettacom would incur an expense of at least $5,000. The building would have to be extended 26 feet, and furnished with six new amalgamating-barrels and three settlers, with all other necessary articles. To increase the amalgamating department of the Mettacom Mill where it now stands would, I consider, be an injudicious outlay of money, and I would here like to make a few remarks on the subject, although I am satisfied the matter has been fully and clearly laid before you in Mr. Prideaux's letters. The Mettacom Mill is situated about five miles north of the city of Austin and your Lander Hill mines; and in reality the same distance from the principal ore-producing mines of the district; consequently the hauling of custom ores, as well as your own, is quite an important item, as it has cost $5 per ton to deliver all ores reduced at the Mettacom Mill for the last two months; still the delivery of ore is not so important an item as the delivery of wood, which costs at the Mettacom $2 per cord in the most favorable season for hauling, and from 4 to 5 per cent. in winter more than to deliver in Austin. Besides these, there are many other smaller disadvantages. I would therefore recommend the removal of the Mettacom Mill machinery into your Empire building at Austin. This Empire building is a substantial brick and stone structure, in fact, the finest in this place, although it is a conceded fact here that the Mettacom is the best ten-stamp mill in the State. This is due, however, only to the superiority of its machinery, and this same machinery can be taken from the Mettacom and put into the Empire building in as good shape as it is at present, and in a very short time, at an outlay of money not to exceed $15,000, which would include the necessary increased amalgamating capacity. This done, you would have a mill second to none on the Pacific coast, and in which you could reduce your own ores at a low figure, and make a good profit on all custom ores. I think it safe to say that, with your mill in Austin, $30,000 per month in bullion could be produced. Taking into consideration the cost of increasing the amalgamating capacity of the Mettacom where it now stands, which will have to be done before the mill can be worked profitably, and the cost of moving the machinery into the Empire building, there is really but $10,000 difference, and the difference in favor of the Empire in the hauling of wood, ore, and salt would be at least $55 per day. In erecting the cylinder furnace in the Mettacom mill I made many improvements on the one first built here, and after three months' careful attention to the working of the furnace in the Mettacom on various kinds of difficult ores to treat, I have noted a further improvement that can be made, but as it is the furnace is a complete success in every respect. In conclusion, I would state that with your mill in Austin you would have one of the finest milling and mining enterprises in this section of the country.

JOHN HOWELL.

According to the latest information I have obtained, the board of directors have not approved of the plan for the removal of the mill to Austin, but it is the intention to increase the amalgamating capacity of the works and to do custom work, besides working the ore from their own mines.

The principal operations in this district have been those of the Manhattan Company, which has worked in its mill not only the ores of its own mines, but a great deal of custom ore from other Reese River mines and from Lander Hill. A most important purchase of mining property was made in July by this company, putting into the hands of the Manhattan almost the entire control of Lander Hill. The property bought was that of the Reese River Consolidated Company, comprising about forty locations, some of which (e. g., Whitlatch Union, Whitlatch Yankee Blade, Savage, Wall, Isabella, Camargo, &c.) have been extensively and productively worked. The Reese River Consolidated Company was in some way entangled in the affairs of the First National Bank of Nevada, which failed, and the mines were sold by the sheriff, for $60,000, to the Manhattan Company. The following is the list of mines enumerated in the deed: The Apollo Ledge; Blue Ledge; Black Ledge; Camargo; Congress Independent; Jo Lane; Eclipse; Whitlatch Yankee Blade; Wall and Isabella; Beard and Seaver, original location; Beard and Seaver, (both on Union Hill;) Hornet; Erie Ledge, two locations, 1,200 feet each; Chicago; Harker; Honest Miner; Union No. 2, first southerly extension; Union No. 2, first northerly extension; Yosemite; Silver Cloud; Governor Seymour; Isabella; Wall Ledge; Monitor Ledge; Nevada Ledge; Peerless Ledge; North Star, second extension west; Gale & Beckwith Company; Yosemite Ledge; Jefferson Ledge, Pleasant Company, first southeast location; Jefferson Ledge, Madison Company, first location northwest; Jefferson Ledge; Madison Ledge; Sally Davis Ledge; Oregon Ledge, Wall Company, first westerly extension; Oregon Ledge, second westerly extension; Southern Light Ledge, first westerly extension; Diana Ledge; Eclipse Ledge, first northerly extension; Pride of the East Ledge; Savage Ledge; Consolidated Union Tunnel.

This purchase assures the future of the company, by giving it undisputed titles (in many cases United States patents) to a large number of locations, covering the best part of Lander and Union Hills. The narrowness of the Lander Hill lodes and their perplexing "faults" and "slides" necessitate, on the one hand, the opening of much ground to keep up the production of ore, and, on the other hand, have led to perpetual conflicts of proprietors, arising from encroachments upon a neighbor's ground in the prosecution of energetic drifting and stoping. It has long been foreseen that only a consolidation of ownerships could permit the efficient and economical exploitation of the undoubtedly rich ore-bodies of this part of the district. This consolidation is now measurably complete. A few mines, such as the Buel North Star, of the Pacific Company,) the Florida, Troy, and Plymonth (?), are in other hands, but the Manhattan Company is the owner of a large surface of area, and can conduct its operations henceforward with increased security, economy, and success.

The exclusive right to use the Stetefeldt furnace in this district has given to the Manhattan Mill an advantage of which no opposition has been able to deprive it.

The operations of the Manhattan Mill during 1871 were as follows: Of the ore from the company's mines there were reduced 1,537 tons, yielding $387,580, an average of $252.16 per ton; of custom ore from Belmont and Reese River, 3,513 tons, yielding $596,043, an average of $169.66. Total ore treated, 5,050 tons, yielding $983,623, an average of $194.77 per ton. The company's ore was mined mostly on contract, involving a percentage of ore to the miners, besides a certain sum in wages. The estimated cost of extraction was $150,581, or over $98 per ton; but this is largely made up of drifting rather than stoping. The cost of taking out ore from the narrow veins of Lander Hill is large, but not so large as those figures would indicate. They properly include operations which have opened up much new ground. The additional cost of dead work during the year was $91,109, and the cost of milling the 1,537 tons alluded to was $41,416, or $26.95 per ton. Total cost, $283,106; profit on Manhattan ores, $104,474. The cost of milling 3,513 tons of custom ore was $88,376, or $25.16 per ton; the profit (about 10 per cent. on the amount paid to customers in returns or in purchase of the ore outright) was $49,060. The aggregate statement of earnings, as taken from the company's books, is as follows:

From company's ores, 1,537 tons	$104,473 61	
From custom ores, 3,513 tons	49,059 97	
Mineral Hill Milling Company, profit	6,085 14	
American Mining Company, proceeds of machinery	4,596 25	
Appreciation of currency on hand December 31, 1870	34 66	
		$164,249 63

From this are to be deducted the following items:

Bullion freight, reclamations, &c	$19,838 29	
Exchange account	909 14	
Mill repairs	1,652 53	
Taxes	4,633 21	
Interest account	3,125 51	
Fire insurance	2,507 85	
Expense account	15,298 79	
Bullion stolen	3,730 06	
Sundry losses	790 82	
		52,486 20
Leaving as net earnings, (coin)		111,763 43

A dividend of 5 per cent. in coin, amounting to $19,375, was declared in February, 1871, and a similar dividend was declared in January, 1872. The sum of $70,419.38 was spent during the year in mill improvements, $31,338.33 in the acquisition of mines, real estate, &c., and the erection of buildings. This sum, added to the dividend of February, 1871, makes $89,794.38, or $21,969.05 less than the earnings of the year. The surplus of December 31, 1871, was $104,982.07, and this was, therefore, increased by the sum mentioned, so that the company entered upon 1872 with a surplus on hand, in supplies, bullion, and coin, of—

Supplies	$82,785 79
Bullion	45,496 64
Coin	47,832 59
	176,115 02
Less indebtedness to individuals	49,163 90
Net surplus, (coin value)	126,951 12

Eureka district has witnessed extraordinary developments during the year, and stands now third in rank of the silver-producing camps of

Nevada. A general description of its resources and works has been given in my preceding report. On this occasion I shall, therefore, only trace the general progress.

The most important company operating in Eureka district is still the Eureka Consolidated. During most of the year four and sometimes five furnaces have been in blast. The newest of these furnaces are built according to original plans by Mr. A. Arents, the metallurgist of the company, and are essentially a combination of the Rachette and Piltz furnaces. Their construction and mode of operation are described at length in another part of this report, in the chapter on Metallurgy. The company's operations during the year, and the results of mining and smelting, as well as the prospects on October 1, 1871, are very clearly set forth in the report of the superintendent, Mr. W. S. Keyes, M. E., and that of the secretary of the company, which I reproduce here in full :

In the middle of January last the present superintendent assumed general charge of the mines and works of the company. At that time the works consisted of three small furnaces, entirely inadequate to the true capacity of the mines. The ground then owned by the company consisted of the Buckeye, 600 feet; the Kohinoor, 600 feet; the Mammoth, 400 feet; the Savage, 600 feet; the Champion, 400 feet; the Sentinel, 600 feet, and the Roseland, 600 feet. The Buckeye and the Kohinoor had been patented under one application, likewise the Champion, and the remainder were held in accordance with the local mining laws.

Work was prosecuted in but three places, viz, in the old Champion, which was nearly worked out; in the south and middle shafts of the Buckeye, in all of which not more than two months' supplies of ore were in sight. The superintendent immediately instituted vigorous prospectings, and was fortunate in proving the continuance of the old bodies and in discovering new, particularly in the ground purchased during the present administration.

The claims now owned by the company are : 1. Buckeye (and Kohinoor ;) 2. Savage; 3. Nugget; 4. Sentinel (and Roseland ;) 5. Champion ; 6. At Last ; 7. Margaret ; 8. Lookout ; 9. Triangle ; 10. Elliptic ; 11. Lupita; 12. Mammoth—in all, fourteen original locations.

Certificates of patent, which in case of controvery may be used in lieu of a patent, have been received for the Buckeye (and Kohinoor,) Savage, Sentinel (and Roseland,) Champion, Lookout, and Mammoth, viz, for eight original locations. The Elliptic application is now in course of publication, and will be concluded in October.

Patents of the United States will soon be forwarded from Washington, the Department having recently sent on for proofs of citizenship of the company's officers, which have been returned thither by your superintendent. The claims are held and patented both as ledges and square locations.

The extreme length of all the claims from the northwest boundary of the Lookout to the southeast point of the Triangle is 1,680 feet, the mean width over 800 feet, thus embracing very nearly the entire surface of the hill.

To a better understanding of the company's properties, a word or two in reference to the formation may not be out of place. Ruby Hill, whereon all the claims are situated, is an isolated spur from the main range reaching down toward the valley. The underlay or foot-wall of the ledge is a hard quartzite. On this we find a clean clay selvage of from two to four feet in thickness. Immediately upon the clay is found the lowest quartzy stratum of ore, rich in gold, moderate in silver, and almost barren of lead. Above this we find the true vein-matter, 300 to 350 feet in thickness, consisting of decomposed and compact oxides of iron, in and through which occur the true ore-bearing strata, varying in size from a few inches up to 35 and 40 feet. These, of which we have certainly eight, although frequently thinning out, have never in a single instance disappeared, thus furnishing a clew to follow which may at any time open out into a paying body. Superimposed upon the vein-matter we find a capping of altered lime-rock, thin on the southwest slope of the hill, and quite thick on the opposite side. Piercing the crust of lime-rock, the irony vein-matter has in every instance been reached, the Lookout shaft alone excepted, where the overlay seems to be exceptionally thick. The vein is of the character known as a "bed-vein ;" it may also be viewed as a regular contact-vein, having a limestone hanging-wall and a quartzite foot. The underlay dips to the southeast, at an angle of 38 to 40 degrees on the southern end of the claims, and 80 degrees near the center of the hill. The latter number approximates the dip of the limestone on the northeast slope of the hill, and probably represents the true dip of the vein itself. All developments tend to establish the correctness of this hypothesis.

The general course or direction of the vein is north 52 degrees west. In a word, the mines of the Eureka Consolidated Mining Company are not limestone deposits, and bear no resemblance to such. This character of vein is not uncommon either in Europe, Mexico, or South America.

The mines have been opened by ten principal shafts and inclines, one tunnel with side drift and numerous prospect holes: 1. Lookout shaft; 2. Champion incline: 3. Savage incline; 4. Windsail shaft; 5. Quartzite shaft, on Savage, near the road; 6. North shaft of Buckeye; 7. Middle shaft of Buckeye; 8. South shaft of Buckeye; 9. Keyes or main shaft on Buckeye; and 10. Sentinel shaft.

The various shafts and inclines have reached depths of 75 to 170 feet perpendicular. The exact figures, with the relative depths below the crest of the ridge, may be seen on the maps. Ore is at present being extracted from the Champion and Savage inclines, (all of which will soon be hoisted through the Windsail shaft, with which a connection has already been made;) from the north drift of the north shaft of the Buckeye; from the Sentinel shaft, and recently from both wings of the tunnel. The ground between the south and middle shafts, and extending to the main shaft, has been mostly worked out, as also the old Champion chamber. Streaks of ore, however, continue onward from both places, which may at any time be followed.

Between the north and middle shafts of the Buckeye, in the old workings, we have still 1,000 tons and upwards of fine smelting-ores, which can only be extracted with facility through the main shaft. This will be done as soon as connection is made between the west drift from the main shaft and the middle shaft. In this way we shall be enabled to withdraw the ore-bodies near the surface, over which, at present, the ore-teams are obliged to pass.

The present width of ore-streaks is, in the Savage and Champion inclines, 4 to 6 feet, all with a general tendency nearly perpendicularly downward: north shaft of Buckeye, 6 to 15 feet; Quartzite shaft, 4 feet; Sentinel, 4 feet; left branch of tunnel, 10 feet, undoubtedly a different body from the Sentinel, and dipping at a very high angle; right branch of tunnel, probably same body as in the other wing, entirely across the face of the drift; Windsail body, 2½ to 3 feet, and dipping to meet those of the Savage and Champion.

In Savage and Champion inclines I estimate fully two months' supply; in north shaft of Buckeye, old workings of Buckeye, Quartzite shaft and Sentinel, at least two months' supply of ore-reserves. In new ore-body developed in tunnels, assuming it to be the same, and of like character, 100 feet long between the tunnels, 10 feet thick, and 70 feet high, (which will not carry the ore to the surface,) we can reasonably count upon upward of 5,000 tons, or, say, two months' supply—in all, with approximate certainty, at least six months' supply.

The widths of the ore-streaks heretofore enumerated refer to the ends of the present drifts, and may be expected to continue. For future supplies we may count upon the following :

1st. The continuation of the present Richmond ore-body, the largest single mass ever found on the hill. This body dips toward the company's Lookout ground, and cannot be distant from the dividing line more than 30 feet. The Lookout is an older location than the Richmond, or any other on that portion of the hill. The Lookout shaft is now down 80 feet, and is being hurried forward night and day. We cannot expect to reach the ore-level inside of an additional 80 feet.

2d. Ore-body recently struck in the north drift from the main shaft; samples of the streak assayed rich in lead, $129 in silver, and no gold. The vein-matter, at a depth of 170 feet from the surface, commences 12 feet from the bottom of the shaft, and extends a distance of 115 feet, and not yet through. We are daily expecting a large body.

3d. The continuation of the ore-body struck in both ends of the tunnel.

4th. The as yet undeveloped wide stretch of ground from the Sentinel shaft through the Mammoth ground quite to the Windsail shaft, the whole of which shows on the surface a very large iron outcrop, with patches of ore.

As a general rule, the surface ores are the richer in lead. As depth is attained, we find a general diminution in the percentage of this metal, and an increase, if anything, in the amounts of the precious metals. To this, however, the tunnel-body and that in the north drift from main shaft are plain exceptions. This decrease in lead would seem to indicate the future necessity of milling at least a portion of the ores raised. At the beginning of the present year the ores became all at once of a more quartzy character; changed, however, in the course of two months to their former condition, and the same may possibly take place with our present deep workings. At all times we have had and shall continue to have some ores more suitable for milling than smelting.

The cost of hauling ore to furnaces is, for 60 tons per day, $2 per ton; for any amount above that quantity, $1.87½.

The cost of mining will be seen in the reports of the home office. The superintendent does not receive the bills of supplies for the mines, of which the accounts are kept in San Francisco.

From January 1 to September 30, inclusive, there was raised from the company's mines a gross amount of 14,985 tons, 1,315 pounds.

The smelting-works of the Eureka Consolidated Mining Company are most eligibly situated in the lower portion of the ravine, wherein is built the town of Eureka. They consist of three old and two new furnaces, one blacksmith-shop, with two forge-fires, and a complete set of tools, anvils, &c.; two boilers, one engine of 40 horse-power, one largest-size Blake's rock-breaker, four large Sturtevant blowers, which supply all the furnaces and both forge-fires; also a complete equipment of tools, tanks, trucks, shafting, pipes, pumps, &c., for the thorough working of the ores.

The company owns, in addition, one stone office, containing apartments for superintendent, clerk, and metallurgist; assay office, and adobe addition used as a store-room; also a stable on the opposite side of the ravine, and away from the furnace-fumes, with stalls for five horses.

The company owns likewise a fine water-privilege, which is housed in, from whence the water is conducted underground to the distributing-tanks adjoining the works. A small earth embankment has been thrown up in front of the works to catch the surface water from all the other furnaces above us, which water flows into a settling-tank, from which a 6-inch pump lifts the water so collected into the general distributing-tanks.

The capacity of the furnaces is, Nos. 1 and 2, (old furnaces,) 16 to 18 tons per day each; No. 3, (old furnace,) 18 to 22 tons per day; Nos. 4 and 5, (new furnaces,) each 35 to 45 tons per day; No. 5 has already smelted as high as 52 tons in twenty-four hours, the ore having been particularly favorable; that is, for all five furnaces, 120 to 146 tons per day, according to the smeltability of the ores charged.

Consumption of charcoal per ton, dependent upon, first, the moisture, and second, the percentage of quartz in the charges, from 30 to 45 bushels per ton of ore—on the average, say, 35 bushels.

The costs of smelting are left to the home office, for the same reason heretofore stated, viz, that supplies, tools, castings, &c., purchased in San Francisco do not appear upon the books at Eureka. Number of tons reduced from January 1 to September 30, inclusive, 14,951 tons, 1,315 pounds. Number of tons of bullion produced during the same interval, 2,550 tons, 1,402 pounds. Number of tons of ore to produce one ton of bullion will average, therefore, very nearly 5¾. Price of charcoal has varied between 28 cents and 30 cents per bushel. Price of wood, $6 to $6.25 per cord. Freight on bullion to Central Pacific Railroad has varied from $12.50 to $17 per ton; is now fixed by contract at $13.75.

In all smelting operations the question of fuel is one of vital importance, the cost of charcoal alone consumed in the company's works being the largest single item of expense incurred in the production of the metal. Already the nut pine, the only wood suitable for coaling, has been cut off within a radius of ten miles of Eureka. With every year the price per bushel of charcoal must increase, and in view of the probably increased consumption in the immediate future, your superintendent has the honor respectfully to suggest that steps be taken to test the feasibility of obtaining coke from some of the mines of the Rocky Mountains.

An attempt to use gas coke in one of the company's furnaces failed for the reason that the blast used was not of sufficient force to penetrate the heavy mass of compact coke and ore. To enable us to do this, there will be required a powerful engine of upwards of 100 horse-power, and a double-cylinder blast. Experiments on a small scale have shown that some at least of the Rocky Mountain lignites may be coked.

Charcoal now costs $40 per ton. Coke, your superintendent believes, can be made and delivered at the works at Eureka for $32 per ton; the smelting power of coke compared to charcoal is as 8 to 5, and therefore, could coke be employed, there will result a saving to the company of over one-third of the present outlay for fuel.

This question may possibly be left in abeyance until the succeeding spring and summer, at which time the enhanced price of charcoal will necessitate its solution.

Wood supply is of minor importance; the price will probably rise to $7 per cord.

As a pendant to the charcoal question, and in order to hold in check the grasping aspirations of the coal-burners, it may not be unadvisable to contemplate the possibility of erecting a mill for amalgamation.

Should the ores in depth assume a slightly increased percentage of quartz, it would be desirable to reduce them by the process of amalgamation; and in order not to proceed too hastily, a few tons of our quartzy ores might be worked in a mill. The Metropolitan Mill, within a few yards of the company's works, will within two months offer an opportunity to definitely test the question.

The superintendent would further respectfully suggest that he be authorized to secure the refusal for the company of two additional plots of mining ground on Ruby Hill, adjoining the Lookout on the north and west, and the Triangle on the south and east.

Very respectfully, your obedient servant,

W. S. KEYES,
Superintendent.

The following is the Secretary's report:

From sales lumber, &c..	$146 85
From rent of scales...	12 75
From exchange on coin-drafts...	1,923 62
From proceeds 2,038 tons base bullion refined.....................	619,275 67
From product 1,430 tons base bullion at refining-works, Newark, New Jersey, and *en route*...	504,800 00
	1,126,158 89

For construction and improvements:

Cost of furnace No. 3................................	$7,268 61	
Cost of furnaces Nos. 4 and 5.......................	26,523 53	
Cost of new boiler, &c., complete....................	2,428 96	
Cost of rock-crusher.................................	1,866 50	
Cost of one Excelsior pump and fittings, all complete......	775 24	
Cost of buildings and other improvements...............	6,359 58	
		$45,222 42

Mine account:

For labor..	58,967 00	
For hauling ore to furnaces..........................	35,695 72	
For candles and oil..................................	2,061 27	
For powder and fuse.................................	1,111 87	
For lumber and timbers..............................	5,284 21	
For freight on supplies..............................	686 57	
For tools, nails, hardware, rope, &c..................	1,933 00	
For blacksmithing...................................	152 40	
For coal and borax..................................	64 43	
For iron and steel..................................	384 42	
For chest carpenters' tools..........................	50 00	
For purchase mules, horses, and harness..............	534 75	
For barley, oats, and hay............................	237 30	
For surveys, register's fees, &c......................	234 25	
For incidentals.....................................	79 00	
		107,512 19

Smelting account:

For labor..	112,841 55	
For coal...	255,761 50	
For wood and coke..................................	7,016 95	
For castings and foundry work.......................	5,087 23	
For iron, steel, and metal...........................	3,349 13	
For hardware, nails, tools, rope, &c.................	3,463 62	
For oil and candles.................................	879 18	
For freight on castings and supplies..................	4,379 09	
For blacksmithing...................................	3,091 13	
For tinsmithing.....................................	1,387 20	
For sandstone and fire-brick.........................	3,830 58	
For hauling slag and clay............................	939 00	
For lumber and poles................................	3,485 67	
For belting-lace, leather, and packing................	1,006 60	
For lamps, brooms, &c..............................	98 25	
For galvanized iron and gas-pipes....................	1,011 89	
For horse hire......................................	214 22	
For purchase horse..................................	240 00	
For barley, oats, and hay............................	620 37	
For cotton drilling, (cover for coal).................	287 68	
For chemicals and paint.............................	402 78	
For incidentals, repairs, &c.........................	297 65	
For insurance......................................	1,788 20	
		411,479 47

General expenses, Eureka:

For salaries superintendent and officers................	8,644 54	
For books and stationery............................	291 75	
For express charges, revenue-stamps, newspapers, franks, &c.	1,411 41	
For traveling expenses superintendent................	143 50	

For barley, oats, hay, and horse hire, &c	$1,185 43	
For purchase horse, wagon, and harness	1,272 00	
For hardware, buckets, brooms, incidentals, &c	487 27	
For assays bullion and ore	85 00	
For purchase Oertling assay balance	250 00	
For attorney's fees	2,270 00	
For services of porter	276 00	
For chemicals	55 45	
For printing and advertising	283 50	
For taxes	1,043 84	
For applications United States patents	646 00	
For notaries' and county clerk's fees	61 60	
For surveys, recorder's fees, &c	1,262 65	
		$19,669 94
Expense, San Francisco:		
For salary of officers and employés	4,821 50	
For office rent	1,040 00	
For books and stationery	535 59	
For printing, advertising, franks, newspapers, &c	219 05	
For services of porter	194 00	
For traveling expenses of president, &c	1,796 00	
For assays of bullion	201 75	
For notaries' fees, revenue-stamps, &c	257 60	
For telegraphing, incidentals, &c	471 89	
For express charges	228 91	
For coal and fuel	56 00	
For report on mine and furnaces	1,000 00	
For gas-fixtures and repairs of office furniture	48 76	
For city and county taxes	14 85	
		10,885 89
Office fixtures:		
For San Francisco office—counter, desk, safe, carpets, &c..	540 00	
For Eureka office—safe, counter, desk, &c	802 37	
		1,342 37
Interest, &c.:		
For interest on overdrafts, &c		6,120 85
Mining properties:		
For purchase of mining ground		10,499 42
Freight, refining, &c.:		
For transportation and refining-charges on bullion		158,305 05
Dividends:		
Nos. 1 to 5, inclusive, paid to stockholders		235,500 00
		996,027 60
Balance of net earnings over all expenditures		130,131 29
		1,126,158 89

RECAPITULATION.

Receipts and disbursements from July 7, 1870, to September 30, 1871.

Receipts:	
For sales of material	$159 60
For exchange on coin-drafts	1,923 62
For proceeds of 2,038 tons of bullion refined	619,275 67
For value of 1,420 tons of bullion at works, Newark, and *en route*	504,800 00
	1,126,158 89
Disbursements:	
For construction and improvements	$45,222 42
For mine account	107,512 19
For smelting account	411,479 47
For general expense, Eureka	19,669 94
For expenses, San Francisco	10,885 89
For office fixtures	1,342 37

For interest, &c...	$6,120 85
For mining properties	10,489 42
For freight and retiuing-charges on bullion	158,305 05
For dividends paid stockholders.....................................	225,000 00
	996,027 60
Balance of net earnings over expenditures...........................	130,131 29
	1,126,158 89

RESOURCES AND LIABILITIES.

Resources:	
Inventory of supplies at Eureka...................................	$6,463 87
Charcoal on hand	38,041 92
W. S. Keyes, superintendent	2,744 36
Sundry book accounts	912 94
1,430 tons of base bullion.......................................	504,800 00
	552,983 09

Liabilities:	
Overdrafts...	$85,760 19
Drafts against bullion shipments.................................	224,062 14
Bills payable..	50,000 00
Superintendent's drafts, outstanding and not presented...........	17,820 80
Book accounts, (not due)	642 88
	378,326 01
Net resources, September 30, 1871................................	174,657 08
	552,983 00

Cost of extracting ores.

Expense of extracting and hauling to furnaces 18,847 tons of ore is.......	$107,512 19
Less supplies on hand per inventory	3,100 00
	104,412 19

Or $5.52 per ton.

Cost of smelting ores.

Expense of smelting 18,825 tons of ore is.............................	411,479 47
Less coal and supplies on hand.......................................	42,425 79
	369,053 68

Or $19.60 per ton.

Eighteen thousand eight hundred and twenty-five tons of ore reduced, produce 3,468 tons of base bullion, or 5.75 tons of ore produce 1 ton of bullion, at a cost of $135.76.

W. W. TRAYLOR,
Secretary.

Later in the year, and during the first months of 1872, the company discovered extraordinarily large and valuable bodies of ore in the Lawton tunnel. My correspondent writes in regard to these in March : "I have just visited the stopes connected with the Lawton tunnel, and am now fully convinced that the new discoveries are indeed immensely valuable and extensive, though they hardly come up to the extravagant estimate made by the local newspaper." *

Ruby Hill is a spur of the Diamond Range. Its general trend is north-northwest to south-southeast. The old openings of the Eureka Consolidated, as well as those of the Richmond and Tip-Top, are on the western, the new ones on the eastern slope. The strike of the ore-body is nearly east and west, and its dip about 45 degrees to the northeast. For this reason ore was first discovered on the western slope of the hill,

* The estimate referred to was, I think, something like $20,000,000 worth of ore in sight.

H. Ex. 211——12

where the vein crops out. The main or Lawton tunnel, the mouth of which is toward the town, (on eastern slope of hill,) is now in over 600 feet, and passes 120 feet to the north of the Keyes shaft, between it and the Windsail shaft. At its end it is in ore. The first ore was met with about 300 feet from the mouth of the tunnel in the K K claim. The shafts mentioned above are connected by galleries, and from the main tunnel runs a short side tunnel into the Sentinel grounds. The main object of the tunnel is, therefore, to transport through it all the ores from the works connected with the two shafts named and from the Sentinel claim. From the mouth of the tunnel the ore is to be transported by means of a narrow-gauge railroad, which is to run along the Jackson grade to the smelting-works. On the back trip the trains are to bring water and supplies to the mines, where the erection of powerful hoisting-works is contemplated as soon as stopes are opened below the level of the tunnel.

The Keyes shaft is now 175 feet deep, and serves as the main hoisting-shaft for the old works. These are to the largest extent in a broken quartzite which crops out below the Nugget and Savage, on the western slope of the hill, and, to judge from its dip, can only be reached at greater depth on the eastern slope. The so-called cap-rock of the ore-bed is limestone. The ore-body itself, though it exhibits a certain regularity, is neither a cross-vein nor a contact-vein, and I cannot give it any better name than that of ore-bed or zone. Horses of a broken limestone, with every stratification, are frequently met with; also cavities with splendid druses of wulfenite, (molybdate of lead,) calcspar, aragonite, quartz, and, lately, malachite and azurite. The approach to the vein-matter is first distinguished by a yellow color of the first dense, afterwards broken limestone, next by a stronger impregnation of pulverulent brown and yellow iron-ore and stripes of the first. Finally, the ore-body proper, brown iron-ore, with impregnations and bands of mimetite, carbonate of lead, massicot, or lead-ocher, &c., is reached. While on the western slope, besides the yellow mimetite, (Buckeye,) large masses of solid carbonate of lead, with the so-called "black carbonate" (which is probably a new mineral) and little galena, (Champion,) were found. The ores encountered on the eastern slope in the iron-stained masses, which are poorer in lead, are principally highly argentiferous galena and "black carbonate" in lumps and nests of often over a hundred-weight.[*] For this reason there is now much more base bullion produced now one ton of lead, while formerly it required 10 to 12 tons.

The Marcellina, belonging to another company, has only spurs from the ore-body of the Sentinel. Opposite the Marcellina, and divided from it by the gulch, is the Carson mine, and adjacent to this are the mines of the Phenix and Jackson Companies. The latter has been prospecting since February, 1871, but has so far found no ore. The K K has large masses of ore in sight, galena and "black carbonate." It is expected that this claim will consolidate with the Carson and other mines, and, with the furnaces of the Buttercup Company, will be soon thrown into one company, to be incorporated in San Francisco. The time for redemption of its property, accorded to the Buttercup Company, is drawing to a close. The Eureka Consolidated intends to tear down during the next summer the two remnants of former times, furnace No. 1, (built under Buel,) and No. 3, (built by Liebenau,) and to

---

[*] The "black carbonate" above referred to is no carbonate at all, but more probably boulangerite, or a new mineral analogous in composition and origin to stetefeldtite. I have seen no analysis as yet.

replace them by two new and larger furnaces, so that the capacity of the smelting-works will then be 200 tons per day. No. 2 was built in December, after the pattern of the new ones, (see article on "Metallurgy" in this report,) but has square corners, and is a little smaller. It has four tuyeres, of $3\frac{1}{2}$ inches mouth, works very well, and smelts 30 tons per day.

At the Richmond smelting-works they are still building energetically. These works will undoubtedly be the best and most perfect in the State, but it remains to be seen by the results of the future whether these very large expenditures are justified.

The Phenix smelting-works, with their two furnaces, are still idle, because the machinery for the hoisting-works of the Adams and Farron mines has not yet arrived, and the mines can therefore not be worked.

At the 15-stamp Lemon Mill (formerly the Metropolitan of Shermantown) Mr. John Howell is now putting up a White's cylinder roasting-furnace, and it is expected that the mill will soon be in working order.

Besides the Adams Hill Company there is now another company located on Adams Hill. This is the Star Consolidated, a San Francisco corporation, which has bought several mines on the quartz belt of that hill. They have beautiful horn-silver ores, but as to the quantity I am not informed. All the signs point to an enormous industrial increase during the coming year, especially if capitalists should take up the Prospect Hill mines.

The Richmond property has been transferred to English hands during the year, and the new company is still building on it very extensive smelting-works. The product of this property during the year appears in the appended statement of the product of the district.

The monthly product of the Eureka Consolidated Company during the year ending December 31, 1871, was as follows:

Product of bullion.

Month.	Tons.	Pounds.
January	152	273
February	134	1,092
March	244	1,155
April	255	1,983
May	177	560
June	341	269
July	520	722
August	426	1,673
September	297	600
October	355	500
November	264	250
December	2	500
Total	3,172	652

The average contents in gold and silver for the whole yearly product may be safely set down as $250 per ton. Adding $100 per ton for the lead, we have a gross value of $1,110,314.10.

Until the end of May only the three smallest furnaces were alternately running, (two running at a time, while one was being repaired.) Since then four have been working at a time, while one was standing idle. In January, 1872, the consumption of the Eureka Consolidated smelting-works, with four furnaces running, was 142 tons of ore and 4,000 bushels of coal per day. The production was about 15 tons of base bullion, containing $230 in gold and silver. The ores were then rather poor in lead and the precious metals.

The quantity shipped in December was so small, because all the furnaces, except No. 4, were blown out in November, and No. 4 on December 2. This was done on account of a strike of the teamsters, whose contract had expired, and who refused to haul more ore at the old prices. I am not informed how this difficulty has been overcome, but at latest accounts smelting was again going on.

The following is the total production of Eureka, as transmitted to me by Mr. O. H. Hahn, M. E., my correspondent at that place:

Number of furnaces.	Names of companies.	Bullion shipped.		Gross value, (gold, silver, and lead.)
		Tons.	Lbs.	
5	Eureka Consolidated Company	3,172	652	$1,110,314 10
1	Richmond Mining Company	1,012	....	354,038 23
1	Phenix Mining Company	222	1,483	71,278 06
2	Jackson Mining Company	212	934	77,550 43
2	Buttercup Mining Company	596	....	208,658 00
1	Roslin Smelting Company	450	....	213,750 00
1	Tilton Smelting-Works	....	....	....
12		5,665	1,074	2,035,588 90

NOTE.—The number of tons shipped by the companies marked (†) is estimated.

In explanation of the high value of the bullion shipped by the Roslin Company, I should say that Mr. J. M. Robertson, the manager, gives the following values per ton of bullion and the amount shipped during the total running time of six months, in 1871:

Amount shipped	450 tons.
Average value per ton in silver	$250 00
Average value per ton in gold	125 00
Average value per ton in lead	90 00
Total value per ton*	465 00

For a detailed description of the mines belonging to the companies besides the Eureka, I have no data on hand at present, but I may say that all the base-metal mines in the district have nearly the same characteristics, and vary less in the classes of ores occurring in them, or the mode of occurrence, than in the size of the ore-bodies. Notes on this subject will be found in a subsequent chapter on the smelting of argentiferous lead-ores in Nevada, Utah, and Montana.

Mineral Hill district has been the scene of much activity during the year. The Mineral Hill Silver Mining Company, limited, an English corporation, has invested largely in the mines of the district, and owns the principal works. The mines lie in limestone, and are not on veins, but on irregular deposits, some of which are very large. These ore-deposits are, however, principally situated on a well-marked ore-channel, which is about 300 feet wide. On this are situated the Giant, Rim-Rock, and Live Yankee, claims in which there seems to be a continuous body of ore of from 22 to 30 feet in width. The greatest depth reached in the Giant in the fall was 120 feet. The ores contain a considerable amount of base metals. They consist of argentiferous galena, anti-

* Silver and gold are here calculated at full value, and lead at the price obtained in San Francisco. Selby & Co. have paid, generally, $1.15 per ounce for silver, and 90 per cent. of the assay value of the gold.

monial ores, oxidized copper-ores, cerargyrite, (chloride of silver,) mendi-
pite, sulphide of silver, &c. But only on the western side of the hill the
ore is sufficiently lead-bearing to make reduction by smelting desirable.
So far this has not been done. The ores of the upper mines are beneficiated by a preparatory chloridizing, roasting, and amalgamation. The
English company has a 15-stamp mill and a 22-ton Stetefeldt furnace.
The working results are reported to be 92 per cent. of the assay value,
and the average fineness of the bullion 0.750. The same company commenced in the fall to build another mill with 20 stamps, a second Stetefeldt furnace, and all the late improvements. After the completion of
this mill the company's works will have 35 stamps, 18 pans, 2 roasting-
furnaces, and a capacity of 50 tons per day.

The total production of Mineral Hill district for 1871 is given by Mr.
Valentine, of Wells, Fargo & Co., as $701,014.

I am indebted to Mr. S. O. Clifford, the county assessor of Lander
County, for the following statement of the bullion product, reported to
him, of the mines during 1870 and 1871. The amounts given for the first
two quarters of 1870 exceed those given in my last report, and the
cause of the discrepancy is the fact that in last year's report many
small mines were not included. The total product for two years, of
Lander County, as here given, is not as large as the total of shipments
from Lander County during 1871 alone. The assessor says in his letter
that his returns are only from two districts, Reese River and Eureka,
while there are a number of others in the county, the most important
of which is Mineral Hill; but even adding this would not make good
the discrepancy. For comparison, I insert here both statements:

Bullion statement of Lander County, Nevada, for the years 1870 and 1871.

	No. tons of ore.	Pounds.	Gross yield.
Quarter ending March 31, 1870	1,457	1,230	$228,896 83
Quarter ending June 30, 1870	2,390	396	119,681 63
Quarter ending September 30, 1870	3,203	1,700	360,709 26
Quarter ending December 31, 1870	4,858	1,166	407,755 43
Quarter ending March 31, 1871	6,341	402	419,477 81
Quarter ending June 30, 1871	8,716	1,429	424,204 56
Quarter ending September 30, 1871	10,762	1,368	639,455 30
Quarter ending December 31, 1871	10,174	1,616	615,276 24
			3,215,457 06

S. O. CLIFFORD,
County Assessor.

*Statement of bullion shipments from Lander County during the year 1871,
as reported by the general superintendent of Wells, Fargo & Co., in
San Francisco.*

From Austin	$965,536 17
From Eureka	2,173,105 56
From Mineral Hill	701,014 00
	3,839,655 73

In former reports I have explained why the returns of the mine-
owners to the county assessors are, in nearly all cases, far below the
actual values of the ores extracted, and in last year's statements a cer-

tain class of ores is not reported at all, as of too low grade, according to the new law, to be subject to taxation.

NYE COUNTY.

I have no returns from this county of an exact and detailed character, the reason being that the very estimable and courteous citizens who promised to furnish them have not kept that promise. I can, therefore, only say in general terms that the Belmont and Mammoth districts in this county have maintained some production, the principal (or only?) mills running being the Canfield Mill at Belmont and the Ellsworth at Mammoth. The shipments from Belmont amounted to $268,903. The principal producing mine was the El Dorado South, which has been described in former reports. It is rumored that this very valuable mine has been sold to a San Francisco company. The incline is reported to be now 375 feet deep, with levels 100 and 80 feet, respectively, both ways on the vein. The vein is from 4 to 12 feet wide, and has thus far produced about $400,000, the ores worked having yielded, it is said, an average of $150 per ton. This high average indicates that only first-class ores have been worked.

The Arizona mine, which is said to have been consolidated in one property with the El Dorado South, is opened to a depth of about 200 feet by a tunnel, and about 200 feet of drifts on the tunnel-level. It is said that a ledge of high-grade ore, 2½ to 6 feet thick, is developed throughout this distance. The ore worked has averaged, according to report, $275 per ton.

Another very promising mine at Belmont is the Monitor, which, according to the Reese River Reveille of March 4, 1872, had yielded, since June, 1871, 240 tons of rock, averaging $503, or a total of $120,732.

At Mammoth there is beginning to be a revival of activity, partly due to the starting of the Ellsworth Mill. The following account is from the Reese River Reveille of February 27, 1872:

This mill was completed last summer by a company organized at Bridgeport, Connecticut and is running exclusively on custom ore. It has ten stamps and a Stetefeldt furnace, is under the supervision of Mr. Kustel, the well-known assayer, Mr. F. W. Smith being general agent. They have adopted the Manhattan Company's schedule of prices, and give complete satisfaction to their customers. Their sources of ore-supply are Ione, ten miles distant, Belmont, eighty miles, and San Antonio, sixty miles. No mining worth mentioning is being done at Mammoth. The cost of hauling from these various places is, from Ione from $5 to $6 per ton, and from Belmont and San Antonio $30 per ton. During the past few months some 40 tons of ore from the Liberty and Potomac mines of San Antonio have been worked at Mammoth. Mr. Roberts does not know the exact yield, but it was quite satisfactory. From Belmont the Monitor and Arizona mines have sent several lots; the last, consisting of 36¼ tons of Monitor ore, yielded at the rate of $462.17 per ton.

The camp of Ione, ten miles distant, is, however, the principal source from which the mill must obtain its ore. This is an old camp, and has passed through many vicissitudes. It has had many ups and downs, at one time promising to be a leading district, and anon all but deserted. Two mills have been erected there, but for causes too numerous to mention here they did not prosper. They were first closed and then dismantled, and it did look for sometime as if Ione had yielded up the ghost. There are good mines there, however, as we know of our own knowledge, having done some prospecting there ourselves in early times. They are small, averaging in the neighborhood of one foot in thickness, but sufficiently rich to pay well. Mr. Roberts tells us that the satisfactory returns given by the Ellsworth Mill have infused new life to the place, and that owners of mines are returning to work them. The principal mines upon which work has been resumed are the Indianapolis, Shoo Fly, Pleiades, and Stonewall. The first of these is down 280 feet, and works 26 men, principally on dead work. They take out about three tons of ore per day, which pays all expenses and a little over. The amount of ore now on the various dumps and at the mill ready for reduction is not less than 400 tons. This looks encouraging, and we hope that the mine-owners of Ione will show a reasonable degree of enterprise, and not allow the mill at

Mammoth to want for ore, for they cannot afford many more failures. The little town is looking quite lively ; it has two saloons, two billiard-tables, one store, one restaurant, one livery-stable, all doing a good business, and it only rests with themselves to make it as prosperous as it was in 1864 and 1865.

The Manhattan Mill at Austin has worked a good deal of the Nye County ores during the year. This is one reason of the high yield per ton. Only very rich ores would pay for transportation to Austin.

Lida Valley district.—This district has attracted some attention during the latter part of the year, especially in Austin, which seems to have been the source of supplies for the new camp. The Reese River Reveille of March 4, 1872, contains the following correspondence from that point:

Lida Valley mining district was organized on the 29th of last August. It embraces an area of one hundred square miles, with Scott's Springs as the center. This is the most prominent and affords the greatest amount of water of a series of springs in the eastern part of a small basin lying between the Palmetto range of mountains on the north and west, and Mount McGruder on the south and east. The valley was christened "Alida" by Colonel D. E. Buel, some eight years ago, when on his memorable trip through the Death Valley country in search of the mystical "Breyfogle" lode, but either from a corruption of the word or by common usage the district is known as "Lida."

Rich mines were known to exist here before the organization of the district, and subsequent developments have proven their value beyond a peradventure. Since about the first of last December Messrs. Hiskey and Walker have purchased from Messrs. Scott, Black & Co. their water-right and mill-sites, together with the Cinderella mine, and at once commenced active operations for the development of the ledges and the improvement of their property generally.

There are now about thirty persons in the district, and thirty-two claims on record. The mineral-bearing belt is about one-fourth of a mile in width, and extends from the valley between Mount McGruder and Gold Mountain on the east to and including Palmetto district on the west, a distance of fifteen miles. The country-rock or formation is diversified by slate, limestone, porphyry, and granite. The ledges run in an easterly and westerly direction, and the ore is chiefly chloride with galena as the principal base metal. As to the value of the ore, I am not yet prepared to speak advisedly in general terms. Messrs. Hiskey and Walker are shipping ore to their mill at Deep Springs, Esmeralda County, for reduction, the result of which I will give you when known. A lot of ore from the Cinderella mine, recently worked at Columbus, yielded a fraction over $500 per ton. Other mines of recent discovery "rank" that ore by 100 per cent.

The editor of the above paper, to whom several specimens from the new mines were sent, says the ores are similar to the rich green-stained decomposed surface ores of some of the mines on Lander Hill, and predicts that the Lida Valley mines will receive wider attention during the present year.

WHITE PINE COUNTY.

I am indebted to Mr. A. J. Brown of Treasure City for a report on White Pine district. His notes, together with other information received during the year, are embodied in the following pages:

Operations in the free-metal mines have, on the whole, been vigorously prosecuted. But nothing has been done during the year in the smelting line except twelve days' run of the Alsop furnace, and there are no indications that smelting will be resumed prior to the advent of the Eastern Nevada narrow-gauge railroad. The prospects for the early building of that road are not flattering. Some mining property has been sold to English companies during the year; but it is well known that but little money has really changed hands; not more than $60,000 in this district can really be traced. The vendors of property generally receive the major part of the payment in paid-up shares of the company. The properties belonging to London companies are the East Sheboygan, Ward Beecher South, Earl, North Aurora, South Aurora, Eberhardt, Idaho, and Great Western.

The prospects of the district appear to be slowly but surely improving. Most of the miners are beginning to realize that the "spar-vein" is the guide-board that points to deeper bodies of ore if any exist, and of that there can be but little doubt, for the formations to the west of Treasure Hill, including White Pine Mountain, comprising granite, silurian, and Devonian rocks, all contain mineral-bearing veins. For most of the information in regard to the "spar-vein" Mr. Brown acknowledges his indebtedness to Mr. D. H. Barker, civil engineer and surveyor, who has made several maps of the vein and accompanying ore-deposits. In the description of the mines the words "vein," "deposit," and "ore-channel" are indiscriminately used for the same thing.

The most important mining operations in White Pine district during the past year have been mainly carried on along the supposed north and south "ore-channel," extending through Treasure Hill, from the Mammoth and East Sheboygan mines at the extreme north end, to the Mazeppa at the extreme south end of the hill. Early in the year a large vein of calc spar was discovered forming the western or foot-wall of the South Aurora, Ward Beecher, and some of the other important mines on the east side of Main street, Treasure City. Later explorations have established its continuity along the whole length of the hill from the O. H. Treasure mine to the South Aurora. At the south end of the last-named mine an apparent break occurs in the vein, but it makes its appearance again near the north wall on the west side of the Eberhardt, and continues thence south along the ridge nearly to the Mazeppa mine. This spar-vein in itself may possibly be of little or no importance as a mineral-bearing one, but taken in connection with the character of the ore-bodies existing along the eastern side of its course, it is important, and worthy of a careful description. This spar has a general north and south course, and dips to the east at an angle varying from 33 degrees near the surface to 38 degrees in the deepest workings. As might be expected, it is often faulted, and generally from east to west. It does not correspond with the strike and dip of the country-rock, its dip being at an angle of 65 degrees. Its structure is banded, i. e., it is made up of narrow bands or layers running parallel to the walls. In thickness, it varies from 10 to 40 feet. Although there are numerous masses and veins of spar in the district, none of them bear any resemblance to the one described. They are generally limited in extent to a few hundred feet at most, lie mostly flat, and correspond with the dip and strike of the country-rock, while the one forming the foot-wall of the "main-ore channel" is persistent in its course for a known length of nearly two miles, and does not coincide with the strata of the country-rock, which it cuts, on the contrary, all along its course. The mines along its eastern side, although partaking of the irregular deposit character common to limestone formations, form an almost continuous "ore-channel," and all, with the exception of the South Aurora, are proving persistent in depth. Those lying on the west side were merely superficial horizontal deposits, corresponding with certain limestone strata. Belonging to these superficial deposits are the mines on Chloride, Bromide. and the other flats on the west side of Main street, which have been mainly exhausted and abandoned. A number of east and west vertical fissures appear to cross the main north and south "ore-channel" at right angles, and, although themselves poor in mineral, they seem to have wonderfully enriched the main channel at the points of intersection. Of these east and west fissures, two are well known, and have often been described, viz, the Eberhardt and California. Another is found 800 feet south of the Eberhardt, crossing from the Grant and Colfax, through the Eureka and Indianapolis mines, and developing in the last-named mine an immense body of low-grade ore. Another smaller one, 400 feet still farther south, passes through the El Dorado mine. But perhaps the most important and most productive is that crossing the Ward Beecher, and forming, with the main vein the massive bodies of rich ore developed in that mine.

Mines east of the spar-vein.—Commencing at the north end of Treasure Hill, the first of the important mines situated on or near the main ore-channel is the Mammoth. During the first two years following the discovery of the district, extensive explorations were made on this mine, and something like $60,000 were expended without discovering any bodies of ore extensive enough to reward the owners for their time and expense; but during the present summer work, was commenced on the large croppings situated 600 feet north of the old works, and an extensive body of good ore has been developed. This ore-body, as exposed in the different works, is 10 feet in thickness and 300 feet in length. The deepest shaft is in good ore at a depth of 50 feet from the surface. The vein courses east and west, and dips to the north at an angle of 50 degrees to the plane of the horizon. The hanging-wall is arenaceous shale, and appears to be smooth and well defined. The foot-wall is the ordinary lime stone of Treasure Hill, and is not defined, but gradually blends with the quartz of the vein. Two hundred and fifty tons of ore are on the dump.

The Miner's Dream is situated about 500 feet east of the new works on the Mammoth, and is probably a part of the same vein or ore-channel. It was accidentally discovered by the superintendent of the Mammoth on the 16th of October of the present year. He was on his way to Hamilton, when he noticed a bunch of quartz, a piece of which he broke off and had assayed. The result was $109 per ton in silver. The ledge, as exposed in an open cut, is eight feet thick. Some very rich rock has been mined, but the explorations are quite limited, and the unfavorable weather will be likely to prevent developments during the winter. The vein has the same casings as the last.

The East Sheboygan, situated immediately east of the Mammoth old works, was located early in 1869. Considerable work was done at the time, and some of the ore was worked with fair results. But the owners were not able to develop the property, and it has lain idle until the present summer. It is now the property of an English company, with head-quarters in London. Since the 1st of September explorations have been vigorously prosecuted, and with the most satisfactory results. Several shafts have been sunk on the course of the vein, and are connected with each other by drifts, thereby exposing a body of ore at an average depth of 50 feet from the surface, 300 feet in length, and 10 feet in thickness. Specimens of ore taken from the present depth are almost solid coin-silver; but the average mill value of all that has been extracted to the present depth does not exceed $35 per ton. The quality of the ore has so far appeared to improve with the increase of depth, and the mine bids fair to become one of the best properties in the district. The vein has a north and south course, and dips to the east at an angle of 30 degrees. Like the two last named, it is a contact-vein, with a hanging-wall of slate and a foot-wall of limestone. The vein is well defined, both walls being smooth and well marked. Fifty tons of selected ore, recently worked, gave an assay value of $78 per ton. Two hundred and fifty tons of second-class, worth from $25 to $30 per ton, are on the dump. This mine is most conveniently situated for cheap working. A mill, situated 100 yards below the present works, can receive a full supply of water from the pipes of the White Pine Water Company at a trifling expense, and the ore can be dumped from the mine directly on the battery-floor, if necessary, by means of a chute, thereby saving the expense of transportation by teams, which usually costs from $3 to $5 per ton.

O. H. Treasure.—This mine was the first discovered on Treasure Hill, and has been so often described that its locality is too well known to require any further description in these pages. During the first two years that it was mined it was generally supposed, from its location at the point of contact of the lime and shale, to be a contact-vein pitching to the west, but the explorations of the past year have disproved the theory. It is now known to dip to the east at an angle of 38 degrees, thus corresponding in dip and course with all the mines situated on the main north and south ore-channel. The greatest depth attained at the present writing is 160 feet from the surface. At that depth the vein is found to be from 10 to 30 feet in thickness. It contains low-grade milling ore, which is very much mixed with lime and spar, and requires careful sorting. A tunnel is projected and is already driven in 56 feet. It starts from the O'Neil grade on the east slope of Treasure Hill, and will eventually explore the mine to a perpendicular depth of 500 feet below the croppings. The distance to be run will be something less than 400 feet, as the east slope of the hill is very precipitous. The probability is that, on this account, much of the deeper explorations along this channel will eventually be carried on by means of tunnels. The prospects of this mine have materially improved during the past year. We find the returns for the first quarter to amount to only 64 tons, and the gross yield to $1,310.72, while those for the quarter ending September 30 are 1,162 tons, giving a gross yield of $48,540. The company have purchased the Big Smoky mill, and have thoroughly overhauled it preparatory to working the ore from the mine.

Silver Wave is the next mine south of the last named. It has been explored to a depth of 170 feet from the surface. The vein of ore is very large, but generally too poor to pay. Some small bodies of good ore have been encountered, but not enough to cover the expense of exploration. The future of the mine is not encouraging. The vein has the usual north and south course, and dips to the east at the same angle as the last, of which it is a known continuation. Two hundred and ninety-nine tons of ore have been extracted and milled during the present year, giving a gross yield of $10,831.63.

The Edgar, situated 500 feet south of the last-named mine, was prospected during 1870, by a perpendicular shaft, to a depth of 140 feet, and by a drift run in 30 feet east from the 100-foot level. Nothing was found, however, to encourage further expense, but the developments elsewhere along that line during the past year induced some parties to lease the property. They went to work some time in September, and have continued the 100 foot-level 40 feet further east. At a distance of 40 feet from the shaft they encountered a large body of excellent ore, through which they have continued the drift 30 feet without finding the end in that direction. The ore, as taken from the mine, yields $50 per ton. The present yield is 10 tons per day, which can, without doubt, be increased to 50 tons per day when the mine is properly opened. The vein appears to course north and south with the usual easterly dip. It is impossible to form any estimate of the size and importance of this newly discovered orebody, but it is evidently one of the largest and deepest yet found on Treasure Hill.

Portage is situated 300 feet further south. A shaft was started from the surface on the east wall of the spar-vein, with the expectation of finding the "ore-channel," which was supposed to be there from the developments made further south. A small quantity of low-grade ore was found 47 feet from the surface, but the property became involved in litigation with the Ward Beecher Consolidated, and work has been suspended since September.

The Ward Beecher Consolidated, 200 feet south of the last named, had a good body of ore during the summer, but this is now exhausted, except 20 feet, which are involved in litigation with the Ward Beecher South. An incline is, however, being sunk through a brecciated mass of black spar and quartz, with fair indications of ore. Two thousand four hundred and twenty-seven tons of ore have been extracted and worked during the year, giving a gross yield of $61,976.59.

The Ward Beecher (English company) is situated next south, and its ore-body connects with that in the last-named mine. The Ward Beecher includes within its works the Autumn No. 2, Red Rover, Montrose, and Colfax locations. No description can give any very clear idea of the underground workings of this mine. It is, perhaps, the best-managed piece of mining property in White Pine district. Every change in the appearance of the mine, as well as every fault and slip, is carefully noted and taken advantage of in the exploitation. The Earl portion of this mine was worked quite extensively and made considerable stir during the summer and autumn of 1869, but the ore-body apparently gave out, the mine was abandoned, and remained idle during the whole of 1870 and until June of the present year. About the last of that month the present owners commenced work in the old Earl chamber, and soon discovered that a slip or fault had occurred, the upper part of the vein having slid down the hill. A drift was accordingly started east from the old works, which encountered the main ore-body 30 feet from the starting-point. This part of the mine is now known as the Ladies' chamber. The body of ore in this chamber has been opened by shafts and drifts to a perpendicular depth of 122 feet from the surface; its greatest breadth, as far as known, is 150 feet from east to west. Its length from the Autumn chamber, with which it connects on the north, is something over 200 feet, and its southern limit has not been found, although a drift has been extended from the chamber 50 feet south, toward the Risdale chamber in the North Aurora mine. It is scarcely probable that so large an ore-body will be found to extend unbroken through the 600 feet of virgin ground that separates the two mines, although it is the general impression that they will finally connect. The broken and disturbed character of the surface limestone fully warrants this conclusion; in fact, small quantities of mixed limestone and ore have been found in several shallow shafts sunk along the line. But a small portion of the immense ore-body exposed in this chamber has been extracted. The open space is 70 feet long, 40 feet wide, and 55 feet high. The Philpotts chamber is situated between the last described and the Ward Beecher Consolidated, with both of which it forms an ore-connection. Work was commenced on this portion of the mine during the summer of 1870, and 1,331 tons, giving a gross yield of $35,000, were extracted during the last quarter of that year. From the 1st of January to the 1st of October 9,706 tons were extracted, giving a gross yield of $496,223.64. The ore-body in this chamber was 35 feet in thickness from east to west and 200 feet long, the greatest depth reached being about 116 feet from the surface. The greater part of the deposit has been worked out to that depth, but the ore in the bottom of the works is as good as ever, though the body is somewhat narrower than it was nearer the surface. The ore-body exposed in this mine, taken in connection with that in the Ward Beecher Consolidated, forms a continuous "ore-channel," about 500 feet in length from north to south, and the southern limit is not yet found. Several thousand tons of ore are broken and ready for hoisting, and the quantity exposed in the different workings is enormous; it cannot be less than 25,000 tons, even if the chimney should be found to terminate in length and depth within 10 feet from the present limits. The hoisting for this mine is done by means of a 20 horse-power engine, and the quantity delivered daily at the surface is 80 tons.

The North Aurora, situated next south, was worked extensively during 1868 and 1869, but remained idle during 1870, and was supposed to be exhausted. The explorations of the past summer have developed a large body of excellent ore only 4 feet below the east end of the tunnel belonging to the old works. This ore-body is situated about 100 feet north of the old works, and has been sunk upon to a depth of 60 feet from the surface. A drift has been driven north through good ore 100 feet, and another east 30 feet, without encountering the line-wall. The quantity of ore exposed in this chamber is estimated at 15,000 tons. Seventy tons of ore are daily shipped by tramway since the 1st of November.

The South Aurora, lying next south, has been actively worked since 1868, and has yielded a large quantity of ore, but it is now apparently exhausted. Prospecting is, however, being vigorously prosecuted, but so far without success. The greatest depth attained in this mine is 225 feet from the surface. The present indications for deeper ore-bodies are not as favorable as might be desired. At the greatest depth the material encountered was brecciated limestone and spar with some quartz. The amount of ore extracted from the South Aurora mine during the year 1871 is given by Mr. B. N. Lilienthal, the chemist of the Stanford Mill, where the ore was worked, as 5,765$\frac{1779}{2000}$ tons, which yielded $148,804.60 in fine bullion.

The mines above named are all supposed to be on the great north and south ore-channel, and form a continuous chain of locations nearly one mile in length. From the Hidden Treasure to the South Aurora there are only two noticeable breaks in the continuity of the ore. The ore-body early found in the Hidden Treasure continues through the Silver Wave and to the Edgar, a total distance of about 1,000 feet. The most of this ore, however, is of too low grade to pay at present. From the Edgar south to the Ward Beecher Consolidated, traces of ore have been found near the surface, but the only shaft in that distance has been sunk only 60 feet, and has probably stopped at least 40 feet short of the depth of the main channel. The second great ore-body extends from the Ward Beecher Consolidated 500 feet south into the north part of the North Aurora ground. South of this lie 600 feet of unprospected ground. Several shallow shafts, however, have shown traces of ore near the surface, and the prospect is good that ore will be found extending south to the Risdale shaft, the locality of the Aurora deposit. The third and last ore-body formerly extended from the Risdale shaft in the North Aurora to the O'Neil grade, a total distance of 600 feet, but the South Aurora seems now exhausted.

The Eberhardt has been but little worked during the present year. A new prospecting-shaft has been sunk to a depth of 180 feet from the surface, without encountering a new ore-body. Only 500 tons of ore have been extracted from the old works and reduced during the year.

The Indianapolis is situated 800 feet south of the Eberhardt. It has one shaft 80 feet deep, and a drift from the bottom 15 feet east in low-grade ore. A drift has also been run 20 feet east at a depth of 30 feet, and another north 60 feet from the same level, all in ore worth $25 per ton.

The Sharp mine was discovered in February last, while grading the foundation for station 20 on the tramway. Several tons of good ore have been mined, but the title is disputed, and the property will probably have to lie idle for some time.

The Grant and Colfax and Eureka have been worked but little during the year. Very extensive bodies of low-grade ore are exposed.

The Bourbon has been worked most of the year on lease. It was sold early in the summer to a company located at Erie, Pennsylvania.

The Genesee has been worked to a depth of nearly 200 feet from the surface. The shaft is still sinking.

The Noonday has one shaft 140 feet deep, in which the owners claim to have encountered a well-defined vein of ore, 8 feet thick.

The Iceberg, south, is yielding considerable ore at a depth of 30 feet from the surface.

The General Lee has yielded some good ore during the year.

The Pocatillo has a tunnel running west on a vertical vein. It is in about 300 feet toward the Ward Beecher.

The Virginia has been worked during most of the season, and 287 tons of ore, giving a gross yield of $14,523, have been extracted.

From the Silver Plate, situated northwest of Hamilton, 115 tons, giving a gross yield of $4,768.33, have been extracted and reduced during the year.

The Great Western, situated west of the Eberhardt, is owned by an English company, who have sunk two shafts, 80 feet deep, in limestone, with the vain hope of finding the west extension of the Eberhardt ore-body.

The Caspian has an incline 190 feet deep, a drift west from the bottom of the incline 60 feet, and one 40 feet in spar, with some quartz. The owners are still driving west toward the summit of Treasure Hill.

The Asbury, situated in the cañon east of the Eberhardt mine, has a tunnel 400 feet in length. The main tunnel is under contract to be run west 100 feet further; there are also two cross-drifts of 100 feet each in length to be run.

The Featherstone is situated west of Hamilton. It has been worked considerably during the year. Only 30½ tons of ore, giving a gross yield of $1,005.50, have, however, been extracted.

The Blair and Banner, Mahogany Cañon, has been worked on lease during most of the year. It has yielded 173½ tons of ore, worth $11,881.40.

In the Glazier a good body of fair ore has been exposed during the year. Work is suspended for the present.

The Caroline, Mount Ophir, has been worked quite extensively, and partly on lease. The vein is small, seldom exceeding one foot in thickness, but contains exceedingly rich ore, which is, however, refractory. The course of the ore-channel is north and south, and it appears to stand vertically.

A late communication of Mr. Brown informs me that there is but little that is worthy of note in mining affairs since he sent his report in November. The unusually severe weather experienced ever since the 20th of December had almost wholly suspended mining operations. A limited amount of prospecting work had, however, been carried on in localities where the conditions were favorable, and some new bodies of good ore had been brought to light.

The most important developments had been made in the Silver Plate mine, situated about one mile northwest of Hamilton, in the low foothills, a short distance east of the Truckee mine. The formation in this locality bears a strong resemblance to that near the Hidden Treasure, Sheboygan, and other mines at the north end of Treasure Hill. The ore is found lying between a limestone foot-wall and a slate or shale hanging-wall. Both walls are smooth and well defined, and lightly striated. The deposit has the appearance of being a sheet deposit, but probably further developments will disprove the theory, as it has in all the mines on the hill occupying a corresponding position. The ore-body exposed in the present workings is something over 100 feet in length, and from 4 to 7 feet in thickness, most of it of very fair grade, 75 tons lately milled having yielded at the rate of $880 per ton.

A new and apparently large body of excellent ore has recently been encountered in the Ward Beecher Consolidated, 30 feet east of the old works and at a slightly increased depth.

Ward Beecher South has materially improved during the last month. A new chamber has been opened 100 feet south of the Ladies' chamber. Its present dimensions are 60 feet in-length by 30 feet in width and 40 feet in height. The ore-body developed in this mine, taken in connection with the Earl and Ward Beecher Consolidated, of which it forms a part, shows a continuous ore-channel over 700 in length, by an average thickness of about 60 feet, depth unknown, at 160 feet perpendicularly from the surface. This is the greatest depth yet attained. The ore yields as well as nearer to the surface, and from appearances may continue in that direction indefinitely. This deposit has yielded something over 25,000 tons of ore since it was first discovered in the Earl chamber, of an average value of something over $40 per ton, or, in round numbers, $1,000,000. The present yield is 50 tons of $50 ore daily.

The Aurora North has developed the finest body of ore at present worked in the district. The mill assay for the last month has run from $78 per ton to $128; present yield 40 tons per day; greatest depth attained in ore, 73 feet. The ore from the two last-named mines is transported to the mill by tramway at a cost of 65 cents per ton. The cost of mining per ton is estimated at $7 and milling at $8.

I have been furnished with the following detailed description of the Stanford Mill, and with an account of the mode of working, and the results obtained up to the middle of November, by Mr. B. N. Lilienthal, the chemist of the establishment.

The 30-stamp Stanford Mill, designed and built under the supervision of William H. Patton, esq., in 1869, at Eberhardt City, White Pine County, Nevada, consists of one main building, 58 by 164 feet, to which the engine and boiler building in the shape of an L, or a wing 38 by 42 feet, is attached.

The main building is subdivided as follows:

58 by 24 feet, ore-house.
58 by 16 feet, drying-room.
58 by 48 feet, battery-room.
58 by 60 feet, pan-room.
58 by 16 feet, retort and melting room.

The fall of the mill between dump-boards and tail-race is 44 feet, divided as follows:

8 feet between dump-boards and ore-house floor.
8 feet between ore-house floor and drier.
2½ feet between drier-battery and floor.
13¾ feet between battery and pan-room floors.

12 feet between pan-room floor and tail-race.
—
44 feet, total fall.
==
The wing is subdivided as follows:
42 by 16 feet, engine-room ;
42 by 22 feet, boiler-room ;
and is built so as to bring the crank-shaft of the engine level with the cam-shaft.

The ore-house has a capacity of 350 tons when filled to the level of the dump-boards. There is a niche 8 by 10 feet in the center of its lower side for a Varney and Rix rock-breaker, the mouth of which is level with the ore-house floor. The drier, 52 by 10 feet, is divided into two equal portions, having each its own fire-place (6 by 2 feet in the clear) and chimney. Each has four flues, built of common brick, which are covered over with cast-iron plates, 36 by 30 by $\frac{3}{4}$ inches, joined at their ends by countersunk bolts and flanges. The straight dry-crushing battery of thirty stamps is divided into six batteries, each pair having one cam-shaft in common. The battery is a knee one, and has nothing peculiar in its construction. The stems are of 3-inch turned iron, placed 10 inches from centers, and weigh 750 pounds mounted, viz :

Stem	286
Boss	230
Shoe	120
Tappet	114
Total weight	750

The batteries make 98 drops of 8 inches per minute, and discharge on both sides, the stamps rising in the order 1, 4, 2, 5, 3. The screens have an inclination of 13 degrees from the perpendicular, No. 40 (1,600 meshes to the square inch) being used on the front and No. 30 wire screen on the back side. The batteries are fed by C. P. Stanford's self-feeder, which does its work satisfactorily. Double-armed cams are in use, constructed after an evolute of a circle, the distance between centers of cam-shaft and stem being $4\frac{3}{16}$ inches.

When a stamp drops 98 times per minute, the time during which one rises, drops, and is at rest, is—

$$t = \frac{60}{98} = 0.612 \text{ second.}$$

The time (t_1) of rising, by construction is $t_1 = 0.263$ second.
The time (t_2) required in falling 8 inches is—

$$t_2 = \sqrt{\frac{2h}{g}} = \sqrt{\frac{2 \times \frac{2}{3}}{32.166}} = 0.204 \text{ second,}$$

showing the time (t_3) of rest to be

$$t_3 = t - (t_1 + t_2) = 0.612 - (0.263 + 0.204) = 0.145 \text{ second.}$$

Rittinger, in his Aufbereitungskunde, gives as an empirical rule that the stamp requires 0.2 second rest. But the friction in a California battery is less than in a German one.

In addition, we find, by construction, that at the instant the stamp touches the mortar, the highest point of the ascending cam is $3\frac{3}{4}$ inches below the tappet, allowing a sufficient modulus of safety.

The theoretical horse-power required by the battery, when making 98 drops of 8 inches per minute is, calling—

$n = 98 =$ number of drops per minute,
$w = 750 =$ weight of a stamp in pounds,
$h = \frac{2}{3} =$ drop in feet,
$m = 30 =$ number of stamps,

$$x = \frac{n \times w \times h \times m}{33000} = \frac{98 \times 750 \times \frac{2}{3} \times 30}{33000} = 44.54 \text{ horse-power.}$$

Fifty-five tons of ore per day for thirty-one successive days is the best record of the battery; and forty-six tons of ore per day for fifty-four successive days is the worst record. (The stems had been worn so as to prevent the proper fitting of the guides.) In the first case we find that one horse-power per twenty-four hours crushes—

$$x = \frac{55}{44.54} = 1.235 \text{ tons.}$$

In the second case—

$$x^1 = \frac{46}{44.54} = 1.032 \text{ tons.}$$

A set of shoes and dies lasts about five months, and a set of cams about fifteen months. The pans are placed at right angles to the battery, in two rows of eight each, 15 inches below the level of the pan-room floor. They are the common flat-bottom pans, with steam-chamber, built by H. J. Booth & Co., $4\frac{1}{2}$ feet in diameter, 32 inches deep, making 57 revolutions per minute, and holding 25 cwt. of dry pulp. Five feet lower, to each two pans, is placed one Belden settler, with wooden shoes, $7\frac{1}{4}$ feet in diameter, $2\frac{1}{2}$ feet deep, making 11 revolutions per minute, and discharging the amalgam through a siphon. The settlers have each four plugs placed respectively 6, 12, 17, and 22 inches from the top. The lowest plug is only removed during the clean-up. Five and a half feet lower are the agitators, one to each two settlers, $6\frac{1}{4}$ feet in diameter, $2\frac{1}{2}$ feet deep, making 17 revolutions per minute. They discharge into the tail-race, and can be run down by means of plugs. Two Knox pans, 4 feet in diameter, are used to clean the amalgam. There are four retorts, 14 inches in diameter and 6 feet long, each set in a furnace separate and independent from all the rest. Each of the two melting-furnaces is capable of holding a No. 50 graphite crucible. A set of pan-shoes and dies lasts about four months.

The machinery is driven by a 140 nominal horse-power horizontal engine furnished with Scott & Eckart's governor and cut-off, built by H. J. Booth & Co., San Francisco, who also constructed all of the other machinery of the mill. Steam is furnished by three tubular boilers, 52 inches in diameter, 16 feet long, and each containing 51 3-inch tubes. The boilers are in one bank, with no dividing walls.

The ore is principally chloride of silver in silicified limestone. An analysis of an average of South Aurora pulp, worked during six months, gives the following composition:

$Si\,O_3$	=	49.600
$Ca\,O\,Co_2$	=	48.808
$Fe_2\,O_3$	=	0.600
$Al_2\,O_3$	=	0.400
$Mg\,O$	=	A trace
$Ag\,Cl$	=	0.192
$H\,O$	=	0.400
		100.00

The ore is hauled to the mill by teams and unloaded on the dumping-floor, where it drops 8 feet into the ore-house. Thence it passes through the breaker, where it is reduced to egg-size, caught in an apron, and then distributed over the drier by means of wheelbarrows. Here it is turned until it is dry, then shoveled into wheelbarrows, and emptied into the hoppers of the self-feeders. The ore leaves the battery as very fine pulp, a mechanical analysis giving:

Water	0.400
Metallic iron	0.006
Silver	0.145
Remained on No. 40 sieve	0.025
Remained on No. 60 sieve	1.978
Remained on No. 100 sieve	16.150
Passed No. 100 sieve	81.296
	100.000

As the pulp is discharged from the battery it is caught in cars, which, when full, are run out, weighed, and charged into the pans in quantities of 20 to 25 cwt., according to the rapidity with which the battery furnishes it.

Method employed in working the ore.—The pans to be charged are filled partly with water; the pulp, 20 pounds of salt, and ¼ pound of commercial cyanide of potassium, (containing 55 per cent. K Cy, c. p.,) to each part, are added; the whole is thinned down to the necessary consistency with water. (When the shoes and dies are new, about 200 pounds of amalgam and quicksilver remain undischarged in the pan.) The muller is then let down and left to grind for four and a half hours, then 250 pounds of quicksilver are added, and the grinding is continued for one hour. The muller is then raised so as to give the amalgam a chance to collect. Seven and a half hours after charging, sixty pounds of quicksilver are added, and the mass is thinned down with water. Eight hours after charging, the contents are run off into the settler, and the pan is ready for a new charge. Should, at any time during the charge, the quicksilver appear in bad condition, it is remedied by adding a small piece of cyanide of potassium.

The settler receives the contents of two pans, and all the additional water it will hold. The arms are kept revolving, the amalgam sinks to the bottom, collects at the siphon, and escapes through it into a tub. Six hours afterward the first plug is drawn, and a stream of water turned into the settler, the surplus, with the suspended sands, escaping into the agitator. Seven to seven and a half hours after the settler receives the charge, the second and third plugs, respectively, are drawn. The settlers are cleansed of the deposited sands every forty-eight hours. They are recharged in the pans with the pulp. The agitators receive the tailings from the settlers. They pass merely through these, so that the suspended sands and amalgam may have an opportunity to settle. A small stream of water runs continually into the agitator. The remainder run through the tail-race to the tailings-pile, where they are settled. The agitators are, once in twenty-four hours, relieved of the deposited sands which are recharged into the pans. These sands usually assay 50 per cent. of the ore-value.

The amalgam which passes through the siphons of the settlers is strained through No. 5 canvas strainers, and then carried to the cleaning-pans. There it is diluted with quicksilver and water, and a small piece of cyanide of potassium is added, stream is turned on, and the muller is allowed to revolve for a few hours. If any iron shows itself, it is removed with a magnet, and then the amalgam is strained as dry as possible.

When 1,000 to 1,200 pounds of amalgam have accumulated, it is retorted and melted in the usual manner.

Sampling.—Every half hour the sampler goes around the battery, catching some of the pulp as it falls into the cars. This is placed in a box, the contents of which are thoroughly mixed every twelve hours. From this an average sample is taken, which is assayed. The tailings-sample is taken from the agitator half an hour after the second plug is drawn from the settler, dried, and assayed, and if it shows over 20 per cent. of the pulp-assay, the ore is not amalgamating well, which must be remedied.

The work of the mill during the past year can be seen from the following table:

Date.	Tons worked	Bullion produced.	Percentage obtained.	Loss of quicksilver per ton in lbs.
January 17 to February 17	1,706$\frac{141}{111}$	$56,202 47	84.84	2.20
May 1 to May 31	1,567$\frac{1808}{888}$	35,972 90	85.00	1.83
July 6 to August 3	1,327$\frac{383}{385}$	42,797 34	85.80	1.53
August 6 to September 6	1,504$\frac{756}{900}$	64,733 68	88.20	1.68
September 20 to November 13	2,480$\frac{1822}{2000}$	67,766 99	85 60	1.33

County assessor's returns of ore worked in White Pine district for the quarter ending March 31, 1870.

Name of mine.	Number of tons.	Pounds.	Gross yield.	Remarks.
Aurora South	3,299	800	$95,734 70	
Aurora Consolidated	1,609	1,570	33,032 09	
Alta	157		1,570 00	Smelting ore.
Andrew Jackson	2	1,000	62 00	Do.
Butter Cup	4	401	133 08	Do.
Burning Moscow	9	500	762 58	
Binghamption	28	1,750	1,047 58	
Bounty	10		192 50	Smelting.
Badger Hill	7	500	123 25	Do.
Bismuth	13	1,500	206 25	Do.
Chloride Flat Consolidated	1,750		59,762 84	
Constitution	14		1,032 65	
Cliff	10	914	446 80	
Chloride Flat	22	174	992 97	
Chihuahua	24	1,000	490 00	Smelting.
Cadiz	154	1,000	1,725 50	Do.
Crescent	53	1,500	337 50	Do.
Cohalep	13		130 00	Do.
Cordoza	76		680 00	Do.
Dell	3	370	158 90	
Don Juan	10		80 00	Smelting.
Eberhardt	880		34,011 36	
Earl	307		614 00	
Elko	10	1,500	193 50	
Erie Company	18	1,500	337 50	Smelting.
Empress Josephine	22	1,000	450 00	Do.
Fletcher Mining Company	15	1,500	1,849 18	
Frazier Company	61	1,000	552 00	
Germania	20		160 00	Smelting.
Hoosier State	6		48 00	Do.
Hemlock	93	600	1,959 30	Do.
Iceberg	35	1,580	1,018 47	
Imperial	178	1,000	3,593 00	Smelting.
Jennie A	15		225 00	Do.
J. C. Hill	18	896	662 11	

County assessor's returns of ore worked in White Pine district, &c.—Continued.

Name of mine.	Number of tons.	Pounds.	Gross yield.	Remarks.
Lockport	26	1,000	397 50	Smelting.
Mazeppa	8	1,660	255 64	
Nelson	1	55	126 29	
O. H. Treasure	464	1,000	27,334 08	
Owego	137	830	11,274 50	
Port Wine	4	....	104 00	
Pinto	13	....	130 00	Smelting.
Ralbei and Steele	22	....	330 00	Do.
Stockholm	7	....	210 00	Do.
Silver Wedge	81	1,000	2,005 30	
Summit and Nevada	148	1,725	6,214 18	
Sage Brush	97	1,250	2,390 76	
Silver Star Consolidated	37	....	3,953 97	
Snow Drop	37	140	1,375 29	
Tom Paine	32	1,240	715 73	
United States	7	500	94 25	Smelting.
Wabash	178	500	6,114 83	
Montgomery	9	630	302 26	
Manhattan	16	1,930	596 97	
Mineral Point	22	1,000	270 00	Smelting.
Mollie Stark	128	....	1,024 00	Do.
Miser's Dream	35	....	280 00	Do.

County assessor's returns of ore worked in White Pine district during the quarter ending June 30, 1870.

Name of mine.	Number of tons.	Pounds.	Gross yield.	Remarks.
Aurora South	2,833	100	$133,982 67	
Aurora	568	275	15,220 27	
Autumn No. 2	41	1,307	2,604 17	
Alta	96	....	576 00	Smelting.
Burning Moscow	26	205	4,234 70	
Butter Cup	5	428	201 86	
Banner State	91	222	2,872 47	
Bourbon	11	....	352 00	
Baldy Green	11	805	300 90	
Blood & Co.	19	1,000	1,300 94	
Bismuth	89	1,000	1,342 50	
Blue Cloud	12	....	120 00	Smelting.
Big Treavare	3	1,000	87 50	
Chloride Consolidated	433	....	10,439 63	
Chloride Flat	86	1,867	4,401 05	
Clyde	14	....	539 00	Smelting.
Chaparell	51	1,000	1,249 00	Do.
Cadiz No. 2	71	....	863 75	Do.
Cadiz No. 1	21	1,000	215 00	Do.
Chihuahua	9	1,000	190 00	Do.
Caroline	13	....	130 00	Refractory.
Cream City	17	1,000	350 00	Smelting.
Derby	10	....	1,600 00	Do.
Davis	11	240	417 24	
Delmonico	11	170	763 95	
Double Eagle	17	85	545 03	
Dickinson	11	....	132 00	Smelting.
Eberhardt	186	1,485	3,809 53	
Empire	2	560	668 00	
Eunice	60	....	1,326 40	
Emerady	12	922	619 14	
Elko	81	1,000	1,874 50	Smelting.
Empress Josephine	16	1,000	330 00	Do.
Fletcher	7	1,312	1,599 78	
Fay	7	1,500	310 00	Smelting.
Foney	13	500	132 50	
Frank Ruland	7	1,000	105 00	
Genesee	64	743	2,360 55	
Great Valley	74	....	740 00	Smelting.
Hartwell	15	125	1,047 98	
Hemlock	229	205	6,128 49	
Iceberg	18	1,550	496 80	
Imperial	110	500	2,010 00	Smelting.
Jennie A	28	1,000	350 00	Do.
Kingley	15	500	122 00	
Mazeppa	1	775	714 45	
Mammoth	78	1,000	1,707 38	
Miser's Dream	69	....	707 25	Smelting.

County assessor's returns of ore worked in White Pine district during the quarter ending June 30, 1870—Continued.

Name of mine.	Number of tons.	Pounds.	Gross yield.	Remarks.
Montezuma	37	1,000	450 00	Do.
O. H. Treasure	2,472	....	119,425 27	
Owogo	33	1,452	1,016 76	
Ohio State	6	935	213 37	
Post Hole	158	1,750	3,119 44	
Progress	9	461	508 39	
Promontory	24	....	463 50	Smelting.
Roman Empire	19	1,000	243 75	
Summit and Nevada	758	1,500	23,629 72	
Sago Brush	24	859	793 00	
Silver Wedge	84	915	1,704 25	
Snow Drop	53	390	5,367 20	
Sierra Pasco	5	995	211 30	
Silver Star	2	1,164	373 99	
Stonewall	19	1,500	938 25	
Seymour No. 2	82	1,000	1,806 75	
Haratoga	4	1,780	235 52	
Saunders	3	566	218 61	
San Pedro	10	955	542 72	
Spanish	4	1,940	185 80	
Stamboul	4	1,441	264 77	
Smith, J. R	....	1,550	119 05	
Silver Brick	9	1,000	133 00	Smelting.
Soto	11	1,000	105 00	Do.
Trench	104	....	4,552 00	Refractory.
Do	12	852	447 43	Do.
Virginia	5	1,875	216 48	Do.
Wabash	292	685	3,919 93	
Wilson and Grasstree	12	....	300 00	Smelting.
Wamebails	10	....	150 00	Do.
Wagnicallicanian	8	....	160 00	Do.
Virginia	2	....	129 60	

Assessor's report of ore worked in White Pine district during the quarter ending September 30, 1870.

Name of mine.	Number of tons.	Pounds.	Gross yield.	Remarks.
Aurora South	2,889	. 150	$1,112,363 68	
Aurora	218	....	5,446 19	
Aurora Consolidated	20	....	100 00	
Autumn No. 2	52	462	1,833 63	
Burning Moscow	69	1,143	6,049 69	
Bourbon	66	99	2,120 22	
Bullion Hill	27	804	1,438 71	
Belmont	1	1,442	140 55	
Bamboo	32	825	1,621 77	Smelting.
Bismuth	11	750	182 00	Do.
Blue Cloud	41	....	451 00	Do.
Bowie and Brown	11	....	130 00	
Chloride Flat	61	117	4,814 51	
Cheshire	2	900	901 70	
Copiapo	7	1,540	415 27	
Carolina	43	750	433 75	Refractory.
Colfax	13	500	202 50	
Campardon	24	....	192 00	Smelting.
Delevan	10	1,568	451 16	Do.
Eberhardt	116	....	7,535 74	
Empire	9	1,264	1,126 71	
Eclipse	22	1,687	1,171 43	
Elko	23	1,000	470 00	Smelting.
Farewell	19	1,815	749 35	Do.
Falloy	2	1,429	255 33	Do.
Frank, F. M	....	108	378 00	
Frasier	48	....	384 00	Smelting.
Fay	37	1,000	1,425 00	Do.
Great Basin	30	100	1,051 75	
Genesee	321	1,000	7,604 57	
Gilkey	3	1,492	329 43	
Gorrilla	13	116	425 77	
Garibaldi	4	818	210 60	
Gregory	8	1,250	279 90	
Hartwell	10	438	304 78	
Harrington	4	1,380	183 06	
Iceberg	27	1,400	664 75	

Assessor's report of ore worked in White Pine district, &c.—Continued.

Name of mine.	Number of tons.	Pounds.	Gross yield.	Remarks.
Imperial	41	500	625 00	Smelting.
Indian Chief	11	1,131	765 78	
Jewett	5	292	419 57	
John Bull	10	500	143 50	Smelting.
Jennie A	22	380	266 28	Do.
Kit Carson	2	1,302	101 85	
Little Bilk	4	590	204 98	
Live Yankee	19		323 00	
Miller	8	1,887	335 82	
McCormick	4	1,814	238 63	
Martin		1,000	134 14	
Maria	7	824	518 09	
McBride	1	142	161 61	
Miser's Dream	20		350 00	Smelting.
Mollie Stark	101	1,500	1,017 50	Do.
Maria Louisa	10	1,000	105 00	Do.
Madison	63		353 50	Do.
Mapulesna	7	500	116 00	Do.
Nevada Star	16		360 00	Do.
Nevada	195	1,057	7,295 70	
O. H. Treasure	1,333		33,324 63	
Othello	2	1,371	261 00	
Pogonip	9	125	303 79	
Piermont	4	1,218	639 02	
Summit	28	947	848 65	
Silver Wedge	6	54	438 28	
Snow Drop	47	700	1,022 83	
Saratoga	13	652	836 60	
Spanish	15	1,090	738 84	
Scott	6	1,418	359 19	
Silver Plate	15	380	501 02	
Salazar	2	672	231 14	
Sheboygan	6	383	311 81	
Truckee	67		6,495 23	
Ward Beecher	651		17,520 00	
Winnebago	17		246 50	Smelting.

Assessor's returns of ore-worked in White Pine district during the quarter ending December 31, 1870.

Name of mine.	Number of tons.	Pounds.	Gross yield.	Remarks.
Aurora Consolidated	238	663	$9,470 46	
Aurora	49	1,928	1,160 10	
Aurora South	3,914	1,000	126,751 31	
Ballarat	5	1,935	388 60	
Bismuth	10	1,120	147 84	Smelting.
Black Diamond	3	008	370 77	Do.
Bowie and Brown	9	1,550	121 19	
Burning Moscow	33	1,900	935 14	
Cadiz	15	1,810	159 05	Smelting.
Charter Oak	13	278	851 00	
Cheshire	2	1,950	153 69	
Chihuahua	10	56	200 56	Smelting.
Copley	4	1,728	229 63	
Dubuque	19	588	192 94	Smelting.
Eberhardt	330	225	12,101 92	
Eclipse		1,800	408 00	
Enterprise	17		1,324 64	Smelting.
Farewell	11		110 00	Do.
Fay	15		450 00	Do.
Grant Valley	13		130 00	Do.
Hemlock	27	594	1,127 10	
Iceberg North	4	1,943	145 41	
Iceberg and Indiana	9	1,352	523 47	
Indiana	21	750	760 62	
Maryland	11	1,328	1,149 55	Pinto district.
Do	31		4,250 10	Do.
Mammoth		656	119 16	
McNevin	2	1,760	455 64	
Mineral Point	18	116	160 58	Smelting.
Miser's Dream	33	1,270	470 89	Do.
Mollie Stark	44		440 00	Do.
Mono	7	1,947	581 58	Do.
Noel	7		425 95	Do.

Assessor's repor of ore worked in White Pino district, &c.—Continued.

Name of mine.	Number of tons.	Pounds.	Gross yield.	Remarks.
Noonday	25	585	641 16	
O. H. Treasure	996	....	20,940 90	.
Pocotillo	1	244	178 65	
Slerm Pasco	10	1,125	425 97	
Seymour and Darby	. 14	760	571 00	
Sheboygan	.5	750	192 01	
Silver Plate	62	1,321	1,661 75	
Silver Wave	15	....	409 50	
Snow Drop	36	....	903 50	
Snow Drop South	129	233	3,402 08	
Summit and Nevada	248	1,839	9,613 27	
Toll Road	8	1,458	78 00	Smelting.
Trench	10	....	670 00	Refractory.
Do	5	1,669	439 23	Do.
Truckee	36	1,865	3,210 00	
Uncle Sam	8	....	192 00	Smelting.
Ward Beecher	1,457	569	80,565 78	
Do	55	....	6,266 15	
Waterloo	9	885	165 24	
Yosemite	29	1,192	205 96	

The smelting-ore was in all cases sold on the dump, and those mines here marked "smelting" constitute less than one-half of those from which ore was extracted for that purpose. To get at a correct estimate of the ore worked in this district it will be necessary to add about one-quarter for small lots and low-grade ore not taxable.—A. J. B.

Statement of ore worked in White Pine during the quarter ending March 31, 1871, (taken from assessor's books.)

Name of mine.	Tons.	Pounds.	Gross yield.	Cost of extraction.	Remarks.
Aurora South	1,553	200	$50,756 00	$14,754 45	
Aurora Consolidated	14	1,351	1,464 40	703 86	
Banner State	80	1,000	5,847 00	2,775 50	
Black Jacket	15	1,386	1,709 40	....	
Belmont	323	1,800	3,391 50	651 80	
Bourbon	27	365	2,123 25	1,094 82	
Charter Oak	20	704	582 50	293 50	
Compromise	4	1,320	915 52	....	
Dahlgren	1	800	199 35	35 00	
Dominion	2	1,005	180 26	50 00	
Empire State	3	100	365 61	....	
Emerald Isle	3	796	262 00	524 75	
Genesee	148	1,250	5,366 00	1,411 94	
Do	21	1,500	727 00	200 00	
Do	500	500	4,875 00	....	
Iceberg South	17	500	1,346 50	135 25	
Hemlock	11	100	828 50	181 55	
Maryland	39	....	3,194 80	812 50	Pinto district.
Mazeppa	0	100	623 00	476 00	
Metropolitan Mill	548	....	2,304 87	....	Tailings.
McCormick	....	1,739	234 44	30 00	
Mountain Chief	9	75	411 70	....	
Nettie McCurdy	6	1,405	975 00	....	Warm Spring district.
O. H. Treasure	64	....	1,310 72	320 00	
Oakland	1	1,000	174 00	100 00	
Pope	2	305	235 61	....	
Summit and Nevada	24	1,787	1,092 82	128 93	
Silver Wave	79	1,180	2,044 44	795 00	
Silver Plate	30	1,265	973 66	1,181 50	
Stonewall	331	1,900	3,497 25	....	
Smith	1	1,250	195 66	100 00	
Sentinel	1	28	138 25	....	
Swansea Mill	900	....	4,506 34	....	Tailings.
Truckee	16	200	1,308 12	·463 00	
Trench	14	100	4,350 00	1,689 60	Ore refractory.
Treasure Hill Mining and Melting Co.	1,208	....	17,274 00	....	
Uncle Sam	6	1,279	1,716 70	....	Argentiferous galena.
Virginia	224	1,546	11,649 98	1,123 86	Refractory.
Wabash	9	1,000	794 50	104 50	
Ward Beecher	904	649	99,485 25	7,328 86	

Statement of ore worked during the quarter ending June 30, 1871.

Name of mine.	Tons.	Pounds.	Gross.	Net.	Remarks.
All Around.............................	1	619	$214 82	$107 41	
Bourbon................................	124	1,235	2,355 31	471 06	
Bowie..................................	16	1,331	907 03	366 81	
Huffman................................	3	1,000	270 62	108 25	
Caroline...............................	2	178	1,090 55	545 27	Refractory.
Chloride Flat..........................	501	1,625	13,683 35	2,730 67	
Eberhardt and Aurora..................	2,072	....	165,060 00	85,125 51	
Flyan..................................	....	604	347 00	173 50	¶
General Lee............................	4	1,255	137 67	....	
Hamilton...............................	8	854	903 70	451 89	
Iceberg North..........................	2	1,404	106 93	42 77	
Lew Morgan............................		1,500	273 85	136 92	
Lucky Boy..............................	3	1,300	287 75	115 10	
Manhattan Mill.........................			2,213 00	....	Tailings.
Montgomery............................	6	990	263 43	105 38	
Metropolitan Mill......................	1,110	....	4,460 63	....	Tailings.
Mazeppa...............................	1	1,411	1,101 45	550 72	
Noonday................................	6	115	250 00	100 00	
O. H. Treasure.........................	107	1,185	7,223 57	2,889 43	
Do..................................	126	1,790	507 58	....	Sold.
Padding................................	1	1,978	279 00	139 50	
Powell.................................	1	782	308 65	154 32	
Piermont...............................	446	....	4,395 97	439 60	Piermont district.
Republic...............................	3	1,540	673 60	336 80	
Sheba Mill.............................	726	....	6,500 00	....	Tailings.
Silver Pinto...........................	45	550	1,644 87	673 94	
Silver Wave............................	215	550	8,554 69	3,421 87	
Smith..................................	18	1,410	1,498 88	599 55	
South Aurora..........................	1,065	1,575	22,287 65	4,475 93	
Summit and Nevada....................	31	1,643	1,736 68	1,096 68	
Treasure Hill Mining and Melting Co.	1,040	....	5,616 00	....	Tailings.
Trench.................................	30	383	7,714 37	3,657 18	Refractory.
Truckee................................	43	1,411	2,856 94	1,134 77	
Uncle Sam.............................	13	996	1,518 15	759 07	
Virginia...............................	62	1,940	2,040 93	816 37	Refractory.
Ward Beecher Consolidated...........	227	....	5,873 55	1,767 55	

Statement of ore worked during the quarter ending September 30, 1871.

Name of mine.	Tons.	Pounds.	Gross.	Net.	Remarks.
Alceon.................................	12	1,000	$860 00	$344 00	
Black Jacket...........................	13	1,000	411 50	164 60	
Blair and Banner......................	92	1,750	6,014 50	2,417 80	
Smoky Mill............................	498	....	3,486 00	....	Tailings.
Caroline...............................	5	911	1,893 78	830 06	Refractory.
Chloride Flat Consolidated............	163	1,700	8,129 50	3,250 80	
Copper-Silver Glance..................	9	1,500	334 25	141 70	
Crown Point...........................	1	300	250 00	125 00	
Daisy..................................	12	1,400	613 00	245 00	
East Sheboygan.......................	12	....	538 00	215 20	
Eberhardt and Aurora.................	5,874	....	201,680 33	99,436 80	
Do..................................	304	....	....	....	Tailings.
Featherstone...........................	30	1,000	1,005 50	402 20	
O. H. Treasure.........................	1,102	300	48,540 00	19,416 00	
Metropolitan Mill......................	540	....	1,860 12	....	Tailings.
Mofitt.................................	3	500	411 60	205 80	
Piermont...............................	538	....	7,450 08	1,491 34	Piermont district.
Sheba Mill.............................	369	....	1,349 00	....	Tailings.
Silver Pinto...........................	38	1,500	2,110 00	844 00	
Silver Wave............................	5	100	232 50	93 00	
South Aurora..........................	1,171	500	28,623 00	5,724 60	
Summit and Nevada....................	279	650	5,865 80	1,181 87	
Truckee................................	26	300	1,940 00	776 00	
Trench.................................	24	360	6,674 00	3,297 00	Refractory.
Ward Beecher Consolidated...........	2,258	....	56,103 04	13,353 04	

County assessor's returns of ore worked in White Pine district for the quarter ending December 31, 1871.

Name of mine.	No. of tons.	Pounds.	Gross yield.	Remarks.
Caroline	20	400	$5,101 00	
Eberhardt and Aurora	7,135	....	145,643 00	
East Sheboygan	15	400	1,075 60	
Edgar	377	....	6,972 25	
Iceberg	334	320	17,946 62	
Manhattan	49	1,000	1,530 00	
Noonday	8	1,800	792 50	
O. H. Treasure	600	1,800	12,790 87	
Do	47	....	327 85	Tailings.
Oasis	170	....	1,139 78	Do.
Piermont	236	....	5,099 80	Piermont district.
Pocatillo	70	1,800	2,684 40	
Swansea Mill	600	....	3,300 00	Tailings.
South Aurora	2,075	1,500	47,135 90	
Silver Plate	41	100	1,588 75	
Silver Stone	6	1,500	1,350 00	
Truckee	7	1,500	682 50	
Trench	61	1,200	10,165 91	
Ward Beecher Consolidated	150	....	6,370 00	
Manhattan Mill	137	....	12,577 00	Tailings.
Do	250	....	3,625 00	Ore purchased.
	7	....	1,000 00	Name of mine not given.

NEW DISTRICTS IN EASTERN NEVADA.

Several new mining districts have attracted attention during the year, but little has been done in these as yet beyond the location of claims and preliminary testing of the ores. Mr. A. J. Brown, who has visited these districts, sends me the following report:

The Schell Creek mines are situated in the Schell Creek range of mountains, about seventy-five miles northeast of White Pine district, and eighty miles south of Toano, the nearest station on the Central Pacific Railroad. The first discoveries of mineral-bearing lodes were made in the early part of June, 1871, about four miles north of the Schell Creek station on the old overland stage-road, and the Schell Creek mining district was organized. The formation containing the ledges is dolomite, overlying granite. The strata are very near horizontal in this part of the district, having a slight dip of only 6 or 8 degrees to the east, which is evidently caused by the eruption of masses and dikes of the rocks at the western base of the range. The dolomite appears to be entirely destitute of organic remains.

Very few of the mines are at all developed, but what little has been done has proved highly encouraging for their future value.

The McMahon ledge is situated low down, near the west base of the range, and very convenient for cheap working. The croppings are about 1,000 feet in length and 14 feet in thickness, and lie entirely in dolomite. The ledge is opened by a surface cut, exposing a face of ore, 20 feet in length by 8 inches thickness, that will average $75 per ton. A few tons of assorted ore, worked at the Big Smoky Mill, Hamilton, yielded $360 per ton. The gangue in this vein is principally quartz, intermixed with a small quantity of calc-spar. The ore is mostly silver-copper glance or stromeyerite, but black sulphurets, horn-silver, and some native silver occur also.

The Woodburn is situated about 300 yards above the McMahon, and near the summit of the range. This is a parallel vein, and is also incased in dolomite, the ore and gangue being similar to the last named. In fact, all the ledges situated wholly in the dolomite contain ore of the same character, and differ from the main mineral belt in containing a considerable percentage of copper, while the main belt is absolutely free from any base metal except a slight trace of antimony. The Woodburn crops to the surface for a·distance of 200 feet, and shows good ore the whole length. The thickness of the vein is 8 feet; it is opened by a surface cut, from which several tons of ore have been taken that will work well into the hundreds.

The Summit is situated half a mile north of the Woodburn, and has the appearance of being a layer or " shut vein." The foot wall is dolomite, and the hanging-wall a thin bed of argillaceous shale underlying quartzite. Its dip and strike correspond with the strata of the country-rock. The croppings are 600 feet in length and 50 feet in thickness. The ore is almost entirely black sulphuret, very evenly distributed through a quartzose gangue. Some of the ore is very rich in silver, and assays of it run high into the thousands. One ton of ore worked at the Big Smoky Mill yielded $395 per ton, but the average of the vein is probably not over $50, and not over one-fifth of the whole vein can be worked at a profit.

There are several other promising mines in this part of the district, but as nothing has been done toward their development, we can form no estimate of their character or importance.

The Queen Spring mines are situated three miles south of the overland road, and are separated from the North Shell Creek mines by a hill of porphyry three miles in length. This hill is traversed by dikes of greenstone and greenstone-porphyry, but no mineral-bearing lodes have been discovered on it. On the next hill south of the porphyry the dolomite again makes its appearance, and with it an immense quartz outcrop, one and a half miles in length, and from 50 to 150 feet in thickness. The dolomite occupies here the east side of the range from the base to the summit, and dips to the west at an angle of about 25 degrees. It forms the foot-wall of the main vein or belt, the hanging-wall being quartzite. The vein here has the same general character as the Summit, before described. The mineral combinations, as well as the quartz-gangue, are identical, and its position in regard to the dolomite is the same.

The Gem, situated at the extreme north end of the belt, is the oldest location in this part of the district. It is opened by a surface cut 20 feet in length, exposing a good body of milling ore.

Silver Chariot is opened by a shaft 20 feet in depth in a mass of high-grade ore. A few tons milled yielded $188 per ton. About 25 tons of the same class are on the dump.

El Capitan is located on a mass of croppings 600 feet in length, 60 feet in breadth, and 30 feet high. Little work has been done, but samples broken from the croppings assay from $20 to over a hundred per ton, and there are probably several thousand tons of fair milling ore in the outcrop alone.

The Sweepstakes is opened by a shaft 15 feet in depth, all in good milling ore. A hundred tons now on the dump will work $70 per ton.

The Excelsior is opened by a surface cut. A good face of ore is in sight.

The War Horse is opened by a shaft 30 feet deep, and by a drift which is 50 feet in on the vein. The ore is mostly low grade.

On the Fairy Bell there is no work done as yet, but a mass of croppings 10 feet thick and 50 feet in length. Prospects very well.

The Nutmeg is opened on the surface by a cut 60 feet in length and 8 feet deep, showing good ore everywhere. A shaft is sunk 12 feet in a body of black sulphuret ore. Two hundred tons on the dump are worth $80 per ton.

On the Storm ledge but little work has been done, but there is some ore of good quality on the dump and in sight in the outcrop.

All these mines are located on the main belt, and yield sulphurets and horn-silver ores. There are slight traces of copper, but no lead-ores.

Silver Queen lies to the east of the main belt, and is incased in dolomite. The vein is 8 feet thick, the ore is stromeyerite, with scales of native and horn-silver.

The San Francisco, incased as the last, carries the same ore. The vein is 10 feet thick, and opened by a surface cut, exposing a large body of good ore.

The Le Bross has dolomite casing, lies near the main belt, and the ore partakes of the character of both the above-mentioned systems of veins. It is opened by a short tunnel, 20 feet in length. A few tons of good ore are on the dump, some of which yielded $95 per ton at Big Smoky Mill.

South of the Nutmeg mine the dolomite is gradually replaced by a dark-colored calcareous shale, and the main belt becomes poorer on the surface. Nevertheless the whole ground is located, and may possibly improve when developed. Several well-defined veins have been discovered near the eastern base of the range, about four miles south of Queen Springs. These veins are found partly in a highly siliceous, stratified rock, and partly in the limestone near it. They crop out boldly for a distance, varying from 1,000 to 4,000 feet in length, but all correspond with the strata of the country-rock in dip and strike.

The Home Ticket is the oldest location in this part of the district, and has been worked quite extensively, developing a good body of ore, 5 feet in thickness, and worth $75 per ton.

The Austin ledge is a late location, and has not been worked. Samples taken from the surface assay from $382 to $612 per ton. The ledge is 6 feet thick and well defined.

The Elephant is incased in quartzite, and is 12 feet thick. Much of the ore shows well in black sulphurets, and some of it is so rich that the heat of an ordinary blacksmith-forge completely coats the surface with globules of pure silver.

The Ruby Hill mines are situated nine miles south of Queen Springs, near the summit of the same range, and are evidently a part of the same mineral belt, as the dolomite is again found on the east side of the vein, forming its foot-wall. A large mass of greenstone, one mile in length and more than 1,000 feet across its greatest breadth, has split the vein here into two branches, one of which follows the dolomite along the east side of the greenstone, while the other runs along the calcareous schists on the west. The deposits in this part of the district are very rich, some of the ore being literally one mass of black antimonial sulphurets of silver; but the disturbed and broken character of the deposits must necessarily detract much from the value of the mines.

The Cow and Calf is the first location made on this hill. Very little work has been done. The vein is 10 feet thick, and several tons of good ore are on the dump.

The Ferret near by shows a vein 6 feet thick, or perhaps more properly a chimney of ore 6 feet thick, for both mines are evidently on the same outcrop.

The Columbia is opened by a surface cut, and shows a face of ore 20 feet in length and 6 feet in thickness. Twenty tons are on the dump, most of it very rich in silver.

Silver Wreath is located on a mass of wonderfully rich croppings, some of which assay as high as $18,000 per ton, and several tons now sacked will work $800 to $1,000 per ton.

Ratlers' Pay and Ramblers' Luck are near by, toward the east, and in the same mass of croppings. They show the same class of. mineral in abundance.

Silver Bluff has a chimney of ore 10 feet in thickness, and several tons on the dump will work well into the hundreds.

Lookout has a chimney 100 feet in length and 30 feet· thick. Several tons of rich ore are on the dump.

The five last-named claims are all located on rich chimneys of ore in the same outcrop. The whole croppings taken together are 600 feet in length and 200 feet wide on the surface. This belt of mineral has been traced four miles south of Ruby Hill, and many of the locations show promising indications of future value.

All the localities so far described are collectively known as the Schell Creek mines. They form one of the most promising as well as one of the most extensive mineral belts known in Eastern Nevada. The whole western base of the range, to within a short distance of the mines, is evidently of volcanic origin.

Warm Spring district is situated twenty-two miles east of Queen Springs, in the Antelope range of mountains. So far as known there is only one mine in this district.

The Nettie McCurdy was discovered in the summer of 1870, and is opened by a tunnel or drift along the ledge, 40 feet in length, and by a side drift across the ledge of 14 feet, the thickness of the vein. The ledge is incased in limestone and slate. The gangue containing the ore is a sandy quartz. The ore contains considerable lead. The length of the ore-body on the surface is 200 feet. Eight tons worked at the Big Smoky Mill yielded from $141 to $278 per ton. One hundred tons of second-class ore now on the dump are estimated to be worth $8,000.

Piermont district is situated about twenty-four miles south of Schell Creek, at the eastern base of the same range, and was discovered and organized in 1869. The formation is argillaceous slate, alternating with graphitic slate and quartzite. The mines are either incased in argillaceous slate, or between it and the graphitic slate or the quartzite, and correspond with the strata of the slates in dip and strike. There are several locations in the district that promise well if properly and systematically developed. The Elephant, Latrobe, and Spear mines have all been tested by shafts sunk to a depth of 35 to 50 feet in large bodies of fair milling ore. The ledges are large and well defined, and wood and water are very abundant in the immediate vicinity of the mines. The Piermont mine is located at the very base of the range, on the south side of Piermont Cañon, and convenient for cheap working. It has been quite extensively prospected during the past year, and is now opened by a tunnel, run to the ledge through 100 feet of country-rock, and a drift thence along the hanging-wall 150 feet in length. For convenience the ledge has been divided into sections of 200 feet each, which are numbered, from the north to the south, from 1 to 6. The first section, in addition to the tunnel, has two shafts, one 36 feet deep and the other 45. Section 6 is opened by a shaft 36 feet deep, and by drifts from the bottom along the lode 25 feet each way. The vein north of this station is badly broken up, and considerable difficulty has been experienced in working that part to advantage; but the mine is yielding 20 tons of ore per day, worth $40 per ton. The ledge varies from 3 to 8 feet in thickness. A fine new 10-stamp mill belonging to the property

204 MINING STATISTICS WEST OF THE ROCKY MOUNTAINS.

is situated 200 yards from the mouth of the tunnel, and a car-track connects it with the mine, by which the ore is delivered at the mill for less than $1 per ton. The whole expense of mining the ore and converting it into bullion is estimated at $10 per ton.

WHITE PINE MOUNTAIN.

The mines on the west side of White Pine Mountain were the first discovered in White Pine district, but during the excitement attendant on the discovery of the rich ores of Treasure Hill they were neglected until the building of the new Monte Christo Mill in June last. Since then the prospect of having a good mill, convenient of access, and one adapted to the reduction of the refractory ores found in that locality, has induced several of the mine-owners to start work on their mines; and in nearly every instance developments have given satisfactory evidence of valuable mines.

The Trench mine has been opened by two shafts, each 50 feet deep, and by drifts from the bottom of the shafts, 50 to 100 feet in length, in all aggregating about 350 feet, and running mostly through bodies of high-grade ore. No ore has been extracted from the mine beyond. that necessarily excavated in running the drifts. This, amounting to about 100 tons in all, has been worked, giving a gross yield of $32,000 or $320 per ton, one-half of which has been profit. The vein is in a thin bed of dolomite overlying shale, and is very irregular, varying in thickness from 1 foot to 8 feet. The ore is argentiferous galena.

The Bald Eagle is a well-defined vein incased in an altered shale. The granite formation is only a few yards distant from the works, and the ledge passes into it. The ledge is only 3 feet thick on top, but at the bottom of the incline, 120 feet from the surface, its thickness has increased to 10 feet, nearly all of which is fair milling ore. Ten tons of assorted ore yielded $138 per ton. There are 50 tons of first-class and 150 tons of second-class ore on the dump.

The Philadelphia is a parallel vein. The incline is 110 feet deep, and the ledge 4 feet thick. Assorted ore works $124 per ton.

The Badger State is incased in granite. The vein is 16 feet thick, and opened by an incline 30 feet deep. The ore is nearly pure sulphide of antimony.

The Uncle Sam is opened by an incline 56 feet deep, which shows a ledge 5 feet thick. The ore is argentiferous galena. The average value per ton is $68, while some of the richest ore yields $320 per ton.

Mills in White Pine.

Name of mill.	No. of stamps.
Manhattan	24
Dayton	20
Big Smoky	20
Swansea	10
White Pine	10
Sheba	10
Oasis	10
Stanford	30
International, (new)	60
Staples	8
Monte Christo, (new)	20
Two other small mills, 5 stamps each	10
Number of mills, 13; number of stamps	232

· The Dun and McCone Mill, 10 stamps, was destroyed by fire; the Henderson, 5 stamps, was removed to Schell Creek, and the Chicago, 10 stamps, to Pioche. The Metropolitan, 15 stamps, is now being removed to Eureka. Moore and Barker, 8 stamps, has been dismantled.

HUMBOLDT COUNTY.

Mining operations in Buena Vista and other districts of this county have been tolerably active during the year. Many old mines have been re-opened with greater skill and economy, and with better results than formerly; but the lack of capital has in many cases retarded operations.

The leading mine of Buena Vista district in 1871, as in preceding years, was the Arizona. This is, indeed, one of the leading mines of the West, and merits attention by reason of its curious vein-formation, as well as its extent and value. The ledge is quartz, carrying distributed particles or bunches of sulphuret of silver, argentiferous lead and . copper ores, and antimonial ores, with occasional chloride of silver and native silver. It is inclosed in calcareous slate, forming the crest of a ridge of siliceous porphyry. The vein appears to follow the stratification of the slates, yet has all the other characteristics of a fissure. The outcrop runs partly around the hill, and two companies at least, the Manitowoc and the Arizona, at one time held locations upon it, and worked their mines toward each other in depth, one location showing a strike of N. 30° W., and a southwest dip, while the other had a strike of N. 15° W., and an easterly dip of 35 degrees; that is, on the supposition of a single vein, the vein at these two points had two dips, approaching each other. In fact, the ore-seams came together in depth; and, in the lawsuit which arose between the companies, it was held to be established that this portion of the ledge constituted a fold or basin, on the opposite slopes of which the two companies were working. The Manitowoc and Arizona being now consolidated, this question has ceased to have a legal bearing; but it is highly important in its relation to the still problematical true course and dip of the whole deposit, as I shall presently show.

The principal opening of the consolidated mine is a tunnel, which starts from the north face of the hill, and cuts, about 150 feet from its mouth, the foot-wall of the vein, or the bottom of the "basin." The axis of this basin dips southward, and the tunnel, if continued in a straight line, would pass through the vein; but it is forked at the point of intersection, and two horizontal drifts are run, one on each side of the fold. These branches are called the east and west tunnels, and separate from one another at the rate of about one foot for every four feet of running length. At the distance of 1,000 feet from the fork they are about 275 feet apart. The east tunnel underruns the old Manitowoc ground.

On the supposition of a "basin," it must be concluded that the general course of the ledge is east and west, and the dip southward; but it is still possible that the old Arizona, with its eastern dip, is the main ledge, and the Manitowoc a spur of it, ascending to the surface. The theory that the two actually cross each other, and that their continuations will be found below the "fold," lacks proof. But I noticed, in September, 1871, a spur shooting downward from the foot-wall on the eastern side, which closely resembled, in dip and other characteristics, the Arizona vein as shown above, on the west side. And I infer, with some reason, from the fragmentary reports of operations which have since come to hand, that this promises to be a strong and regular vein. ·

The west tunnel had reached, in October, a distance of about 1,100 feet from the forks. Here a displacement of the ledge was met with, and, having a large amount of ground open, the company postponed continuing the tunnel for the present. As many breaks have been found in the ledge, and all insignificant in extent of displacement, no special importance attaches to this particular one at the end of the west tunnel. The east tunnel seems to have lost the ledge in broken ground, some 350 feet from the forks. It has been continued to 800 feet without certainly finding the ledge proper. This is beyond the workings of the Manitowoc.

During the year, while the west tunnel was driven ahead, the ledge has been stoped along its upper and western side, toward the north, (i. e., toward the entrance of the mine,) for about 150 feet on the gentle rise of the ledge, and at the other end for about 10 feet. There is, therefore, a stope 1,100 feet long, 10 feet high at one end, and 150 feet high at the other; only, instead of standing steeply, as in most veins, it is not very far removed from a horizontal position. The workings look, in this respect, more like a coal-bed than like a metal-mine. The above estimate of 1,100 feet of breast for stoping makes no account of the few and small barren spots caused by little "jumps," or displacements of the vein. The ground stoped out is filled mostly with waste, the necessary passages and dumping-spaces being left open. The average thickness of the ledge is $2\frac{1}{2}$ feet, the range being from 10 inches to 5 feet. The hanging-wall is very strong, and there is usually a convenient clay "gouge" between ore and slates.

On the east side, and below the tunnel, the ledge has been worked out to a varying depth. At about 85 feet from the forks a passage is cut through, following the "sag," from the west to the east tunnel. The lowest point is about 20 feet below the tunnel-levels. A similar passage between the two tunnels, a few hundred feet farther south, would have to go much deeper.

For about 100 feet southward from the lowest point referred to, quartz has been found, going downward to the east, and removed. And again, at various points in the tunnel, rich rock has been removed to the depth of 10 or 20 feet when discovered. For exploring the ground east of the west tunnel, it is reported an incline was sunk at a point about 900 feet from the forks, to the depth of 120 feet, on a uniform good ledge, dipping about 20 degrees east, with only one slight displacement.

The yield of the ledge is remarkably regular in quality. The first-class ore is worth $500 per ton and upward; the assay value of the mill-ore about $60; and the yield by first process (Washoe amalgamation) about $33. There are large quantities of tailings, variously estimated in value at $250,000 to $1,000,000.

An important economical improvement of the year has been the construction of a tramway from the mine down to the bottom of the cañon below. The loaded cars, carrying about one ton of ore each, descend by gravity, and pull up the empty ones. The descent of 1,900 feet occupies about five minutes. The saving is $2\frac{1}{2}$ miles of bad road in the distance, and about $1.25 per ton in the cost of transportation. The amount of ore extracted from the Arizona during the year is about 7,000 tons.

The Arizona Association operates two stamp-mills,* and a pan-mill for the treatment of tailings. The stamp-mills have ten stamps each, of about 700 pounds, dropping 8 inches. At the lower mill the rate of

---

* One is owned, I believe, by another company, called the Silver Mining Company.

running was 96 drops per minute.   This was perhaps too fast, considering the extreme fineness (No. 40 wire) of the screens.   The capacity of each mill is about 15 tons daily, or 1.1 tons at the lower, and about 1.3 tons at the upper, per horse-power developed at the stamp.   The tailings-mill will be mentioned below.

The North Star mine has been worked during the larger portion of the year.   The ledge is much disturbed and broken up by porphyry, which surrounds and underlies the ground.   The irregularity thus occasioned is the cause of the expensive and difficult working of the mine, as much prospecting is required and much quartz has to be moved, which needs much picking to obtain the required grade for milling or shipping ores.   Thus far the works have not reached 100 feet depth. The mine is in limestone.   The amount of ore extracted during the year is about 650 tons, of which about 7 tons have been selected for shipping-ore, the balance being treated at the Pioneer Mill, and yielding about $17,000.

On the Henning, (an old claim, about a mile south of the Arizona), work was commenced in July, and has been prosecuted without interruption since.   A tunnel has been run for about 150 feet on the ledge. The outcrop had spots of rich mineral; but the amount has been increasing steadily as the tunnel is driven into the hill.   A few tons of the ore taken from near the mouth of the tunnel were worked by the Pioneer Mill, but, being poor, assayed only $25 per ton.   The mine is in calcareous slate, similar to the Arizona country-rock.

On the old Peru, now the Agamemnon, a drift has been run on the ledge in the upper tunnel for about 150 feet.   Several tons of ore, rich in mineral, have been selected for trial, and await now more favorable weather for removal.   Assays of pieces run from $45 to $500 per ton, with a good percentage of gold.   A contract has been let to run 75 feet of drift and 75 feet of incline on the ledge for the ore to be taken out during the work.   The mine is situated about half a mile southeast of the Arizona, in a deep ravine.   The outcrops of the ledge can be traced for several hundred feet, crossing the hill east of the Arizona.   It is in the metamorphic rocks of the country, quartzite and slates.

The Eclipse is located on the main range of hills west of the district, at a short distance from the summit, and is in calcareous slate, which caps, with the limestone, the hill above.   The tunnel driven to intersect the ledge was constructed during the year, and a good-sized, mineral-bearing ledge was found.   Drifts were run, and sufficient quartz was taken out for a fair trial at the Pioneer Mill; but the result was not very favorable.

The prospecting of other claims has not been carried far enough to render them worthy of special mention.

The three stamp-mills in the district have been at work with little interruption during the year.   The Pioneer has worked the old tailings in its reservoir, the North Star rock, and the tailings of about 2,000 tons of Arizona rock worked last year.   The Arizona and the Silver Mining Company's mills reduced, until the end of October, rock from the Arizona mine.   Since then the latter has worked the tailings lying in its reservoirs, and the former has worked rock three-quarters of the time and tailings the rest.

During May and June the Arizona Association built a tailings-mill with six Varney pans, to work the tailings from the Arizona and the Silver Mining Company's mills.   It is built between these two mills, and the tailings are brought to it in sufficient quantities by a one-horse wagon.   An engine and boiler give the necessary motive-power, yet a turbine wheel is connected with the machinery, worked by a 45-foot fall. It is the intention to work it with this power, when the water of the cañon is in sufficient quantity for the purpose. The mill works from 25 to 30 tons of tailings in twenty-four hours. The mere reworking of the tailings did not give the satisfaction expected, and an Akin furnace has just been completed to roast the tailings. The furnace is said to have worked well; but enough drying-surface had not been provided, so that the required quantity of tailings could not be dried in twenty-four hours for the run of the whole mill. This, together with the poor success of a machine for breaking up the tailings, led to the stoppage of the furnace, after four or five days' running. Another difficulty may be the fumes of mercury from the tailings while drying. The chlorination attained to 84 per cent. by the first working; but it cannot be said that this is the best result possible, for the furnace was not run long enough to regulate the working thoroughly.

The bullion produced by the mills is as follows:

From Arizona rock, by two mills, about	$225,500 00
From Arizona tailings, worked at different times by the three mills, about	72,000 00
The Pioneer Mill shipped from rock about	12,500 00
The Pioneer Mill shipped from tailings about	36,000 00
Total product about	346,000 00

There were shipped from the Arizona mines about 170 tons of select ore to San Francisco, netting $78,000.

Monthly shipments of bullion.

For January, 1871	$32,788 62
For February	23,966 76
For March	28,422 55
For April	24,664 87
For May	23,096 21
For June	31,478 21
For July	35,535 78
For August	30,199 43
For September	27,870 26
For October	35,737 43
For November	27,061 07
For December	26,000 00
Total	346,821 19

Star district.—One of the earliest and most furious mining excitements in the State of Nevada was that which followed the discovery of the Sheba mine and the organization of the Star district, ten years ago. For a considerable period this cañon was the scene of an enthusiastic and busy industry; Star City became a flourishing mining town, with two hotels, an express office, daily mails, a telegraph-line to Virginia City, and a reported population of more than a thousand inhabitants. Yet in 1868, when I first visited the district, a decline as sudden and rapid as its rise had left of all this prosperity and promise no trace except the empty houses of the town, the abandoned mining-works, and the daily mail and the telegraph and express offices, which had not yet been removed. The collapse had been quick and apparently complete; yet a study of the deserted district led me to report at that time to the Government (see my Report on Mining Statistics rendered January 18, 1869, page 127) that it would certainly sooner or later receive attention again. This opinion was based partly upon evidences presented by the district itself, partly upon inferences from its history. It was notorious that the whole community had followed the fortunes of the Sheba mine. Scarcely any other in the district, except the De Soto, and perhaps one or two minor enterprises *on the same belt with the Sheba*, had been productive of anything more than assays and prospects. When the ore-bodies of the Sheba were exhausted, and the expensive-prospecting-works of that company failed to disclose any continuation or repetition of them, the ruin of Star City was inevitable and immediate. But in this very circumstance was the ground of hope for the future. The successful re-opening of the Sheba mine would be certain to recall the prosperity which its close had driven away. Moreover, the evidence presented by the mine itself and its surroundings was sufficient to justify the expectation (which has since been realized) that the main deposit, outlying chambers of which had formed the basis of former operations, might be found by more careful and rational search. This point will more clearly appear from a description of the district. •

Star City is about twelve miles north of Unionville, at present the principal town of Humboldt County. The Star district lies on the eastern slope of a range of mountains, the highest peak of which (Star Peak) has an altitude of about 11,000 feet above sea-level. The deep, wide, and tortuous cañon runs from the summit of the range easterly to the broad, dry valley, and collects the waters of a considerable area into Star Creek, which carries 70 miners' inches in the dryest summer months and swells to 200 or 300 inches in the early summer, after the snows begin to melt. This stream flows across the outcrops of the metalliferous veins, which mostly course parallel with the range. The *débris* and gravel are perhaps 20 or 30 feet deep in the bottom of the cañon; and although the mainly argentiferous character of the ore-deposits of the district gave little hint of alluvial gold, yet a company of enterprising prospectors have recently commenced gulch-mining, and are reported to have opened very profitable, though limited ground. This occurrence of gold placers below the outcrops of silver-mines is not unprecedented. It will be remembered that the Comstock ledge in Nevada was discovered by following up the gulch deposits to which it had given rise. That vein has always since produced a considerable quantity of gold associated with its silver; but less auriferous silver-mines may still, in the course of time, have given rise to accumulations of gold in placers. The two metals habitually occur together in nature.

In ascending the cañon, the sections of successive strata may be observed, showing a general north and south course, and a dip westerly, into the mountain. They consist of alternating quartzite, limestone, shales, and slates. The first formation crossed above the foot-hills is a metamorphic quartzite, gray when freshly broken, but changing on exposure to brown. The next layer above is a gray limestone, dipping west about 60°, and bedded in layers of a few inches thickness, the whole group being about 100 yards thick. Overlying this is a large development of black limestone, in which several silver-bearing deposits have been discovered. This is succeeded by gray quartzite, and the latter, in turn, by black slate. Between the two formations last mentioned is a belt or channel, more than 200 feet wide horizontally, of bluish-gray silicious and calcareous rock, characterized by the prevalence of small threads of crystallized quartz, and (on all the cleavages)

H. Ex. 211——14

coatings of talc. It was in this channel that the Sheba deposit was discovered, a few feet only from grass, and close under the hanging-wall, bounded, in fact, by the black-slate roof above, and by a well-defined horizontal floor in the veinstone below. The first and largest chamber having been worked out, one or two smaller ones were found close by, likewise lying upon the same floor and close to the hanging-wall. About $125,000 were extracted from these bodies, and very large sums were expended in the search for further deposits. Quite naturally, the floor which formed the lower limit of those already exploited was supposed to be a foot-wall, and it was expected that one would be again found, if found at all, hugging the hanging-wall as before. Consequently, all explorations were confined to this part of the belt, with the exception of a single cross-cut, which was run about 200 feet towards the underlying quartzite. At that time it was by no means certain that the whole belt between the quartzite and the slate belonged to one vein, and must be considered as veinstone. On the contrary, two other parallel claims were located on the surface, within these limits, upon. small stringers or threads of ore, of no real independent importance.

The futile exploring-drifts of the Sheba Company presented one indication which might have led, and did finally lead, to the discovery of the true nature and position of the vein. I refer to the fact that the numerous small stringers of quartz, and sometimes of ore, encountered in the veinstone mass, dropped away towards the distant east or foot-wall. This fact appeared insignificant at the time, in comparison with the encouragement given, by the large bodies already found on the hanging-wall, to further explorations in that part of the channel. These explorations were expensive, thorough, and utterly barren of results; and the company suspended operations (as so many companies do) just when the negative evidence accumulated would have led them inevitably to continue their search, if they had continued it at all, in the right quarter.

After several years of abandonment, the property came into the hands of its present proprietors, and work was resumed under the direction of Mr. Samuel Stewart, superintendent of the well-known Arizona mine at Unionville, Mr. Richard Nash having immediate charge of the operations. The latter, who was for many months in the employ of the old Sheba Company, and possessed perfect familiarity with the history of their operations, has made good use of the experience of the past. After some further exploration along the hanging-wall, resulting in nothing more than the discovery of small outlying pockets—mere remnants of the old bonanza—he resolved to follow the indications of the slips and stringers in the veinstone, and to look for a main body further east than any workings had previously done. Accordingly he drove a tunnel northward into the belt from the cañon, and cross-cutting towards the foot-wall, considerably under the old works, struck, at a distance of about 650 feet from the tunnel-mouth, a large, well-defined, and rich vein.

It is now perfectly clear that the numerous stringers and threads of quartz traversing the channel of veinstone drop to the foot-wall, upon which they unite to form the vein. The bodies found in the old Sheba mine were within the wide channel, but far outside of the ore-vein now disclosed. Such outlying chambers are of frequent occurrence in ore-deposits of this character.

The vein recently opened has been exposed by a drift for about 100 feet in length. Stoping has been carried on for about 25 feet above this drift, and a winze has been sunk below the drift about 15 feet.

The minimum width of quartz and ore is 20 inches, the maximum 9½ feet, by measurement in the roof of the stope. This large width constitutes a bulge in the vein about 16 feet in horizontal length. At the bottom of the winze there is a four-foot vein. The average width of the ore-vein throughout these exposures is 4½ feet. At the north end of the opening it is somewhat scattered, but recovers itself beyond and below. At the south end there is a clay seam cutting and heaving the vein. At the time of my visit (September 6) a short cross-cut was in progress, to strike the vein beyond this break; and a characteristic band of black rock, which everywhere accompanies the vein-walls, and which has been seen nowhere else in the mine than in that position, had already been struck in the cross-cut, leaving no reasonable doubt of the immediate vicinity of the ore. The vein where it is cut off has its maximum width, and the cross-seam is so thin and smooth, and gives so "clean" a cut, (without broken ground,) as to indicate but a very small heave.

The structure of the vein is perfectly regular and normal in appearance. Vugs and combings occur in it frequently, as well as a banded structure, which promises excellently for both permanence and quality. The clay partings are thin, and the foot-wall is hard and well defined. Although so many stringers are dropping in from the hanging-wall, to swell the dimensions and value of the vein, none have ever been found to penetrate the foot-wall.

The course of this vein by compass is north 11° west by south 11° east. The dip varies from vertical, and in one place, 70° east, to 60° west. The immediate gangue is quartz, and the ore is principally fahlerz and argentiferous zinc-blende. The first-class ore, comprising apparently about 5 per cent. of the whole, is worth from three hundred to five hundred dollars per ton. The second-class ore (judging from its appearance as compared with what I remember of the old Sheba ore) should be worth fifty to seventy-five dollars.* The mill of the Sheba Company, formerly situated at the north of the cañon, was sacrificed and lost during the pecuniary embarrassments and confusion of an interregnum; but a small water-mill has been erected in the cañon, containing five stamps and a small number of concentrating-machines, for the purpose of crushing and concentrating the ore thrown aside as unprofitable during the former workings. This mill is run at a small profit, in spite of the great waste of silver involved in the use of wet concentration upon ores containing brittle compounds of silver.

These ores consist of sulphurets of silver, argentiferous fahlerz, and zinc-blende, the last mineral having been found by separate assay to carry a high value in silver.

The true method of treatment would be to roast with salt all the ores, of every grade, in the Stetefeldt furnace, the Brückner cylinder, or the ordinary reverberatory, and then extract the silver by close amalgamation in barrels or pans. I entertain no doubt that the whole mass of the vein described can be thus treated at a handsome profit. The ore extracted during recent prospecting operations has paid all expenses, no assessments having been levied since the re-opening of the mine.

So long as the two real walls (the slate and the quartzite) inclose so wide a channel, there is of course a possibility that the ore-vein may scatter through the intermediate space. But all indications prove that

* The lowest assay in quantities of ten tons, of the first-class ore shipped from the mine since re-opening, has been $562 per ton. The second-class ore has run $69 and upward. My figures are, therefore, very low.

this will not be the case at depths below the present tunnel-level. The vein having once regularly formed on the foot-wall, by stringers dropping eastwardly to that wall, may be expected to stay there in depth. The slips of the rock also dip eastward. The clay cross-seam already alluded to courses north 45° west and dips 84° east. From all these indications, as well as from the appearance of the ledge itself, which is far more persistent and regular than anything hitherto discovered in the Sheba ground, it is impossible not to believe that, though there may be outside ore-bodies in the west, there will not cease to be, from present workings downward, a strong vein on the foot-wall.

A decisive corroboration of this view is furnished by the circumstance that the two walls of the wide channel are drawing nearer together in depth. This may be clearly seen in the underground workings, and its effect will be to exclude gradually the barren ground which now occupies so large a space in the channel, and to secure, in all probability, a compact and reasonably uniform ore-bearing vein.

The facilities for working this mine are good. The company owns the whole water-right of Star Creek, and has a wagon-road, constructed at great expense, down the side of the cañon. The distance to Mill City on the Central Pacific Railroad, is twelve miles, and the cost of hauling ore by teams $5 per ton. Little timber is required in the mine, and the extraction of the ore for some time to come may be cheaply carried on through the tunnel and stopes now open. Exact estimates of cost are at present impracticable, since a large portion of the work hitherto has been preliminary, and the era of regular production is but just beginning.

The old works on the east have been worked on tribute by four Cornish miners for three months with good success. The mineral being scattered, is liable to "make bunches;" and this is a safer way for the owners and more profitable to miners of some experience.

The concentrating-mill connected with the mine was run for a short time in the spring. Although the quartz was well separated from the mineral, the result was not satisfactory, the probability being that the fine particles of silver-ores floated away to some extent on the water.

The De Soto was worked but little during the year. About a hundred tons of ore were sent to the Reno Mill for reduction. The mine having been bonded to an English company, this was done to make a trial of the working of the ore. The work gave satisfactory results. It is not known yet whether the sale has been effected.

The Yankee Blade has been worked during the year. The result is not yet very satisfactory. This mine is in calcareous slate.

In Star Cañon fair results being anticipated from gulch-washing, last summer dams were made to obtain a head of water to wash the stream-deposits. The water failed at the end of the summer, before final results were obtained, and the work waits for the wet season.

In the same district, in Bloody Cañon, about four miles south of Star Cañon, an antimony-ledge has been worked this fall, and about 100 tons shipped to San Francisco. The little demand for the metal here, and the danger of an over-stocked market, make the business somewhat precarious.

Central district.—Mining operations have been carried on during the greater part of the year, mostly in prospecting and testing the rock from the different ledges. The ledges contain, in many cases, besides silver, a large per cent. of gold. Those specimens seen by the writer are quartz with galena, a little zinc-blende and antimony, or their decomposed products.

The little stamp-mill of Philip Muller, in the district, has been at work when necessary to make trials of a few tons of rock at a time for prospectors. The owner has also been experimenting on a little roasting-furnace of his own make during the year, with varying results.

Sierra district.—A mill has been put up for dry amalgamation, according to A. B. Paul's patent. The first experiments are said to be very satisfactory by the superintendent of the mill, Mr. Charles D. Smyth. People await the final result with much interest. The mill has been placed in Dun Glen Cañon, above the town, where the Old Lang Syne mill used to be. The Lang Syne mine has been worked for a few months to extract the necessary quartz for the new mill. The Tallulah mine also has been running its tunnel to reach the ledge about 150 feet below the old works. The Auburn mine has also been worked in a small way. For want of means to work their quartz, rich in gold, the owners, who work the mine themselves, have stamped and extracted by hand the gold from the rich pieces to obtain the means to prosecute their work.

Echo district.—The Butte mine was worked until last September, when the mill of the company at Rye Patch was burned down. The work done during the first part of the year was on the incline, sinking deeper on the ledge, besides the removing of the ledge between two drifts run north of the incline. This was the main source from which the mill was furnished with ore. Last July the company, after purchasing a tunnel about 500 feet in the hill, well located to reach their ledge, began work on it. They will be about 130 feet vertically below the present mouth of the mine when the ledge is reached by the tunnel.

The work on the old Alpha mine was also resumed last fall, and a good deal of ore picked from the dumps has been shipped to the Reno mill for reduction.

The mill at Rye Patch was burned, by accident it is supposed, last September. A new one, stronger and better adapted to the wants of the place, has been built, and will probably be ready to run early in 1872.

Antimony in Nevada.—For the following interesting description of some remarkable antimony-mines, I am indebted to Mr. William L. Faber, a metallurgist of scientific training and practical experience.

The mines are situated in Humboldt County, twelve miles south of Battle Mountain, a station on the Central Pacific Railroad, five hundred and forty miles east of San Francisco, and three hundred and sixty miles west of Ogden and Salt Lake City.

There are two parallel veins, about 100 feet apart, one of which has been prospected. Both crop out for over a mile, commencing at the top of a ridge, where the Mountain King shaft has been opened, running downward about 1,000 feet in a distance of 1,500 feet north, where they cross a cañon or gully, and thence rise on the opposite ridge, where another shaft, the Columbia, has been sunk to a depth of 93 feet, at a point about 80 feet above the cañon.

The Mountain King shaft is 15 feet deep, and exhibits, from surface to bottom, and in the bottom, a continuous vein, two feet thick, of solid sulphuret of antimony. The vein is perpendicular, and has well-defined, regular walls, clearly cutting the country-rock.

In the Columbia shaft the vein is not so regular or well defined, but still contains, in a width of four feet, fully two feet of solid ore, sometimes in a body, sometimes divided in two or three strings by intervening horses. From the excavation 3,750 cubic feet of rock were removed, which furnished 150 tons of clean ore, being at the rate of one ton per 25 cubic feet. Of this ore 50 tons have been removed and sold or used, while 100 tons are on the dump.

The ore from the Columbia shaft is an intimate mixture of sulphuret and oxide of antimony, quite free from any other mineral or metal. A careful analysis of a fair average sample of the ore, rough-dressed, resulted as follows:

Moisture	2.82
Alumina, (clay)	1.58
Silica	12.62
Antimony	62.28
Bismuth	6.63
Sulphur	15.31
Oxygen, (calculated)	4.06
	99.30

A specimen from near the surface, at the Columbia shaft, assayed 3 ounces of silver per ton; one from the bottom of the shaft, 20 ounces; and one from the Mountain King shaft, 19.5 ounces.

Mineralogically, the ore consists of—

Blue sulphuret of antimony	56.71
Yellow oxide	29.09
Quartzose gangue	14.20
	100.00

These mines are peculiarly interesting on account of the singular purity of the ore, since the absence of lead and copper greatly facilitates the production, by the simplest reduction process, of an excellent quality of regulus of antimony. This was fully proved by an experiment which I made at the works at Battle Mountain Station, smelting the ore in a reverberatory, with native alkali as flux, and fine coal as the only reducing-agent, and thus producing, in one operation, from crude ore, metallic antimony, several tons of which were shipped to New York and sold to consumers for various puposes. All who have used it pronounce it equal to the imported refined regulus; although some of the pigs were not quite free from sulphur.

The details of the smelting process Mr. Faber does not wish to make public, but that it is inexpensive appears from the following facts: The furnace, of a capacity to hold only a ton of melted ore, could work this off in twelve hours, producing 800 to 1,000 pounds of metal from 2,000 pounds ore. Fifty per cent. was the best yield obtained, and this only when the furnace worked to perfection. The consumption of coal was 1 ton in twenty-four hours, the fine coal sifted from which was used as the reducing-agent, mixed with the melted ore.

The loss of metal by volatilization was quite insignificant, and nearly the whole loss was due to the difficulty of keeping the antimony out of the slag. Several times, from an error in the working or in fluxing, nearly all the metal was scorified, nor could it be reduced again from the slag, at least in the reverberatory.

Mining the ore costs, by contract, $2 per ton. Hauling from the mines to the station, by job teams from a livery-stable, costs $4 per ton. Freight to San Francisco, $10 per ton; to England, via Cape Horn, say $15 per ton, meaning always the short ton of 2,000 pounds. The ore is worth in England from £12 to £15 per miners' ton, of 2,352 pounds, equivalent to about $50, to $62 for 2,000 pounds. The regulus is worth from 12 to 14 cents per pound.

From these figures it is apparent that a mine of base metal exclusively—for the silver in these antimony-ores is too insignificant to be regarded—in the Pacific States, may return quite handsome profits, and be more desirable property than mines of silver or gold not strictly first class.

Abstract statement of the assessment-rolls of the proceeds of mines of Humboldt County, Nevada, for the year 1870.

Names of owners.	Description and location of mine.	Quarter of year.	Number of tons extracted.	Value per ton.	Total value.	Remarks.
Nevada Land and Mining Company, (limited) Fall & Temple	Alpha mine, Echo district; 2,000 shares. Arizona mine, Thorne Vein district, 900 feet; original location of Arizona mine.	1 2 3	165 670 1,780 28¼ 1,970 41	$95 96 11 34 35 50 304 00 29 30 300 60	$17,623 10 7,397 80 63,190 00 8,892 00 55,554 00 10,000 00	Ceased producing in 1870. In the latter part of the year 1870, Fall & Temple united with the Silver Mining Company, forming an association known as the "Arizona Association."
Silver Mining Company	First south extension of the Arizona mine; 1,200 shares.	1 2	7 567-2000 47 9 1254-2000	310 07 38 84 228 80	2,258 39 1,731 48 2,241 16	
Arizona Association	Arizona mine	3	770 274½ 1,535	33 14 281 23 33 91	25,517 80 7,163 82 52,051 85	From original Fall & Temple's ground.
	Original and first south extension		85 250-2000	302 02	25,574 20	From extension Silver Mining Company's ground.
G. W. Fox	Little Giant mine, Battle Mountain district; 1,400 feet in original location.	1 2 3 4	57 30 500-2000 120 10 825-2000 30 20 10 13	132 41 93 33 142 21 50 63 130 00 22 82 49 60 53 62	7,547 37 1,669 93 17,085 20 527 08 3,900 00 530 20 496 00 684 06	
Butte Mining Company	Butte mine, Galena district; original location; 1,200 shares.	1 2 3 4	42 70 59 531-2000 10 150-2000 68 138-2000	168 00 131 44 165 67 175 25 46 33	7,056 00 9,200 00 9,818 51 1,705 64 3,289 09	
Union Series Company	Union Series mine, Union district; 900 feet.	1	19	90 00	1,710 00	
Shiloh Mining Company	Shiloh mine, Galena district; 1,800 shares.					
S. H. Knowles & Co.	White mine, Battle Mountain district.	2	13 77 1304-2000	100 00 90 00	1,300 00 6,988 68	Original 2,000 feet.
C. Howell & Co	Central Pacific mine, Relief district.	4 2	5 60	190 00 62 00	950 00 3,720 00	Original 2,000 feet.
W. Smith & Co.	Butte mine, Echo district.	3 2	22½ 10	76 80 50 00	1,615 43 500 00	Original 2,000 feet.
Lott & Co.	Trenton mine, Battle Mountain district.	3	14	100 00	1,400 00	Original 1,400 feet.
Golden Age Company	Golden Age mine, Central district.	3	5 600-2000	272 00	897 60	
Total					362,996 47	

Abstract statement of the assessment-rolls of the proceeds of mines of Humboldt County, Nevada, for the year 1871.

Names of owners.	Description and location.	Quarter of the year.	Number of tons extracted.	Value per ton.	Gross value.	Actual cost of extracting.	Actual cost of transportation to place of reduction.	Actual cost of reduction or sale.	Total cost.
Arizona Association	Arizona mine, (original and extension.)	1	2,560	$29 35	$73,149 29	$15,630 00	$10,630 00	$29,750 00	$56,050 00
		2	...,000	448 61	21,130 21	1,461 57	857 33	706 87	3,965 07
		p	2,003	29 59	65,400 00	18,620 00	5,450 00	19,300 00	43,370 00
			291	29 71	39,517 00				
		4	2,300	422 54	12,042 33				
			300		25,000 00				22,180 80
			1,067		31,476 00				3,200 00
			*38		19,243 00				36,432 00
			13,680		40,480 00				751 43
J. S. Clark & Bro	Locomotive mine, Central district	1	11 1300-2000	104 24	1,214 40	233 00	168 93	349 50	2,161 17
		2	20 250-2600	88 30	1,706 00	1,412 80	245 00	503 37	
Pioneer.	North Star mine, (original,) Buena Vista district.	1	1,100	8 25	9,074 00				
		2	1,950	9 32	11,742 00				
		3	1,980		10,689 68				
			110	18 00	1,980 00				
White Mining Company	White mine, Galena	4	76 1000-3000	159 89	12,159 02	8,536 75		2,879 82	11,416 57
		2	43 340-2000	160 68	6,918 78				
Butte Mill and Mining Company	Butte mine, Butte Cañon, Rye Patch.	4	40 325-2000	131 50	4,960 71	1,667 15	417 00	3,092 00	5,460 35
		2	76 1716-2000	131 08	10,068 39	5,335 00	1,244 25	4,433 00	5,064 30
		1	106 1430-2000	38 45	6,382 85				5,086 15
		4	8334	43 95	24,306 82				11,012 25
P. Mellor & Co.	Golden Agr, Central district	1	40		4,242 73				
		2	9	55 55	500 00	450 00	18 00	997 00	1,830 83
									675 00
Battle Mountain Mining Company, (limited.)	— mine, Battle Mountain district.	1	255 835-2000	46 62	12,450 63	10,679 75	3,306 05		13,987 80
		2	244 630-2000		12,312 16	11,717 25	3,398 98		15,115 53
		3	322 116-2000		15,006 39				16,856 35
Nevada Butte Mining Company	Butte mine, Galena								
Sheba Company	Sheba Mine	1	1,600	12 00	19,200 00				
		2							

Abstract statement of the assessment-rolls of the proceeds of mines of Humboldt County, &c.—Continued.

Names of owners.	Description and location.	Quarter of the year.	Number of tons extracted.	Value per ton.	Gross value.	Actual cost of extracting.	Actual cost of transportation to place of reduction.	Actual cost of reduction or sale.	Total cost.
Little Giant Mining Company......	Little Giant......	3							$650 00
		4	8¼		$2,541 75				
		1							
		2			1,490 74				
		3	10 92-2000						1,163 00
		4			519,488 38				

* Shipping-ores.   † Tailings.

ELKO COUNTY.

Cope district.—I have not yet been able to visit this district personally, nor could my deputy do so. For this reason I cannot speak as intelligently of the situation of its mining industry as I should wish to, especially as, from correspondence, I must come to the conclusion that this district, as well as the neighboring one of Bull Run, promises to become quite important in the near future.

It appears that during the year the development of the mines in Cope has satisfactorily progressed. The Argenta and Excelsior mines have been worked with great vigor during the summer, in order to prepare them for the extraction of large quantities of ore in the fall. By that time it was expected to have the mines sufficiently opened to give employment to forty or fifty miners in extracting ore. The Independent, El Dorado, and Monitor, have also been energetically worked.

There was only one dry-crushing and roasting mill in Mountain City in the summer, and as the ores carry large quantities of base metal, and cannot, therefore, be worked to advantage by the wet-crushing mills, which were first foolishly erected, this one mill (P. F. Davis's, formerly Vance's) was continually overcrowded with work. Finally, Mr. Norton resolved to add roasting-furnaces to his 10-stamp mill, but whether this programme has since been carried out I do not know.

In *Bull Run district* several mines have been worked throughout the year, and the ore has been brought to Mountain City for reduction. About one hundred miners are reported to have been engaged here in mining in the summer.

Lone Mountain district is situated twenty-eight miles north of Elko. Its name is derived from the position of the mountain, rising alone from the plain, but which properly belongs to the chain on which Mineral Hill, Railroad, Cope, and Bull Run are located, further south. Although this district is a very promising one, there is no work being done this winter. Several mines will, however, be opened in the spring. There are three formations of rock running through the mountain, namely, limestone on the east, and granite and slate on the west. The mines now located are mostly in the limestone, though the most important are situated between the granite and slate. The most promising is the Paulina. It contains 800 feet, and is situated between the granite and slate; its course is north and south; dip 50°. A shaft is sunk near the center of the claim, 50 feet. The vein is 4 feet wide, very regular, and the ores are steadily improving in descending. The average yield of the ore in the bottom of the shaft is $120 in silver per ton, and 25 per cent. lead. This claim is also opened on its northern extremity; it is traced for a mile. Several locations to the north promise well, and one to the south, owned by parties in San Francisco, has a shaft sunk on it 50 feet deep, showing the same characteristics of the ore as Paulina. The Paulina was located by Messrs. Lowe, McKenzie & Smith, who sold one-half of the mine to Messrs. E. V. Robbins and J. W. Hussey. The same parties own also the Monitor, located in the limestone. A shaft has been sunk 40 feet, and some stoping is done on this lode. Its width varies from 6 inches to 4 feet. The ore taken out is sold for $20 per ton, on the dump. It was hauled to Elko, and there smelted. The road from Elko is very good, affording plenty of water and grass nearly the whole year round. At the foot of Lone Mountain there is a stream of water that, in its lowest stage, furnishes 100 inches. If ores needing concentration, or milling-ores, should be found hereafter, this water would be extremely valuable. On the southern end of the mountain is a very strong vein of iron, containing some copper and silver. A shaft has been sunk on it 20 feet, and a drift run across the vein for 16 feet; but the hanging-wall has not been reached. From indications on the surface, the vein is supposed to be 40 feet in width. The time must soon come when this vein will be valuable for its iron, especially as a flux for quartzose-lead ores.

Railroad district.—Mr. J. W. Hussey, of Elko, has kindly furnished me some notes on Railroad district, the substance of which, together with other information, is embodied in the following:

The principal mines are situated on the eastern slope of Bunker Hill Mountain, the highest peak in the district. Two spurs putting out from the main range form a horseshoe, or crescent, and within this crescent the principal labor has been done. Bunker Hill Mountain is 9,050 feet above the level of the sea, and mines have been discovered within 200 feet of the summit. The mine most developed in this horseshoe is located on one of the principal spurs, and is called the Last Chance, No. 2. Its course is nearly in accordance with that of the main range northwest and southeast, and cutting the spur on which it is located at right angles. The claim is 600 feet in length, and is developed by an incline, sunk to the depth of 90 feet, following the foot-wall. The hanging-wall was not reached until, at the depth of 60 feet, a drift was run across the lode, which was found to be 13 feet wide. A tunnel was run, cutting the lode at the depth of 90 feet, and communicating with the incline. The pitch of the lode, to the depth of 60 feet, is at an angle of 37°. At that depth it changes to 60°, which it now is. The principal ores, to the depth of 60 feet, were galena and carbonate of lead, with occasional streaks and spots of sulphuret, and the oxides of copper diffused through the lead-ore. At this point red and black oxides of copper came in strongly on the foot-wall, to the width of 8 feet, and, as the incline descended, native copper, in considerable quantities, was found diffused through the red oxides. The copper-ores continued for 20 feet in depth, when they gave way to galena and carbonate of lead again; but the ore was richer than found above. Whilst the upper ores will yield from $30 to $150 per ton in silver, those found underneath the copper yield from $50 to $300. The pure galena, which is concentrated in the center of the vein, yields upward of $200 per ton in silver. At the depth of 90 feet the lode is found to be 21 feet wide, 17 of which is good smelting-ore. The walls are limestone, very smooth and regular, with bodies of spar lining either wall. This is a characteristic of the mineral-veins of this district; the sides are usually lined with spar, and just before reaching a vein in a tunnel large bodies of spar are encountered in the country-rock. A shaft is now being sunk in the tunnel on the vein, 15 feet from the foot-wall, which was down 20 feet in November, finding the same ore at that depth as encountered in the tunnel. This mine was purchased of the original owners about one year ago by E. V. Robbins, of Chicago, Illinois, and by him sold to J. W. Hussey, acting as agent for a company of New York capitalists, under whose directions recent developments have been made. The mine is now in condition to yield 40 tons of ore per day. Developments are still going on, as probably the company will not erect furnaces before spring. This is the largest deposit of lead-ore ever found in this vicinity, and bids fair to become as celebrated as the famous deposits at Eureka.

Easterly of the Last Chance, and 150 feet farther down the spur, are situated the Lone, True, and Red Jacket mines. These locations are in close proximity, and supposed to be on the same lode. They produce galena and carbonate of lead, though, up to this time, of a lower grade than the ores of the Last Chance. These mines were bonded about six months ago by the New York company, of which Mr. Hussey is the agent. In prospecting the same a tunnel, known as the Hussey tunnel, was run into the mountain 137 feet, supposed to cut the veins 100 feet deep. About 100 tons of good ore have been taken out. Some of the ore sent to San Francisco yielded 55 per cent. lead and $47 in silver. The vein now being worked is 7 feet wide, 3 feet of which is solid galena, of a little higher grade, both in lead and silver, than that sent off. Mr. Hussey intends pushing the tunnel further into the mountain, as there are indications in the end of the tunnel of striking another deposit.

Still farther to the east is the Tripoli, owned by W. J. Raveston and some San Francisco capitalists. This lode is opened by a cut and a shaft sunk to the depth of 30 feet. Some of the richest ore ever found in the district, assaying as high as $1,500 per ton in silver, has come from this shaft. The lode is from 3 to 6 feet wide. Some good milling ore, containing sulphurets of silver, has been taken from this mine.

Just east of this lode are several locations of some promise, among them the Otto, Republic, and Mayflower, which mostly contain lead-ores, though the last named contains copper of a high percentage.

Adjoining the Last Chance on the west is the Humboldt, now incorporated in San Francisco under the name of the Highland Silver Mining Company, and two tunnels are now being driven into the lode; but up to this time no favorable results have been reached.

Southerly of the Humboldt, and well up on the side of Bunker Hill, is located the Web-Foot, owned by the same parties as the Tripoli. A tunnel has been run on the vein 30 feet, and a shaft sunk to the depth of 25 feet. Some of the ore taken out has been sent to San Francisco, and yielded $90 in silver and 45 per cent. lead. There was some ore in the bottom of the shaft, in November, that would yield $200 in silver. As the hill is very steep here, it is the intention of the owners to cut the lode at greater depth with a tunnel in the spring.

Northwest of this is the Shoo-Fly, owned by E. V. Robbins. Some very good ore has come from this vein. There were 30 tons of ore on the dump, in November, that would yield 45 per cent. lead and over $100 in silver per ton. A tunnel has been run on the vein 40 feet. The first body of ore struck is exhausted, and the tunnel is now being extended into the hill, with strong indications of speedily striking another body. The Rhino is a location 150 feet northwesterly of the Shoo-Fly. The ore resembles that of the Shoo Fly somewhat, but there is not enough work done to develop the characteristics of the vein. On top of the ridge, just above this mine, is located the Bunker Hill. Ten tons of red oxide of copper, through which native copper was generally diffused, was taken out here and sent to San Francisco. The vein is not well defined. Just below the Shoo-Fly are located the Bullion and Bullion Extension, both of which have afforded considerable quantities of galena, rich in lead, and containing from $40 to $50 in silver per ton. Work on both has been very irregular, and the owners have been somewhat discouraged through selling their ores and the purchasers failing to pay. A tunnel was recently started to strike the Bullion Extension 200 feet deep, which will be 250 feet long. After penetrating the hill 95 feet, and coming to hard rock, the tunnel was stopped. Still northerly and westerly are located several mines of considerable promise, if they were in the hands of capitalists who could carry on the work until furnaces were erected, making a market for the ores. The Dally contains galena with some copper. The Pine Mountain has afforded some very good copper-ore, containing about 830 in silver. The ore known as copper-glance is found in this mine. The Nevada, close by, belongs to the Highland Silver Mining Company. It is a very large lode, but not sufficiently concentrated for present profit. A spur puts down from the main range at this point, forming the westerly side of the Crescent. The first mine of importance on this is the Sweepstakes, owned by Brossomer, Norton & Co. It carries mostly copper, in the form of black and red oxides. A shaft has been sunk 35 feet, and a drift run from the shaft 20 feet westerly. Considerable of the ore now coming out yields native copper, and assays 30 per cent. Near this are located two mines belonging to Mr. Ritchie, of Philadelphia, known as the Snow Drop and Orphan Boy. Both seem to be on strong lodes, containing some lead, but mostly copper. On this spur the lodes are well defined, and below the mines just named they contain more silver and less lead. When better developed they will probably prove to contain good milling-ore. Among the mines affording ores of this description are the Mountain View, Haskell, Pixly, Highland, Little Emma, and Black Warrior, the last three being locations on the same lode. Between the two horns of the Crescent is a small hill, which seems to be a slide from the main range, but, on penetrating, it is found to be composed of limestone and sandstone, and to contain probably the largest body of carbonate of copper ever found in the State. The mines are called the Ella and Jensen, and are owned by A. J. Roulstone, J. W. Hussey, and Messrs. Lynde, Hough, and Thurman, of San Francisco. The vein has been penetrated by several shafts, and cut by two tunnels. One of the tunnels cuts the vein 30 feet deep, where it is 20 feet wide. The ore is shipped as it comes from the mine to San Francisco, where it has been found to contain from 22 to 28 per cent. of copper. One shaft is sunk 25 feet below this tunnel, and the same ore is found. The course of the vein is northerly and southerly. The owners were shipping 80 tons per month in November, but from that time forward expected to ship 200 tons per month. On the southerly side of the mountain there has been less work done, yet there are some mines that promise well. The most promising are the following : Red Bird; contains galena and carbonate; shaft sunk 40 feet deep; vein from 6 inches to 4 feet wide. The galena in this mine is of high grade. Owned by the Chase Brothers. On the same range is the Last Effort, owned by Messrs. Hall & Houghtalin, which also contains galena of a high grade. Still farther to the south is the Lyon, a lode 13 feet wide. The vein-matter is filled with streaks of rich galena, assaying over 8150 per ton in silver. These streaks are from 1 inch to 4 inches wide. A shaft has been sunk 30 feet in the vein, and carbonate ore is coming in at the bottom. The mine is owned by Messrs. Morgan, Peyton & Co. Near this is a promising mine of carbonate of lead, called the Walla-Walla Chief. Still farther south are the Wormer and Rising Sun, the ores of which resemble those of the Walla-Walla Chief, and contain some iron. On the western slope of Bunker Hill a very good mine has been opened this summer. It is called the W. S. Lee, and is owned by Messrs. Armstrong, Gillette & Piott. This vein has been stripped on the surface for a length of 400 feet. The ores are silver and lead, though containing less lead than those on the eastern slope. A shaft is sunk to the depth of 90 feet, in which there is ore all the way. The vein is from 18 inches to 4 feet wide; in the bottom of the shaft it is 4 feet. The course of the vein is northerly and southerly; the dip 50°. The walls are lime-stone. Cuts have been made on the surface of the vein, showing it equally as good as found in the shaft. Forty tons of ore have been shipped to San Francisco, which yielded $100 in silver. Upward of 100 tons of like character were on the dump in November. Near the W. S. Lee, and south of it, are the Pine Creek and True Blue. These mines, as far as developed, show rich lead and silver ores, which con-tain also a considerable percentage of copper. The owners of the True Blue are prosecuting work on the mine, meeting, so far, with good success. The shaft is only 12 feet deep, but yields a high grade of ore. There are not many mines on this side of the mountain that are opened to any extent. Access to them has been difficult, until this season, when Messrs. Armstrong, Piott & Co. made a road that leads to Palisades, which is thirteen miles distant.

The Union Copper mine, owned by C. M. Grout & Co., has a better surface-showing than any mine in the vicinity. The shaft is 10 feet deep; width of lode not ascertained. Copper-ores, yielding 40 per cent. of copper and $80 in silver, have come from the shaft. The claim contains 2,000 feet, and the croppings are 30 feet wide. The Mountain Boy is also a promising location.

Coal near Elko.—Reports have reached me that twelve and one-half miles north of Oreana, and about twenty miles northeast of Elko, sev-eral coal-veins, which had been discovered a year or more ago, have been further prospected. A vein of coal 4 feet wide is reported to have been first struck in a shaft at the depth of eight feet from the surface. Since then a tunnel has been run for 200 feet, cutting, in the space of 100 feet, five distinct veins of different width, amounting in all, it is reported by the local paper, to 60 feet in width of coal, resembling very much the best of what is known in Elko as the Rocky Mountain coal. The tunnel cuts the veins at no point at a greater depth than 15 feet, as it has been run simply for the purpose of prospecting the veins and to ascertain the dip, &c., so that permanent works may be commenced.

The discovery of workable seams of good coal in the vicinity of Elko, from which point it might be shipped to the neighboring base-metal districts, including Eureka, is of the highest importance. But I am in-clined to believe that these coal-beds are neither of the thickness ascribed to them, nor of a quality such as would warrant their use in blast-furnaces. It is undoubtedly lignite; and this western lignite will unfortunately break up into small cubes as soon as it is exposed to heat. So far this kind of coal has been found unsuitable for metallur-gical use. For the reverberatory it is possible that it may be found convenient for use in stair-grates; and in the blast-furnace, if it can be used at all, it must be used in pressure-furnaces, such as Mr. Bessemer has proposed for the melting of iron in cupolas.

LINCOLN COUNTY.

No single district in Nevada, outside of the Washoe country, has pro-duced as much bullion, during 1871, as Ely district, in Lincoln County. Last year's operations, which, in themselves, produced results sufficiently grand, were insignificent in comparison to what has been done this year.

Stimulated by the astonishing success of the two leading mining companies in this district, numerous other companies have been formed during the year, and nearly all of them have so far succeeded. The Meadow Valley Mining Company, and the Raymond and Ely Mining

Company, have taken the lead in magnitude of operations and bullion product.

Mr. Aug. J. Bowie, jr., mining engineer, has lately made a very complete report on the Meadow Valley mine and mill, and with his permission I insert it here as the best which has ever been published.

The Meadow Valley mine is situated in Pioche City, the county-seat of Lincoln County, Nevada, distant one hundred and fifty miles from Hamilton, or two hundred and seventy-three miles from the Palisades, (station on the Central Pacific Railroad.)

The property at present worked, and known by the name of the Meadow Valley mine, consists of a series of claims located on the vein, adjoining one another, and aggregating a total length of 2,315 feet.

The formation of the hill in which the vein occurs is quartzite. Including the two branches, the vein has been traced fully one mile and a quarter.

The quartzite formation in this country is entirely novel ;* in fact, v. Cotta, in his work on Stratification, records but three instances of important mines in which quartzite appears, viz, Przibram in Bohemia, Poullaouen in Brittany, and Hiendeleneia in Spain.

The general course of the Pioche vein (as it is called) is northeast and southwest, but its direction varies in general as well as in special cases. The size of the vein is irregular, being from 1 inch to 9 feet. The dip varies ; it is sometimes very flat, and then quite vertical.

In the west end of the Meadow Valley mine, the vein dips in shaft No. 3, with an average angle of 81° ; while in No. 7, on the east end, the average dip is 61°. The texture of the vein is strong and adheres to the encompassing walls. The size of the vein is occasionally designated by a border of galena. The gangue is quartz, and is very much decomposed. The silver occurs chiefly as a chloride, combined also with carbonates and sulphurets of lead. No arsenic or antimony has been as yet detected in the ores. The vein in the east end of the mine contains also carbonates of copper and copper silver-glance.

The peculiarity of veins in quartzite is the number of stringers, or feeders, attached to the main vein. So far, at Pioche, none of them have been followed to any great length.*

As to whether the forks† at the east end of the mine will again meet, it is impossible to state. One fact especially worthy of notice is the encroachment of the vein in shaft No. 7 on the Burke ground.

The ledges on the surface are 630 feet apart ; at a perpendicular depth of 440 feet, in No. 7 shaft, the distance between the 440-foot level and the Burke level is 420 feet. Of course it is impossible to foretell whether they will unite in depth.

The annals of mines in quartzite lead us to expect that the Pioche vein will continue to a considerable depth ; but the continuity of a vein must not in any way be confounded with its richness or productive capacity in precious metals.

A mine developed only 400 or 500 feet deep has not acquired a sufficient depth to warrant a positive opinion as to its future development. At present the prospects of the Meadow Valley mine are very flattering.

The mine is worked through three shafts. The Extension shaft on the west end is perpendicular, and first cuts the vein 150 feet from the surface. At a depth of 307 feet (bottom of shaft) a cross-cut to the vein shows it to be only 18 feet distant to the south. By means of a horse-whim at the mouth of the shaft the ore is hoisted to the surface. A double engine has been bought, and will be erected here in the course of the next two months. The shaft is small and is closely timbered. It has two compartments, one of which is used for a ladder-way. In case of need the shaft could be enlarged on the south side. The first hundred feet of timbering is poor. The only work at present prosecuted here is the driving of a cross-cut to the vein for the purpose of opening the 300-foot level, now being driven from shaft No. 3.

The Discovery shaft is situated 31 feet from the Pioche monument. It has been sunk to the 200-foot level on the vein. It is well timbered in sets, and could be divided into double compartments. Since the purchase of the Extension shaft, it is no longer used, and will in all probability be converted into an ore-chute.

No. 3 shaft is situated 228 feet to the northeast of the Discovery shaft. This is an incline shaft, and has been sunk to a depth (vertical) of 400 feet. It is the main working shaft of the mine, and all the levels have been regulated from it. It is well timbered and has double compartments. An engine at the mouth of the shaft, with sufficient capacity to sink 800 feet, hoists the ore and waste to the surface.

---

* Feeders have made their appearance in the 200-foot level, No. 3 shaft, and near the Black shaft, 63-foot level.

† Raymond & Ely and Meadow Valley branches.

Descending No. 3 shaft to the 63-foot level, the ore in the stopes on both sides (west to Discovery and east 98' 8") to surface has been nearly exhausted. Forty-five feet from the surface a thin seam of quartz made its appearance in the shaft, on the hanging-wall side. At first it was only about an inch in size, but as the shaft descended the seam widened. Down to the 63-foot level the horse was all taken out. The vein at this tunnel-level measured 8 feet. Drifting east 33 feet, a bifurcation of the vein became apparent. Drifts were immediately started on the forks, and on both the north and south forks through to the company's line the vein has been traced. On the north fork the distance traced was 1,705 feet; south fork, distance traced 570 feet.

The amount of ore now remaining in the stopes above the 63-foot level approximates 150 tons, of a value of $130 per ton. West of the Discovery to Extension shaft there are 69 tons—value 138.16 per ton. From the 63-foot level to the 120-foot level the shaft is sunk in the vein on the foot-wall side. For 20 feet below the 63-foot level the horse was extracted, and 80 feet from the surface the distance between the hanging and foot walls, (across the horse) was 23 feet. This immense space left open between the walls has since been filled up from the shaft to the hanging-wall.

The horse is composed of quartz. For the first 40 feet it was quite solid, but in depth it appears to partake more of the nature of quartzite, the quartz commencing to disintegrate and to be discolored by iron or ferruginous substances. On the 120-level (foot-wall) the shaft passed through a large body of rich ore. The vein of the shaft measured 5 feet, and assayed $230 to $250 per ton. A drift driven east from a winze sunk from the 63-foot level to the 120-foot level, distant 66 feet from shaft, west, disclosed the bifurcation of the vein to the west of the main shaft; and on the opening of the 120-foot level the bifurcation was found to exist 48 feet to the west of No. 3 shaft.

East of this last-mentioned shaft, on the 120-foot level, for 110 feet the ore is stoped out to within 23 feet of 63-foot level, and under foot for a depth of 50 feet. the vein continues 56 feet further east.

One hundred and sixty-six feet from the shaft, east, the vein is broken and has been thrown north, and is badly contracted. The drift beyond the disturbance has been run along the foot-wall 70 feet in the vein. The ore in this drift is very poor and lies in bunches. Average samples gave $54.95 per ton. The vein in the present end of the drift only shows a trace of metal. The hanging-wall drift shows a small vein of low-grade ore which is of little value.

There are several cross-cuts connecting the foot and hanging wall veins.

At the point of dislocation there is a cross-cut from one wall to the other, showing the forks to be 50 feet apart. The hanging-wall drift extends 28 feet northeast beyond the cross-cut. Forty-two feet from cross-cut, west, on the hanging-wall vein, there is a chute to the upper level, and a stope 20 feet long has been worked.

At shaft No. 3 the distance from foot to hanging wall is 19 feet. West of this shaft to the central winze the stopes have been exhausted to the upper level. From the winze to the Extension shaft the stopes remain unworked. The amount of ore remaining developed above the 120-foot level to the 63-foot level is 823 tons.

Descending a winze, west 63 feet from the main shaft on the 120-foot level, 15 feet, at 93 feet east of winze the bifurcation again shows itself, the vein attaining a size of 9 feet, with an average assay value of $121.63 per ton. A detailed description of the stopes is here unnecessary. Suffice it to say, that between the 120-foot and 200-foot levels there are 752 tons of ore, averaging, assay value, $160.20 per ton. The 200-foot level is connected through the Extension shaft, a distance of 261 feet. On the east side of No. 3 shaft the level has been opened on 62 feet. A cross-cut at shaft from foot-wall, across the horse, shows the vein on the hanging-wall at a distance of 32 feet. Drifts have been started on the hanging-wall vein, running at present 22 feet north 80° east, and 9 feet south 78° west. The vein does not look at all encouraging.

From the 200-foot level there are, besides the main shaft, two winzes connecting it with the 280-foot level. The winzes have opened up a considerable quantity of good ore; the greater part of which can be cheaply mined. Size of the vein—varying from 1½ to 2¼ feet.

Descending the main shaft, a fine ledge is seen all the way from the 200-foot to the 280-foot level, averaging from 1½ to 3 feet. The 280-foot level has only been driven 53 feet east; but west from No. 3 shaft, the level has been run 251 feet (nearly its entire length) in excellent ore. The present face of the drift is only 10 feet from the Extension shaft. A grand average of twenty samples taken from along the bottom of the 280-foot level, with an average vein of 20 inches, showed a value of $178.98 per ton. Total amount of good ore developed between the 200-foot and 280-foot levels, from No. 3 to Extension shaft, is 2,275 tons;* average assay value, $169.55 per ton.

Seventy-five feet west of main shaft, on the 280-foot level, a winze connects with the 360-foot level. The winze is sunk in the vein and the ore is good. Immediately west

---

* This includes a small quantity of ore estimated in sight, east of No. 3, between the levels.

of this winze, on the 360-foot level, the recently reported strike of $300 ore was made.

Descending No. 3 shaft to the depth of 340 feet, a small cross-cut made from the shaft on the hanging-wall side in the vein, which at this point seems to dip flatter into the hill, exposes 1 foot of fine ore. A sample from it assayed at $643.28 per ton. This same body of ore again makes its appearance in the shaft at the 360-foot level with an average size of 6 inches, assaying $411.34 per ton. The total size of the ledge in the shaft at this station is 18 inches.

From the main shaft along the 360-foot level, west to winze, the vein is small and bunchy but rich, and the stopes will doubtless open well.* The Extension shaft on the west end, No. 3 shaft on the east end, together with winze from 280-foot to 360-foot levels, and the 360-foot level itself, as far as opened, permit an estimate to be made of the number of tons of ore above this level, between the aforesaid points. Total amount of ore in sight, 2,514 tons. Average size of vein, 1¼ feet. Average value of ore per ton, $178.98. These last figures ($178.98) are taken from the average assay value of the ore in the bottom of the 280-foot level. The recent discovery of richer ore on the lower level will probably increase the value of the ore in the stopes above the 360-foot level.

The main shaft has been sunk 40 feet below the 360-foot level, but is not in the vein. The vein from the 360-foot level appears to 'change its dip, and, to preserve a uniform incline in the shaft, it has been sunk regardless of the vein, at least for the present. Returning to the 63-foot level, 404 feet east of the bifurcation, following the North Fork, is the Old Black shaft. The stopes overhead of this level through to the New Black shaft have been exhausted. West of the Old Black shaft, above the 63-foot level, for a distance of 100 feet, the stopes are worked out. A cross-cut in the 63-foot level, 100 feet west of shaft, through to Boyd shaft, has exposed an 18-inch vein of rich ore on the south side of the drift, assaying $200 per ton. The distance through this cross-cut to the hanging-wall vein is 109 feet 6 inches.

Between the Old Black and Summit shafts the ore above the 63-foot level is nearly all stoped out. In the bottom of the 63-foot level, between the New Black and Summit shafts there is a fine body of first-class ore, extending, it is supposed, down to the 120-foot level. * Its size is unknown, as the 120-foot level is not as yet under it; the body itself is dipping rapidly west. At present, this stope is furnishing daily 10 tons of first-class ore. The mine assay of these stopes, October 26, was $339.12.†

The ore from the Black shaft stopes contains a large quantity of manganese, and has always been very rich. Selected ore, worked from the old stopes, has reached the value of $720 per ton. The driving forward of the lower levels will develop all this ground, which, up to date, is unprospected in depth.

Ascending 32 feet through the chute at the end of the cross-cut, (100' 9'' west of Black shaft,) on the hanging-wall vein, the ground acquired by the compromise between the Meadow Valley Company and the Washington and Creole Company is reached.

This ground, formerly in the possession of jumpers, was badly gutted at the eastern end, but fortunately their shaft on the west ground, now owned by the Meadow Valley Company, did not strike the vein.

Although considerable ore has been extracted from this claim, there still remain, between the 63-foot level and the surface, 432 tons, of an average value of $152.30 per ton. In depth the prospects are favorable.

The Summit shaft has been sunk 400 feet, (vertical.) It was sunk on the vein, and, excepting the first 30 feet, has always been in poor ore. For the last six months no work has been done in it. It is timbered and has two compartments. The present indications of the vein in the bottom of the shaft are poor. Nineteen and one-half feet east of the Summit shaft, on the 120-foot level, there is a cross-cut from the foot to the hanging-wall vein. Two hundred and fourteen feet south of this level the vein under the Creole ground was struck by the cross-cut and found to be very poor. Returning to a point in the cross-cut, 28 feet 9 inches north of the vein, and running through the quartzite 105 feet, the vein under the Boyd stopes has been again found. Although not as yet developed to any extent, the mine assays of October 26 showed the bunches of ore to be worth $109.90, $144.44, and $113.04 per ton. West of the Summit shaft, (foot-wall vein,) the 120-foot level has been driven 150 feet. A raise is now being made to connect this level with the stopes already described in the bottom of the 63-foot level.

Four hundred and sixty feet northeast of the Summit shaft, along the 120-foot level, (foot-wall vein,) is the Receiver's shaft. This shaft was sunk perpendicularly 31 feet to this level. The immediate cause for sinking it was to prevent jumpers from claiming the ground. Simultaneously with its sinking, jumpers sank a shaft on the Pioche vein.

---

* Since writing the above this level has been pushed forward 114 feet west of the shaft, the vein measuring 30 inches, and assaying $329 per ton.

† Samples are taken every two days from all the stopes and drifts in the mine.

From the 120-foot level the Receiver's shaft was sunk on the vein to a total depth of 200 feet, and at the same time the lodge was traced through on the 120-foot level to the Summit shaft west, and to the Challenge shaft east, in order to prove the vein. Between the Summit and Receiver's shafts the vein is distinctly traced, and, although somewhat irregular at the latter, farther west it is a bold, hard ledge, varying in size from 1 foot to 3½ feet, assaying from $25 to $32 per ton. None of this ground has been at all prospected, and, no doubt, in depth, good ore will be found. The 200-foot level of No. 7 shaft connects with the Receiver's shaft.

Between the Receiver's and Challenge shafts, on the 120-foot level, very rich ore has been extracted. In the Receiver's shaft there are several places which indicate good ore, but no drifts have ever been run to explore the ground to the west of it, into which, probably, the ore-chutes from the Challenge ground have passed. Forty-five feet east of the Challenge shaft the 120-foot level connects with a shaft, which in turn connects through a drift with No. 7 shaft. In order to decrease the expense of hoisting the ore, and to facilitate its transportation, a tunnel was driven north from Receiver's shaft, 227 feet through the country-rock, corresponding with the 120-foot level. This is called the Tunnel east of 120-foot level.

As soon as the raise from the cross-cut on the 120-foot level to the Boyd stopes is open, all the ore from that portion of the mine, which is now conveyed west, will come out through this tunnel. At present everything east from the Black shaft passes through it, and is dumped with the ore from No. 7 shaft* into the ore-house situated on Meadow Valley street, in front of the Meadow Valley Company's office.

No. 7 shaft is an inclined shaft sunk on the vein to a vertical depth of 500 feet. Average dip 51°. The shaft has two compartments.

A steam-engine, (similar to the one at No. 3 shaft,) situated at the mouth of the shaft, does the hoisting for this portion of the mine. The first drift opened from this shaft was at 97 feet below the surface. East seventeen feet on the same level the lodge first made its appearance. All the rich ore on this level east for 140 feet, and west to Receiver's shaft, has been stoped out. The length of the rich stope on this level east was 79 feet, and west 78 feet. The ore remaining above it assays as follows: East of shaft, $32.97; west of shaft, $35.12 per ton. Size of the vein: east one foot, west two feet. The ore on the west side runs as low as $12.56 to $10.99 per ton. From the first to the second level the distance is 77 feet 2 inches; dip 48° 30'. West from the shaft on the 200-foot level, for 45 feet, the stopes are worked out. A small pillar of low-grade ore, 68 feet long, follows the stope, and then for a distance of 24 feet it is worked out to the level above. Under foot, at the end of this stope, there is a chute to the level below.

East from the end of the last-mentioned stope, for 60 feet, the ground is solid overhead; then follows a stope 35 feet long and 29 feet high.

Twenty feet farther west the drift terminates in quartzite. Value of ore remaining in stopes, from $9.42 to $15.70 per ton. For 44 feet under foot, on the east side of shaft, (200-foot level,) it is solid; overhead for 100 feet it is stoped out; immediately adjoining is a barren streak 44 feet long.

The first chute of fine ore succeeds this barren spot, and is opened for 13 feet, then unworked for 39 feet, and the ore again exposed for 43 feet. Average size of the vein, 2 feet. Average value of the ore $193.82 per ton. Between the 200-foot and 100-foot levels 815 tons of ore may be estimated as in sight.‡

The ore-chute mentioned east of the shaft, on the 100-foot level, crossed the shaft at the 200-foot level. From the 200 to the 280-foot level, it is 74 feet; dip, 58° 30'. The drift west has been driven 76 feet; the vein is poor, and averages two feet. The ore-chute is 50 feet long; average value, $7.80 per ton.

In the east drift on this level, 66 feet from the shaft, there is a chute from the level above. Fifty-five feet from here there is a winze being sunk, and overhead there is a stope.

One hundred and thirty-seven feet in the drift beyond the first stope there is a break in the vein. At 160 feet a second stope has been commenced, but is raised only a short distance. Adjoining this stope a fine ledge makes its appearance, extending 26 feet; the ledge is hard and rich. For 63 feet beyond the raise the ledge is left on the foot-wall, and appears pinched until it reaches the third stope, which is opposite the cut made by the Pioche Company. This stope has been opened for 24 feet, and shows a fine vein 3 feet in size. The amount of ore exposed by the stopes on 280-foot level is 454 tons; value per ton, $163.44; average size of vein, 2½ feet. It is to the east of this same stope, above and below it, that the Pioche Company overran the Meadow Valley Company's line, and extracted 150 tons of ore, valued at $250 per ton.

To the east of the Meadow Valley Company's line, in the Pioche Company's ground,

---

* That is, ore extracted from the eastern end of the mine.
† The position is indicated on the chart.
‡ Although this ore is not opened by the 100-foot level, there can be little doubt of its being found when the level is pushed forward and the stopes raised.

a winze has been sunk 100 feet lower than this level, and excellent ore has been found pitching from their winze into the Meadow Valley mine.

This is a most favorable prospect for future developments in the lower levels, east from No. 7 shaft. From the 200-foot level to the 440-foot level, No. 7 shaft descends 210 feet, on an angle of 52° 15′. The 360-foot level has not as yet been started. The 440-foot level, on the east side of the shaft, is in 40 feet. As yet no rich ore has made its appearance, nor is it expected for the first 150 feet or more.*

The ledge in the drift now assays $9.42 per ton. The shaft is 30 feet below this level; † the vein appears to be dipping flat into the hill, and is not in the bottom of the shaft.

General observations.—On account of the narrowness and peculiar character of the vein, and other considerations, shafts Nos. 1, 3, and 7 should be kept 200 feet ahead of the work. Levels should be regularly opened and driven. I would particularly recommend that the 360-foot level be connected throughout the entire mine. The probabilities are that this level will open up valuable unprospected ground between shafts Nos. 3 and 7, and afford greater facilities for working the various ore-chimneys from the east, all of which are pitching west, and in depth will be found far beyond the plane of their original boundaries. The only profitable manner of extracting the ore in this mine is by overhead-stoping. The character of the ore and the formation of the encompassing walls render underhand-stoping unprofitable.

The store-house at mine is well supplied. The present stock on hand amounts to $12,933.28. (*Vide* mine store-house report, November 1, 1871.)

It is absolutely necessary for the company to keep a large supply of hardware and materials on hand, as there are no firms in Pioche, with sufficiently assorted stocks, with which the company can deal. The recent fire showed the importance of the company having its own store-house well supplied.

The mill.—Everything at the mill is in good condition. The scarcity of water has prevented the use of the concentrators. By the middle of November the new engine will be in position. The additional two pans, *en route*, when in place, will increase the productive capacity of the mill about 180 tons per month.‡

The Horn pans I did not examine, but am informed by Mr. Forman that they are in excellent order, and, so far, no amount of wear by action of the chemicals, excepting on two of the pans, is at all observable. In future, the bottoms of the shells should be cast thicker than those now in use.

The store-house at the mill contains a large stock of chemicals, quicksilver, castings, &c. (*Vide* store-house report, November 1, 1871.) Value of present stock on hand, $46,678.09; supplies *en route*, about $25,000.

The assay office at the mill should be immediately removed from its present position. Its proximity to the mill endangers the latter. For the sum of $1,800 a brick office with a tin roof can be built. One mile below the mill, in Dry Valley, the company owns an interest in a large spring which is said to run 200 inches of water.§ The water, by analysis, proves to be the purest found in the neighborhood. The office at the mine is 962½ feet higher than the spring, and distant from it ten miles. I would call the special attention of the board of trustees to the possibility of bringing this water to the mine and erecting the mill in Pioche City. At a cost of $85,000, 10 inches of water can be brought into town from this spring.

The ore-hauling account for the coming year will, in all probability, exceed $100,000.

In case the company should deem it advisable to bring the water into town, (the cost of which can only be determined after a thorough examination of the country,) the old mill could be converted into a tailings-mill. There are on hand at the mill, in the reservoirs, 14,000 tons of tailings, averaging $33.75 per ton;∥ and 2,000 tons of slum, averaging $106.76 per ton.

RECAPITULATION.

Location of mine, Pioche City, Lincoln County, Nevada.

Length of claim, 2,315 feet.

Formation in which the vein occurs, quartzite.

The gangue is quartz, decomposed. The silver occurs as a chloride, combined likewise with carbonates and sulphurets of lead. (*Vide* report.)

Size of vein, from 1 inch to 9 feet.

---

* The drift is in 114 feet, assaying $56 per ton. (November 14, 1871.)

† The shaft is now 78 feet below the 440-foot level. (November 14, 1871.)

NOTE.—If the present (monthly average) high grade of the ore is reduced to a lower standard, say $120 per ton, the mine will be worked to a greater advantage, and it will prove in the end more profitable to the stockholders.

‡ Since arrived, and now running. (November 22, 1871.)

§ Measured by civil engineers.

∥ Average value of tailings, estimated from monthly mill reports from 15th July, 1870, up to date, (October 31, 1871,) shows $33.96 per ton.

Average size is from 2 to 2¼ feet, irregular.
Course of vein, from northeast to southwest; dip southwest. (*Vide* report.)
The present prospects of the mine are good.
No. 3 shaft, 400 feet deep, vertical measurement; No. 7, 500 feet deep, vertical measurement; lowest level at east end of mine, 440 feet; lowest level at west end of mine, 360 feet.

Ore in sight in mine.

West end.

	Tons.	Gross value.
Above 63-foot level	219	$27,032 04
63-foot level to 120-foot level	893	123,444 76
120-foot level to 200-foot level	752	120,470 40
200-foot level to 280-foot level	2,275	385,726 25
280-foot level to 360-foot level	2,514	449,955 72
Boyd stopes	432	65,793 60

East end, No. 7 shaft.

	Tons.	Gross value.
100-foot level to 200-foot level	815	157,963 90
200-foot level to 280-foot level	454	69,661 76
Ore at mill, November 1, 1871, (value, $140 per ton)	1,500	210,000 00
Totals	9,854	1,610,047 83

In reservoirs, November 1, 1871.

Tailings, 14,000 tons, at $33 75 per ton		$472,500 00
Slums,... 2,000 tons, at $106 75 per ton		213,520 00
Totals... 16,000		686,020 00

Supplies at mine, November 1, 1871	$12,933 28
Supplies at mill, November 1, 1871	46,678 08
Supplies in *transitu* for mine and mill	25,000 00
Total	84,611 36

Total cost of mining, milling, taxes, &c., estimated from accounts of 1870 and 1871, $44.11 per ton.
Average bullion yield for same period, $105.34 per ton, or 73.4 per cent. of gross value of ore worked.
Average bullion yield for August, September, and October, 1871, $114.78 per ton.

The following is the very complete biennial report of the superintendent and secretary of the company for the fiscal year ending July 31, 1871:

PIOCHE, NEVADA, *August* 5, 1871.
To the president and trustees of the Meadow Valley Mining Company:

GENTLEMEN: Herewith I hand you statement of operations at the company's mine and mill for the fiscal year ending July 31, 1871.
During the company year just ended there have been extracted from the mine 16,500 tons of ore obtained from the following sections:

	Tons.
Tunnel west of 63-foot level	7,703
No. 3 shaft	3,511¼
Black shaft	1,349¼
Summit shaft	374
Discovery shaft	223¼
No. 7 shaft	2,335
Tunnel east of 120-foot level	913¼
Total	16,500

You will observe from the above statement that tunnel west has been the most productive section, having produced nearly one-half of all the ore extracted during the year; but the ore above that level (the 63-foot level) is now nearly exhausted, and hereafter the principal supply must be hoisted through shafts Nos. 3 and 7.

The development of the mine has been greatly retarded by our inability to do the necessary amount of hoisting with the whims, which have been taxed nearly to their utmost capacity hoisting ore to keep the mill supplied, and consequently but comparatively little sinking and drifting could be done below the tunnel-levels.

During the month of July steam hoisting-works were erected upon shafts Nos. 3 and 7, and are now in successful operation, obviating that difficulty, and we are now able to proceed rapidly with the sinking of shafts, drifting, &c.

No. 7 shaft has attained a depth of 364 feet. The average dip of the vein in this shaft is 53° to the south.

The east drift 200-foot level in this shaft is being driven on the ledge, and in very fine ore. Average samples taken from the face of the drift for the past week show an average assay of $216.66 per ton.

At 280-foot level east drift is being driven forward as rapidly as possible to cut the ore-chute that we have left under the track-floor in 200-foot level.

Station at 360-foot level is now being opened.

Summit shaft has been sunk to a depth of 369 feet, and, being upon a portion of the mine that has been comparatively barren from the surface, we have stopped work upon it for the present. The average dip of the vein in this shaft is 58° to the south.

No. 3 shaft has been sunk to a depth of 285 feet, and the work of sinking is being vigorously prosecuted. The average dip of the vein in this shaft is 77° to the south.

The 280-foot level, which is just being opened, promises well for a large amount of first-class ore. The west drift from this shaft is being driven in the ledge, and in very fine ore. Average samples taken from the face of the drift for the last ten days give an average assay of $205.67 per ton.

Winze No. 2, 160 feet west of No. 3 shaft, is being sunk from the 200-foot level to connect with the 280-foot level, and is now down 13 feet, showing first-class ore.

From present indications the section lying between Discovery shaft and No. 3 shaft, and extending from 200-foot level to the 280-foot level, will produce a very large amount of first-class ore.

The ledge throughout the mine below the 200-foot level contains much more picking, ground than it does above that level, and I anticipate a large reduction in the cost of extraction.

From careful measurements made, I estimate the amount of ore developed, that will mill $80 or more per ton, to be 6,758 tons, and in the following sections of the mine:

	Tons.
No. 7 shaft, above 280-foot level	1,476
Tunnel west, above 63-foot level	464
Washington section	701
No. 3, below tunnel-level	3,767
Black shaft	350

In this estimate I have not included the body of ore lying between No. 3 and Discovery shafts and the 200 and 280-foot levels, nor any body of ore that is not fully developed and its full extent known.

But little work has been done during the year upon the outside mines owned by the company, except the necessary amount required to comply with the mining laws of the district.

In several of these mines the indications are quite as promising as they were at the same depth in No. 7, and I would recommend the sinking of shafts upon them to the depth of 100 feet or more to develop them.

During the past three months the company's mill at Lyonsville, Dry Valley, has been thoroughly overhauled and repaired, the old pans having been replaced by new ones of the most approved pattern, and three new boilers have been added, making the mill complete in all of its appointments, and in much better condition for doing good service than when it was first built.

During the first few months the mill was run much time was lost for want of an adequate supply of water. The ditch for supplying the mill was dug across a sand-flat, and the water was absorbed and evaporated during the summer, and the ditch frozen up during a portion of the winter.

This has been remedied by building a flume to take the place of the ditch, and covering it with earth to a sufficient depth to prevent freezing during the winter.

During the year just ended 17,458 tons of ore have been received, and 16,172 tons reduced by the mill.

The necessary reservoirs have been constructed for catching tailings, and no tailings are now allowed to run to waste.

I would respectfully renew my recommendation relative to the building of a cupola, and making our shoes and dies at the mill. There are now 100 tons or more of old castings at the mill, and I am confident that a saving of 50 per cent. upon the present cost can be made by using this iron and making them at the mill.

Your attention is respectfully called to the tabular statements herewith, which will give you a more comprehensive idea of the details of operating the company's mine and mill.

Yours, truly,

CHAS. FORMAN, *General Superintendent.*

TABLE No. 1.—*Condensed statement of cost of production, &c., for company, year ending July 31, 1871.*

ORE STATEMENT.

	Tons.	Lbs.
Ore on hand August 1, 1870	378	1,659
Ore received from Hanchett & Rutherford	482	1,758
Ore received from Barnes & Scott	95	1,890
Ore extracted by company during year	16,500	977
Total	17,458	6,284
Ore worked by company during year, plus moisture deduction	16,171	1,535
Ore on hand August 1, 1871	1,286	749

COST PER TON.

Extracting	$14	20.6
Prospecting, improvements, and incidentals	10	30.7
Reduction	19	60.4
Total cost per ton	44	11.7
Average yield per ton of ore worked	105	34.

WORK OF ASSAY OFFICE.

Number of troy ounces of bullion before melting	2,288,782.63
Number of troy ounces of bullion assayed after melting	2,170,352.28
Average loss in melting, per centum	5.4
Number of ore assays made	1,890
Number of bars made	1,328

TABLE No. 2.—*Statement showing the amount of ore extracted and cost of extraction for the company year ending July 31, 1871.*

Months	ORE EXTRACTED																SUNDRIES		Incidentals	EXTRACTING ORE	
	Tunnel west		Shaft No. 3		Block shaft		Summit shaft		Discovery shaft		Shaft No. 7		Tunnel east		Total		Officials	Office expenses		Labor	Material
	Tons	Lbs	Tons	Lbs	Tons	Lbs	Tons	Lbs	Tons	Lbs	Tons	Lbs	Tons	Lbs	Tons	Lbs					
1870																					
August	632	1,551	448	1,504					28	1,760					1,326	815	$813 50	$383 25	$1,594 14	$14,719 74	$1,767 78
September	755	40	10		277	1,840	14	565	196	2,315					1,229	195	283 18	758 00	2,740 63	24,298 00	3,882 92
October	893	1,030			329	160	72	1,725			6	1,320			1,247	1,620	616 66	125 50	2,693 85	16,173 00	2,568 72
November	1,062	901	59	863	332	1,318	137	575			137	435			1,464	764	616 66	105 50	2,493 22	14,802 00	2,394 41
December	1,055	1,057		1,940	174	1,705									1,534	1,062	616 67	376 95	2,375 83	17,865 75	3,758 14
1871																					
January	850	563	409	150	79	1,700	131	1,495			433	940	61	340	1,904	848	616 66	618 50	481 69	20,998 00	3,374 98
February	453	184	272	990	16	144					130	456	16	1,716	973	1,114	994 43	422 00	2,746 01	11,366 34	3,800 00
March	391	1,068	415	624	29	446	17	755			340	684	250	133	1,193	538	866 66	349 83	*17,083 56	15,694 60	3,041 73
April	343	1,838	478	1,936	47	812					370	900	301	1,330	1,514	738	866 66	598 30	15,436 91	16,091 75	3,213 55
May	416	1,188	510	78	31	1,124					301	1,023			1,420	1,883	866 66	419 77	1,358 30	15,974 00	2,319 75
June	361	1,063	460	379	30	1,688					348	91	301	1,330	1,275		866 66	227 40	1,249 66	15,051 75	2,684 36
July	327	250	476	1,096									173	600	1,325	37	866 66	781 07	1,249 99	15,957 50	2,393 92
Total	7,783	59	3,511	859	1,349	1,117	374	115	225	1,075	2,235	409	913	1,356	16,500	977	8,691 06	5,139 47	40,298 70	198,282 43	36,130 97

*$15,000 tax.   †$4,387 09 bullion tax.

TABLE No. 2.—*Statement showing the amount of ore extracted, &c.*—Continued.

Months	PROSPECTING AND DEAD WORK				IMPROVEMENTS				Grand total	RELATIVE COST				Total cost per ton	Material on hand
	Day labor.		Contract system.		Day labor.		Contract system.			Sundries.	Extraction.	Prospecting and dead work.	Improvements.		
	Labor.	Material.	Labor.	Material.	Labor.	Material.	Labor.	Material.							
1870.															
August	$5,276 30	$1,061 21			$35 00	$697 94			$25,688 86	$1 60.7	$12 41	$4 78.5	$0 45.6	$19 46.8	$34 38
September	1,165 00	1,500 00	$6,629 26	$198 75	641 00	700 00			35,942 73	3 05.8	22 91.8	2 16.7	1 09.3	29 93.6	5,914 65
October	3,383 50	400 00	6,336 50	161 45	1,383 00	500 00	$500 00	$1,100 00	36,268 75	2 75.3	15 01.9	5 50.2	2 79.2	29 06.6	9,302 35
November	2,307 50	600 00	3,479 67	112 36	1,297 00		400 00	500 00	28,989 24	2 19.2	12 46.5	3 84.4	81.7	19 31.8	9,669 22
December	1,752 50	1,200 00			159 00	800 00			22,496 07	2 19	14 09.8	4 26.4	62.4	21 17.6	12,768 59
1871.															
January	2,072 75	1,800 00	1,480 61	92 96	928 75	1,000 00	148 00		33,443 10	90.7	12 70.8	2 96	1 08.5	17 56	12,034 03
February	2,699 00	1,745 96	3,490 65	196 17	585 50	1,000 00			29,056 06	4 27.7	15 57.1	8 36.2	1 68.8	29 84.5	7,558 65
March	2,414 75	2,251 90	5,699 75	23 75	1,165 25	2,139 31	401 50		32,135 39	*15 53.8	15 70.1	9 54.5	1 10.8	43 69	5,162 53
April	4,542 00	1,465 50	1,505 60		1,283 00	897 32	275 05		31,166 84	14 76.5	12 74.9	4 96.3	1 62.0	29 58.3	6,044 68
May	5,596 25	1,600 00	3,430 64	559 79		119 15			134,664 91		12 50.4	7 24.9	08.4	24 59.6	6,552 62
June	4,496 00	1,524 35	2,883 77	600 00		113 14			29,816 09	1 79	12 68.8	6 90.9	08.2	21 65.9	5,552 63
July	4,332 50	3,415 68	2,061 56	443 02	1,221 00	2,588 61			35,306 71	2 18.7	13 85	7 73.8	2 87.1	26 64.6	11,025 65
Total	41,037 05	17,784 90	33,046 21	2,388 15	7,637 50	10,460 47	1,724 55	1,600 00	404,480 75	3 29.2	14 20.6	5 71.3	1 30.2	24 51.3	11,025 65

† $84,387 09 bullion tax.

* $15,000 bullion tax.

TABLE No. 3.—*Statement showing the amount of ore on hand, worked, and cost of reducing for company year ending July 31, 1871.*

Months.	ORE.								TIME.								AVERAGE ASSAY PER TON.		AVERAGE ASSAY OF TAILINGS.	
	Received.		Worked.		Moisture deducted 4½ per ct.		Balance on hand.		Running by pans.		Lost by pans.		Running by stamps.		Lost by stamps.		Gold.	Silver.	Gold.	Silver.
	Tons.	Lbs.	Tons.	Lbs.	Tons.	Lbs.	Tons.	Lbs.	D's.	H.m.	D's.	H.m.	D's.	H.m.	D's.	H.m.				
1870																				
August	1,689	119	1,161	400	70	15	451	1,704	29	5 30	1	18 40	29	19 00	1	12 00	$5 03	$117 04	$3 50	$47 98
September	1,241	1,455	1,027	1,300	55	1,759	610	903	26	6 00	3	18 00	28	12 53	1	11 05	7 57	170 71	3 11	43 30
October	1,253	1,715	1,071		56	846	736	1,072	28	13 18	5	10 42	29	2 37	1	21 33	7 13	165 52	3 05	49 30
November	1,464	764	1,017	900	63	1,794	1,177	1,142	23	3 18	6	20 42	29	8 28	1	15 32	7 28	153 98	3 96	37 05
December	1,534	1,082	1,460	1,987	68	107	1,182	110	26	16 50	4	7 10	27	23 41	3	0 19	5 80	151 12	2 98	34 38
1871																				
January	1,904	848	1,503	1,300	85	1,399	1,477	159	25	13 45	5	10 15	29	4 30	1	10 30	8 44	123 43	3 49	37 97
February	1,510	148	1,449		67	1,906	2,430	509	21	17 43	6	8 17	25	0 18	2	23 42	5 04	115 90	2 52	29 10
March	1,087	300	1,312	638	44	443	2,098	1,818	24	22 30	6	1 40	28	8 19	2	13 41	6 02	133 70	2 54	30 63
April	1,722	1,247	1,007	674	68	277	1,745	1,914	27	18 40	1	3 30	27	10 47	2	13 13	5 71	116 70	2 17	18 53
May	1,409	728	1,161	1,798	63	849	1,930		29	3 45	3	20 15	28	3 59	2	20 8	5 43	132 67	2 27	25 33
June	1,771	1,863	1,608	1,531	61	1,833	1,615	780	27	15 18	2	8 42	27	23 59	2	0 21	4 63	127 70	1 63	21 36
July	1,255	37	1,614	606	59	1,251	1,286	749	26	21 37	2	2 23	29	21 34	1	2 26	5 45	159 90	2 03	26 40
Total	17,458	284	15,395	1,064	778	471	1,280	749	316	13 54	46	10 6	309	12 30	25	11 30	5 96	137 25	2 90	32 28

TABLE No. 3.—*Statement showing the amount of ore on hand, worked, &c.*—Continued.

Month.	Average yield of bullion per ton.	Average percentage extracted.	Total amount of bullion produced.	TOTAL COST.					RELATIVE COST.				Total cost per ton.	Average loss quick-silver per ton.
				Hauling ore.	Labor.	Materials.	Improvements.	Grand total.	Hauling.	Labor.	Material.	Improvements.		Lbs.
1870.														
August	$66 41	54.4	$77,100 63	$5,876 73	$6,516 49	$7,989 51	....	$20,382 73	$4 75	$5 96.7	$6 45.3	....	$16 47.5	3.77
September	114 69	64.3	114,488 33	5,146 52	4,478 68	8,385 27	$1,669 08	19,679 55	4 75	3 84.9	7 92.4	$1 54	16 10.3	4.76
October	151 60	87.9	166,011 06	5,355 90	4,684 87	7,373 54	7,316 08	24,720 29	4 75	4 15.5	6 54	6 49	21 53.5	2.39
November	133 06	62.5	133,387 62	5,145 89	4,680 80	6,629 19	8,510 97	25,025 55	4 75	4 32	6 11.9	7 01.1	23 10	2.40
December	103 65	81.8	151,731 20	7,267 72	6,137 72	16,328 79	7,365 14	37,119 37	4 75	4 01	10 66.9	4 62.5	24 25.4	5.88
1871.														
January	102 46	70.7	154,097 78	7,549 17	6,237 90	14,160 07	6,813 73	34,730 89	4 75	3 92.5	8 90.9	4 98.7	21 87.1	4.48
February	93 80	77.6	136,047 65	7,305 53	6,637 57	13,047 58	573 00	24,537 68	4 75	4 36.9	6 67.2	37.7	16 16.8	2.03
March	95 60	68	124,786 01	6,453 06	4,925 90	13,047 43	655 50	25,084 21	4 75	3 62.5	9 50	47.5	18 35	4.38
April	82 17	67	62,746 06	5,108 98	4,597 50	10,817 70	1,606 25	22,322 43	4 75	4 56.5	10 05	1 62.1	20 74.6	5.65
May	102 71	74.4	119,108 57	5,345 27	5,588 25	10,673 86	5,439 90	27,724 63	4 75	4 68	10 37.1	4 63.3	24 61.6	3.41
June	97 90	74	153,498 31	935 92	5,588 25	11,683 58	4,435 36	20,891 21	4 75	3 54.3	7 11.4	2 64.8	17 85.5	2.06
July	119 96	73	193,665 37	7,951 70	5,983 37	9,921 10	2,022 70	25,877 87	4 75	3 57.2	5 92.4	1 50.8	15 45.4	3.66
Total	105 34	73.4	1,612,899 14	76,340 79	65,500 55	128,532 64	46,682 63	317,056 61	4 75	4 05	7 94.8	2 83.6	19 60.4	3.77

TABLE No. 4.—*Statement showing the work of company's assay office for company year ending July 31, 1871.*

Months	Number of bars made.	Weight in troy ounces.		Average per cent lost in melting.	FINE BULLION. Fineness.		Average value per ounce.	Value.		
		Before melting.	After melting.		Gold.	Silver.		Gold.	Silver.	Total.
1870.										
August	38	153,986.48	149,340.58	3.4	.00099	381	$0.51.02	$3,089 27	$74,015 09	$77,103 36
September	78	129,924.29	121,308.60	5.0	.00099	699	92.84	2,962 43	111,524 77	114,487 19
October	92	136,582.25	146,501.00	7.0	.00085	843	1.11.00	2,872 10	159,770 08	162,642 18
November	96	141,299.00	130,114.50	7.9	.0012	785	1.04.00	3,199 08	132,188 54	135,387 62
December	119	205,907.00	193,493.00	5.1	.00155	668	78.50	3,887 21	147,843 99	151,731 20
1871.										
January	194	206,909.00	193,537.50	6.2	.0005	587	77.13	2,922 67	147,074 13	149,297 00
February	101	176,589.30	166,069.30	4.8	.0009	584	77.32	3,115 57	126,959 06	130,074 63
March	88	154,674.00	147,931.70	4.3	.0007	602	79.32	2,118 56	115,237 30	117,356 18
April	63	108,680.00	102,827.00	5.1	.0009	586	77.63	1,914 14	77,391 56	79,305 70
May	93	155,680.00	147,128.70	5.3	.0008	573	76.05	2,564 73	109,940 88	112,505 60
June	117	203,974.00	193,145.30	5.2	.0007	580	77.21	2,951 95	146,191 42	149,143 37
July	130	234,492.00	221,170.30		.00067	625	81.14	3,112 30	179,473 00	182,535 20
Total	1,180	2,029,569.03	1,917,684.98	5.4	.00089	619	82.01	34,009 69	1,527,609 84	1,561,619 53

TABLE No. 4.—*Statement showing the work of company's assay office, &c.*—Continued.

Months	BASE BULLION.						Grand total.	Number of ore assays made.	Cost of office.
	Number of bars made.	Weight in troy ounces.		Average per cent. loss in melting.	Silver fineness.	Value.			
		Before melting.	After melting.						
1870.									
August	----	----	----	----	----	----	$77,103 36	280	$330 00
September	----	----	----	----	----	----	114,487 19	156	680 77
October	----	----	----	----	----	----	106,011 46	65	744 40
November	----	----	----	----	----	----	135,387 62	67	675 40
December	11	10,461 60	10,001 30	2.3	151	$3,360 68	151,731 20	97	680 00
1871.									
January	15	24,368 00	23,732 00	1.3	150	4,800 78	154,097 78	99	720 56
February	12	21,210 00	20,405 00	3.5	224	5,972 72	136,047 65	101	690 00
March	17	29,797 00	29,261 00	2.5	196	7,423 83	124,786 01	114	640 18
April	9	16,040 00	15,435 00	2.7	172	3,440 80	62,746 50	140	644 00
May	25	44,577 00	43,412 00	2.6	193	6,832 97	119,338 57	201	747 42
June	28	49,482 00	49,531 00	1.8	134	8,314 84	157,498 21	249	760 64
July	31	54,108 00	52,760 00	1.4	163	11,080 17	193,665 37	279	820 75
Total	148	259,313 60	252,667 30	2.8	164	51,280 99	1,612,000 52	1,890	8,353 32

TABLE No. 5.—*Statement of materials purchased and consumed at the mine for the company year from August 1, 1870, to July 31, 1871.*

Articles.	Quantity received.	Value.	Quantity consumed.	Value.	On hand July 31, 1871.	Value.
Lumber ....feet..	237,356	$12,728 05	231,350	$12,368 05	6,000	$360 00
Hewn timber ....feet..	216,560	7,580 04	188,510	6,318 69	28,050	1,262 25
Round poles...............feet..	25,791	2,186 20	24,301	2,000 20	1,400	126 00
Fuse ....feet..	102,500	1,165 22	86,700	889 16	25,800	276 06
Charcoal ....bushels..	13,130	4,810 81	12,780	4,688 31	350	122 50
Candles ....boxes..	1,577	7,906 14	1,417	7,102 04	160	803 20
Powder ....kegs..	881	4,112 33	834	3,861 23	47	240 10
Barley ....pounds..	54,264	3,326 00	53,964	3,305 00	300	21 00
Hay ....pounds..	59,128	1,375 64	55,608	1,314 04	3,520	61 60
Assorted iron ....pounds..	20,648	3,366 44	20,314	2,638 03	6,334	728 41
Assorted steel ....pounds..	14,025	3,251 13	10,003	2,366 29	4,022	884 84
Rope....................pounds..	2,952	730 28	2,515	638 77	437	100 51
Nails...................pounds..	12,030	1,603 28	9,290	1,280 28	2,749	323 00
Borax ....pounds..	344	149 34	94	46 84	250	102 50
Stone-coal ....pounds..	2,790	136 60	2,790	130 60	....	....
Shovels................number..	276	452 68	225	381 37	51	69 61
Assorted handles.........number..	1,205	551 64	1,030	490 10	175	61 54
Wheelbarrows ....number..	24	334 60	24	334 60	....	....
Wheelbarrow-wheels......number..	12	47 40	6	23 70	6	23 70
Tamping-bars ....number..	101	216 04	83	180 04	18	36 00
Car-wheels ....sets..	11	459 24	7½	336 74	3½	122 50
Axes ....number..	71	128 51	49	92 76	22	35 75
Picks ....number..	408	788 94	384	740 94	24	48 00
Sledges ....number..	161	752 82	128	631 38	33	121 44
Powder-cans............number..	46	61 76	46	61 76	....	....
Lard-oil ....gallons..	90	166 85	44½	83 60	45½	83 25
Coal oil ....gallons..	136	161 66	112½	107 87	53½	53 79
Files ....number..	194	190 58	135	81 43	59	49 15
Screws ....gross..	88	136 24	78	120 24	10	16 00
Axle-grease.............cans..	74	44 86	30	17 36	44	27 50
Padlocks ....number..	15	8 00	12	5 60	3	2 40
Water-barrels ....number..	58	278 05	40½	190 55	17½	87 50
Water ....gallons..	62,367	2,363 90	62,367	2,363 90	....	....
Nuts ....pounds..	259	37 42	99	13 42	160	24 00
Bolts ....pounds..	1,225	249 67	1,053	231 07	172	18 60
Hinges ....number..	50	41 75	44	37 25	6	4 50
Brooms ....number..	16	14 45	6	6 12	10	8 33
Brick ....number..	16,500	270 44	16,500	270 44	....	....
Wagons ....number..	3	693 00	1	153 00	2	540 00
Horses ....number..	9	2,115 00	1	115 00	8	2,000 00
Shingles ....M..	52,500	513 00	52,500	513 00	....	....
Harness ....sets..	5	320 00	2	120 00	3	200 00
Lanterns ....number..	24	14 34	24	14 34	....	....
Bellows ....number..	4	108 00	4	108 00	....	....
Shears ....number..	9	174 80	8	156 80	1	18 00
Office-fixtures	....	1,800 00	....	800 00	....	1,000 00
Giant powder ....pounds..	130	148 00	85	100 75	45	47 25
Giant-powder caps.............	6	7 50	5	6 25	1	1 25
Track-nails ....pounds..	470	122 20	154	40 04	316	82 16
Wood.................cords..	38½	300 30	6	46 80	32½	253 50
Sundries	....	10,937 60	....	10,368 64	....	568 96
Total		79,389 34		68,363 69		11,025 65

Receipts and disbursements of the Meadow Valley Mining Company for the biennial term ending July 31, 1871.

DISBURSEMENTS.

For mine properties:

Purchases made by company in extinguishment of adverse claims, &c....... .... $250,589 76

Law expenses contesting adverse claims. $32,679 65
Traveling expenses of law-agents.......   6,288 76
 ——————   38,968 41
   —————— $289,558 17

For construction and improvements at mine:

Cost of the company's mining-works at Piocho......   43,546 88
Freight of the materials and machinery from San Francisco 1,256 65
 ——————   44,803 53

For construction and improvements at mill:		
Cost of company's works at Lyonsville	$171,717 10	
Freight of the materials and machinery from San Francisco	42,349 22	
		$214,066 32
For mining:		
Opening company's mines, explorations, and extracting ores, as per Exhibit No. 2		474,738 41
For milling:		
Reduction of ores at company's mill, as per Exhibit No. 2		337,396 10
For miscellaneous expenditures:		
Freight to San Francisco on bullion yield to January 1, 1871..................		23,352 70
Discount on bullion yield to date....................		25,660 23
State of Nevada taxes on bullion..................		20,340 48
Exchange for coin equivalent on superintendent's drafts, fiscal year of 1870.....		3,688 18
Interest on company's notes and overdrafts at bank, fiscal year of 1870		5,614 34
Insurance premium on mill property.................		2,531 00
Telegrams to and from company's works.............		1,192 35
Office salaries for the term:		
To consulting engineer, secretary, and fees of trustees for attendance at board meetings......		8,098 33
Incidental expenses at San Francisco for the term..		6,374 35
Office furniture.....................................		1,123 55
For dividends:		
From No. 1 to 5, inclusive, paid to stockholders.....	329,585 00	
Remaining in special fund unclaimed...............	415 00	
		330,000 00
Total disbursements for the term..................		1,788,738 04

UNLIQUIDATED BALANCES.

Charles Forman, general superintendent.—Current balance of account to date		418 48	
Bullion *in transitu.*—Credited per contra to bullion yield..		72,831 67	
Bills receivable.—Unmatured balance of $50,000 due company in settlement of adverse mining claims....		25,000 00	
Cash.—Balance in hand to new fiscal year:			
Petty cash..........................	$250 00		
General cash........................	93,801 89		
		94,051 89	
			192,302 04
			1,981,040 08

RECEIPTS.

From assessments on capital stock:			
No. 1, levied June 5, 1869....	$15,000 00		
No. 2, levied August 14, 1869...	30,000 00		
No. 3, levied October 9, 1869......	15,000 00		
No. 4, levied December 4, 1869......................	15,000 00		
No. 5, levied January 29, 1870......................	30,000 00		
No. 6, levied March 26, 1870........................	45,000 00		
No. 7, levied May 11, 1870..........................	60,000 00		
		$210,000 00	
From bullion yield of company's mines:			
As per Exhibit No. 2.................................		1,671,465 66	
From rents:			
Of company's boarding-house, at mine..............	474 33		
Of company's boarding-house, at mill..............	612 16		
		1,086 49	
From sales of materials:			
To mining contractors and others...................		3,127 45	
Total receipts for the term.....................		1,885,679 60	

UNLIQUIDATED BALANCES.

Superintendent's drafts :
  On San Francisco office, unpresented................ $59,453 09

Bills payable :
  Unmatured balance of $100,000, paid to Rutherford
  & Co., to extinguish adverse claim to "No. 7 mine" 20,000 00

Unmatured book accounts :
  60-day bills for mining and milling supplies........ 15,907 39

        $95,360 48

        1,981,040 08

        T. W. COLBURN, *Secretary.*

SAN FRANCISCO, *August* 17, 1871.

A statement of the gross proceeds of bullion from the mines of the Meadow Valley Mining Company, and cost of production and reduction of the ores yielding the bullion, being for the biennial term ending July 31, 1871.

Dr.

MINING DEPARTMENT.

Explorations and dead work............................ $68,864 80
Labor in extracting ores............................... 281,259 87
Mining supplies...........................$75,516 59
Freight from San Francisco on supplies....... 14,999 18
        90,515 77

Contingent mining expenses.......................... 17,989 71
Salaries for the term................................. 16,108 26

    Total expenditures in mining department 474,738 41
Deduct amount of inventory of supplies on hand at date, 11,025 65

        $463,712 76

MILLING DEPARTMENT.

Ore-transportations from mine to mill, (10 miles)........ 79,613 87
Chemicals and other supplies...................$79,081 01
Quicksilver................................... 57,294 06
Freight from San Francisco on supplies....... 37,265 84
        173,640 91

Labor in reduction of ores........................... 71,901 80
Contingent milling expenses.......................... 4,589 60
Salaries for the term................................. 7,649 92

    Total expenditures of milling department...... 337,396 10
Deduct amount of inventory of supplies on hand at date, 37,273 15

        300,122 05

MISCELLANEOUS.

Freight on bullion yield to January 1, 1871............. 23,352 70
Discount on bullion yield to date.................... 25,660 23
State of Nevada taxes on bullion.................... 20,340 48
Exchange on general superintendent's drafts on San Francisco..................................... 3,688 18
Interest... 5,814 34
Insurance on mill property, premiums................. 2,531 00
Telegrams.. 1,192 35
San Francisco incidentals............................ 7,497 90
San Francisco office salaries for the term............ 8,098 33

        98,175 51

    Total expenses................................ 862,011 22
Net profits for the term............................. 813,668 38

        1,675,679 60

CR.

DULLION.

By proceeds of experimental workings of ores from the company's mines prior to construction of company's mill...	$22,272 89	
By proceeds of shipment of 67¼ tons of ore to England from company's mines, per Barron & Company........	36,293 63	
By proceeds of company's reduction-works at Lyonsville, as per tabular statements of general superintendent, for the current year ending July 31, 1871.................	1,612,899 14	
Total gross proceeds of bullion for the term....		$1,671,465 66
By rents of boarding-houses for workmen..............	1,086 49	
By sales of materials............	3,127 45	
		4,213 94
		1,675,679 60

T. W. COLBURN, *Secretary.*

SAN FRANCISCO, *August* 17, 1871,

The Raymond and Ely mine has greatly increased its operations since it has been incorporated in a company, as will appear from the following reports of the president, superintendent, and secretary :

FIRST ANNUAL REPORT OF THE RAYMOND AND ELY MINING COMPANY FOR THE YEAR ENDING DECEMBER 31, 1871.

President's Report.

To the Stockholders of the Raymond and Ely Mining Company :

It is with satisfaction that I submit my annual report in connection with the reports of the other officers, embracing all the business transactions of the company for the first fiscal year ending December 31, 1871.

At the time the company was organized a year ago, the facilities for the reduction of the ores were limited to a 10-stamp mill; but, on further explorations of the mine, ores in abundance were discovered, sufficient to supply double this number of stamps. Therefore, your trustees increased the capacity of the mill to twenty stamps.

These stamps have been regularly employed since their completion, producing, during the four months ending December 31, 1871, the unprecedented amount of $906,219.25 in bullion.

I consider this large production of silver from only twenty stamps unparalleled, for the same time, by any mining enterprise on this coast, if not in any part of the world.

Continued and systematic prospecting of the mining grounds of the company has developed still larger and richer deposits of mineral, warranting a further increase of mill capacity. Hence, on the 10th of October last, a contract was entered into for the construction of a 30-stamp mill, which is now almost completed. With this additional facility for working the ores, the production of bullion must be largely increased, even should the ores decrease in value; and I see no reason why dividends of not less than $5 per share, or even of a larger amount, cannot be paid for many months to come.

Notwithstanding the great success which has attended the operations of the company during the first year, with most excellent prospects for its continuance in the future, yet to attain this promise it is necessary that the mines should be well explored far in advance of the immediate daily consumption of ore.

To do this expenditures must be incurred, the benefits from which will be derived in the future.

The extraordinary receipts of bullion from the mines have excited the cupidity of covetously disposed persons, who have endeavored to embarrass the operations of the company and to depress the value of the stock ; and even with a lingering hope to possess the exceedingly valuable properties of the company. In every instance disappointment has followed their efforts.

So far as title to the mines of the company is in question, it is my opinion that if the company is not secure in the possession of its mining ground, then I know of no property in the district in which they are located that has a good and reliable title.

Every precaution to protect the property from destruction by fire has been taken, by

II. EX. 211——16

having good fire-apparatus and water in and around the buildings; besides, the mills and works are covered by insurances to the amount of $115,000.

For a complete and comprehensive summary of all operations for the past year at the mines, their future prospects, the quantities of ores extracted, and from what mines they were taken, the cost of mining and milling, the average yield of metal per ton, as also all kinds and sorts of material consumed at mine and mill, as well as a careful estimate of the quantities of ores now developed as a reserve for future extraction, I refer you to the full and able report of your superintendent, Mr. C. W. Lightner, and ask a close examination of it.

All items of receipts and disbursements during the year just closed are to be found in the report of the secretary, as also the assets and liabilities of the company. From this report it is readily determined for what purpose the resources of the company have been employed.

The report possesses sufficient merit for your careful attention.

ALPHEUS BULL.

SAN FRANCISCO, *January* 16, 1872.

Superintendent's report.

PIOCHE, NEVADA, *January* 9, 1872.

Alpheus Bull, esq., President Raymond and Ely Mining Company, San Francisco, California :

MY DEAR SIR : The amount of ores extracted from the different mines belonging to the company, during the past year, is as follows :

From the "Vermillion mine," tons	77¼$\frac{858}{1000}$
From the "Burk mine," tons	2,764$\frac{785}{1000}$
From the "Creole mine," tons	2,689$\frac{453}{1000}$
From the "Panaca mine," tons	6,708$\frac{795}{1000}$
Making a total of tons	12,239½$\frac{691}{1000}$

Of which we sent to the company's 20-stamp mill, at Bullionville, (distant ten and a half miles,) 10,574$\frac{784}{1000}$ tons, leaving on hand at the different dumps in Pioche 1,707 tons, distributed, as per table sent.

In addition to the ores remaining at the mines, we have about 125 tons of second-class "Burk," "Creole," and "Vermillion," of low grade, and about 200 tons of "base metal" at the "Burk," which, by the present mode of reduction at hand, would not bear working, but which may hereafter be of some value to the company. These I have not included in the statement of property on hand.

We have worked at the company's 20-stamp mill 10,372½$\frac{818}{1000}$ tons of ore, a portion of which I found on hand there on the 6th of January, 1871, but which has been replaced by that sent during the year. The tables show a gain of ore at the mill of 16½$\frac{848}{1000}$ tons.

The product of all the ores worked has been in gross, (assayed value here,) $1,361,-628.76, showing a yield, per ton, in bullion, of $131.27.

The large amount of base metals contained in the ores, together with imperfection in the machinery, which has at all times been overtaxed, will account for our percentage of working not being over 74.27 of the assayed value of the pulp. The tailings, however, have all been saved, and lie convenient for working, and may in future be a fund for the company to work upon.

In the "Vermillion" mine we cleaned out the shaft, (sunk 90 feet,) and drifted west from the bottom 40 feet. We also ran the old 40-foot level 10 feet farther, and stoped out a few feet in the bottom. The vein shows ore, but not in sufficient quantity or of a quality profitable to work at present.

In the "Burk" we have sunk the main shaft to the depth of 386 feet, (on the incline;) but little ore has been found below the 170-foot level, the body found in the upper portion of the mine gradually working itself westward in its descent until it gave out in the broken ground lying between the west incline and the eastern workings of the "Creole." The ground has been thoroughly explored to the east on the 72-foot level, and fair quantities of ore are to be found, but of a low grade. The fissure has never been lost sight of in sinking, and the nature of the ground, changing at the lowest depth, gives promise of ore existing in it.

In the "Creole" mine we stopped extracting after finding the large body of ore in the "Panaca," and have persevered in sinking and working for its development. The main shaft is down 383 feet, and, at this depth, we have cut a body of ore 3 feet in width, of excellent quality. There is still left in the upper levels of this and the "Burk" mine a large quantity of ore of medium grade. The new body cut at the bottom looks well, and will make the "Creole" as fair a producing mine as it was during the early part of the year.

The "Western Extension" (acquired during the year by purchase) has a good body of ore developed in the 200 and 300-foot levels. As we have but 87 feet to drive to connect the main working level of the " Panaca" with the 200-foot level of this mine, we may expect a good supply from this source.

At the " Panaca," the work of sinking the main shaft and drifting on the 223-foot level is progressing as rapidly as possible. At the 100-foot level we have connected with the old " Panaca " shaft on the west; and have run eastward 100 feet from the shaft, where we have a winze 74 feet deep, showing ore for 40 feet. On the 223-foot level we have run 134 feet west and 197 feet east from the shaft, making 331 feet in all, driven upon this level, the whole showing ore ranging from 2 to 6 feet in width, of fine quality. The face of the drift west is in a cross-heave, which cuts off the ore temporarily. The face of the drift east, going toward the " Western Extension," is still in good ore.

I shall refrain from making an estimate in tons and value of the ores developed in any of these mines, owing to the varying nature of the veins in this district, but will express my own confidence in having at the least three months' supply for fifty stamps in this mine alone above the 223-foot level. Before that time expires the shaft will be down to the 323-foot level, and indications are favorable that this, when reached, will exceed in extent and production the present lowest working level.

The whim-houses and whims over the "Burk" and "Creole" are in perfect condition, and the new steam hoisting-works, over the " Panaca," are very complete and effective.

The old 20-stamp mill is in poor condition, and requires a thorough repairing—the boilers threatening, and the pans and shafting worn and out of line. A week's repair, at least, with a full force, will be needed, to enable it to work with safety and economy.

The new 30-stamp mill will be, when completed, one of the very best in the State. Delays have occurred in the transportation of the most important part of the machinery ; but I have every assurance that the stamps will drop during the present month.

Hereafter, our bullion will be assayed at our own office, as we have now all the means for doing it in the proper manner.

Herewith I hand you tables, inventories, and memoranda, which will enable you more clearly to understand our past operations.

With our improved facilities for extraction and reduction, and the healthy appearance of our mines, we may hope for a very prosperous year's work.

Very respectfully,

C. W. LIGHTNER,
Superintendent.

Statement of receipts and disbursements of the Raymond and Ely Mining Company for the year ending December 31, 1871.

RECEIPTS.

From bullion yield of the mine		$1,361,628 78
From water-works		569 10
From mining supplies :		
Sale of supplies at Pioche		337 50
From milling supplies :		
Sale of supplies at Bullionville		2,180 91
From mill salaries ;		
Weighing ores, &c., for other mines		733 28
Superintendent's drafts :		
Advised but not yet presented		30,521 45
Total		1,395,970 57

DISBURSEMENTS.

For purchase of property and claims :		
Purchase of lands, claims, and titles	$108,798 75	
Purchase of water-works	2,000 00	
Law expenses	13,349 32	
		$124,148 07
For mining :		
Wages paid to miners	178,476 49	
Supplies for the mines	33,640 29	
Freight on supplies	2,791 19	
Contingent mine expenses	5,781 57	
Salaries of the superintendent and clerk	8,240 00	
		228,929 54

For improvements at the mine............................		$35,689 32
For milling:		
Wages paid to employés................................	$36,684 66	
Supplies for the mill...................................	59,101 37	
Freight on supplies....................................	6,734 99	
Salaries of the superintendent and clerk of the mill.......	6,148 00	
Contingent mill expenses..............................	10,487 26	
Ore-hauling from the mine to the mill...................	47,376 59	
		166,532 87
For improvements at the mill:		
Including the increasing of the 10-stamp mill into twenty		
stamps, &c.......................................		31,643 91
For the new 30-stamp mill:		
H. J. Booth & Co., for the machinery	46,053 50	
Freight on the machinery to Bullionville.................	19,377 42	
Labor and materials for the erection....................	24,051 68	
		89,482 60
For taxes:		
On real estate and personal property....................	1,477 98	
On bullion yield of the mines..........................	9,085 09	
		10,563 07
For dividends..		615,000 00
For discount on bullion yield..........................		21,642 77
For miscellaneous expenditures:		
General expenses.....................................	1,741 38	
Office expenses......................................	1,489 04	
Office furniture......................................	676 85	
Office salaries	2,792 50	
Fees of trustees for attending meetings of the board	830 00	
Interest and discount................................	127 97	
		7,657 74
Balance of cash on hand		64,680 68
Total disbursements.............................		1,395,970 57

ANDREW J. MOULDER,
Secretary.

SAN FRANCISCO, *January* 16, 1872.

ASSETS AND LIABILITIES.

Property and stores on hand December 31, 1871.

Improvements:		
At "Panaca" mine	$20,286 37	
At "Burk" mine	6,000 00	
At "Creole" mine....................................	2,400 00	
At "Western Extension" mine.........................	800 09	
Thirty-stamp mill	89,482 60	
Twenty-stamp mill...................................	56,500 00	
Water-ditch	2,000 00	
Pump-house	500 00	
Blacksmiths' house..................................	500 00	
San Francisco office furniture	676 85	
		$179,145 82
Stores:		
At "Panaca" mine	$6,248 45	
At "Burk" mine	5,427 79	
At "Creole" mine	562 96	
At "Western Extension" mine.........................	134 30	
At 20-stamp mill.....................................	47,059 17	
At warehouse and magazine...........................	5,776 98	
		65,209 64
Ores on hand:		
1,707½⁰⁰⁰ tons, as follows:		
1,330 tons from the "Panaca," estimated as of milling		
value at $150 per ton	199,500 00	
351½⁰⁰⁰ tons of different qualities, chiefly from the "Burk,"		
"Creole," and "Vermillion" mines, returned as of milling		
value at $60 per ton..............................	21,108 39	

2C tons from the "Western Extension" mine, at $125 per ton	$3,250 00	
		$223,858 39
Toll-roads:		
Work done on roads from the "Panaca" ground over the mountain, and improving old roads around the "Burk" and "Creole" mines........................		1,583 09
Cash on hand .....:..		64,680 68
		534,477 62

LIABILITIES.

Superintendent's drafts, advised but not yet presented................................	$30,521 45

Statement of the working of mines of the Raymond and Ely Mining Company for the year ending December 31, 1871.

Months.	ORE					DETAILS OF COST.					
	Extracted.		Sent to mill.		Balance at mills.	Extraction.		Prospecting and dead work.		Improvements.	
	Tons.	Pounds.	Tons.	Pounds.	Tons.	Labor.	Materials.	Labor.	Materials.	Labor.	Materials.
1871—January	553	1,331	553	1,351	......	$2,383 12	$1,806 80	$1,370 13	$1,806 80	$2,171 04	$1,806 80
February	516	1,040	516	1,010	......	3,026 75	939 08	2,132 00	616 44	2,261 00	1,426 69
March	839	381	619	381	230	6,443 73	1,865 98	2,977 39	1,473 64	1,162 88	1,133 47
April	724	1,171	798	903	230	7,317 50	1,064 60	7,324 49	1,491 11	1,110 00	1,569 98
May	652	607	680	607	302	8,719 00	1,563 63	7,380 00	1,233 68	419 00	775 11
June	969	1,957	969	1,957	195	10,236 25	1,840 30	6,868 00	1,698 60	205 50	357 75
July	908	1,784	868	1,764	235	8,618 50	1,323 16	7,270 00	1,179 06	1,120 33	248 57
August	975	429	975	429	235	6,685 35	1,755 41	7,449 00	1,408 92	1,229 25	1,132 74
September	1,272	1,025	937	1,025	570	5,806 45	1,674 95	7,272 59	1,732 93	2,353 50	1,615 69
October	1,845	730	1,155	731	1,220	7,303 00	2,992 17	12,583 59	1,860 55	1,143 00	1,192 70
November	1,462	1,573	1,187	1,573	1,495	7,542 00	2,848 79	6,061 80	1,716 19	2,085 00	4,183 62
December	1,395	1,024	1,183	1,024	1,707	11,081 25	3,175 15	7,959 96	2,164 23	1,925 00	5,864 66
Total	12,239	1,072	10,314	785	1,707	85,161 60	24,469 92	78,551 09	18,437 23	17,284 69	21,308 81

CONTINGENT.

	Expenses.		Total cost.	Total cost per ton not extracted.
	Mining expenses.			
1871—January	$506 50	$790 00	$14,660 10	$26 48
February	127 00	550 00	11,078 96	21 45
March	185 60	550 00	15,792 67	18 68
April	355 75	550 00	21,376 73	29 36
May	171 12	550 00	20,749 73	22 59
June	48 25	550 00	20,744 65	22 50
July	147 15	550 00	20,456 96	21 70
August	588 57	800 00	21,222 68	17 30
September	766 36	850 00	23,404 71	15 75
October	469 74	850 00	23,947 68	17 74
November	618 98	850 00	25,947 68	17 74
December	789 25	850 00	33,809 85	24 93
Total	4,793 57	8,940 00	257,246 68	21 01

WHENCE EXTRACTED.

Mines.	Extracted.		Shipped.		Balance.
	Tons.	Pounds.	Tons.	Pounds.	Tons.
Panaca	6,708	045	5,196	045	......
First class					192
Second class					1,170
Third class					140
Creole	2,680	1,964	2,589	1,264	100
Burk	2,714	108	2,729	108	35
	77	1,655	17	1,655	60
Vermilion	10,229	1,072	10,514	785	1,707

Statement of the working of the 20-stamp mill of the Raymond and Ely Mining Company, at Bullionville, for the year ending December 31, 1871.

Months	ORE Received Tons	Pounds	ORE Worked Tons	Pounds	Balance Tons	Pounds	TIME LOST By pan Days	Hours	Minutes	By stamps Days	Hours	Minutes	Assayed value of pulp in silver	Bullion per ton produced in silver and gold	Percentage of working
1871—On hand	355	1,091													
January	553	1,351	529	1,921	359	1,520		15		5	9		$129 00	$76 56	63.5
February	516	1,040	475	788	401	90	2	16	30	3	9	13	145 07	137 24	91.8
March	494	989	480	788	415	311	1	17		5	1	51	109 13	104 86	96.1
April	726	903	530	636	611	378	1	13		6	5	44	97 66	79 30	83.0
May	820	450	958	498	493	330		17		10	21	7	102 03	67 77	66.7
June	969	1,957	994	306	468	1,981	3	21		1	18	13	77 91	57 37	72.6
July	868	1,764	1,056	34	311	1,711	3	11		1	2	46	100 14	89 68	89.4
August	675	429	1,042	1,136	344	1,004	3	6		5	4	35	132 04	130 45	88.0
September	937	1,025	918	389	263	1,677	8	21		7	19	56	289 94	189 21	63.4
October	1,155	731	1,149	1,392	269	1,016	3	18		3	56		356 74	231 83	65.0
November	1,187	1,573	1,131	1,398	325	1,191	4	18	20	9	3		312 38	181 18	58.0
December	1,183		1,157	508	351	1,013	4			7	22		257 84	144 77	58.6
Total	10,389	1,536	10,372	1,619	351	1,613	46	14	50	63	15	7	170 74	131 97	74.37

Months	Total bullion, silver and gold	Hauling ore	COST OF WORKING Labor	Materials and implements	CONTINGENT Assaying and office expenses	Salaries	Total	Cost per ton
1871—January	$40,485 08	$3,542 46	$2,099 58	$3,808 80	$210 85	$367 00	$9,049 69	$17 11
February	63,293 97	2,118 91	7,496 37	3,451 33	652 68	443 00	8,562 29	18 01
March	50,371 99	2,009 23	4,094 85	3,487 58	778 19	494 00	10,705 05	22 47
April	43,563 92	2,942 30	4,517 65	3,631 83	353 15	480 00	10,144 76	22 50
May	63,936 36	3,391 90	3,302 38	7,111 69	674 74	571 00	14,981 71	15 97
June	57,646 12	3,024 42	3,305 70	7,217 55	1,060 71	430 00	15,932 38	16 03
July	91,818 33	3,634 61	2,785 45	7,450 99	595 08	450 00	14,903 11	14 52
August	139,563 26	4,567 64	2,832 10	7,560 04	775 59	491 00	16,225 37	15 57
September	174,226 46	4,687 56	2,885 35	4,965 95	846 98	480 00	15,468 84	18 46
October	266,521 30	5,790 24	4,306 75	8,346 79	1,243 95	650 00	10,938 41	27 30
November	205,046 64	5,938 53	4,509 15	8,170 87	630 50	650 00	20,329 72	17 57
December	107,547 64	5,917 56	5,185 55	8,189 15	1,297 37	650 00	19,689 45	18 31
							21,212 63	
Total	1,301,590 33	47,376 39	41,528 08	73,344 57	9,109 77	6,148 00	191,921 90	18 50

Property tax, 1871.
Bullion tax, June 30.

Average loss of quicksilver for the year, 2.55 pounds per ton worked.

C. W. LIGHTNER, Superintendent.

The Pioche mine first came into general notice during the spring of the year. It has since contributed largely to the bullion product of the district.

A correspondence in the White Pine News, of August 8, gives the following particulars in regard to this mine:

The ledge varies from 2 to 7 feet in thickness. In one place it even winds to 12 feet. It is incased in quartzite. Its strike is east and west, with a dip of 48° south into the ledge down for 200 feet. There is also a shaft 200 feet deep, and from the bottom of it a tunnel is run 50 feet to the ledge. The ledge is then drifted on both ways from the end of this tunnel, and from this depth two inclines, 115 feet apart, follow the ledge. One is 58 feet deep, the other 40. A continuous body of ore is disclosed, from the surface to this depth. From 10 tons of first-class ore, two average assays were made, giving $857 and $1,005 per ton. From points where I saw large bodies of the same, July 8, ultimo, I took samples of first-class ore, assays of which, since made by myself, give $1,315.81 and $1,297.26. Also two assays of second class give $812.54 and $907.95 per ton.

The following description of the mine appears in the News of August 23:

Forty-four feet from the surface comes First station, drifted on east and west along the ledge. Here a large body of ore was extracted, and a large body remains in sight, but of a base quality, which was immediately left, on striking freer ore below, 11 feet deeper. Next comes Second station—drifted on westwardly. This is also in base ore. Ninety-five feet below Second, comes Third station. Here drifts east and west have been run along the ledge, in good free ore. Fifty feet below Third, comes Fourth station, and between these two stations the principal work of the mine has been done. Here the ledge showed to perfection, from 4 to 7 feet wide, and the ore of a free and excellent quality. Extensive stopings were carried on here on the ledge, several hundred tons of good ore extracted, and several hundred tons of ore yet remain in sight of this point. Then prospecting commenced below the fourth level, by means of two winzes, 115 feet apart, along the incline of the ledge. The ledge appears to gradually and surely improve as depth is made. Here it is straight, compact, of an even width, with smooth and perfect walls. The workmen were enabled to sink some 60 feet on each winze before the air became too foul. Here they drifted east and west, again on the perfect ledge, and now are pushing vigorously to connect the two winzes, in order to establish a current of air. However, a blower has been put in to bear on these winzes, which makes comfortable work in these quarters, only somewhat slow hoisting with buckets and windlass. Here, also, a little stoping has been done, to prove the ledge, and many a dividend is reposing quietly here in sight. All the stations down to the Fourth are connected with the main shaft by adits; the tunnel on Third station being 55 feet long, on account of the ledge pitching southwardly, and the adit-level on the Fourth station being 72 feet long. All the stations are also connected by a series of chutes and winzes along the ledge. On arrival and placement of steam-hoisting works it is the intention to prosecute the sinking of the shaft to a great depth at a rapid rate. Everything points downward to the location of hidden treasure, unprecedented in extent and richness. Although the ore throughout is known to be somewhat base, yet the company's mill worried out of it, last month, 76½ per cent. of pulp assay, and from 168 tons of mixed ore turned out $40,000 worth of bullion—over $35,000 fine, and $4,000 base. The Stetefeldt furnace, in connection with the mill, is nearly completed, and when finished but a very small percentage of these rich assays will slip through. Three hundred tons of rock from this mine yielded $54,000. The Stetefeldt furnace has since been completed.

The Alps is a new mine. According to the latest news it looks encouraging, and will probably add considerably to the bullion product of the district for next year. The shaft has been sunk to the depth of 115 feet, and a drift to the east has been run about 120 feet, showing a body of ore about 2 feet wide. The average assays from the stopes and dump are $165.54. There is also a winze sinking below the first station, showing a fine body of first-class ore, about 3 feet wide, assaying $222.61. The company have built a mill, and intend to commence crushing ore early in January. There are about 200 tons of ore on the dump.

A number of other mines have been acquired by corporations during the year, and have commenced operations energetically. They are at present not sufficiently advanced to warrant special notice.

The mills in this district being, of necessity, located a considerable distance away from the mines on account of the necessary water for milling, the cost of transportation is at present rather high and onerous. The building of a narrow-gauge railroad from Pioche to Bullionville has, therefore, been considerably agitated. In its argument for this scheme the Ely Record estimates the following as the amount of ore which the below-mentioned mines would furnish monthly as freight to the railway company:

Since the necessary steps have been taken by some of our leading men to secure the construction of a narrow-gauge railroad from Pioche to Bullionville, the public will be interested to know what benefit the road will be to the community. Among the benefits to arise will be the great saving in cost of transportation of ores from mines to the mills. During the years 1872-'73, the mines in Pioche will yield ore as stated herein, and perhaps much more than the present estimate. The Raymond and Ely will ship to the mills 2,000 tons per month, or 24,000 tons per annum. At present rates, $6 per ton, this costs $144,000 yearly. The same can be shipped on the railway for $2 per ton, total expense, $48,000, thus saving the nice little sum of $96,000 per annum.

The Meadow Valley Company, by using the mill at Dry Valley to work up the large quantity of tailings now on hand at the mill, would be large gainer by building a new mill at or near Bullionville, where good water can be had in abundance. Then the company could ship as large a quantity of ore as the Raymond and Ely is estimated to ship. The saving in freight would be $96,000.

The Pioche Company will ship 800 tons per month, 9,600 per annum, at a cost of freight, at present rates, of $57,600 per annum; at railroad rates, $2 per ton, would save $38,400.

Washington and Creole Company will ship 300 tons per month, 3,600 tons per annum; present rates of freight, $21,600; railroad rates, $7,200; saving $14,400 per annum.

The Bowery will ship the same amount as the Washington and Creole, and the probability is this company will ship much more than that amount. The saving to this company will be at least $14,000 per annum.

The Alps mine will ship 260 tons per month, 2,400 tons per annum, costing at present rates $14,400; railroad rates, $4,800; saving $9,600 per annum. Thus we have a total saving from these mines alone of $283,200 on the freight of ores, without reckoning the saving on freight of wood, timber, &c. Many other mines will ship large quantities of ore to the mills, which must be taken to the valley for reduction.

The savings in the above amount will be greatly increased by the new mines that are opening out in that district. This mining camp is already sufficiently developed for the undertaking of such a work, and the benefits that would be derived would aggregate sufficiently in two years' time to pay the whole cost of building one.

For the quarter ending December 31 the bullion shipments aggregate $1,203,542.83. This is the result of the crushing of seventy-five stamps, and is equivalent to $16,074 per stamp. This showing is far superior to any other district in Nevada.

The total shipments of Ely district, during 1871, are given by Wells, Fargo & Company, as $3,982,228.

CHAPTER III.

IDAHO AND OREGON.

I make no apology for the meagerness of this chapter. The fact alluded to in the letter introductory to this report, that it is impossible with the means at my disposal to secure full, prompt, and reliable information every year from every part of the vast region covered by my work, is a sufficient explanation. Idaho and Oregon, being comparatively outside of the great advance of the mining industry, resulting in other States and Territories from the completion of the Pacific Railroad, and the extraordinary discoveries of new and rich deposits of the precious metals, were more neglected last year than other parts of the country. I am sorry that any part had to be passed by; and I am grateful to the few and scattered correspondents in these districts who have made it possible for me to say something, though less than it deserves, of the condition and progress of their mining industry. I estimate the total product of Idaho at $5,000,000, and of Oregon and Washington at $2,500,000.

Owyhee district.—Of the history of this district, during 1871, but little is to be told. In the Golden Chariot mine the ore began to grow poorer in March, the seventh level proving poor, and, since that time, the ore has not averaged $30 a ton. Within the first two weeks in December, however, the vein was struck on the eighth level, and good ore found; and, about the same time, rich ore was found in the fifth level south, some 350 feet from the shaft. A $5 assessment levied in September depressed the stock to $4 and $5 a share; and there was another assessment delinquent December 23, of $3. In the face of this last assessment the stock jumped to $30 on the news of good ore, but fell again to about $21. By the 1st of February the mine will be able to deliver out 1,000 tons a month, and the indications for pay-ore are very favorable.

In the Poorman no developments have been made; the ore worked being from the old dumps, and from tributors.

The developments on the Oro Fino have not been satisfactory, there having been found no body of pay-ore in the lower works of the mine. From the upper levels, (the old levels,) however, there have been taken out and worked in part some 2,000 tons of fair ore; 800 tons of Oro Fino ore were lying in the yard at the Owyhee Mill in December.

The Mahogany Company has collected one $3 assessment, and has now one of $2 delinquent on the 19th. There has been but little ore worked lately; but the mine is being opened in good shape, and by the 1st of March 800 tons per month will be taken out, and the mine will be in condition to keep up this production for several months.

Some prospecting work is being done in the Ida Elmore mine, but no body of ore has been found as yet.

In Flint district about a dozen persons have worked just hard enough to take out 196 tons of ore. It is very rich, and is from a great many different ledges.

On the top of Florida Mountain has been found a vein of very rich ore, the bullion giving about $3 per ounce. The ore milled has averaged $100 a ton; but the title is under a cloud, and figures are kept very secret. In the Empire mine, also, some very rich ore has been found.

It is worked by tributors, and the ore milled has paid about $250 a ton.

Since December 1 of last year Mr. J. M. Adams, mining engineer, has been in charge of the Webfoot, Ida Elmore, and Owyhee Mills. He has made a good many improvements in the Owyhee Mill. He has, for instance, decreased the speed of the pans to 55 and increased the speed of the battery to 90 drops a minute, and now he can work 45 tons of Golden Chariot ore a day without any trouble. Two years ago the mill could not average 30 tons a day on the same ore, and now 45 tons are worked with less fuel than was previously used in working 30 tons. The settlers and pans have also been considerably altered, and there is a great improvement in the work done, and in the saving of quicksilver. There is now a railroad into all the slum-yards, and, when running on slums, $450 a month is saved thereby.

The following statements, furnished by J. M. Adams, esq., give the production of gold and silver of Owyhee County during the year from July 1, 1870, to July 1, 1871. It will be observed that the aggregate exceeds largely that of the previous year. Most of the figures are taken by my correspondent directly from the books of the different companies, and when he has been obliged to estimate amounts he has taken pains to be rather below than above the facts:

Statement No. 1 of the bullion production of the mines in Owyhee County, Idaho Territory, for the year July 1, 1870, to July 1, 1871.

Name of mine.	Number of tons.	Yield per ton.	Total product.
Golden Chariot	13,751	$55 36	$761,273 96
Poorman	928	19 53	18,126 64
Oro Fino	958	28 09	26,968 00
Slums and tailings	870	14 36	12,494 39
Mahogany	1,126	50 08	56,390 00
Ida Elmore	3,242	26 67	86,490 00
Ledges in Flint	190	177 56	34,802 00
Ledges in Silver City, sundry	1,289	66 42	85,620 20
Placers			58,000 00
Total	22,360		1,140,105 19

Statement No. 2 of the bullion production of the mills in Owyhee County, Idaho Territory, for the year July 1, 1870, to July 1, 1871.

Name of mill.	Name of mine.	Number of tons of ore.	Yield per ton.	Total product.
Owyhee	Golden Chariot	8,254	$64 37	$531,370 48
	Poorman	928	19 53	18,126 64
	Oro Fino	176	51 24	9,019 00
	Slums and tailings	870	14 36	12,494 39
		10,228		
Webfoot	Mahogany	372	40 00	14,880 00
	Oro Fino	335	30 50	10,917 00
		707		
Ida Elmore	Ida Elmore	3,242	26 67	86,490 00
	Mahogany	254	65 00	16,510 00
	Sundry	39	87 48	3,411 00
	Oro Fino	190	30 00	5,700 00
	Golden Chariot	3,097	53 80	166,649 35
		6,822		

Bullion production of the mills in Owyhee County, Idaho Territory, &c.—Continued.

Name of mill.	Name of mine.	Number of tons of ore.	Yield per ton.	Total product.
War Eagle	Golden Chariot	1,100	$32 24	$35,466 68
	Oro Fino	57	12 19	693 00
	Sundry	350	28 56	10,000 00
		1,507		
Cosmos	Oro Fino	200	6 38	1,277 00
	Golden Chariot	1,300	21 37	27,187 45
	Mahogany	500	50 00	25,000 00
	Sundry	600	70 37	42,208 70
		2,600		
Black's Mill	Ledges in Flint	196	177 56	34,802 00
Arrastras	Sundry	300	100 00	30,000 00
Placer				58,000 00
Total		22,360		1,140,105 19

Report of the Golden Chariot mine for the year ending February 1, 1871.

The receipts and disbursements for the year ending February 1, were as follows:

RECEIPTS.

From bullion	$598,625 62
Discount on bills	917 82
Merchandise sold	44 50
Premiums	2,852 23
Slime sold	2,333 96
Total receipts	604,778 13

DISBURSEMENTS.

Dividends to stockholders	$115,000 00
Labor account	152,938 69
Milling account	110,842 15
Supplies	43,169 94
Bills payable	24,043 97
Hauling ore	22,165 88
Freight on bullion	11,867 61
General expenses	10,303 42
Freight on supplies	9,132 79
Miscellaneous items	44,360 07
Total disbursements	543,824 52
Cash February 1, 1871	60,953 61
Total	604,778 13

At the commencement of the last fiscal year, the company owed about $25,000, which, as will be observed, has been paid; in addition to which, $115,000 has been returned to stockholders. Dividends were resumed last September, when $20,000 was paid, followed in October by $25,000, and in November by $30,000. There was no dividend in December, but in January $40,000 was paid, equal to a monthly average of $31,000 since the resumption of dividends. There were 8,404 tons of ore crushed during the year, yielding $71.23 per ton. The total expense of reducing the ore to bullion, including labor, supplies used and on hand, freight on supplies, hoisting, hauling, and milling, was $40.51 per ton.

The company have no liabilities. Their assets, on the 1st of February, were as follows:

Supplies on hand	$17,301 63
House, engine and machinery	15,000 00
Ore on hand, 600 tons	35,000 00
Due from Cosmos Mill	1,500 90
Cash on hand	60,953 61
Total assets	129,755 24

The company paid a dividend of $70,000 on the 10th of March. On February account $128,729 had been received.

Report of the Golden Chariot for the year ending February 1, 1872.

The secretary's report for the year shows the following items:

RECEIPTS.

Cash on hand, February 1, 1871	$60,954
Assessment	80,000
Bullion production	396,653
Miscellaneous	5,393
Overdrawn in bank	4,084
Total	547,084

DISBURSEMENTS.

Dividends to stockholders	$130,000
Golden Chariot Stull Company	20,234
General expenses	11,921
Milling	133,280
Hauling	24,039
Labor	143,871
Supplies	48,205
Freight on treasure	7,167
Miscellaneous	26,548
In hands of superintendent	1,819
Total	547,084

The disbursements for miscellaneous purposes embraced assaying, exchange, freight on supplies, machinery, interest, and San Francisco office and other expenses. The total assets aggregate $74,903, against $19,084 liabilities.

The following is a statement of the gold-dust and bullion (coin value) shipped by Wells, Fargo & Co. from Silver City during the year ending December 31, 1871:

	Dust.	Bullion.
January	$2,790 00	$150,375 39
February	2,136 63	147,564 46
March	2,500 84	93,270 47
April	2,990 00	44,819 64
May	3,480 50	86,170 06
June	4,310 00	66,438 86
July	6,470 00	92,773 58
August	4,350 00	77,285 23
September	3,640 50	53,781 15
October	4,903 58	46,734 51
November	4,517 66	32,218 02
December	3,038 67	44,803 00
	45,128 38	936,234 37
Total		981,362 75

This amount exceeds the treasure shipments for 1870 considerably.

During the fall accounts reached me of the discovery of valuable mines at South Mountain, about twenty-five miles south of Silver City. According to the reports of the Idaho papers, the mountain on which the mines are located is quite as steep, though not so rocky and probably not so high, as War Eagle. On the northern slope of the mountain a magnificent stream of water, fed by numerous never-failing springs, wends its way through a deep and well-timbered gulch to the foot-hills and plain below. The principal mines hitherto discovered are contained in a zone of half a mile in width from north to south, and about three miles in length from east to west, near the summit of the northern slope of the mountain. The lodes run east and west, and dip to the south. They are from 18 inches to over 100 feet in width, and are embedded in a species of limestone. The ore is argentiferous galena, and also contains pyrites of iron and a small quantity of gold. A large number of assays have been made, ranging from $20 to $40, and even higher, in silver per ton. On the west side of the gulch above mentioned is what is known as Mineral Hill, on which are located the following-named mines: Cottonwood, Yreka, Yellow Jacket, Independent, Old Mortality, Narragansett, and Connecticut, most of them well defined and showing a rich quality of ore. North of Mineral Hill about a quarter of a mile are the Saint Croix and Saint Lawrence mines, and a mile and a half farther down the gulch is the Wide West. The Golconda, Galena, and Original run across the gulch near its head. The latter is 120 feet in width, and has been traced and located for nearly three miles. The Golconda is narrower than the Original, but is decidedly the finest looking mine that we saw in the district. A tunnel has been commenced on the ledge, opening up a solid mass of ore from 2 to 3 feet in thickness. Around the run of the mountain, and on the slope east of the gulch, are the Summit, Warn, Arvica, Mona, Scandia, Imperial, and other mines, which prospect well and yield satisfactory assays. Extensions have already been taken up in all directions, and new discoveries are being made every day. Most of the locators are from Silver City and vicinity.

It is one of the most favorable localities for mining purposes in this part of the country. But little labor will be required to get out immense quantities of ore. The gulches furnish splendid locations for furnaces and mills, with plenty of wood and water for all practical purposes.

A town has been laid out at the intersection of the two gulches at the base of Mineral Hill, and christened Bullion City. About fifty people were there in October, and more were going in every day. Wagons can be taken within 300 yards of the camp by going by the way of Camp Three Forks.

The Boisé Basin.—No important changes have taken place in the placer-mining interest of this section during the year. The supply of water held out until the end of September in the most important localities; but about this time the placer-mining season came to an end. The general results are reported to be quite as good as during the previous year. Detailed accounts, however, have not reached me at the time of this writing.

A few quartz mines have been worked during the year on Granite Creek, but the principal work in prospecting for and opening new quartz mines was done after the placer mines had shut down. Quite an excitement as to quartz claims was reported in the vicinity of Granite Creek during the last quarter of the year, and many new claims have been located and partly opened by shafts. A new quartz mill was started about that time by Mr. T. S. Hart, which crushed ore from the Sawyer ledge. Enough ore was reported on hand to run the mill for the next ten months. The Gold Hill mine and its 25-stamp mill have been in operation as during the previous year, most of the time, and, I am informed, with satisfactory results.

The United States assay office at Boisé City, which was to have been put in operation in July of last year, is unfortunately not yet organized; but there seems to be no doubt that the impediments heretofore in the way are now removed, and that the office will shortly be in working order.

Warren's Camp and Northern Idaho.—My correspondent, Mr. Richard Hurley, who has for several years past furnished information on the above portions of Idaho Territory, writes to me in November that there has not been any change of note in the working and the production of Northern Idaho.

An article lately published in the Idaho Statesman gives a short history of the mines of these districts, and points out the causes why quartz mining has not assumed greater proportions. Its moderate tone and the entire absence of that flight of fancy, which unfortunately characterizes so often communications from western mining districts, entitle it to confidence.

Early in 1861 the attention of the masses was first attracted to the Oro Fino district; in the summer of said year the Elk City district became an attraction. In August of the same year the first discoveries were made in the Florence basin; and early in 1862 the Warren's district was first made known; since which time the Miller's Camp, Palouse, Gnat Creek, Moose Creek, Newsome Creek, Clearwater Station, the bars of Salmon and Snake Rivers have had their attractions, all of which camps have been worked to a greater or less extent ever since their first development. The summer of 1862 witnessed the presence of the largest immigration to the Florence district, and the remarkable yield of that district in 1862 and 1863 is generally well known to the whole country.

In 1863 the discovery of rich mines in the Boisé Basin caused much of the larger portion of the miners then in the northern districts to drift south of the Salmon range, so as to be among the first to select the best locations in the new district, and the early disclosed richness of the new district and its continued prosperity held them fast, till, with few exceptions, the last tie which bound them to the northern districts had been severed. New developments and new enterprises, combined with the continued success of the first discovered camp of Boisé, have bereft the northern camps of the requisite mining population and mining capital and skill essential to the full development of the mining resources of the north. Especially has this been the case in reference to quartz mining.

But nevertheless there has been an annual product of these northern placer mines, from the period of their first discovery up to the present time, by the labor which has remained, that we think will equal, for each day's labor, that of any other district in the Territory.

As an evidence of this, in none of the camps has the price of day labor of white men in these mines fallen below the sum of $5, and yet in the majority of cases the employer has made a profit upon said labor.

Within the past two years Chinese labor has been introduced into several of these camps, and in that of Oro Fino district has proved highly satisfactory to the owners of claims.

Respecting quartz, it seemed to possess no attractions in these northern districts until the summer of 1865.

During that summer and fall several ledges were discovered in the Florence district, and many claims were located, and some of them prospected during the following winter, but with unsatisfactory results. In the summer and fall of 1866 about one hundred distinct ledges were discovered in the Warren's camp and immediate vicinity, the surface prospects of many of which were highly satisfactory.

But few miners then in the district were conversant with quartz, and the imagination of many was greatly excited as to their richness. But nothing more than surface prospecting of these ledges was done till the fall of 1867. During that fall, and the early part of winter, several arrastras and two 5-stamp mills were constructed, and during the winter and spring of 1867 about 1,500 of ore were extracted from seven ledges and reduced for free gold, the average yield per ton ranging a little more than $37. But all this ore had been extracted and milled by inexperienced men, and under the most embarrassing circumstances, from the want of capital necessary to perform what is termed the "dead work" in quartz mining, and these embarrassments had the tendency to produce contentions and strife among the operators, and to destroy confidence in traders and others upon whom they were more or less dependent for credit to push forward their enterprises. The camp has not yet recovered from these embarrassments, though much is still being done by way of developing the fact that many of these leads can, with capital to properly open them, be made productive of profit to the investment. Upward of $125,000 in gross have already been taken from the few leads which have been worked, and one lead is now being worked with success.

During the past two years several leads have been discovered in Elk City and Oro Fino districts, where surface prospecting is fully equal to those of the Warren's Creek. In Florence camp one 5-stamp mill was erected, which developed the fact that the quartz of that district is good, though the veins are not numerous, and are smaller than those of Warren's. The millmen became embarrassed, and the mill is at present idle.

These districts are so isolated from the great thoroughfares of travel, along which men of capital pass in making their tours of observation, that none visit these districts except upon a special mission for the purpose, and these kind of missions are fewer than angels' visits.

In conclusion it is fair to presume that these northern districts, if they do not receive the immediate attention of mining capitalists, yet their prospects warrant, yet it is to be hoped they will constitute a reserve of rich mining territory for Idaho, when other portions have become exhausted.

According to all accounts the water has held out unusually well in the northern placer mines during the year.

In regard to Warren's camp mining enterprises, during the past year, my correspondent says:

The Rescue quartz mine is worked now by a new company, known as the Rescue Mill and Mining Company. It is paying well; they take out on an average of $2,100 a week. As I attend to the company's outside business, I know that there is $1,000 a week clear profit. The company have a 10-stamp steam-mill on the ground. The mine is worked by an incline, about 200 feet deep. The best ore is in the bottom of the lower level. All the ore worked averages about $22 to the ton, the fineness of the gold being about .680. I assayed from the ledge since September 4, 1871, $19,673.37. Nearly half that time they had only five stamps running. The machinery is very imperfect, as the company bought first a small 5-stamp mill, and afterward rented a 5-stamp battery from parties here, that have a 10-stamp water-mill.

The Rescue is the last ledge discovered in this camp. There are some 250 recorded, but very little work has been done on them. Still parties all over North Idaho are very much encouraged with the present prospects of quartz in this camp. The price paid for labor in the mine and mill is from five to six dollars a day, so it is easily seen what an advantage it is to men, who would otherwise have to lie idle all winter, to get employment here. It has been tried to work some ledges as good, if not better than the Rescue, but the owners had to pack the ore on mules to an arrastra, and pay from four to eight dollars a ton freight, and about $20 dollars for crushing. About a ton in twenty-four hours could be crushed, and so people came to the conclusion that it would not pay.

Early in the spring of 1872 Mr. R. Hurley sent me the following estimate of last year's production of the placer mines in Northern Idaho, which I insert as the best obtainable:

Warren's camp	$160,000 00
Florence camp	100,000 00
Oro Fino camp	200,000 00
Elk City	100,000 00
Clearwater Station and Newsome Creek	180,000 00
Salmon River	40,000 00
Snake River	15,000 00
Moose Creek, east of Oro Fino	30,000 00
Other small creeks, about	50,000 00
	875,000 00

Eastern Oregon.—The number of workmen in the mines (principally placers) of Eastern Oregon has decreased perhaps 10 per cent. during the past year; but the increased facilities for working, such as hydraulic pipes and derricks, have made up the deficiency, and the yield is probably not far from that of former years.

Cañon district and Dixie have fallen off slightly in their yield, as compared with 1870, while Olive and Burnt Iron districts have increased fully enough to make up this deficiency, and Granite and Elk districts have about held their own. Camp Watson, or Spanish Gulch, has been added to the list, with a fair yield, hydraulic mining having paid well in that locality during the year. The quartz of that district, which is supposed to be rich, still lies untouched. There are no quartz-mills running in Grant County, the Prairie Diggings Mill having been closed, in the course of the summer, for reasons unknown to me. I do not anticipate activity in quartz mining so long as the placers hold out, and communications are so difficult and expensive.

An effort was made near Cañon City to penetrate the cement underlying the gulch diggings, and commonly called bed-rock. The shaft was sunk nearly 300 feet, and abandoned without reaching the real bedrock.

It is believed that this region will soon have communication with San Francisco, by way of Winnemucca and the Central Pacific Railroad, a much shorter route than the present one via Dalles City and Portland. At present the gold product goes to San Francisco via Portland. The high express and insurance rates over this route, particularly between Cañon City and the Dalles, favor the carriage of gold-dust in private hands, and its transmission in small packages through the mails. Hence I have made a larger allowance, over and above the express shipments, than does the superintendent of Wells, Fargo & Company, in his statement, published in the appendix. I am indebted for most of the foregoing information to Mr. W. V. Rinehart of Cañon City.

The districts east of the Blue Mountains, being near the great overland Boisé stage-road to Umatilla, are more favorably situated as regards communications, and this fact has led to some activity in quartz mining, with promising results. Mr. E. W. Reynolds, of Baker City, has favored me with a few notes on the condition of affairs early in the present year.

The Virtue Gold Mining Company (on the old Ruckel mine) are making satisfactory progress. The 10-stamp mill, at Baker City, has been refitted with new pans and new machinery. The mine has also been greatly improved; a fine steam-engine has been erected at the long tunnel-level (see my Report of 1870, page 231) for hoisting and pumping purposes, and the Rockafellow lode will be developed in depth. The rock crushed thus far, by the new company, maintains the average of former operations, (say $40 per ton.) Mr. Joseph Potthast is superintendent, and Mr. David Morrow has charge of the mine.

The Eagle Canal Company's ditch, in the Kœster district, is completed, and that neighborhood, which contains a good deal of excellent placer ground, will be actively worked during this year.

The Olive Creek and Rye Valley silver mines are reported to give very encouraging returns for prospecting, and to have attracted some attention from San Francisco capitalists.

Western Oregon.—From the districts of Jackson, Josephine, and Douglas Counties, once famous, and still to some extent productive, I have been unable to obtain, for the last two years, any trustworthy details.

CHAPTER IV.

MONTANA.

The greater part of the mining districts of this Territory have been personally visited, during last summer, by Mr. Eilers, my deputy, or myself, the former having traveled over the western portion of Montana, while I devoted myself to an examination of the eastern counties.

When it is considered with what difficulty and expense communication, travel, and transportation are maintained between the Territory of Montana and the rest of the world, it seems marvelous that any one should come there or stay there at all. The route by the Missouri River boats and Fort Benton is tedious and precarious, by reason of the low water, which stops navigation always before it is desired, and frequently before it is expected to do so. The only other route now employed is the road which leaves Corinne, Utah, on the Central Pacific Railroad, passes near Fort Hall, in Idaho, and, crossing the vast basaltic plains of the Snake River, enters Montana by Pleasant, Sheep Creek, and Beaver Head Valleys. With four hundred and fifty miles of hauling to be represented, as well as the railway transportation, in the prices of all imported articles, (among which must be included many of the necessaries of life,) Montana is heavily weighted in the race with other Territories; and the fact that she maintains prosperity, and is increasing in permanent population and sober industries, points to extraordinary natural resources.

First and fundamental are the agricultural capacities of Montana. A region which does not produce its own food must carry on every other industry at a fearful disadvantage. I know that the wonderful State of Nevada will be quoted as an example of prosperity, based almost exclusively upon mining; but this illustration really supports my proposition. It is not true, by the way, that there is no agriculture in Nevada; still less that an extensive agriculture may not hereafter arise in the valleys of that State. The sterility of the sage-brush country is an exploded superstition. The land lacks only water, and irrigation has already, in many places, produced wonders. But granting that Nevada has been hitherto, and will long continue to be, devoted chiefly to mining, and that food, as well as other supplies, has been imported into the mining districts, it is a notorious fact that this condition of affairs has crippled the mining industry from the beginning. The profits of the mines have been much smaller than they would otherwise be; a scanty and wandering population has made labor both dear and hard to control; and, finally, the net gains of the industry have mainly gone out of the State, leaving behind as "improvements" stamp-mills, cheap temporary houses, and holes in the ground. The railroad, the steady growth of agriculture, and other causes, will doubtless improve this state of affairs; but thus far, it must be acknowledged, Nevada has bled at all her veins without gaining a healthy life from such phlebotomy. The southern part of Idaho belongs to the same category.

In traveling by the stage-road northward from Corinne, no sooner is the Montana boundary passed, than nature assumes a different face. The sage brush gives way to nutritious and abundant bunch-grass; the vast, arid *mesas* are succeeded by lovely valleys; and instead of the barren brown ranges of the South, appear the pine-clad summits of the Belt and Rocky Mountains. Abundant streams of clear, pure water traverse the fertile bottoms; and though, by reason of insufficient rain-falls at certain seasons, irrigation is a necessary part of agriculture, yet the means of effecting it are ample and easily available.

Corn is not cultivated with much success, and fruits have been raised by a few enterprising and skillful horticulturists only; but all grains and roots flourish amazingly. The heaviest wheat-ears I have ever seen were harvested this year in the valley of the Stinking Water, or Passameri.

The grasshoppers have been, for three years past, the most pestilential enemies of the Montana farmer. But this season they have disappeared. I found them in great numbers on the Union Pacific Railroad, in the neighborhood of Laramie, careering westward'in fiendish glee, and whitening the air with their hosts; but they were too late to do much harm, even in Utah; and meanwhile Montana has escaped them altogether. To offset such occasional scourges as this, the ranchman of this Territory has the certainty of high prices for his product. At times flour is worth $26 a barrel, and oats are selling at over $2 a bushel. These are unusually high prices, though not quite so bad as those of early days, when, in one of the first winters of the placer miners here, the Mormon wagoners demand $80 for flour, per sack. The usual price of oats is $1 per bushel, or about 3 cents per pound, and with all the growth of the production during the last few years, the supply has never yet exceeded the home demand. Probably there is no other region in the United States at present where such inducements are held out to farmers as in this Territory. Many immigrants are coming in now, in the good old-fashioned way, with their teams and wagons, and wives and babies, to locate in the valleys of Montana. But it is a long and tedious journey; and, at the end of it, one is shut out from the world. Make Montana as accessible by rail as is Utah or Colorado, and the tide will come in grandly.

Another hinderance to agriculture, which the railroad will remove, is the danger from hostile Indians. This does not at all affect the greater part of the fertile districts of the Territory. It is mainly in Gallatin Valley that the settlers suffer. During the last summer the Sioux of Sitting Bull, a noted outlaw chief, not under treaty with the United States, and mustering, it is said, a thousand braves, made a sudden descent, for stock-stealing purposes, in the region referred to, and got away with some 300 head of horses. They killed two or three persons in an incidental way, and successfully escaped to their mountain fastnesses. Without the facilities of transporting and concentrating troops, which a railroad gives, it is almost impossible to follow up and catch these bands, to say nothing of maintaining such a police as to prevent their depredations. The Sioux and Blackfeet are perhaps the most numerous and warlike of our red enemies. I am satisfied that the problem of dealing with them, like the minor problem of the Apaches in Arizona, will be settled finally by railroad, and in no other way.

It is to the stock-raiser, even more than to the farmer, that robbery, whether at the hands of Indian or white, is a frequent source of loss, and the raising of cattle and horses is pre-eminently the business for which large portions of Montana are fitted in a most remarkable degree. The bunch-grass, which grows here in such luxuriance as to lose, in some places, its characteristic distribution in bunches or clumps, and to cover the whole surface with continuous pasture, is already famous as a nutritious and fattening food for stock. Cattle and horses are turned out upon it at all seasons, even in the winter, and improve in condition while grazing. This grass dies early, but retains its nutritious properties all winter. It thus constitutes a standing hay—only it is much better fodder than hay, and almost like grain in its effect. To be more exact, I might say that to pasture a horse on bunch-grass is like giving him plenty of good hay, with regular and liberal feeds of grain. There are a good many horses in the Territory now, but the breed has hitherto been poor. Now more attention is given to breeding; and in a a few years this Territory will furnish, I am convinced, a strain of serviceable blood, worthy of the great advantages nature has bestowed upon the stock-raiser here.

The grass to which I have alluded makes excellent beef also; the price during my visit was 25 cents a pound. The herds in some of the valleys amount to 5,000 or 6,000 head. There is a great demand still for oxen as well as cows; and Montana is importing cattle, as well as receiving into her ample grazing lands the stock of other States and Territories. The value of the dairy products of Montana is already over $500,000 annually; but that is only a feeble beginning. Like everything else here, except gulch mining, this business is in its earliest infancy. The Territory contains 23,000,000 acres of agricultural and 69,000,000 acres of grazing land; and these vast areas are merely dotted here and there with the cabins of perhaps 5,000 ranchmen, the rest of the population being gathered in the mining towns and camps.

There is as yet not much sheep-raising; but every wool-grower will see that this must be a country excellently adapted to that business. But there is at present no home market for wool, because there are no home manufactures. However, the Territory is not yet ten years old; and everything cannot be done at once. When the time comes the mountains stand ready to offer abundant water-power and lumber.

All the industries I have mentioned will start into vigorous life when the railroad shall have opened the way to the civilized and commercial world; and behind them stands the great mining industry, the extent of which, even at the present time, is quite astonishing for a Territory so isolated as Montana, and which must grow into vast proportions as soon as cheap communication with the outside world is established.

The amount of the gold and silver production of Montana is usually underestimated by the San Francisco statisticians. Mr. Valentine, superintendent of Wells, Fargo & Co.'s Express, in a statement which will be found in the appendix, gives the invoiced shipments of bullion for 1871 at $4,060,929, and adds to this sum but 20 per cent. for "other conveyances." This is certainly far too little. The proportion of bullion privately carried, and the undervaluation of the express shipments are always dependent upon the rate of express charges and insurance. To the circumstance that this rate is high in Montana, is added the facility for private shipments of ore, bars, and dust by the numerous empty returning freight-wagons. The Montana agents of Wells, Fargo & Co., themselves, (who must be supposed to know the facts more familiarly than the San Francisco superintendent,) have repeatedly declared that the invoiced express shipments are about half the actual product.

The following extract from a letter of William F. Wheeler, esq., United States marshal of the Territory, dated December 16, 1871, presents the case as clearly and as closely as it can be done, and corroborates my own personal observations:

I have procured in person, from the four principal places of shipment, the value of the dust and bullion sent away by express. The result is as follows:
From January 1 to December 1:

From Helena	$2,140,000
From Virginia City	630,000

From Deer Lodge	$890,000.
From Bannack	150,000
Estimate for December:	
From Helena	250,000
From Virginia City	100,000
From Deer Lodge	120,000
From Bannack	20,000
Total silver in refined bars, 1871	125,000
Base bullion (silver and lead) shipped by freight to Corinne, 50 tons, valued at $300, silver, per ton	15,000
Crude ore shipped by freight, 200 tons, valued at $50 per ton	10,000
	4,450,000
To this amount should be added the amount in the hands of Chinese, who usually sell only what is necessary for their current wants, and take it away when they go out of the country in parties; the amount retained by a class of men who believe the price will be higher—hoarders; the amount taken out of the country by individuals and parties of men who, this year, have availed themselves of going together in numbers of fifteen or twenty on the "fast-freight line," which takes them from here to Corinne in eight or nine days, for mutual protection, and to save express charges on their gold	3,600,000
	8,050,000

I have submitted these figures to many of our leading bankers, merchants, and miners and not one has said I have placed the amount too high. Some of them believe the yield has been ten millions. The express agents think that I should double the amount shipped by express. The bankers and merchants agree that the yield this year is larger than it was last year, because their business proves it. The year has certainly been a prosperous one for Montana. The next year promises to be still more prosperous than the present. Much more snow has already fallen than fell all of last winter. This shows that we may expect an abundant supply of water next year, which is all we require for a season of great prosperity, because the bulk of our gold is the product of placer mines.

WESTERN MONTANA.

In Beaver Head County the old placer mines of Grasshopper Creek are still worked to a considerable extent. Their palmiest days are, of course, gone by, and the rude methods of pan and rocker have long been replaced by sluice and hydraulic mining, and even these methods have not yielded very satisfactory results during this year. White labor is still very high, and it is evident that diggings must be very rich to enable the employer to pay $5 or $6 per day to his hands. At the same time there is such an unreasonable enmity of white miners against the Chinese, that those of the latter who came into Bannack in the spring to engage in mining were notified by the whites to leave at once. Now the majority of the inhabitants of Bannack would be glad to welcome them back, but they have engaged in other parts of the Territory, and for this year, at least, Bannack will certainly remain a dull camp as far as placer mining is concerned.

At the time of my visit there was only some sluice mining going on at claims on the bar opposite the town. I must add here that Bannack is situated on the north side of Grasshopper Creek, and that the productive ground which has been heretofore worked reaches from one and a half miles above the town (west) to about six miles below it. A great part of the bed of the creek is practically worked out, and that part which is still untouched cannot be worked at present, on account of the costly raising of the gravel through shafts, water being at the same time very troublesome. But on the bars, though the majority of the rich claims here, too, are exhausted, there remains still a large field, especially for the time when labor shall have become reasonably low. Four

. miles below town the White's Bar Ditch Company and the Cañon Ditch Company have had two hydraulics at work since the middle of May, the former company employing eleven, the latter four men. The gravel on the hill-side is here from 12 to 20 feet deep, the portion nearest to the slate bed-rock alone carrying gold. This auriferous stratum is found from 1 to 6 feet thick. The White's Bar Ditch Company has worked pretty regularly with eleven men since the middle of May, as above mentioned. Up to the 28th of July they had, however, only taken out $466. At the same time they expected to have twelve or fifteen hundred dollars in the fine dust closest to bed-rock, which is reserved for cleaning up when water becomes scarcer later in the season. So far, they had only worked a week and a half in cleaning bed-rock, and the returns for the last week had not come into town.

The Cañon Ditch Company had employed up to the 22d of July only four men, who made the necessary repairs on the ditch, which leaked in many places. Since then they employed fourteen men, washing off dirt. There had, of course, no clean-up been made up to the 28th.

The Pioneer Ditch Company have employed four to five men up to July 1, and since then only two. I am told that they have probably taken out $700 or $800 this year, but I could not obtain any definite information on this point.

The Spring Gulch Ditch Company, for which the four partners constituting it have done the work principally themselves, is reported to have taken out $5,000 this season. The gold from the Bannack diggings is, on an average, .955 fine. Since the 22d of July three companies of Chinese, about fifty in all, have commenced work several miles below Bannack. There is nothing known as to the yield obtained yet, but their diggings are thought to yield not above $2 per hand per day. Indeed, Mr. L. Newman, the agent of Messrs. Wells, Fargo & Co., at Bannack, a gentleman who has the best opportunity for correct observations, thinks that there are but very few claims in Bannack and vicinity which will exceed the above yield.

At Horse Prairie Gulch, twenty-eight miles southwest of Bannack, the Yearing Brothers are reported to employ between thirty and forty men in their hydraulic claims, and Merrill's, Hyde's, and several smaller claims employ about thirty more. About $11,000 have been brought into Bannack from this locality up to the end of July. The gold is found .900 to .910 fine.

The two principal buyers of gold in Bannack, Messrs. Wells, Fargo & Co., and Mr. Ike Roe, had bought, up to the time above mentioned, from the Bannack, Horse Prairie Gulch, and neighboring placers, $65,000 worth of gold. This amount has, of course, been considerably increased during the latter part of the year. Mr. Ike Roe writes me in December that he alone had bought $60,000 up to that time. Information from Wells, Fargo & Co., applied for some time since, has not reached me at the time of this writing, but I feel sure that they cannot have bought less than $40,000 during the year. This makes the yield of gold of this district less than any previous year, and the reason for this is to be sought in the falling off of the placer yield.

The Dakota is the best-known quartz vein in the immediate vicinity of Bannack. It occurs about a mile below town, in the hill north of Grasshopper Creek. It is, unfortunately, divided up into a great number of very small claims, of which Nos. 1 to 3 and Nos. 6 to 10 inclusive, have been worked the most. No. 6 is better opened, and has had more work done upon it than all the rest, although the entire length of the claim is only 100 feet. It belongs now to W. C. Hopkins alone, who has bought out his former partner. The claim was worked for several years previous to 1870, when, for the first time, it lay idle during almost the entire year. About eight weeks previous to my visit to the district, work was recommenced on the mine, and it has since kept the mill busy.

The vein being, as above stated, best exposed in Dakota No. 6, the following statements refer more particularly to that claim. The vein strikes nearly east and west. It lies between a granite dike on the south and crystalline limestone on the north, and is, consequently, a contact-vein. The dip of the granite wall being quite irregular, sometimes to the south and more frequently to the north, that of the vein is also varying, and the first exploring-shafts which followed the vein closely are therefore rather crooked. The width of the vein is also very variable, being from a few inches to 15 feet. The horses in the vein, which occur quite often in the larger portions of the vein, are always limestone, never granite. The great bulk of the ore is a very dark-colored quartzy brown iron-ore, while around and in the limestone horses, down to a depth of over 260 feet, sheets and threads of green carbonate and soft black oxide of copper are always found. In a large bonanza, only lately discovered, quite near to the surface, these sheets of copper-ores are quite thick, from half an inch to an inch, and they completely envelop the limestone boulders, which lie very loosely in the brown ore. The greatest depth reached on the whole vein is 310 feet, in a shaft on No. 6. A long tunnel, which exposes numerous cavities filled with beautiful crystals of calc-spar, is also driven on this claim across the limestone into the vein, striking it at a point where it is at least 15 feet wide, 165 feet below the surface. For 20 or 25 feet before reaching the vein the limestone becomes quite brecciated and loose. From the bottom of this tunnel a shaft is sunk 145 feet deeper on the vein, which, for its entire depth, stands in a large mass of soft brown ore, showing free gold quite frequently. The existence of the large body of fine ore lately found quite near to the surface was unknown to the owner, both its extremities in the longitudinal direction being hidden by walls of dead matter, so that he had thought the whole intervening space was filled with the same. All the ore down to the depth now reached is very rich in oxide of iron, quartz being present in a much smaller proportion. No water has been reached in the shaft yet, and iron pyrites occur, therefore, only in small bunches on the lowest levels. But whenever the water-level is arrived at, the pyrites will undoubtedly be found in the lode to be very solid, an occurrence which, so far, has in most gold-veins not proved to be very favorable for the richness of the veins in depth, and which is, to say the least, a great impediment to the extraction of the gold by milling. The ore worked at the mill at present yields only $12 to $15 per ton, much gold being lost on account of imperfect machinery.

The Wadham vein is located on the opposite bank of Grasshopper Creek, high up toward the summit. It runs northeast and southwest, and dips northwest. The width of the paying portions of the lode is from 3 to 4 feet, as exposed in several shafts from 50 to 100 feet deep. The ore has a redder color than that from the Dakota, and contains from $200 to $240 per cord, (of six tons.) The mine has been opened to a depth of 125 feet, and the ore is worked in three arrastras, by Mr. Wadham himself, and in the R. T. Hopkins Mill of five stamps. A good deal of the ore is slightly copper-stained.

The Saint Paul is higher up on Grasshopper Creek, just opposite Bannack, and on the same bank as the foregoing. The vein lies between syenite and talc-slate on the hanging, and quartzite on the footwall. Nearest to the vein on the hanging-wall generally lies a two-foot band of syenite, which, however, thins out often and disappears, permitting a one-foot layer of talc-slate, which lies above it, to form the wall. The vein has a general northeast and southwest course, and dips first near the surface, almost imperceptibly, but after a depth of 10 or 20 feet has been reached, very steeply toward the northwest. The first feature is undoubtedly the result of the erosion of a steep ravine which runs along the vein and near to it, and which caused the upper softer portion of the vein to tip over into and across it, after the supporting rocks were mostly washed away. The vein, which is shown in the different prospecting-tunnels and shafts to be from 3 to 6½ feet thick, is distinctly divided into two layers, which differ materially in appearance and composition. The upper layer, from 1 to 3½ feet thick, is an excellent highly ferruginous gold quartz, in which free gold is frequently visible ; it is often quite soft for a considerable distance, but sometimes quite hard, yet porous. The latter kind of ore shows quite as much free gold as the former. The lower layer, from 2 to 3 feet thick, is a whitish, red-spotted, decomposed material, which may have been syenite; but it always contains gold in the red spots, though it is not near as rich as the upper layer.

The claim of Mr. George Brown, which is the only one worked, contains 350 feet. He has opened it by six or eight tunnels and small shafts, in all of which the lode appears very regular as soon as the upper portion, which, as mentioned above, is tipped over toward the ravine, has been passed through. Six hundred tons of ore from this claim have been worked, the greater part in the New York and Montana Company's Mill, and the remainder by Mr. Brown himself in the N. E. Wood Mill, which he has lately rented. But he can only work three tons in twenty-four hours when the ore is soft, and does not extract more than half the assay value of the ore, which is reported to be from $24 to $28 per ton. The claim is excellently located for tunneling on the vein, as a depth of about 250 feet can be reached by starting in the bed of Grasshopper Creek.

The following notes on the other claims along the Dakota were furnished me by Mr. W. C. Hopkins, of Bannack.

Dakota No. 5, west.—Owned by a New York company. The top of the claim is stripped of surface material about 15 feet in width, and to a depth of about 25 feet. Ore seemed to be in pockets near the surface. At the present depth the vein is 3 feet in width, and dips to the west. The ore is of the same character as that heretofore described from No. 6.

Dakota No. 7, west.—Two shafts, each about 80 feet deep, with drift connecting, are on this property. A large amount of ore has been taken out of this claim, which was all milled, yielding as well as other ores on the lead. The claim has been in litigation for the past two years, but a settlement has now been made, and the owner will recommence work as soon as the weather will permit.

Dakota No. 8, west.—There are also two shafts on this claim, about 80 feet deep. Ore commenced being taken out at a depth of about 20 feet. At a depth of about 60 feet one of the shafts broke through into a cave, about 20 feet in depth, and extending nearly the whole length of the claim, or about 100 feet. This opening dips to the west, and its bottom was covered with "burnt" quartz containing considerable free gold. The hanging-wall is limestone, full of small pockets showing calc-spar crystals, and stalactites and stalagmites of great beauty. No attempt has been made to sink below this cave, the owners preferring to wait developments on No. 6.

Dakota Nos. 9, 10, and 11, west.—Same number of shafts as on the foregoing claims. The ore is apparently of the same character, but does not occur in the same quantity as in the other claims. The vein still shows a heavy dip to the west.

Dakota No. 12, west.—The shaft on this is over 100 feet in depth, and shows hardly an indication of ore. The lead does not seem to extend much farther west than No. 11.

The Dakota lode possesses some peculiar characteristics. The Discovery claims are in a ravine. The richest claim is No. 6 west, which is on the west slope of a hill about 150 feet above the Discovery claim. No ore is found west of No. 2 west until No. 5 west is reached. There is so much water found in No. 2 west, which is on the east slope of the hill, as to impede working, while in No. 6, on the western slope, at a depth of 300 feet, or nearly 100 feet lower than the shaft in No. 2, there has been no water, and the bottom of the shaft is perfectly dry. No ore is found east of No. 2 east, from Discovery, and none is found west of No. 11 west as yet.

Estimate of cost of mining and reducing ores in Beaver Head County, Montana Territory, reported by Walter C. Hopkins, December 31, 1871.

Population of Bannack, 500; population of Argenta, 400; wages of first-class miners, $4 to $5; wages of second-class miners, $3; wages of surface laborers, $3; cost of lumber, $50 to $100 per 1,000; cost of mining-timber, usually 75 cents per stick of 20 feet; cost of common powder, $6.50; cost of giant powder, $6.50; cost of quicksilver, $1.25 per pound; cost of freight from Corinne, Utah, 3 cents per pound; cost of fuel, $8 to $10 per cord; cost of 10-stamp mill, California pattern, including freight, erection, &c.—none in this section; cost of 20-stamp mill, freight, erection, &c.—none in this section; minimum mining cost per ton of ore—no estimate, work being generally done by the day; average yield of ore : gold, $12 to $20 per ton ; silver, from $30 to $100 per ton, at mine.

The mills of Bannack are nearly all old and imperfect, and sadly out of repair.

The N. E. Wood Mill, a Bullock crusher and grinder, is not at all fitted for working hard ores, and even with soft ones has a very small capacity, as above mentioned.

The Walter C. Hopkins Mill, belonging to the owner of Dakota No. 6, has twelve stamps, and at the time of my visit the batteries leaked badly, so that one could not be kept in operation at all, and, shortly after, the mill had to be shut down altogether on account of the giving way of the battery foundations. Besides the stamps, there are two dolly-tubs and a settler in this mill. It is the only mill in the vicinity of Bannack which is driven by steam. [Information derived from Mr. Hopkins himself late in the year renders it probable that this mill will be fitted with steam-pipes, pans, and settlers very soon, the object being to fit it for the working of the Blue Wing silver-ores by the Washoe process, and to make it more effective for gold-ores.]

The R. T. Hopkins Mill, a little 5-stamp affair, with an arrastra attached, is the only one which has been running very regularly of late years. It crushes ore from the Wadham, and is said to work nearly as close as Mr. Wadham's arrastras, which are lower down the creek.

The New York and Montana Company's Mill of twenty-four stamps was idle and closed up at the time of my visit, on account of litigation, so that I could not even get into the building, which has a solid and substantial appearance, quite in contrast to the shanties covering the other mills.

Thomas W. Wood & Sons were erecting in the latter part of the year a 12-stamp steam-mill on Taylor's Creek, three miles from Bannack. This mill is intended to amalgamate the Blue Wing ore, after a preparatory chloridizing roasting. I am not informed of its details.

In the neighborhood of Bannack are three smelting-works, or what appear to have been intended as such. Two of them into which I had access bear evidence that the builders had not the slightest idea about metallurgical operations; the third was closed up. All of them are, of course, idle, there being no lead-ores in the vicinity, and even if these were present such works could never be conducted profitably.

About three miles north of Bannack is the Blue Wing district, which contains silver-ores in deposits in limestone. All of them are generally narrow, and the widest rarely exceed 3 feet in width. But the ores, argentiferous galenas and fahl-ores, are very rich, assaying from $125 to $150 per ton. The Blue Wing and Bostwick's mines were worked to a small extent during the summer, and the ores were sold to the smelting-works at Argenta.

For the following remarks on this district, which reached me only just in time to be incorporated in this report, I am indebted to Mr. W. C. Hopkins, of Bannack.

Blue Wing district is situated about three miles from Bannack, on the divide separating the waters of the Grasshopper from those of the Rattlesnake, on a spur of what is known as the Bald Mountain range. The belt of lodes is about three miles in length by two in width, and the ores are mainly amalgamating-ores.

The Blue Wing lode was discovered in 1864, and gives name to the district. It has been developed to a greater or less extent for over 1,500 feet. There are several shafts and tunnels upon it, particularly upon the Discovery claim, of 1,000 feet. The deepest shaft is down about 80 feet, and from this drifts or tunnels were run. The crevice averages about 3 feet in width. The lead is not yet well defined, being still in broken rock. The ore shows occasionally native silver, and is almost a pure amalgamating-ore, which now readily commands $65 per ton at the dump-pile of the mine. The general direction of the crevice is east and west.

The Huron is situated about one mile from the Blue Wing, on an opposite bluff of the same range. It has also been developed for about 1,000 feet in length. The ore is of the same general character as that from the Blue Wing, but richer, and commanding readily $100 per ton on the dump. The main shaft is about 80 feet deep, and exposes a 3-foot crevice, easily worked, and showing every indication of being a permanent lead. Native silver is often found. The general direction is east and west. The owner has a large amount of ore now out, ready for sale to either smelting or amalgamating works the coming season.

The Wide West is an exceptionally rich lead, from which a large amount of ore was taken out in 1865 and 1866. The crevice is about 3 feet wide, and at the start seemed to be partially closing in. It is owned by a New York company, but has been practically abandoned since 1866 on account of monetary difficulties of the company. The receiver of the company proposes to re-open the mine the coming season. The general direction is east and west.

The Kent, a lode with a crevice of about 7 feet, has been developed for about 600 feet. The ore is different from that of other leads in the district, being soft, of a reddish color, as if stained by iron, easily mined, containing some gold, and worth $25 to $30 per ton at the dump.

The Brick Pomeroy is situated about one and a half miles from the Blue Wing, on the same range, and is a very large lead, the crevice being about 7 feet in width. It has been developed for several hundred feet, and has upon it several shafts, all of which give ores commanding about $40 per ton at dump-pile. The ore shows occasionally native silver, and is of a different character from that of other mines in the district, inasmuch as it contains quite a large amount of argentiferous galena. This lead promises, so far as can be judged from present developments, to prove a permanent one. Of other leads there are prominent the Bright Silver, Black Hawk, Sibley, Silver Rose, Milton, Whopper, Bonaparte, John Wesley, Victory, Highland, Black Hawk, No. 2, Charter Oak, Del Monte, Sherman, Centreville, and Puritan. All of these have shafts upon them, and many of them are also prospected by tunnels. From all of them considerable ore has been taken, and upon most of them the various owners have been at work this winter, throwing out a large amount of ore for summer consumption. The value of these ores is as yet unknown, but being of the same general character as those from the Blue Wing, Huron, &c., they will probably be of about the same value. Most of these leads run east and west, but some of them are cross leads, and run northerly and southerly.

The smelters purchase only the richest class of ores from this locality. As they are mainly amalgamating-ores, the expenses required for the fluxes are too great to render a profit possible except from the rich ores. The consequence is that quite a large amount of low grade ores is on the hands of the miners, and will probably remain there until proper amalgamating-works are erected near by to work them.

Argenta district is at present the only silver district in Montana in which the ores are beneficiated on the spot. They are treated by smelting, though by far the greater part of the Montana silver-ores are really amalgamating-ores. The works do not treat Argenta ores only—the production of the district being far less than the capacity of the smelting-works—but also nearly all the silver-ores which are, at the present time, mined throughout the whole Territory.

In some respects the location of these works was well, in other respects very badly chosen. When the mines of Argenta were first discovered, seven or eight years ago, the mineral deposits, which are nearly all located on the limestone hill north of Argenta, displayed lead-ores on top, some of them very rich in lead, and most of them with a satisfactory percentage of silver. True to the usual mode of developing mining districts in the West, several parties rushed to the conclusion that the ores of these mines must be smelted, though there was no mine opened to a greater depth than 25 feet. Works were consequently erected, first by the Saint Louis and Montana Mining Company, afterward by A. M. Elsler, and still later by Messrs. Tootle, Leach & Stapleton, so that Argenta now boasts six blast and two cupelling-furnaces. But, unfortunately, it was soon discovered that in all the lead and silver deposits occurring in the limestone, which comprise ninety-nine hundredths of the Argenta mines, the ore became continually poorer in lead with increasing depth, though it retained, in most cases, its original percentage of silver. This soon stopped one of the works entirely, while the others could work only from time to time, at long intervals, whenever a sufficiency of lead-ores had accumulated to permit of a short campaign. Smelting, under these unfavorable circumstances, to which the high price of charcoal, of labor, the poor quality of accessible fire-proof material, and the costliness of smelting large quantities of fluxes must be added, could, therefore, not be very profitable, especially as the reduction of the litharge and subsequent shipment of the lead were out of the question. Even since the railroad has come within less than four hundred miles of Argenta the lead cannot be shipped to advantage, except in the winter, when, on account of the scarcity of return freight to Corinne, this may be done at a cost of $20 per ton. But even at Corinne lead has a value of only 3½ cents per pound, thus leaving 2½ cents per pound as a margin for reduction from ores poor in lead and for losses in smelting. The consequence is, of course, that the smelting-works cannot pay anything for the lead in the ores, and that the cost of smelting, as well as freight and the profit of the smelting-works, must come out of the contents in silver. This, and what is still worse for the miner, quite arbitrary buying rates on the part of the furnaces, has permitted the mining of silver-ores in Montana to dwindle down to a small fraction of what it actually should be. Only the richest ores are now mined in especially favored localities, and the production of silver in Montana for this year will, under the most favorable circumstances, not exceed $150,000, whereas it should be, according to the abundance of the ores in the Territory which are available even at present, not less than $2,000,000.

As I mentioned before, by far the greater part of the now known silver-ores of Montana should at present not be treated by smelting at all, but by chloridizing-roasting and amalgamation. This process is not only far cheaper than smelting, but it is also, under the circumstances, less affected by fluctuations in the quality of the ore as furnished by the mines, especially when the Stetefeldt furnace or the Brückner cylinder is used for roasting, since in these apparatus a varying percentage of lead will have little influence on either the cost or the perfection of the roasting.

In the summer Messrs. S. H. Bohm & Co., who own the largest smelting-works of Argenta, acquired by purchase a vein which occurs in the granite immediately on the bank of Rattlesnake Creek. This vein carries very good lead-ores, carbonates, phosphates, and molybdenates on the surface, and galena in the lower part of the shafts. The owners are now hard at work to open this vein so as to get adequate stoping-ground and reserves as soon as possible. The lode is, on an average, as far as exposed, 3 feet wide, and contains an ore-seam of 12 to 15 inches, which is very solid and free from gangue. This vein, which is christened the Ferdinand, and the Eaton and Legal Tender—deposits in the limestone above, which both contain very good smelting-ores in nests—are probably sufficient to deliver all the lead-ores needed for the present capacity of the smelting-works, and to extract the silver from the quartzose-silver ores bought from other districts. But if in these mines also the percentage of lead should decrease as much in depth as it has done in all the others of the district, the furnaces of Argenta will, indeed, be in a poor location; for, besides the existence of these plumbiferous ores in the immediate vicinity, there is nothing in the location to recommend it, except a very good water-power and the neighborhood of a marl-bed, which furnishes a good hearth for the cupelling-furnace. Charcoal and iron-ore have to be bought at high prices, and a good lining for the blast-furnaces is not at hand, granite being used for this purpose in default of something better.

And, in a commercial point of view, Argenta will certainly never amount to anything, as no railroad is ever likely to touch it. I have hinted above that, in years hereafter, when lead shall have acquired a certain value in Montana, the amalgamating process, which would be so advantageous at the present time, will not be in place for the beneficiation of the silver-ores of the Territory. But nature, which has so bountifully supplied this Territory with the ores of all the metals, has also here furnished the means of introducing a rational process for the extraction of the precious metals from the ores under discussion. I refer to the existence of a vast amount of sulphureted-copper ores in various parts of the Territory. Some of these deposits are located close to a natural line of railroad, *i. e.*, near the low Deer Lodge Pass in Butte district. It is probable that even the main line of the Territory, the North Pacific Railroad, will run through this pass and down the valley of the Deer Lodge.

These copper-veins carry ores very free from gangue, principally yellow sulphurets and peacock-ore, both mixed with iron pyrites. Near the surface, however, these minerals are oxidized and converted into carbonates, oxides, and silicates. We have here, then, the true basis for the extraction of silver from the quartzose ores of the Territory by smelting. It is true the extraction by means of lead is a much less complicated process, but the use of copper sulphurets will prove far more reliable, because the adequate supply of the latter is assured. It has these further advantages, even at the present time, that copper has a commercial value in the Territory, while lead has none, and that in the case of copper-smelting no buying of iron-ores is required for the purpose of fluxing the quartz of the silver-ores, while the use of the Montana lead-ores involves a heavy outlay in this direction. The reply might be made here, that the argentiferous galena-ores could be dressed up to a high percentage of lead before smelting, thus rendering unnecessary a great portion of the fluxes which are now required. But, unfortunately, this cannot be done with economy, as the friable character of the silver-ores associated with the galena makes dressing unprofitable. The ore, after dressing, would probably be poorer in silver than before, much of the soft floating fahl-ores, sulphides, and antimonious ores having gone with the tailings.

Thus, as the case stands, the extraction of the silver from the Montana ores by smelting, will, in time, be the only rational one, except in districts, if such should ever be found, which would by themselves be able to furnish true silver-ores enough to supply amalgamating-works continually. But in the smelting-works copper, not lead, must be looked for to play the *rôle* of the necessary medium.

On the hill-side above the lead mines at Argenta there are a few placer mines. One of the gulches is worked by hydraulics, and employed in the summer four men. The yield of this claim was at that time reported at $40 per day, and there is no doubt that work could be continued until late in the year, as the ditch delivered an abundance of water. This ditch belongs to Mr. Kingsley, and takes its supply from Rattlesnake Creek, about four miles above the town of Argenta.

The districts which principally furnish ores for the Argenta smelting-works, outside of the Argenta district, are the Blue Wing, (already described,) Moose Creek, and Vipond districts. Both the latter are as yet little developed, but contain rich ore.

Vipond district, especially, promises to furnish a great deal of rich silver-ore in the future. It is situated in the Big Hole country, an exceedingly rugged part of Beaver Head County, about forty-eight miles from Argenta. The ore must be packed eight miles, to the Big Hole River, where it is transferred to wagons and hauled to Argenta.

It was mentioned in last year's report that a large area in this district is literally covered with float quartz. During the year active prospecting for the ledges has been going on, and a good many have been located and partially opened.

For the following detailed description of Vipond district, I am indebted to Mr. P. Knabe, mining engineer, of Red Mountain City. This district lies about fifty-five miles northeast of Bannack City, and is bounded east by the Big Hole River, north by Wisdom River, and south by Cañon Creek. From each of these streams it rises suddenly to perhaps 1,000 feet, forming then a plateau which is intersected toward the streams by deep cañons, while in the west it is limited by high mountains. A large area of this plain is covered thickly with glacial detritus, while the banks of Cañon Creek are lined with gigantic moraines of an ancient glacier. This renders prospecting extremely difficult in many places.

The formation of the country is dolomitic limestone; the mineral deposits are invariably silver-bearing.

The first mine was located by the Vipond Brothers in the fall of 1867, but not until the summer of 1871 did this locality become the field of vigorous prospecting. Consequently developments are yet very limited, though prospects are very good.

In enumerating the different mines, commencing from the west, the the following are to be mentioned:

The Miwanotack appears to be a net-work, or a system of lodes or pockets. In four shafts ore was found from 1 to 4 feet wide, an assay of which yielded as high as $410 per ton, average about $200 in silver, besides about 40 per cent. of lead. Sixty-five tons of ore were treated at the Bohm smelting-works in Argenta.

The Forest, apparently a pocket, 4 feet wide. The ore resembles greatly that of the Miwanotack, both in value and general character. The minerals observed in this claim are quartz, carbonate of lead, galena, blue and green carbonate of copper, silver-copper glance, horn-silver, native silver, and a greenish-yellow substance, consisting of an oxidized mass of lead, copper, and arsenic, and rich in silver. Thirty tons of ore hauled to the Bohm smelting-works.

The Gray Jockey, a deposit about 12 feet wide, and supposed with some reason to be a true fissure-lode, is explored by a shaft 20 feet deep. Its dip is perpendicular, or nearly so; the strike is northeast and southwest. Mineralogically the ore resembles that of the two former mines; it is, however, not so rich in lead and silver. A selected sample assayed $173 per ton.

The Onyx, running northeast and southwest, is 15 feet wide, and shows large croppings of quartz. It is developed to a depth of only 6 feet, and shows white quartz, galena, carbonate of lead, and compounds of copper, arsenic, and lead. A selected sample assayed $162 per ton.

These four mines are located within only one-quarter of a mile from each other.

The Juno is about half a mile distant, in a westerly direction. It is developed by a shaft 36 feet deep. The deposit, 4 feet wide near the surface, terminates in the bottom of the shaft in a mass of decomposed limestone. A sample of the ore, which resembles also that of the foregoing mines, assayed $143 per ton.

The Mammoth, running northeast and southwest, is located one mile northwest from the Juno. It is developed by a shaft 25 feet deep, exposing a body of ore of 4 feet at the surface, and but 5 inches in the bottom of the shaft.

A considerable quantity of ore is contracted for to be treated at the Bohm smelting-works in Argenta.

About three miles northeast of the Miwanotack is situated what is known as the Quartz Mountain. It is a ridge running southeast and northwest, about two miles long and one-half mile across, sloping gently toward the southeast. From the middle to its southwestern base its summit is crowned by large outcroppings of a number of silver-bearing deposits; hence the name of the locality. The formation is also limestone, its strata dipping 30° toward the south. They are intersected by all the deposits mentioned in the following. The general character of the deposits of this portion of the district varies materially from that of the foregoing mines. They are vertical or nearly so, and not only, as stated already, intersect the strata, but appear also to have caused the latter to be dislocated. The walls are not always well defined, but a seam, the sides of which are striated planes, runs in the middle of the deposit, and the richest ore is always found in close proximity to it. These facts seem to indicate that the deposits of Quartz Mountain are true fissure-lodes. The gangue of the deposits in question consists of quartz and heavy spar, which latter mineral is entirely wanting in all the mines mentioned heretofore. The next five mines to be cited are parallel to each other, running northeast and southwest, and from fifty to one hundred yards apart from each other.

The Bismarck, the first northwest, is located on the summit of the mountain, and about 500 feet above its base. A shaft 14 feet deep disclosed a body of ore, from 1 foot to 3 feet wide, an average sample of which assayed $422 per ton, and a selected sample as high as $1,169. The following minerals were observed: native silver, silver glance, silver-copper glance, horn-silver, and ruby-silver ore. By amalgamation retort was obtained of .992 fine.

The North Star, a few feet below, is explored by a shaft 5 feet deep. It is about 3 feet wide. A sample of ore assayed $167 per ton.

The Humboldt is near by. It is developed to a depth of 6 feet, and shows a body of ore 5 feet wide, a sample of which assayed $162 per ton. Carbonate of lead is the predominating mineral besides quartz; but all the minerals found in the Bismarck occur here also.

The Aurora, a deposit about 3 feet wide, is explored by a shaft 15 feet deep. It shows all the minerals observed in the Bismarck, also some carbonate of lead. A selected sample assayed $1,451.

The Lone Star, or Pettingill, the ownership of which is in dispute, is cropping out 10 feet high, and is 8 feet wide. The ore resembles that of the Humboldt, and assays as high as $200 per ton.

On the south eastern end of the ridge three more notable deposits are located. These appear to run east and west, and parallel to each other. They stand perpendicular.

The Argyle is 10 feet wide, separated from the northern wall by a clay casing. The ore occurs in irregular bunches in the gangue, which is composed of quartz and heavy spar. Among others, especially galena, carbonate of lead, silver-copper glance, and native silver are met with. It is developed by a shaft 20 feet deep, and eight tons of first-class and twelve tons of second-class ore were obtained from this opening. A selected sample assayed $656 per ton.

The Banner and the Handy Andy are recent locations near the Argyle, which have as yet not been subjected to any investigation; they look, however, promising.

The facilities for erecting reduction-works are ample in this district.

With very little cost a good road can be built to the Deer Lodge road, a distance of only twelve miles. Timber of the best quality abounds, and the streams referred to already, and several springs, may be made available for water-power.

In Deer Lodge County the placer mines have furnished by far the bulk of the gold-product of the year. In the richest and most important gulches water has held out longer than usual, and fair amounts of gold have been taken out. Among these German Gulch and Yamhill are especially noteworthy. On some of the head gulches of Moose Creek (ten or twelve miles south of Silver Bow,) several companies have been sluicing during the greater part of the summer. They are reported to have made an average of $10 per day to the head. The extensive placers of Silver Bow have, on the contrary, not done as well as usual during the season; in fact, only four or five claims of the hundreds in this locality were reported to me as having paid wages. The whole product up to August was given as only about $50,000, a sum ridiculously small for so large a field. The same ill success has attended operations at Butte and Rocker, about two miles above Silver Bow. These three districts suffer in common from an inadequate supply of water and insufficient height of the ditches heretofore constructed. There are here thousands of acres of gravel-ground, which cannot be worked at all for that reason at present. To remedy this state of affairs Messrs. Humphrey & Brother have undertaken the construction of a tunnel through the main range for the purpose of furnishing an inexhaustible supply of water for these rich placers. It is one of the most gigantic enterprises ever undertaken in Montana, and one that will add largely to the wealth of Deer Lodge County.

The copper mines in the vicinity of Butte City, which were mentioned in a previous report, have so far not been worked. There is very little gold and silver in the ores, and to use them for the extraction of the silver from the quartzose-silver ores of neighboring districts has as yet not been thought of in Montana.

There are many Chinese in the three foregoing camps, and every year numbers of new comers are added.

German Gulch has had a prosperous season. In July nine companies of white men were here engaged in mining, most of whom had been using the abundance of water to the greatest advantage by washing off the heavy top earth. Some of them had commenced washing up, and some large clean-ups had been made. Chinamen have purchased mining ground in German Gulch during the last year to the amount of $61,000, and yet there is no perceptible falling off in the number of white men. Two new claims were opened above Dr. Beale's ground, which, up to the present season, was the uppermost claim worked.

Nine miles westward, over the range by way of a rugged trail, French Gulch is reached. This is a tributary of the Big Hole. About seventy-five whites and some twenty-five Chinamen have been engaged mining here, within a circuit of four miles. The principal companies are Birmingham & Co., of French, Wier & Co., and Lynch, Garrett & Co., of Fenian Gulch, Brunell & Co., of the swamp claim at the head of First Chance, and Allen & Co., of the French Gulch Bar. Some of the companies mentioned have a large extent of rich ground. Several other parties of French and Spanish miners have been working, with fair compensation for their labor. But, on the whole, French Gulch has hardly done as well this year as previously. Leaving French in the direction of Deer Lodge, the gulch and bar mines of Golden, McMinn,

McGraith & Co., thirteen miles distant, and a little below Brundy's old mill, are met with. These parties own the water they use, and have an unlimited quantity of good ground; they were running a hydraulic, and had cleared off over four acres of bed-rock at the time above mentioned. They were still running eight-hour shifts, and intended not to make any general clean-up until near the 1st of August. These mines paid an average of $9 per day to the hand last season, and as they are being worked to much greater advantage this year, they will, without doubt, pay proportionately better wages.

Fredrickson is pleasantly situated on a high flat, commanding splendid views of the Deer Lodge Valley, and the south side of Powell's Peak. Good paying mines are known to exist in Antelope, Spring, Prairie, and Dry Gulches, and some sixty men are employed in them all. A deep gravel-channel, containing some gold, runs through the bar on which the town stands, and parties were engaged in the early summer with hydraulics in opening it up, with the most favorable indications of good pay. Should this bar prove to be good, the reputation of the Race-Track Diggings will be established.' Prospecting was going on at several other points in this vicinity. The receipts of the Miners' Ditch Company for water were at that time over $100 a day.

At Highland, in the southeastern corner of the county, the Only Chance Company has been running three arrastras. The mine produces a large quantity of first-class ore, and there is no doubt that the gold-product from this mine during the present season is much greater than that of former seasons. The Nevins Company has been running two arrastras. Their ore is paying well, and the lode shows a fine body of quartz. Trainor, Conovan & Co. are reported to have struck a very rich deposit at the head of their flume in Highland Gulch, which was said to yield from 50 to 75 cents to the pan. Charles Wunderlich, who leased the Langworthy Flume Company's ground, was doing remarkably well. Five companies were working in Basin Gulch.

Henderson Gulch, a tributary of Flint Creek, I am informed, has yielded fairly during the season, as have also the placers in the vicinity of Blackfoot. Georgetown, on a small branch of Flint Creek, on the foot of the western slope of the ridge, on the eastern side of which the Cable mine is located, has been little worked during this year. At the time of my visit to this locality only two parties, of two or three men each, were at work sluicing. The town is deserted and dilapidated.

Pioneer Gulch, situated fifteen miles northwest of Deer Lodge City, is an affluent of Gold Creek. The diggings are bar diggings. Three white companies were working five hydraulics here in August, and four companies were sluicing in French Gulch, a tributary of Pioneer. Three more hydraulics were at work on Wilson Bar, two miles below the town of Pioneer. Several companies had already, at that time, been obliged to stop work on account of the scarcity of water. The gold from French Gulch is the best in this vicinity. It brings $19 to the ounce, while the gold from the other localities is rated at $18 to $18.75. So far no goldbearing quartz-veins have been found at the head of Pioneer or the other gulches emptying into it. The bed-rock underlying most, if not all, the claims is a calcareous shale.

The camp was discovered in 1861, but was for several years abandoned until the Pioneer Company commenced working by hydraulic in 1867. There are a great many very expensive ditches in this vicinity and about Yamhill. About eighty white miners and one hundred and fifty Chinese have been working in Pioneer the last season. Yamhill and Pike's Peak are about four miles nearer to Deer Lodge City than the camp just mentioned. The second of these is the oldest camp, and, at the present time, nearly worked out. It is situated in the ravine along the foot of the hill on which the new camp of Yamhill is located. No white miners work now in Pike's Peak, but about fifty Chinese are at work there, using second water, for which they pay 10 cents per inch. The Chinese wages in this place are about $50 per month. They are exclusively employed by their own countrymen. At Yamhill about one hundred and eighty white men were employed at Pilgrim Bar and Gold Hill, while on the other side of the hill twenty-five more men were employed on Dry Gulch. Wages of white miners are here $5 per day without board. The auriferous deposit on this hill lies in the deep bed of an ancient river, the channel running north and south, and crossing the present ridge at a very sharp angle, so that it finally meets the valley in which the old Pike's Peak claims are located. Its existence was first discovered in this place last year while the bars of the small creek were being washed. Near the valley this channel is about 1,200 feet wide, and here are the best claims—Smith, Boyd & Co.'s, Bell's, and Hagan & Co.'s. Next to them are Chinese. The size of a claim is 200 by 600 feet, but one company may own several claims. The depth of the gravel in the claims near the valley is about 25 feet to the bed-rock. Higher up the channel is narrower, from 500 to 600 feet wide, and the gravel is in some places 70 feet deep. The bed-rock is indurated clay. There is this peculiarity noticeable about the gravel, that wherever it is composed of quartzite and quartzite slate it pays well, but when it is largely composed of granitic rocks there is little or no gold found in it. About twelve men are employed in every claim, (night and day shifts,) and a week's clean-up produces usually from $1,000 to $1,600. In one of the upper claims three men had been killed by the caving of the high bank a few weeks previous to my visit. The Rock Creek ditch, carrying 2,500 inches of water, furnishes most of these claims, and all the upper ones are supplied by it. The charge for water is 25 cents per inch per twenty-four hours, which is certainly a high charge. Still, all the claims pay exceedingly well. Exact statistics promised me by Mr. D. L. Irvine, the secretary of the Ditch Company, have not yet arrived, but it is certain that this camp has produced more gold this year than any other placer field of the same area in Montana. There are few quartz mines worked at the present time in this county; the Atlantic Cable mine, at Cable City, and the Philippsburgh mines being the only ones on which work has been done to any extent during 1871.

Cable City is situated forty-five miles southwest from Deer Lodge City, in the Cable range, a spur of the main chain of the Rocky Mountains. The town owes its existence to the discovery of the Cable lode, which raised sufficient excitement at the time to cause the whole eastern and western slopes of the divide between Hot Springs and Flint Creek to be prospected for other gold-veins. Many were indeed found, as the great number of costeaning pits, especially on the western slope, attest. But the ore cannot have been sufficiently rich, for none of these veins have been worked beyond an inconsiderable depth.

The Atlantic Cable mine is located almost on top of the divide before mentioned, and on the Hot Spring Creek slope. The vein lies in a zone or dike of crystalline limestone, which is incased by granite. This dike runs northeast and southwest, and dips northwest at the surface. The ore-vein has so far been rather irregular in dip, strike, and width, but it has always been found very large, too large, in fact, for convenient timbering, wherever it has been worked. The ore is a soft, highly iron-stained and porous quartz, which frequently contains the decomposed ores of copper. At the time of my visit there was no work going on at the mine, beyond pumping, the mine having been flooded a short time before by a ditch which runs across a portion of the outcrop. A large amount of gold has been taken from this mine, as will be seen from the annexed statements by Mr. Cameron, but the owners have derived no benefit therefrom, and furthermore the mine. with all its shafts and galleries, is in a worse condition now for profitable working, than it would be if it had never been touched. In fact there can probably not many mines be found, even in this country, which will surpass this one in badly-planned, irregular, expensive, and dangerous workings. It is very difficult now to secure the upper portion of the vein (part of the large, old ore-chambers having caved in completely) so as to render the lower workings perfectly safe and secure against too great an influx of water. But the very large cost of hoisting the ore might easily be remedied by the sinking of a new working-shaft, or by the completion of the tunnel at right angles to the lode, now in progress of excavation. The ore seems to have been encountered in several very large bodies, which have been taken out entire. In some portions of these were found a great many limestone bowlders, similar to those spoken of in connection with the Dakota lode near Bannack. Mr. Aiken, the superintendent and one of the owners, has furnished the following data : There are four shafts sunk upon and near the vein, only two of which struck it. One of the latter is 148 feet deep, and had ore in all the way except a few feet near the top. At the depth spoken of the vein suddenly contracted, the small seam remaining dipping to the northwest. This was not followed farther. At a depth of 82 feet in this shaft there was a level run towards the northeast for 300 feet. There was an ore-body mined out here 80 feet deep and from 45 to 55 feet wide. Below the depth of 85 feet the vein was full of loose bowlders, lying in the soft ore. When the depth of 148 feet was reached in this shaft, the wide and high ore-chamber in the level above, which had been poorly timbered, caved in for a length of 250 feet. This happened during the time that Nowlan and Plaisted worked the vein. After this the mine remained idle for some time. The present working-shaft, like one or two others, was originally sunk as a prospecting-shaft. It is located in the granite, below the limestone, and was first sunk to a depth of 90 feet. From here Mr. Aiken drove a tunnel in the limestone along the granite wall 450 feet long toward the ore left in the 300-foot level mentioned above. He found here a body of ore 11 feet wide and 100 feet long, on which he stoped downward away from the shaft and found water. He timbered the opening made, and followed the vein a short distance northward, where the increasing number of loose bowlders in the ore stopped his progress. There was no other course left now, but to sink the shaft lower in order to drain the stopes. Pumping and hoisting machinery was then put up, the shaft was sunk 20 feet deeper, and this drained the stopes for a distance of nearly 500 feet through the loose vein-matter without any other communication being established. Later the shaft was sunk still 40 feet deeper, and a drift was run in northwest for about 20 feet, which carried it to loose material and drained all the ground above this level effectually. An incline was then sunk from the upper level, near its northeast end, to a depth of 60 feet. This encountered an ore-body 30 feet deep, from 55 to 60 feet wide, and extending about 150 feet farther north than the old stopes, which was rapidly removed. This ore-body pitched to the southeast. In the lowest part of the incline the water was very troublesome. The main shaft was then sunk about 60 feet deeper in granite, and from here a drift was driven toward the incline. This struck limestone in 20 feet, and not until it struck a seam at a distance of 95 feet from the shaft did it drain the incline, which was subsequently sunk somewhat deeper, the ore-body retaining its width. It was stopped on account of many loose boulders, which were found in the bottom. The main shaft is at present 240 feet deep, and from it a cross-cut was being run in the summer, which was then in 125 feet, and it was expected that in 25 feet more the vein would be reached. This drift will have 45 to 55 feet of stoping-ground above it in the greater part of the mine. The ore near the surface from this mine was so decomposed and rich that Messrs. Aiken, Stowe & Pierson, the original owners, could before they had a mill, wash out $50,000 in a very short time in the creek below.

The following statements in regard to this district, and especially the Cable mine, were sent to me by Mr. Cameron at the end of the year:

Return of the production of gold and silver in the Moose Creek mining district, Deer Lodge County, Montana Territory, for the year ending December 31, 1871. Reported by D. Cameron, book-keeper for S. Cameron & Co.

Mill, Hanauer; owner, A. Hanauer; location, Cable City; mine, Atlantic Cable; gold lode; number of tons of ore, 1,646; average yield, $22.38; total product, $36,839.45; time of running, 66 days; average number of stamps running, 20; whole number of stamps in mill, 20; power, steam, 40 horse-power.

REMARKS.—The Atlantic Cable mine was discovered in 1867 by Mr. Alex. Aiken, and soon afterward passed into the control of W. Nowlan, esq. This gentleman, being engaged in the banking business, in Helena, probably did not pay that attention to the timbering and working of the mine which it required and deserved, and the consequence was that it was worked very expensively and timbered badly, which finally resulted in its completely caving in, in consequence of which it was closed up for a period of ten months. At the expiration of this time it was re-opened by S. Cameron & Co., who are now working it. Notwithstanding, however, the many drawbacks to a successful working of the mine, in the way of extremely bad management and prolonged and complicated litigation, this mine has produced, since its discovery, about $400,000.

The company commenced running a tunnel from the bottom of the hill in the fall of 1870, and have run it 600 feet. This tunnel when completed will be about 1,000 feet, and will tap the mine at a depth of 300 feet from the surface. The Miners' and Mechanics' Tunnel Company commenced running a tunnel, in order to tap this same lode, three years ago. This tunnel is now in a distance of 1,000 feet, and when completed will be about 1,400 feet in length, and will tap the mine at a depth of 375 feet from the surface.

List of live mining claims in Moose Creek mining district, Deer Lodge County, Montana Territory, on the 1st day of January, 1872. Reported by D. Cameron, book-keeper for S. Cameron & Co.

Name, Atlantic Cable mine; owner, S. Cameron & Co.; character, lode; course, northeast and southwest in depth; dip, southeast; dimensions of claim, 2,200 feet, and the shaft, which is 240 feet deep, is sunk in the center of the claim; country-rock, granite; vein-matter, crystallized limestone, feldspar, and iron-ore, which contains occasionally lumps of iron and copper sulphurets; ore, decomposed quartz and free gold; value per ton, $22.38, average value during the year.

REMARKS.—The quartz from this mine is conveyed up an incline of 75 feet by means of a windlass, thence through a tunnel, 500 feet, to the shaft, up which it is hoisted by means of steam hoisting-works; engine, 20 horse-power. The mine is drained by a double-acting force-pump, which is, however, of too small a size for an emergency. In consequence of mine being flooded last spring, from a water-ditch, it was found necessary te stop operations about three and a half months, and a further delay of several months was occasioned in consequence of the necessity of running a new tunnel, 95 feet lower down, which will greatly reduce the mining cost of ore.

The average, $22.38, as given above, although correct for the time specified, is not a correct average value of the ore taken from the mine. One thousand seven hundred and thirty-two tons of ore, taken from said mine in November and December of 1870, yielded an average value of $28.41 per ton, which is not more than a fair average value of the ore produced by the mine, so far.

Estimate of cost of mining and reducing ores in Moose Creek district, Deer Lodge County, Montana Territory. Reported by D. Cameron, book-keeper for S. Cameron & Co., January 1, 1872.

Population of district, 150 persons; wages of first-class miners, in summer, $4.50; in winter, $4 per day without board; wages of second-class miners, in summer, $4; in winter, $3.85 per day, without board; wages of surface laborers, mechanics, same as first-class miners; laborers, same as second-class miners; cost of lumber, $50 per 1,000 feet; cost of mining-timber, $250 per 1,000 feet, running measure; cost of common powder, $7.50 per keg, delivered; cost of giant powder, $1.40 per pound, delivered; cost of quicksilver, $1.25 per pound, delivered; cost of freight from Deer Lodge, from 1¼ cents to 2½ cents per pound; cost of fuel, wood delivered at mill, $3.25 per cord; cost of 10-stamp mill, California pattern, including freight, erection, &c., about $10,000; cost of 20-stamp mill, freight, erection, &c., about $20,000; minimum mining cost per ton of ore, $4, exclusive of hauling, which costs 90 cents per ton; mine from which this is reported, Atlantic Cable mine; character of rock at that mine, decomposed free-gold quartz, with small quantity of copper and iron pyrites; depth of mine, 240 feet; maximum mining cost, per ton, $7; minimum reduction cost, $4; name of mill, Hanauer; number of stamps, twenty; character of process employed, copper plates; maximum milling cost, $5; average mining cost per ton, $5.50; average milling cost per ton, $4.50; average yield of ore for the year, $22.38 per ton.

REMARKS.—Average width of the vein, from surface to present depth, is 60 feet, although the granite walls which incase the vein-matter are about 200 feet apart, but on the southwest side there is a deposit of crystallized limestone of 140 feet in width.

On the slope of the hill, which is diagonally cut by the Cable vein, and for more than one thousand yards below, it is probably the richest placer ground in this part of Montana. In view of the softness of that vein, and the thick layer of detritus which covers the whole slope, it is astonishing that nobody should have before thought of embarking in the enterprise of washing down the gravel. It was only during the last season that Mr. Conrad Kohrs, a shrewd business man from Deer Lodge, secured the right to the whole ground, and brought water to it from a distance by a ditch which cost him $10,000. This was not finished until late in the summer, and was especially delayed by the necessity of laying a considerable distance of flume where the ditch crossed the Cable lode twice on the same hill. The hydraulics could therefore be run only eight weeks before the first snow-falls, about October 1, effectually closed the work for the season. At the same time the ditch gave way in several places, but this has since been repaired, and no further trouble in this direction is anticipated. During the eight weeks of actual working twenty-two thousand cubic yards of ground were washed, which yielded $18,000, or 81 cents per cubic yard of free gold. Besides this over 70 tons of float quartz from the Cable lode, in much of which free gold is plainly visible to the naked eye, were picked up at the tail-race. The ground so far has been 15 feet deep to granite bed-rock. Later in the season Mr. Kohrs has sold one-half interest in the water-right for $27,000.

From the few placer camps lying still further north in this county I have not received satisfactory information up to the time of this writing. In the fall, while I was in the Territory, the water-supply was reported short, and the yield up to that time not as good as the year before. The same was reported to me from the placer mines on Cedar Creek, in Missoula County, which before that time had enjoyed a short-lived prosperity. The gold was said to occur very pockety, and a great many miners had left the diggings.

In the *Flint Creek silver district*, which at one time created so much excitement, work has been just sufficiently prosecuted to keep alive an interest in those mines. Mr. Cole Saunders, who is largely interested here, has kindly furnished me with the following statement in regard to this district.

Flint Creek district.—Situated twenty-five miles above the mouth of Flint Creek is what is known as Flint Creek mining district. It is one of the most promising silver-quartz camps in Montana. It was discovered and brought to notice in the spring of 1866 by a party of prospectors under the leadership of Charles W. Frost, and the district organized, although a prospector named Horton had previously visited the place.

The original locations that gave life to the place, and caused a stampede of fifteen hundred persons, were the Comanche, Poor Man's Joy, Comanche Extension, Cliff, Speckled Trout, Kitty Clyde, and Hope. Rich silver-ore being exhibited in Helena and other points in the Territory, attracted the attention of practical men from Washoe and other silver-mining places on the Pacific coast. The specimens of ore exhibited by Frost being so extremely rich, caused the wildest excitement and a general rush to the new "Silverado." The Saint Louis and Montana Mining Company erected a 10-stamp mill in 1867, supplied with Wheeler pans and all the adjuncts of a complete Washoe mill. The rich croppings of the Hope and Comanche were run through, and yielded from $40 to $100 per ton. From a disposition on the part of the owners of the mill to acquire the most valuable mines of the camp, a system of "freeze-out" was commenced that speedily blasted the name of the richest quartz camp in Montana, and crushed the hopes of the fortune-hunters who had flocked there. The mill having been closed down was taken as the best evidence of its "failure," and Philipsburg became deserted except by the original discoverers, who had remained firm in their "first love," believing that time and railroads would cause a recognition of the value of their mines, and that capital would be offered for the manipulation of the same.

' For the past three years developments have been pushed steadily by the limited force of men who remained. The Eastern Comanche, Poor Man's Joy, Speckled Trout, and Franklin have been more or less developed, showing bodies of rich ore varying in thickness from 6 inches to 10 feet.

The Eastern Comanche has been stripped on the surface a distance of 1,200 feet, showing a continuous body of ore. Three shafts have been sunk in it, the deepest being 75 feet. A tunnel is now being pushed on the Discovery with gratifying results, showing a body about 8 feet thick that is believed will mill $50 per ton.

The Cordova also shows good indications of a strong vein.

The Poor Man's Joy has been opened on the surface for several hundred feet, and ore of extreme richness was shipped to Newark, New Jersey, and to Berlin, Prussia. The owners have never reported the yield of the ore shipped, (20 tons,) but it is believed to have been at least $500 per ton. The same ore yielded $100 per ton in the Saint Louis Mill without selection, (49 tons.) Antimony and lead prevailed in this mine to such an extent that the ordinary mill process failed to save 30 per cent. of the assay, and further working was suspended.

The Speckled Trout mine has been opened by a shaft on the Discovery of 85 feet in depth. At 50 feet a level was run east 60 feet, and at the bottom of the shaft another level was run east 70 feet. The ore-seam averages in width from 2 feet to 10 feet, and the latest workings in the mine show a heavy body of rich ore that cannot be surpassed in any country. One hundred and fifty tons of this ore have been placed in the hands of the First National Bank of Helena for shipment during the past summer. It was sent to Reno, San Francisco, and Swansea. Results, as far as heard from, leave handsome dividends above the cost of shipping, though five hundred miles of land transportation in wagons from Philipsburg to the railroad is a heavy tax on shipments of ore. Two kinds of ore are found here, one being similar to the Comstock ore, containing sulphurets and chlorides of silver, and scarcely a trace of lead with quartz and spar as gangue. These ores have been worked practically to 70 per cent. of the fine assay, but owing to the difficulty of having ores worked in the Saint Louis Mill, except at ruinous prices for crushing, but small quantities have been treated. The second class of ore spoken of is what is known as "base," containing galena, manganese, arsenic, antimony, with small quantities of copper and sulphurets of silver, being similar to many of the ores of the Reese River country in Nevada. An attempt to work them over in blast-furnaces was made last year by Mr. Cole Saunders, who, in conjunction with other parties, organized what is known as the Cole Saunders Silver-Concentrating Company. They erected two furnaces, the blast being furnished by a Sturdevant fan. A Dodge crusher to crush the ores, and, in fact, all the other appliances necessary to make a complete smelting-works were procured. It was found after starting that the fluxes of iron and galena, that had been calculated on, could not be had in quantities to keep the works running; and after a number of ineffectual attempts to run the furnaces the process was pronounced a failure, and smelting as a "business" was abandoned. Not in the least discouraged, this company leased their mines and works to the Imperial Silver Mining Company, Colonel J. J. Lyon, superintendent, and the furnaces have been removed in the last four months, and a 5-stamp mill for dry crushing has been erected. This mill is now complete, and has been running for the last month (December, 1871) with gratifying success, or, to express myself more correctly, at least it appears a success, as bullion is being produced, and the works are steadily running. The ore is crushed dry and roasted in reverberatory furnaces and chloridized, about 10 per cent. of salt being used in the operation. It is then amalgamated in Freiberg barrels. There is no doubt this is the true way to work the ores of this camp. The expense is great, but the success is certain. Salt can be delivered here in the summer months for 6 cents a pound; wood $4 a cord; labor $3.50 and $4 per day, currency.

It is believed the following figures cover the expense of running a 5-stamp mill and working the ores as above stated:

2 engineers, $3.50 and $4 per day	$7 50
2 feeders, $3.50 per day	7 00
3 roasters, $4 per day	12 00
1 extra man	3 50
1¼ cords wood for engine, $4	6 00
1½ cords wood for furnace, $4	6 00
10 per cent. alt on 3 tons, being 600 pounds per day, at 6 cents	36 00
Lights and oils	1 00
Loss 2 pounds quicksilver to ton, or on 3 tons, 6 pounds, at $1	6 00
Daily expense	85 00

The works having a capacity of 3 tons per day, this shows an average cost of reduction of $28.33 per ton. A 10-stamp mill would reduce the expense per ton at least one-third.

The above figures show about the cost of running in the Imperial Mill as now constructed, but by adding any of the improved roasting-furnaces, now in use in Nevada, the expense could be very materially decreased. The Speckled Trout ore is being worked, at present, but as the returns are not made public they cannot be exactly stated. Rumor places their yield at $100 to $150 per ton.

The "leaching process," as explained in Küstel, will have a practical trial here this winter, as Colonel Lyon at present is preparing vats, &c., necessary to make the experiment. Numerous tests in a small way have proved successful, and it is believed the same results will be met with on a larger scale.

James A. Brown, of Deer Lodge City, has a number of men at work this winter prospecting ledges in the camp with a view of leasing the Saint Louis Mill or of erecting a new one in the spring. As far as developed the ledges show well, and several thousand tons of rich ore are already on the dumps and in the levels ready for hoisting.

Another season, it is confidently expected to show a flourishing camp with weekly shipments of silver bullion that will rival the successful districts of Nevada and Colorado.

EASTERN MONTANA.

Under this head will be considered the mining districts around Virginia City, in the Jefferson Valley, near Helena, and between Helena and Bozeman.

With regard to the valley of the Upper Madison or Fire Hole River, and the country immediately surrounding the Yellowstone Lake, it is sufficient to say at this time that they do not promise anything to the miner except sulphur, fire-clays, and natural cements. An account of this region, prepared after personal examination, is omitted from this report on account of the more detailed description about to be published by the Government in the report of Dr. F. V. Hayden, United States Geologist, acting under orders of the Secretary of the Interior.

I can say little about the placer mines. They continued to be actively worked through the season, and their product was not less than in former years; but there is little to note, beyond changes of ownership, in addition to what has been fully set forth concerning the different gulches in my former reports.

Alder Creek.—As is usually the case in the immediate neighborhood of very productive gulch diggings, the quartz-mining industry around Virginia City has been developed slowly and with many failures. Several causes may be adduced for this almost universal phenomenon. Alluvial mining attracts a population usually without capital and not specially experienced in the very different requirements and risks of quartz mining. Not realizing the complex nature and amount of the expenses attending the extraction and reduction of ores, the gulch miner is apt to be over-sanguine in his estimate of the value of veins, and to underrate the difficulty of working them to permanent profit. Moreover, in isolated districts like those of Montana, there is frequently a surprising ignorance of what has been done elsewhere in the way of determining the best machinery and processes; and the miner frequently wastes his time and money in experiments which have long ago been rendered unnecessary. Again, the cost of freight operates strongly to encourage the adoption of all sorts of patent machines, on account of their cheapness and portability. Another serious trouble is the high rate of wages, coupled with the irregular supply of labor. Miners are not inclined to work steadily at any one place. They take employment when the gulches are dry, or when bad luck has left them without the means to go "prospecting." As soon as the favorable season or the accumulation of a little money permits them to try their luck again, they are off for new fields. It is very difficult to maintain under such circumstances a regular industry like deep mining. Finally, there has been, in Montana, at least the usual proportion of wild investment by eastern capitalists, reckless mismanagement by incompetent or dishonest agents, and plundering by everybody of the non-resident owners, who seem to be considered "fair game" in many mining districts; and it should be added that the locators of veins here, as everywhere in the West, attach an exaggerated value to undeveloped property—an error which the folly of capitalists has done much to encourage.

Some of the mountain districts around Virginia City will doubtless become in time the seats of productive industry. A few good mines have been already developed, though sadly expensive failures, from one or more of the causes just enumerated, have been too frequent.

The rock in Alder Gulch is mainly feldspathic gneiss, in which hornblende, mica, graphite, and garnets occur. Lava (of which there are large overflows in the neighborhood) crosses in several places. At the very head of the gulch, the gold in the gravel is traced up to the edge of the apparently overlying limestone of Bald Mountain, and it has been asserted that the auriferous channel actually runs under the limestone; but this is, *à priori*, unlikely, and not supported by proof. The limestone abounds in fossils, (Devonian?) and is indubitably older than the agencies which carved the present water-channels, and filled them with gravel and boulders. There is no reason to doubt that the source of the gold in Alder Gulch was the auriferous veins in the gneiss and slates of the mountains. That the gulch has proved enormously rich, while the veins are but moderately so, is quite in accordance with the universal law of the concentration of gold in alluvial deposits. The shape of the gulch, its tortuous windings, its intersection of the rocks at various angles, and the great denudation of the surface, have co-operated to concentrate in a comparatively small space a large amount of the precious metal.

Summit district, about eight miles up Alder Creek from Virginia City, contains several mines and mills. The Oro Cache is the leading mine. It is situated on Grant Hill, near the head of Alder. The vein runs into the hill, north 10° east by compass, dipping 65° to 70° west. It is variable in width, ranging from a mere seam up to 4 feet as maximum.

The quartz is bluish and whitish gray, associated with some feldspar and garnets, and carrying free gold and some iron pyrites. It yields by mill process $16 to $60 per ton. The mine has been irregularly opened, and worked rather for immediate results than permanent convenience and productiveness—a policy which it is now proposed to reform. The principal useful workings at the time of my visit in July were two drift-tunnels from the face of the hill. The upper tunnel is 600 feet long and about 100 feet below the top of the hill. The first 380 feet are on the vein, and the available ore above the tunnel has been worked out for this distance. The vein is lost in the last 120 feet, and it is not worth while to hunt for it here, as the lower tunnel will follow it with greater certainty, and open up twice the amount of ground per running foot. The extraction of ore was carried on in July through the upper tunnel, by means of winzes and stopes below the level. One winze, 120 feet from the tunnel-mouth, was over 80 feet deep; another, 340 feet from the tunnel-mouth, was 30 feet deep. The vein between them is somewhat pinched (at *P.*) A few hundred tons of ore, standing in the stopes connected with these winzes, constituted at that time the available reserves; but the completion of the lower tunnel was expected to open at once a considerable amount of ground. This tunnel, beginning 100 feet further south than the upper one, was in 320 feet, having already passed under the first winze

and stopes. A connection was soon to be made, and the ore extraction carried on, without hoisting, through the lower tunnel. The ore was transported by wagons, over a winding grade, two miles, to the mill in the valley 1,000 feet below; but it was intended to build an incline from the tunnel-mouth down the face of the hill, directly into a side gulch, and deliver the ore at the bottom, only a short distance, by a good road, from the mill. The daily product was then three wagon-loads, or between 6 and 9 tons.

The ore is worked in the Excelsior Mill of Mr. John How, who is also a principal owner of the mine. The mill has fifteen stamps of 600 pounds

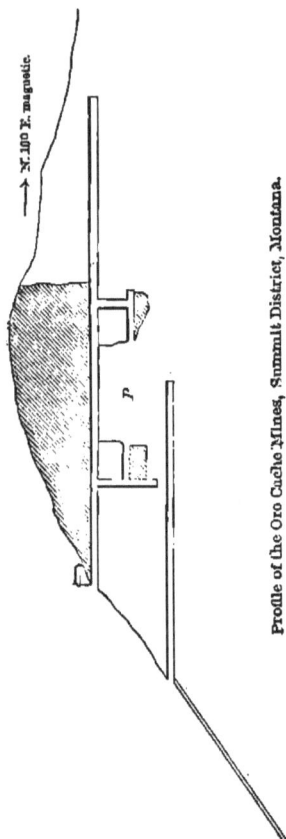

Profile of the Oro Cache Mines, Summit District, Montana.

Scale, 200 ft. to the inch.

N. 100 E. magnetic.

each, drop 8 to 12 inches, rate about 60 per minute. The average capacity is 15 tons daily, or about 1.1 ton per horse-power developed. The
collars are cast on the stems, a bad arrangement, as it prevents any
alteration of the drop, which consequently remains unalterable, except as
it is irregularly changed by the wear of shoes and dies. Amalgamation
is carried on in the battery and upon copper-plated aprons. Blankets
are used, and the blanket-washings are treated in two arrastras. The
tailings are collected in a reservoir, and an attempt has been made to
work them in a Wheeler pan. The experiment was unsuccessful, probably because the settler used was much too small. Ten stamps of the
mill were running on Oro Cache ore, and the rest on second-class rock
from the Kearsarge.

Other veins in Summit district are the Kearsarge, (ore at present low
grade,) Keystone, (not then vigorously worked, but highly spoken of,)
Nelson, Polar Star, &c. The Excelsior Mill was the only one running;
the Hawkeye, (Postlethwait's,) with fifteen stamps, and the Lucas,
an expensive 20-stamp mill, were idle. Southmaid's Mill, an alligator-
crusher of some kind, ran a short time and broke the machinery, which
has never been repaired.

A couple of miles below Summit, in Alder Gulch, is a curious water-
power derrick which deserves mention. The power is transmitted more
than 40 feet by means of a rope from a small overshot wheel to the
pulleys of the crane, which serves to lift the heavy boulders out of the
gulch, and thus afford access to the gravel and bed-rock.

On the opposite side of Grant Hill from the Oro Cache is *Spring
Gulch*, the scene of the operations of the New York and Montana (Colonel McClure's) Company. It is believed that the Oro Cache vein extends through the hill to this gulch. The company referred to spent a
great deal of money, and is said to have extracted $60,000 from Oro
Cache ore. But it failed disastrously, and the property has been sold
under execution. It consists of a village of deserted houses, and a mill,
containing two Chilian grinders and (originally) sixteen heavy iron cylinders for amalgamation. A portion of the cylinders were afterward
removed to the Connor Mill in Brown's Gulch.

Mr. Christinot has opened in Spring Gulch a vein 4 to 6 feet wide,
parallel with the Oro Cache. The quartz was expected to yield at least
$10 per ton, and to be worked in the mill above mentioned.

All the veins opened in Summit and Spring Gulch districts appear to
have the same strike as the country-rock (gneiss)—say, north 10° or 20°
east.

Brown's Gulch, entering the main valley several miles below Virginia
City, from the south, presents veins of a different direction, (north 57°
to 60° west;) the course of the country-rock I did not determine. The
most important of these veins carry silver, with some gold. The Pacific
Ledge (strike north 57° west, dip 70° north) has a maximum width of
10 feet, pinching in places to 3 feet and less. It carries fine antimonial
and sulphuret ores of silver, with specimens containing native silver, in
wire and leaf form. There is apparently no galena. The claim of Mr.
Johnson and his associates, 1,100 feet on the ledge, is opened with two
cross-tunnels, one 70 feet below the other, and a shaft on the ledge connecting with the latter. A good deal of ground (considering the great
width of the vein) is available for stoping. The total product up to my
visit in July had been about 250 tons. The owners being without capital, were working in a small way, treating the ore without selection, in
a small pan, by raw amalgamation. The pan was set in the gulch and
run by water-power, being charged in the morning, and left to itself

until the men returned from work. It worked about 2 tons per week, extracting $60 per ton. As this method of treatment does not reach anything but the native silver and free gold, (there being probably no decomposition of the sulphurets or antimoniurets,) of course a large part of the value of the ore is thus wasted. An arrastra, formerly used, extracted $70 per ton. Some 6 tons of selected ore, to be shipped in sacks, via San Francisco, to Swansea, was estimated to contain 300 ounces of auriferous silver, worth $2 to $3 per ounce. The owners preferred this course to that of sending the ore to Argenta, where there are smelting-works, because at the latter place the prices paid for ore are not so favorable to the miner. The freight account would stand about as follows : To Corinne, by empty returning freight-teams, $20 to $30 per ton ; from Corinne to San Francisco, via Central Pacific Railroad, $15 to $17 per ton in car-loads of 10 tons. Freight from Brown's Gulch or Virginia City to Argenta, at least $20 per ton, by special teams.

This statement affords a striking illustration of the advantages and commercial effects of a railroad. The Pacific Railroad, 400 miles distant, controls and cheapens the southward freights, because it is the route by which supplies enter the Territory. When the Northern Pacific shall have entered Montana, all the freight and supply trains will move from the line of that road into the various mountain districts, and, returning " empty," will afford cheap transportation for ores to centers of reduction.

I do not mean to say, however, that such ores as those of the Pacific mine, just described, necessarily require smelting. On the contrary, their freedom from galena indicates their suitability for amalgamation ; but they must certainly be first subjected to chloridizing roasting. This may be done in reverberatories, in the Brückner cylinder, or in the Stetefeldt furnace. It is a curious proof of the lack of intercommunication on such matters among our mining districts, that while ores of exactly this nature have been successfully reduced for years in Eastern Nevada, they should have been considered hopelessly refractory here. One attempt was made to roast them with salt, in connection with a small mill, (How's,) the mine-owners offering half the gross proceeds as the price of reduction. The mill contained five stamps, two Wheeler & Randall pans, (one of which has since been removed to the Excelsior Mill at Summit,) one settler, and, in a shed outside, the reverberatory furnace. This is small and badly constructed. The chimney is not bigger than a stove-pipe. An inspection of a heap of " roasted " ore in the yard showed that the work was ignorantly performed, and sufficiently explains the failure in which this experiment resulted.

The Black Ledge, near the head of this gulch, is said to be wider than the Pacific, but not so rich. I did not enter it, there being no one about at the time of my visit. A huge, misshapen shaft, intricately but insecurely timbered, and a tunnel from the shaft-mouth into the hill, are the visible workings. The quartz on the dump resembled gold rather than silver ore, and seemed of low grade.

Hot Spring district.—This district, once a scene of much active and speculative mining, has passed through a period of re-action, and is now reviving again. The most noted mine now in operation is the Red Bluff, owned by Mr. J. J. Lown. It is described as follows by Dr. A. C. Peale, of Professor Hayden's party, who visited it in July, 1871 : Dip, north ; strike, east and west ; width, 2 to 7 feet ; country-rock, hanging-wall, " gray granite ;" foot-wall, gneiss ; two shafts, 100 feet apart, respectively 105 and 110 feet deep, connected by a drift, extending 45 feet beyond the second shaft, getting below water-level. The ore is principally red jasper, with the particles of metallic gold disseminated through it, and plainly visible. Below this jaspery ore, from which the lode received its name, occur galena and pyrites. Dr. Peale obtained also specimens of blue chalcedony and semi-opal. Approaching the hanging-wall the ore assumes a porphyritic *habitus*, with large masses of bright red jasper. The ore had averaged, for six months preceding July, $60 per ton. Eight men were employed in the mine, at $3 per day.

Iron-Rod district is situated in the mountains on the west side of the Jefferson Valley. At Iron-Rod Station, on the stage-road, there was a Bullock arrastra mill in process of erection at the time of my visit. It had formerly been running at Granite Gulch. Near by is the Stevens & Trivett 12-stamp steam-mill, which I found idle, on account, it was said, of litigation affecting the title to the Iron-Rod mine also.

The Iron-Rod vein runs north 67° east and dips 56° southeast, following apparently the inclosing granite or gneiss. At the depth of 150 feet the dip grows steeper. The vein varies in width from a mere seam to 4 feet. It has been worked more or less for a distance horizontally of 1,100 feet. The diagram shows the principal works. The deepest shaft is an incline of 350 feet; another shaft, 700 feet further northeast, is 150 feet deep; and 300 feet further northeast a drift begins, running southwest into the hill 250 feet. This was run in connection with a lawsuit, to establish the continuity of the vein, which splits on the surface northeast from the second shaft mentioned, but re-unites at slight depth. Northeast of the deep shaft, and between the depths of 100 and 300 feet, (at *PP*,) the vein is pinched—for how long a distance horizontally is not known, as the ground between the main shafts has not been explored. The ore contains limonite, galena, and some pyrites—a partly decomposed and indurated quartzose pyritic ore. It yields in the mill from $25 to $100 per ton, in gold .550 fine.

The Clipper, owned by Porter, Mant & Lehmer, is about one-fourth of a mile northwest of the Iron-Rod, with a similar course and dip. The ore carries much free gold, (.670 to .675 fine,) yielding in mill as high as $75 (average $45) per ton, with gangue of quartz and associated limonite, galena, &c. The country-rock is white quartzite and gneiss.

The diagram shows the workings, with the exception of the stopes, which I could not inspect, as no one was in the mine at the time of my visit. Ore was obtained, it was said, from the bottom level of the main shaft. The tunnel in the northeast exposes a vein having similar course and character as that in the shafts. As shown in the workings the Clipper crevice is 4 to 6 feet wide, and the pay-streak varies between 3 inches and 3 feet.

The Glen's Falls, supposed to be an extension of the Clipper, to the

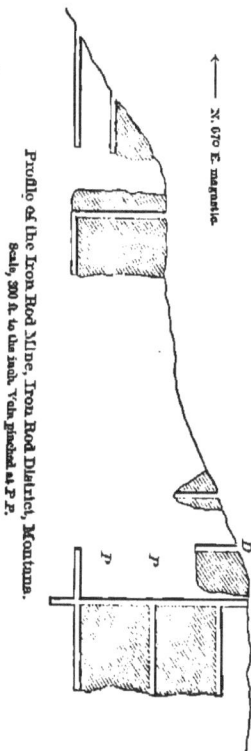

northeast beyond a small gulch, has a shaft 48 feet deep, and a level from the gulch west of the shaft, 125 feet long. Good ore, yielding as high as $100 per ton, has been found in pockets.

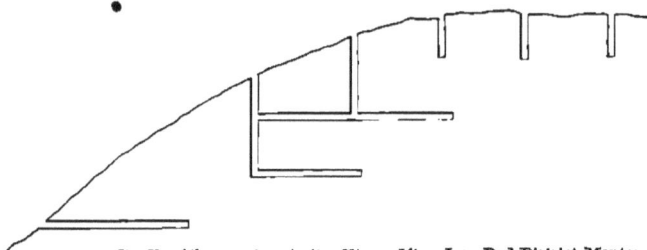

Profile of the openings in the Clipper Mine, Iron Rod District, Montana. Extent of stopes not known. Scale, 250 ft. to the inch.

The Pinchbeck, northwest of the Clipper and parallel with it, has a shaft 85 feet deep, and a level of perhaps 125 feet, running northeast from the shaft. The ore is good, but the vein is narrow, ranging from 3 inches to 1 foot, (in pockets.) Operations are said to have about paid expenses.

Northeast of the Pinchbeck are several mines, temporarily or permanently abandoned, such as the Queen of the West, (shaft 10 feet only,) the Flint-lock, (low grade; $10 per ton; vein 4 feet,) and the Tolin, or Saulsby silver mine. The latter has been worked with more energy than wisdom or success. Two shafts, 100 and 125 feet deep respectively, and a level between them, have resulted in the production of about 30 tons of ore, of which 3 tons have been hauled away, and the rest remains at the abandoned mine. The dumps give no evidence of value in the ore.

The Nugget lode, three-quarters of a mile west of the Clipper, strikes north 73° east and dips 55° northwest. The vein is opened by two levels, the upper one about 50 feet under the summit of the hill, and 125 feet long, and the other 300 feet long, 130 feet below the summit. By reason of the slope of the hill, the lower level is not far beyond the upper at its remote end, though so much longer. A shaft 50 feet deep, beginning on the hill-side below the mouth of the upper level, connects with the lower. The vein varies in width from a seam to 4 or 5 feet; in the upper levels from 2 to 4 feet; general average in the reserves about 2½ feet. The ground opened and not yet stoped contains about 450 tons; the ore is ferruginous quartz, carrying free gold .760 fine, and easily worked in mill, with a reported average yield of $30 per ton.

The Morning Star and the Bedford are two promising lodes, carrying ferruginous quartz with free gold. A small lot of the ore was under treatment at the time of my visit, in a new 6-stamp mill of Tripp & Ainslee, erected in the open air by the side of the road and on the river-bank, between Iron-Rod and Silver Star, about two miles from the latter place. This mill was built at Saint Louis, and is run by water-power. The stamps weigh 550 pounds, and drop 8 inches 24 to 48 times per minute, according to the supply of water. Capacity 4 tons daily; efficiency probably about 1.25 tons daily per horse-power developed. The amalgamation was performed chiefly in the battery, where a large surface of copper-plate was exposed; the tailings were run directly into the Jefferson River. The worst feature in the construction of the mill is the fact that the tappets are driven tight on tapering stamp-stems, so that their position cannot be changed, either to alter the drop or to maintain the average drop when the wear of shoes and dies would increase it too much. The mill was expected to take Nugget ore on a custom contract, at $6 per ton for working.

Silver Star district.—The Morning Star and Bedford lodes above alluded to are properly situated in Silver Star district, which borders the Iron-Rod district on the north. The mountain range continues northward along the west side of the Jefferson Valley, and several other districts between Silver Star and Helena have been more or less developed. It seems to be a prevailing characteristic that the higher parts of the hills contain gold-bearing veins, in granite, gneiss, &c., while below, along the foot of the hills, limestone crops out, with silver-bearing deposits. The extent of the argentiferous deposits of our Western States and Territories in limestone has been already demonstrated to be very great. I need only allude to the mines of Utah, and to White Pine and Eureka in Nevada, to show that the productiveness of such deposits has been unsurpassed, except by the Comstock lode, in our history; and I am convinced that similar developments will be made in the limestones of Montana. Hitherto, however, these silver mines, particularly where they carried galena, have been but little worked. Possibly the town of Silver Star got its name from one of these discoveries. At present it derives its support and its fame from a gold mine, the celebrated Green Campbell.

This mine is two and a half miles west of the town. The vein courses north 85° east and dips 42° south. The country-rock is gneiss, changing in places to an indurated slate, or "bastard trap," as the miners call it. A bed of limestone is seen a quarter of a mile below in the cañon, which must underlie, not very deeply, the gneiss; but the vein seems nowhere to touch the limestone, and is probably in the main conformable to the dip and course of the country. The ore and the walls are much decomposed and highly ferruginous. The best ore is stained with copper. Much heavy timber is required in the mine; but the expense of extraction is in all other respects extremely light, as no part of the workings has yet penetrated below the zone of decomposition. The average width of the vein in the whole mine is 10 feet, and of the paystreak (which meanders curiously from side to side in the crevice, and is only to be distinguished, in the general reddish and yellowish mass of crumbling material, by an initiated and practiced eye) about 8 feet. The average yield of the ore in mill has been $17 per ton; the best ore has yielded $30 per ton.

As will be seen by the diagram, the mine is open to the depth of 120 feet, where it is drained and ventilated by a long tunnel, which serves also for transportation. This tunnel is only shown in the sketch where it is on the vein. Coming from the southeast, it intersects the vein obliquely at X. Between this point and the mouth of the tunnel is a

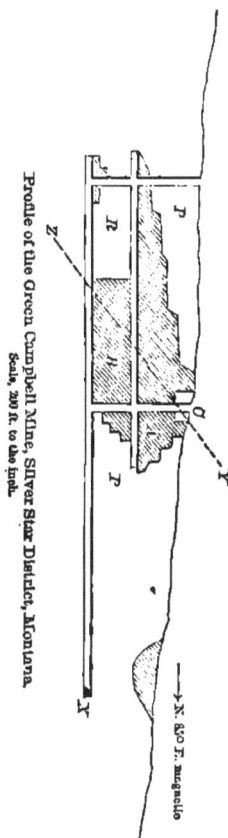

Profile of the Green Campbell Mine, Silver Star District, Montana. Scale, 200 ft. to the inch.

distance of some 800 feet; and outside the tunnel-mouth there is a trestle-work for 300 feet further, carrying the track to the dumps. This arrangement secures cheapness of extraction down to the tunnel-level. But it is now necessary to extend the workings below this level, and it is unfortunate, in view of this fact, that so long an adit was constructed so near the surface. It should be borne in mind, however, that in this soft material the cost of drifting is not nearly what it would be in solid rock; and possibly the tunnel has paid for itself already. The available reserves at the time of my visit were in the block R, where the crevice was 8 feet and the pay-streak 5 to 7 feet wide—rather less than the average. At A and B, the pay-streak had been 16 feet! And generally in the direction of the line YZ the vein had been so good, both in dimensions and in yield, as to lead to the belief that this line is the axis of a body or chimney of rich ore, on each side of which the vein, as at P and P, contracts and is impoverished. It is, therefore, intended to sink an incline on the vein from a point in the tunnel-level, 75 feet west of the main shaft C, and to rise vertically from the same point to the surface. Through this incline the deeper levels will be operated, the tunnel being still used to a great extent, though the new vertical shaft will serve as an upcast and man-way, and in the transmission of power, if for no other purposes. Its cost will be small.

The ore is hauled by teams down a well-graded road, constructed by the company, two and one-half miles to Silver Star, where it is treated in the Green Campbell steam stamp-mill. This mill has ten stamps of 650 pounds, dropping 7½ to 8½ inches, 65 to 75 times per minute. Capacity, 17 to 20 tons daily, the latter of very soft material. Efficiency, according to the average lift and speed, 1.84 tons (for soft material, 2.17 tons) daily per horse-power developed. This is high efficiency, (1.50 tons being very good;) but in fact all the rock here treated is more or less decomposed. The mill is, indeed, however, an excellently constructed one, as is shown by the fact that at the time of my visit (July 30, 1871) it had run steadily for thirteen months almost without repairs. Besides the stamping-machinery there are four Horn's pans, not now in use. Amalgamation is effected by means of plates and loose mercury in the battery, and copper-plated aprons in front, with one plated sluice-box for the escaping pulp, and one or two settlers. Most of the gold is caught in battery and on the aprons. The average yield, as already stated, has been $17 per ton. Mr. Thomas J. Johns, the superintendent of the company, deserves great credit for the intelligence and prudence with which its affairs have been conducted.

In Little Prickly Pear a few companies have been at work during the year. Most of the work was done below what was once the town on the bars in Trinity Gulch. From seventy-five to one hundred miners have worked on Silver Creek and bars. The bar above Silver City has especially paid well, and there seems to be on the whole no falling off in yield. There is plenty of ground yet for years to come.

The Big Prickly Pear mines are now worked and owned entirely by Chinamen, and Montana City contains but half a dozen whites. The main creek was expected to be prospected with a drain, below town, in the fall, and, as Chinamen are making good pay in the creek two miles above, hopes were entertained that it would also prove good below, and that Montana City would again be populated by a considerable number of whites. A Mr. Hall has several bars of tin, weighing a couple of pounds each, reduced by himself from ore picked from the gravel at the adjacent bar. This ore is very pure, and similar to the float-tin found in Durango, Mexico. Some prospecting has been done for the ledge that this ore is evidently derived from, but so far without success.

The Jefferson Silver mines have been worked to a very small extent during the year, although they carry ores of fair value. The principal difficulty heretofore has been to reduce the ores successfully on the spot. Smelting has been tried because the ores contain much galena, but they are, at the same time, too quartzose to work well without large quantities of fluxes, and the local small furnaces had no success. In the latter part of the year, however, since the Helena smelting-works have been in the course of construction, a great impetus has been given to silver mining in this region, and another year will probably witness good results. The Helena works, which receive their ores from different parts of the Territory, will have no difficulty in smelting the Jefferson ores, as they can mix the ores suited to make a proper charge.

Unionville.—I am indebted to Mr. J. C. Ricker for valuable notes on mining operations in this district, bringing the history down to a much later date than that of my visit in July. The principal mines at Unionville are located upon the Whitlatch Union lode, a description of which, with a diagram of the underground workings in 1869, will be found on pages 287-'89 of my report of 1870. With one or two exceptions all these mines were worked more or less during 1871, and with satisfactory results, one noteworthy feature being a decided improvement in the value of the quartz in depth. I will here briefly recapitulate the general characteristics of the lode, as given in a former report. It strikes about north 85° east, crossing at right angles Oro Fino and Grizzly Gulches, the two forks of Last Chance, (the gulch upon which Helena is built.) The country-rock near the surface is coarse granite, passing at a depth of between 100 and 200 feet into fine-grained syenite. The vein is irregular and much disturbed just under the alluvium, but soon assumes a dip of 35° to 40° north into the hill, which it retains to a depth of 250 feet. Below this level it runs over 100-feet horizontally, and in some places even rises at an angle of 5° to 20°. It varies in width from 1 to 20 feet, (averaging perhaps 6 feet,) carrying in the upper levels chiefly white quartz with an occasional greenish tint, and a very small percentage of iron sulphurets; in the lower levels a more bluish quartz with a rapidly increasing proportion of sulphurets. Runs upon select ore have yielded as high as $65 to $80 gold per ton; but the general average has been from $15 to $25. It is the uniform distribution of gold throughout the vein, the consequent steady value of the quartz, extracted, and the small amount of dead work required, that have placed this vein at the head of the profitably productive quartz mines in the Territory. It is unfortunate that so valuable a piece of mining-ground is divided into so many claims. A consolidation of proprietary interests long ago would have done away with many items of expense and secured the continuous working of all the claims, many of which have been forced, by trouble from water, lack of capital for necessary machinery, or dead work, &c., to lie idle for considerable periods.

The Hendrie mine, comprising the Discovery claim No. 1, and half of No. 2 east, was worked with fair success during the first part of 1871. The Discovery incline was extended about 80 feet, and now reaches nearly 500 feet in depth. Steam hoisting-works were erected, and upwards of 2,000 tons of ore were extracted and reduced, yielding from $17 to $22 per ton. In the lowest level on the Discovery claim the ledge is 2½ to 6 feet wide, carrying white quartz, with occasional greenish tint,

H. Ex. 211——19

and gold, free from base metals, or nearly so. Work was suspended in May, probably on account of the trouble from the water, the means of removing which are inadequate. There is now about 75 feet of water in the incline.

The IXL mine (called the Ricker in former reports) commenced in January with the work of opening ground, and preparing to put on a large force of men in the fall and winter. The incline was sunk over 100 feet during the year, and is now more than 500 feet deep. From the 400-foot level it maintains a pitch of about 45° to the bottom. At 30 feet from the bottom a level was run 125 feet west, and the vein was stoped upwards to the 400-foot level. About 40 feet is the present height of the stope. The breast exposes a body of very valuable ore—the richest on the hill. The vein varies in this mine from 4 to 6 feet in width, the walls being of fine-grained syenite, very hard and flinty. The ore retains all the characteristics of the Union lode, white quartz, with occasional green stains, and the gold free and coarse. The water is hoisted from the mine by steam-power, with a sheet-iron tank or car, containing about 125 gallons. The amount of water hoisted in twenty-four hours is about 800 gallons, which keeps the lowest level dry. At the time of my visit the production was about 400 tons of ore per month; the amount extracted during 1871 is reported as upwards of 4,000 tons, which has yielded, according to Mr. Ricker, "full one-third more per ton than any previous run made from this mine." I infer that it yielded $25 to $30. A level has been run 30 feet east from the bottom of the incline, which exposes a large seam, 3 to 5 feet wide, of similar character to the ore in the western stope. The mine is comparatively well off as regards reserves; none of the Whitlatch Union mines can be said to be models in this respect, one reason being that, on account of the shortness (horizontally) of their claims, each added section opened in depth represents less ground made accessible than it ought to. In other words, there are more inclines on the vein than mining requires, either for exploring, ventilating, pumping, or hoisting. This is a result of the divided ownership.

The National Mining and Exploring Company owns claim No. 2 west, amounting, by the last settlement of boundaries, to 233 feet on the vein. The incline, located about 20 feet from the western line of the claim, is down 550 feet. Three levels have been run east, each from 125 to 150 feet long, and mill-rock of good quality has been extracted from the stopes, probably ranging between $20 and $25 per ton. The company's mill reduced in 1871 upward of 4,000 tons. There is a good vein, 3 to 9 feet wide, exposed in the lowest level, (about 500 feet down.) No work was done for the last two months of the year, except running the pump which drains the next claim west, and partially the IXL mine on the east. This is an 8-inch plunger to 400 feet, and a lift from the bottom of the incline to that point. It is run ten hours out of twenty-four. The mine has good steam hoisting-works. The peculiar variations of the vein in dip are shown in this and the Columbia mine, adjoining. For the first 250 feet it dips about 36° north, and then changes to 18° north. The dip in the Columbia is given below.

The National Company's Mill has 20 stamps' weight, estimated, 760 pounds; drop, estimated, 10 inches; rate, 70 per minute; capacity, about 17 tons per twenty-four hours; efficiency, 0.6 tons daily, per horsepower developed by the stamps.

The Columbia Company owns the next claim, No. 3 west. The incline is down 630 feet, though the vertical depth is not so great as that of the

National incline, because the dip of the ledge is remarkably different. For 300 feet from the surface it is about 32° north; then, for 150 feet, it pitches only 5° or 6°, and finally assumes a dip of 29°, which is maintained to the bottom. These curious variations in dip, accompanied with some curving of the strike of the ledge, led to embarrassment in determining the boundaries of the claims underground, since the usual method of surveying at right angles across the local strike at the surface boundary, and continuing the line down the local dip at that point, here operated so as to cut out one claim in depth altogether. The matter was settled by compromise, as, under the present confused and incomplete mining laws, it must be settled, if at all.

The Columbia mine has been in constant operation during the year, and over 7,000 tons of ore have been extracted and reduced at the company's mill. In the lower levels the vein sometimes attains a width of 10 or 12 feet, and the ore is very rich in this locality, yielding from $27 to $32 per ton. The steam hoisting-works are handsome and substantial; in fact, the machinery of the mine and mill of this company is considered the best in Montana.

The Columbia Mill contains thirty stamps in 6 batteries; weight, 690 pounds; drop, 7 to 8 inches; rate, 78 per minute; capacity, 37.5 tons per twenty-four hours; efficiency, 1.24 tons per horse-power developed by the stamps. Amalgamation is effected in battery and upon outside plates. Twenty of the stamps are provided with Hungerford concentrators, one to each battery; and there are two large Wheeler pans for sulphurets. The tailings are exposed in the yard for one winter, and then charged again into the batteries with the ore. It is found that a considerable amount of tailings can be run through in this way without affecting the capacity of the mill for fresh ore, which seems to indicate an ample discharge. The screens are Russia slit, (not punched with round holes.)

The Owyhee claim, on the Parkinson lode, a western continuation of the Union fissure, is now owned by the National Company above mentioned. It comprises 300 feet in length, and is supposed to be very rich. An incline commenced in the early autumn was 180 feet deep in December, and was expected to strike the vein and pay-ore at 250 feet, when it was intended to erect steam hoisting-machinery and a 40-stamp mill, with all the modern improvements. Substantial buildings are already at the mine.

The Parkinson claim, adjoining this on the west, and owned by Messrs. Whitlatch, McClure & Argyle, has been but little worked during the year, probably on account of a prospective sale, for which Mr. Whitlatch has been negotiating in England, it is said, with John Taylor & Sons. One run of ore, made in the fall, yielded $25 per ton. The mine is regarded as a very good one, and I presume active operations were resumed in January.

The Park lode is about half a mile further west, and is either the same or a parallel fissure. South and east of the Park lode is the Evelyn, owned by Messrs. Cartright & Harvey. This mine was discovered more than a year ago, and has been considerably worked in a somewhat irregular way. The vein varies in width from 1½ to 6 feet, strikes apparently northwesterly, and dips 35° northeast. Upwards of 5,000 tons of ore have been extracted and worked in the Diamond City Mill (ten stamps) in Grizzly Gulch, yielding from $15 to $22 per ton. As no blasting is required, the cost of extraction has so far been very low. It is reported as not exceeding $1.50 per ton.

Collecting the foregoing statements as to the Unionville district, we may estimate its production during 1871 as follows :

Mine.	Tons.	Estimated yield.
Hendrie	2,000	840,000
IXL	4,000	100,000
National	4,000	80,000
Columbia	7,000	200,000
Evelyn	5,000	80,000
Owyhee, Parkinson, and other veins	800	20,000
Total product	22,800	520,000

Last Chance Gulch.—The workings in this gulch, half a mile below Helena, on the so-called ancient or red-gravel channel, were mentioned at length in my report of 1870, (p. 283.) To that account I will add but a few remarks, based on later personal observation. This red gravel, which is found underlying the cement or false bed-rock of the gulch, is supposed to be an older gulch, running east and west, or nearly at right angles across Last Chance. It is 35 to 45 feet in thickness, and extends 800 feet along Last Chance Gulch. The theory that it is a pot-hole or basin in the bottom of that gulch, filled with auriferous material while the stream flowed over it, is disproved, first, by the color of the gravel, which is red granite, with large boulders, while that of the overlying channel is gray limestone; secondly, by the abundance of water, invariably struck by sinking the red gravel, which indicates that it is, itself, a water-course; thirdly, by the tracing of the red gravel by shafts west of the present gulch. The long drain-tunnel, intended to strike the extension of this deposit, is proceeding slowly, for lack of capital. The most westerly shaft which has struck the red gravel, to my knowledge, is that of Colonel Keeler, two miles from the gulch. Meanwhile the 800 feet width of the claims in Last Chance (which, it should be remembered, are located, as upon that gulch, without reference to any theory of a cross-channel) has been rapidly exploited. The present owners of the upper claim (200 feet) are Taylor, Rumsey & Co.; of the second, Mr. Williams; while the two lower claims were purchased last summer by a Chinese Company, whose rights, it appears to me, under present territorial laws, must be held by precarious tenure. The legislature of Montana is reported to have passed a bill prohibiting aliens from acquiring or maintaining any titles. Whether this sort of law is constitutional or not, it is certainly destructive of the interests of the community, as may be shown in numerous instances where the Chinese have purchased, for cash, claims which white men could no longer afford to work, and have proceeded to make them productive, at a smaller profit to themselves than to the Territory. Besides being bad policy, this course toward the Chinese is rank dishonesty. The men who are glad enough to sell their old and worn-out diggings to these patient and frugal strangers join in the cry that the Chinese will overrun the land, and propose to eject them from the property they have paid for. It seems to me there is no harm to society in that kind of usurpation of soil which consists in buying it and paying for it; and, at all events, those who, by virtue of laws of their own manufacture, excluding foreigners from rights of original location, get the land in the beginning for nothing, and then, after having skimmed it of its richest treasure, sell it at their own price, and pocket the money of the purchaser, are no better than highway robbers, if they conspire thereafter against the title they have transferred.

I must candidly say, however, that while the legislation of Montana is outrageous, and there is a great deal of silly and wicked talk of the same tone among certain classes of the population, I have heard of only two or three cases of actually perpetrated injustice. Practically the Chinese have not been persecuted with violence, nor directly robbed, except in a few instances. But there is certainly danger, under the recent law I have mentioned, that they will suffer serious wrongs. It is indeed a ludicrous illustration of human stupidity—a Territory full of whites, every man of whom owns mining ground which he is anxious to sell, deliberately destroying their own market by excluding all the purchasers of second-class ground. In every enlightened community the producer takes pains to find consumers for his refuse and otherwise worthless material; in Montana three thousand men, whose ambition is to save money and buy cast-off claims—the refuse of wasteful white mining—are considered as interlopers, and kicked out of doors. One thought is calculated, however, to provoke anger rather than laughter, namely, that the ignorant and narrow-minded authors of this folly themselves have no rights to the mines except such as the United States has generously granted them, in consideration of the benefit to the whole country from the proper development of its resources. They are wasting the bounty of the people, and attempting to exclude from its benefits those who would economically utilize what they squander.

To return to the claims in Last Chance Gulch. The red gravel in Thompson's claim has panned, from the best layer or streak, as high as 50 cents to 81 per pan. In the Williams claim 20 cents is high. A great deal of stuff in both claims has much less value, as the material is spotted and streaked with strata of varying value. But the work would doubtless continue to be profitable for a long time had not the cost of dead work been progressively and enormously increased by the system of extraction adopted. Of course there would in any event be much expense involved in stripping the bed and pumping out the water. But all this has been, I think, more than doubled by the lack of forethought in the disposition of tailings and boulders, which have been heaped mountain-high on the top of ground which was afterward to be worked. Hence the removal of thousands of tons of this waste material has become necessary, sometimes twice over; and there is a considerable amount of ground (as, for instance, the east side of the Williams claim) which might be profitably worked but for the vast heaps piled upon it, and which will perhaps never be attacked, in consequence of these accumulated artificial hinderances, causing extra trouble and cost beyond the expected gain. A description of the method of extraction here pursued is given in my report of 1870. A wooden tower is erected in the gulch or on the bank, and the gravel is hoisted by steam upon inclined planes to the top of it, then dumped into the sluice-boxes and washed over the riffles. The inclined planes radiate to different parts of the claim, and can be easily moved, to suit the convenience of working, without disturbance to the hoisting-machinery in the tower. The dump of boulders, strippings, (overlying cement, sand, &c.,) and tailings (gravel from the sluices) accumulates around the tower, causing the permanent evil already alluded to.

For an enumeration and description of other gulches in Lewis and Clarke County I refer to my former reports, the statements of which I do not wish to repeat, while at the same time I have not heard of new or remarkable developments in the county during the year past.

Cedar Plains or Radersburgh district.—This is one of the most active and promising gold-quartz mining districts in Montana. The importance of its placer mines (gulch and hill diggings) has never been very great, though a limited number of men still find profitable occupation in that business within the district. Radersburgh is situated fifty-five miles from Helena, on the road to Bozeman. The mines are in the hills west of the town, in what used to be called Cedar Plains district. The principal discoveries were made as long ago as 1866, and the veins have been developed mostly without capital—a circumstance which is highly creditable to the energy and perseverance of the owners. It would scarcely have been practicable, but for the decomposed nature of the vein-material above the drainage line, and the freedom with which it can be amalgamated in mill.

The successful operations of quartz mining have doubled the population of the district within the year. The inhabitants now number six hundred, about one hundred and fifty of whom reside at Keatingville, a camp near the mines, which dates its existence from the spring of 1871.

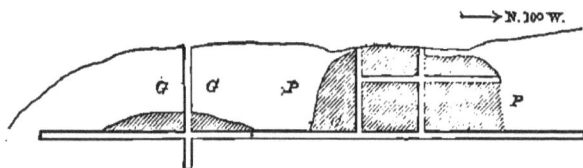

Profile of the Keating Mine, Radersburg, Montana.
Scale, 250 ft. to the inch.

The Keating lode is perhaps the most celebrated. The above diagram sufficiently shows the workings. The course is north 10° west, except for 125 feet, beginning 75 feet north of the whim-shaft, during which it courses north 60° west; but at 200 feet from the shaft resumes its usual strike. The dip varies from 65° west to vertical. The average width of the vein where it is worked is 20 inches. At C a branch of ore shoots off to the southeast and east, which has been supposed to be the left-hand lode mentioned below. At G G the vein is good, at P P it is pinched. The yield of the ore is reported at $20 per ton. This lode was discovered, in 1866, by John Keating and David Blacker, its present owners, and worked with arrastras till the spring of 1870, when the Postlewait Mill was purchased, and has been running with fifteen stamps almost constantly since June of that year. The stamps weigh about 680 pounds, and drop 8 inches 60 times per minute. Capacity 20 tons per day; efficiency, $\frac{33,000 \times 12 \times 20}{680 \times 8 \times 60 \times 15} = 1.72$ tons daily per horse-power developed. This high efficiency is due to the softness of the quartz. The mill has a low battery, with amalgamated plates inside and out, and one arrastra for the tailings from the settler. It is run by steam, and was built in Burlington, Iowa. The yield of the ore may be roughly estimated at $20 per ton, and the production of the mill at $10,000 per month. The country-rock of this mine is an altered magnesian-argillaceous rock; the vein itself is decomposed, quartzose-ferruginous matter, running into sulphurets in depth.

Near the Keating lode, but coursing north 80° west, and dipping 65° south, is the Left-hand, at one time supposed to be a spur of the Keating, but now held as a separate crossing vein. Its course cuts the Keating near the point C in the diagram. This vein had been opened, at the time of my visit in August, by two shafts, respectively 30 and 35 feet deep, and a level 125 feet long between them. Its average width was

20 inches, widening in places to 3 feet. The ore was decomposed and partially indurated iron-oxide. The last run of the rock had yielded $40 per ton.

By information dated in December, I learn that the lode is worked to the depth of 50 feet, and that the main shaft is down 130 feet, showing a 2-foot vein of rich ore. The aggregate yield of the mine to that date is given at 600 tons, and the average yield at a little over $35 per ton, or $22,000 in all. The main shaft will be sunk to 160 feet, and levels run at that depth for regular extraction. The work done hitherto, though temporarily profitable, has been scarcely miner-like. Two shafts within a hundred feet, two levels in a depth of 50 feet, and the whole vein "gouged out," leaving an empty crevice, is the appearance of the mine in its first stage. The great promise of this property will certainly justify more systematic work. Indeed, operations would probably have been more regular from the beginning, but for the litigation which rendered it desirable, on the one hand, to develop the independent character of the vein as soon as possible, and, on the other hand, to get as much money out of it as possible, while its title remained in doubt.

The average yield for the first 25 feet in depth is said to have been $21 from one shaft and $35 from the other. It has been worked in the Sample Mill, a steam custom-mill with twelve stamps, 500 pounds, dropping 8 to 12 inches, and at the time of my visit, when very soft rock was being worked, only 34 times per minute. On hard quartz it is said to run at 60 drops. This very soft material was from the Allen mine, (see below,) and the capacity at the rate given was 10 to 12 tons daily, or an efficiency of 2 tons per horse-power daily. The Left-hand lode is said to keep one arrastra also running. Messrs. Clancy and Davis, the owners of this mine, have three other partially developed lodes, the General Washburn, Don Juan, and Morning Star, of which the first promises to be very rich, according to the indications of 60 tons of ore extracted from a 70-foot shaft, which yielded $16 to $22 per ton.

The Don Juan is said to have a 2-foot crevice of dark ores yielding $20 to $36 per ton. The Morning Star has been opened to the depth of 40 feet, shows an 18-inch crevice, and prospects well. Other lodes worthy of mention are the Pennsylvania, discovered September, 1871, 2-foot crevice, dark ore, $15 to $20 per ton; the Northeast, discovered January, 1871, crevice 2½ feet, ore yellow, much free gold, $30 per ton; the Rica, (Ferguson, Gonn and Smith,) one foot, decomposed yellow ore, prospecting well, traceable for 2,000 feet; and the Hidden Treasure, (A. Campbell,) 2-foot crevice, dark yellow ore, free gold, $20 per ton.

Messrs. Keating and Blacker, the owners of the Keating lode, possess also the Ohio and Leviathan. These are situated near the Keating Mill, about half a mile up the gulch, above a little settlement called Keatingville, two or three miles from Radersburgh. The Ohio courses north 38° west, and dips 77° southwest. The average width is 4 feet, in the bottom of the shaft 5 feet, (but here it has passed into sulphurets, which are not at present worked.) The ore is decomposed and ferruginous in the upper levels, and carries calc-spar in fine crystals—a circumstance which I did not notice elsewhere. The walls are white decomposed argillaceous rock above, and probably greenstone below. The vein crops out boldly in a bluff by the side of the road opposite the mill. It has been opened with a drift from the road, 50 feet northwest, and beyond this on the top of the hill a shaft 35 feet deep, showing a handsome 8-foot lode. No stoping has been done at this point. On the southeast side of the road a shaft is down 180 feet, (August, 1871,) and a drift has been run 150 feet southeast and 40 feet northwest, at the depth of 75 feet, from which the vein has been stoped out to the surface. The bottom of the whim-shaft is in sulphurets. So is the bottom of a shaft at the mouth of the tunnel-drift first alluded to. This latter shaft is 65 feet deep, but its mouth lies 30 feet lower than that of the whim-shaft. The Ohio ore has averaged, according to Mr. Keating, $25 per ton. There seems to be still a good deal of the decomposed ore accessible; but, sooner or later, this mine, and all others in the Radersburgh district, will have to face the sulphurets in depth. At present, these are universally avoided. There are no facilities for working them, and I could not even learn that they had ever been assayed or tested in any way for gold.

The Leopard lode has a 1-foot crevice of dark-brown ore, shown in a 20-foot shaft. It prospects well. The Twilight is 2 feet wide, yielding a light yellow ore. It has been developed to the depth of 60 feet, was very rich near the surface, but is not now worked.

The Iron-Clad lode, giving its name to a small settlement of a dozen houses, is situated in the foot-hills, about two miles west of Radersburgh.

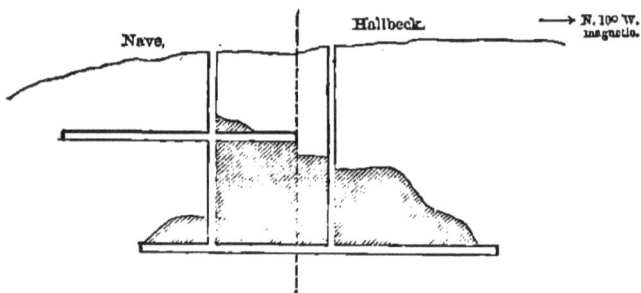

Profile of Workings on the Ironclad Vein, Radersburg, Montana.
Scale, 100 ft. to the inch.

The course is south 10° east, and the dip, taken at the mouth of the whim-shaft, about 70° west; below, 67° north. The average width is 3¼ feet. The country-rock is magnesian, (talcose slate?) and the vein follows apparently its dip and course. The ore is bluish-gray quartz, containing disseminated sulphurets; in the upper levels, ferruginous decomposed quartz. The sulphuret zone in this vein begins about 40 feet from the bottom of the present workings. Very little of the gold in the sulphurets is saved in the mill; but the tailings have been saved for future treatment. At the time of my visit, a 6-stamp water-power mill at Hot Spring was at work on the Nave ore, and the Sample Mill has treated some of Hallbeck's ore. The larger part of the quartz extracted remains for the present on the dumps. The owners contemplate the erection of a new mill in the spring. The vein has been superficially traced for 3,000 feet. The discovery was made in 1866 by John Spears, but the greater part of the lode is held by other parties, of which the most prominent are: Charles G. Hallbeck, Discovery, 200 feet; Jacob Nave, No. 1 south, 1,000 feet; J. F. Allen, No. 1 north, 206 feet. The diagram shows Nave's and Hallbeck's workings. At each of the two shafts shown there is a horse-whim. Water is raised in barrels, amounting, in August, on Nave's claim, to 20 barrels, at 33 gallons, or 660 gallons daily. The yield of the quartz is $20 to $25 per ton.

The depth of workings on the Iron-Clad, according to my latest advices, (December,) was: 130 feet in Nave's claim, No. 1 south; 180 feet, in Hallbeck's Discovery claim; 188 feet in Allen's claim, No. 1 north now owned by Terrill and Merritt; and in the Allen shaft itself, 204 feet.

The Mammoth, owned by Cleaveland and Naves, has a 4-foot crevice of dark-brown ore, yielding $14.80 per ton. Depth of working, 30 feet. The Robert E. Lee has a shaft 60 feet deep; is from four inches to one foot wide, and yields light-yellowish, soft ore, milling $40 to $90 per ton.

The Vanderbilt, owned by Kerwin and Baier, is developed to the depth of 50 feet, showing a 12 to 14-inch crevice, with brown quartz, carrying some visible free gold. Thirty-five tons milled $15 per ton.

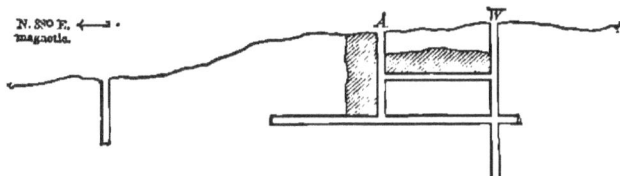

Profile of the Allen Mine, Radersburg, Montana.
Scale, 100 ft. to the inch.

The Allen lode, about half a mile east of the Iron-Clad, was discovered by Mr. J. F. Allen, who owns 800 feet of it. The diagram shows his workings up to August 5, 1871. All those were accomplished in two months, the vein being remarkably soft, and the walls, especially the hanging-wall, good. There are thin clay partings and no slides of rock. Only two blasts had been found necessary in the drifts. The course of the vein is north 28° east, and the dip 70° northwest for the first 30 feet, and 65° below. The width is, at northeast end of lower drift, 4 feet 8 inches; between air-shaft (A) and whim-shaft (W) above lower drift, 3 feet; at bottom of whim-shaft, 3 feet; average width, 3 feet. The whim-shaft is 270 feet northeast of the Discovery. The vein carries sulphurets, no doubt, in depth, but they are not yet reached, and the ore resembles that of the Iron-Clad above the sulphuret zone. The amount of quartz extracted up to August was 300 or 400 tons. The last run of 80 tons in the Sample Mill yielded $13.55 per ton. It is supposed that the softness of the material caused a loss of gold. This mine was in October under negotiation for sale to a New York company at $27,500, cash, for 800 feet. The owner proposed erecting a mill immediately, whether the sale should be effected or not. The width of the ledge, and the cheapness and ease of extracting the rock, render it a very promising mine.

The Congress lode (A. Campbell & Co.) has a 2-foot crevice of dark-brown ore, assaying $57 in silver and $45 in gold, per ton. Depth of workings, 33 feet. The John Spear lode, developed to the depth of 20 feet, shows a 3-foot crevice, carrying much galena, and assaying $90 in silver and $11 in gold, per ton. The Live Yankee, Moore Campbell, and a number of other promising outcrops, have been but slightly prospected.

South of Cedar Plains district, high up in the mountain-range separating Crow Creek from North Boulder, large quantities of galena "float-rock" have been found, and it is expected that rich silver-bearing lodes will be discovered in that neighborhood during this year. Indeed, some lodes of galena have already been found which assay 860 silver per ton.

The facilities for the reduction of ores have increased with a rapidity almost but not quite equal to the progress of extraction in this district. The following mills were in operation in December :

Keating and Blacker's, fifteen stamps, steam, cost $20,000; Davis Mill, twelve stamps, steam, cost $12,000; Allen's Mill, six stamps, steam; How & Wood's arrastra, with two stamps; and Nave Brothers' Mill, five stamps and arrastra, water-power. The combined capacity of this machinery is sufficient to crush 56 tons of quartz every twenty-four hours. The product for the year 1871 was about $200,000, one-half of which was from ores of the Ohio and Keating lodes, crushed in the Keating and Blacker Mill.

The placer mines of this district have yielded well during the past year, and some new and very remunerative discoveries have been made. Mr. Thompson discovered a lower channel on Keating Gulch which yielded $5,000 in one week, with only five workmen employed, and which is by no means yet exhausted. The same channel was afterward struck at a depth of 60 feet, about 1,000 feet below Thompson's claim; and it is not improbable that it may extend along Keating Gulch a distance of two or three miles, in which event this district will receive a new impulse to its prosperity. In Mountain Gulch a deep channel was recently discovered on Mr. Harvey's claim, which proved to be very rich, and from which 145 ounces of gold were taken during the few days which it was worked before cold weather put a stop to mining operations. This discovery will probably lead to the development of a very extensive gold deposit underlying the false bed-rock upon which the gulch has been hitherto worked. The yield of the placer mines during the past season has been not less than $50,000.

The ditches, three in number, which provide the water for the working of these placer mines, are owned by Mr. Quinn. Their combined length is about twenty miles, and the cost of their construction was about $30,000. Water will be supplied in future at such rates as will justify the working of very extensive deposits which have heretofore lain idle.

We have already seen that the yield of this district, which is only about three miles square, was, during the year 1871, $250,000. The indications are that these figures will be increased to $500,000 during the current year.

Indian Creek district.—About eight miles north of Cedar Plains district, and situated among the foot-hills of the Crow Creek Mountains, is Indian Creek mining district, supporting the towns of Saint Louis and Springville, located respectively in the upper and lower portions of the mines.

The placer mines of this section were discovered in 1866, since which time they have been worked to only a moderate extent on account of scarcity of water. Indian Creek is worked for a distance of about four miles, gives employment, while water lasts, to about one hundred men, and pays about $7 per day to the hand. The aggregate length of water-ditches is fifteen miles. The yield of gold for the year 1871 was in the neighborhood of $50,000. Some rich bars in this district have paid from $20 to $50 per day to the hand.

Among the quartz mines are the following:

The Diamond Ledge is 13 to 15 feet in width, and is sunk upon to the depth of 95 feet. It yields a decomposed reddish quartz, 40 tons of which, worked in an arrastra, paid $10.25 per ton. It has the appearance of being a "chimney;" is owned by Messrs. Lewis & Reece, of Pittsburgh, Pennsylvania. Blacksmith Ledge, 2 to 4 feet wide, reddish quartz, in a line with the Diamond Ledge, and perhaps a continuation of the latter; has a shaft 75 feet deep, and is actively worked by its owners, Messrs. Foster, Ross & Co. Jaw-Bone Ledge, 4 to 5 feet wide; well-defined crevice, producing hard, brown quartz, of a very regular character, showing some fine free gold. Fifty tons in an arrastra yielded $15 to $33 per ton. Combined length of shafts and levels, 250 feet. Has been worked to a depth of 60 feet. Its owners, Messrs. Kerwin, Paschley & Baier, contemplate the erection of a small mill at an early day. A number of other ledges in this district prospect well, but are undeveloped. Both this district and that of Cedar Plains are very advantageously situated on the main stage-road between Helena and Bozeman, and on the borders of one of the most productive agricultural sections of Montana. Their supplies of merchandise can be obtained from Helena in from one to two days, while the products of the farm are abundant in the immediate neighborhood.

CHAPTER V.

UTAH.

During the past year the mining resources of this Territory have been rapidly brought before the public, and the influx of prospectors, miners, and speculators has been very great. Capital, too, has, in many instances, found its way into mining and smelting works; but, on the whole, it may be asserted that few mining regions have in so short a time acquired an importance like that of Utah with the aid of so little capital.

There is perhaps not so much excitement in Utah as there has been in many new mining districts on far less foundation; the cause being that there is little opportunity for persons without capital to engage in large and profitable enterprises. But the owners of claims hold them at enormous figures, a sure indication that the thing is overdone, whatever may be the actual basis.

Many new districts have been organized and prospected within the past year, and some of them are regularly shipping ore. A few mines, and among these especially the celebrated "Emma," have, thus far, furnished the principal basis of actual business, as well as the stimulus for sanguine operations. The advantages of this Territory as a mining field, under existing circumstances, it is true, are inviting.

First among the advantages of the situation must be reckoned the presence of a large agricultural population in the Territory. Utah will not have to import food to supply its mining population; and this secures reasonable prices of supplies and abundance of labor. The stories told about the cheapness of mining labor in Utah are, however, exaggerated. The Mormons take from one another very low wages. The standard is annually fixed, I am informed, by the church authorities; and I believe it was this year $1.50 per day. But they take all they can get from Gentile employers, and, moreover, few of them will work as miners; so the wages of this class of labor are $2.50 to $3.50 per day, even in the districts nearest to Salt Lake City. The prices paid for hauling ore, on the other hand, are very reasonable, considering the distances. Most of the teamsters are Mormons.

Another advantage is the facility of railway transportation for ore and base bullion from Salt Lake to the East and West. In this respect it is true the miners and smelters are dependent upon the railway companies. During the summer all shipments of ore were paralyzed by a new and enhanced schedule of freights. Only the Emma Company, which had a contract with the railway, at low rates, (running till September, I believe,) was able to go on. The rates were subsequently reduced, though not to the former point. They were then $18 per ton for ore and $20 for bullion, from Salt Lake to Omaha. But aside from these fluctuations, it is evident that without the railroads the mines of Utah would not have been successfully developed. Even for those ores which are smelted in the Territory most of the charcoal is brought by rail from Truckee, in the Sierra Nevada, (though a considerable amount is burned in the Wasatch Mountains, and in piñon districts further south.) The Truckee charcoal can be had in large quantities at 25 cents per bushel; the Utah charcoal costs 22 to 30 cents, but is frequently inferior in quality, while the supply is precarious. This price is far higher than it should be; but the fact is that the charcoal-burners here will abandon the business unless they can make as much money as others.

Again, the character of the ores in the Utah districts is such that they can be either shipped at once to foreign reduction-works, or smelted into argentiferous lead on the spot, and shipped in that form, or, finally, both reduced and separated in the Territory. Some of the mines in the western range furnish milling ores, which are treated by the Washoe system. The majority, however, contain galena, carbonate of lead, with gangue of ferruginous dolomite or quartz and admixtures of antimony and arsenic.

If we inquire, however, how these favorable conditions are utilized, we find much to criticise and lament. The metallurgical industry here is conducted in a sadly careless and ignorant manner. There are now in the Territory some nineteen or twenty furnaces, mostly small shaft-furnaces. The aggregate production in July was, however, only about 15 tons of base bullion daily—a proof that the furnaces are run very irregularly, as an inspection of the works also sufficiently shows. Ignorance of the nature and proper treatment of the ores is one reason. The furnaces are run so badly that salamanders are of frequent occurrence. In fact, I do not believe the average campaign exceeds a week.

The Messrs. Robbins, aided by their skillful metallurgist, Mr. Rüger, have erected excellent works for the treatment of galena-ores, comprising a large reverberatory roasting-furnace, with a smelting-hearth underneath. These works are not yet fairly in operation; and there appears to be some difficulty in obtaining a regular supply of the kind of ore for which they are calculated, not so much because there is a lack of galena among the ores produced, as because the miners do not like to sort them out, and prefer to mix all their ores together, and ship them in that condition. The Robbins works are, moreover, interesting, as the only ones, so far as I am aware, employing the coal of the Green River field as fuel. Experience thus far has shown that the coal is not of uniform quality, even when coming, nominally at least, from the same mine. The best of it will probably be a moderately good fuel for the reverberatory processes. The shaft-furnaces employ charcoal.

The loss of lead and silver in the shaft-furnaces is very great. It may almost be said that the bullion produced is not richer in silver than the ore. I feel sure that in many cases, moreover, half the lead is lost in the slag, or up the chimney. The astounding ignorance of the smelters may be illustrated by a circumstance which I personally noticed. At one of the principal works I saw heaps of hematite-iron ore and of limestone, which I found were brought there (the hematite at a cost of $17 per ton) to mix with the silver-ores and make a "slag." As the ores on the floor contained plenty of iron, and more than plenty of lime, it may easily be understood that what this addition really made was not a slag but a salamander.

Yet in spite of these most evident losses, the smelting-works in general are reported as paying high prices for ore—prices, in many cases, which would exclude the possibility of profit, even under good management. The only explanation I can offer is, either that the proprietors of the furnaces are losing money without knowing it (a thing which may easily occur to persons inexperienced in the smelting business) or they are running the furnaces at a loss, with the intention of selling mines on the strength of alleged favorable results. I fear that in many instances the latter is the true explanation. Certain it is, if anything in metallurgy or mathematics is certain, that the smelting-works now in operation cannot be legitimately making money, operated as they are, and paying what they do for ores.

One great obstacle is too often in the way of the free development of the best mines in the Territory. The blighting curse of litigation rests upon almost every good mine in the older districts. The mining laws are vague and bad, and the Territory is infested with unscrupulous jumpers and black-mailers. There were, in the summer, at least three parties fighting over the Emma mine; and the Emma was by no means a solitary instance. Lawsuits were springing up all over the Territory— a new crop of mischief from the indolence and neglect of the Federal Government with regard to the mining law. For my part I am more and more thoroughly convinced that the men to whom the United States is virtually giving away its mineral lands are not the proper persons to regulate the tenure of their titles. One would scarcely say to his best friend, "Here, take my property on your own terms!" and the penniless speculators in mining claims can least of. all be trusted to make the laws defining their own rights. In my opinion the United States law, which declares all citizens entitled to mine upon the public domain, gives them no right to any dog-in-the-manger titles. The object of the law is to develop the mines, not to help a few individuals lock them up and demand high prices for them. A discoverer has the right to occupy and improve; this is properly his only right; it is all he can sell; and until the mine is purchased of the United States nobody can have any rights in the premises which abandonment or neglect to improve does not defeat. In Utah there were many mines discovered some years ago and abandoned. Now a second crop of discoverers has come, and the old ones have returned in swarms to claim their "rights."

The product of 1871 in gold and silver may be estimated as follows:

Express shipments of gold	$221, 262
Express shipments of silver	130, 175
Ore shipped by railroad, 10,806 tons, at $150	1, 620, 900
Base bullion by railroad, 2,378 tons, at $175	316, 150
Allowance for undervaluation by express	11, 513
	2, 300, 000
To which may be added for the value of the lead contained in ore and bullion	500, 000
Giving a total of	2, 800, 000

The shipments of ore and bullion from Salt Lake City in the different months were as follows:

	Ore, tons.	Bullion, tons.
1871—January to May 30	3, 984	269
June	842	100
July	2, 819	410
August	1, 123	370
September	554	420
October	220	370
November	624	350
December	440	90
Total	10, 606	2, 378

Mr. E. P. Vining, general freight-agent of the Union Pacific Railroad, informs me that 8,880 tons of ore and 2,185 tons of bullion were shipped from Utah eastward over that road during the year. This would leave for shipments over the Central Pacific Railroad westward 926 tons of ore and 193 tons of bullion.

The material for the following pages has been furnished by notes from Professor Blake, Mr. Fabian, Captain Stover, O. Hahn, Mr. Heffernan, and by those made by my deputy, Mr. Eilers, and myself during our visits to the different districts in the Territory.

The mining districts of Utah which have attracted most attention during the year are in the region east and south of Great Salt Lake. They are laid out in the two principal ranges of the Wasatch Mountains, and thus lie in two longitudinal belts, one on the east of the valley of the Jordan, the other on the west. Commencing near Corinne, at the northern and eastern end of the lake, the districts succeed from north to south in nearly the following order:

Cache County.—Dry Lake, Logan, Millville, Mineral Point.

Box Elder County.—Willow Creek.

Morgan County.—Weber, traversed by the Weber River and by the Union Pacific Railroad.

Davis County.—Farmington, Centreville.

Salt Lake County.—Hot Spring, New El Dorado, Big Cottonwood, Little Cottonwood, American Fork, Snake Creek, Uintah, Silver Fork.

Utah County.—East Tintic, Mount Nebo or Timmins, Spanish Fork.

Tooele County.—Tooele, Stockton, Ophir, Lower district, Camp Floyd, Osceola, West Tintic. In the same, or Oquirrh, range, but on the eastern slope, are the West Mountains, or Bingham district, facing the Big Cottonwood, and the Little Cottonwood on the opposite side of the valley. Lake Side district is directly west of Great Salt Lake. The island in the lake is known as Church Island district.

The geological structure of the Wasatch Mountains is intricate and interesting. The principal range east of Salt Lake has a broad exposure of gray granite rising into peaks generally whitened by snow. This rock is flanked on the east by an immense thickness of quartzite and limestone strata, the last mentioned generally holding the lead and silver-bearing veins. In these rocks Professor Blake obtained numerous fossils, principally encrinites and one species of Archimedes. He refers the formation to the Lower Carboniferous or Devonian period. In the upper portion of the principal cañons or valleys, cutting the range transversely, there are distinct and well-marked traces of former glaciers. At the head of Little Cottonwood Cañon the granite is worn by glacial action into smoothly rounded summits—*roches moutonnées*—and part way down the valley the ancient terminus of the glacier is marked by a moraine stretching across the valley from side to side, except where it is cut through by the existing stream.

In the higher portions of the Wasatch, especially east of Salt Lake City, in the valleys of the Big Cottonwood and the Little Cottonwood, there is an abundant supply of timber, consisting chiefly of fir, pine, and cedar. The quaking ash grows in great luxuriance along the streams lower down.

During the winter snow falls in great quantity, and accumulates to a depth of from 6 to 20 feet or more in some places, and where it is sheltered from the direct rays of the sun it remains through the summer or until September and October. The melting of this vast accumulation of snow gives a constant supply of pure water, and forms rapidly-flowing streams, valuable not only for saw-mills and smelting-works but of far greater importance for the irrigation of the valley lands. Both of the Cottonwood Creeks are thus utilized, and are two of the most important sources of water for the irrigation of the valley near Salt Lake City.

The unrestrained use of the timber by the miners and the Mormons is working its speedy destruction. It is not only cut freely for the mills, but for making coal, and immense quantities are annually consumed by forest-fires, the result of carelessness or neglect, or of willful determination to destroy. It is abundant now, but in a few years will be very scarce and valuable. The Mormons have for many years obtained their supply from the Big Cottonwood, and it is said that the first discoveries there of mineral deposits were made by the lumbermen.

The discovery of metalliferous deposits in this region, inaugurating the series of discoveries leading to the formation of the several districts named, may be said to date back to the year 1863. The first silver-lead veins were found in the Oquirrh range, in Bingham Cañon, and in the mountains bordering Rush Valley. In October, 1862, the United States volunteers from California, under the command of General Connor, arrived in Utah and established the post known as Camp Douglas. Many experienced California miners were in this command, and naturally enough took every favorable opportunity for prospecting the hills and valleys of that vicinity. In the fall of 1863 Lieutenant Weitz and a small party discovered the outcroppings of a lode in Bingham Cañon, and from that time prospecting was energetically prosecuted. In 1864, when Company L was stationed at the Government reserve in Rush Valley, many lodes of argentiferous galena were found. In the same year many locations were made by parties of persons emigrating from the Western States to California, and about this time the town of Stockton was laid out. The land was "taken up" for the purpose by General P. E. Connor, J. F. Rogers, Joseph Clark, and J. J. Johnson. The first mining district was organized in December, 1863, and was named "West Mountain Mining District." It embraced the whole of the Oquirrh range of mountains lying west of Jordan Valley. At a meeting of the miners, held June 11, 1864, this large district was subdivided, the eastern slope of the mountains retaining the name of "West Mountain Mining District," and the western slope was called the "Rush Valley Mining District."

Rush Valley district.—In the years 1865 and 1866, when the troops were mustered out of service, a great many men turned their attention to the recently-located veins, and Stockton soon grew to be a mining town of considerable importance. About fifty new buildings were erected, and the trade in supplies for the miners was very brisk, though the cost of every article was enormous. Transportation of supplies from the Missouri River to Salt Lake City at that time cost 25 cents per pound. This heavy expense upon all tools, together with the great difference between gold and currency, and the large profit asked on goods by the traders, made mining a very costly occupation in those days. A shovel, for example, cost $5; steel, $1.40 per pound; powder, $1.50; sugar, $1.25 per pound; coffee, $1.50; tea, $4.50; and other articles in proportion.

In the year 1864 the West Jordan Mining Company was incorporated under the laws of California, and work was commenced on a lode of that name in Bingham Cañon. A tunnel was run into the hill about 40 feet. The work was continued until the great expense, amounting to $60 per running foot of tunnel, caused it to be suspended.

The first furnace for smelting the ores was built by General Connor and others, associated under the title of the "Pioneer Company," at Stockton, in the year 1864. About the same time a company, called the Rush Valley Smelting Company, was formed at Camp Douglas; the stockholders were the twenty officers then stationed there. Other furnaces were built soon after by General Connor in the reverberatory form. They were made of adobes and sandstone, and lasted long enough to prove that the ore could be successfully smelted, but also that the materials used in construction were not suitable, for the furnaces soon burned out and were left to ruin. There were three of these furnaces of large size, and one of them had a flue 150 feet long. The ruins of these and of other furnaces are still to be seen, deserted, upon the hill-sides near Stockton.

Smaller blast or draught furnaces were built by each of the following-named parties, Finnerty, James, Gibson, Nichols, Brain, Warren, and Davids, and all smelted more or less of the ores. The first cupola-furnace was built by Mosheimer, Johnson & Co., but for want of sufficient blast did not work very successfully. One cupelling-furnace was built by Stover & Weberling.

In 1865 the Knickerbocker and Argenta Mining and Smelting Company was organized in New York, and disbursed $100,000 in the purchase of claims, machinery, and material for furnaces, buildings, &c. The superintendents were Captain C. B. Dahlgren, Colonel J. G. Cooper, and J. M. Forbes. This company suspended operations in 1866, and their property was sold by the sheriff, by the order of the county court, to pay the indebtedness of the concern. This large expenditure of money, without any satisfactory result or profit, was inevitable considering the cost of all materials necessary to carry on smelting-operations. To mine and smelt argentiferous lead ores at a profit was simply impossible under those conditions, and it did not become possible to conduct this business profitably until, by the construction of the Pacific Railroad, supplies and transportation were cheapened. Since the completion of the road these old, abandoned, or neglected claims and furnaces are invested with new interest and value.

The discovery and location of the principal lodes of this district were made during the years 1864, 1865, and 1866. There were upward of five hundred locations recorded. Not more than forty of these were in what is now called the Ophir or East Cañon district. The first discovery in that district was in 1864, and was called the "Subjugation." This was soon followed by the location of the "Wild Delirium," Saint Louis, Mountain Gem, (with extensions,) Pollock' IXL, Metropolitan, and others. At Stockton, the earliest discoveries were named the Eureka, Quandary, Potomac, Great Basin, Great Central, Silver King, Last Chance, New York, Silver Queen, Lady Douglas, Mineral Hill.

No discoveries of any importance have been made in this district north of Dry Cañon since the year 1866, with the exception of a few locations made at Soldier Cañon recently.

The Rush Valley district was divided, in July, 1870, into three districts, the north end being named the Tooele, and the part taken from the south end of the original, the Ophir district. The central portion retained the name of Rush Valley district, and is about seven miles square. The town of Stockton is beautifully situated upon the slope of the foot-hills of the mountains, facing Rush Valley Lake, a sheet of water six miles long and three miles wide. From well-known and understood causes this district has not attracted as much attention as others during the general enthusiasm and excitement attending the discovery and opening of mines in other districts, although it is believed that many of the lodes are equal in value to some of the best in the Territory. Instead of increasing in population, the town of Stockton is retrograding; there are not as many buildings and residents in it now as it had three years ago.

H. Ex. 211——20

For most of the foregoing details regarding the early history of this district, I am indebted to Captain David B. Stover, who has been familiar with the progress of mining there and in other parts of Utah since the first discovery in 1863. The following are special details of several of the principal claims.

Great Central, first northern extension, 1,000 feet; located in August, 1864; vein 3 feet wide; assays $100 per ton in silver, and 45 per cent. lead. The development in April, 1871, consisted of an open cut 30 feet long, with a pit 10 feet deep at each end. Ten tons of ore had been shipped. Owners, Connor, Stover, Butler, and Kean.

Bolivia, 1,000 feet; located in 1865; ore an argentiferous galena, assaying $40 to the ton, and giving an average of 35 per cent. of lead. Seventy-five tons of ore have been shipped, and there was about the same quantity on the dumps. The development consists of a shaft 132 feet in depth; vein represented as 3 feet wide. Owned by Connor, Stover, Brown & Co. Wood and water plenty.

Eureka: Located in 1864, and work commenced in the same year; vein 4 feet to 6 feet wide; average assay value in silver $75; lead 60 per cent. Opened by a tunnel 140 feet long run in to cut the vein, and by an excavation on the surface equal to 100 cubic yards. The tunnel had not reached the vein in July, 1871. Only about 20 tons of ore had been taken out up to that time. Owners, Connor, Stover, Church & Co.

Hard Times: The claim was located in 1864, and is 1,200 feet in length; vein 18 inches wide, in limestone; ore is a hard, green-stained carbonate. Assays average $100 to the ton, but they have reached as high as $550. Native silver in small scales has been found in this lode. Opened by two shafts, each about 12 feet deep. Several tons of ore were taken out. Owners, Connor, Stover, Benson & Co.

Lady Douglas, 1,700 feet; located in 1865; vein, 2 feet wide; ore, galena and carbonate of lead; assay value about $50 to the ton in silver and 40 per cent. of lead; opened by a cut of 20 feet in length, and by several pits, each a few feet deep; also by another cut 25 feet in length. This vein is said to have been traced for 1,000 feet; about 3 tons of ore have been taken out. Owned by Butler, Kean, Stover, and others.

Exchange: Location 800 feet in length; vein said to be 6 feet wide and the ore to assay $30 per ton in silver, and 48 per cent. of lead; opened by a pit 8 feet deep; work commenced in March, 1871; owned by Butler, Rice, Chase, and Kean.

Elizabeth: 1,200 feet located; vein 18 inches wide; average assay of silver, $50 per ton; lead, 40 per cent.; opened by a shaft 12 feet deep, and by an open cut 25 feet long; work commenced in 1866; twenty tons of ore have been taken out by the owners, Messrs. Carle, Stover & Co.

Quandary: Length of claim, 1,200 feet; vein, 3 feet wide; average assay, $45 in silver, and 50 per cent. of lead; shaft 110 feet deep; work commenced in 1864; resumed in 1871, and 10 tons of ore taken out; owners, Chase & Co.

Pendleton: Length of location, 1,200 feet; vein, 2 feet wide; average assay, $55 in silver per ton, and 40 per cent. of lead; opened by a shaft about 60 feet deep; work commenced in 1864; about 10 tons of ore have been taken out; owned by Stover and Butler.

Last Chance: Length of location, 1,000 feet; vein, about 2 feet wide; averages in silver $100 per ton, but samples have assayed as high as $1,400; average yield of lead, 40 per cent.; opened by a shaft 35 feet deep, and by several cuts on the surface across the vein; quantity of ore, about 5 tons; owned by Nevitt, Stover, Connor & Co.

Rush Valley: Location, 1,200 feet long; vein, 2 feet wide; ore, an argentiferous galena, assaying $40 in silver, and 30 per cent. of lead; shaft 8 feet deep; work commenced in 1864; owned by Connor, Chase, and Gibson.

Silver King—first extension west: Location, 1,000 feet; vein, 4 feet wide; average assay, $30 per ton in silver, and 55 per cent. of lead; one shaft is 165 feet, and one 40 feet deep; work was commenced in 1865. This was owned by the Argenta Silver Mining Company. Sixty tons of ore have been taken out. It is now owned by Dilidine, Stover, and Benson.

Potomac: Location of 1,200 feet made in 1864; vein, 3 feet wide; shaft 100 feet deep; 25 tons taken out; assay value is about $55, silver, and 60 per cent. of lead; owners, Bayliss, Kerr, Benson, and Stover.

Constitution: Location, 1,000 feet; made in 1864; vein, 1 foot wide; ore assays $50 per ton in silver, and 60 per cent. of lead; shaft 15 feet deep; and 10 tons of ore have been taken out; owners, Church, Stover, and others.

Pleasant Hill: Location, 1,100 feet; vein, 2 feet wide.; average assay in silver $35 per ton, and 55 per cent. of lead; worked by an open cut 30 feet long and a shaft 12 feet deep, in 1865; twenty tons of ore were taken out. Relocated under the name of "Grand Cross" in 1870. It is now opened by a tunnel 130 feet long and a shaft 30 feet deep. More than 100 tons of ore have been taken out since 1865. Some of this ore was shipped to San Francisco, and some has been worked at Simon's Furnace. Owners, Payne, Paxton, and others.

Silver King: Location, 1,000 feet; vein, 5 feet wide; ore, argentiferous galena and carbonate of lead; average assay value $40 per ton in silver, and 50 per cent. of lead; opened by two shafts, one 200 feet deep, and the other 50 feet. There are also a tunnel 150 feet long connecting with one of the shafts, and drifts equal to 100 feet more in length. Work was commenced in 1865; 300 tons of ore have been taken out, most of which was sent to San Francisco. A part was worked at Simon's Furnace. Owners, Gail, Connor, and others.

Defiance: Location, 800 feet; vein, 4 feet wide; assays $40 per ton, and 40 per cent. of lead; opened by a tunnel 65 feet long; work commenced in December, 1870; only about 2 tons have been taken out; owned by Butler, Rice & Co.

John Adams: Location, 1,000 feet; vein, 2 feet wide; assays $40 in silver, and 35 per cent. of lead; opened by a shaft 10 feet deep, and by cuts in the surface; worked in 1865 and reworked in 1870; only 2 tons of ore taken out; owned by Butler, Kean & Co.

New York Lode: Location, 1,000 feet; vein, 18 inches wide; the ore is a hard carbonate of lead, some of which has assayed as high as $3,000; the average is about $125 in silver, and about 40 per cent. of lead; shaft, 45 feet deep; owners, Nichols, Stover & Co.

Saint Patrick: Location, 1,400 feet; vein, 3 feet wide; ore assays $40 in silver, and 60 per cent. of lead; shaft 90 feet deep; work commenced in 1871, soon after the discovery; probably 100 tons had been taken out in July, 1871. The ore is smelted at Simon's Furnace.

Legal Tender: Location, 1,000 feet; vein, 3 feet wide; average assay $50 in silver, and 60 per cent. of lead; 100 tons taken out; work commenced in 1871; owned by True, Tierpan & Co.

Putnam: Location, 3,000 feet; vein, 3 feet wide; average assay $60 in silver, and 40 per cent. of lead; opened by cut and shaft, the latter 60 feet deep; owners, Delamater, Wells & Co.

Tucson: Location, 1,000 feet; vein, 3 feet wide; ore assays $40 in silver, and 55 per cent. of lead; opened by a shaft 100 feet deep; owned by George Berry & Co.

Ophir or East Cañon district.—This district was formerly, as mentioned above, a part of the Rush Valley or Stockton mining district. Horn-silver and other rich silver-ores were found August, 1870. The Silveropolis mines have since yielded from $50,000 to $75,000 of base bullion. Some of the most prominent claims in April, 1871, were the Tampico, Mountain Lion, Mountain Tiger, Petaluma, Zella, Silver Chief, Defiance, Virginia, Monarch, Blue Wing, Silveropolis. It is now probably the most productive district of the Oquirrh range. Mining is prosecuted with energy and success, and mills and smelting-works are in full operation. Up to April, 1871, over five hundred locations had been recorded. One furnace was in operation, and two more were erecting. Ophir City, the center, was a thriving town of one thousand to twelve hundred inhabitants, and it is increasing in importance daily. It is located in a cañon leading from the center of the range to the open plain or valley south of Rush Lake, and is accessible by carriage-road from Salt Lake City in one day. Stages run daily, passing around the south end of Salt Lake through Rush Valley and the town of Stockton.

There are two distinct groups of mines; the one north of the town affording an abundance of galena and pyritous ores of low grade, while the other, upon Lion Hill, south of the town, yields a richer class of ore, the decomposed portions of which can be successfully treated by the ordinary mill processes.

This Lion Hill, or rather mountain, (for it rises abruptly some 2,000 feet above the town,) is noted for the quantity and richness of the silver-ore it has already yielded, and for the number of silver-producing claims located upon it. It is formed of a great mass of limestone strata, which here rise in one grand anticlinal curve. The edges of these strata show in an almost vertical wall along the valley which cuts directly through them transverse to the axis. The silver deposits crop out between the upper layers of rock, which there pitch to the eastward at an angle of about 20°. The outcrops are upon the brow of a very steep descent, facing the west, and overlooking the valley and the cañon below the town. The new mill erected there by the Walker Brothers, of Salt Lake City, is directly in sight below, and is not more than a mile distant. But as the descent is too abrupt for a roadway the ores are not sent directly down, but are carted in the other direction along the ridge, descending gradually by the bed of the ravine to Ophir City, and thence a mile further down the valley to the mill.

The ores of Lion Hill are chiefly the soft, ochery, and earthy-looking mixtures resulting from the decomposition of argentiferous galena and other argentiferous minerals containing antimony and arsenic. An examination made by Mr. Blake of a portion of ore taken from the Rockwell claim showed that the silver existed in it in the form of chloride, so that it could be easily and cleanly worked in a mill without preliminary roasting. But there are also large quantities of carbonate of lead and nodules of antimonial galena, which require roasting or smelting in order to liberate the silver contained in them.

Most of the ore in these claims is soft enough to be taken out without the use of powder, the pick and shovel sufficing to detach it. It is wheeled out through tunnels to the surface, and is there packed in canvas sacks for shipping or for sending down to the furnaces or mills. The veins lie at such an angle upon the hill that they are readily and cheaply opened by tunnels, and do not require expensive shafts and machinery for hoisting or pumping. There is as yet no water of consequence in any of the excavations. Apparently it freely finds its way downward through the crevices in the limestone, so that no difficulty from that source need be expected in depth.

It is proposed to run a tunnel directly through the mountain at a depth of about 2,000 feet below the principal claims. The survey of McLaren's East Cañon tunnel shows that the height of some of the principal points and claims above tide is as follows:

	Feet.
Silver Chief mine	8,500
Mountain Tiger and Zella	8,650
Summit of Lion Mountain	9,100
Monarch and Virginia	8,775
Horn-Silver Hill	9,225
Vallejo and Occidental mines	8,850
Mountain Lion mine	8,620

Silver City, or the valley just below it, is found to be nearly 6,675 feet above the sea.

Probably the largest amount of ore has been shipped from the Mountain Lion claim. The Occidental, Tiger, Rockwell, Zella, and Silver Chief are all prominent claims, and have yielded notable quantities of good ore. The last-mentioned claim is reported to have had, on the 1st of August last, from 250 to 300 tons of excellent " chloride ore" upon the dump, and to be opened by a tunnel 120 feet long, with a drift 40 feet long to the northward, showing a fine body of ore extending north and south.

Mountain Tiger, Rockwell, and Zella.—These three contiguous claims, after having been successfully worked separately, have recently been consolidated, and are being systematically developed by tunnels and shafts, under the superintendence of Mr. Mark Daly. Some details regarding the production and value of the ores of these claims will serve to give a general idea of the value of the ores from other claims in the immediate vicinity.

The greater portion of the work has been done upon the Mountain Tiger claim. It is opened by a tunnel and open cut 150 feet in length, following the ore for a part of the distance, and ending in an inclined shaft reaching under the outcroppings of the Rockwell claim, to connect there with a shaft sunk from the surface. At the bottom of this incline, in August last, a small oven-like cavern was opened into, and the floor was found covered with a yellowish, earthy deposit, which, though not very promising in its appearance, contained at the rate of over $2,000 per ton in silver. It appeared to be a mixture of chloride of silver and oxide of antimony.

Upon the Zella claim the ore came up to the surface of the ground, and has been excavated for about 75 feet in length and to a depth of 15 to 20 feet, for most of this depth under the overhanging wall of limestone. The thickness varied from 2 to 5 feet, and the ore was very rich. Both this claim and the Mountain Tiger have yielded large quantities of ore. Some of it has been worked in the Pioneer Mill of Messrs. Walker & Brothers, and some has been shipped. The value of the ores on the dump-piles in August last was estimated at $135,000 by the superintendent, and the total of ore out and in sight in the three claims at over $360,000. On the 30th of August last the shaft in the Petaluma workings (a portion of the Mountain Tiger claim) was 30 feet deep, and showed ore all the way.

In regard to the value of these ores, as raised, there is abundant and satisfactory information. Large sales have been made from time to time, the value being ascertained by average sampling and careful assays. Most of the lots of ore offered for sale were purchased by Mr. C. T. Meader, of Salt Lake City, at from 63 to 65 per cent. of the assay value. The value of several lots is shown in the following table:

Returns of sampling and assay of nine lots of ore from the Mountain Tiger claim.

Quantity in pounds.	Moisture.	Lead per cent.	Ounces of silver per ton.	Value of silver per ton of 2,000 pounds.
4,989	3.50	0.00	166.21	$214 89
24,443	17.00	....	97.68	176 28
22,671	6.50	15.00	64.15	82 94
4,587	4.50	3.00	53.94	69 74
4,135	9.75	9.50	104.97	135 72
7,000	8.50	5,02	189.54	245 05
8,546	9.03	5.00	93.31	120 64
27,322	9.00	10.20	164.75	213 00
10,825	5.00	7.00	104.98	135 72

Returns of sampling and assay value of eleven lots from the Zella claim.

Weight in pounds.	Moisture.	Lead per cent.	Ounces of silver per ton.	Value of silver per ton of 2,000 pounds.
3,567	7.00	2.5	198.28	$256 36
7,872	10.00	9.0	121.01	156 45
16,995	12.50	11.0	163.29	211 12
6,016	10.50	9.0	209.95	271 44
10,859	9.05	19.0	199.45	257 86
5,432	18.00	14.0	166.21	214 89
2,395	7.00	7.0	330.96	427 89
2,520	4.50	Trace.	67.06	86 71
13,753	18.00	5.0	105.85	136 85
7,112	18.00	3.0	99.14	128 18
3,810	14.50	6.0	116.64	150 80

An average sample of the ore standing in the Tiger claim in August, yielded at the rate of, silver 185 ounces, value $239, per ton of 2,000 pounds. An average sample of the Zella claim previously taken gave $138.86 as the value in silver.

Silveropolis, Tampico, Occidental, and other claims.—About a mile beyond the Mountain Tiger claim, on Lion Hill, in the direction of Camp Floyd, there is a group of claims from which a considerable quantity of rich chloride-of-silver ore has been taken. At the Tampico a great open cut along the slope of the hill exposes the ends of curved strata of lime-stone for 250 or 300 feet. The limestone is hard and flinty, and irregu-larly seamed with masses of dark-colored calc-spar, some quartz and heavy spar. Good ore is found in the midst of these seams, but not in any clearly defined or regular vein.

At the Silveropolis claim horn-silver was found interstratified with limestone or filling a seam parallel with the strata. After a few hundred tons of good milling ore had been taken out of an open cut the seam " pinched" and work was discontinued at that place. But tunnels to cut the stratum in depth have been commenced. The prospect for striking a prolongation of the deposit is very fair. The strata at this place trend southwest and dip at about 45° to the southeast. Nearly $50,000 worth of silver ore was taken from the cut, and much of the rock yielded from $400 to $600 per ton. The ore was very free from base metals, but some nodules of galena were found toward the bottom of the cut. The limestone upon which the chloride of silver rested is much veined with white calc-spar, and the same mineral was found with the ore.

Some of the ore from this locality, worked at the Pioneer Mill, gave bullion .996 fine; it evidently was chiefly chloride uncontaminated by other metallic compounds and reducible by the mill process. Ore from the Tampico also gave very fine bullion. One lot of 24 sacks of ore from the Tampico assayed $1,116.88 per ton, and another of 22 sacks $934.63 per ton. The average of the pile as selected was over $800. Six sacks shipped abroad assayed at the rate of $4,104.72 per ton.

The Pioneer Quartz-Mill, East Cañon, or Walker's Mill, already referred to as situated one mile below Silver City, at the mouth of the cañon, is one of the best for its size upon the Pacific coast. It was made by a branch of the establishment of W. J. Booth & Co., of San Francisco, and has all the latest improvements and modifications on the old-fashioned mills of two or three years ago. There are one Blake's rock-breaker, three 5-stamp batteries, six grinding and amalgamating pans of one ton capacity each, three 7-foot settlers and two large retorts for the bullion, set in brick-work. The engine is a fine piece of workmanship, nominally 70 to 80 horse-power, with cylinder 16 inches in diameter and 30 inches stroke. The boilers are 48 inches in diameter and 16 feet long, with 40 tubes in each. The engine-building is 40 by 30 feet, and the main building 40 by 60 feet. The ore-dump platform above the rock-breaker has a capacity of 1,200 tons. The large Fairbanks scales will take a load of 20 tons.

The batteries are arranged for dry crushing. The discharge is from both sides, and the powdered ore is received in two tram-wagons standing under each side of the mortar. The stamps are run as fast as from 80 to 90 lifts per minute with double cams; the drop is 8 to 10 inches. The fifteen stamps crush about twenty-five tons a day. The ore is delivered at the mill by ox-teams at $4 per ton from the mines. Wood for fuel is delivered at $4 per cord.

This mill was started in August last, under the superintendence of Mr. Lathrop Dunn, and worked several small lots of ore from different claims. The silver bars first turned out were remarkable for their purity and fineness, ranging from .991 to .996 fine. The first four bars weighed and assayed as below:

	Ounces.	Fineness.	Value.
Silveropolis claim	769.75	.991	$986 28
Silveropolis claim	850.00	.996	1,094 63
Tampico claim	1,024.75	.991	1,313 00
Occidental claim	770.00	.993	988 60

There were also a few arrastras at work in the cañon. One owned by J. D. Lomax is run by a water-wheel 18 feet in diameter. He uses also a Wheeler pan for amalgamation. The works of the Ophir Mining and Smelting Company are owned by Colonel Weightman & Co. There is one furnace in operation, and a new one is being built. The former is 14 feet from the top to feed-hole; 30 inches in diameter at top, 40 in the middle, and 30 at the bottom. They have a 16 horse-power engine and a No. 7 Sturtevant blower. About two tons of bullion are made a day. There are three tuyeres to this furnace. The fire-bricks are made 8 inches thick. It takes twelve to go around the furnace.

In Ophir district are also Fawcett's "patent" furnaces. These are draught-furnaces, in which the draught is produced by conducting steam into the stack above the charge, which draws air in rapidly below through a large number of openings or tuyeres. The idea is old, and has long been given up as impractical in older mining countries. The steam is furnished by a 40 horse-power boiler for two large iron stacks lined with fire-brick. The furnace burned out a very short time after it had been started without producing any bullion. Subsequent trials resulted no better. Schofield, Abbey, Drake & Co.'s furnace, in the same district, has two stone stacks, boshes lined with quartzite, of the same capacity and construction as that of Jennings, near Salt Lake City, which will be mentioned hereafter. There is a 20 horse-power engine, a Gates crusher, and a No. 7 Sturtevant fan. Some bullion has been made here, but in the latter part of the year the works were idle.

In the fall the Brevoort Mining Company built a steam stamp-mill at the lower end of East Cañon. This is the patent of J. W. Forbes, of La Porte, Indiana, and cost $2,500 in the East. It is known as the "automatic steam-battery," and the mortar is oval in shape, discharging from a No. 40 screen, on all sides. There are two stamps, the stems of which are 6½ feet high, and 65 pounds of steam are required to run them. They use two engines, one of 10 and one of 30 horse-power. This mill is run on the same principle as Wilson's stamp-mill, so well known on this coast, the stamp-stems acting as pistons to two vertical cylinders, so that the force of the steam is thrown directly on the stamp. This company also have five improved Varney pans, and one Farnham & Warren patent pulp-grinder, with a capacity, it is said, of 1 ton an hour. The grinder is 30 inches high and 3 feet in diameter, grinding the pulp three times over, and then discharging it into the amalgamating-pans.

According to an estimate furnished me by Mr. James Heffernan, of Corinne, smelting in this district cannot be profitably conducted under the present circumstances, and especially with the present high prices of ore. It is based on ore from the Velocipede mine, which contains about 30 per cent. of lead with 30 ounces of silver per ton, and of which it takes for the present smelting operations 5 tons to make 1 ton of bullion.

Cost:

5 tons of Velocipede ore, at $30 per ton	$150 00
Cost of smelting, at $35 per ton	175 00
Cost of transporting 1 ton of bullion to Salt Lake City...............	10 00
Total cost of 1 ton of bullion, not including interest, &c............	$335 00
Proceeds:	
1 ton of lead bullion, at 3½ cents per pound	70 00
5 tons of ore yielding 30 ounces, total 150 ounces, of which 80 per cent. is saved, 120 ounces, at $1.15 per ounce........................	138 00
Total value of bullion, per ton	208 00
Loss	127 00

This seems high, but I am satisfied that it is not far from the actual facts.

CLAIMS NORTH OF OPHIR CITY.

The claims located upon the north side of the cañon, just above Silver City, yield ores not so well adapted for milling as for smelting. These ores consist chiefly of a mixture of galena and iron pyrites, both in small crystals, and they occur in beds from 2 to 5 feet thick apparently inter-stratified with the rock, and dipping into the mountain at an angle of 20 to 30 degrees. The rocks are thinly stratified limestone with slaty partings, passing into calcareous shale. The upper portions of the beds of ore, at and for several feet below the surface, are much decomposed, giving an ochery, yellow and greenish mass, easily mined and smelted. Among the claims producing such ores are the Silver Shield, Velocipede, Hidden Treasure, General Grant, Burnett, Cooley Sevier, California Antelope, Wild Delirium, &c.

Burnet: Claim 1,200 feet; one mile from Ophir; ore as above described; selected samples assaying about 48 per cent. of lead and $60 per ton in silver. The sulphurets are found in mass about 20 feet below the croppings.

General Grant tunnel: Located to intersect several lodes—the Harriet, Seymour, General Grant, Blue Monitor, Lola Montez, and others. The ores of these lodes are worth about $45 per ton in silver, and contain 20 per cent. of lead. The tunnel was commenced July 16, 1870. In April, 1871, $2,000 worth of work had been done on it and the claims. Upwards of 300 tons of ore had been taken out, one car-load of which was shipped to Liverpool.

Blue Monitor: Claim 3,000 feet; work was commenced in July, 1870, and a large amount of ore taken out. It assays about 60 ounces of silver per ton and 15 per cent. of lead. Owned by W. W. Angel, William Traus, L. W. Clark, and others.

The Raymond Smelting-Works are located at Ophir City, East Cañon, fifty miles from Salt Lake City. Operations commenced March 1, 1871; building 40 by 80 feet; steam-power; one blast-furnace; capacity 10 tons of ore per day; on an average, 3¼ tons of the ore smelted produced one ton of metal. Forty tons of base bullion were produced up to April 30, 1871; value per ton, $257. Another blast-furnace was then erecting. Charcoal is used as fuel. The average assayed value in silver per ton was 131 ounces. I am indebted for these details to Mr. S. A. Raymond, superintendent of the works.

Camp Floyd district.—This district adjoins Ophir or East Cañon district on the south. It is comparatively new, and the claims are not yet much developed. The outcroppings of veins are well defined, and are not so high up on the mountain as at Lion Hill. The winters are not so severe, and it is claimed that miners can work in open claims during the season. Among the principal claims are the Sparrow-Hawk, Silver Cloud, (reported to have been recently sold to an English company;) Mormon Chief, and the Grecian Bend. There is also a vein affording cinnabar of low percentage.

Sparrow-Hawk: This claim, opened during the summer by Mr. McMasters, shows a considerable body of shaly quartz, of a dull, bluish-gray color, and coated with films of chloride of silver. The vein is marked by very heavy quartz croppings. The thickness at the open cut, from which most of the silver-bearing ore has been taken, is about 50 feet, but it is irregular. A large portion of these croppings is apparently quite free from ore in paying quantity. In August last there was a large pile of ore on the dump, estimated at 100 tons of first-class and the same quantity of second-class ore.

Grecian Bend: This claim is a short distance beyond the Sparrow-Hawk, and may be a prolongation of the same vein. This, and the Mormon Chief claim beyond it, are adjoining claims, each having 2,000 feet upon the lode. They are characterized by an enormous outcrop of quartz stretching up the side of the mountain for a mile or more. It rises from 20 to 50 feet or more in height, and has an irregularly broken, precipitous face. It pitches into the hill at an angle of about 20°, and the general direction of the cropping is 10° south of west. This outcrop is in general quite hard and compact, and gives little indication of being ore-bearing, though the color is dark, and it much resembles the quartz at the Sparrow-Hawk claim, where chloride of silver has been found. Very little work has been done on either of these claims. With the exception of two small pits upon the lower edge of the croppings the mass of the vein is untouched, and awaits vigorous work, conducted upon a liberal scale, to break into the rocky mass and show whether it is ore-bearing or not. At one of the excavations there are some small streaks of ore, which, it is said, assay well for silver. At that place the ledge appears to be split up into several layers, but all of them are conformable to the strata of shaly limestone above and below. The indications are sufficient to justify the expenditure of some money in prospecting the ground, especially along the contact of the quartz with the wall-rocks.

West Mountain Mining district, Bingham Cañon.—Among the numerous claims in this district may be mentioned the Buel and Bateman mines, sold to an English company during the summer, the Vespasian, worked by Kelsey & Sons, the Silver Jane, Kenosh, Winnamuck, Washington, Spanish, and Equi.

Kenosh lode: Claim, 1,600 feet; "vein" said to be 34 feet wide; ore assays 35 ounces silver to the ton, and 60 per cent. of lead. One thousand dollars' worth of work had been expended up to April, 1871, in shaft and tunnels. Work was commenced in July, 1870. Some 350 tons of ore had been taken out and 100 tons sold. Fifty tons of base bullion, valued at $300 per ton, were shipped to Chicago.

Winnamuck mine: Two thousand feet located; "vein" varies in width from a foot up to 16½ feet. The ore is argentiferous galena and carbonates. Work was commenced on this lode in 1864, and the vein is opened by an incline shaft 300 feet deep, and two drifts 70 to 80 feet long and 72 feet apart. Some 1,200 tons of ore had been taken out up to April, 1871. At present the mine belongs to Messrs. Bristol & Dagget, who have worked it very successfully during the year. As their furnace (a large one of the Piltz pattern) is located immediately below the mine, so that the ore can be directly run down to the charge-floor by means of a chute, which commences at the mouth of one of the tunnels, and, as a part of the ore from this mine (carbonate) contains much more oxide of iron, and is richer in lead and silver, than is usual in the Bingham mines, these gentlemen prosecute their business under somewhat more favorable circumstances than the other mine and furnace owners. In fact, as far as I was able to ascertain, they are the only successful mine-owners in the district, a fact due partly to the above causes and partly to their superior intelligence and the good tact which caused them to employ an accomplished metallurgist to build their furnace and to start it running successfully. These gentlemen have also worked the ore from the Spanish mine, a great part of which was galena, and of medium richness in silver, in their furnace.

The mines and works most frequently mentioned during last summer and fall have, however, been those formerly owned by Messrs. Buel & Bateman, and transferred in the summer to the Utah Silver Mining and

Smelting Company, limited, an English company, which is reported to have paid the high price of $450,000 for the property. The mines are located at the head of Bingham Cañon, and the claims cover several hills completely by being staked out on imaginary veins running in all conceivable directions. In reality, however, there appear to be no veins here, but irregular pockets in quartzite, which carry quartzose lead ores, very poor in lead and silver. The claims located are the Dartmouth, Bullion, Portland, Sturgess, Warrior, Allison, Chance, Onesimus, and Belshazar, each containing 1,200 feet. At the time of the writer's visit to the property, which was shortly after the transfer to English hands, the Portland was principally worked, as it carried the best ore. But this ore contained, I am informed, only from $4 to $30 in silver per ton. The ore-body opened in the tunnel was extremely irregular and much mixed with gangue. The ore here consisted principally of carbonate and leadhillite. As widely differing reports had been made by an American and by two English mining engineers—the American having condemned the mines as not worth working, while the Englishmen figured up large prospective profits for the purchasers—the writer took some samples of the ores then being smelted to ascertain their real value. These samples assayed as follows:

	Lead.	Silver per ton.
1. Silicious and argillaceous iron-stained carbonate of lead	4.6 per cent.	1.21 ounces.
2. Ochreous carbonate	25. per cent.	10.93 ounces.
3. White ore, (mixture of carbonate and leadhillite)	57. per cent.	12.10 ounces.

The last sample was taken from a small pile of picked ore, and came from the Portland, where it occurred in patches.

The extraordinarily unprofitable smelting operations of this company, as well as the successful ones of Bristol and Dagget, are mentioned in another part of this report, under "Lead-smelting in Nevada and Utah."

The English company commenced, immediately after their purchase was consummated, to erect a large furnace of the combined Piltz-Raschette pattern, the capacity of which is 45 tons a day. This costly furnace, as well as equally costly prospecting operations, swallowed up the original working capital of the company very soon, and in December it was reported that prospects here were very discouraging. According to still later news, however, the company had raised a new working capital, and a new ore-body had been discovered in the Warrior.

The Oro claim is situated in Markham's Fork, and is opened by a shaft about 70 feet deep. The Washington is a location about 2,000 feet southwest of the Vespasian claim, opened by a shaft about 80 feet in depth in September last. The Vespasian is about a mile and a half west of the Oro claim. Shaft, in September, about 60 feet deep, and reported as showing a vein of ore about three feet in thickness. The Silver Jane, about one-quarter of a mile beyond the Vespasian in a northwesterly direction, has a shaft about 70 feet in depth.

There are numerous other claims in this district, but in all of them the ores are poor in silver as well as in lead. Concentration, which seems to me the only rational method for working these mines successfully, has not yet been attempted.

Gold-placers.—Placer gold has been found and worked for in the lower parts of Bingham Cañon for the past four or five years. The claims are numerous, and much work has been done. Costly preparations have been made during the past summer to reach the bed of the old channel under a considerable depth of earth and gravel. At Mason M. Hill's claim, located in the cañon, two miles above its mouth, expensive machinery for hoisting and pumping has been erected. In October last the shaft was about 80 feet deep, but the progress of the work was retarded by the great influx of water, which the Cornish pumps, then erected in the shaft, could not master. In November, however, Mr. Hill procured a compound propeller pump, Shaw patent, of a capacity of 1,000 gallons per minute, from Philadelphia, and by this means it was expected that he would be able to continue sinking his shaft, and reach bed-rock. Heretofore the only claims worked in the cañon had been the bars on the hill-sides, and in several instances old river-channels on top of some of the spurs coming down to the main cañon. These have generally paid very well, and in some cases extraordinarily large strikes have been made. During the last season a company of foreigners, Italians or Spaniards, were reported as having been especially successful. They had taken out of their claim, which is located on the top of a considerable hill close to Bingham City, over $50,000 in a very short time. The total yield of the Bingham placers, since they have been worked, is given by those best informed as over $500,000, over $100,000 of which is last year's product. The gravel in Bingham is little washed, and consists mostly of angular fragments of quartzite. The supply of water in the cañon is insufficient, except in the early spring, when the melting snows furnish an adequate supply for the hydraulics and sluices for a few months.

Bed-rock in the lower part of the cañon has never been reached, and, in view of this, Mr. Hill's enterprise is of great moment to the future of the placer-mining interest.

According to information which reached me at the end of the year, gold-bearing quartz-veins had been discovered in one of the side cañons, but I am without any particulars.

Tintic district.—This district is in Tooele County, about seventy miles southwest of Salt Lake City. Both East and West Tintic districts are reached by Concord stage-coaches from the city, and they have the great advantages of accessibility, with plenty of wood and water, and very' mild winters, without snow, stock being kept unhoused all winter. The name is supposed to be that of an Indian chief formerly living in that valley. West Tintic first attracted attention as a mineral region in December, 1869. The first discoveries were made by Messrs. Stephen B. Morne, Peck, Hyde, and others. They found the outcrops of the now famous Sunbeam lode. This is now one of the principal veins in the region, and it extends far enough to permit a great many locations along its course, some of which have been opened to a considerable depth. Among other important localities are Eureka Hill, the Mammoth, and the Armstrong copper claims, the Shoebridge, the Martha Washington, Black Dragon, Gray Eagle, Highlander, Swansea, Argenta, Diamond, Evening Star, North Star, and James Bird. There are many more from which much is expected.

Eureka Hill: This is by far the most prominent and best-known mining locality in the district. It has been successfully worked during the year, and has produced large quantities of rich smelting-ore. The formation is stratified limestone, uplifted nearly on edge. The veins are upon a projecting spur, with a rounded, elongated surface. Pits sunk, from 2 to 10 feet in depth, almost anywhere upon this hill, reach argentiferous ore, much of it highly charged with horn-silver. Galena, and its derivatives by decomposition, are abundant, and some of the ores contain considerable quantities of copper, enough to produce brilliantly-colored specimens of green and blue carbonates. The ore-deposits appear to follow the vertical stratification of the limestone, and to occur in irregular masses. They are not in all cases confined to one bed or division of the rocks, but crop out in many nearly parallel irregular veins, sometimes connected, without doubt, by cross-courses or seams cutting across the strata. The contiguity of the outcrops, and the possible intersection of the veins below, have led to much difficulty and litigation among the numerous claim-holders. The principal part of the ore from this locality is carted to smelting-works at Homansville, a few miles distant. They are known as the Utah Smelting and Milling Company's works. These works commenced operations June 17, 1871, and at the time of the visit of the writer had been running sixty days. The company have two furnaces, but had been running only one of them at a time, partly because of the scarcity of workmen, most of them being required to put the mine, the Scotia, in such working order that there should be no lack of ore for both furnaces. It was expected in the month of August that both furnaces would soon be in full operation.

In the sixty days from starting there had been run out 2,849 bars of silver-lead, averaging 121 pounds to the bar, or in all $172_{\frac{3}{100}}$ tons, with an average value of $210 per ton. The furnace was running upon a mixture of ores from Eureka, the Scotia, and other places. A very considerable amount of arsenic and antimony was evidently present, for the fumes of the former pervaded the atmosphere around the works. Iron-ore is used to mix with the charges. Good charcoal is delivered at the works at 16 cents per bushel. The blast is obtained by Root's blower, worked by a portable steam-engine, built in Chicago. The product ranges from 44 to 50 bars a day.

Other smelting-furnaces have been erected at Diamond City, and have run upon ores from the Shower mines and on other ores obtained by purchase.

The Mammoth copper claim is a remarkable deposit of ore in limestone, cropping out upon the western slope of a hill facing the broad and well-wooded valley of Tintic, and only a few miles south of Eureka Hill. It is opened upon the surface by a broad cut, a cross-cut, and a shaft to a depth of about 170 feet. This last is irregular in its direction and dip, but follows the mass of the ore. A cross-cut at the bottom of the shaft, 52 feet long, has not reached the limits of the ore-mass in either direction. This ore is, much of it, very ferruginous and poor in copper, but there are masses and seams of rich, dark-colored ore, mixed with green and blue carbonates of copper. The undecomposed ore occurs not only in amorphous masses, but in bladed crystals several inches in length, radiating through the greenish vein-stone. This mineral contains sulphur and arsenic, and is probably the species enargite. The secondary ores resulting from its decomposition are very highly colored, and give beautifully-variegated masses of green and blue carbonate, besides masses of silicate of copper. Over 150 sackfuls of black oxide and green carbonate of copper were taken from one of the open cuts at the surface. Considerable quantities of ore from this and adjoining claims have been shipped to Swansea. Most of it has been sold to dealers in ores at Salt Lake City.

There is doubtless a large amount of ore remaining in these irregularly-formed deposits, which, without forming a regular vein between well-marked walls, appear to extend along a certain belt or zone following the stratification of the limestone. Similar ore appears again in the adjoining hill, at the Armstrong claim, from which large amounts have been shipped.

The percentage of copper in the ores from these claims varies with the care taken in selecting. From 10 to 50 per cent. may be regarded as a possible range for the ore in shipping quantities. A very considerable quantity probably will not run over 8 per cent. The value of silver is reported to be from $20 to $100 per ton. Some 700 tons of ore had been extracted and shipped up to April 1, 1871, 300 tons of which were sent to purchasers in Salt Lake City.

A tunnel is now being run into the hill, in a northeasterly direction, so as to intersect the Mammoth claim in depth. This tunnel will be about 500 feet long and 243 below the croppings. On the 17th of August last this tunnel had been run 117 feet, 66 feet of which was open cut. The ore-mass trends about northwest and northeast, and dips to the north and east at about 45°.

The Martha Washington claim is about two miles from the Mammoth, and carries silver and lead, without much copper. It has a distinct vein-structure trending northeasterly and dipping westerly about 70°. In thickness it ranges from 2 to 4 feet. The ore is quartzose, rusty, spongy, and appears to be the result of the decomposition of argentiferous and antimonial galena. It is reported to assay from $40 to $150 in silver per ton. From 40 to 50 tons were upon the dump in August last. The incline shaft, from 40 to 50 feet in depth, was so full of carbonic-acid gas that it could not be entered with a burning candle.

The North Star location is 900 feet in length, and is about one mile south of the Mammoth. The vein is in limestone, with a gangue of heavy spar, and is said to be 14 feet wide, and cuts the strata vertically. A tunnel is being run in to cut the vein at a depth of 75 feet. This ore is reported to carry gold and silver, and to assay from $20 to $200 per ton. Owners, Messrs. Congar, Loomis, Oakley and Carter.

Black Dragon: Location 2,200 feet in length, and on the same range as the North Star. The vein is reported to be from 4 to 10 feet wide, in limestone, carrying carbonate of lead and ferruginous matter, assaying, on an average, about $50 per ton. Owners, Messrs. Moore, Peck, McCurdy and Morgan.

The Sunbeam is a remarkably well-defined vertical vein, cutting through hard, porphyritic rocks, and having distinct and hard croppings for about one mile in length. These croppings are quartzose, and they stand from 2 to 5 feet above the surface. In many places they are much divided up by intermediate masses of rock. It may be called a thin vein, for the ore-bearing portion rarely exceeds 6 to 12 inches in thickness, although in places the croppings indicate a much greater breadth. The ore is a decomposed galena, giving oxides and carbonates of lead rich in silver. The assays have ranged from $32 to $848 for silver, according to the samples. On this claim there was one shaft of 130 feet, and another of 50 feet in depth. It is opened in many other places, and considerable quantities of ore have been taken from several of the claims, chiefly from Congar's, the O. K., and Moore and Peck's.

The Shoebridge claim is located and opened on the top of a high hill beyond the Sunbeam. It is remarkable for the unchanged condition of the ore at a moderate depth, the sulphurets being found there in their full brilliancy without any rusty or ocherous ore. The vein is opened by a shaft, from which drifts have been run part way down, and at the bottom, 116 feet from the surface. The vein runs nearly north and south, and is vertical. It has a good clay gouge along the walls, and is 3 to 4 feet thick in the widest part, about half way down the shaft. The ore is extremely interesting, inasmuch as it consists of a mixture of the rare mineral enargite with iron pyrites. This mineral is a compound of sulphur, arsenic, and copper, containing about 47 per cent. of the metal and some silver. It is taken out in large, brittle, black masses, which are easily broken up after exposure, and show numerous cavities lined with small rhombic crystals characteristic of the species.

Big Cottonwood district.—The first location made and recorded in this district was in June, 1870. Nine locations were recorded in that month, 9 in July, 18 in August, 17 in September, 4 in October, in all of which it is claimed that galena-ore was in sight. The veins vary in thickness from 6 inches to several feet, and the ores assay from $25 per ton to several hundreds, according to samples. The Davenport, Theresa, Wellington, Highland Chief, Wandering Boy, Antelope, Prince of Wales, Congress, Lone Star State, Rock Island, Beckwith, Marfield, Hidden Treasure, Cooper, Scott, Read and Benson, and the Ophir, all had ore in sight in April, 1871. The Davenport, Theresa, Wandering Boy, Marfield, and Prince of Wales, had each yielded some ore for shipment already in the fall of 1870.

There were over two hundred locations recorded in this district up to April, 1871. Its southern boundary line joins upon the northern edge of Little Cottonwood district along the summit of the mountain, above the Emma mine, the Savage, Montezuma, and the Flagstaff. The claims near this dividing line send their ores down into Little Cottonwood Cañon.

The Davenport mine is one of the most notable of the district, and has produced a large amount of ore during this summer. The claim is nearly upon the dividing line of the two districts at the top of the hill. It is opened by an incline shaft following the dip of the vein. The ore is similar to that from the principal claims of Little Cottonwood, and is excellent for smelting. It appears to be the result of the decomposition of argentiferous antimonial galena, for carbonate of lead and the earthy-looking oxides of lead and antimony are abundant. The developments made up to the close of the season were favorable to the extent and richness of the mine. Considerable ore has been sent down the Big Cottonwood Cañon to the Hawkeye works.

Gold is reported to have been found at the head of the Big Cottonwood, high up among the granitic and gneissic rocks.

In November, 1871, the district contained over six hundred and twenty-five locations.

Little Cottonwood district.—Leaving the great highway about seven miles south of Salt Lake City, the road to Little Cottonwood crosses for five or six miles the low foot-hills formed by the detritus washed from the mountains. These are not generally cultivated, and present to the traveler little but the dusty gray sage-brush, except in the immediate vicinity of the streams. Along the base of the steep mountain-faces runs the bench or old water-line, indicating the former line of the great fresh-water basin, a small portion of which is now occupied by Salt Lake. This bench can be distinctly seen, even at a great distance, and can be followed for many miles along the mountains.

The Wasatch range at this point consists of upturned strata of sandstone, quartzite, slate, limestone and granite, the latter apparently of sedimentary origin, like the rest. Whoever undertakes to explain the geology of the Cottonwood Cañons on the theory that the granite is eruptive, and forms the central mass of the upheaval, will be involved in serious contradictions, and will be obliged to twist the facts considerably to fit them to this notion. The general course of the strata is nearly northwest and southeast, bearing rather towards east and west, and thus crossing obliquely the geographical axis of the range, which is nearly meridional, as well as the course of the cañons, which is, windings apart, on the whole, east and west. Thus in riding through the cañons, one may observe on both sides the successive strata, the edge or outcrop of each one on the southern side being farther up the cañon (east) than its continuation on the northern side. The dip is usually about 60° northeast.

The scenery in the Cottonwood Cañons is both grand and lovely. The Big Cottonwood Cañon is wild, precipitous, narrow, and tortuous. At twenty different points, as one rides along the bank of the stream, the rock-masses before and behind seem to close up, and leave neither inlet nor outlet for the tumbling waters. But the reckless river, getting used to this sort of thing at last, plunges boldly toward the apparently impenetrable barrier, and lo! a narrow fissure, unseen before, opens around some jutting crag, and the flood surges through, to enter another *cul-de-sac* and escape again by a hidden outlet. Those who have admired the Devil's Gate, in Weber Cañon—the most romantic point on the Union Pacific Railroad—will understand, from the hint which that one spot gives, what must be the picturesque effect of this cañon, which is crowded full of such surprises. Weber Cañon, in fact, has in a feeble degree many of the characteristic features of the Wasatch scenery; but Big Cottonwood excels it in every particular. The vast overhanging peaks and cliffs on either side, rising 3,000 feet or more above the road; the musical brooks that pour down their steep gorges, now leaping in cascades, now burying themselves, to re-appear as cool, clear springs; the stately forests of pines and aspens; and, last touch of beauty, the stains and patches of brilliant color from innumerable wild flowers that cover acres and acres of the mountain sides with pure white and delicate blue, and bright yellow, and fiery red and imperial purple hues—these elements all combined, and viewed through the marvelous lens of the spotless upper air, present a picture impressive and inspiring beyond words. Each of our great mountain-systems, the Alleghany, the Rocky Mountains, the Wasatch, the ranges of the great inland basin, the Sierra Nevada, and the Coast Range, has its peculiar style; but it seems to me, after seeing them all, that the Wasatch unites more completely than any other (unless it be the Cascades of Oregon) the softness, beauty, and luxuriance of the East, with the sublimity and solemn grandeur of the West. In these respects the cañon of the American Fork perhaps surpasses those of the two Cottonwoods; but as I have not yet had the pleasure of a personal acquaintance with it, I cannot speak with so much enthusiasm concerning its scenery.

When it is added that Big Cottonwood cañon is not more than fourteen miles from Salt Lake City, and that even this short distance will soon be traversed by rail, so that the tourist can leave the cars almost at the very mouth of the cañon, it will be evident that this remarkable scenery is destined to become well known and loved by thousands of travelers. Inded, I do not doubt that Salt Lake City, so interesting on many accounts, will be a great resort of pleasure and beauty seekers henceforward.

At the mouth of Little Cottonwood Cañon are Colonel Buell's reduction-works, comprising two Piltz furnaces. The location is magnificent, affording a fine water-power, excellent dumping-grounds, &c. The furnaces were both idle during my visit, one being just ready to start and the other being choked with a huge salamander. Half way up the cañon I passed the works of Jones & Pardee—one-shaft furnace, half strangled with a salamander, but smelting bravely and persistently on to the last gasp. At the head of the cañon, about nine miles from the mouth, is the town of Central City.

Little Cottonwood district includes the valley of the Little Cottonwood, and extends to the summit of the mountains on each side. The lower part of the cañon is walled by granite remarkable for its uniformity in structure and grain, its large masses, and the ease with which it breaks into rectangular blocks for building. The stones for the foundation of the new Mormon temple at Salt Lake City are obtained here. Higher up the cañon the granite is overlaid by metamorphic sandstones, slates, and limestone strata, extending to and beyond the summit. The strata show in the most distinct and striking manner, whole mountains being cut through so as to give splendid natural sections. It is in the limestone rocks, at about the middle of the series, that the principal ed-posits of ore occur. The claims are chiefly upon the north side of the valley, on the slope of the mountain facing the south. The celebrated Emma claim is about half way up, and above it are the Savage, Montezuma, and Flagstaff, besides many others more or less opened and developed. The first legitimate location in the valley, according to Dr. O. H. Congar, was made by Mr. Silas Brain, in August, 1865. This and other locations were bought by Dr. Congar, in connection with the New York and Utah Prospecting and Mining Company. One thousand feet of each of the North Star, American Eagle, and Morning Star claims were sold by this company to Mr. Bruner, of Philadelphia. Attempts were early made to construct and work furnaces, and, after some unsuccessful trials, Dr Congar, in September, 1866, succeeded in producing about 3,000 pounds of silver-lead, worth about $300 per ton, in silver.

In 1869 Mr. J. B. Woodman sunk a shaft a short distance below the North Star claim, and followed indications of ore until he suddenly opened into the immense deposit now known as the Emma mine. At the present time the claims, shafts, tunnels, and open cuts upon the hill may be counted by hundreds, and several claims beside the Emma have produced large amounts of argentiferous ore.

The Emma mine is one of the most remarkable deposits of argentiferous ore ever opened. Without any well-marked outcroppings, there was nothing upon the surface to indicate the presence of such a mass of ore except a slight discoloration of the limestone and a few ferruginous streaks visible in the face of a cut made for starting the shaft. Some of the earliest locators in the cañon assert, however, that in the little ravines below this shaft large masses of galena, some weighing over 100 pounds, were found upon the surface and in the soil. After the discovery of the deposit, by means of the shaft, a tunnel was run in so as to intersect it in depth. This tunnel extends in a northwesterly direction, and is 365 feet long. It intersects the ore-mass where it was about 60 feet long and 40 feet wide, measured horizontally. From this level, called the first floor, ore has been mined above and below until an excavation or chamber has been formed, varying from 20 to 50 feet in width, and from 50 to 70 in length, and 77 in height above the tunnel-level, and 50 in depth below.

In August last a portion of the ore below the tunnel-level was still standing, but the mine had produced from 10,000 to 12,000 tons of ore, assaying from 100 to 216 ounces of silver per ton of 2,000 pounds, and from 30 to 66 per cent. of lead, averaging about 160 ounces of silver, and

II. Ex. 211——21

from 45 to 50 per cent. of lead. The total value of this ore, at the cash price paid for a large part of it in Liverpool, £36, or $175 in round numbers, was about $2,000,000.

This ore was extracted at comparatively little cost. Most of it was stoped from below upward, and was delivered by chutes into the cars upon the tramway laid in the tunnel. In general the ore was soft and easily excavated by picks and shovels, without the aid of gunpowder. It consisted chiefly of ferruginous and earthy-looking mixtures of carbonate and oxide of lead, oxide of iron and of antimony, mixed with nodules of galena. It appears to have resulted from the decomposition of argentiferous galena, and other sulphureted and antimonial minerals, containing silver. The ore may be said to be without gangue, and does not require hand-sorting or separating by mechanical means from worthless vein-stone. This ore was shoveled up and put into sacks for shipment without any other delay or expense. The larger part was shipped overland by railroad to New York, and thence by steamer to Liverpool.

The walls of the excavation are very irregular, but consist of a hard, white dolomitic limestone. The ore-mass appears to conform to the stratification, and to have a general northwesterly direction, dipping to the northeast. The extent of the ore-mass in the direction of its length had not been fully ascertained at some of the levels when I visited the mine in July, though in most of the floors it had all been taken out, and the form of the excavation may be taken as marking, in a general way, the limits of the main body. A peculiar brecciated mass of dolomitic limestone accompanies the ore, and may be regarded as vein-matter, for nodules of galena are found isolated in its midst as well as small patches of soft earthy ore disconnected with the main body. The limits of this ore-bearing breccia are not yet ascertained, and prospecting-drifts to the northwest along its course may reach other bodies of rich ore. Late reports from the mine (in November) state that such masses have been found at the end of drifts run in from the fourth floor. It is, of course, not to be expected that the extraordinarily rapid and cheap production of the past season should continue unchanged in the future. Such masses of ore will be found to vary and to be pinched in size, as is already evident in the mine. But where such enormous deposits occur the miner is justified in following for great distances in length and in depth the merest threads or seams of ore, which may lead to other heavy deposits. Up to the close of the season favorable for teaming and shipping the ore, the attention of the company had been chiefly directed at the mine to the extraction and shipment of ore. There was little time to give to prospecting ahead for future development. This part of the intelligent miner's duty has been neglected until recently.

The main ore-chamber is well timbered throughout with a framework of squared and mortised timbers set at regular intervals of 4 feet from center to center, in the same manner as practiced at the mines on the Comstock lode, Virginia City. The vertical space is divided into "floors" or levels of 6 feet 8 inches each. There are eleven floors above the tunnel and eight below.

The circumstances attending the sale of this mine in England do not require discussion here. I have elsewhere declared my opinion that the price obtained (£1,000,000) was not justified by the appearance of the mine as I saw it; but I must frankly confess that the very important discoveries reported since have greatly enhanced its value. If the danger of litigation has been successfully arrested, I do not doubt the company will produce a great deal of ore at a handsome profit—probably sufficient to justify their large capital.

The North Star is a claim of 1,200 feet, a short distance northwest of the Emma or Woodman shaft. It is opened by an incline shaft along a seam of ore marked by a strong ferruginous, gossan-like outcrop. Very little ore has been found in depth. It is one of the oldest locations on the hill, and some years ago a furnace was erected near by to smelt the ores, but it had no success. The claim is now worked under the superintendence of Mr. Bruner.

The Western Star claim is 800 feet in length, and located west of the North Star. About 37 tons of ore were taken from near the surface. The first lot of 10 tons not being well selected, failed to pay the expenses by about $15 per ton. The second yielded a net profit of about $33 per ton, and the last about $44 per ton. In April, 1871, the owners had about 75 tons of ore on the dump, assaying about $60 in silver and 40 per cent. of lead. The shaft at that time was 90 feet in depth.

The Monitor and Magnet claim is 2,400 feet long. Considerable quantities of ore have been sent away from this claim with good results. For a time it supplied Woohull's furnace.

The Black Prince location (1,700 feet) is about 200 feet south of the Monitor and Magnet. In April last there was a pit 15 feet deep upon this claim.

The Caledonia, about 800 feet west of the Emma claim, 1,600 feet, is opened by an incline shaft and a tunnel driven in about 300 feet and crossing "two well-defined veins" and several small leaders, "producing an excellent quality of ore."

The Cincinnati, at the eastern end of the North Star, had, in April, 1871, a tunnel of 200 feet in length driven to intersect the vein.

The South Star claim, 500 feet, had a shaft 100 feet deep, and showing some good ore for smelting. This is about 2,000 feet northwest of the Emma.

The Morning Star is about 200 feet west of the South Star, and has a shaft 175 feet deep. Has produced some good ore for shipment.

The Flagstaff is situated about 3,000 feet northwest of the Emma, and has 1,400 feet in the claim. Up to April, 1871, over 80 tons of ore had been shipped from this mine. The highest yield per carload of 10 tons was $120 for silver, per ton, and 58 per cent. of lead. In April there were estimated to be about 400 tons on the dump, assaying about $60 in silver and 45 per cent. of lead. In August the excavation had reached a depth of about 260 feet on the dip of the vein. There were two shafts connected below by drifts some 35 feet in length serving to "block out" some of the ore-ground and show its value. The vein follows the stratification of the limestone rocks above and below it and pitches into the mountain at an angle of about 45°. Like all the veins of that vicinity there is no out-crop of quartz, or any other evidence, along the surface, of the existence of a vein below. The ore is soft and earthy, and poorer in silver at the surface than it is in depth. The quantity of carbonate of lead increases toward the lower workings, and in general the ore is richest on the foot-wall. The vein varies in its thickness, but both of the walls appear well defined. The thickness of the ore will probably average 2½ to 3 feet. In quality the ore is excellent. It sold during the summer on the dump for $70 to $85 per ton. Mr. Webster, of the firm of Webster & Lewis, of New York, purchased 41 tons of the rich lead-bars run out from this ore, and found it soft and easy to work, containing only 1 or 2 per cent. of antimony, and from 104 to 110 ounces of silver, and 1½ ounces of gold per ton.

Messrs. Lewis, Johnson & Co., of Salt Lake City, (sampling-mills,) sampled and assayed 116 sacks of this ore, weighing 9,277 pounds, and found it to contain per ton at the rate of 0.70 ounce of gold, equal in value to $14.46, and of silver 57.57 ounces, worth $74.42, and 55 per cent. of lead. Moisture 3 per cent. A parcel of 10 tons was sold in San Francisco, in November, 1870, to Cross & Co., for James Lewis & Son, Liverpool, which assayed, on arrival, 73½ ounces of silver per 2,240 pounds, equal to 65½ ounces per 2,000, equal, in value, to $85 per 2,000 pounds, and 60⅞ per cent. of lead.

The following results were obtained by Major Meader in Salt Lake upon four separate lots, amounting in the aggregate to about 380 tons :

Moisture.	Per cent. of lead.	Ounces of sil- ver to ton of two thousand pounds.	Ounces of gold to ton of two thousand pounds.	Value of sil- ver.	Value of gold.	Total value of gold and sil- ver.
3 to 5 per cent........................	58. 50	57. 30	00. 58	$74 45	$11 98	$86 43
3 to 5 per cent........................	50. 82	00. 21	00. 82	77 85	16 94	94 79
3 to 5 per cent........................	53. 00	71. 00	00. 73	91 79	13 08	108 87
3 to 5 per cent........................	51. 00	53. 54	00. 73	71 81	14 88	86 69
Average value per ton........	....	....	....	....	....	93 09

Savage: This claim comprises 1,400 feet, and is located high up on the hill-side, about 1,500 feet above the Emma, and a few hundred feet east of the Flagstaff. It is opened to a depth of over 230 feet by a single inclined prospecting-shaft following the vein, and without any side drifts. The ore shows near the entrance of the incline as a rusty, gossan-like mass, or vein, cutting the beds of limestone vertically. A few feet below the surface, within the incline, the thickness of the vein overhead is about 3 feet. It pinches up at a point lower down, and toward the bottom of the incline opens out again to a vein from 2 to 3 feet wide of rich ore, yellowish and rusty in color, and in places streaked with green stains of copper. Quartz vein-stone is found at the bottom of the mine, and it is hoped that this will prove to be a continuous regular vein formation.

The ore is soft and earthy, much like that from the Emma and other claims. It is rich in silver and lead. The mineral, wulfenite, is found disseminated in small, thin crystals throughout the vein.

The Montezuma is about 90 feet west of the dump of the Savage. The vein is vertical, or nearly so, like the Savage, and extends apparently from 3 to 5 degrees west of north (magnetic.) The croppings are rusty and rather hard, but below the ore is softer and richer in silver and lead. The country-rock is a hard, black limestone. This vein, like the Savage, is opened by an incline to a depth of 240 feet. This incline follows the ore, and its direction is about north 40° west. The vein may be said to average, where opened, 2½ feet in thickness. Some 200 tons of ore had been shipped up to July, and about the same quantity remained upon the dump.

The ore from both the Savage and the Montezuma is sacked at the mouth of the incline, and then lowered down the side of the mountain upon a wooden tramway 1,285 feet long. The loaded car, in descending draws up the empty one, laden at times with water and supplies for the mine. The movement is controlled at the top by a friction-band upon the shaft of the drum around which the rope passes. At present all the ore is raised from the mines by a hand-windlass only, no hoisting-machinery having been erected.

A line for a tunnel has been surveyed, and 86 feet had been excavated up to August last. When completed it will be from 250 to 300 feet in length, and will serve as an outlet for the ore from both claims.

The Hiawatha is about 300 feet west of the Montezuma; claim 3,000 feet. Ore from this claim assays $100 in silver and 40 per cent. of lead. The Bell is about 700 feet west from the Montezuma; shaft about 150 feet deep; produces good ore. The Gopher claim, 1,000 feet, has a shaft 60 feet deep. The Lilliwah is a claim located about 100 feet southeast of the Gopher. The Revolution is about 1,000 feet east of the Montezuma; shaft 150 deep in April, 1871. The Stoker is 600 feet beyond the last; shaft, in April, 1871, 75 feet deep.

From this last claim it is only a short distance to the Rock Island and the Davenport, just over the divide, and inside of the line of Big Cottonwood district. South of these claims we find the Lavinia, the Grizzly, Pocahontas, Idaho, Lincoln, and Diamond. Still lower, and beyond the Emma tunnel, are the Chicago, a claim of 1,600 feet, the Ohio, and the Relief.

Several extensive tunnel enterprises are under way. One of them, by the Emma Hill Tunnel Company, has been run 350 feet into the solid rock of the mountain. This is located at a point about 300 feet west of and below the Emma shaft. About 1,000 feet farther east is the mouth of the Little Cottonwood tunnel, and in the other direction, and at about the same distance, the Utah tunnel has been started.

Professor B. Silliman, of New Haven, has made some interesting investigations to determine the composition of the ores occurring in the Wasatch Range, and more particularly of those in the Emma. With his permission, I insert here his remarks on the subject:

The ores of the mines thus far opened in the Wasatch Mountains are largely composed of species resulting from the oxidation of sulphides, especially galenite and antimonial galena, with some salts of zinc and copper, all containing silver, and rarely a little gold. Iron and manganese ochers occur in considerable quantity in some of them; but the process of oxidation has prevailed very extensively, so that the ochraceous character of the ores is the striking feature of most of the mines in this range.

The great chamber of the Emma mine, which is an ovoidal cavity measuring, so far as explored, about 110 feet vertical by about 80 by 110 feet[*] transverse, was found to be filled almost exclusively with epigene species, the product of oxidation of sulphides, and capable of removal without the aid of gunpowder for the most part. The study of this mass reveals the interesting fact that it is very largely composed of metallic oxides, with but comparatively small proportions of carbonates and sulphates. Fortunately I am able to present an analysis of an average sample of 82 tons (= 183,080 pounds) of first-class ore from the Emma mine, made by James P. Merry, of Swansea, April, 1871, which is as follows, viz:

Silica	40.90
Lead	34.14
Sulphur	2.37
Antimony	2.27
Copper	0.83
Zinc	2.92
Manganese	0.15
Iron	3.54
Silver	0.48
Alumina	0.35
Magnesia	0.25

[*] The horizontal section of this chamber is greatly overrated by Professor Silliman. The size given heretofore is the correct one, and actually measured.—R. W. R.

Lime 0.72
Carbonic acid 1.50

                                                                           90.42
Oxygen and water by difference 9.58

                                                                          100.00

The quantity of silver obtained from this lot of ore was 156 troy ounces to the gross ton of 2,240 pounds.

This analysis sheds important light on the chemical history of this remarkable metallic deposit, and will aid us in the study of the paragenesis of the derived species. It is pretty certain that all the heavy metals have existed originally as sulphides, and we may, therefore, state the analysis thus, allowing 8.52 sulphur to convert the heavy metals to this state:

Silica 40.90
Metallic sulphides 52.60
Al, .35; Mg, .25; Ca, .72; Mn₂; Mn, .20 .... 1.52

                                                                           95.02
Water, carbonic acid, and loss 4.95

This calculation assumes that the sulphides are as follows, viz:

Galenite 38.69
Stibnite 3.30
Bornite 1.03
Sphalerite, (blende) 3.62
Pyrite 5.42
Argentite 0.54

                                                                           52.60

This statement excludes the presence of any other gangue than silica, and considering that the ores exist in limestone, the almost total absence of lime in the composition of the average mass is certainly remarkable. The amount of silica found is noticeable, since quartz is not seen as such in this great ore-chamber, nor, so far as I could find, in other parts of the mine. The silica can have existed in chemical combination only in the most inconsiderable quantity, since the bases with which it could have combined are present to the extent of less than 1½ per cent.; nor do we find in the mine any noticeable quantity of kaolin or lithomarge, resulting from the decomposition of silicates, nor are there any feldspathic minerals. It is most probable that the silica existed in a state of minute subdivision diffused in the sulphides, as I have seen it in some of the unchanged silver-ores of Lion Hill, in the Oquirrh Range.

The absence of chlorine and of phosphoric acid in the analysis corresponds well with absence of the species *cerargyrite* and *pyromorphite*, of which no trace could be found by the most careful search among the contents of the mine. The miners speak of the "chlorides," and the unscientific observers have repeated the statement that silver-chloride is found in the Emma mine, but the ores indicated to me as such are chiefly antimonic ochers.*.

The general (perhaps total) absence of the phosphates of lead in the Wasatch and Oquirrh Mountains, so far as explored, is a striking peculiarity of the mineralogy of these ranges. On the other hand, the absence of chlorine in the mines of the two Cottonwoods and the American Fork is in striking contrast with the constant occurrence of cerargyrite (horn-silver) in the Oquirrh and also in the southern extension of the Wasatch. I have sought in vain for a trace of this species in the districts of the Wasatch just named, and the occurrence of pyromorphite is extremely doubtful.

Molybdic acid, however, exists pretty uniformly disseminated in the mines of the Wasatch in the form of *wulfenite.* Although it occurs in minute quantity, it is rarely absent, and may be regarded as a mineralogical characteristic of the districts of the two Cottonwoods and of the American Fork. For this reason a few particulars will be in place here.

Wulfenite is found associated with calamine, (smithsonite,) cerusite, malachite, az-

---

* There exists generally among the mining population of the central Territories of the United States a distinction between *horn-silver* and *chloride of silver*—an error arising, as I am persuaded, from supposing the ochraceous ores to be chlorides not so perfectly developed as to be sectile.

urite, and more rarely alone in little cavities in the ochraceous ores. In the Emma mine vugs or geodes are occasionally found lined with botryoidal, apple-green calamine, rarely crystallized, often brownish and sometimes colorless, but invariably associated with wulfenite. The calamine incloses and covers the crystals of wulfenite, which form a lining of considerable thickness. The wulfenite is in thin tabular crystals of a yellow color, resembling the Carinthian variety of this species. The crystals are very brilliant and perfect, but quite minute, rarely two or three millimeters in width, and not over one millimeter in thickness, often less. They are quite abundant in this association, no piece of the calamine which I have seen being without them. They sometimes, but rarely, penetrate through the globules of the calamine so as to show themselves on the upper surface of that species. But the calamine has obviously formed botryoidal masses around the wulfenite, a crystal of this species being often seen forming the nucleus of the calamine globules.

These facts are of interest in the paragenesis of these epigene species. The order of production has obviously been, first, the cerusite resting on ochraceous iron, manganese, and other metallic oxides; next, the wulfenite crystals were deposited upon and among the crystals of cerusite, and lastly came the calamine, crystalline at first, and, as it accumulated, becoming fibrous and amorphous, completely inclosing and capping the other species.

Wulfenite occurs also in this mine, as likewise in the Flagstaff, the Savage, and Robert Emmet, without the calamine, but never, as far as observed, without cerusite and other carbonates. In the Savage, masses of cerusite, with various oxides, are interpenetrated by the tabular crystals of wulfenite.

Although wulfenite forms a very minute factor of the entire ore-mass in these mines, by the law of mineral association it may be considered as the characteristic species of the ores of these districts, occurring in the maguesian limestones. So far as I am informed or have observed, wulfenite has not been hitherto found in any of the other mining districts of Utah; but by the same law it may be reasonably looked for whenever deposits of epigene minerals are explored in the same geological and mineralogical relations in the Wasatch range of mountains.

The oxidizing and desulphurizing agency which has acted upon the great ore-mass of the Emma mine, whatever it was, has performed its work with remarkable thoroughness. A careful study of its action discloses some other facts of interest in the paragenesis of species. From the appearance of numerous large blocks of ore, forming solid boulders in the general mass, a concentric arrangement is easily recognized. On breaking these masses across, the fresh fractures disclose a dark center which consists almost entirely of decomposed sulphides, composed chiefly of cerusite blackened by argentite and metallic silver in a pulverulent form. This dark center, chiefly of cerusite, is often pseudomorph of galenite in its fracture. Next is usually a zone of yellowish and orange-yellow antimonial ocher, cervantite, often quite pulverulent, at times only staining the cerusite; then follows a narrow zone of green and blue copper-salts, malachite, azurite, cupreous anglesite, with, rarely, wulfenite; then follows cerusite, sometimes stained with antimony ocher, and not unfrequently associated with wulfenite; outside all are the iron and manganese ochers. This concentric arrangement I have observed in a great number of cases; and the above order of species, while not invariable, is believed to reflect accurately the general arrangement. Well-crystallized species, as mineralogical specimens, are rare in this great mass; but the following may be recognized as its chief components:

Galenite, sphalerite, pyrite, jamesonite (?), argentite, *stephanite*, boulangerite (?), *antimonial galenite, cervantite*, mimetite (?), limonite, wad, kaolin, lithomarge, *cerusite*, anglesite, linarite, *wulfenite, azurite, malachite, smithsonite*. Those most abundant or best crystallized are in italics. This list can no doubt be extended as opportunity occurs for the more careful study of the ores, the great mass of which, amounting to many thousand tons, have gone into commerce without passing under any mineralogical eye.

American Fork.—This district adjoins Little Cottonwood on the southeast. Among the principal claims are the Miller, Pittsburgh, Wyoming, Kentuck, Alpine, Silver Glance, Waterloo, Emeline, Conqueror, Champion, Chelsea, Castor, Terrible, Mary Ellen, Live Yankee and Silver Tie.

The Silver Glance is opened by two cuts, each about 25 feet long, a tunnel 120 feet long, and an incline shaft 117 feet long, with a drift from it 35 feet long. There is a large pile of ore on the dump.

The Miller is a very large deposit of ore, on which much work has been done. It is opened by two tunnels and an incline, the largest tunnel being 60 feet in length, following a solid body of ore all the way from the surface to the heading, where it showed a vein 19 feet in thickness in June. At that time a portion of the mine had caved in, owing to defective timbering, but even when seeing it under this disadvantage the body of ore exposed to view appeared very large. Two miners, with pick and shovel, could easily keep two wheelbarrows busy running out ore to the dump, where five men were employed sacking it up, after which it was slid down a ravine to the gulch below. From twelve to fourteen tons per day was the usual quantity taken out when worked in this way. When we consider that this ore will run from $90 to $180 per ton, and that the quantity which can be taken out when the mine is properly opened for working will be only limited by the means of transportation and reduction, we can hardly comprehend the true value of this mineral deposit. The surrounding mountains are thickly covered with a heavy growth of pine, sufficient to furnish fuel in abundance for many years to come.

The Wyoming lode is situated about 400 feet above the Miller, on the very crest of the mountain, and is supposed by many to be the same vein. In June a short drift had been run in on the vein from the east face of the hill, at the end of which a shaft had been sunk to a depth of 20 feet, disclosing a vein of ore 8 feet in width. It shows nearly the same character as the Miller, being carbonate mixed with galena, assaying at the rate of $122 per ton.

The Kentuck lode is nearly a mile further north on the same mountain. A shaft has been sunk to the depth of 25 feet from the surface, the ore-vein being about 4 feet wide in the bottom. The mineral is coarse galena and honey-combed quartz.

The Alpine mine, within a few hundred feet of the Miller, has developed a large body of ore, thought to be equal to that of the Miller. The company had about 100 tons on the dump in August. The Alpine has been pierced and laid open by a tunnel run into the side of the mountain something over 100 feet, which is well timbered, and the mine is in good condition to deliver much ore.

The Pittsburgh mine showed, in November, in an incline 43 feet deep and a cut of 12 feet, a large body of solid lead-ore, which was reported to assay $60 per ton in silver, and 69 per cent. of lead. The quantity in sight was estimated at 5,000 tons, which, it was expected, could be brought to the surface, as soon as the new tunnel should be completed, at the small cost of $1 per ton.

The Champion is opened by a shaft 60 feet deep, and by two tunnels each 100 feet deep. Galena-ores.

The Mary Ellen claim shows a mixture of galena, iron pyrites, and copper pyrites, and is said to contain gold.

The Sultana smelting-works, consisting of three furnaces, were built to work the ores of the Miller mine. They are at the junction of Miller and the main cañon. Other works are to be erected at the mouth of the cañon, and refining-works at Lehi. The foundation for a quartz-mill has been laid.

The district has two recording offices, and over 500 claims had been recorded up to November, 1871.

Lucien district.—This district is about one hundred and twenty-five miles west of Ogden, nearly on the line between Utah and Nevada, and six miles south of the Central Pacific Railroad. It was reported, in October last, that Messrs. Buel and Bateman had purchased the Zecoma claim in this district, a claim which carries galena, said to assay 65 ounces silver per ton, and a high percentage of lead, and that they would speedily erect a furnace for smelting the galena-ores which it yields. About 100 tons of galena are reported to have been previously shipped West. There are also several copper-claims in this district, from which a few tons have been shipped to San Francisco.

Saint George district.—Up to April, 1871, no claims had been opened and worked in this district, though many locations had been made.

Star district, Beaver County.—Up to April, 1871, the total amount of ore shipped from this district would not exceed 100 tons, and this from various locations. One lot worked at Ely & Raymond's Mill yielded $197 per ton in silver. This was from the Taylor mine. One shipment to San Francisco netted $250 per ton, and one lot sampled in Salt Lake City yielded $288 per ton.

Hamilton district.—A new district, southwest of Camp Floyd, was organized in September last. The quartz-croppings are reported to contain galena and some gold.

Parley's Park, a district about thirty miles east of Salt Lake City. The principal mine here is the Piñon, in which a large body of galena and carbonate has been struck. Assays are reported to yield from 30 to 250 ounces of silver per ton. The owners, Lowe & Co., are reported to have contracted to deliver 20 tons a day to the new smelting-works to be erected at Ogden.

List of furnaces and mills in Utah in the latter part of the summer, 1871.

Ophir, furnaces, 5; mills	2	Salt Lake, furnaces............	4
Stockton, furnaces..............	2	Bingham, furnaces	2
Tintic, furnaces................	2	American Fork, furnaces.......	2
Cottonwood, furnaces..........	3	Corinne, furnaces..............	1

For the following estimate of working expenses of a 15-stamp mill, with Brückner roasting-cylinders, in Utah, I am indebted to Mr. L. Huepeden, the agent of Mr. Brückner:

Estimate for 15-stamp mill in Utah, employing two Brückner-cylinder furnaces, 13 feet by 6¼, (in clear.) Capacity of 20 to 30 tons.

[All these figures are high.]

Labor:		
Engine, 4 cords, at $5, ($4)......................................		$20 00
Wood:		
Furnaces, 4 cords, at $5, ($4).....:.............................		20 00
Two engineers, at $5 and $4...............................	$9 00	
Two crushers....................................	8 00	
Foreman, (night shift)............................	5 00	
Head amalgamator.............................	5 00	
Two amalgamators, at $4...............................	8 00	
Two roasters, at $5...................................	10 00	
Two roustabouts, at $3 50............................	7 00	
Assayer, $6; smelter, $5................................	11 00	
Clerk, $4; superintendent, $20.......................	24 00	
		87 00
Salt, 8 per cent. on 20 tons, at ¼ cent........................		16 00
Loss of iron...		15 00
Loss of quicksilver, 50 pounds...............................		40 00
Oil, grease, &c...		3 00
Charcoal and chemicals, (assays)...........................		5 00
Wear and tear, 20 per cent., and interest 10 per cent., ($30,000)...............		30 00
Office expenses..		4 00
Total on 20 tons, at $12 per ton...............................		240 00

Tin.—In the fall an effort was made to create an excitement on account of alleged discoveries of vast deposits of tin in the vicinity of

Ogden, and many poor miners were actually allured into spending considerable time and money in that district. But the "tin mines" were speedily disposed of by scientific men as a gigantic fraud on the public, and in time, I believe, to prevent serious losses by inexperienced men. The following letter of Dr. F. A. Genth, of Philadelphia, to the editor of the United States Railroad and Mining Journal, shows the nature of the "tin mines" and "ores:"

DEAR SIR: In the Philadelphia Ledger of the 25th ult., and several other newspapers, appeared the following from Salt Lake City, under date of October 23: "The *tin* mines of Ogden are enjoying increased attention. The governor and a large party went to-day to visit them. An experienced miner and expert from Cornwall, England, reports them wonderful, and that vast quantities of ore in sight at the Star of the West, the pioneer discovery, will average 20 per cent. of fine tin. He says these discoveries are destined to work a revolution in the tin trade of the world. New discoveries are being made daily, and another claim has been bonded for $200,000."

These are certainly wonderful discoveries, and, judging from the character of the "ore," there can be no doubt that it exists in vast quantities. It was my good fortune, already over one month ago, to receive some of that which had been sent to Washington. About fourteen days ago I received a second lot for examination, and was also favored with a visit of one of the owners, who brought larger lumps, and showed some bars of tin which had been melted from the ore, and also some copper which had been tinned with the product of such smelting operations. My specimens are *undoubtedly authentic.* They consist of a rock, composed of white feldspar, (probably albite,) hornblende, and a small quantity of quartz. The albite and hornblende are present in variable quantities, sometimes the one, sometimes the other predominating. Ocular inspection did not show a trace of *tin;* concentration of the heavier portions, by grinding the rock and washing off the lighter, and the chemical examination of the heaviest, did also show not a trace of *tin.* A very careful analysis and a crucible assay showed likewise total absence of *tin.* The specimens which I received are, therefore, *no tin-ore* at all, but syenite—a granite in which mica is replaced by hornblende. As undoubtedly strong efforts will be made in the East to dispose of these "*valuable tin-mines,*" I consider it my duty to make this plain statement of my experiments, and hope that all the newspapers which have helped to circulate the report about these *great tin discoveries* will now correct their error.

Yours, very truly,

F. A. GENTH.

CHAPTER VI

ARIZONA.

In spite of the unsafe condition of this Territory during the year, on account of Indian depredations, material progress has been made in the discovery and partial development of new mining districts, while in the older districts quartz mining has been carried on in the same mines as were worked last year.

The principal new discoveries were made in the Bradshaw, the Hualpai or Sacramento, and in three new districts which were organized in the western foot-hills of the Pinal Mountains. The latter have been named the Holsted, Pioneer, and Nevada districts. There were also valuable discoveries made in the immediate vicinity of Prescott, these new silver mines being located only four to five miles southeast of the town, in the heavy pine timber of the Sierra Prieta.

Mr. John Wasson, the surveyor-general of Arizona Territory, has kindly furnished me with a summary of mining operations in the Territory during 1871, which follows in full:

UNITED STATES SURVEYOR-GENERAL'S OFFICE,
Tucson, Arizona Territory, November 28, 1871.

SIR: In accordance with a request made by Mr. A. Eilers, of date October 23, and which I received November 14, I at once addressed letters of inquiry to different parties throughout the Territory, whereby I hoped to get more exact information regarding the progress of mining during the present year than in any other manner, but as yet only a few responses have reached me, and but one of them of much value.

The annexed statement, marked A, is furnished me by O. H. Case, deputy surveyor of mineral lands in Yavapai County, and is as nearly correct as probably any one could have prepared on short notice.

In consequence of a scarcity of water, mining operations in that county during the present year have been almost wholly confined to developments of old and the discovery of new mines, and in these respects hopeful progress has been made. In many letters received from well-informed men, resident in Yavapai County, all speak more encouragingly regarding the future of mining in that region than was common one year ago. The unusual drought of the past two years has rendered placer mining generally impossible; yet much of the country in the vicinity of Prescott is known to contain placer-gold in fine paying quantities, which can easily be made available with a supply of water for washing.

The discovery of very rich mines in Bradshaw mining district last January, diverted the attention of many miners from moderately-paying mines. While numerous discoveries or locations have been made in that locality, the Tiger lode is the chief one. A shaft is down on it 100 feet, and the drifts and cross-cuts run show the width of the vein to be 31 feet. The development is going forward, and San Francisco operators are spending their money on it. Early in this year a renewal of prospecting was made in Hualpai or Sacramento district, which lies in Mojave County, and near the Colorado River. The greatest activity and interest prevail there at this time. Over five hundred miners have recently gone there, and more are going and taking means and experience with them. All doubts of the richness of these mines and the abundance of ore are removed. Deputy mineral-surveyor, O. H. Case, writes me, under date of the 20th instant, that the owners of the Clinton, Niles, and Jones lodes had three tons of silver-ore recently worked in San Francisco, which yielded $8,400; also that Clark & Co. had worked at the same place four tons of Keystone ore, which had given in silver $10,720. Nothing is said of any yield in gold. From various letters written by men of experience, and who are expending their own time and means in those mines, I am led to believe that the recent discoveries in Hualpai district are of vast value, and will, as one writer says, make that section one of the best for mining on the Pacific coast.

In the month of July a party of nearly three hundred men was formed for exploring the Pinal Mountains, which, I believe, wholly lie in the new county of Maricopa. This party was commanded by Governor Safford, and guided by a man named Miner.

Owing to the latter's misrepresentations of his knowledge of the country, the time and supplies of the company were exhausted in a fruitless search for pretended discoveries made in 1862; but the ramblings of the expedition undoubtedly led to the late discovery of quartz-veins of much promise in the Pinal Mountains. Three districts were formed early in this month, viz, Holsted, Pioneer, and Nevada. The Silver Queen lode, in the Pioneer district, is reported best, so far as developed. Twenty-four hundred pounds of the ore are now on the way to San Francisco for accurate test. Assays running from hundreds into thousands per ton are reported. Much interest is manifested by men of all occupations, and nearly all are trying to get some interest at the expense of previous earnings. The ledges are but a few miles from the Gila River, and wood and water are reported quite convenient and abundant.

Other discoveries which promise well have been made in many parts of the Territory, but mostly where it is unsafe for small parties to go, or stay when there. Instances of this kind are within my knowledge, and citizens of Tucson are interested. Several locations, a few miles southward, the owners have endeavored to prepare for patent, but the Apache Indians have frequently stolen or destroyed their tools and improvements; and laborers cannot be induced to carry on work at any price. With few exceptions, locators are intensely anxious to positively secure their mines, because of the prospect of early railway communication. But danger prevails everywhere except near the Lower Gila and Colorado Rivers. Every act of locators exhibits honest faith in the great value of the quartz-veins of Arizona.

I addressed a letter to a gentleman at Wickenburgh for statistics of the Vulture mine, the largest producing one in the Territory, but no reply is yet received. In a friendly letter from the superintendent of the mine, of date November 14, he says it "still holds its own, and the prospect of finding water in the mine for milling purposes is very flattering. I am sinking a shaft inside the mine, and am down with it 310 feet from the surface. I struck water at 295 feet, and it has steadily increased as the shaft deepens. I think 75 feet more sinking will find water enough for a 20-stamp mill. * * I have an unusual quantity of good milling ore in sight."

Should water for milling purposes be found in the Vulture, I think Mr. Eilers will assure you that it will then be one of the most valuable mining properties in America. As it is, water for the men, mining, cooking, &c., has to be hauled fifteen miles, and the ore a like distance for reduction.

Extensive work has been reported at various times as going on in Castle Dome district, situated about sixty miles above Arizona City, and near the navigable waters of the Colorado. Many shipments of ore have been made to San Francisco for sale and reduction, and I am informed that satisfactory returns have generally been secured. A patent has recently been applied for of the Flora Temple lode, in that locality, and much interest is manifested in securing title to mining property there.

A large vein of coal, and of valuable quality, has been discovered near Camp Apache, in Yavapai County. The coal has been used by smiths with entire satisfaction. Another vein of great size, and reported good quality, was located some weeks ago about sixty or eighty miles east and north of Tucson, in Pima County.

Generally speaking, nearly every prospecting expedition which makes a determined effort finds new mines of the precious metals or other minerals, which have every indication of permanency and value.

Now, they cannot be occupied and developed, except where many are found near each other. It is certain death for half a dozen or a dozen men to attempt to carry on work beyond the direct influence of military camps or large settlements.

Very respectfully, your obedient servant,

JOHN WASSON,
Surveyor-General.

Exhibit of producing mines in Bradshaw, Big Bug, Hassyampa, and Lynx Creek mining districts, Yavapai County, Arizona, on the 1st day of January, 1872. Reported by O. H. Case, United States deputy surveyor.

Name.	Owner.	Character.	Course.	Dip.	Dimensions of claim.	Country-rock.	Vein-matter.	Ore.	Average value per ton.	Mills.	Product for the year ending July 1, 1869.
BRADSHAW MINING DISTRICT.											
Tiger	Tiger Company	Vein	N.E. & S.W.	S.E.	1,000 ft.	Granite	Quartz	Galena and silver ores.	$41 00	None	14 tons worked in San Francisco, $10,374.
Del Pasco	Jackson & Co	Vein	N.E. & S.W.	S.E.	1,000 ft.	Porphyry and slate.	Quartz	Gold	70 00	6 stamps	$7,428.
BIG BUG MINING DISTRICT.											
Big Bug	Gray & Hitchcock	Vein	N.E. & S.W.	Vertical	2 ft. wide	Greenstone and slate.	Brown quartz	Gold	21 50	10 stamps	
HASSYAMPA MINING DISTRICT.											
Sterling	Sterling Company	Vein	N.E. & S.W.	S.E.	8 ft. wide	Greenstone and slate.	Quartz	Gold	100 00	10 stamps	Pursued by Indians, May, 1871.
Davis	C. C. Bean & Co	Chlorides and sulphides.	N.E. & S.W.	S.E.	6 ft. wide	Quartzitic slate.	Quartz	Silver	54 50		3½ tons worked. $191 10.
Benjamin	Noyes & Curtis	Chloride.	N. & S		20 in. wide	Quartzitic slate.	Quartz	Silver	1,000 00		
LYNX CREEK DISTRICT.											
Vernon	C. Y. Shelton	Gold	N.E. & S.W.	N.W.	18 in. wide	Granite	Quartz	Gold	150 00	Arrastra	$5,000.
Pointer	William Pointer	Gold	N.E. & S.W.		4 ft. 9 in. wide	Greenstone.	Quartz	Gold	40 00	Arrastra	$1,000.

The Tiger lode and the new developments in the Hualpai district have formed the prominent features of the mining news from Arizona during the year. Of the former, the most extravagant descriptions have reached me. The vein was discovered towards the end of last year, but the value of the ore was first ascertained during the first quarter of 1871. Several assays, which ran very high, created an immense excitement. A town was laid out in the immediate vicinity, which filled rapidly with people, and actual prospecting of the ledge was energetically prosecuted. The vein is reported to have been traced for over six miles, and for a large part of this distance claims have been staked out, most of which have, by this time, been more or less tested, by shafts and drifts. The shaft of the Discovery Company (Moreland & Co.) is the deepest, and they have taken out the most ore, some of which has been sent to San Francisco, and yielded nearly $750 per ton. The assays of samples from this vein, which have been published, run all high, and many of them exceed $2,000 per ton, while I have seen none which gave a yield below $58. The following are a few, all of them having been made in Virginia City, Nevada: $1,008.75; $736.50; $22.50; $122; $1,742.63; $1,953.40; $2,547.90; $58.13; mixture of ores, $1,318.69. Also, $1,804.69; $1,586.31; $972.22; $1,305.40; $2,990.69; $8,028.47. It is reported that San Francisco capital has come to the aid of the Discovery Company, and that thus it has been enabled to continue in its course of sinking and drifting, (one drift is said to be in over 80 feet on the vein,) while nearly all of the other claims on the lode have been lying comparatively idle.

The silver veins discovered near Prescott, during the fall, are reported as narrow, but exceedingly rich. In regard to yield per ton, it is asserted that the Bismarck, Cornucopia, and Homestake ores fully equal those from the Tiger. I am not informed as to the erection of reduction-works to treat the ores of the silver mines.

The Del Pasco, which was mentioned favorably in last year's report, has been further sunk upon during the year, and the 5-stamp mill, erected near by, has been at work a small part of the time. In the fall, the supply of water, which is at no time very abundant, became so scanty that for two months the mill could only run three and four hours per day. In an aggregate run of one hundred and ninety-two hours during that time, 42 tons of ore were crushed, which yielded $2,500, or an average of $59.52 per ton, considerably less than the ore worked before that time used to contain. In November the mill had to be shut down, as the water gave out completely. Meanwhile some water had been struck in the mine, and, in sinking deeper, this increased so much, that about the end of December a sufficient supply was reported to run the mill fourteen hours per day.

In regard to the Big Bug Mill and mines, I have not received the information which I have requested of the superintendent. But it appears from other information that the company has not worked regularly during the year.

In the famous Vulture, at Wickenburgh, an important change has taken place. This is the striking of water at a depth of about 300 feet. The sinking goes on energetically, with a hope of a sufficient increase in the supply of water to at least obviate the necessity of hauling the water for blasting purposes, and for the use of the men and animals, a distance of fourteen miles from the Hassyampa River. Even this would be a great saving, and it is not impossible that, in the course of sinking, enough water should be found for crushing purposes. The work at the mine and mill has been constantly going on during the year, with the exception only of the few days needed for repairs at the mill. The ore in the lower level is reported much richer than usual, and many tons of it have been worked. From information received, I conclude that the yield of this mine in gold during the year has been somewhat larger than last year, but official information, for which I have applied, has not yet come to hand.

There are still various schemes spoken of for the purpose of effecting a reduction in the cost of handling and crushing the ore. The plan of conducting water from the Hassyampa River to the mine appears to have been dropped, and, in conformity with the prevailing idea of the times in regard to cheap transportation, a narrow-gauge railroad from the mines to a point on the Gila River, and a transfer of the mill from Wickenburgh, to that point is now spoken of. I am not familiar with the distance from the mines to the nearest point on the Gila, nor with the peculiarities of the route, but judging of what can be seen of the country to the west from the road between Wickenburgh and Phenix, I should think the undertaking to be not only feasible, but even easy of accomplishment, at a moderate cost. It is highly probable that the road can be laid over an almost level mesa for nearly the entire distance, and there will certainly be no mountains to cross.

An important discovery has been announced as having taken place in the Weaver district. This is the Sexton lode, a vein which is considered almost as valuable as the Vulture. No thorough test of the vein has, however, as yet been made, and the value of the ledge is so far only deduced from its size on the surface, and a test made of several tons at the Vulture Mill.

The placer-mining interest in Central Arizona has, according to all accounts, suffered severely from the protracted drought, and no more gold has been extracted in this manner than during the preceding year, which was also remarkable for an extraordinary scarcity of water.

Hualpai or *Sacramento district*, which, it will be remembered, was favorably spoken of in last year's report as a district in which mines might be profitably worked at the present time, has greatly gained in importance by new discoveries and the developments made during the year. While in the fall of 1870 there were not a dozen white men in the district, there are now nearly five hundred men reported to be at work there. A great many new discoveries have, of course, been made, the Cerbat Range having been prospected north and south of the veins which are mentioned in last year's report. Most of the later discoveries carry argentiferous-lead ores, like those described in my last report; but there have also been located several veins which carry amalgamating-silver ores, and at least one lode, the Vanderbilt, which carries a heavy percentage of gold. The work done during the year has principally been prospecting, and on many claims shafts of from 15 to 60 feet have been sunk, and selected lots of ore have been shipped to San Francisco for experiment. Several of these shipments have given very flattering returns.

Mineral Park and Parkerville are two new settlements in the district, in which a large number of veins have been discovered. In fact, the whole Cerbat Range seems to be filled with veins from its southern to its northern extremity. A mill is reported in the course of erection, and Mr. W. J. Fee, an intelligent gentleman, who was the first to enter the district after the abandonment of the old Sacramento district in 1866, is about erecting a smelting-furnace. From personal reports of my deputy, Mr. Eilers, concerning the ores of the Cerbat Range, it is evident to me that the smelting process is the one for which they are best adapted.

There is no doubt that another year will witness important results of mining in this district.

Concerning the production of gold and silver in the Hualpai mining district, Mojave County, Arizona, for the year ending December 31, 1871, C. A. Luke reports as follows: There is one steam 5-stamp mill in the district, located at Silver Park, and owned by Meacham and Hardy. This mill is not fully completed, but will be soon. The ores are at present shipped to San Francisco, and yield from $200 to $800 per ton. Over 100 tons have been shipped. One shipment of 20 tons yielded $800 per ton.

Exhibit of producing mines in Castle Dome mining district, Yuma County, Arizona, on the 1st day of January, 1872. Reported by George Tyng.

Name.	Owner.	Character.	Coarse.	Dip.	Dimensions of claim.	Country-rock.	Vein-matter.
Buckeye	Butterfield Bros	Lode.	N. W×S. E	87	800 feet...	Thick-coarse granite.	Talc and clay.
Flora Temple	Polhamus & Gunther	.do ..	....do	60	2,000 feet.		Don't know what it is.
Castle Dome	Barney & Tyng.	.do ..	....do	87	400×15...		Talc and clay.
Extension of Castle Dome	......do	.do ..	....do	87	2,000×15.		Do.
Don Santiago	......do	.do ..	....do	90	1,600×4..		Fluor-spar.

Name.	Owner.	Ore.	Average value.	Mills.	Product for the year ending July 1, 1871.
Buckeye	Butterfield Bros	Argentiferous, galena, and lead ores.	$70 00	Ore shipped to San Francisco.	250 tons.
Flora Temple	Polhamus & Gunther		60 00		400 tons.
Castle Dome	Barney & Tyng		77 00		600 tons.
Extension of Castle Dome	......do		77 00		
Don Santiago	......do		60 00		30 tons.

Many other claims not worked. Deepest shaft in the district is on the Castle Dome mine—160 feet—showing much ore. Mexican laborers. Hoisting all done by hand-windlass. Whims, &c., to be erected from proceeds of ore now *en route* to San Francisco and Truckee. Freights from Colorado River near mines to San Francisco, $12.50, coin, per ton. Ore assays (in bulk) 50 per cent. to 65 per cent. lead, 7 ounces to 182 ounces silver per ton. Galena-ores average 60 per cent. lead, 35 ounces silver, and sell for $60 coin, about.

The Castle Dome district has been organized since 1863; worked only by prospectors without capital until October, 1870. Not $300 invested January 1, 1872, in buildings or other permanent improvements. Parties now working are cautious, and work no lodes that do not pay their way from the top down.

In the Castle Dome district and about Gila City mining has been carried on most of the time during the year. From the former locality the lead-ores are shipped to San Francisco, and at the latter the capacity of the stamp-mill for working the gold-ores is reported to have been increased. It is also reported that a stamp-mill is being erected by Mr. Booger, near La Paz, to work the ores from the Constantia, a vein which is described in my last report.

On the whole, the immediate prospects for mining in Arizona are more favorable at the close of 1871 than the year before.

CHAPTER VII.

NEW MEXI·CO.

My limited means have not enabled me to visit this Territory during the year, or even to keep a paid agent there. For the information I have received from that quarter I am principally indebted to persons residing there, who take sufficient interest in the mining resources of the country to undertake the trouble of communicating information gratuitously. Among these, thanks are principally due to Mr. R. B. Willison, surveyor-general of the Territory, Dr. Hilgert, Messrs. M. Bloomfield, Eugene Goulding, and A. H. Morehead.

In the *Moreno mines* the placers have been worked with moderate success during most of the year, as will appear from the following statement. The Aztec mine has also been worked, and produced bullion. The Montezuma has only been prospected, and the mill has remained idle:

Exhibit of producing mines in Moreno mining district, Colfax County, New Mexico, on January 1, 1872. Reported by M. Bloomfield.

Name.	Owner.	Character.	Course.	Dip.	Dimensions of claim.	Country-rock.
Willow Creek......	Six companies.......	Placer ..	Gravel	......	2 miles by 300 feet.	....
Moreno Creek......	Three companies	...do ...	...do	......	½ mile by 300 foot.	River diggings.
Grouse Gulch	Seven companies ...	...do ...	...do	......	2 miles by 300 feet.	....
Humbug Gulch.....	Five companies......	...do ...	...do	......	1½ miles by 300 feet.	....
Last Chance	One company........	..do ...	...do	......	2,000 feet by 300 feet.	....
New Orleans Flat...	One company........	...do ...	...do	......	1,500 feet by 300 feet.	....
Sundry other claims.	Sundry persons	...do ...	...do	......	....	....
Aztec Mine	Aztec Mining Company.	Lode...	N. W. and S. E.	......	3,000 feet...	Slate.
Montezuma	Maxwell Land-grant and Railway Company.	...do ...	...do	......	3,000 feet...	Granite.
Chester	Graham, Dimick & Co.	...do ...	...do	......	3,000 feet...	...do ....

Name.	Owner.	Vein-matter.	Ore.	Average value.	Mills.	Product for the year ending Jan. 1, 1872.
Willow Creek......	Six companies	....	....	Unknown	....	$40,000
Moreno Creek......	Three companies	....	....	...do	....	40,000
Grouse Gulch	Seven companies ...	....	....	...do	....	60,000
Humbug Gulch	Five companies......	....	....	...do	....	25,000
Last Chance	One company	....	....	...do	....	10,000
New Orleans Flat...	One company	....	....	...do	....	5,000
Sundry other claims.	Sundry persons	....	....	...do	....	10,000
Aztec Mine	Aztec Mining Company.	Quartz ...	....	...do	15 stamps..	Unknown.
Montezuma	Maxwell Land-grant and Railway Company.	...do	....	...do	30 stamps.	Idle the whole year.
Chester	Graham, Dimick & Co.	....do	....	...do	25 stamps.	Unknown.

REMARKS.—Aztec mine idle since October, on account of water in mine. Montezuma Mill idle; but twenty five or thirty miners at work the whole year developing a lead of 30 inches of quartz, which will probably pay from $15 to $20 per ton. Chester Mill ran several weeks in May and June, but failed to pay expenses.

A company had been formed in January, 1872, to work the Moreno Creek by machinery, and it is thought that they will succeed beyond doubt. There are some five or six miles of good mining ground, which have been secured by this company, and which they will divide to run about three claims to the mile. The plan is to employ steam-shovels to dig and hoist the dirt; the dirt to be washed on the surface, or rather in sluices, some 8 or 10 feet above the surface of the ground. This does away with the inconvenience occasioned by the slight grade of this creek, which is only about one in one hundred. The ground has heretofore been worked entirely by hand and wheelbarrows, a process which is expensive and slow.

This creek averages about 50 cents or 15 grains of gold to the cubic yard. The company calculate they can work the ground for 25 cents the cubic yard, and work 400 yards per day. Their enterprise seems entirely practicable, and may be the means of increasing the production of gold by $300,000 per year. The general prospects of the district are good, owing to the heavy snows of last winter.

From *Silver City* I am informed that the district still suffers on account of a lack of capital invested in the mines. Still it is reported that about $90,000 worth of silver slabs have been shipped from there since the mines were discovered. Most of this was smelted out by Mexicans in their primitive way. No mills or smelting-works are as yet erected. In July it was reported that about sixty miners were at work here taking out ore from the various silver-lodes. The Sophia lode, owned by the Spring Hill Mining Company, had a shaft 25 feet deep, and showing a splendid vein of rich ore. The vein is from 4½ to 5 feet wide. The Reinhart lode, owned by Mr. William Kronig and others, was 8 feet wide at a depth of 16 feet, and very rich ore. The Colfax lode showed 3 feet of good ore at 10 feet in depth. The Great Eastern lode, owned by the Eureka Mining Company, showed a splendid vein of rich ore 4 feet wide. The Abbey lode, and a great many others, showed also good ore.

There has been a great drawback on these mines by reason of the miners being compelled to lie idle on account of not even having proper tools to work with. As a general thing they came to the district without means. In January, 1872, there were over three hundred lodes located, some of which have shafts on them of the depth of 60 feet.

It is asserted that none of the lodes, by practical tests, produce less than $50 to the ton. The area covered by these silver-lodes is about thirty miles square, and it is believed that these mines will, at some day, whenever capital comes to the country, prove far richer than is now supposed.

From *Pinos Altos* I have no reliable data as yet, beyond a general estimate of the product, which has been small.

There has been some activity again in the Organ Mountains during the year, but lack of capital is here also in the way of a speedy development of the mines.

Exhibit of producing mines in Organ mining district, Doña Ana County, New Mexico, on January 1, 1872. Reported by A. H. Morehead.

Name.	Owner.	Character.	Course.	Dip.	Dimensions of claim.	Country-rock.
Aztec............	J. Freudenthal, H. Lesinsky, A. H. Morehead, F. Blake, et al.	Lode....	N. E. and S. W.	E.	3,000	Granite casing.
Stephenson	H. Lesinsky, W. F. Shedd, et al.	...do	....do	E.	3,000	....do ....
Cuebas	J. Freudenthal, G. E. Blake, et al.	...do	....do	E.	· 3,000	....do ....
Bennett	N. V. Bennett, Shedd, et al.	...do	....do	E.	3,000	....do ....
El Quibedo	W. H. Graham, James Foley, et al.	...do	....do	E.	3,000	....do ....
Rosalia..........	M. Lesinsky, Joseph Lesinsky, et al.	...do	....do	E.	3,000	....do ....

Exhibit of producing mines in Organ mining district, &c.—Continued.

Name.	Owner.	Vein-matter.	Ore.	Average value.	Mills.	Product for the year ending Jan. 1, 1872.
Aztec	J. Freudenthal, H. Lesinsky, A. H. Morehead, F. Blake, *et al.*	Lead, antimony.	Silver, small per cent. of gold.	$125	Furnace.	$4,000
Stephenson	H. Lesinsky, W. F. Shedd, *et al.*	Quartz, lead	Silver	340	...do	6,000
Cuebas	J. Freudenthal, G. E. Blake, *et al.*	Lead	Lead	$16, silver and lead.	...do	Not worked.
Bennett	N. V. Bennett, Shedd, *et al.*	Lead, &c ..	Silver			New discovery.
El Quibedo	W. H. Graham, James Foley *et al.*		Lead			Do.
Rosalia	M. Lesinsky, Joseph Lesinsky, *et al.*	Iron, antimony.	Silver			Do.

REMARKS.—The Stephenson is the oldest mine, having been discovered about thirty years ago, and worked successfully for about twenty years. Upon the breaking out of the rebellion in 1861 labor was suspended, and since that time but little has been done. Work will begin again shortly, and every indication shows that it will prove remunerative to the owners. There have a great many new discoveries been made within the past six months, but I am unable to give the names of the mines, &c. The locations, however, are in the Organ district. The bullion produced has been extracted by the crude Mexican method. There are no regular smelting-works or mills.

CHAPTER VIII.

COLORADO.

Want of funds has compelled me to investigate the progress which the mining industry of this Territory has undoubtedly made during 1871, with less detailed care than was my intention. Although the principal mining districts of Colorado are not included in the field allotted to me by congressional resolution, I have so far managed every year to record the developments made; and the processes used for the extraction of the precious metals have received their due share of attention. This last year my attention was necessarily directed to other fields which had not before been personally visited by me, and in using the greater part of the small amount appropriated by Congress for my work, in that direction, only enough has remained in my hands to furnish in this report a general outline of what has been accomplished during the year in Colorado.

As in former years, Messrs. Jacob F. L. Schirmer, assayer of the United States branch-mint at Denver, and J. H. Jones, agent of Wells, Fargo & Co., have kindly furnished me with an estimate of the product of Colorado for 1871. Their intimate acquaintance with the mining industry of the Territory, and their knowledge of the shipments made, entitle this estimate to the highest confidence, and I accept it, therefore, as my own:

Gold and silver product of Colorado Territory during 1871.

Shipped by express	$2,820,000
In private hands, (estimated)	140,000
Shipment in matte	923,000
Shipment of ore	500,000
From southern mines	130,000
From northern mines	50,000
Used by manufacturers	100,000
Total	4,663,000

The most accurate estimate which I could obtain of the product of the previous year gave the yield of the whole Territory for that year as $3,675,000. There is therefore an increase of very nearly a million of dollars.

This increase has been apparent on every side during the last year throughout the Territory in more extended mining, milling, and smelting operations; and it is clear that the industry is now looked upon as legitimate business more than ever before. The time of wild and extravagant speculation, with undeveloped properties and "processes," has passed away in Colorado, and if all the signs do not deceive me, an era of steadily progressive industry has at last fairly been inaugurated.

In Gilpin County many of the older claims, which had been idle for several years, have been taken up again. It is noteworthy that the larger number of these have been leased by miners, who in almost every instance have made good wages, and in some cases small fortunes. They have generally done vastly better than the companies who preceded them, working the same veins.

On the Kansas lode, for instance, in Nevada district, there were, in the latter part of the year, nine claims worked, the majority of which turned out exceedingly well, so that the lode produced at that time much more gold than ever before. Wheeler & Sullivan, who work the "First National" claim on this lode, were furnishing 40 stamps of the New York Mill with ore. Richards & Co. were deepening their shaft on the Ophir Company's claim, and were in good ore all the way. The Garrison claim was worked by Wolcott & Co., who were both drifting and stoping, and took out good pay ore. Ira Easterbrooks, who took the lease of the Mead claim off the hands of William Lyon & Co., was drifting and stoping in his mine, which presented a better appearance than ever before. D. L. Southworth was sinking on Waterman's claim, the next one east. Mr. Root was obtaining rich ore from his claim on the Kansas, near Boston Mill. On the second claim east of Root's, Andrews & Sullivan were raising large quantities of fine-looking ore, which was being crushed at Lake's Mill, on North Clear Creek.

The English Kansas Gold Mining Company were doing well on their claims purchased some months ago of J. F. Hardesty.

Another claim, between Root's and that of Andrews & Co., was being worked in a small way, and the ore hoisted by a windlass.

The second-class ores from this vein assayed, at the time spoken of, about 1½ ounces of gold per ton, and the yield in the mills was given to me as 4½ to 5 ounces per cord. First-class ores brought, at Hill's works, about $100 per ton.

The Kent County lode, about 400 feet above the foregoing, was also very actively worked during last year. Mr. Eilers visited, in the fall, the Ætna Company's ground, 1,000 feet in length, which was leased by Messrs. Nichols, Roe, Fisher, and Mitchel. These men started to work the mine with almost nothing, but were doing exceedingly well at the time of this visit. The mine was 320 feet deep, and the force employed in three levels and on the surface was twenty-seven men. The principal stoping was done between the first and second levels, 100 and 160 feet from the surface, both east and west from the hoisting-shaft. Here a great mass of very rich ore had been found sticking to the hanging-wall, which had been overlooked in the old stopes on account of a thin sheet of slate which separated it from the ore on the foot-wall. About 30 tons of ore were hoisted per day, a considerable portion of which was first class, bringing at the smelting-works from $95 to $126 per ton. The second-class ore averaged about 7 ounces per cord. This claim is very well opened. The levels are all connected by winzes, and sinking for reserves was steadily kept up. In the rich ground spoken of, the vein was over 5 feet wide, containing the usual mixture of iron and copper pyrites, blende, and quartz.

This claim connects by drifts with a neighboring one leased by the Bradley Brothers, so that in both of them the ventilation is very good. The latter claim is 450 feet long, and is opened to a depth of 200 feet, but at the time of Mr. Eilers's visit it was idle, on account of litigation with the owners of the Ralston Company lode, which appears to be a feeder to the Kansas. There were a large number of claims worked on the lode which could not be visited.

The Prize is situated opposite the foregoing. It strikes northeast and southwest, and dips steeply to the southeast, like the two last named. It is opened for a length of 900 feet, in two claims of respectively 400 and 500 feet. The latter belongs to the Cornwallis Company. Neither of these claims has been opened long, but the capacity of each, last fall, was 25 tons per day. First-class ore contained $94 in gold and silver, about 8 ounces of which was silver. Second-class ore assayed 6 to 10 ounces of gold per cord. The Iron Ram is situated in the same neighborhood, and can deliver about 10 tons per day. The best ore assays from $50 to $80 per ton. On the California, situated on the same hill-side with the Kent County, the deepest shaft was down 740 feet, in the fall. Another one was 360 feet deep; and there were several others of less depth. The vein is from 2 to 3 feet wide, and dips and strikes very nearly parallel to the Kent County and Kansas. First-class ore from it is reported to contain $150 per ton in gold and silver. For such ore Professor Hill pays $75; and I am informed that one claim on this vein (Harper's) can furnish 12 tons of it per week. Stalker's claim is reported to furnish nearly as much. If this is really so, these claims contain exceptionally rich ore. (See my last report.) The first-class ore of this vein contains about 4 per cent. of copper, and the second-class 1¼ per cent. For second-class ore, which is said to assay, on an average, 2 ounces of gold and 10 to 15 ounces of silver per ton, Professor Hill pays $37 per ton.

Of lodes in other districts, the Illinois Central, Burroughs, American Flag, and Gunnell, (the latter since the middle of summer,) have been actively worked, and produced much ore. The Gregory, Parmelee, Briggs, Bates, Bobtail, and Fisk were all idle in the fall and full of water. The Fisk and Bobtail will be the first lodes drained by the Bobtail tunnel, driven by Mr. Rogers and others. This tunnel was in 300 feet in the fall, and 150 feet further were expected to bring it to the Fisk, which it will strike at a depth of about 450 feet below the outcrop.

The principal mines in the vicinity of Central City have been described very fully in my last report, and, as no very important changes have taken place, it is unnecessary to go again over the same ground.

The following statistics of the average contents of ores in gold and silver, from different districts in the vicinity of Central City, have been kindly furnished me by Mr. A. von Schulz, whose facilities as assayer in Central City enabled him to take his averages from a large number of samples assayed within the last year:

	Per ton, in coin.
Gregory district:	
Milling ore, average of 72 samples..............................	$35
Smelting ore, average of 35 samples	169
Nevada district:	
Milling ore, average of 56 samples.............................	35
Smelting ore, average of 32 samples	127
Illinois Central district:	
Milling ore, average of 31 samples.............................	33
Smelting ore, average of 9 samples.............................	126
Russel gulch:	
Milling ore, average of 59 samples.............................	37
Smelting ore, average of 23 samples	112
Central City district:	
Milling ore, average of 22 samples	27
Smelting ore, average of 8 samples	87

Enterprise district:

Milling ore, average of 25 samples............................   35

Eureka district:

Milling ore, average of 17 samples............................   41

Lake district, (two miles southeast of Central City:)

Milling ore, average of 12 samples............................   24

From notes submitted to me by Colonel G. W. Baker, of Central City, I reproduce here the following statistics, showing the relative values of gold and silver in Colorado pyritous gold ores. According to 77 assays, made by the territorial assayer at Central City during the last two years, and certified by him, the average contents of these samples, which represent a great number of veins from different districts in the vicinity of Central City, in gold and silver, are: gold, $61.93; silver, $37.30. The average of silver is here, however, larger than usual in the pyritous gold ores of Colorado, because some of the ores assayed are more properly silver than gold ores. In the following 28 of the above assays the usual proportion of gold and silver present in the gold-ores is more accurately given:

	Average.	Aggregate.		Average.	Aggregate.
1	$49 60	$15 08	15	$74 41	$12 22
2	70 27	13 26	16	27 90	14 04
3	84 72	21 71	17	74 41	52 26
4	90 94	23 92	18	22 73	15 08
5	41 34	19 76	19	33 07	9 55
6	47 53	18 07	20	165 36	37 44
7	29 97	10 14	21	39 27	24 31
8	29 97	9 36	22	80 61	11 05
9	20 67	26 13	23	103 35	38 03
10	50 98	14 56	24	39 27	5 72
11	33 07	11 18	25	59 94	10 01
12	86 81	14 04	26	29 97	6 11
13	37 21	15 93	27	49 60	18 34
14	28 93	7 54	28	41 34	5 33

Taking 237 assays of second-class and 102 of first-class ore which were made in the territorial assay-office somewhat over a year ago, and averaging those made from ores of particular districts, the contents of gold and silver vary for the different districts as follows:

Milling ores, (second class.)

Number of assays.	District.	Average per ton of ore.	
		Gold.	Silver.
29	Gregory	$24 10	$11 37
54	Nevada	22 51	12 85
30	Illinois Central	19 93	13 39
55	Russel	20 07	17 14
29	Central City	17 30	10 60
21	Enterprise	8 47	27 05
17	Eureka	29 42	12 02
9	Lake	6 31	18 60
237			

Smelting ores, (first class.)

Number of assays.	District.	Average per ton of ore.	
		Gold.	Silver.
34.....................	Gregory	$138 98	$30 32
32.....................	Nevada	99 30	37 62
19.....................	Russel	50 98	61 90
9.....................	Illinois Central..	86 39	40 57
8.....................	Central City	63 61	23 44
102			

In the assays made by Mr. A. von Schulz about the same time, the proportion of gold and silver contained in the ores of the various districts, and obtained in the same manner as above, appears as follows:

Milling ores, (second class.)

Number of assays.	District.	Average per ton of ore.	
		Gold.	Silver.
31.....................	Gregory	$25 14	$6 69
83.....................	Nevada	30 59	12 50
9.....................	Illinois Central..	32 33	25 00
14.....................	Russel	16 90	27 11
12.....................	Central City	28 09	28 56
3.....................	Eureka	20 14	18 21
151			

Smelting ores, (first class.)

Number of assays.	District.	Average per ton of ore.	
		Gold.	Silver.
72.....................	Gregory	$158 72	$18 84
32.....................	Nevada	95 50	21 48
104			

Mr. Burlingame, the territorial assayer, and Mr. A. von Schulz both certify that the above assays from their books were made from ores from the following number of different lodes in the districts named: From Gregory district, from 46 different lodes; from Nevada district, from 51 different lodes; from Illinois Central district, from 17 different lodes; from Russel district, from 64 different lodes; from Central City district, from 34 different lodes; from Enterprise district, from 36 different lodes; from Eureka district, from 14 different lodes; from Lake district, from 20 different lodes.

It is seen that the proportion of gold and silver in the ores varies considerably in the averages from the two assay-offices; but they seem to establish the fact that there is really *more* silver in the Colorado gold-ores than was assumed in Mr. Reichenecker's article in my last report. Taking the average of 428 assays made in the territorial assay-office and in that of Mr. A. von Schulz, at Central City, we find the proportion of gold to silver in Colorado gold-ores: gold, $22.56; silver, $17.51 per ton of ore.

Accepting the last statement, which is taken from the greatest number of assays, as the one which probably comes nearest to an average, there would be to $100 gold in the ore $77.62 silver. The gold bullion of Colorado contains, however, according to Colonel Baker, only $20 silver to every $100 in gold.

The great loss of gold, silver and copper in the Colorado stamp-mill process I have discussed in my last report. It is there shown that the actual saving is only about 40 per cent. of the assay value of the ore. From the above it appears that the loss of silver, considered for itself, is much greater, as, of $77.62 silver, only $20 are actually saved; the loss is, therefore, $57.62 in silver to every $100 in gold saved, or 74.2 per cent. of the original contents of silver in the ore.

The above assays, being of small samples only, may not represent the true proportion of gold and silver, but the figures obtained from them must evidently not be very far from the truth. In this case it is evident that concentration and smelting of the ores would be far preferable to the present mode of working these ores, even if we accept, in this case, the maximum loss of brittle silver-ores in concentration, i. e., from 20 to 30 per cent. In this connection a process of beneficiating ores containing both gold and silver, and little galena, by means of smelting, as practiced in South Austria, is worthy of the higest consideration, and I draw attention to it for the benefit of those most directly interested. Though practiced under conditions somewhat dissimilar from those in Colorado, it seems to me that it would fulfill, in that country, the requirements of the times. The process was witnessed and described by Mr. John A. Church, E. M., who kindly permits the use of his article. It will be found in full in a subsequent chapter of this report.

Professor Hill's smelting-works at Black Hawk have been enlarged during the summer by one smelting-furnace, a reverberatory of the same pattern as the two older ones. This, in itself, is proof that the mines in the vicinity of Central City furnish much more ore than before, for it is well known that the manager of these works is very cautious, and not apt to invest additional capital unless he is certain of a steady supply of ore. Nevertheless, the old and, I am sure, unreasonable dissatisfaction among miners, who have for years joined in the cry for more beneficiating-works and higher prices for the ore, has not been allayed, so that Professor Hill has been finally obliged to defend publicly his policy and his works. I reprint his letter here, because it contains many valuable statements; it was addressed to the Central City Register:

Articles frequently appear in the papers of Central City, similar to yours of a few days ago, under the head of "California mining." Such articles, so far as they exert any influence, are calculated to convey false impressions. Besides, there are persons who, either from ignorance or designedly, are constantly making false statements about the comparative cost of treating ore in this and in other places.

It is stated in the article referred to that, in California, stamp-mill rock, worth $9 per ton, and sulphurets, worth $15 per ton, can be treated with profit. How are they treated? The stamp-mills in this county treat ores for $3 to $5 per ton. If it is claimed that the mills here do not save so large a proportion of the gold as they save in California, it must be remembered that saving gold, when it is in combination with almost every base metal and sulphuret that is known, is a very different thing from saving gold when it exists free in a quartz matrix. The writer also says that in this county gold and silver ores worth from $60 to $75 per ton, and rich in copper, are not available at the present time, and thinks it is a shame that such ores cannot be worked here with profit.

Ores containing gold and silver to the value of $60 per ton, in currency, and allowing (as the writer says they are rich in copper) that they contain 8 per cent. of copper, are worth here $36 per ton, without any cost beyond that of delivery at the smelting-works. If ores worth $8 to $15 per ton in California can be mined and reduced with profit, and if in Colorado ores worth $36 per ton, over and above all costs of reduction, cannot be treated with profit, it would seem that the great difference between California and Colorado was in the expense of mining and not in the expense of reduction.

It would be instructive for those who complain of low prices of ore in Colorado to inquire what prices are paid in other places.

The four principal smelting-works for the treatment of gold and silver ores in the United States, and outside of Colorado, are at San Francisco, Reno, Omaha, and Newark.

At San Francisco they charge from $50 to $100 (gold) per ton for smelting gold and silver ores. I have not any printed scale of prices paid by this establishment, but derive these figures from the company's statement of returns made to persons who have sold them ore.

At Reno, according to their published scale of prices, they charge, for ores assaying $80, $46 per ton; for ores assaying $100, $50 per ton; for ores assaying $200, $70 per ton; for ores assaying $500, $105.

At Omaha they have treated but little ore, but they have recently made offers to miners in this county to treat their ores for $100 per ton, and guaranteed 100 per cent. of the gold and 95 per cent. of the silver.

At Newark they charge $50 per ton for the lowest grades of ore, when they work for gold and silver, and advance, with the increased richness of ore, to over $100 per ton.

I have before me the account of sales of two lots of ore from Nevada, Gilpin County, of about five tons each, which were sold in the open market in Swansea, Wales, to the highest bidder. One of these lots assayed $136 in currency; price paid, $95; charge for treatment, $43 per ton. The other lot assayed $170, currency; price paid, $122.96, currency; charge for treatment, $47 per ton. These prices were paid for the gold and silver alone. The copper, 6½ per cent. in the former case, and 10 per cent. in the latter, was paid for at the regular market rate, which, of course, gives on the copper a fair profit to the smelter. I have the statement of the purchasers of these ores, that they are the most complex of any ever offered at their works, and this will in part account for the charges for treatment, which, for Swansea, are very high. In Swansea, coal suitable for smelting costs from 50 cents to $1.50 per ton, labor about 75 cents per day, and the best fire-bricks, $12 per thousand.

The Boston and Colorado Smelting Company are treating ores, of which the assay value of the gold and silver, estimated in currency, is $50, $100, and $150, at a cost to the miner of $35, $40, and $45, respectively; that is, for ores which contain $50 per ton, currency value, all over $35 is paid to the seller, and for ores containing $100 per ton, also all over $40 is paid to the seller, and so on. For intermediate grades a *pro rata* charge is made.

This company also pays for the copper $1.50 for each per cent. on the dry Cornish assay, which is the assay on which all copper-ores are sold.

No one who is acquainted with the facts will deny that the ores of Colorado are the most complex which are worked on this continent, containing, as they do, mixed with the sulphurets of copper and iron, large quantities of the sulphurets of antimony, arsenic, zinc, and lead, and a refractory gangue. Neither can any one deny that the actual costs of all the principal elements employed in smelting, viz, fuel, labor, fire-bricks, and iron, are more than double here what they are east of the Mississippi River, and much higher than they are in California.

It must be remembered that the charge of $35 to $45 per ton not only covers all the first cost of smelting, including calcining, crushing, and bringing the ores to a liquid state, without the aid of foreign fluxes, and often a second calcination and melting to bring the matte to a workable condition, but also all the costs of separating and refining the metals, and all losses which must inevitably occur when so many different processes are employed.

The statement that ores which contain $8 to $15 per ton can be treated in California with profit may be true. It is equally true that the greater part of Colorado ore is treated at $3 to $5 per ton. But the statement, that there is any place in the United States where ores are smelted at so low a cost to the miner as in Colorado, is not true.

The above statements about the charge for smelting in other places have been derived either from the company's printed scale of prices, or from their own reports of prices paid to the sellers of the ore. If any one doubts their accuracy, I will make this proposition, viz, if there is any place in the United States where ores similar to the ores in this county are smelted, or treated on a large scale in any other way, so as to realize 95 per cent. of their contents in gold, silver, and copper, at a lower charge than is made by this company, I will at once reduce our charges to the same terms. If any one thinks that low-grade ores can be smelted in Colorado for less than $35 per ton, and their full value realized, where pine wood costs $5 to $6 per cord, common labor $3 per day, and skilled labor $5 to $6 per day, and fire-bricks $130 per thousand, he has only to try it. He can soon be convinced by an experiment.

It is often represented that the capacity of the Boston and Colorado smelting-works is limited, and that they are now overstocked with ore. In answer to this, I will say that this company provided a large surplus capital, which could be employed, if needed, and it was supposed when the company was organized that before this time it would employ twelve furnaces. It is now working three smelting-furnaces. All that is required to double or treble the present capacity of the works is a production of ore to justify it. When any one fails to find a market here for his ore, at the terms above stated, it will be time to represent that we are overstocked and have ceased to buy.

The smelting-works of Mr. West, at Black Hawk, have, as far as I know, been idle during the greater part of the year. I am not informed whether the reason for this is to be looked for in a failure of the process (shaft-furnace) or in a combination of commercial conditions, which are very apt to affect new works where old ones are already successfully established.

Gulch mining on Clear Creek has been followed by a few men during the year with good results. S. S. Chambers & Co.'s claim, about 200 yards from the junction of North with South Clear Creek, is reported to have yielded over $5 per day per hand.

The owners of this claim have been working it for nearly five years, but last year was the first which paid them something like wages; and this year their work appears to have been quite profitable.

Dr. F. Page's claim is the next one above the foregoing. It has also been worked, and, according to report, with satisfactory results.

Blake & Co.'s placer claim is about half a mile above Page's. This company have nearly a quarter of a mile of fluming, and their ground is very rich, some of it having yielded at the rate of over $2 per cubic yard.

Above this claim are Hamilton's, J. W. Fries's, Pitman & Wiley's, Huggel & Co.'s, and Alexander Cameron's. The latter has been very extensively worked, but has not paid as well during the last season as in former times. There is much unoccupied ground between the claims above mentioned.

In Clear Creek County the principal features, in the way of advance in the silver-mining industry, are the re-occupation of Argentine district and a considerable development of the argentiferous-galena mines in the vicinity of Idaho, as well as the discovery of several extraordinarily large and rich silver mines near Georgetown. But the industry has received a serious loss by the burning, during the latter part of 1871, and in the early part of 1872, of the three principal silver-mills in the county, viz, the Washington Mill, at Georgetown, the Baker Mill, at Bakerville, and Stewart's reduction-works, at Georgetown. It is reported that all these works are to be rebuilt soon, but meanwhile the product of the county is lessened to the extent of the capacity of these mills, and miners are very much inconvenienced in regard to a market for their ores. Mr. Stewart had introduced into his new mill the Airey furnace, an apparatus belonging to the same class as the now celebrated Stetefeldt furnace. There are some features which distinguish it from the latter, but it is questionable whether these features add anything to its usefulness. The two principal ones are:

1. The fixing to two inner opposite sides of the furnace of a series of iron plates, which can be set at any desired angle, and are intended to retard the fall of the ore. It has been shown by the working of the Stetefeldt furnace that this improvement is uncalled for and undesirable—uncalled for, because the short time of the fall of the ore suffices to chloridize it up to 92 and even 96 per cent. of the assay; and undesirable, because these plates must be very rapidly destroyed by the acid-fumes developed in the furnace, and they are therefore a source of very large expense for repairs.

2. The position of the fire-places and flue. In the Airey furnace the upper fire-place is situated at the top of the stack, whence the flame goes down in the furnace, meeting the one from the lower fire-place about midway, the two draughts abutting squarely against each other.

It is an old rule in metallurgy never to let two draughts meet directly from opposite directions, because they are weakened, if not altogether destroyed. This was actually the case at the Airey furnace, when the writer visited it last summer, and, as there was a poor draught in the furnace, the heat was not sufficient for a successful roasting. This was, 1 believe, afterwards corrected; at least the results of chlorination were said to be better, but the precise results reached in the furnace, before the conflagration of the mill, have not yet become known. Aside from the question of draught, this arrangement of fire-places suffers from the same objections as affect the Whelpley and Storer furnace, viz., that the ore is exposed to the highest heat in the commencement of the roasting, when it should be subjected to the lowest, and at the end of the operation to the lowest, when the temperature should be the highest. For the roasting of the inevitably large amount of dust there appears to be no provision in the Airey furnace. It must, apparently, pass over and over again through the furnace.

The Stewart reduction-works treated, from July, 1870, to July, 1871, 1,390 tons of ore, of which the assay value was $201,700, and the yield $180,785. From July 1 to November 1, 1871, the quantity treated was 500 tons, and the yield $58,811.

The works of Palmer & Nichols treated, from March 26 to July 1, 1871, 228 tons of ore, assaying 38,493 ounces of silver, and from July 1 to November 1, 1871, 203½ tons, assaying 36,136 ounces of silver, making a total of 431½ tons, with 74,629 ounces. The yield has been about 86 per cent. of the assay, and the concern has shipped, up to November 1, $84,000, coin value. Adding 80 or 90 tons treated previously to March 26 by Huepeden & Co., (mainly second and third class ore from the Brown and Terrible, and estimated at $77 yield per ton,) we have over $90,000 as the product of these works during ten months of 1871. The average yield per ton of ores treated by Stewart has been, for the year ending July 1, 1871, $130; for the four months ending November 1, 1871, $117; for the whole sixteen months, $126. The average yield of ores treated by Palmer & Nichols has been, for three months ending July 1, 1871, $189; for the four months ending November 1, 1871, $198; for the whole seven months, $194. This high yield is due to the fact that much rich surface ore has been brought in from small mining operations to these works. Stewart's supply was mostly from the lower grades of Terrible ore, with which he kept ten stamps (half his capacity) running for the greater part of the time.

Dibbin's reduction-works in East Argentine district (one Brückner cylinder) are reported to have shipped from $1,200 to $1,500 per week since middle of July.

On the whole the prospects for the silver mines of Colorado are encouraging. When we sum up the yield of the ores treated in the Territory for the year 1871 and the reported shipment of ores, it will be seen that the production considerably exceeds that of former years, though, of course, the extravagant expectations of the sanguine ones have not been realized.

Passing by the older mines, which have been repeatedly mentioned in my former reports, I add here a brief description of some mines near Idaho which have come into favorable notice during the year, and of the new discoveries in the vicinity of Georgetown.

Mines in the vicinity of Idaho.—The Queen is situated in the gneiss, about a mile and a half north of Idaho, on a hill-side about 800 feet above the town. The lode is worked by different parties, and opened by six shafts, from 60 to 80 feet deep. It is from six to eight inches wide, runs northeast and southwest, and dips about 33° northwest. The ore at the depth reached is mostly decomposed, but there are patches and streaks of white iron pyrites found in the vein. The decomposed ore is brown and yellow quartz, reported to assay 111 ounces of silver per ton, while the pyrites contain from 75 to 80 ounces per ton. There were twelve men (six drills) at work at the time of the writer's visit, but owing to the narrowness of the crevice and the hardness of the rock only 4 to 5 tons of ore could be taken out per week. There were as yet no levels or stopes. In one of the shafts occurs a solid layer of zinc-blende, with little galena, 3 inches thick, which was said to assay $6000 per ton. The claim is 1,500 feet long, and portions of it have been leased by Captain Hall, William Hill, and A. Morgan, the owners, to different parties. The ore is shipped to Professor Hill's smelting-works at Black Hawk, four and a half miles distant, the freight being $5 per ton.

The Franklin, near by, is a vein from 6 inches to 2 feet wide, and carries more galena than the foregoing. The main shaft on it is 120 feet deep. The ore contains about $40 to $50 per ton in silver. Much money has been uselessly spent on the mill belonging to this mine. First, furnaces were erected to smelt the ore, but the blende interfered, though the ore contains less of it than usual in Clear Creek County ores. Next, the mill was erected and reverberatories were used for chloridizing the ore. But the enterprise failed both financially and technically, the latter principally because there was no reliable supply of water in the gulch where the works are located. Now it is reported that the company intend to lease a mill on Clear Creek, (Buford Mill,) where the erection of an Airey furnace is said to be contemplated. This mine is better opened than the rest, and has a capacity of 3 tons per day.

The Seaton mine has been frequently mentioned in the Colorado papers on account of its rich ores. Mr. Eilers visited two of the claims on this lode, No. 7 and No. 2, east of the discovery.

Seaton No. 7 is owned by Captain Dean, and is a 100-foot claim. The depth of the mine is 120 feet in one and 81 feet in another shaft, and the width of the ore-seam is here from 6 to 12 inches. The ore assays from $50 to $400 per ton. There were only two drills at work in the fall, and 3 tons a week was the yield of the mine. To a depth of about 50 feet in both shafts the ore is a brown decomposed material, but lower down about 12 inches of quartz, carrying blende and galena in the proportion of 3 to 1, and some fahlore, are found. This class of ore is reported to assay from 100 to 273 ounces of silver per ton.

The Seaton No. 2 east of the discovery, owned by Lewis & Co., is 700 feet long on the vein. The deepest shaft is 270 feet deep. The main vein runs from northeast to southwest, and dips from 48° to 50° northwest. It is from 3 to 12 inches thick. It was originally worked for gold, and during that time the deep shaft was sunk. This is now full of water up to within 70 feet from the surface. A second vein has been found lately, which intersects the original one at right angles to its dip, the pitch of the new discovery being about 30° to southeast. This is from 8 to 18 inches wide, and is characterized by frequent faults in the plane of the dip; the continuation of the vein at each fault being found from 1 to 2 feet lower than the vein above. In the southwestern part of the claim, the plane of this new vein crosses the plane of the original one at a depth of about 30 feet from the surface, but toward the northeast the line of crossing sinks steadily, until, in the Dean shaft, the new vein is found at a depth of 110 feet. It is worked by under-hand stoping, and a large chamber reaching from shaft to shaft is already opened. The capacity of the mine is from 6 to 8 tons per week. The ore is a mixture of zinc-blende, galena, iron pyrites, and fahlore in a quartz gangue, the blende predominating. There is much more fahlore in this than in any other mine in the district, which I have seen, and the average contents of silver are very high.

The Santa Fé runs parallel in strike to the foregoing vein, and lies higher up on the mountain. Its dip is less steep. The claim, of 500 feet in length, is owned by R. B. Griswold. There are five shafts on it, which, commencing from the northeast end of the claim, are respectively 78, 35, 25, 40, and 25 feet deep, and all of them are located within 200 feet on the vein. Nevertheless the grades of ore raised from these shafts differ widely. In the first shaft southwest of the discovery, for instance, the first-class ore, as selected for shipment, assayed 25 ounces of silver and one-half ounce of gold per ton, and was rich in lead. In the next shaft to the southwest the ore contains more iron pyrites; is also rich in lead, and contains from 2 to 4 ounces of gold, and from $125 to $140 of gold and silver per ton. Ore from the following shaft on the southwest contains between 50 and 60 ounces of silver per ton, and 2½ ounces of gold. There is, as yet, no communication between the shafts underground, and all of them are being sunk deeper. About six tons of ore are taken out weekly, by six men.

There are a number of other mines in this vicinity which hold out fair promises for the future, but all of them are, like those described, not yet well developed, and can, therefore, furnish little ore at present.

I must mention, however, here a mine which, in connection with the often-talked-of smelting-works in this vicinity, or a little higher up on Clear Creek, is of great importance. This is the Edgar. It is situated north of and opposite the old Whale mine, which has been described in former reports, but is now idle. The Edgar is by many considered a continuation of the Seaton lode, though with what right, considering the distance between the two mines, I cannot determine. Its strike is northeast and southwest; its dip northwest. The lode is from 6 inches to 2 feet thick, and carries very solid galena and fahlore. There are two tunnels driven on the lode and connected by a shaft. The mine is sufficiently opened to be able to furnish about 3 tons per day. First-class ore is reported to assay, on an average, 80 ounces of silver and one-half ounce of gold per ton, and contains seldom less than 45 to 50 per cent. of lead. Some of the ore mined has assayed as high as 165 ounces of silver per ton. This ore contains very little zinc-blende, which, in nearly all the other silver mines, is the predominant mineral, and it will therefore be of the highest importance to the smelting-works which are said to be on the point of erection in this region. Mr. Eilers was informed that one or two other mines, lately opened in this vicinity, carried ores very similar to those of the Edgar, as far as the contents of lead are concerned, but he was unable to visit them.

Among the new mines in the vicinity of Georgetown, which have turned out very satisfactorily from the commencement of operations, the Pelican is the most important.

The Pelican is situated on Sherman Mountain, on the east bank of Cherokee Gulch, and about three miles west of Georgetown. The ground on either side of the gulch rises at a rapid rate, and affords a fine opportunity to gain great depth from the surface as the work of drifting east and west on the vein goes on. There were, at the end of the year, four adits on the vein, aggregating in length 420 feet. The depth obtained from the surface in the gulch was 85 feet. The greatest depth obtained from the surface in the lower west adit was about 200 feet. This adit is cut in the gulch by a tunnel about 100 feet in length.

A tunnel 100 feet in was being driven 300 feet in length, to cut the vein 110 feet from the surface. When this tunnel shall have been completed perfect drainage of the mine will be secured. A winze was being sunk on the vein from the lower adit to intersect the line of the long tunnel where it will cut the vein. By the 1st of June, 1872, the mine is expected to be in condition to yield 20 tons of rich milling ore a day. The lower, middle, and second adits are connected by air-shafts or winzes. The mine is well timbered. Everything is well arranged about the mine to secure perfect drainage, safety for the workmen, ventilation, and speedy and cheap delivery of the ore at the mouth of the tunnel and adits. Very little stoping has, so far, been done on the mine.

The Pelican is, compared with others in Clear Creek County, a large vein. The walls are well defined. The breadth of the vein between walls averages about $3\frac{1}{2}$ feet, from 12 to 20 inches of which contain rich ore. The vein-matter consists of argentiferous galena, copper-ore, fahlore, blende, and sometimes native silver. Some of the assays made from ore of this vein are exceedingly high, as the following list shows:

Selected specimens of ore—

No. 1.. 2,823 ounces silver.
No. 2.. 1,700 ounces silver.
No. 3, fine gangue........................... 1,782 ounces silver.
No. 4, blende .... 428 ounces silver.
No. 5, containing much gangue............... 808 ounces silver.

From the books of the mine and mill certificates the following information is derived :   .

From the 7th of February, 1871, to November 18, 1871, a little over nine months, the Pelican has yielded the following amounts of ore:

One hundred and twenty-four and a half tons of second and third class ore, treated at Palmer & Nichol's and the Stewart Silver-Reduction Company's Mills, gave a yield of 32,280 ounces of silver. The average yield of this lot of ore was 240 ounces per ton of 2,000 pounds. There were at the mine at the end of the year 200 tons of second and third class ore, worth, at a low estimate, 50,000 ounces; 30 tons of first-class ore at the mine and at Silver Plume Mill, worth 510 ounces per ton, equal to 15,300 ounces; 27 tons shipped to Swansea, Wales, 510 ounces per ton, equal to 14,770 ounces; 4 tons sold to Frank J. Marshall, 572 ounces per ton, equal to 2,288 ounces; a small lot, 1 ton, sent to Professor Hill for treatment, 294 ounces.

Total product in tons, $386\frac{1}{2}$; total product in ounces in silver, 113,932.

Total cost of mining $386\frac{1}{2}$ tons of ore, transportation, milling, incidental expenses, &c., 10,000 ounces. Therefore, 113,932—10,000= 103,932 ounces. This leaves consequently a clear profit of over $130,000 in coin.

The amount of ore in sight in the mine is large. As yet only a little ground has been stoped out.

The mine can deliver, with sixteen miners, four tons of ore per day, and the owners think they can keep up this rate of production throughout the next year from the present reserves. All the ore must be packed down the mountain at a cost of $2.50 per ton, and transportation to the mills costs $3 more.

The Fletcher is another new discovery of great promise. It is situated on the Bald Creek slope of Democrat Mountain, and very rich ore, some of which yielded $799 per ton, has been found near the surface. The same may be said of the Elkhorn and Maine lodes. In regard to the developments and yield of the latter, I take the following information from the Georgetown Miner of February 8, 1872.

The vein, a large one, with well-defined walls, is situated on Sherman Mountain, a short distance above the village of Silver Plume. The matrix of the vein is feldspathic rock and quartz. The pay streak is argentiferous galena, interspersed with gray copper, ruby silver, and iron and copper pyrites. The country-rock is granite. In some portions of the mine there is a solid ore-deposit of 2 feet in breadth. We have frequently examined the Maine, and have never seen a barren spot in the mine.

The yield of the mine since its discovery in May, 1871, about eight months, has been as follows:

Ore treated at Stewart's, 29 tons, 566 pounds, yielded	$13,350 32
Shipped to Hill, Black Hawk, 100 tons..................	19,350 00
Sold to W. Bement 13 tons.............................	1,509 00
Out at the mine 80 tons...............................	15,480 00
Total yield of mine..............................	49,689 32
Cost of work done on mine, &c....................	11,000 00
Net yield of mine	38,689 32

The above figures were obtained from W. T. Reynolds, one of the owners of the mine, and a personal examination of the records of the mine shows that they are substantially correct.

All the above calculations are coin value. The average yield of 222 tons and 596 pounds, total product of mine in eight months, is a little over $223 per ton.

The various tunnels in the vicinity of Georgetown, such as the Burleigh, Marshal, Lebanon, Eclipse, have been steadily driven ahead, and some of them are reported to have struck good veins. There is, however, so far as I am aware, no regular mining going on through the tunnels on any of the new lodes discovered.

West Argentine district.—For notes in regard to this district I am indebted to Mr. A. Wolters, M. E., formerly of Bakerville. The first discovery of silver-bearing lodes was made in 1864 in East Argentine, and the ensuing excitement drew of course a large number of prospectors to this and the adjoining district of West Argentine. Hundreds of lodes were thus discovered in 1865; only a few, however, were being worked to any extent. Owing to several circumstances, the chances in favor of profitable mining were so small that all work was abandoned after the fall of 1866, with the exception of that on the Baker and Belmont lodes, and even these, though both of them were undoubtedly well-defined large fissure-veins, were worked at a loss to the owners, from the following reasons:

1. Great ignorance of the character and qualities of silver lodes prevailed amongst the owners of the lodes, and the waste of large sums of money in foolish experiments was the consequence.

2. Until the fall of 1867 there was no market for silver-ores in the county, and when one was established at that time, the reduction-works of Garrott, Martine & Co. charged the enormous amount of about $75 mill-fees per ton of ore.

3. The location of the lodes was very unfavorable. Situated as they were at a distance of ten miles from the nearest market for ores, and in places where even the construction of a mule-trail could not be thought

of, the cost of mining and transportation alone was sufficient to kill any mining enterprise.

Considering these circumstances, to which is to be added the generally prevailing opinion that the district was inaccessible during from six to seven months of the year on account of snow, it can certainly not be surprising that the camp was deserted, and so remained until the summer of 1870. At this time one-half of the Stevens lode was purchased by the Crescent Silver Mining Company, of Cincinnati, and vigorous development soon proved it to be a paying vein. When this fact became generally known, a few prospectors made their appearance and discovered some very promising lodes, prominent among which were the Dresden, Bismark, Muscatine, Pocahontas, Mountain Lion, Walter Scott, Worcester, and Wayne County. At the same time the Baker Company had struck two very large deposits of ore, and were mining at a profit for the first time. Besides this, they had without any trouble succeeded in keeping up communication with the mine during the whole very severe winter of 1869–'70 with only a single team, and had thus furnished conclusive evidence that the lodes were not inaccessible on account of snow. Moreover, they had finished their large and well-planned reduction-works at Bakerville, from two and a half to three and a half miles from the mines, and put them in operation the 3d of September, 1869. They at once reduced mill-fees to $39 per ton, and thus offered all possible inducements to the mine-owners to go to work and develop their property.

All these circumstances combined led to an increased activity in the summer of 1871, an activity hardly surpassed by that exhibited in 1865. Work was resumed on the west half of the Stevens, the Coney, Democrat, and Fortunatus, all once abandoned as non-paying, and now leaving their owners a liberal margin for profit. Prospectors flocked in day by day, and scarcely a week passed without one or more rich discoveries. The Fourth of July, Fifth of July, Minneapolis, and General Moltke justly caused more or less excitement by their rich deposits of fahlore, stephanite, and silver-glance.

Though most of the lodes worked were only discovered this very summer, or late in the fall of 1870, and though up to this spring the Baker Company had not received a single ton of custom-ore from West Argentine, they had in July about 40 tons to their 15 tons of Baker ore, with a fair prospect of seeing the production doubled in August. At this time, however, a great drawback was experienced by mining operations in the burning down of the Baker Mill. This caused an additional expense of $8 per ton for transportation of the ore to Georgetown, forced the miners to go to town to attend to their business, and stopped operations on nearly all the lodes owned by employés of the Baker Company, because the latter became unable for some time to pay their men. Other lodes, which had been worked at a profit so long as there was a home market for their ore, were abandoned, because they could not stand the extra expense of hauling the ore to Georgetown; and thus there were only ten lodes worked in November against about twenty in July.

Another obstacle in the way of vigorous development is the location of many of the lodes on the excessively steep and rocky western slope of McClellan Mountain, which rises at an angle of 38°. Nearly all are very difficult of access without wire cables or tunnels. The former are not erected in most instances, because the lodes are owned by workingmen, who possess no capital for the purpose. In regard to running tunnels, the facilities are better than in any other place in the county. As mentioned before, the mountain rises at an angle of 38°, and the

II. Ex. 211——23

lodes cross the mountains, giving the much-desired chance of running in on the vein. But, unfortunately, the lodes as a rule pinch up toward the base of the mountain, and therefore the tunnels would have to be run a couple of hundred feet before they could be expected to pay. This expense the owners are unable to undergo without the aid of capital; hence, to make the district an active and profitable one, capital is indispensable.

In regard to quantity as well as quality of lodes, West Argentine is equal to any other district in the county; and with some money to open the mines properly, they certainly offer great facilities for cheap working. Nearly all of them may be opened by adits, which is certainly the cheapest way if 75 feet in depth are gained for 100 feet of tunneling; and then the fact that the lodes occur here more or less concentrated in groups in a small area, gives a chance to work quite a number of them by one adit and shaft. This is an item of the highest importance, as no one can deny that a large percentage of failures in mining operations is owing to the fact that they were not carried on upon a sufficiently extensive scale. Whilst half a dozen lodes, each worked by itself, with its own adit, shaft, hoisting-machinery, and superintendent, are very apt to turn out compete failures, there may be a certainty of success if all six are worked by one party, with only one shaft, one adit, and one engine.

In their general character the West Argentine lodes are very similar to those of Reese River district, Nevada. They are all true fissure veins from 2 to 10 feet wide, and averaging probably from 3 to 3½ feet. The ore-streaks are narrow but rich, averaging probably from 3 to 4 inches, and assaying from $250 to $500 per ton. Frequently pockets of fahlore are met with, especially near the surface, and the galena-bearing veins often carry from 12 to 16 inches of solid and rich ore. There are two distinct systems of lodes, one bearing nearly northeast and southwest, the other nearly north and south, and dipping considerably to northwest and west respectively. The gangue matter is quartz, feldspar, and fluorspar; the country-rock is granite. As in the veins of Reese River district, the ore occurs as a rule in pockets, mostly united by thin seams of ore, sometimes only by a small selvage, and it is therefore absolutely necessary to develop the lodes to a considerable extent before stoping is commenced, in order to have always large reserves on hand.

Though the lodes of this district are all worked at altitudes above timber-line, the tunnels by which they ought to be developed can nearly all be started either in the timber or at a level with it. The timber itself is splendid and abundant. Quail Creek furnishes sufficient water-power during six months of the year; and 50 feet head can be obtained almost anywhere with a flume of 500 feet in length. Nutritious grass grows in unlimited quantity above timber-line, affording excellent pasture for stock. The Baker Company's well-kept wagon-road has reduced the cost of hauling ore to Bakerville to from $2.50 to $3 per ton.

The following is a list of those veins which either promise to be valuable, or by actual development have been shown to be so:

1. The Baker, owned by the Baker Silver Mining Company, of Philadelphia, runs northeast and southwest; dips northwest. Opened to a depth of 320 feet; worked by three adits, 187, 212, and 420 feet long, all connected by a shaft, extending to a depth of 168 feet below the third level. The mine is in good ore above the first level. The ground between the first and second levels is entirely worked out, and between the second and third levels nearly so. Though nine-tenths of all the ore found in the lode were on the foot-wall, the deep shaft has been sunk on the hang-

ing-wall, and no ore was found beyond a depth of 60 feet.   This summer a cross-cut was started towards the foot-wall, which, at a distance of 10 feet from the hanging-wall, struck a vein of solid ore $2\frac{1}{2}$ inches in width, and assaying $650 per ton.   Here, again, as in all instances when true fissure-veins have been sunk upon to greater depth, it has been proved that the ore continues downward, though of course varying in richness as well as in thickness, and sometimes pinching up entirely for some distance.   In the stope above the first level there is a vein of ore, varying from 2 to 10 inches, yielding ore of $200, (mill assay.)   The stope between the second and third carries from 1 to 6 inchesof $130 mill ore.   Below the third level no ore has been taken out, except in sinking the shaft.   The driving of levels preparatory to taking out ore was just contemplated when the company's mill was destroyed by fire.   This accident stopped operations for a considerable period.   Next year a deep tunnel, gaining over 400 feet depth below the third level, is to be started to facilitate cheap working.

2. The Stevens.   One-half owned by the Crescent Silver Mining Company, of Cincinnati, the other by Frank Dibbin.   Both parts are worked; and, considering the amount of development, they furnish a large amount of ore, about one-half of which is a very pure galena, containing from 55 to 60 per cent. of lead, and from 160 to 230 ounces of silver per ton, whilst the other half is a decomposed ore, of ferruginous character, worth from 100 to 120 ounces of silver per ton.   The lode, which runs nearly due north and south, dipping west, is in an almost inaccessible location, but this difficulty was efficiently and also cheaply overcome by the construction of a wire tramway between 700 and 800 feet long, with only two supports at the higher end, whilst the lower 500 feet are without any.   The cable is one-half inch thick; the buckets, made of No. 8 sheet-iron, are capable of holding about 150 pounds of ore each; and the two supports are formed of 4-inch gas-pipe, let into the solid rock and fastened there by pouring molten lead around it.   This very cheap and effective arrangement, planned by Mr. Kurtz and executed by Mr. Lowe, works to perfection, and could be employed at any place on McClellan Mountain where the construction of heavy timber supports would cause too great an expense.   The Stevens cable, the first one of this kind put up, cost less than $2,000, and, at the present reduced prices of labor and materials, it would not cost more than $1,300.   The lode is worked by adits connected by a shaft.   It will hereafter undoubtedly be found advantageous to run a cross-cut tunnel of about 600 feet in length, which will intersect both the Lindell and Stevens at a depth of nearly 450 feet, and do away with hoisting and the whole cable arrangement.   The shaft must, of course, be sunk to that depth to secure the proper ventilation.   The crevice is $2\frac{1}{2}$ feet wide; the ore-vein from a couple of inches to over one foot.   The lode crops out for a distance of several hundred feet.

3. The Lindell, owned by F. Watson & Co., is a vein of the same character, running north and south and dipping west.   It is 3 feet wide, with a streak of galena, and crops out for a considerable distance.   The galena is probably even richer in lead than that in the Stevens, but it is rather poor in silver, though in one place some has been found containing 113 ounces per ton.   The lode runs parallel with the Stevens at a distance of about 50 feet; is easy of access, but is not now worked.

4. The Coney, owned by Smith, Graves & Co., runs northeast and southwest, dipping northwest.   The crevice is 6 feet wide, with a heavy streak of decomposed gangue matter interspersed with sulphurets and fahlore, yielding milling ore of 60 to 150 ounces per ton.   The lode is

opened by three shafts, the deepest one being between 50 and 60 feet deep. No well-defined and solid walls had been found in November.

5. The Democrat runs northeast and southwest, dips northwest, and has been worked during the fore part of this summer under a lease, furnishing some very rich galena-ore, but not enough to yield a profit. The lode is in a bad locality, and the expense of working it without a tramway or deep tunnel is excessive.

6. The Fortunatus, owned by an eastern company, has been worked under a lease by Wolters & Bechtel. It strikes northeast and southwest, with steep dip northwest. An open cut has been made 22 feet deep, and a drift started from the bottom. The ore-vein averages about 4 inches the whole distance down, assaying in different places from 80 to 240 ounces per ton, and averaging 100 ounces. Small pockets of sulphuret of silver are frequently met with, yielding ore assaying at the rate of $1,400 per ton. Though, so far, all the work has been done in frozen ground, the 3½ tons of ore taken out have paid more than expenses. The crevice is 3 feet wide.

7. The Argus lode runs northeast and southwest, and dips slightly northwest. There is one shaft, 10 feet deep, which shows about 2 inches of good ore, assaying over 300 ounces per ton, and a well-defined crevice 3 feet wide.

8. The Sonora, owned by the Sonora Silver Mining Company, strikes northeast and southwest, and dips northwest; worked by an adit about 200 feet long, and a cross-cut 25 feet long; adit run in, not on the vein, but on a large white mass of rock mistaken for the vein. Thirty feet northwest of the tunnel a vein of fine-looking mineral crops out, about 2 inches wide, to which no attention has been paid. Not now worked.

9. The Richmond. Course, northeast and southwest; dip, northwest; crevice, 7 feet wide; worked by adit, between 150 feet and 200 feet long, from the base of the mountain; no ore struck yet; not now worked.

10. The Tunnel. Course, northeast and southwest; dip, northwest; worked by shaft 80 feet deep; small vein of ore on the hanging-wall, 1 inch thick, assaying 70 ounces per ton; crevice, 4 feet wide.

11. The Proteus. Course, northeast and southwest; dip, northwest; worked by an adit 70 feet long, started on a good-looking vein of sulphuret-ore 2 inches wide, which, at a short distance from the mouth, turned into a very pure, fine-grained galena, some of which is said to have yielded over $1,000 silver per ton. At this point work was abandoned.

All these lodes are old discoveries made in 1865 and 1866. There are others, such as the Savage, Black Hawk, Jackson, Hampton, &c., about which I have, however, no information.

Among new discoveries are the following:

1. The Dresden, discovered in 1870, owned by Isaacs, Wolters & Bechtel, one of the most promising veins in the district; course, northeast and southwest; dip, slightly northwest. The vein crops out for a distance of 700 feet, and has been opened at three different places, respectively 250 and 400 feet apart, and showing in all a well-defined crevice from 3½ to 5 feet wide, and a vein of decomposed galena and zinc-blende from 1 to 8 inches thick, averaging probably 3 inches, and worth 240 ounces silver per ton. From the bottom of the discovery-shaft a drift has been run for a distance of 12 feet, showing at the head a 2-inch vein of pure sulphurets, worth $700 per ton. In running this drift, over 2 tons of ore were taken out, which netted, after deducting $96 for mining, $60 for milling, and $20 for hauling to Georgetown, $43

above expenses. Four hundred feet below the discovery another 10-foot shaft has been sunk, showing two veins of ore, one 1 inch, the other 2½ inches thick, assaying, respectively, 136 and 123 ounces silver per ton. Sixty feet west, and parallel with the Dresden, is—

2. The Bismarck, owned by the same parties, running northeast and southwest, and dipping slightly northwest; opened by shaft 11 feet deep, and showing a crevice 4 feet wide and a pay-streak of 4 inches of decomposed argillaceous material, impregnated with fahlore, zinc-blende, and galena, worth 73 ounces per ton. About 150 feet above the discovery-shaft there is a large "blow-out," 40 feet wide, showing five distinct crevices, two of which are mineral-bearing. Some 100 feet below the shaft the ore-vein pinches up to one-half inch of solid zinc-blende, assaying $1,300 per ton. The mineral crops out for over 400 feet, and the crevice is well defined at the base of the mountain, where it comes down a nearly perpendicular cliff 50 feet high. The rocks project here about 20 feet on each side of the crevice, forming the best natural tunnel-site in the district. It is contemplated to work both lodes, and several others close by, by means of a tunnel run in on the Bismarck, and cross-cuts run both ways to the other lodes.

3. The Fourth of July, owned by an English company, who bought, for $10,000, one half, which they are developing now; the other half is owned by the discoverers, E. Riley & Co. Crevice 2 feet wide, running north and south, and dipping east. It is opened in several places by small shafts, from 3 to 6 feet deep, for a distance of from 700 to 800 feet, showing in every one a well-defined crevice, with a streak of quartz, from 1 to 5 inches wide, containing more or less fahlore, carbonate of copper, and fluorspar. At the discovery-shaft a pocket of solid fahlore, from one-half to 2 inches thick, was found, assaying $2,800 per ton. This lode was the first rich discovery made this season, and the excitement which the "big strike" caused gave a great impetus to prospecting, followed by the discovery of several other lodes of the same character, i. e., carrying only silver-ore proper, green and blue carbonate of copper, and fluorspar, mixed with more or less quartz. Among these discoveries the following three take a leading position.

4. The Minneapolis, discovered and owned by F. Smith & Co.; course, northeast and southwest; dip, northwest; worked by adit 60 feet long; crevice, 3 feet wide; mineral streak, 2 to 5 inches, consisting of fahlore, silver-glance, carbonate of copper, and specks of galena. Value of ore from 250 to 1,000 ounces per ton.

5. The Fifth of July, owned by A. Wolters and Charles Myers; course, northeast and southwest; dip, about 75° northwest; worked by shaft 20 feet deep; crevice, 3¼ feet wide; ore-vein, 1 to 4 inches, containing fahlore, silver-glance, and carbonate of copper. Value of ore from 124 to 882 ounces per ton. The ore taken out paid a profit over all expenses, though worked in frozen ground at an elevation of nearly 13,000 feet.

6. The General Moltke, discovered and owned by A. Wolters; bearing, northeast and southwest; dip, slightly northwest; cropping out for a distance of 400 feet; crevice, 3¼ feet wide, with a streak of quartz from 4 to 16 inches thick, carrying fahlore and fluorspar; worked by shaft 12 feet deep; pay-streak 3 inches, with several small streaks of an intimate mixture of fahlore and fluorspar, aggregating about ¾ inch of solid mineral, worth 630 ounces per ton. One other lode of this kind was recently found near the General Moltke, and still another one in the vicinity of the Fourth of July, but I have no data in regard to them.

7. The Grunow, owned by William Mendenhall; course, north and south; dip, slightly west; crevice, 3¼ feet wide; cropping out for a couple

of hundred feet; worked by shaft about 20 feet deep; ore-vein from 1 to 6 inches, containing quartz, with galena, fahlore, and carbonate of copper and copper pyrites, assaying 300 ounces.

8. The Wayne County, owned by J. Mavis, P. Petersen, and P. Beauregard; course, northeast and southwest; dip, northwest; crevice, 4 feet wide; mineral streak, 2 to 3 inches of very rich decomposed galena, with fahlore; worked by adit 12 feet long.

9. The Worcester, owned by the same parties; bearing and dip the same as the foregoing; shaft sunk 10 feet; crevice, 2½ feet wide; ore-vein, 1½ to 3½ inches of decomposed galena and zinc-blende, with fahlore to a depth of 6 feet; then 4 inches quartz, with a small quantity of galena and carbonate of copper.

10. The Pocahontas, owned by R. Wood & Co.; course, northeast and southwest; dip, northwest; worked by adit 50 feet long; crevice, 2½ feet wide; ore, galena.

11. The Brooklyn, owned by R. Wood and William Mendenhall; worked by adit 20 feet long; just coming into ore; crevice, 3 feet wide; ore, galena.

12. The Muscatine, owned by R. Wood & Co.; bearing, northeast and southwest; dip, northwest; worked by adit 20 feet long; crevice, 3½ feet wide; ore, galena.

13. The Essex, owned by Isaacs, Wolters & Bechtel; course, northeast and southwest; dip, northwest; shaft, 10 feet; crevice, 3 feet wide; ore-vein, 1 to 3 inches of galena and zinc-blende; worth 60 ounces per ton.

14. The Dickey, owned by Dickey, Crocker & Kinread; shaft, 12 feet deep, showing 2 inches of galena and zinc-blende; milling 240 ounces per ton.

15. The Growler, owned by the same parties; ore-vein, 3 to 4 inches; shaft, 12 feet deep. Both veins are not worked on account of location.

16. The Mountain Lion, owned by J. Williams and A. Bechtel; bearing, nearly north and south; dip, west; crevice, 5 feet wide; shaft sunk 10 feet. There are several small pay-streaks, aggregating about 6 inches, worth 160 ounces per ton.

17. The Goslar, owned by A. Wolters and A. Bechtel; course, northeast and southwest; dip, northwest; worked by adit 12 feet long; ore cropping out 100 feet above; no ore in adit; crevice, 4 feet wide.

18. The Praga and—

19. The Slovan, both owned by J. Shimmel; course, northeast and southwest; dip, northwest; crevice, 2½ and 3 feet wide; both showing 1 inch mineral on foot-wall, containing fahlore, galena, carbonate of copper, and zinc-blende, with rich pockets.

In November, work was going on on the following ten lodes: Baker, Stevens, Coney, Fortunatus, Fourth of July, Dresden, Minneapolis, Wayne County, Pocahontas, Brooklyn, all the rest lying idle from one or another of the reasons already given. After the Baker Company's Mill shall have been rebuilt, work will probably be resumed on as many more; but to make the district as productive as it could be and ought to be, from $200,000 to $300,000 ought to be expended in proper development.

The bullion shipment from Clear Creek County during 1871 was, according to the Georgetown Miner, as follows:

By the Stewart Silver-Reducing Works	$239,528 60
By the Palmer & Nichols Silver Works	100,002 49
By the Brown Company	25,845 00
By the International Company	26,125 70

By the Baker Company $4, 509 55
Gold from Empire and Idaho, (alluvial washings) 20, 000 00

    Total in bullion 416, 011 34
    Ore shipment estimated at 453, 035 00

                                            869, 046 34

The Stewart reduction-works treated 1,801 tons, which produced an average of value per ton of $135.

The Palmer & Nichols works treated 528 tons, which average $182.66 per ton.

In Boulder County, the Grand Island district has attracted the most attention. But although nearly three years have now passed since the first rich discoveries, there is to this date no mine developed so as to insure a steady supply of ores, with the single exception of the Caribou. This is the mine that created so much excitement in regard to the district, and which caused the discovery of a host of others. Many of the latter, it is true, promise, at the slight depth to which they have been opened, to become as good mines as the Caribou; but 30 or 40 feet shafts are not enough to reveal the true nature of a mine or the ore in it. The deepest mines, setting aside the Caribou, are the Idaho and Boulder County, the shafts on these being respectively 45 and 50 feet deep. My opinion is that very many of these mines will turn out to be very rich and valuable, but at the time of Mr. Eilers's visit they were so little developed that it was impossible to form an intelligent opinion as to their value. There has been, so far, very little capital brought into Grand Island district, and this, coupled with the high cost of transporting ore to Black Hawk, the nearest market heretofore, has prevented as extensive developments as the mines appear to justify. I am confident, however, that now, since the splendid mill of Mr. Breed, at Middle Boulder, has gone into operation, the district will be rapidly developed, whether foreign capital offers its help or not.

The Caribou mine (claim 1,400 feet on the vein) has been described, as it appeared last year, in my previous report, to which account I have little to add. When Mr. Eilers saw the mine in October, Mr. Breed, the new owner, was continuing with energy to sink the shaft. The main shaft was 205 feet down, and the one 110 feet to the east of it, 115 feet, the two being connected by drifts. The vein was throughout the shafts, drifts, and stopes from 3 to 5 feet wide, though in one place it had bulged out to a much greater size. This, however, continued only for a few feet. The vein does not show well-defined walls. A hanging-wall especially can never be recognized, while a foot-wall shows itself in spots, being there separated from the vein by a very thin selvage. But in the greater part of the workings there was nothing found to define the vein sharply. The gangue is a very hard and tough quartz. There is sometimes only one pay-streak, from ½ to 4 or 5 inches wide, and in these cases this is exceedingly rich. At other times there are a great many thin seams of high-grade ore running through the vein. But very rarely is the quartz interspersed with silver-bearing minerals throughout its width, or even its greater part.

There is, on the whole, as it would appear to the eye, much barren gangue in the vein; but the owners assert that this "third-class ore" assays $60 per ton. The solid ore-streaks contain sulphuret of silver, stephanite, silver-copper glance, sulphuret of copper, a little galena, and zinc-blende. The capacity of the mine at the time of Mr. Eilers's visit was about 30 tons of second-class ore and less than one ton of first-class daily. The $60 ore was, at that time, thrown aside to be worked in Mr. Breed's own mill in the future. Only the ore between the surface and the first level, 50 feet deep, was stoped out entirely. The second level is driven 100 feet from the top, reckoning from the mouth of the engine-shaft, and 80 feet from the mouth of the east shaft. It connects the two shafts, and is also driven toward the west, where it will eventually connect with the west shaft, now only 60 feet deep. West of the main shaft the ground between the first and second levels was stoped out for a length of 35 feet, and some underhand stoping below the level had also been done here. East of the main shaft, between it and the east shaft, a winze had been sunk from the first to the second level, and a small portion of the ground was stoped out above the second level near the east shaft. The third level in the main shaft was started 180 feet from the top, and had been driven 20 feet to the east and 15 to the west. In the east shaft the third level started from the shaft in its bottom, 115 feet from the top, so that it would not connect with the third level of the main shaft. The level had been driven 25 feet to the east and 30 feet to the west. In the latter portion a chamber about 20 by 25 feet had been stoped out. There was ore visible in every part of the vein exposed, and the reserves were quite large. The first-class ore yielded from $500 to $700 per ton; the second-class from $150 to $200.

Mr. Breed was erecting at Middle Boulder a splendid mill, with four Brückner cylinders, which has since been finished and put in operation. Much delay was caused by the breaking of castings during the earlier part of its running, but it is now reported in working order, and several heavy shipments of bullion have been made.

The hill above and below the Caribou is covered with a complete network of veins, and a great number of locations have been made. But all these veins must, as yet, be considered undeveloped, though small lots of ore from many of them have been shipped to Black Hawk, which generally yielded well. Some of these veins, especially the Perigo, appear more like gold than silver veins, but most of them carry rich silver-ores—so far all decomposed.

A Mr. Kearsing erected during the fall a reverberatory furnace in the town of Caribou, in which it was his intention to smelt ores from the district. It is a pity that such mistakes should still be made at the present time. The Grand Island district contains very few smelting-ores, and none which are free enough from quartzose gangue to be smelted, without enormous loss, in reverberatories. The latest information from that quarter is, that these smelting operations are a failure.

In the Ward district mining is reported to have been carried on quite actively. I am, however, unable to speak, with positive knowledge, as neither Mr. Eilers nor myself could visit the district. In June the Caribou Post reported the following:

On the Ni-Wot hill a crowd of busy men, repairing old buildings and preparing for new ones, give the appearance of reviving prosperity. Adjoining the mill, buildings, one hundred and twenty feet long in all, are projected to accommodate the chlorination-works of Mr. Richardson, excavations for which are now being made. He will use four Brückner cylinders at first, and add others as the business increases. Mr. W. M. Tobie will run the stamp-mill for Smith & Davidson. That wonderful mine, No. 10 west, on the Columbian, discloses an 8-foot crevice in the west drift. This, and a 4-foot crevice on the Benton, are thought sufficient to feed forty-five stamps. Carson, Long, and others are working the Nelson; Mitchell, Williams, Mooney, McDonald, the Benton; Crary, Benson, and others, the Columbian. The miners in Ward have lately gone down into the Columbian shafts, in which the surface quartz has once been worked out, and here they ran a cross-cut for parallel veins. Some of these side veins are found larger and richer than the one originally worked. In this way thousands of dollars will be obtained this season. The width of the Columbian lode at the surface is not yet determined. The gangue which fills the vast space between the walls yields, under stamps, about two ounces of gold per cord. This is easily mined, and may be obtained in such quantities that, if all the stamps in Colorado were put at work on it, the supply would not soon be exhausted. The iron and copper pyrites contain from three to four ounces of gold per ton, and fifteen to twenty ounces of silver per ton, and this gold is diffused, with remarkable evenness, through the ore the whole length of the vein. None of the rock which fills the vast space between the walls is entirely destitute of gold. About 20 per cent. of the assay is saved by the stamp-mills. The lode has yielded, up to the present time, not far from a quarter of a million of dollars, and, in obtaining this, it is probable that not less than three-quarters of a million have been run down the creek and lost to the commerce of the world.

GOLD HILL.—H. Fullen is working a 7-foot crevice on his White Rock lode. The best of his ore—solid mineral, iron and copper pyrites—runs 14 ounces per cord, and it averages so well that he is making a net profit of not less than $100 per day. He is proposing to put up another stamp-mill with steam-power, so as to run next winter.

FOUR-MILE.—Six or seven parties are now working the bed of this creek for gold, and are doing well. Some of the claims worked are well up the creek. It is the opinion of the miners that the coarse or shot-gold obtained is washed down from the head of the creek, which forms on the eastern slope of the bald mountain between this place and Ward. This place (the head of the Four-Mile) is as unknown and unappreciated as Grand Island was two years ago. Without doubt there is a gold area there as yet undiscovered. Rich specimens of gold-ore have been brought in from that locality.

In July the same paper brought additional correspondence from this district:

THE WARD MINING COMPANY.—Ames, Dixwell, and associates, of Massachusetts—E. K. Baxter, of Central, agent—are opening the Volcano, Belfast, and York lodes, and will put a whim on the Manhattan. Their 20-stamper, now run by Mitchell & Williams, is to undergo thorough repairing in view of continuous work. This mill is supplied with percussion concentrating-tables, which, on account of the siliceous character of the Ward ores, work admirably. There is already quite an accumulation of concentrated tailings. This company propose the erection of smelting-works. They have 600 consecutive feet on the Princeton, an easterly extension on the Columbian vein, and the same number of consecutive feet on the Manhattan, besides as much undeveloped property, amounting in all to above 6,000 lineal feet. Their main shaft on the Princeton is 250 feet deep. Their mill-building is a costly and substantial structure, 85 by 50.

NI-WOT.—Sam. Graham is driving forward the construction of Richardson's chlorination-works. Smith & Davidson are running the 50-stamper to great profit. Their main shaft opens into the mill, and is now 220 feet deep. In the bottom the converging mineral-seams are nearly united. It is thought that 10 feet more will unite them in one solid crevice. We are told their purpose is to sink 30 feet deeper, and then run a level each way. The ore above this level will be broken down by overhand stoping, while the main shaft is continued downward. The mine is now in condition, the superintendent reports, to supply Richardson's works in full. The mine will be worked by running levels, and disclosing the ore in advance of present requirements, in order to insure a full supply for larger operations. It has been in an unsafe condition, but is now being timbered and made secure. The mill will also be overhauled and put in perfect order. It is reported that Davidson & Smith are negotiating a sale of a third interest to Mr. Gill, of Denver. It looks as if there would be a stubborn fight for the title to this property.

The Celestial is one of Deardoff's old discoveries, newly opened, and now worked by Benson & Long. The ore prospects for $200 per cord, and there appears to be a great deal of it. The situation, at the head of Spring Gulch, is most convenient. The quartz goes to the James Creek Mill, the most northern stamper in Colorado. Mitchell & Williams, at the depth of 60 feet on the Benton, have passed through the surface quartz to iron and copper pyrites, which run well under stamps.

I am not informed whether the chlorination-works above mentioned have been completed and are in operation.

In Summit County, the placer-mining season of 1871 has not been as prosperous as heretofore, owing to the small amount of snow that fell during the winter, and also to the scanty rain of the summer. The supply of available water has been much less than in average years, and, as a necessary consequence, the amount of bullion produced has been less than usual. Still, the yield of gold per hand per day is reported as nearly one half ounce, and the total shipments of gold from the county are given as 3,700 ounces. Considerable new placer-ground has been discovered and developed; many new ditches have been built, and some companies have made very extensive preparations for next season. Although there were not as many companies at work in French Gulch as the year before, a fair share of placer-mining has been done here. According to a correspondent of the Central City Register of July 5, the following work was going on at that time:

George Day was running two flumes, 600 and 500 feet in length, respectively. He was working tén men, had considerable ground stripped, and was averaging about $10 per day to the man. Calvin Clark was working fourteen men, had in about 1,500 feet of flume, had a No. 1 derrick, and considerable ground sluiced off ready for shoveling, and, as heretofore, was averaging about $10 a day to the man. J. Todd was working four men, had in about 800 feet of flume, and was in good pay. The Badger Flume Company, owned by Rood, Clark, Eyser & Co., William McCartney, superintendent, had in about 500 feet of flume and were working three men. The mine so far had not paid expenses, but they expected to reach bed-rock inside of 200 feet. The Grant Flume Company, owned by Iliff, Pollock & Co., had in 1,700 feet of flume, and were working eleven men. They were running the flume and also side drifts, all of which prospect largely. On Stilson's Patch, west side of French, Mr. Sissler was taking out good pay. Mower & Hays, by means of a tunnel and shaft, developed some good ground in the Patch last winter, having struck dirt which averages 40 cents to the pan. Pearce & Co. (late J. McFadden) intended to start up in a few days. J. Johnson had taken up some old ground in the vicinity. Jeff. Davis and Lilian Patches, on the west side of French, near the head, have yielded immensely, but owing to several reasons they were not worked this season, except by C. H. Blair, who had in about 50 feet of flume, was working in two places, employed seven men, and was obtaining fair pay. Two men were working ground on shares which belongs to Calvin C. Clark, and were making it pay. George M. Clark & Co. were employing two men in opening some new ground, which prospected well. Rippey & Co. were taking out good pay in Webber Gulch. Fred. Dorl and others were booming in Gibson, with good results.

French Gulch is about five miles long, and, with Stilson Patch, has about 17 miles of ditches, 6,700 feet of flume, five hydraulics, and in July had a population of 165.

About the same time, Gold Run was worked by the following companies, who were all averaging about $10 per day to the man: Solon Peabody was working twelve men; Moffat & Shock six; Blodgett & Mayo six; Catel six; Walker six; and John Nolan seven. Buffalo Flats, situated at the lower end of Gold Run, were worked by George Mumford, who employed fourteen men, was running four flumes, and taking out good pay. Gold Run and Buffalo Flats are covered by two and one-half miles of large ditches, and use seven hydraulics. Delaware was being worked by Stogsdill & Twibell, who worked five men, and were taking out their usual good pay. Andy Delaine was working four men, and expected to realize better than last year. Delaware has about six miles of ditches and two hydraulics. Galena was worked by two companies. Riland, Coatney & Roby were working ten men, were running two flumes, and expected soon to make clean-ups similar to last season. Messrs. Roby & Co. were working the upper portion of the gulch, employed a number of men, and were doing well. Galena is covered by a five-mile ditch and uses two hydraulics. Georgia, Humbug, and American Gulches were owned by six companies, and considered the richest in the county, as they yielded from one to two ounces a day to the man, with a few inches of water. Eli Young & Co. were running a bed-rock flume in the Swan, near the mouth of Georgia, for the purpose of striking the pay-streak in each.

In Illinois Gulch William McFadden was working six men. He averaged $10 per hand per day—more than in previous seasons. In Salt Lick Gulch the yield was satisfactory. Toward the end of August, T. H. Fuller & Co. had finished their extensive preparations in Mayo Gulch and commenced working by the booming process, which gave them good results. They were, however, at the same time constructing a ditch from Indiana Gulch, which they hoped would give them sufficient water for ground-sluicing during the next season. At the same time Greenleaf & Co. were mining extensively in Utah Gulch. They were building a ditch from the Blue River to the head of the gulch, and expected by this means to do the largest placer-mining business in the county during the next season. In Hoosier Gulch, in the extreme southeastern end of the county, Bemrose & Co. have been mining with good results, their ground, an old channel, being very rich.

Many new lodes have been discovered during the year in the county, especially in Ten-Mile district, but the principal work in lode-mining was done by the old companies mentioned in my last report. Prominent among these stand the Boston Silver Mining Association and the Saint Lawrence Silver Mining Company.

The Comstock, the property of the Boston Silver Mining Association, was reported, in August, in shape to furnish 20 tons a day, and 1,500 tons of ore were on the dump. The company employed 100 men. A substantial tramway was constructed from the mine to the new mill, which was under construction. It is to have a capacity of 20 tons per day, and will include smelting-furnaces for the beneficiation of the galena-ores, while the greater portion of the ore is to be roasted and amalgamated. There is a 100 horse-power steam-engine at the works. The mine was, at the time mentioned, 260 feet deep, and about 1,200 feet of stoping ground were exposed.

The Saint Lawrence Silver Mining Company has also been energetically at work. Their Silver Wing mine is about 500 feet above the works on Glacier Mountain. A tunnel was being driven, in August, which was expected to be in 200 feet by the 1st of September, and which will give 200 feet of stoping ground. The crevice is 5 feet wide, and the vein of solid mineral about 17 inches in width, which produces about one ton of ore to the foot advanced in the tunnel. The ore assays from 30 to 180 ounces of silver per ton, and contains brittle silver, zinc-blende, antimony, gray copper, and galena. The ore has, so far, increased in quantity as the tunnel progressed.

About 300 feet northward is the Napoleon lode, in which a tunnel is also being driven, which will be as long, and open as much stoping ground as that in the Silver Wing, when the contract for running it is completed. The ore is similar to that of the Silver Wing, but gives a higher assay. A track covered with sheds will connect the Napoleon with the Silver Wing. At the tunnel-entrance of the latter commodious ore-houses are being built for the reception of the ore from the two lodes, and from here a double-track tramway will be laid, on which the ore will be conveyed through the ore-houses to the rock-breaker. The ore will then pass from the rock-breaker on to the drying-floor, which will be heated by the escape gases from the furnace. From this it passes to the stamps, and is then conveyed by two endless-chain conveyances into the weighing-hopper. After weighing it is dumped into the receiving-hopper at the base of the furnace. The ore is then raised by an elevator to the feeding-hopper at the top of the furnace. After roasting and chlorodizing it is drawn from the base of the furnace and conveyed to the cooling-floor. After cooling it is passed into the concentrator, then into the amalgamation-pans, after which the amalgam is retorted.

The main building is 30 by 50, contains one of Howland's 10-stamp rotary batteries, two of Wheeler & Randall's amalgamating-pans, (all cast iron,) settler, and retorts. The furnace-building will be 35 by 40 and 50 feet high, ore-house 20 by 40, and the blacksmith and tool shops will be adjacent. The works will be operated by a 50 horse-power engine; their capacity will be 10 tons per day, and next spring another battery of stamps and two additional pans will be put in, which will double the capacity. The Airey furnace, conveyances, &c., will be similar to Stewart's works at Georgetown, and the furnace will be constructed by the same men who built that of Mr. Stewart. By the arrangement above described it will be seen that most of the labor will be performed by simple mechanical agencies and machinery.

The works were expected to be completed in September, but they were not ready to start at the end of the year.

Of other mines which have become well known during the year, the Chautauqua, Register, Tiger, Coley Extension, and Walker should be mentioned. They are, however, not nearly as well developed as the mines of the two companies above spoken of.

The completion of the reduction-works in the early future will undoubtedly do much for the further development of the quartz interests of the county, which have so far principally suffered from want of a market for the ores.

In Lake County the placer-mining interest has suffered from the same causes which affected Summit. In California Gulch, a tributary of the South Arkansas, the most work has been done, and a few men were at work as late as October. Since the discovery of gold in this gulch, it is estimated to have yielded over two and one-half millions of dollars. The yield this year has not been as large as usual.

On the Arkansas, below Granite, some placer mining was carried on during the last months of the year, when the low stage of the water permitted the working of dirt from the bed. Between forty and fifty men were employed there, as late as December, in "rocking." The yield is reported at $2.50 to $8 per hand per day.

Of veins, the Printer Boy, Pilot, Five-Twenty, American Flag, and Berry Tunnel have been the main objects of attention.

The Printer Boy was discovered in June, 1868, by Messrs. Smith & Mullen. For a year at least very little attention was paid to it by the discoverers and owners; but during their absence other parties jumped it and took out several thousand dollars. This drew the attention of the owners, and a suit of ejectment was commenced. Litigation in this, as in hundreds of other cases, brought the lode more into notice. Since Messrs. Paul, Smith & Co. (now the Philadelphia and Boston Gold and Silver Mining Company) proved the property as theirs, a main shaft has been sunk 190 feet, and a boundary shaft 78 feet deep. Between the two shafts a level has been run 450 feet, over which is a stope of ground 60 feet in depth. The greater part of this was worked out the present summer. One hundred and forty-five cords of ore from this stope, treated at the Five-Twenty Mill, gave an average yield of 18 ounces per cord. In November the mill (one battery) was running on wall-rock that yielded from 3 to 6 ounces per cord. At the mine it was estimated that there were at least 250 cords of waste or wall-rock in the dump pile. During the winter, Mr. Cooper Smith, the mining foreman, intends sinking the main shaft 100 feet deeper, making it 240 feet in depth, and then drifting north 450 feet to the boundary shaft, making a stope of ground 100 feet in depth and 450 feet in length. Two whim-houses have been built during the past season. The excellence and durability of the work on the surface, combined with the safety and neatness with which the mine is timbered, and the manner in which the mine is being opened, are very flattering to the skill of the managers. This company intend to put up a mill of their own next year, which is to be located in Iowa Gulch, and driven by water-power.

Adjoining this, on the north, Messrs. Breece & Co. are working their mine. The main shaft, 130 feet in depth, carries a crevice of pay-ore 6 or 8 inches in width. In the drift running south, 18 feet from the shaft, is a crevice of pay-ore from 6 to 10 inches in width. In the breast of the drift running north there was, in November, an inch of rich gold-ore. In this mine the gold is found in pockets that yield from 5 up to 1,000 ounces.

East of the Printer Boy Mr. John Hoover discovered a lode the last summer, which he christened the American Flag. The first ore treated gave a yield of 8 ounces per cord. In the bottom, 58 feet from the surface, the crevice has split. On the foot-wall the pay is 4 or 5 inches in width, and about the same on the hanging-wall, a horse, 4 feet in width, being between the pay-streaks.

The Five-Twenty, Printer Boy, and American Flag, and Berry Tunnel lodes are in granite, as also is the western wall of the Pilot. Overlying the granite, about 50 feet from this wall, is a stratum of limestone. From here to the Mosquito Range this limestone overlies the whole country, with here and there ledges of schist and granite breaking through it.

Probably next to the Printer Boy in richness is the Berry Tunnel lode, owned by Captain S. D. Breece. A tunnel 100 feet in length has been driven on the vein, the breast of which is 40 feet from the surface. Work has been suspended for several years, no attempt having been made until within the past year to introduce a process for reducing the sulphuret-ores of this locality. Careful assays show that this ore contains a large percentage of gold, silver, and copper. The tunnel is now badly caved in. Within a hundred yards of this lode, to the westward, the limestone makes it appearance.

The Pilot is now opened by the main shaft and three levels, 50 feet of stoping ground being between each two of them, and between the first and the surface. About 20 tons of rich gold-ore have been beneficiated, and much galena is out awaiting the erection of reduction-works.

From Park County, I have only some information in regard to the new silver discoveries on Mounts Lincoln and Bross. They were first discovered late in July or early in August, and the extraordinary richness of the ores soon raised a great excitement. Many prospectors hastened to the scene, and the location of an immense number of claims was the consequence. The ore occurs in limestone, which here covers the whole country, evidently in deposits, not veins. Not many developments have been made, the time of the miners having principally been spent in prospecting.

On Mount Bross the Moose, owned by Myers, Plummer & Dudley, is opened on the surface about 400 feet in length, and in depth about 20 feet. The vein is about 2 feet wide, and the ore averages, by assay, $460 per ton. The company were preparing to ship 30 tons of ore to Swansea, Wales, in November; cost of shipment will not exceed $70 per ton. This company also own the Dwight, which is developed similarly to the above, and contains about the same grade of ores. Ten tons from this will be shipped, making 40 tons in all. The Fairview, owned by Myers & Plummer, has been traced several hundred feet. The depth is 10 feet, which shows two veins, one of galena 2 feet wide, and the other of honeycomb quartz 18 inches wide. The latter prospects well in gold. The Tar Heels, owned by Burroughs & Co., though only sunk 6 feet, assays $800 per ton. The Park Pool Association owns numerous veins on both mountains, all of which contain good ore. This association was organized by Judge Stevens. Messrs. Safford, Sykes & Co. own four on Mount Lincoln; one of them, the Muskox, contains ore equal to any yet found; two assays recently made by Professor Schirmer are reported as yielding $473.80 and $1,326.50. A. M. Janes owns seven veins on Mount Lincoln, two on Mount Bross, and three near the head of Blackskin Gulch, which assays up to $700 per ton.

The discoveries have been pre-empted as "lodes," "ten-acre lots," "one hundred and sixty acres," and "fifteen hundred feet square," thus showing that nobody is certain in which form the mineral bodies occur. Mr. Stevens, I am informed, started the "acre" method, and called it "placer-ground."

The advantages of the district are an abundance of wood, coal, water, hay, and hardy grains and vegetables. Hay is delivered near the mines at from $20 to $25 per ton, and could be contracted for $15, less than half the Central price. The South Park is an extensive and natural hay country. Vegetables are cheap, brought in from the lower part of the Park. The coal mines on George Licner's ranch, ten miles easterly from Fairplay, are spoken of as excellent in quality and especially valuable to coke. Quartzville, one mile southwest from Montgomery, at the base of the range, is to be the supply-point to the mines, from which a wagon-road will extend to the workings. The place selected for the reduction-works is at the junction of the Quartz Gulch and the Platte River, where another town is probable. In November there were not more than 6 inches of snow, while there was a foot or more at Montgomery, and 18 inches at Breckinridge; but this is undoubtedly exceptional. Usually towns in that section, close up under the range, are accessible at all seasons of the year.

The Moose Company intend to prosecute their work through the winter. They have commenced to tunnel the mountain about 800 feet below the lode, and anticipate no inconvenience from wintry weather, and, indeed, the difficulties of mining in the high altitudes are not as great as popularly supposed, providing the mines are inclosed. It is a common remark among miners that they prefer the cooler and more even temperature of the high mountains to the sometimes hot and sometimes cold climate of the valleys.

There will probably be a great rush to the new mines early in the spring, and no doubt much litigation will result from the mixed methods of location.

CHAPTER IX.

WYOMING.

In my last report I mentioned the extensive coal-field of Wyoming, without giving any detailed description of the mines. Since then the business of coal-mining along the Union Pacific Railroad has assumed such large proportions, and the lately-developed base-metal mines of Utah render the existence of mineral coal in that region so important for the extraction of the metals from the ores, that I have considered it my duty to examine this subject closely. My deputy, Mr. A. Eilers, who was charged to make the field-examinations, was freely assisted in his endeavors by the superintendents of the two principal coal-mining companies, Mr. Thomas Wardell, of Rock Springs, and Mr. Charles T. Deuel, of Evanston.

Coal has been discovered in many localities, from a point 100 miles west of Cheyenne, to Echo Cañon in Utah. Three coal-beds have been principally worked, at Carbon, at Rock Springs, and at Evanston. The geological horizon of these beds in relation to each other has not been definitely determined, but from the general westerly dip of the strata, it is inferred that the Carbon coal is the lowest of the three, and the Evanston bed the highest. A still lower one has lately been opened, immediately at the western slope of the Black Hills. When the local disturbances, which are of frequent occurrence, shall have been better studied than is at present the case, this order of superposition may possibly be found different.

The Wyoming Coal and Mining Company, which has the contract to supply the Union Pacific Railroad, works mines at all three of the above-named points, and the Rocky Mountain Coal and Iron Company, which supplies the Central Pacific road, works the same bed in three different places at Evanston.

The Carbon seam, one hundred and forty miles west of Cheyenne, is opened immediately by the side of the railroad-track by a shaft 70 feet deep. Like all the Wyoming coals, this coal is a lignite, but very compact, and full of resinous matter, which. being finely distributed throughout the bed, is also often found in translucent patches very similar in appearance to amber. Before November of last year this coal was extensively mined, but the unfortunate fire which broke out in the bed at that time has closed the mine. Spontaneous combustion is reported to have been the cause of the conflagration, and it was found impossible to stop it speedily. It was finally extinguished, after the pillars had been burned out, by the caving of the overlying strata. The coal is 8 to 10 feet thick, and the mine was being re-opened in the summer, and has, no doubt, resumed active operations by this time. The coal is found to be the best for gas purposes west of the Missouri. The following analysis of the coal is furnished by Mr. Wardell: water, 6.80; ash, 8.00; volatile, 35.48; fixed carbon, 49.72.

Rock Springs lies in the midst of the Bitter Creek desert. The coal mines are about a mile east of the station, and close to the track. A little village has sprung up here, most of the houses being owned by the company, and inhabited by the miners. There is no sweet water in the vicinity, and for domestic purposes it is therefore brought by rail from Green River, fourteen miles farther west. There is, however, a sulphur-spring about a mile from the mines. The valley of Bitter Creek, a dry stream in the summer, is here entirely underlaid by a seam of coal 10 to 12 feet thick, and a smaller one of about 1 foot above it. Toward the east the same seams appear in a small hill of 8 or 10 acres, some 150 feet above the valley. All around this hill the coal is exposed to view, except in a narrow strip of about 200 feet in width, where the bed connects with the portion running under the valley. The strike of the rocks is here nearly north and south, the dip 5° to 8° slightly south of west. At present coal is only extracted from the hill by pillar-work, but an incline has been sunk on the same bed, close to and under the railroad-track, about a half mile west of the hill. The mouth of this incline is intended to be the point to which the coal from under the valley, and that from under the slope of the hill, is to be brought, the work being done by the same engine.

The Rock Springs coal is very firm, and full of resinous matter. It leaves very little ash, and does not fall so easily to pieces, on exposure to the air, as other Wyoming coals. It is a good gas-coal, and well fitted for steam purposes, and for use in reverberatory furnaces. For the blast-furnace it is not applicable, as it does not coke, and splits up into small angular fragments on exposure to the heat. The daily supply from the mines is about fifteen car-loads, or 150 tons; but in winter this production is increased to 200 tons a day. The floor of this bed is a coarse white sandstone, on which lie 8 to 9 feet of very clean coal. Next comes a seam of slate, from 1 to 3 inches thick, and above this are from 3 to 3½ feet of coal, overlaid by an arenaceous shale. The coal is only removed up to the band of slate, which, together with the coal above it, furnishes a very good roof. An analysis of this coal gives: water, 7.00; ash, 1.73; volatile, 36.81; fixed carbon, 54.40.

At Evanston the coal-bed now worked measures at least 22 feet in thickness, and in the middle mine of the Rocky Mountain Coal Company the thickness of the vein is even 26 feet. The mines are three miles northwest of Evanston, in Bear River Valley. The bed shows, in the different inclines sunk upon it, a dip of from 20° to 26° northeast. Commencing from below, the strata exposed in the Wyoming Coal Company's mine show the following order and thickness:

	Feet.	Inches.
Soft shale	4	0
Indurated calcareous clay	4	0
Calcareous and argillaceous spherosiderite	0	8
Indurated calcareous clay	4	6
Coal	0	4
Argillaceous limestone	2	0
Slate, with thin seams of coal	3	0
Coal, with 9 inches slate in the middle of the bed	4	0
Shale and slate	10	0
Dark bituminous fire-clay	5	0
Slate, with coal-seams and impressions of leaves	2	0
Coal	8	0
Hard coal	0	3
Coal	0	4
Slate	0	3
Coal	2	0
Slate	0	5
Coal	2	0
Slate	0	2
Coal	8	0

		Feet.	Inches.
Slate		0	4
Coal		0	10
Sandstone		1	0
Shale		4	6

The incline in this mine is 850 feet long. Galleries are driven both ways, at 150 feet below the surface, and again 192 feet below this point. The upper one was, in the summer, driven in on a level 600 feet to the northwest, and 450 feet to the southeast. From the galleries, oblique ascending gangways are driven on the upper side every 50 feet, and from these the chambers start right and left, the breasts being 20 feet, and the pillars 18 feet wide. Ventilation is as yet satisfactorily maintained by a small furnace in the mine. The pump is in the bottom of the mine, and the steam is conducted to it from above.

The Rocky Mountain Company has opened the bed on the same hill by three inclines, from the middle one of which all the coal was raised in the summer, the others being not yet quite prepared for work. It was expected that by January 1, 1872, the machinery on all three inclines would be in running order. Their combined capacity will be 1,000 tons per day. On the middle mine the company have erected an excellent 60 horse-power hoisting-engine, built by Booth & Co., of San Francisco, with which they hoist ten cars at a time. The incline is straight to a depth of 270 feet, where a fault in the vein was encountered, letting down the coal abruptly 8 feet. A slight curve in the incline was here necessary, from which it is sunk again in a straight line to a total depth from the surface of 486 feet. With the exception of a slight variation in the size of the working chambers and pillars, this mine is worked on the same plan as that of the Wyoming Coal Company. The lower seam, which is alone taken out at present, as in the Wyoming Company's mine, is here 9½ feet thick instead of 8 feet, as above given in the enumeration of the strata. The coal and slate above make a firm roof, if the chambers are not over 18 feet wide. It is intended to mine in the future workings the upper 8-foot bed. For this purpose one side-gallery has been run obliquely across the coal-bed on the lower side of a main gangway until it struck the roof of the upper seam. The sandstone roof seems to be sufficiently strong. This mine delivered 150 tons per day.

The northern incline was in 120 feet, and two gangways were started from it 85 feet from the surface. The coal-bed is here also 26 feet thick. Machinery for this mine was on the ground, and soon to be erected.

The incline on the southern mine, nearest to that of the Wyoming Coal Company, was down 386 feet, and two sets of levels were driven in right and left for some distance. The mine was, however, not worked yet, the engine not being in position. The work in all these mines has been done very neatly and accurately. On January 1, 1872, the depth of the three inclines on the Rocky Mountain Company's property was reported to me by the superintendent, Mr. Deuel, as follows: No. 1, 386 feet; No. 2, 512 feet; No. 3, 290 feet.

The Rocky Mountain Company employs mostly Chinese, a sufficient number of English and American miners being only retained to train the former. The Wyoming Company employs English, Scotch, and American miners at Evanston, and Scandinavians at Rock Springs. Wages vary from $1.50 to $2.50, with board. The Evanston coal is clean, and exhibits almost no stratification, while cross-seams are extremely numerous, so that undercutting is carried on at a disadvantage, and the production of a vast amount of slack is the consequence, which is filled in on the lower side of the main gangways so as to level them. The coal, and especially the slates, containing much iron pyrites, and the layers of slack often being from 4 to 5 feet thick, there is great danger of spontaneous combustion; and the Wyoming Company intends, therefore, to hoist in future the greater part of the small coal and burn it on the surface.

The Evanston coal analyzes, according to Mr. Thomas Wardell, as follows: water, 8.58; ash, 6.30; volatile matter, 35.22; carbon, 49.90.

The following are tabular statements of the coal mined and shipped by the Wyoming Coal and Mining Company, and by the Rocky Mountain Coal and Iron Company, since the mines were started:

Coal mined and shipped by Wyoming Coal and Mining Company, from August 1, 1868, to December 31, 1871, in tons.

Months.	1868.			1869.			
	Carbon.	Point of Rocks.	Rock Spring.	Carbon.	Point of Rocks.	Rock Spring.	Evanston.
January..........................	....	....	....	2,360	1,330	1,287	808
February	....	....	....	300	676	1,218	68
March...........................	....	....	....	757	647	763	....
April............................	....	....	....	1,430	180	333	....
May.............................	....	....	....	1,049	101	386	....
June............................	....	....	....	541	597	957	....
July.............................	....	....	....	921	946	1,377	....
August..........................	650	....	....	3,481	348	1,591	....
September	1,010	....	....	3,421	378	1,539	....
October.........................	1,400	....	....	4,038	61	1,972	....
November.......................	1,900	....	....	6,977	160	2,487	204
December	1,600	1,830	365	5,853	aband'ed	2,913	886
Total...................	6,560	1,830	365	30,428	5,426	16,903	1,966
Average number of men employed	119			163			
Distribution of coal.	*Tons.*			*Tons.*			
Union Pacific....................	8,735			53,733			
Other parties	....			990			
Total...................	8,735			54,723			

Months.	1870.			1871.		
	Carbon.	Rock Spring.	Evanston.	Carbon.	Rock Spring.	Evanston.
January	6,801	3,479	1,400	2,235	4,639	2,629
February	4,522	1,887	1,552	3,010	4,423	1,240
March	3,293	967	1,608	3,200	4,582	1,681
April	3,677	605	1,276	1,528	2,564	1,708
May.............................	2,689	1,389	639	2,104	1,159	1,670
June	3,355	1,541	408	2,349	3,102	1,531
July.............................	3,685	976	285	2,635	2,025	1,004
August..........................	5,080	886	726	2,790	3,080	1,266
September.......................	4,973	1,417	952	3,504	2,706	1,418
October..........................	5,534	2,009	667	3,713	3,868	2,034
November.......................	4,016	2,642	838	2,904	3,685	2,883
December	4,334	2,348	1,991	1,594	3,757	2,009
Total	51,519	20,390	12,447	31,687	40,498	21,163
Average number of men employed........	233			240		
Distribution of coal.	*Tons.*			*Tons.*		
Union Pacific	80,003			89,034		
Other parties	4,353			4,314		
Total.......................	84,356			93,348		

Statement of coal mined by the Rocky Mountain Coal and Iron Company, at Evanston mines, Wyoming Territory, in the years 1869, 1870, 1871, and the first quarter of 1872.

1869 and 1870.

	Tons.
Mined from October 24 to December 31, 1869	2,473
Mined from January 1, 1870, to December 31, 1870	18,187

1871.

	Net tons.
January	4,655
February	5,871$\frac{7}{20}$
March	6,726
April	4,212$\frac{12}{20}$
May	1,219$\frac{5}{20}$
June	2,179$\frac{5}{20}$
July	3,317$\frac{19}{20}$
August	3,859$\frac{5}{20}$
September	5,483
October	6,043$\frac{8}{20}$
November	4,493$\frac{11}{20}$
December	5,808$\frac{17}{20}$
Total	53,869$\frac{7}{20}$

1872.

	Tons.
Mined in January	8,481$\frac{7}{20}$
Mined in February	7,596$\frac{14}{20}$
Mined in March	8,856$\frac{8}{20}$

The Seminole mines.—At various times during the last year accounts reached me in regard to the discovery of gold and silver mines in the Seminole Mountains, situated about thirty miles north of Fort Steele. The best and most reliable account of this discovery and a description of the mines has been furnished by a letter of General Morrow, of Fort Steele, to Dr. Silas Reed, the surveyor-general of Wyoming Territory. It appears from this letter, that as early as 1869 silver mines were discovered in the Seminole Mountains by three miners, who were all subsequently killed by Indians. They were met on their return from their discovery by Lieutenant R. H. Young, with a detachment of soldiers, and gave him some specimens. These samples gave by assay the extraordinarily high yield of $2,000 in silver per ton. In consequence of this a party was organized last June by General L. P. Bradley and Captain Thomas B. Deweese, to explore the Seminole Mountains for the silver mines from which the samples had been brought to the post. The mines were not found, but gold-bearing veins were discovered instead. General Morrow describes the Seminole Mountains and the gold-veins as follows:

The chain of mountains of which the Seminole Range is a section has its rise about the forty-first parallel of latitude, near North Park, in Colorado, and runs in a northerly direction to Fort Fetterman, where it bends suddenly to the west, and then trends a little north of west until it meets the Wind River Range near South Pass.

From North Park to Fort Fetterman, and thence to the point where the North Platte River breaks through, making a grand cañon, the range is known as the Black Hills,

West of the Platte Cañon it takes the name of the Seminole Range, which it retains until it unites with the Sweetwater Mountains, a little east of the one hundred and eighth meridian line.

The mines are situated in the Seminole Mountains, about eight miles west of the Platte.

Geologically these mountains belong to the igneous or metamorphic period, as is shown by the character of their mineral-bearing rock, as contrasted with the succession of later strata reclining against their sides. The highest peak, Bradley's Mountain, is 9,500 feet high, as determined by an aneroid barometer; but the mean elevation of the range is thought to be something less than 8,000 feet. The average width of the range is about three miles.

The character of the mountain-rock indicates that it has come up from a great depth, being highly metamorphosed; but the slopes are not, as a rule, precipitous, and there is hardly any portion of the mountain, in the vicinity of the mines, which does not admit of roads being built without much labor or expense.

The mines, as before stated, are located eight miles west of the Platte, in a group or cluster of elevations, of which Bradley's Mountain is the highest by nearly a thousand feet. The principal deposits thus far found seem to be confined to a single elevation, known in the district as Gold Peak.   *   *   *   *   *   *   *

The country has been imperfectly prospected, and it may be that hereafter the deposits of gold and silver will be found to have a more extensive range than at present ascertained. Many claims, perhaps one hundred, have been located, but the true fissure-veins do not exceed a dozen or fifteen.

The Ernest, the Mammoth, the Break of Day, the Jesse Murdock, the Slattery, the Edward Everett, and several other mines, have well-defined quartz-veins through which gold is disseminated in large proportions. On these and some other claims the work of sinking shafts and running tunnels is being pushed forward rapidly. In all of the above-named locations free gold is found.

It has been thought by some persons that the various fissure-veins in this district are "spurs" from the Ernest lode. In this view I do not concur, for two reasons: first, because the strike of the several veins of fissures does not concur in direction; secondly, because the vein-matter of the several veins is not by any means the same.

In some instances rich copper-colored quartz largely predominates; in others, the quartz is deeply discolored by protoxide of iron. Again, in some of the veins the quartz is almost a pure white, while in others it is greatly decomposed. If anything may be inferred from the dip of the several fissures, this may also be urged against the theory of a single-fissure formation, for I observed that the dip varies in the several mines from almost a vertical to a slope of a few degrees. The dip is not the same in any two veins.

I regard it as quite certain that there are at least a dozen true fissure-veins in the district already developed; and that others will be found hereafter I have no reason to doubt.

I ought to add here that, as a rule, the ledges run parallel, or nearly so, with the axis of the mountain. A true fissure-vein has never been known to give out, though it may "pinch" or be "faulted;" and hence the only question, as it seems to me, in this district, is as to the quality of the ores.

On this subject all that can be stated is, that numerous assays of the ores have been made in Omaha, Denver, and Salt Lake, and in every instance a very large percentage of gold is reported. In several instances the ores have gone as high as $100 to the ton, and in one instance an assay made at the office of D. Buel & Co., Salt Lake, showed $250 to the ton, as reported to the writer by Colonel Buel.

In many of the claims the vein-matter is decomposed quartz, with sulphurets of iron and copper.

The country-rock in the immediate vicinity of the veins is described by General Morrow as micaceous slate and gneiss.

The Sweetwater mines.—These mines have been worked during the year as well as the limited capital here invested would permit. But lately an English company is reported to have bought a two-thirds interest in the Cariso lode, and the whole of the Wild Irishman. They are said to contemplate very extensive mining operations; and all the other mine-owners look forward, of course, with great interest to the developments to be made.

Surveyor-General Reed, who visited the Sweetwater gold mines in August, 1871, gives the following account of the mines and the work going on at that time:

One of the best lodes or mineral belts in the South Pass district commences within a few hundreds yards of South Pass City, and bears off to the northeast. I will mention some of the principal mines upon it. .

The Young America mine is the first one of note, only 300 or 400 yards north of the village, and is situated west of the Cariso Gulch, which was found so rich in placer gold. It is owned by an Ohio company, A. G. Sneath, superintendent. There are two shafts about 80 feet deep, where the vein is about 2 feet wide. The strike of the vein is north 86° east, the lode perpendicular. The ore is said to be worth $40 to $45 per ton of quartz, which is a whitish blue, and carries free gold. This company had a fine mill of ten stamps and a 20 horse-power, which I saw in ruins from fire, in Hermit Gulch, half a mile distant. There is an engine, and good building over the shaft, for hoisting and pumping. .

The Cariso lode is situated upon the hill, east of the Cariso Gulch, about half a mile from South Pass City and a quarter of a mile from the mine just described. Some suppose it to be on the same lode as the Young America, but its strike is north 60° east. It was the first-discovered lode in the district, by H. S. Reedall, in 1867. The party was soon attacked by Indians, and three killed. In the winter following the mining was resumed, and from the croppings of the lode, which they crushed in a hand-mortar, $1,600 in free gold was obtained, and they washed out $7,000 more from the *debris* in the gulch below the vein.   The main shaft is about 210 feet deep, and worked by an engine. Their stamp-mill is on Willow Creek, and run by water-power.

The owner, Mr. Thomas Roberts, has worked the mine with considerable skill and industry, and has made it pay its way, even to the building of a stamp-mill and engine-house, and placing an engine in it. He visited London this summer, by the invitation of some capitalists, who have purchased two-thirds interest, I am informed, for $100,000, and it will now be worked with that energy and skill which will probably result in greatly enhancing the reputation of this important mining district.

The vein-stone, which had been thrown away, and which was found to contain $70 per ton in gold, will now be made to impart its treasure. It holds about $15 per ton of free gold in mechanical combination, and the remaining $55 per ton is probably in the state of sulphuret or other chemical condition, and will have to be extracted by other methods.

The length of the lode is understood to be 3,000 feet, with the discovery shaft near the center; but some of this distance is yet owned by individual parties, in 200-feet claims, and thus there are other shafts than the one the engine is on. Several levels have been run out from the shaft. The dip of the lode is 75° southeast. The average width of ore-streak is 3 feet, between well-defined walls of hornblendic gneiss. The yield of the mine per month is about $5,000 or $6,000, the capacity of the water stamp-mill allowing only about this much. An analysis of the blue sulphuret of iron by Messrs. Johnson & Son, London, gives 3 ounces and 18 pennyweights of fine gold to 2,000 pounds of rock.

The Wild Irishman is supposed to be on an extension of the Cariso lode, upon the crest of the same ridge, several hundred feet northeast. The main shaft is 78 feet deep. The vein is about the same width, and the quartz yields nearly the same per ton as the Cariso. It is owned by the London company before referred to, 1,000 feet on lode; and I am just now informed, while writing, that this is the company that purchased two-thirds of the Cariso mine from Mr. Roberts, and will now work both of these lodes with all the necessary energy and capital.

Mr. Rickard, the superintendent and part owner of both mines, I am informed, will enlarge his operations upon the most approved scale of mining, and will doubtless erect a steam stamp-mill in Hermit Gulch, near the mouth of the tunnel he is running to the Wild Irishman shaft, as the water-mill of the Cariso will not be able to crush half the mineral rock of both mines. They also own the Duncan lode, near by.

The Buckeye Boy is 300 or 400 yards east of the Wild Irishman, at a point of hill on Hermit Gulch. A shaft is sunk, and some drifting done, the material from which indicates a fine vein when fully prospected, as it appeared to be in close proximity to the stratum of gray talcose slate before mentioned. Two industrious miners were the owners, and were at work upon it.

The Carrie Shields lode, situated three-fourths of a mile east of South Pass City, on the north side of Willow Creek, is owned by W. C. Ervin, of South Pass City, to the extent of 1,000 feet on the vein. The strike is northeast, shaft 90 feet, width of vein 2 to 6 feet. The ore yields from $15 to $37 per ton by ordinary stamp process.

I descended into the shaft and found the vein well defined, a good quality of quartz, and I procured some of the decomposed selvage of the vein, which I found quite rich in gold, as shown by washing. I also saw free gold in the quartz, and have no doubt it is a valuable mine. A short tunnel run in from the gulch would intersect the vein about 300 feet below the surface at shaft. The owner is not working the mine this season, and offers it to capitalists for $10,000.

There are numerous other discoveries of gold-veins in the vicinity of South Pass City, with shafts ranging from 20 to 50 feet deep; but as no work is being done on them now, not much could be learned of their yield per ton. These are the Robert Emmett, Nellie Morgan, Golden Gate, Garden City, General Grant, Austin City, &c. Messrs. Thompson & Kimbrough have a prospect named the Tennessee, which, judging from the specimens shown me, promises well.

The Mary Ellen lode has yielded some very rich ore in the croppings, dip 45° north. The hanging-wall consists of slates, the foot-wall of syenite. Some of the ore is reported to have yielded as high as $104 per ton, owing, no doubt, to its contact with the syenite.

The Barnaba, owned by Foster & Co., shows a fair yield of ore, vein 4 to 6 feet wide. It is not worked this season.

Atlantic City, four miles northeast of South Pass City, is situated on Rock Creek, in the midst of valuable mines, and, like South Pass City, has not the population that its advantages and capacity warrant. The gulch diggings in its vicinity yield largely in gold, but the scarcity of water interferes greatly with their proper success. In the bed of Rock Creek, below the village, as high as $100 in gold per day, for each good hand, has been obtained. Upon the north fork of Smith's Gulch, not far from the village, new placer diggings were found this season, which they named Promise Gulch. I found thirty or forty miners at work in them, and they averaged an ounce a day ($18) to each man, with only the water of a small spring, which they used over time and again. Water has since been brought by race several miles, and they now predict that they will obtain $75,000 next season from this gulch.

Wolf Tone lode is situated a short distance above the town, the vein crossing under Rock Creek Branch. It was discovered by the gulch miners working in this creek for placer gold down to the bed-rock, and who there found the vein, which is 2 feet wide, the quartz yielding $40 per ton. Messrs. John Folger, Hughes, and Brennan own 1,500 feet on the vein, which crosses the creek, and is expected to become a valuable mine.

The Buckeye State mine is situated on the ridge northwest of the village, one-half to three-quarters of a mile distant, and is owned by Dr. F. H. Harrison, Edward Lawn, John McCollam, James Forrest, John McTurk, and others, to the extent of 3,000 feet on the lode. It is a good paying mine, and worked with skill and economy, but not to the extent it might be with a larger mill accommodation. Most of the owners work in it themselves, and twenty to twenty-five men were employed at $4 each per day at the time of my visit to it. The main ore pump-shaft is 140 feet deep, and vertical, but cuts the lode at 80 feet in depth. There are only 90 feet of drifts on the lode, 50 feet west and 40 feet east. The width of vein is 2½ to 7 feet, averaging about 4 feet; the strike of the lode north 40° east, dip 60° northwest. They have an engine of 20 horse-power, and 10-stamp mill. The quartz yields $30 per ton. The product, as now worked, is from $50,000 to $60,000 per annum.

The Soles and Perkins lode, owned by Messrs. Perkins, Menifee, Ralston, Taylor, and Logan, has the reputation of being a very good mine, but work is now suspended until an engine and pump can be procured. The shaft is 95 feet, on dip of vein; strike of vein, east; vein, 3 to 4 feet wide. It requires capital to furnish engine for mine and stamp-mill.

The Oriental lode is on the south side of Rock Creek, nearly a mile west of Atlantic City, and owned by Major Horace Holt and Messrs. George B. Thompson, L. Steele, and Peter Haas to the extent of 1,000 feet on vein. The shaft is only 65 feet deep, in which I found the quartz, as well as the selvage matter of the vein, quite rich in free gold. These men work the mine wholly themselves, and they have run in at the base of the hill a fine adit-level, 400 feet toward the vein, which they will intersect at a considerable depth below the shaft. This will doubtless make a valuable mine when fully opened, and the owners deserve great praise for the industry and perseverance they have already shown in developing it. Eleven tons of their quartz, lately crushed, yielded $22 per ton. They found other veins 3 to 4 feet wide along their tunnel, with similar pyritous-gangue rock to that found in the Cariso, which is an excellent indication for the increase of the gold product. They also own a share with Messrs. Jones & Walker in the next 1,000 feet on the southwestern extension.

The Cariboo lode is situated on Rock Creek, above the Oriental, and is owned by Bliss & Co., of California, to the extent of 1,500 feet on the west end, and Cutler & Co., 1,100 feet on east end. The ledge is 3 feet wide, shaft 75 feet deep; rock yields $15 to $20 per ton. They have a 10-stamp water-mill, but are not working this season. The dip of the vein is 60°, and 50 tons of the ore are reported to have yielded $5,000.

The Eldorado mine, formerly owned by Dr. Barr, and now by Mr. Ameritty, of Atlantic, is reported to be a valuable lode; the vein is only 1 or 2 feet wide, but the ore quite rich. The shaft is 120 feet deep. This mine is not worked this season.

The Miner's Delight lode, (west end,) is the richest, perhaps, of all the lodes in this mining district, is situated within the Shoshone reservation, near Hamilton village, four miles northeast of Atlantic, in Spring Gulch, on the north side of the dividing ridge between the Sweetwater and the valley of the Big Horn. The west end, 800 feet, is owned by Messrs. Lightburn, Holbrook, and others. There is a new 60 horse-power engine upon it, sufficient to pump and hoist, and another of 20 horse-power to run a 10-stamp mill which adjoins the engine-house. There are three shafts. The engine-shaft, with three apartments, is vertical, and 150 feet deep. The next is 115 feet, and inclines, with the dip of the lode, about 60° to 70° southeast. The third shaft (whim-shaft) is 85 feet deep. The strike of the gneissic strata is north 40° east. Several drifts have been run on the vein, which averages 3½ feet wide, and carries excellent ore from wall to wall. About 14 inches of it consist of white, transparent quartz, of fine grain, (sometimes of milky and leaden hue,) showing free gold most of the time. The remainder of the lode consists of a selvage of decomposed quartz, next to the wall-rock, of dark, rusty color, and very rich in gold. The width of the ore-streak in the southwest part of the vein varies from 6 inches to 5 feet. The ore, I learn, yields about $40 to the ton on an average.

The Miner's Delight, (east end,) is owned by parties in Tiffin, Ohio, to the extent of 800 feet. I found Mr. Robert H. Morrison, the manager, putting the shafts and levels in true mining order, timbering the shafts and drifts in the best and most approved manner, which, on such a lode, is always the best economy. The whim-shaft is 85 feet deep, and two levels (of 30 and 40 feet) are run each way from the shaft, showing same quality of ore and width of vein as the west end lode. The walls of the lode are smooth and well defined. The lode bends north at its eastern end.

The Hartley lode, owned by the Messrs. Hartley to the extent of 800 feet, is probably on the same vein as the Miner's Delight, which it adjoins on the southwest. The shaft is 100 feet deep, the vein 1¼ feet wide, and drifted upon 100 feet, and the quartz-rock is very rich. But the mine is flooded with water when the Miner's Delight pump does not keep it down, as had been the case the past summer, while the new engine of the Miner's Delight was being set up.

The Peabody lode is southwest of the Hartley, and on the same vein; and is owned by Manheim, Quinn, Frank, Young, Smith, and others, to the extent of 3,000 feet. One inclined shaft is 120 feet deep, the dip being about 45°, and the vein 3 to 4 feet wide. The ore is not as rich as the Hartley, but fair; yields $15 per ton.

Stamp-mills.—Twelve stamp-mills have been erected in this district, carrying about one hundred and sixty stamps, which was double the number required for the small working force and production of the mines. One or two valuable ones were burned, and two were erected on worthless, huge quartz-veins; and one of these is now being taken to the Utah mines.

Gulch-mining.—Gold has been found in nearly every gulch in this district, and some have proved almost as rich as the famous Dutch Flat diggings in California, though of far less extent, the ravines being narrow. But their large yield is the best evidence of the number of rich lodes in this district.

There are six or seven of these rich gulches, which are worked only a small portion of the year, for want of sufficient water, to wit, the Cariso, and Rock Creek, above and below Atlantic, and the Yankee, Meadow, Smith's Promise, and Spring Gulches. The Spring Gulch is just below Miner's Delight, and contains the *débris* of that rich lode. I found thirty to forty men working on it, with only the water pumped from the Miner's Delight engine-shaft. The largest nugget taken from it weighed six ounces. I saw many of an ounce or two in weight, and also saw a lump of gold-quartz, taken from the gravel, as large as a water-bucket, which looked as if it contained a pound or two of gold.

Promise Gulch was discovered this summer, and is a branch of Smith's. It heads up north against the southwest extension of the Miner's Delight lode, on the dividing ridge. It is on the road between Miner's Delight and Atlantic, and I have already made mention of it.

Amount of bullion extracted.—I found it impossible to obtain anything like correct statistics on this point. During the year ending July 1, 1869, the estimate was $155,000 in coin. The product has probably been that much for each of the last two years, and perhaps considerably more, as that is about what is taken from only three of the best mines. This looks like a small amount for so many lodes that yield so well; but it must be borne in mind that it is very little over a year since Indians murdered some of the best young men and miners, within the very center of this eight miles of mines, and killed several other citizens in the valley, not far north of the mines. Miners cannot work and at the same time watch and fight Indians.

This state of affairs has prevented immigration to those mines, and large numbers have been induced to leave and go to the Utah mines within the last eighteen months, where the prospector can pursue his arduous calling, free from the constant apprehension that while he is stooping over his work the arrow of an ambushed savage may pierce his heart.

Fuel for mining purposes.—The question of fuel will become a serious one at no distant period, when capital and experienced labor shall be brought to the energetic development of these mines. Most of the timber for the mines, and lumber for buildings, can be obtained twenty to thirty miles west, in the Wind River Mountains, where there are saw-mills at this time; but fuel for the engines and furnaces, and for domestic purposes, cannot be brought so far except at too great expense.

Coal must therefore be found, and it is possible, and even probable, from what I can hear, that it may be found not far north of the mines—in the "valley," as it is termed.

If it cannot be found, then the next step necessary will be to enlist capital for the purpose of securing the construction of a narrow-gauge railroad from Fort Steele, or Rawlins, via Seminole Gap, and thence up the Sweetwater to the gold mines. This would supply coal from the coal-fields at Carbon, or north of Fort Steele, or from valuable veins that exist not far south of Rawlins. It would also give access to the gold and silver mines of the Seminole Mountains, close to this line of road, which in a short time will exhibit sufficient wealth in mineral products, and so attract public attention and confidence as to command the building of a railroad thus far toward the Sweetwater mines. The whole line would require but very little more grading than for a railroad over an Illinois prairie.

PART II.

METALLURGICAL PROCESSES.

CHAPTER X.

THE SMELTING OF ARGENTIFEROUS LEAD-ORES IN NE-VADA, UTAH, AND MONTANA.

The material for this chapter has been collected in part by my deputy, Mr. Eilers, and myself, by personally visiting the localities and works described. I believe all the works mentioned in the chapter have been visited by one or the other of us; but in many cases the brevity of the time at our disposal prevented us from as detailed an examination as was desirable. The deficiency was remedied by the courtesy of Mr. O. H. Hahn, a metallurgist of skill and experience, well known in the West, whom I requested to furnish me with such data as were at his command. This he has kindly done, and I have incorporated in this chapter his very detailed and admirable article, discussing the principles involved, and giving much information with regard to special operations in Utah and Nevada. I can give Mr. Hahn this general acknowledgment only, since the free use I have made of his materials, interpolating, distributing, and altering, to suit my own views, renders it necessary that I accept the responsibility of this chapter, while I would not deprive any colaborer of his due share of credit. The same remark will apply to the contributions of my deputy, Mr. A. Eilers. Perhaps it would be nearest the truth to say that these two gentlemen are primarily the authors of the chapter, but that their work has been edited after a somewhat arbitrary and self-willed fashion, so that they cannot be asked to adopt and acknowledge it as it stands.

I shall speak here only of such works as beneficiate ores directly in the mining districts. And when I say that more than twenty furnaces exist in Utah, about as many in Nevada, five in Montana, and four in Cerro Gordo, Inyo county, California, it is obvious that a business so extended deserves attention. Wide apart as these different works are located, they have nevertheless to deal in nearly every case with the same or very similar circumstances and conditions, so that, with very few exceptions, virtually the same system of smelting is followed in all these establishments. This is the so-called method of reduction and precipitation in blast-furnaces.

As the principal reasons for the employment of a blast-furnace process, are to be considered: the low percentage of lead in the ores, the high price of the only available fuel, charcoal, and the exorbitant rates demanded for labor. The reasons why the reduction and precipitation process is preferred to a roasting, reduction and precipitation process are the high prices of labor and materials, and the preponderance of oxidized ores over sulphurets, though in some cases the latter are quite abundant.

The weight of these reasons will be better understood when the character of the ores to be treated and the object of the smelting are more minutely stated. The ores are in nearly all cases a preponderating mass of oxidizedlead ores, such as cerussite, anglesite, and leadbillite, in which nests and nodules of undecomposed galena occur. Associated with these are: in Eureka, Nevada, arseniate of iron and arsenical pyrites, hydrated oxide of iron, quartz, and calcareous clay; in Little Cottonwood Cañon and American Fork, Utah, iron oxide, and in some

cases a combination of antimony, the nature of which I have not ascertained; also dolomite and quartz in widely varying proportions; in
Bingham Cañon, Utah, only quartz and comparatively little oxide of
iron, or iron sulphurets; in Cerro Gordo, California, oxide of iron, iron
pyrites, antimonial compounds, copper-ores and, as gangue, carbonate
of lime and quartz. In Argenta, Montana, occur, besides the above-
named lead-ores, pyromorphite, and molybdate of lead. The preponderating gangue of the Argenta ores is quartz, and there is here a larger
proportion of galena than elsewhere in the West. In most of the
localities named, the lead-ores themselves contain sufficient silver to
render its separation from the ore the main object of the smelting; but
in some of the districts, and especially in Montana, the lead-ores serve
only to furnish the extracting-agent for the silver of true quartzose
silver-ores, which at the same time contain a sufficient percentage of
lead to make amalgamation impracticable. They are therefore beneficiated by smelting, although the lead itself has no market value.

As there is more or less sulphur or arsenic present in all these ores, none
of which are submitted to a thorough preparatory roasting, the
formation of matte, or speiss, or a mixture of both, is of course
unavoidable; and as silver has not only great affinity for lead, but
also for sulphur, much of this metal goes with the matte. In
most works the latter is not roasted before adding it to a subsequent charge, if it is at all treated further; and the extraction of
the silver from it is therefore, in this case, only possible after it
has passed the furnace quite often, very little of the sulphur being
driven off in the upper parts of the blast-furnace at each smelting.
In Eureka, a mixture of matte and speiss, the latter predominating, is
formed, the contents of silver and gold in which hardly ever surpass
$12 to the ton; and this amount is at present not considered worth
extraction in that locality. The speiss, or "white iron," as it is there
termed, is therefore thrown over the dump.

The marketable product which the smelting-works produce is argentiferous lead, with the exception of the works at Argenta, Montana,
which cupel the lead and ship the silver only. As a general rule it
pays best in the mining districts to produce argentiferous lead bars or
crude bullion, the contents of which in silver and gold vary from $60
to $500 in the different districts. The main reason for not cupelling the
lead in the West is found in the increased rates and risk of freightage
for bullion; but the separation of the silver and lead, and the refining
of the latter, can also be accomplished at much less cost in the eastern
centers of trade than in the mining districts. There are of course exceptions, as, for instance, in Montana, where the smelting-works are
located so far from the railroad that the price obtained for the lead
would not even cover the cost of smelting and freight, and where only
the silver is therefore shipped, the lead remaining in the furnace-yard
in the form of litharge. Part of this is used over again in smelting
such silver-ores as are naturally too poor in lead; but the greater portion remains to await cheaper reduction and railroads.

A few remarks in regard to the present tendency of metallurgical ideas
as far as smelting is concerned may be here in place.

Formerly the blast-furnaces used for lead-smelting usually had an
oblong rectangular cross-section, the size of the hearth being rarely
larger than 20 inches by 2¼ feet, and frequently they were drawn together at the top. The capacity of such a furnace, with one or sometimes two tuyeres, was about six to eight tons per twenty-four hours.
But of late years essential improvements have been made, the aim of

all of which was a higher production, and less loss of metal in slags and by volatilization. This has been reached by a complete alteration of the shape of the furnaces, by increase of size, and the introduction of proportionately more compressed air through a larger number of tuyeres. In regard to their shape, and the results obtained, two furnaces have come into especially prominent notice. These two are :

1. The *Raschette* furnace. It has an oblong rectangular cross-section and the form of an inverted truncated pyramid. Numerous tuyeres, cooled by running water, are placed in the long sides, in such a manner that the opposite currents of air pass each other.

2. The *Piltz* furnace. It has a hexagonal, octagonal, or circular cross-section, and the shape of an inverted truncated pyramid or cone. Many tuyeres are placed radially around the furnace-center, the breast alone being without them.

Both of these furnaces are furnaces with open breast, and both have the two most important principles in common, *the application of more compressed air in a comparatively smaller space than in old-style furnaces, and a widening of the shaft toward the top.* The first secures a more perfect and rapid combustion, and hence a more rapid fusion; the second causes the smelting zone to commence lower down in the furnace than formerly; the charges, lying firmly upon the slanting sides, force the gases and heat to pass through the whole column above, while the wider section above decreases the velocity of the upward current, and volatilization is to a great extent prevented.

But quite recently experience has taught in this country that a combination of the form of the Raschette and Piltz, so to speak, produces still better results, the capacity of the furnace being thus increased, while the management is less difficult. Last spring two new furnaces were built at the works of the Eureka Consolidated Company, in Nevada. In order to test the comparative merits of a combination furnace and the Piltz, Mr. Albert Arents, the metallurgist of the works, concluded to construct one furnace with elongated hearth, 3 feet wide by 4½ feet in depth, and 6½ feet diameter at the top. The furnace was provided with ten water-tuyeres, four in each side and two in the back wall. The other furnace was a Piltz, of 4 feet diameter in the hearth, 6½ feet at the top, and provided with twelve tuyeres placed radially around the center. Both furnaces were 10 feet high above the tuyeres, and worked admirably, but the combination furnace was found to smelt from one-fifth to one-fourth more ore than the Piltz under the same circumstances. The same experience has been arrived at in the Hartz districts in Germany, where the best results are obtained with a circular furnace of 5 feet diameter at the top, having a hearth 20 inches wide and 3½ feet deep, and seven tuyeres. In this furnace the loss of lead in the slag (although the same charge is used as in the other furnaces) disappears almost entirely, being only ½ per cent., while in the other furnaces it is from 2 to 4 per cent.

The best proportion of the hearth-area to the throat-area may be accepted as 1 : 2½ for a height of from 10 to 12 feet. It is rarely necessary in the western districts to give a greater height to the furnaces. I am at present only aware of the existence of one which exceeds this height. This is situated in Bingham Cañon, Utah, and has to smelt very quartzose ores.

Before proceeding further, I shall give a general statement of the elementary principles of metallurgical operations. It is true that the current text-books on metallurgy cover a good deal of this ground; but I deem it important to introduce this plain and practical *résumé* for the benefit of those who cannot easily consult the best large works upon the subject. Many of these are in foreign languages, others are both dear and scarce, and all, it may be presumed, are more or less difficult of access in our remoter mining regions.

Minerals containing the useful metals in such quantities and in such a chemical combination as to make their extraction profitable, we term "ores," while their earthy portions we designate as their "matrix" or "gangue." In regard to subsequent metallurgical treatment, we can make the following practical classification:

1. *Smelting ores*, viz, ores containing base metals in notable quantities.

2. *Dry ores*, viz, ores containing noble metals and no base ones, or only in limited quantities.

It is my intention to speak here particularly of those pertaining to class 1.

Ores and gangue are always more or less intimately mixed. For the utilization of the metals it is, therefore, necessary to separate them from their gangue by artificial means, which are either of a mechanical or chemical nature. A mechanical separation alone is not sufficient to produce a merchantable product; it can only serve as preparatory to the chemical processes, among which that of smelting will be here specially considered. Smelting is a conversion of solid mineral or mineral and metallic masses into the fluid state by means of heat and chemicals, and the subsequent separation of the metallic from the earthy ingredients by means of their specific gravity. Although there are a great many methods in vogue for utilizing lead-ores by smelting, there are only two which have found application and justly claim attention in the mining regions of the Great Basin: (A) the English process of smelting in reverberatory furnaces, and (B) the blast-furnace process.

The former has some marked advantages over the latter: the possibility of using raw fuel; its exemption from the necessity of using blowing-engines, and the consequent saving of power; an easier control of manipulations, and the production of a lead of better quality in which the precious metals are concentrated. Its general application, however, is greatly impaired by the fact that only comparatively pure ores can be treated successfully. Thus, ores containing a considerable percentage of other metals besides lead, as, for instance, zinc, copper, antimony, &c., or more than 4 per cent. of silica, are unfit for the reverberatory process, silicate of lead, which impedes the process of the operation and gives rise to the formation of rich residues, being formed in the latter case. In the former there is, besides loss in rich residues, also a large one by volatilization. In England the lead-ores subjected to this process contain about 80 per cent. of lead, the gangue generally being carbonate of lime. The English process in its unaltered form can, therefore, only be recommended for pure galenas with calcareous gangue, an ore not often obtained in the western mining districts. To my knowledge there is only one establishment in operation where ores are treated by this process, that of Messrs. Pascoe & Jennings, near Salt Lake City. Another one of this kind, that of Messrs. Robbins, is idle for want of the proper ores. I feel, therefore, justified in omitting to enter upon a more minute description of this process.

Compelled by the high prices of labor, transportation of materials and products, lack of cheap mineral coal, &c., the lead-smelters of the Great Basin have almost unanimously adopted the blast-furnace process of smelting. By its means they are enabled to obtain a salable product in the shortest possible time, and with the least expense, the residues being so poor that they can be thrown away.

To insure success in smelting lead-ores, as all other ores, it is necessary to know their mineralogical character, as well as the chemical properties of the gangue in which they occur. A perfect separation of the ore from its matrix by hand being impossible, and a concentration by water being, in most cases, in the West impossible, on account of the insufficient supply of this liquid, the gangue accompanying the ore must be converted into a fusible compound, termed *slag*. Quartz, we know, is infusible by itself; so is lime; but if we mix both in the proper proportions, and expose them to the necessary heat, the result will be a fusible compound. It has been found by actual experience that not the single compounds of silica and lime, or alumina, magnesia, &c., but double compounds of, say, silicate of lime and silicate of alumina, are the most fusible ones. Replacing one of these bases by alkalies, or the protoxides of the heavy metals, as, for instance, iron and manganese, we increase the fusibility of a slag within certain limits. The fusibility of a slag depends principally upon the proportion of silica to the bases contained in it. Mineral substances which serve to liquefy others not fusible by themselves we call *fluxes*. Under favorable circumstances an ore may contain all the slag-forming ingredients in the proper ratio, but only in a very few instances has nature graciously permitted such a coincidence, as, for example, in Eureka district, Nevada.

According to the ratio between silica and the bases, we discriminate four classes of fusible slags:

1. *Tri-silicates*, in which the silica contains three times the amount of oxygen present in the bases. As there is over 50 per cent. of silica in such slags, they require too high a temperature for their formation to be thought of in lead-smelting.

2. *Bi-silicates*, containing 50 per cent. of silicic acid and 50 per cent. bases, in which the amount of oxygen in the silica is twice as large as in the bases.

3. *Singulo-silicates*, with 30 per cent. silicic acid and 70 per cent. bases, the silica containing as much oxygen as the bases.

4. *Sub-silicates*, with 20 per cent. silicic acid and 80 per cent. bases, the amount of oxygen in the silica being less than that in the bases.

In the latter two the bases are predominant over the silicic acid, therefore they are termed "basic slags," while the first two are termed "acid slags." Chemists have taken the trouble to establish complicated formulas derived from accurate analyses of various slags; but, as they are rarely constant compounds, these formulas have hardly any practical value for the metallurgist; he is content to know the percentage of silica and the quantity of the useful metal which he is endeavoring to obtain. An experienced smelter must be able to draw his conclusions from the appearance of his slag in both the fused and solid states.

The most desirable slag for lead-smelting is the singulo-silicate, or a mixture of bi-silicate with the former, with protoxide of iron prevailing. The singulo-silicates run with a bright-red color, and solidify very quickly with turgescence. The bubbles, after bursting, frequently discharge blue gaseous flames.

These slags have a vitreous, metallic luster, and a higher specific gravity than the bi-silicates, and are, therefore, more liable to entangle metallic particles. If lime and alumina are the prevalent bases, the heat required for their formation is much higher than in the case mentioned before. Such slags are generally pasty, run short, and form incoherent lumps. After solidification they have a honey-combed, stony, or pumice-stone-like appearance, grayish-green color, and radiated, or lamellar-crystalline texture. An earthy singulo-silicate is really almost the least desirable slag for a lead-smelter.

Bi-silicates require a higher temperature, and consequently involve a larger consumption of fuel for their formation than singulo-silicates. They flow slowly like sirup, solidify very gradually, without cracking or bursting, and are not liable to form accretions in the furnace, like basic slags. They appear vitreous after chilling, have a conchoidal fracture, and generally a black color. Being saturated with silicic acid they corrode the furnace-lining much less than basic slags. Their specific gravity is lower and admits of a clean separation of metallic particles; but on the other hand they are apt to take up a large percentage of oxide of lead, and so cause a loss of metal. Furthermore, for their formation it is necessary to have the ore reduced to at least pea-size, which condition is not fulfilled in western smelting-works, where crushers are generally used for breaking up coarse pieces of rock.

Sub-silicates are entirely out of the question, as they are only detrimental. If protoxide of iron is their principal base, they run in a thin stream, like fluid litharge, congeal very quickly, and easily form accretions in the furnace-bottom. Having a high specific gravity, they do not allow a clean separation from the metal. By their corrosive action on the lining, and their tendency to form accretions in the furnace, they shorten a campaign or run to a few days; hence, their production must be avoided.

As *fluxes* the following substances are used:

1. *Acid slags*, for their capability to take up bases, and as solvent agents.

2. *Basic slags*, for their capability of saturating themselves with silicic acid, and as diluting agents.

3. *Iron-stone* is a very efficient agent to slag silicic acid, *i. e.*, quartz, being reduced in the furnace to protoxide of iron, which has a strong affinity for silicic acid, and forms an easily fusible slag. Its price varies in the western districts, according to local circumstances, from $5 to $25 per ton. The best quality for our purposes is hematite or magnetite. Hydrated iron-ores are too easily reduced to metallic iron, and ought to be burned before use. If free from quartz and slag they may be thrown into the furnace in pieces of fist size. Iron-ores are also used as desulphurizing agents.

4. *Soda* is even better than the above as a solving agent for quartz, but it can only be had in a few localities at reasonable rates, the general price being from $60 to $80 per ton.

5. *Lime*, as a partial substitute for iron-stone in solving quartz. It is best used in pieces of pigeon-egg size. From the theoretical standpoint burnt lime would be the best form, but as this is generally in a very fine state, it will partially be blown out at the top of the furnace or roll through the interstices of coal and ore, and thus be prevented from uniting with the silica in the desired proportion. Lime cannot be used by itself as a slagging agent for quartz. Lime-slag is smeary, not very liquid, and deranges the furnace very easily by clogging. The metal separates only imperfectly from it, which is the reason that so much metallic lead is wasted by being thrown away with the slag in some of the limestone districts.

6. *Clay* is only used on a very small scale as a partial substitute for quartz. It must be applied very cautiously, as it often arrives raw at the bottom of the furnace in the shape of dry, incandescent lumps, which stick to the walls and hearth.

7. *Salt* is used by some smelters of Utah who have a very indistinct comprehension of fluxes. Although they allege that it renders the slag liquid, this is an illusion. Any assayer knows that the salt does not enter into a chemical combination with gangues, but forms a slag by itself, which, on account of its lesser specific gravity, floats on the top of the other slag. I noticed slag of this kind at Mr. Easton's furnace, Salt Lake City. Besides its inefficiency upon earthy matters, salt acts injuriously upon the metal by forming volatile chlorides of lead and silver.

8. *Iron pyrites* has been ignorantly used as a miraculous sort of flux. To the skilled metallurgist the effects are obvious, viz, the production of a brittle, sulphureted metal, or of matte, no action upon gangues, and a clogging up of the furnace.

9. *Quartz*, in the form of coarse sand. It is used to furnish the acid for the slag in cases where the gangue of the ores is basic.

In addition to the fluxes enumerated above, I must mention some metallic products occasionally used for various purposes:

1. *Iron*, in the shape of tin scraps, pieces of wrought iron, cast iron, &c., is used to decompose galena, thereby forming sulphuret of iron, (*iron-matte*) and metallic lead. Owing to the high price of iron in Utah and Nevada, it is either replaced by the less efficient iron-stone, or rendered unnecessary by a previous roasting of the ores.

2. *Litharge* was intended to be added to poor lead-ores at Ogden, Dunne & Co.'s works at Eureka, Nevada, in order to prevent the precious metals from being carried into the iron-matte. Owing to the heavy expense of cupelling, and a change of the ore for the better, this purpose was abandoned.

3. *Cinders*, semi-fused matter from previous smeltings to extract the metals.

Fuel.—The only fuel used at present by the lead-smelters of the Great Basin is charcoal, the price of which ranges from 15 to 34 cents per bushel of 1.59 cubic feet, according to locality. The lowest rates are paid at the American Fork and Tintic districts, Utah, where timber is abundant; the highest at Little Cottonwood, Utah, which gets its coal from Truckee, California, by rail, and at Eureka, Nevada. In the latter place the enormous demand has materially influenced the price. The five furnaces of the Eureka Consolidated Company, for instance, consume alone 4,600 bushels daily. The charcoal is chiefly burned from cedar, quaking aspen, mountain mahogany, and nut-pine wood. Nut-pine coal is considered the best, and generally contracted for. The coal-burners make their pits of various sizes, according to circumstances. A pit of 100 cords of green wood burns out in about fifteen or twenty days, and yields from 2,500 to 3,500 bushels of charcoal. The best coal is made about Eureka, Nevada, by experienced Italian coal-burners, the poorest in some places in Utah. The latter is generally made of small timber, and is full of brands and dross. The waste often reaches 15 per cent. As one ton of good, hard coke approximately produces the same effect as 200 bushels of charcoal, it would be a great benefit for the western smelters to use it. But the blast-engines used do not yield a sufficient pressure for a perfect combustion of coke, as experiments at the Eureka Consolidated have shown. The price of charcoal being steadily on the increase, there will be a time when smelters will have to replace the former by coke. It may be reasonably expected that after the Utah Southern Railroad is finished, the development of those fine beds of mineral coal in the southern part of that Territory will tend to the springing up of coke-industry, and so give a new impetus to smelting.

Blast-engines.—The only blast-engines in use in Nevada and Utah are the different sizes of Sturtevant's fan and Root's pressure-blower; the latter, yielding a much higher pressure, is better for lead-smelting, and may possibly compete with cylinder blast-engines, where coke is used in smelting. The only advantages the former have over the latter are their cheapness and the small amount of power they require. A Root's blower No. 8, yielding sufficient blast for three large-sized furnaces, does not require more than twenty horse-power.

Building-materials.—Rubble-stones are used for building the foundations and sometimes the outer casings of furnaces. The latter are generally made of common brick or dressed stone to present a handsomer appearance. Those parts of a furnace, however, which are most exposed to an intense heat and the corrosive action of ore and slag, must be constructed of *refractory* or *fire-proof* materials. Of such we have—

Certain *sandstones,* free from alkaline matter and metallic oxides. A small percentage of iron oxide is less detrimental than alkaline earths or feldspar. An excellent sandstone is found on Pancake Mountain, a series of low hills between the Diamond and White Pine Ranges, distant about twenty-five miles from Eureka, Nevada. Sandstones of the same age—the carboniferous—are also found in the Diamond and White Pine Mountains, but their physical properties, and hence their behavior in the furnace, are different and not satisfactory. The Pancake sandstone has a very fine grit and a light yellow color, and does not crack or fly in the fire after seasoning. Green sandstones of ever so good a quality, and defective ones, viz, such as show flaws or nodules of foreign matter, are not fit to be placed in the furnace. The Pancake sandstone is known to stand for months in a furnace without needing to be replaced. It sells for $20 per ton at the quarry, and $12 additional for hauling.

The coarse-grained reddish sandstones and quartzites of Utah are not to be compared with those before mentioned, and had better be used for outer casings only.

Granite does not often answer the requirements of a fire-proof material, and is mostly used as bottom-stone only. In Argenta, Montana, however, a very quartzose granite is used in the furnaces, and it stands campaigns of three weeks' duration.

Instead of the natural fire-proof stones the majority of smelters use artificial ones, viz, English, Pennsylvania, and Colorado *fire-bricks.* Sun-dried bricks or adobes, molded of various proportions of good clay and coarse quartz-sand, are too expensive, and therefore have gone out of use. They were used in the White Pine Smelting-Works.

The *clay* used about a furnace ought to be refractory, or nearly so, and plastic at the same time. These qualities are combined in the Eureka and the Camp Floyd (Utah) clay; that of Camp Douglas (Utah) is too lean, and that of White Pine (Nevada) almost worthless on account of its large percentage of oxide of iron.

Lean clay serves well enough as a mortar, but is unfit for a great many other purposes, as will be seen below.

Good fire-clay contains from 50 to 70 per cent. of silicic acid, and from 30 to 50 per cent. of alumina.

As a *mortar* for the foundation-walls and the outer casings, a mixture of slacked lime and river-sand is used; for the inside, or lining, however, as for all parts of a furnace directly in contact with heat, a mixture of refractory clay with quartz-sand or ground sandstone has to be used. The clay, of course, must be ground and sifted. Lime-mortar in this instance is unfit for use, as it crumbles off in the heat, and allows the slag in combining with it to creep through the joints.

The annexed sketches show the construction of an improved blast-furnace for smelting lead-ores, such as are now in use in the West, and have given great satisfaction.

Modified Piltz Furnace.—Fig. 1.

The longitudinal section (Fig. 1) is made along the line H Y, in Fig. 2; and the cross-section (Fig. 2) is along the line T V in Fig. 1. A is the shaft of the furnace; B, the chimney; C, the hearth; D, the foundation; E, the bottom-stone; a, the dam-plate; a and b, hearth-plates of cast iron; c, cast-iron pillars, on which the flange d rests; e, dam; f, fore-hearth lying outside of the furnace; g, bridge; h, tymp-stone, or front made of clay; i, breast; k, slag-spout; l, matte-spout, or iron-spout; m, siphon-tap; n, tap-hole; o, lead-well; p, p¹, p², &c., tuyeres through which the blast enters the furnace; q, nozzles, (made of galvanized iron;) r, wind-bags, (of leather or canvas;) s, induction-pipe; t, charging-door or feed-hole; u, throat.

The wall in which the breast lies is called the front wall, the one opposite to this the back wall; the adjoining ones the side walls.

This furnace is called an *open-breasted one*. In foreign countries furnaces with a *closed breast* and without fore-hearth, which have only an opening for the exit of the slag, are often used. Such furnaces are termed "crucible furnaces." Notwithstanding the many advantages they have over the open-breasted ones, they do not permit the detaching of accretions in the furnace, and are, therefore, not suited for our purposes. The mason-work, especially the lower part, of all rectangular furnaces is strongly bound together by 1½-inch tie-rods of wrought iron laid in the outer walls. Each pair of them, lying in the same vertical plane, passes through a wooden, or, better, a cast-iron brace, which is screwed tight to the wall. Round furnaces are tied either by means of iron rings passing around the outside, or by complete shells of boiler or sheet iron.

The height of shaft-furnaces ranges from 8 to 20 feet above the center of the tuyeres. Low furnaces are necessary for basic ores, especially such as carry a great deal of oxide of iron, (White Pine district,) to prevent the reduction of metallic iron. High furnaces are of good service for refractory ores, e. g., argillaceous or quartzose ores, (Bingham Cañon,) and where a bi-silicate slag is desired. In high furnaces a higher temperature is attained with a less amount of fuel than in the low ones. But a low furnace is easier manipulated when deranged than a high one. Where the character of the ores changes frequently a low furnace is preferable. The standard height in this country is 10 feet above the center of the tuyeres. On the top of the furnace is an iron, or, better, brick smoke-stack, high and wide enough to carry off the fumes.

The manner of charging or feeding is of importance, as it affects the working of a furnace materially. Furnaces of small dimensions generally have a feed-hole a few inches above the throat, on that side of the furnace directly opposite the front wall. The proper proportion of fuel, either by measurement or weight, is introduced first, and on the top of that the ore, which may be scattered all over the area of the furnace, leaving an empty space only at the front wall, (Jackson & Roslin furnaces, Eureka; Salt Lake Valley, Stockton, &c.) More capacious furnaces require two feed-holes, which are situated at nearly right angles to the breast, i. e., in the side walls, (Eureka Consolidated Company's and Utah Silver Mining and Smelting Company's new furnaces.) The ore is not spread over the area of the throat, but charged round the tuyere-walls, leaving a core of coal in the center.

To insure regularity in charging, the throat of a furnace is frequently provided with a funnel, the opening of which can be kept closed by a sheet-iron box let down from the top while charging. As soon as it is time to charge the furnace, this box is raised by means of a counterpoised lever, and the charge drops down. After emptying the funnel, the box is lowered again. This arrangement at the same time protects the workman from noxious vapors. Where no condensation-chambers are used, this box runs out into a pipe, which is movable in the stationary smoke-stack, (Richmond furnace, Buel & Bateman's furnaces.) I shall have occasion hereafter to speak more fully about charging.

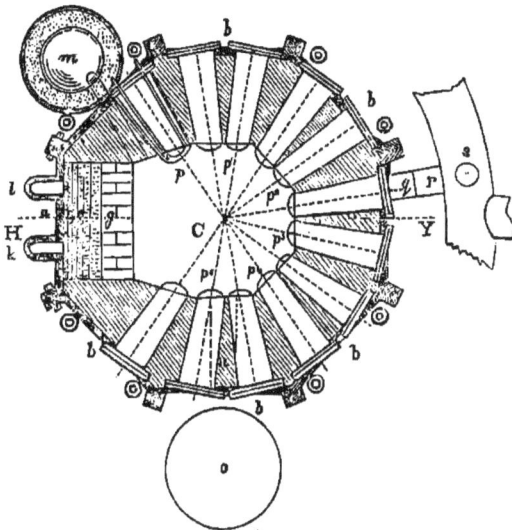

Modified Piltz Furnace.—Fig. 2.

The number of tuyeres and the manner of placing them are really not of so great consequence as is generally assumed, if the proper quantity of air is introduced into the furnace and divided well in the hearth. The majority of smelters in this country place the tuyeres only 6 inches above the level of the slag-spout, and point them downward. This is very faulty in lead-smelting, as it tends to concentrate the heat too far below, volatilizing much metal. Placing the tuyeres too high above the slag-hole is entirely wrong, as in that case the metal in the hearth below cannot be kept sufficiently hot. Before the tuyeres the furnace-temperature is highest. There the separation of the metal from matte and slag, according to their specific gravity, takes place. Below the tuyeres the temperature decreases again. If the tuyeres are, therefore, inserted too high above the slag-spout, the molten masses will stiffen, and even solidify, below. A furnace in White Pine once had the tuyeres three feet or more above the slag-hole. The consequence was a congealing of the fused masses in the hearth, and an entire clogging up of the furnace. The correct way is to place them horizontally, all on the same level, and from 10 to 18 inches above the slag-spout, (Eureka Consolidated Company's and Phenix Company's works.) All vertical dimensions are understood to be measured from the center of the tuyeres. For every $1\frac{1}{2}$ square feet of hearth-area, a tuyere of 2-inch nozzle is required.

Since the introduction of cast-iron or wrought-iron tuyeres cooled by water, the working capacity of lead-smelting furnaces has been greatly increased. Formerly, only sheet-iron, clay, or simple cast-iron ones were in use, giving rise to much inconvenience. In order to protect the furnace-walls from the influence of the reverberated heat, the tuyere had to be provided with a nozzle of clay, or a very acid slag, protruding into the furnace. But, to keep this nozzle or nose of a certain length, and to prevent it from growing or melting off, it had to be constantly watched by attentive and experienced men. During the last century an attempt was made on the Hartz Mountains to increase the production by constructing a large-sized furnace with fourteen tuyeres. It failed on account of the difficulty in keeping the tuyere-walls from burning out. Even the first Raschette furnace, built in 1864, on the Hartz, was provided with sheet-iron tuyeres. But they had to be replaced so often—which always necessitates a stoppage—that it was found expedient to try water-tuyeres, which, indeed, gave entire satisfaction. The best ones in use in this country are wrought-iron ones of the Keyes patent. The lowest point of the hearth is from 36 to 40 inches below the center of the tuyeres, the latter figure being the maximum. If made deeper, the lead will get too cold.

In selecting a furnace-site a great many things have to be taken into consideration in an economical as well as a technical point of view. To answer the latter three conditions are necessary—a sufficiency of water, a spacious ore-floor, and a convenient slag-dump. The lack of one or all of these conditions puts a smelter to great inconvenience, and may even cause a financial failure. After having graded off a suitable location for a furnace at the side of a gently sloping hill, if such a one is convenient, a square or rectangular excavation is made in the ground to receive the foundation. The area is generally 8 by 10 or 10 by 10 feet, the depth depending upon the condition of the subjacent ground. If the same be directly on the bed-rock, as in the instance of the Eureka Company's furnaces, no foundation is required, and a depth of 3 or 4 feet is sufficient to receive the furnace-masonry proper; but if it be moist or in gravel, a depth of from 7 to 14 feet is judicious. The foundation is made of undressed rocks which are laid in lime-mortar, or, better, in cement. The largest ones are used for corners, and the joints must be filled up with spalls. The topmost course, on which the furnace is to be built, ought to consist of dressed stones, well seasoned, and sandstones, if possible. The joints must be perfectly tight. In some instances it is desirable to make provision for draining off the surface-water by arched channels, as the furnace-bottom ought to be absolutely dry.

If the furnace is intended to be provided with hearth-plates, like the one described, those, as well as the cast-iron pillars, are to be put in place now. Then the inside of the hearth-plates is carried up of sandstone blocks 2 feet wide by 1 foot thick, leaving sufficient room for the tap-holes and an open space at the dam-plate. In Eureka, as soon as the mason-work has progressed to 7 inches above the plates, the tuyeres are placed in position and walled in with fire-brick or sandstone. Three feet above the dam-plate the arch over the breast is started and the masonry continued to a level with the top of the pillars. Then the flange which is to bear the upper part of the furnace is put in its place and well bolted to the pillars. The flange is 2 inches thick. The part of the furnace above this flange may consist of inferior sandstone or even common brick, 1 foot or 18 inches thick, as it is less affected by the heat and corrosive action of the ore. About 6 inches or 1 foot above the charging-floor the chimney for carrying off the fumes is started and continued to a height of from 12 to 15 feet, leaving out spaces for the feed-holes 3 feet wide by $2\frac{1}{2}$ feet high at two opposite walls. The chimney ought to have a sufficient opening—say 3 feet—at the top to prevent the smoke from issuing through the feed-holes into the charging-room. The use of sheet-iron smoke-stacks is objectionable, as they always get red-hot in the operation of lighting up and blowing out a furnace, and then rapidly yield to the corroding action of the oxygen of the air.

To keep the mason-work from spreading it is braced by a sufficient number of wrought or cast iron uprights, which are sustained in position by wrought-iron bands passing over them. The latter are bolted together. The first bands are laid round the furnace about 2 feet above the dam-plate, and then follow one another in spaces 1 foot apart. At the Richmond and Winnamuck Company's furnaces the upper part of the furnaces is bound by an iron shell.

Now, the foundation is covered with soil, made firm by pounding, to within 3 feet below the upper edge of the plates, and a track is made for the slag-trucks. After having connected the tuyeres with the water-tank by wrought-iron pipes of convenient size, ($\frac{3}{4}$ or $1\frac{1}{4}$ inches,) the work of seasoning commences. A fire of billet-wood is kept slowly and steadily increasing in the furnace for about two weeks. During that time the bolts ought to be loosened to prevent the stones or bricks from being cracked by the escaping moisture. As soon as the furnace-walls get warm outside and no more moisture is perceptible in the joints, the furnace is ready for use. The fire is withdrawn and the furnace cooled down enough to allow a man to work inside. The bottom-stone, previously put in, is now provided with a thin coating of clay or brasque, (a composition of powdered charcoal and clay in varying proportions,) which is rammed in with a wooden stamper, after wetting it until it just coheres in lumps. The dam is made in the same manner, but of very good fire-clay, and taking care to make it extremely hard. It has a steep pitch toward the bottom. The tap-hole is made by pounding clay into the space left for that purpose and turning a pointed stick on the outside round a central axis, thus circumscribing a cone. The tap-hole may be in the front plate, which is best, or in a side of the furnace. Generally a large furnace has two tap-holes on opposite sides, and at right angles with the front plates. One tap-hole is at the deepest point of the bottom, the other one a few inches above it. Thus the metal may be tapped high or drawn off entirely, according to the circumstances. Mr. Arents, of the Eureka Consolidated works, recently made an attempt to do away with the inconvenient mode of tapping hitherto in use, and his efforts have been crowned with such success that there is not a single furnace in Eureka without this peculiar contrivance, termed the "siphon-tap" or "automatic tap." It consists of a sheet-iron cylindrical shell, which is bolted on to one of the cast-iron plates, in which formerly one of the tap-holes would have been located, and 6 inches below the top of the plate. Through a hole in the side of this shell toward the furnace passes a 3-inch wrought-iron pipe into another hole in the furnace-plate, and obliquely down to the lowest part of the hearth inside. The highest point of the pipe lies in the middle of the shell, and a foot or more below its upper rim. This cylinder is rammed full of fire-clay, the pipe being meanwhile closed by a plug. A basin, 18 inches in diameter, is then cut out and the plug withdrawn. The rim of the basin is on a level about 1 inch lower than the lowest level of the matte-spout, which is from 3 to 4 inches below the level of the slag-spout, so that the two can be drawn off separately. During the running of the furnace the lead stands always as high in this basin as in the crucible inside of the furnace. It is proved by the actual working results, since this improvement was introduced, that—

1. The furnace runs more regularly than before.
2. The lead obtained is purer.
3. "Sows" are prevented.

4. The work of the smelters is lightened.

These results agree entirely with the theories bearing on the subject, and I shall show that a fifth beneficial result might be added, namely, saving of fuel.

When the usual method of tapping a lead-furnace is followed the blast is stopped and the tap-hole in the bottom of the crucible is opened, (sometimes with great difficulty, when metal has cooled in it at a former tapping.) The lead, matte, and slag run out into the kettle, the hole is stopped again with clay, or a mixture of clay and coal-dust, called "stübbe" or "brasque," and the blast is turned on and smelting resumed. With the cleaning of the crucible, building up of fore-hearth, &c., this part of the smelting often takes considerable time, and the temperature in the furnace is reduced, so that much fuel is burned to make up the lost heat. Irregularities in the running of the furnace are frequently directly traceable to this cause; and the first commencement of the formation of "sows" occurs also in nearly all cases during the stoppages, when the small doughy masses of reduced metallic iron have an opportunity to stick to the bottom of the crucible, which is no longer protected by a liquid mass. It is well known to every metallurgist that whenever the foundation is laid for a "sow" it is extremely difficult to prevent its rapid growth; and even if the larger parts are broken or chiseled out at every tapping the iron will continually gain on the smelter.

By the employment of the automatic tap the first formation of "sows" is evidently prevented. Even if there be much iron from the charge reduced to the metallic state, the lumps will not come in contact with the bottom, but will always swim on the lead-bath. Being here exposed to the oxidizing influence of the blast they will be carried into the slag.

Furthermore, this arrangement for tapping carries the molten lead out from the *bottom* of the blast-furnace as fast as the metal is reduced inside. At the same time the lead smelted from the charge above remains in the crucible long enough to give the molten ingredients the required time to re-act upon each other and separate according to specific gravity.

The lead obtained must be purer, because it is taken from the bottom of the crucible, where the purest (heaviest) metal gathers, and because the foreign (lighter) metals, as iron, zinc, &c., are kept longer under the influence of the blast, and thus are mostly oxidized and slagged. The work of the smelters is, of course, considerably lightened, because, in addition to the tapping, the hard work of removing "sows," loosening the charge in the crucible after tapping, &c., is dispensed with.

It seems to me that the invention is one of the first importance for lead and copper smelting. For copper-matte the pipe must be of clay.

When a furnace is blown out the last of the lead is of course drawn off through the lowest tap-hole into a basin of 40 inches diameter and 18 inches depth, at the side of the furnace opposite the automatic tap, in the same manner as this has always been done.

As soon as the bottom is made the breast must be put in. About 6 inches above the front plate a straight arch, called the "bridge" is started of fire-bricks. In accordance with the thickness of the breast desired, they are laid lengthwise or edgewise. At Eureka the breast is made 9 inches thick, although 4 inches would be sufficient. A fire is now started in the hearth, siphon-tap, and lead-well to dry them. In the hearth the fire is continued till it gets red-hot. This is done by filling the hearth with lump-coal and kindling it. After it is all burned down the ashes are withdrawn and a fresh fire is started. These operations are continued till the desired end is accomplished, which generally takes two days.

Ores.—The majority of smelting ores with which the smelters of Utah and Nevada have to deal are galena and the carbonates, sulpho-carbonates, and antimoniates of lead.

According to the quantity and quality of the gangue we may classify them as—

1. Ores containing all the slag-forming ingredients (oxide of iron, silica, lime) in the proper proportions, or *neutral* ores, (Eureka Consolidated Company's mines.)

2. *Basic ores*, with lime and oxide of iron or manganese, and no silica, or not in sufficient quantities, (White Pine, East Cañon, American Fork, Cottonwood, Eureka, Nevada.)

3. *Acid or hard ores*, with silica and clay prevailing, (Bingham Cañon, Stockton, Tintic? Humboldt?)

Provided the slag-forming ingredients alone be present, *galena-ores*, when passed through a blast-furnace, do not yield metallic lead at once, but a mixture of metallic lead with sulphuret of lead, (*lead-regulus, matte,*) and other sulphurets, if such be present—an article that finds no market. In order to produce metallic lead galenas may be smelted with an addition of metallic iron, (5 per cent. or more, according to circumstances,) or after roasting.

In the first case the iron unites with the sulphur of the galena to a sulphuret of iron, called *iron-matte*, and metallic lead is set free. This re-action is, however, not complete, as a considerable quantity of sulphuret of lead, and with it silver, is retained by the iron-matte, necessitating another roasting and smelting operation. The iron-matte being lighter than the lead, floats on the top of the latter, and thus can be easily separated from it after cooling.

The combined *roasting* and *smelting process* is preferable to the *iron-reduction* process. The galena is first roasted in heaps, stalls, or reverberatory furnaces. Roasting in heaps and stalls is cheaper, as the ore may be used in lumps, and no expensive apparatus is required; but it is more tedious and incomplete, and only suited for galenas containing a large percentage of sulphurets of iron or copper. The latter prevent the ore from smelting together and so stopping the roasting process, and their sulphur furnishes the necessary fuel.

The roasting in reverberatories is by all means the best preparation of galena-ores for smelting. In this country it is generally done in small Mexican furnaces, called *galemadors*, (a corruption of " galenadores,") of the shape given in Küstel's " Nevada and California Processes." After the roasting operation is finished the heat is so increased that the ore is converted into a slag, principally *silicate of lead*, which is drawn out of the furnace, cooled, and broken up into large pieces, of convenient size. The agglomerated ore is then passed through the blast-furnace, with the proper quantity of fluxes, (Cerro Gordo, California; Big Cottonwood, Corinne.) The quantity agglomerated is 10 tons in twenty-four hours, at a cost of about 8½ per ton.

Before smelting, the ores ought to be reduced to the proper size. Some of the Eureka ores, yielding a very basic slag, may be thrown into the furnace in any size without disturbing the smelting operations, (Richmond Company's ore.) But siliceous and calcareous ores ought to be reduced to pea-size in a battery. Unless this is done no furnace can be run without sledge and bar.

Ores carrying much oxide of iron, like the White Pine ores, ought to be agglomerated in conjunction with quartz in a reverberatory furnace. Hereby the oxide of iron is slagged, and cannot be so easily reduced to metallic iron by subsequent smelting in a blast-furnace. Metallic iron, not finding heat enough in a lead-furnace to keep it sufficiently fluid to run out with the slag, congeals in the hearth, and forms what smelters term "sows," "bears," "horses," or "salamanders."

Very fine ores ought to be agglutinated by milk of lime, or agglomerated in a reverberatory, as they either escape from the top of the furnace or roll through the charge and arrive raw before the tuyeres, thereby forming nozzles and deranging the furnace.

The *Eureka ores* are principally bog-ores, with argentiferous and auriferous carbonates of lead interspersed. The iron-ore is chiefly in the shape of hydrated oxide of iron; but streaks of pittizite, arseniate, and sulphate of iron are frequent, and phosphates of iron probably occur, although they have not yet been observed. The principal lead-ores are cerussite, mimetite, and galena, in pockets. But wulfenite (molybdate of lead) has been found very frequently in cavities, beautifully crystallized. Owing to the presence of arsenic and sulphur in these ores no reduction of metallic iron need be feared, as this metal is carried off in the shape of a mixed sulphuret and arseniuret of iron, termed "matte" or "speiss," a very fusible compound.

The percentage of gold in these ores decreases with the rise in the percentage of lead. The reverse (Ruby Hill ores) is the case with the silver. There are, however, zones of lead-ore, which do not carry gold at all, and only about 30 ounces of silver per ton, (Bullwhacker mine.) The average contents in lead of the ores delivered at the smelting-works of Eureka are probably about 25 per cent., value of gold and silver varying. Formerly even ores with only 6 and 8 per cent. of lead were smelted in one establishment, along with dry ores. The resulting lead was very rich, sometimes running up as high as $1,500 in gold and silver, but the matte was also rich, assaying about $70 in gold and silver. As matte requires additional expensive operations to extract the useful metals from it, it is, at present, better to make it as poor as possible and throw it over the dump. To do this we have to observe the metallurgical principle—the more lead in the charge the less of the noble metals will go into the by-products.

In the front rank of the works at Eureka, Nevada, are those of the Eureka Consolidated Company, at the north end of the town. This company have five furnaces, of a capacity of 150 tons of ore per day. The motive-power for four Sturtevant blowers, No. 8, an 8-by-10 Blake's crusher, and a 6-inch pump, is furnished by a 40 horse-power engine, with two boilers, one being always in reserve in case of repairs.

Furnace No. 1, having five tuyeres of 2½ inches nozzle, was built after the pattern of the Oreana furnaces, similar to a north-of-England slag-hearth. Its present dimensions are 2½ by 3 feet in the hearth, and 2 by 4 feet at the top, with a height of 12 feet from the center of the tuyeres to the feed-hole. Capacity, 21 tons of ore per day.

Furnace No. 2, with three tuyeres of 3 inches nozzle, is of the same pattern and capacity.

Furnace No. 3 is six-sided, with five tuyeres of 2½ inches nozzle, and otherwise of the same dimensions as the last, except as to height, which is 10 feet. Capacity, 23 tons.

These three furnaces derive their blast from two blowers.

Furnaces Nos. 4 and 5 are octagonal, and have ten tuyeres each, four on each side, adjoining the breast, and two at the back. Each furnace has a blower of its own. Dimensions, 3 by 4½ feet in the hearth, 6½ feet at the top, 10 feet high. Capacity, from 35 to 40 tons per day. The blowers are run at a speed of 2,100 revolutions per minute, and yield air of a pressure of 1 inch mercury. The cooling-water passes from the tank through a 3-inch supply-pipe, and thence through ¾-inch pipes, entering the tuyeres from below. The waste-water passes out at the top of the tuyeres into funnels connected with a 3-inch waste-pipe. The latter leads to a large collecting-tank outside of the building, whence the water, after cooling, is pumped back into the supply-tank. Owing to the inadequacy of the supply-pipe, and the temporary insufficiency of water, only four furnaces can be run at a time. As the water is very muddy, the cast-iron tuyeres are very rapidly destroyed, on account of the accumulation of sediment inside. It is therefore contemplated to use only wrought-iron tuyeres, which, though costing nearly twice as much as cast iron, last much longer.

The nozzles, connected by leather hose, (wind-bags,) with corresponding reducers in the main induction-pipe, are 4 inches diameter at the outer end, tapering down to 2¼ inches towards the mouth. They are pushed tightly into the tuyeres, to prevent the escape of wind. At the outer end or elbow a 1½-inch projection is attached to the central axis of the nozzle. This contains the eye-hole, which is closed with a wooden plug. The latter is removed occasionally, to inspect the condition of the tuyere.

To illustrate the manipulations at a furnace, I will describe them from the commencement of a campaign, viz, from the blowing-in.

The hearth and furnace having been dried in the manner described above, the furnace is gradually filled up to the throat with coal, care being taken to keep it blazing. The fore-hearth and the apertures of the tuyeres are left open during this operation, to facilitate the draught. The filling up takes from four to five hours in a large furnace like Nos. 4 and 5. As soon as the coal at the throat has reached a dark-red heat, the blowing-in proper commences. Previous to putting on the blast, however, the front is put in; that is, the space h under the bridge g is closed up with bricks of stiff clay, rammed in tightly, and reaching a few inches below the dam-plate. Then the fore-hearth is also covered with clay, pounded down tightly. All the tuyeres, except the four nearest the front are closed with clay stoppers; their respective wind-bags are tied up with strings, to prevent the escape of wind. The nozzles of the four tuyeres named are now placed in position, and the blast is allowed to blow with full force for three-quarters of an hour, a long flame issuing all the time from the pipe of the siphon-tap. When the latter is red-hot, the blast is shut off by a cut-off in the main pipe. The clay balls are now removed from the closed tuyeres, and all the nozzles are put into the tuyeres. The blast is turned on again, and the charging commences. About three tons or more of lead are put into the furnace through the feed-holes, in the proportion of two scoops (a 1.2 bushels) of coal to 250 or 300 pounds of lead. This is done to heat the hearth properly, and prevent the accretion of slag or cinders, which might seriously interfere with the good working of the furnace. About 250 bushels of coal are used in all the foregoing proceedings. After all the lead is melted down, the feeding of the ore commences. First, six scoops (1.2 bushels=18 lbs. each) of coal, are scattered over the furnace, three from each feed-hole; on the top of this, but close to the walls, eighteen shovels of fine ore (15 lbs. each) nine through each feed-hole, and four shovels (17 lbs. each) of slag=25 per cent. of the weight of ore. This makes 1 pound of coal to 2.5 pounds of ore, or 3.1 pounds of smelting-mixture.

Every charge is marked by moving a peg on a tally-board for the convenience of the superintendent.

As soon as the lead has entered the pipe of the siphon-tap below, which may be observed by the disappearance of the flame emanating from it up to that time, the basin is covered with live coal and kept so all the time. Simultaneously the clay is removed from the fore-hearth. About two hours after the first charge, the slag, entering from below the breast, rises in the fore-hearth to the level of the slag-spout, viz, 3 inches below the top of the dam-plate. A cast-iron pot of conical shape, 26 inches deep, 15 inches upper, 6 inches lower diameter, is now placed under the slag-spout by means of a truck, and the exit of the slag is urged by detaching the crust along the dam. For the first half hour the slag is somewhat stiff, and only red-hot, from impurities, and from the fact that the furnace has not attained the proper temperature; but in the course of time it increases in fluidity and incandescence. The corners of the fore-hearth have to be frequently cleared from hard accretions, to prevent them from growing. After the lapse of a few hours the blast is shut off and the front removed by means of a sledge and bar, in order to clear the hearth and tuyeres from adhering cinders. If there be a hard crust on top it has to be broken up and pushed out of the hearth. During the stoppage the tuyeres are closed up to pro-tect the workmen from the escaping carbonic-oxide gas, and to economize the heat in the furnace. No new charge is introduced during this time. As soon as the hearth is clear, lump-coal to the amount of about two bushels is thrown over the fore-hearth, and a new front is made. The latter has to enter into the slag; if this is not observed, the blast will come through, not only exposing the furnacemen to lead-fumes, but at the same time chilling the fore-hearth. The front must be made of plastic clay; lean clay does not answer. After having closed the front, the nozzles are adjusted and the blast is turned on again. Hitherto the furnace has been running with a blaze at the top, indicating too high a temperature in its upper portions, which gives rise to great loss of lead by volatilization, and also injures the feeder's health. If the slag is gaining in fluidity, and the tuyeres remain perfectly bright, not even showing the least black ring, two shovels of slag may be replaced by two shovels of ore, but this must be done with the utmost caution, and at intervals of from four to six hours. At last a point will be attained when the blaze at the top disappears and the throat gets perfectly dark, discharging only black smoke. The normal charge has now been reached. In twenty-four hours 180 charges of smelting-mixture are run through with 1,260 bushels of coal. The normal charge for fine carbonate-ore is thirty-four shovels of ore and two of slag, corresponding to 46 tons of ore in twenty-four hours; for coarse ore it is 26 shovels of ore and 2 of slag, corresponding to 35 tons per twenty-four hours. When the fur-nace has been in operation for a week, it will even take more than this for a time, probably because it has then assumed the most favorable shape for smelting.

The furnacemen have to watch everything about the furnace very attentively in order to be always ready to apply the proper remedies. The slag has to run almost constantly while the blast is on. As soon as it becomes smeary and sticky, and emits a spray of sparks, which rise in parabolic curves, the matte-spout is opened and the matte run into a cast-iron pot, lined with clay. It is of smaller dimensions than the slag-pot. Care must be taken to keep wet or even cold tools away from matte or metal to avoid explosion. When the matte ceases, and in its stead slag begins to flow, the matte-spout is closed again with a clay stopper. The matte-spout is 3 inches lower than the slag-spout, and inclined a little towards the outside, while the latter lies horizontal. Thus it is possible to keep slag and matte separate.

Meanwhile the siphon-tap requires some attention. The pipe must be kept clear from accretions by pressing a red-hot bar through it from time to time, because it is very difficult to open it again after it is once closed up. The basin is kept nearly full, and the lead is ladled out as it accumulates. The lead-molds, which are in bar-shape, hold about 120 pounds each. From every fifth bar a sample is taken by means of an iron spoon. The samples from all the furnaces obtained during a certain time (usually twenty-four hours) are melted together to obtain the average sample.

The tuyeres must be kept clear by introducing a bar from time to time to detach obstructions. If there should be any sign of darkening, the charge must be decreased by two shovels, and the result waited for. If the charge be still too heavy, another decrease of two shovels is ordered, until the tuyeres resume their normal condition. If they should at any time get long black nozzles, the blast must be stopped and the hearth cleared out immediately. The reason of this occurrence may be an overcharge, or a preponderance of silica in the charge, i. e., a faulty mixture. If under normal charges ore arrives raw before the tuyeres and the blaze bursts out at the top, an irregular sinking of the charges or their detention on wall-accretions is indicated. These have to be removed. To do this the charge is allowed to descend half way in the shaft of the furnace, and only wood is applied as a fuel. By its blaze the wall-accretions are partly melted down. The balance is removed with chisel-pointed bars, worked through the feed-hole. During this operation the blast is, of course, shut off. Then the furnace is filled up again with coal, and the smelting proceeds as usual. Under ordinary circumstances the hearth is cleaned once in eight hours.

If wall-accretions have increased to such an extent that they cannot be removed without the greatest difficulty; if the charges descend irregularly, in spite of being decreased; and if the furnace-walls show unmistakable signs of destruction, it is advisable to blow out the furnace. The charge is allowed to go down to the tuyeres, the furnace emitting thick lead-fumes and a blaze. As soon as the charge has arrived at the tuyeres, the blast is shut off, and all the loose masses are drawn out of the furnace. Then the tap-hole is opened with a sharp-pointed bar, and the liquid contents of the furnace are discharged into the lead-well or basin, which has been previously heated. The congealed matter remaining in the furnace, consisting of slag ore, etc., is detached with bar and sledge. The breast is only removed when needing repairs. After cooling, which usually takes thirty-six hours, the furnace is freed from wall-accretions, and the injured places are repaired. The hearth and boshes are relined with English fire-bricks, and a new dam and bottom are tamped in. In the siphon-tap only the basin needs repairing. The inside of the tuyeres must be cleared from sediment before they are ready for service again. A furnace, if badly burned out, can be in running order within a week's time.

Furnace No. 5 was lighted up for the first time on May 31, 1871, and made three runs of respectively twenty-six, forty-five, and fifty days, together one hundred and twenty-one days, without ever being repaired. It was only blown out in consequence of repairs not connected with the furnace.

In the three small furnaces the proportion of coal to coarse ore is as 1 : 3.75; to fine ore as 1 : 4.6 by weight. The quantity of ore run through in twenty-four hours is, for coarse ore, 20 tons, for fine, 26 tons, with an average consumption of 33 bushels of coal per ton of ore. The four furnaces constantly in use smelted in the fall of 1871 115 tons of ore per day, yielding from 14 to 24 tons of lead. They consumed on an average 4,000 bushels of coal. The ore assayed on an average $27 in silver, and $35 in gold. The lead-assays have been discontinued as being unimportant. Daily 120 tons of ore were delivered by teams from the company's mines.

The ore is dumped in front of the feeding-floor. The coarse pieces are picked out and run through the crusher, while the fine ore is wheeled directly to the furnaces. Generally charges of coarse and fine ore are given alternately; only when the furnace is deranged the latter are given in preference.

The charcoal is piled up in the open air in a place elevated some distance above the roof of the feeding-floor. Trestle-works with car-tracks connect the latter with the former. Every furnace has a compartment of its own for the coal. It is conveyed from the pile in cars holding 20 bushels, and dumped into chutes leading to the bins. There is always a thirty days' supply, viz, 120,000 bushels, kept on hand. The dross and waste of the coal is about 10 per cent.

The laboring time at a furnace is divided into three shifts of eight hours. The crew for one shift consists—

I. At a large furnace (Nos. 4 and 5) of 1 smelter at $4.50 per day; 2 helpers at $4 per day; 2 feeders at $4 per day.

II. At a small furnace, of 1 smelter, 1 helper, 1 feeder.

Two foremen, at $6 per day, have the immediate supervision of the furnace-hands. They change every twelve hours. Two machinists at $5, under the supervision of a chief engineer at $8, take care of the engine. Besides, there is a blacksmith and outside foreman, and a number of roustabouts adequate to the wants of a large establishment. At the head there is a metallurgist, who reports to the general superintendent.

The products obtained in smelting are:

1. *Silver-lead*, generally called base bullion, with from $250 to $400 in gold and silver, about one-half of the value being gold. As it is not advantageous to treat it any further on the spot, it is shipped to Newark, New Jersey. The expenses of shipping to San Francisco, thence by sailing-vessel to Newark, and the cost of parting, amount to $69 gold.

There were produced at the Eureka Consolidated Company's works, during the time from January 1 to October 1, 1871, 3,000 tons of bullion from 17,000 tons of ore, at a cost of about $39 per ton of ore, all told.

2. *Matte*, or rather a mixture of matte and speiss, that is, sulphurets and arseniurets of iron, with 90 per cent. of iron and from $12 to $15 in gold and silver. It is thrown over the dump as worthless under the present circumstances. Its color is yellowish-white, like that of marcasite, with a blue tint at the surface; its texture is radial; specific gravity 4.02. It is produced in the proportion of 2 to 4 bullion. Sometimes this proportion is larger.

3. *Slag*. It is a mixture of singulo with sub-silicates. It shows only traces of gold and silver by either crucible or scorification assay, and has a specific gravity of 3.6. About 10 or 12 per cent. is used over again, the rest is thrown over the dump.

Its composition is shown by the following analyses:

	I, (thin.)	II, (thick.)	III.
$Si\ O_3$	= 26.12	37.50	30.20
$Fe\ O$	= 52.80	50.70	50.60

	I, (thin.)	II, (thick.)	III.
Pb O · · · · · · · · · · · · ·	= 2.79	8.00	8.70
Al₂O₃ · · · · · · · · · · · · ·	= 5.80	——	3.01
Ca O · · · · · · · · · · · ·	= 12.00	2.80	7.10
			0.90 Mg O
	99.51	99.00	100.51

I and II are analyses made by Arents from recent slags; III by Küstel from former ones.

Wall-accretions, principally sulphurets of lead and arsenic, with about $10 in silver and traces of gold. These are thrown over the dump. They crystallize in small cubes, and have metallic luster and blue color.

Hearth-accretions and *furnace-scrapings*, semi-fused slags, &c., are likewise thrown away.

Dust, assaying about the same as the ore, is a mixture of coal-dust with the finest particles of ore. Its percentage is considerable, but cannot be accurately ascertained without attaching dust-chambers to the furnaces. It would be well for a large company like the Eureka Consolidated to do this.

The principal loss, however, is not in the dust, but in the matte and in some of the slag. The yield of precious metals is 93 per cent. of the fire-assay reduced to dry ore.

The theory of this smelting-process is easily explained. Under the influence of heat the carbonates first lose their moisture and carbonic acid. The remaining oxide of lead unites with the silica present to silicate of lead. The limestone also loses its carbonic acid, thereby becoming a base which has a stronger affinity for silica than oxide of lead. The oxide of iron is reduced to protoxide by means of heat and the reducing power of carbonic oxide from the fuel. The consequence is that we obtain silicate of iron and lime, and oxide of lead, which yields to the reducing action of carbonic oxide, and forms metallic lead. If there is an excess of limestone or oxide of iron, a portion of protoxide of iron, being a weaker base than lime, will remain uncombined, and then will be reduced to metallic iron. Sulphuret of lead in contact with oxide of lead (according to the formula 2 Pb O+Pb S=3 Pb+SO₂) forms metallic lead, while sulphurous acid is disengaged. In the presence of oxides of iron a portion of the latter is reduced to metallic iron, which, in its turn, decomposes with the sulphuret of lead to sulphuret of iron and metallic lead.. These re-actions with sulphur and iron are less complete in the presence of silver than of lead alone, owing to the great affinity between silver and sulphur, which causes more or less silver to remain in the matte or iron sulphuret, though the greater affinity of lead for silver takes the most of the latter into the metallic lead.

Arseniates of lead and coal, acting upon each other in the heat, yield arseniuret of lead, arsenious acid, carbonic acid, and metallic lead. The arseniuret of lead is again decomposed by metallic iron, forming an arseniuret of iron, or speiss and metallic lead.

The next important works at Eureka are those of the Richmond Company, lately passed into English hands, at the southeastern end of the town. They were originally erected by Messrs. Ogden, Dunne & Co., for the purpose of doing custom work. But the scarcity of real lead-ores offered for sale induced these gentlemen to abandon this scheme and consolidate with the owners of the Richmond, a very valuable mine adjoining those of the Eureka Consolidated Company on Ruby Hill.

The ores resemble in their character those of the latter company, being bog-ores, intermixed with gold and silver bearing lead-ores. On an average they yield by fire-assay $40 in gold and silver, and produce, when mixed with about 7 per cent. of quartzose silver ore, (milling ore,) a bullion of $250 per ton, which is shipped to San Francisco for parting, at a cost of $35 per ton. At present the company have a circular furnace of the Piltz pattern, with seven 2-inch tuyeres, mechanical feeder, and siphon-tap, running; but there are two more large furnaces, designed by Mr. Arents, in the course of construction, each of which will reduce 50 tons of ore per day. In addition, there is a German cupelling-furnace of ten feet diameter, a softening or calcining furnace, and a bullion-melting furnace, which are out of use at present. The steam-engine is a vertical one, of 35 horse-power, with one boiler only. It drives a No. 7 Sturtevant blower, a Blake's 10 by 12 inch crusher, a Howland crusher, and a Harrison burr-stone mill. The Blake's crusher is intended for breaking up the coarsest lumps of ore. The Howland crusher and the Harrison mill are only used for sampling-purposes and grinding sand-stone, clay, &c. The former reduces the material to pea-size, after which it goes to the mill to be ground to a fine pulp.

The arrangement for getting coal to the smelting-furnace is substantially the same as that at the Eureka Consolidated works. There is a magnificent ore-floor, built of stone, attached to the works, where the winter supply of ore is piled up. The feeding-floor is spacious, and contains a number of bins to keep different ores separate.

Lately the charge was as follows:

17 large shovels of charcoal - - - - = 90 pounds, about 5 bushels.
24 shovels of Richmond ore, at 15 pounds - - - - - = 360 pounds.
 2 shovels of milling ore, at 12 pounds - - - - - - = 24 pounds.

                                 384 pounds.
 2 shovels of slag, at 17 pounds - - - - - - = 34 pounds.

                                 418 pounds.

This is at the rate of 1 pound of coal to 4.6 pounds of smelting-mixture, or 4.2 pounds of ore, or 26 bushels of coal to 1 ton of ore. There are passed through the furnace 150 charges in twenty-four hours, equal to a capacity of 28.8 tons of ore. The ore, smelted during the month of October, worked $64 per ton. Run, from three to four weeks. The proportion of matte produced along with the bullion is about the same as at the Eureka Consolidated works. The wall-accretions are more troublesome than in the former works, as the mechanical feeder prevents their detachment. The slag is basic, and resembles No. 1, heretofore described.

The works of the Phœnix Company do not present any new feature. They have a small Raschette furnace, of 25 tons capacity, built by Charles Liebenau, running, and another one in the course of construction. The company's mines are in three different ore-zones, viz:

1. That of the bog-ores, (Jackson mine, mines on Ruby Hill.)

2. That of the dry or milling ores, (mines on Adams Hill,) auriferous-silver ores, with little or no lead.

3. That of the lead-ores, (Bullwhacker mine.) The latter are rich in lead, with a moderate yield in silver, and no gold.

The ores of the variety No. 1 are very basic and require an admixture of quartzose material for smelting. This is accomplished by adding the ores of the varieties Nos. 2 and 3.

The bullion yield is four tons per day, assaying $210 in silver and $40 in gold per ton. There is no, or very little, matte produced, owing to the lack of sulphurets in the ore. Wall-accretions do not occur. The furnace looks perfectly clean after blowing out, but the slag is of a more basic character than No. 1 of the Eureka Consolidated Company's works, hence the walls round the tuyeres are very rapidly destroyed. Run, 21 days.

Besides these establishments there are a great many smaller ones in Eureka district, which run, however, only at intervals.

In White Pine district the first impetus to a perfect smelting mania was given, it is reported, by Colonel Charles S. Bulkley. Waiting in vain for the completion of the White Pine Smelting-Works, at a fixed date, he started himself to manufacture a lot of lead, necessary for calking the pipes of the White Pine water-works. For this purpose he built a little brick furnace with a grate inside, in the town of Hamilton, the apparatus being about as high as a German elbow-furnace. Then he purchased several tons of good gray carbonates from the Miser's Dream mine, from which he reduced the lead by throwing it into the furnace, alternating with dry billet-wood. The lead ran into a bowl in front of the furnace. The simplicity of these operations, and the bright shine of the lead-bars produced, which, by the way, assayed $36 in silver per ton, gave rise to the erection of an almost unlimited number of furnaces. "Every miner his own smelter," was the word. Mexicans erected atmospheric or draught furnaces, which, on account of their small cost, were soon copied by the miners, and Welshmen built the more expensive blast-furnaces. But a collapse was soon to come. The small capacity of these furnaces, and the low grade of the lead produced, were out of proportion to the general costliness of the necessaries of life. Other difficulties were associated with these circumstances, and rendered smelting impossible for people of small means. The completion of the Pacific Railroad encouraged other parties to engage in smelting. The first one was a San Francisco corporation, (the White Pine Smelting Company,) who, in June, 1869, built works at an expense of $36,000, with a view of depending entirely upon custom-work. As the business was considered to be very profitable, the Alsop Company and private individuals offered competition, and this was the beginning of the end. One party was overbidding the other in the purchase of ores, to drive their opponents out; finally they had exhausted their resources and ceased work. Just at this time another capitalist stepped in, expending large sums of money for new works. Before fairly getting to work he had to stop, however, because the prices for ores, coupled with the difficulties of smelting them, seriously impaired a financial success.

The smelting ores of White Pine may be classified as follows:

1. *Lead-ores* proper, principally cerussite with occasional nodules of galena and red copper-ore, ($Cu_2 O$,) carrying from $5 to $35 silver per ton. The purer carbonates form solid masses, and have a peculiar gray color; therefore they are called "gray carbonates." The majority of the carbonate-ores, however, are mixed with the oxides of iron and manganese, which give them a black or brown appearance. They are pulverulent, and yield readily to the pick. Both varieties fill cavities in the Devonian limestone, and are confined to a particular branch of the White Pine Mountains, called the base-metal range, (Miser's Dream, Mollie Star, Jennie A., and other mines.)

H. Ex. 211——26

2. *Copper-lead ores.*—They are, according to their chemical composition, a mixture of arseniates of copper and lead, with the carbonates of copper and pockets of galena, and assay on an average $60 per ton. They form either large pockets in the limestone or impregnate the same. For this reason they are not as easily mined as the real lead-ores. There seems to be an abundance of them on the western slope of Treasure Hill, (Elko, Erie, Russian, and Imperial mines.)

As may be inferred from their occurrence, these ores are of a very basic character; the former class being very ferruginous, the latter calcareous. To flux them, clay, clay-slate, and a very siliceous sand from the vicinity of Shermantown were used in default of quartz, which could only be procured with greatest difficulty. Besides, most of the works had no means to crush it. Purely quartzose ores only occur on the White Pine Mountain proper; but the cost of transportation and the high prices compelled smelters to desist from getting them. Occasionally small lots of quartzose silver-ore from outside districts, or quartzose tailings, could be bought, but not enough to avoid those incessant troubles and vexations arising from a want of fluxes. Iron sows were a daily occurrence. Another source of trouble was the lining. The insufficient quantity of quartz added to the ore caused the latter to corrode the lining in order to saturate itself with silicic acid. English fire-brick, pancake sandstone, in fact every kind of lining, would be destroyed in the course of a few days. The commonest lining was a sun-dried composition brick, made at great expense, of kaolin and common clay. But, owing to its not inconsiderable shrinkage, it would soon present to the slag points of attack, which kept the mason busy repairing. Notwithstanding these difficulties, runs were made at the White Pine Smelting-Works and the Alsop furnace of four and six weeks.

The most ridiculous feature in smelting at White Pine was the practice of some smelters to roast or burn calcareous ores of class 2 in a sort of lime-kiln to get rid of the sulphur. Instead of smelting these galeniferous copper-ores in their raw state, perhaps with an addition of galena in admixture with carbonate-ores, with a view to produce a tolerably pure lead and copper matte, those ores were subjected to the above operation, and a mixture of lead and a semi-sulphuret of copper was obtained, which was not salable in San Francisco, and in the East only at a great loss.

The carbonate-ores ought to be agglomerated in a reverberatory furnace with siliceous ores, and then, to enrich the bullion, passed through the blast-furnace with raw copper-lead ores and galenas. The result would be silver-lead of a good grade and argentiferous copper-matte. The latter could be roasted and smelted for concentrated matte or black copper. To insure financial success, however, a company ought to have works of a large capacity in a central location, and own mines of their own. Custom-ore cannot be relied upon, as it takes capital to develop mines so that they can keep a large establishment supplied. This most of the miners do not possess. The furnace at the White Pine Smelting-Works had a capacity of 15 tons per day, and consumed from 26 to 30 bushels of coal per ton of ore. The latter required an addition of from 15 to 20 per cent. of quartz. If quartz-tailings were used, they had to be mixed with clay, and formed into bricks. Raw tailings being very light, and in a fine state of pulverization, are either carried out of the chimney by the blast, or roll through the charge into the hearth without entering into combination with the ore. The bullion produced from the carbonate-ores alone yielded from 18 to 30 ounces of silver per ton; from mixed ores, (carbonates, copper-lead ores, and dry ores,) 130 ounces and upward.

There are many other promising smelting districts in the State of Nevada, but the smelting operations carried on there do not differ materially from those already described.

Most of the lead-ores of Utah differ in this particular from those of Nevada, that the prevailing gangue is quartz. Calcareous ores are, however, also found in considerable quantities in East Cañon, Little and Big Cottonwood Cañons. Bingham Cañon offers the most striking instance of the occurrence of quartzose ores. They lie in a disintegrated quartzite, which intersects a stratified limestone, probably pertaining to the Devonian age. The great bulk of them are the carbonates and sulpho-carbonates of lead, carrying from 15 to 30 grains of silver, with streaks of galena varying in silver contents. A large portion of the ores show traces of gold. Of accessory minerals, small quantities of sulphurets of iron, oxide of iron, and clay-ironstone may be named.

There are at present two smelting establishments in Bingham Cañon, that of Messrs. Bristol & Daggett, and that of the Utah Silver Mining and Smelting Company, limited, both of which work ores from their mines, and also do custom-work. The former is very conveniently located at a hill-side below the mine, belonging to the same parties, the Winnamuck, from which the ore is chuted down on a planked ore-floor, forming part of the housed feeding-floor. The different classes of ore delivered to the works are thrown through a screen; the coarser pieces are run through a Brodie crusher and reduced to walnut-size. Previous to smelting the ores are mixed by weight, so as to produce a bullion of a certain standard.

The company's furnace is a circular one of the Piltz pattern, with eight tuyeres of 2-inch nozzle. It is 14 feet high from tuyeres to throat, 3½ feet diameter in the level of the tuyeres, and 5 feet at the top. The hearth forms a hexagon on the outside, and is inclosed by six cast-iron plates 1½ inches in thickness. The two nearest the dam-plate are provided with slots for tap-holes. The upper part of the furnace, made of brick-work, rests on a cast-iron flange, which is borne by four hollow cast-iron pillars. The part below the flange is of Utah sandstone, 13 inches thick, lined inside with 4 inches of Pennsylvania fire-brick. The motive-power comes from a 10-inch cylinder stationary steam-engine, with 25 horse-power locomotive boiler. It drives a Brodie crusher and a No. 4 Root's blower. The efflux-pipe of the latter is provided with a safety-valve and a wind-gauge, by which the pressure of the blast is measured in inches mercury.

An open bulk-head adjoining the ore-floor holds about 30,000 bushels of charcoal.

The manipulations at this furnace do not differ much from those anywhere else, only in lighting up the proceedings are a little different. After the hearth is heated up sufficiently, a suitable quantity of lead is introduced through the front; then the furnace is filled up with coal in the usual manner. As soon as the coal has reached to within 5 feet below the throat, slag is charged in portions of one pound of the latter to one pound of charcoal. When the charge is in the level of the throat, the blast is turned on. About 1,000 pounds of good, fusible slag, picked out for that purpose, are fed before commencing with light charges of ore.

In the past summer the ores coming to the works for treatment were:

1. Carbonates of lead from the Spanish mine, with from 28 to 30 per cent. of quartz, 55 to 60 per cent. of lead, and $22 silver per ton.

2. Carbonate ores from the Winnamuck mine, with 30 per cent. of quartz, 35 per cent. lead, and about $80 silver and gold.

3. Ferruginous dry ore from Winnamuck mine, with 38 per cent. silica and alumina, 27 per cent. metallic iron, and $65 silver and gold.

4. Same ore, with 45 per cent. silica and alumina, 23 per cent. metallic iron, and 880 silver and gold.

Ore No. 1 was the principal one smelted; occasionally No. 2, which is of the same character, and No. 3, were added; No. 4 was reserved for assorting. The furnace worked well, and without the least difficulty, when the ore was mixed with 40 per cent. of a tolerably pure hematite from Lehi, 20 per cent. of limestone, and 30 per cent. of slag.

The slag produced was stiff, and resembled a bi-silicate. A decrease in the percentage of slag added to the smelting-mixture was always accompanied by evil consequences. The resulting slag in that case was dry, short, and would soon stop running. A diminution of the iron-ore and increase of the limestone also worked unfavorably, and the more so the less oxide of iron was in the smelting-mixture. Pure silicates of lime cannot be perfectly liquefied by the temperature prevailing in a lead-furnace.

The tapping is done at these works in the old manner, by piercing the tap-hole with a bar as soon as the lead has risen to the slag-spout. The tap-hole is just high enough above the bottom of the hearth to leave a suitable quantity in the latter. After tapping, the hearth is cleared from cinders and other accretions.

The production of matte is not noteworthy.

The normal charge was: 5 scoops of charcoal, at 1.1 bushels or 18 pounds = 90 pounds; 15 shovels of lead-ore, at 15 pounds = 225 pounds; 6 shovels of ironstone, at 13 pounds = 78 pounds; (partially Winnamuck ore No. 3;) 4 shovels of limestone, at 13 pounds = 52 pounds; 3 shovels of slag, at 10.5 pounds = 31.5 pounds—total smelting mixture, 386.5 pounds.

The proportion of coal to smelting-mixture is as 1 pound to 4.3 pounds, and to ore as 1 pound to 2.5 pounds; 1 ton of ore to 48.8 bushels of coal.

In twenty-four hours, under a pressure of from 1½ to 2 inches mercury, 140 charges were run through the furnace, corresponding to 27 tons of smelting-mixture, or 15¾ tons of ore, from which resulted 7 tons of lead, carrying between $60 and $80 of silver per ton.

The lead is shipped to Chicago for parting.

The number of hands required was: 3 smelters, at $5 per day; 6 helpers, at $3 per day; 3 feeders, at $3 per day; 2 engineers, at $4 and $3 per day; 1 blacksmith, at $3 per day; 1 coal-receiver, at $2 per day; 4 roustabouts, at $2.50 per day.

Three helpers might be saved by providing the furnace with an automatic tap.

The ores of the Utah Silver Mining and Smelting Company, limited, are of the same character as those of the Spanish mine, viz, very poor and siliceous. At the time I visited these works nine classes were made, for what purpose I did not learn.

The charge was as follows: 2 baskets of coal, at 2½ bushels = 90 pounds; 6 shovels of ore, (chiefly leadhillite) = 90 pounds; 2 shovels of iron-ore = 26 pounds; 2 shovels of limestone = 26 pounds; 2 large shovels of slag, about = 30 pounds—total smelting mixture, 172 pounds.

This is at the rate of 111 bushels of coal to 1 ton of ore, or 58.1 bushels to 1 ton of smelting-mixture. This proportion is exorbitant; but it was all that could be done under the circumstances, the ore being poor in lead, and the iron-ore, though scrupulously assorted, very siliceous. Assuming that 140 charges passed the furnace within twenty-four hours, its capacity would be 6 tons 600 pounds of ore, from which 2 tons 200 pounds of lead resulted. Under a higher pressure of the blast like that at the Winnamuck furnace, the capacity of the latter would probably be attained.

The furnace then running was a six-sided one, with five tuyeres of 2½-inch muzzle, and mechanical feeder. It was supplied with blast by a No. 8 Sturtevant blower. But there was a larger one in the course of construction, an exact copy of No. 5, at the Eureka Consolidated Company's works.

Buel & Bateman's works, at the mouth of the Little Cottonwood Cañon, consist of two circular Piltz furnaces of the same size as the one of the Utah Silver Mining and Smelting Company, limited. They have been running extremely irregularly during the summer, and, as near as I have been able to ascertain from the best authority, at a decided loss. The manipulation of the furnace does not differ materially from that of similar furnaces elsewhere. The ores smelted are those from Little Cottonwood Cañon, which are, with the exception of those from the Emma mine, decidedly basic.

Very good smelting-works have lately been built in American Fork, to smelt the ores from the Miller mine. But at present I know only that they are Piltz furnaces, with automatic tap. The ore smelted here is decidedly basic. In my next report I hope to be enabled to describe these works and the smelting operations carried on there.

There are a great number, probably over twenty, other smelting-works in Utah; but none of these have so far run regularly or with profit.

In Montana, only two smelting-works are in operation now, and both these are located in Argenta, Beaver Head County. A third establishment, a copy of the Argenta works, is building in Helena.

The smelting-works of Argenta, and especially those of S. H. Bohm & Co., are managed as well as can be expected in that locality. Bohm & Co. have had their two blast-furnaces and one cupelling-hearth in blast since May, and Mr. Stapleton's works have also been in operation pretty regularly. A third works, the old ones of the Saint Louis Company, are idle, and have been so for several years.

All three are located a short distance above the town of Argenta, on the south bank of the Rattlesnake Creek. The Saint Louis Company's works were built first, at a time when labor and all the materials for building were at the highest price. A natural tendency to save in the cost of materials is therefore everywhere visible, and it is undoubtedly owing to this that the external appearance of the furnaces and buildings is ungainly, rough, and clumsy in the extreme. The works might, however, have answered the purpose very well, if the inner shape and dimensions of the blast-furnace had been suitable. But neither the wide hearth nor its trapezoidal section could give good results with ores as quartzose as those of Argenta. And, furthermore, it appears from the burnt appearance of the inside of the furnace, from bottom to top, that the ore must have been charged into the furnace in large pieces, and the smelting conducted with the flame blazing out of the top, two very serious mistakes which ought never to have happened. The slag, too, on the dump, shows at once that smelting in reality was unsuccessful, whatever large amounts of silver may have been taken from the furnace during the short time of its running. A part of the slag contains very much lead, (and undoubtedly silver,) while another part is not smelted at all, but was probably pulled out of the hearth with instruments. The German cupelling-furnace, on the contrary, is a very good and substantial structure, and must have done its work well. The establishment is not now in shape to be started up again with little cost, many important parts being entirely missing, and the whole having suffered much from exposure.

S. H. Bohm & Co's works, the old "Elsler furnace," are the next above the foregoing. They consisted, up to August of this year, of two stack-furnaces and a German cupelling-furnace, the blast being supplied by a Root blower, driven by a magnificent water-wheel of 12 feet diameter and 4½ feet breast. The latter supplies also the power for a Dodge crusher. Since August, a third blast-furnace and another cupelling-furnace have been added to the works. The blast-furnaces are the high furnaces, with rectangular section and the same area at the hearth and throat as first introduced into this country for copper-smelting, at Ducktown, Tennessee. The inside height above the tuyeres is 20 feet, the section 24 by 24 inches. The furnaces are necessarily so narrow and high because the ores are extremely quartzose. The lining is a quartzose granite from the neighborhood, which stands the heat about three weeks. There are two common tuyeres in the back of each furnace, which lie horizontally about 10 inches above the upper end of the dam-plate, and have a diameter of about 1¾ inches at the mouth. The smelting is conducted with "noses," that is, the melted charge is allowed to cool locally around the interior openings of the tuyeres, so as to form a nozzle or "nose," protecting the tuyere against heat and chemical action, and at the same time conveying the blast well into the interior of the furnace. The hearth is filled with heavy *stübbe*, or brasque made of charcoal-dust and burnt yellow clay, no white clay being at hand in the neighborhood. This material is reported by the owners to stand very well.

A pressure of about 1 inch quicksilver is intended to be maintained in the blast. The charges vary, of course, considerably, as very different ores are constantly delivered from the mines, but it is intended, and the dump shows that the object is generally reached, to produce a slag ranking between a singulo and a bi-silicate. Rather large amounts of iron-ore and limestone, both from the vicinity, are used for fluxing the great excess of quartzose gangue in the ores; and only from 2 to 2½ tons of bullion are produced from each furnace per day. The charges contain from $80 to $150 silver per ton, and the base bullion produced assays from $250 to $500. Specific statistics in this connection are wanting at present, as the owners have not yet fulfilled their promise of sending them.

The cupelling-furnaces are exact copies of those used in the German lead and silver works. For the hearth, a very good marl is employed, which is found in the limestone a short distance from Argenta.

Stapleton's works, a little higher up the creek, consist of two shaft-furnaces and one cupelling-furnace, all constructed on the same plan as those just described. These works do not run quite as regularly as Bohm's, principally because the ore-supply is precarious.

Such a cause has, so far, not affected Bohm's works; undoubtedly on account of the superior activity of the managers in securing ores from all quarters in advance, and the larger working capital at their disposal. The prices paid by both works for ore to miners are exceedingly moderate, and leave a large margin for profit, although the cost for smelting must necessarily be very high. Charcoal, for instance, costs from 18 to 20 cents per bushel, and labor from $4 to $6 per day, everything else being correspondingly high. The loss of silver in smelting is claimed by Mr. Bohm to be only from 8 to 10 per cent. of the assay value, though there are no arrangements connected with the furnaces

to condense the dust. This is quite possible, as no ores containing much
antimony, arsenic, or zinc appear to come to the works.

The litharge produced in the cupelling-furnaces is, for the greater
part, not utilized at all at present, small quantities only being occasionally required for addition to the charges of the blast-furnaces. The
bulk of it lies in the furnace-yard, awaiting the time when it can be
profitably reduced and shipped.

Much of the lead-ores being carbonates, and such of the galena-ores
as contain a sufficiency of iron pyrites to be fit for open heap-roasting,
being subjected to that process before smelting, there is only an inconsiderable quantity of matte produced, which, being at the same time
poor in copper, is added to the charge without a preparatory roasting.
It would, however, be a better plan to save the matte, until there is
enough on hand to make a roast-heap; as in that case quite an amount
of iron-ore which must now be purchased at the works as flux might
be saved, and the silver and lead would be extracted at once. By the
present method, the greater part of the matte passes the furnace many
times almost unaltered.

The smelting-works in Argenta will undoubtedly do well, if conducted
as at present, as long as there is no competition, either from large
amalgamating-works or from smelting-works using the copper-ores of
Montana for the extraction of silver and gold. But it seems to me that
with either of these they would be unable to compete.

The reasons for this statement may be found in the chapter on Montana, where they are fully explained.

On the whole, very few of the smelting-works in Nevada, Utah, and
Montana have been financially successful, and this in spite of the rich
ores they usually treat. There are two principal reasons for this. The
one is the unprepared state in which the ores are delivered at the
smelting-works, the other the fact that but few works can be found
which are managed by metallurgists who really are what they claim
to be.

In many of the western mining districts, notably in those of Utah and
Montana, there is an abundance of water, which invites a removal of the
greater part of the gangue of the ores by the cheap and effective
means of dressing. Yet this has never been done, though it is so
evident that an enormous saving in fuel and fluxes might be effected
by the removal of the gangue before smelting. It is rarely the case
that in the western lead-ores the true silver-ores are found, which would
occasion much loss in dressing; the silver is, on the contrary, in nearly
all cases, closely allied with the carbonates and galena; and in dressing
such ores very little loss need be feared.

In regard to the other reason, I am sorry to see that mine-owners
will not comprehend that metallurgy is a business which requires long
study and practice, and which cannot be successfully conducted by
those who know neither the theoretical ground on which it is founded
nor the practical details. A noteworthy exception in regard to the
last point are the works of the Eureka Consolidated and those of Bohm
& Co. in Montana. These works are really successful, and might be
more so if a mistaken economy did not prevent a still further perfection.

The Kast furnace.—This furnace, which has given great satisfaction
at the works of Clausthal, Prussia, is essentially a small circular Piltz
furnace, in which the tuyeres are somewhat differently distributed.
Believing that it would be specially useful to works of small capacity, of
which there are likely to be many in our scattered mining districts,

I have procured for the information of American metallurgists descriptions and drawings of the furnace and detailed statements of its working. In respect of economy it surpasses all furnaces with which I am acquainted. The extraordinarily favorable results shown by the following statements are partly due to the favorable circumstances at Clausthal, particularly in the facility with which suitable mixtures of different ores for smelting can be obtained, and the excellent fluxes (copper-slags from the Lower Hartz, and matte-slags and roasted mattes from the works themselves) which are available to the smelter. Since, however, the Raschette furnace, under the same conditions at the same place, loses more lead than the Kast, and has a shorter and more troublesome campaign, it is evident that the Kast possesses an intrinsic superiority.

The particulars of construction are best seen in the accompanying drawings. I have only to add, in explanation, that the previous existence of the large concentration-chambers at the Clausthal works compelled the choice of a certain size, which could not be increased. By the working results of the furnace it has been proved that these chambers are not absolutely necessary, the amount of dust caught in them being exceedingly small.

For the construction of a furnace with four tuyeres and two condensation-chambers, as given in the drawings, the following materials are used in Clausthal:

> 98 cubic feet of sandstone for a sole-stone.
> 763 cubic feet of dressed-sandstone blocks for the outer walls and pillars.
> 23,000 pieces of common brick.
> 200 pieces of chamotte brick.
> 845 cubic feet of rubble-stones for foundation.
> 216 hectoliters* of a mixture of common lime and plaster.
> 36 hectoliters of "leather-lime," (leder-kalk.)
> 32 hectoliters of clay.
> 20 cubic feet of fire-proof sand.

The following materials are of cast iron:

8 plates above the entrance to the side-tuyeres, at $5\frac{1}{2}$ cwt.
1 plate above the entrance to the back-tuyeres, at $6\frac{1}{2}$ cwt.
1 plate above the entrance to the charging-door, at 7 cwt.
3 plates around the fore-hearth, weighing $8\frac{1}{4}$ cwt.
4 water-tuyeres, weighing $4\frac{3}{4}$ cwt.

The conducting-pipes for the blast and wind stacks weigh about 36 cwt.

Besides the above, iron rods and rails, or other heavy bar-iron, are required to bind the furnace and condensation-chambers. The plans followed for this purpose are various. That adopted in Clausthal is sufficiently shown in the drawings.

As far as the iron-work above enumerated is concerned a great saving can of course be effected in places far from founderies by substituting brick arches for the plates intended to support the masonry above, and for the cast-iron conducting-pipes sheet-iron ones can be made to answer. A large saving in the original cost of a furnace is made by omitting the condensation-chambers, which, as I said before, are not absolutely necessary, although it is desirable to have them.

The ores smelted at Clausthal are crushed massive ores, and the dressed ores from the Burgstaedter, Rosenhoefer, and Zellerfelder dis-

---

* One hectoliter = 6,107.4 cubic inches = 2.84 bushels.

tricts. Keeping in view their different gangue and varying contents of lead and silver, they are so mixed in quantities of 1,000 cwt. dry weight, that the average contents in the mixture are: of lead, 58 to 60 per cent., and of silver, 0.1 per cent.; while at the same time the gangues are so proportioned as to furnish an easily fusible slag. Such a quantity of 1,000 cwt. is divided into 20 "charges," at 50 cwt. A "charge" is mixed with 25 cwt. of roasted matte from the first smelting, 40 cwt. of copper-slags from the Lower Hartz, 20 cwt. of matte-slag from the Clausthal works, and 16 cwt. of slag from the first smelting. The last two items vary somewhat, according to whether heavier or lighter gangue is preponderating in the ore.

This charge is smelted in both the Kast and the Raschette furnaces. In the latter, however, 10 additional cwt. of slag from the first smelting is sometimes mixed with the charge.

As seen in the drawings, the Kast furnace is 20 feet high, has a diameter of 3 feet at the tuyeres, and of 5 feet at the top. There are four water-tuyeres, two being in the back wall and one on each side. The distance from tuyere to tuyere is 21 inches, with the exception of the two nearest the front, which are 42 inches apart. It is, however, proposed to put a fifth tuyere into the front wall. The diameter of the nozzles is $1\frac{3}{4}$ inches, the pressure of the blast 10 to 12 lines quicksilver.

The average results of a month's working, reduced to 100 cwt. of ore, which consumed 51 cwt. of coke, were:

59 cwt. of lead, containing 14 to 15 quints* of silver per cwt.
78 cwt. of lead-matte, containing $2\frac{1}{2}$ to 3 quints of silver per cwt. and 7 to 10 per cent. of lead.
Slags with 0.08 quint silver and 0.4 per cent. of lead.

The quantity smelted in twenty-four hours was 63 cwt. of ore, or 190.65 cwt. of charge, equal to about 9.5 tons. One pound of coke carries 6 pounds of charge. The charge is spread over the whole surface of the furnace.

There is hardly ever any trouble in a campaign, the latter being invariably very long, in fact much longer than campaigns have heretofore been made in lead-smelting. There are no accretions, very little dust, and the proportion of fuel used is very small. The yield of the lead is invariably as high as the assay with black flux and iron made of the ore, and is reported to exceed it sometimes.

If I should suggest any improvement at all in the construction of the Kast furnace, in its application to the smelting of western ores, I would propose to straighten out the corners in the inside of the furnace on both sides of the breast. The object of this is simply to make the parts of the furnace immediately behind these corners more easily accessible for the bar and rabble, as in these places, if anywhere, accretions are most likely to occur. It is not to be expected that accretions can be avoided as easily with our western, undressed ores, as with the clean ores of the Hartz.

---

* 1 quint = 5 grammes = 77.165 grains.

CHAPTER XI.

ECONOMICAL RESULTS IN THE TREATMENT OF GOLD AND SILVER ORES BY FUSION.

This chapter was written, at my request, by John A. Church, E. M., of New York City, a metallurgist of much intelligence, to whom I am obliged for the permission to insert here what I think is a very useful and suggestive essay. He desires to make acknowledgment for the information contained in the paper to Dr. Leo Turner, formerly director of the works described, and now at Brixlegg in the Tyrol.

At a time when the treatment of gold and silver ores by fusion, in opposition to the mill-process, is attracting so much attention in this country, it may be useful to consider what is done in a well-conducted foreign works. For this purpose I will ask the reader to accompany me to Lend, in Austria, a small but thoroughly organized establishment. It is situated in the Salzburg Alps, and receives its ore from mines at Rauris and Boeckstein. The former, lying 8,200 feet above the sea, is said to be the highest mine in Europe, some of its openings being made in glacier ice. It was worked by the ancients, who have left the contracted and tortuous workings peculiar to them.

The ore differs in no way, unless in extreme poverty, from countless mines in the West. It consists of gneiss, quartz, and clay-slate, containing the sulphurets of iron, copper, lead, zinc, and antimony, besides arsenical pyrites, gold, and silver. The gold is found in two conditions, free gold and gold alloyed with silver. This alloy for the year 1866 was composed, on the average, of 15.33 gold and 84.67 silver, which gives a specific gravity of 11.28. Mercury has a specific gravity of 13.6, and as the amalgamation of gold by the Austrian method is looked upon as a proceeding entirely mechanical, the separation being effected solely by the superior gravity of gold over mercury, this alloy, which is lighter than mercury, cannot be amalgamated.[*] Such is the lesson of long practice, the free or fine gold being extracted from a part of the ore, at least, by amalgamation, while the tailings are smelted to obtain the alloy. The following table will show the proportion of fine to alloyed gold, and also exhibit the extreme poverty of the ore. To the Rauris and Boeckstein ores I have added those from Zell in the same part of the Alps. The ore from this place is not now worked, the point of poverty having apparently been reached at which the auriferous rock ceases to be an ore.

	RAURIS.	BOECKSTEIN.	ZELL.
	In 2,000 pounds troy ounces.	In 2,000 pounds troy ounces.	In 2,000 pounds troy ounces.
Fine gold	0. 32 to 0. 48	0. 098 to 0. 113	0. 090 to 0. 097
Gold and silver alloy	1. 40 to 1. 47	0. 570 to 0. 660	Unimportant.
Iron pyrites, copper pyrites, galena	8 per cent	4½ per cent	Unimportant.
Value of silver and gold in American coin	$13 49 to $16 92	$5 91 to $8 40	$1 86 to $2 00

[*] See Rittinger's *Aufbereitung*, (ed. 1867, page 469.)

As in 1866 Boeckstein delivered 63 per cent. of the ore, and Rauris 37 per cent., the average value per ton for the year was $10.16,* or 0.004 per cent. gold, and 0.034 per cent. of silver. This does not include the value of the copper and lead, which form, respectively, 2 and 1 per cent. of the ore. The former is extracted; the latter is not sufficient to supply the waste of the process, and lead has to be bought for the works. Even in Europe these ores are considered extremely poor. I am not aware that ores from veins so poor as these have ever been worked in America, but if they have they must have owed their value to the fact that the gold was all fine, and could be amalgamated.

TREATMENT OF THE ORE.

The ore is first sorted to six varieties for the furnace, and one for amalgamation. The former comprise quartzose ore, rich, medium, and poor, compact pyrites, galena and antimonial ore.† The ore sent to amalgamation is the poorest kind of pyritiferous rock. It contains merely traces of pyrites, and is amalgamated, because in that process it undergoes concentration.

Amalgamation.—The ore for amalgamation is crushed under stamps of 220 pounds weight, (total,) through sieves of 1.5 millimeters, (0.06 inch,) the battery-box having a sieve on each side to secure the most rapid discharge of the slime. Two methods of treatment are employed for the slime : first, it is first concentrated, and then amalgamated, or, second, it is first amalgamated, and then concentrated. With ore that contains much pyrites the former is best; with ores very poor in pyrites, the latter.

Amalgamation takes place in pans, there called " mills." They are 24 inches in diameter at the top, 16 inches at the bottom, and 9 inches high, and made of cast iron, one-eighth to three-sixteenths inches thick. They are not directly conical, but the side forms a step 3 inches wide. In this pan mercury is poured an inch deep, and a wooden block shaped like the pan, and 1 to 1½ inches less in diameter, is suspended over it. The upper part of this block is hollowed out like a hopper, with its discharge in the center, and the under side has small pieces of sheet iron placed radially in it, and which just clear the mercury. When this block is revolved, and a stream run into the hopper-like depression on its upper surface, the slime is carried over the mercury from the center to the circumference of the pan, the whole apparatus acting like a " centrifugal" pump. This is the Austrian gold-mill so often described.‡ Great care is taken to prevent too rapid a motion of the stream, which would not allow the gold time to settle and would carry off the mercury. Twelve to thirty-two revolutions a minute is the speed given, depending upon the fineness of the ore, thickness of the slime, and amount of gold present. These mills extract by one operation 75 per cent. of the fine gold, and 15 per cent. more by repeating the process. Each mill passes about one ton of ore in twenty-four hours. Compared with blankets this system does not appear to present any advantages in the first handling of the ores; but I should think the Austrian mill might be substituted with gain in the place of many other amalgamating arrangements now used after the blankets. Compared with the Colorado methods these mills

---

* Unscientific as the method is, I feel compelled to give these values in American coin, since that is the only expression known to the workers in our mines.
† A collection exhibiting these ores, and a full suite of furnace-products, can be seen at the School of Mines, of Columbia College, New York.
‡ Rittinger's is the best account. See his *Aufbereitung.*

extract 20 per cent. more* than the Colorado amalgamators, though this yield necessarily depends upon the proportion of silver in the gold. They require little watching, except when used immediately after the stamps, when the accumulation of gold might require their cleaning up every two or three days.

Smelting.—For four years the ores delivered for fusion were in the following proportions:

	From Rauris.	From Boeckstein.
Quartzose ore	6.50	24.11
Compact pyrites	0.06	0.48
Sulphuret of antimony	1.41	0.41
Slime from amalgamation	28.03	38.00
	37.00	63.00

About 66 per cent. of the smelting ore has, therefore, been amalgamated.

From 70 to 75 per cent. of the ore is worthless rock, and this must be removed before adding lead, which would suffer serious loss if charged with so much quartz. The operations are, therefore, as follows:

1. Fusion for raw matte.
2. Roasting of raw matte in stalls.
3. Fusion (without lead) for a more concentrated matte.
4. Roasting of second matte in stalls.
5. Fusion with lead.
6. Cupellation of rich lead.

The first fusion.—Eleven years' experience has proved that the most efficient slag is one approaching the composition of a bi-silicate. The following is an average analysis:

Silica	51.02
Alumina	2.16
Oxide of iron	19.75
Lime	15.40
Magnesia	8.57
As. Mn. Cu. } Zn. S. (by dif.) }	3.10
	100.00

Each year the matte resulting from the previous year's fusion with lead is roasted, analyzed to ascertain the amount of oxide of iron present, and charged in the first fusion as a flux for the quartz; or, if containing above 35 per cent. of copper, it is treated for copper.

The furnace is not new, and contains none of the late improvements, but it does good service. Its dimensions are as follows:

Height	24	feet.
Diameter of hearth	3	feet.
Diameter of boshes	4.5	feet.
Diameter of throat	2	feet.
Number of tuyeres	2	
Pressure of blast	$\frac{3}{24}$ to $\frac{1}{2}$ inch of mercury.	

---

* See Mr. Hague's Report on Mining Industry of the Fortieth Parallel.

From 100 to 120 bushels of charcoal are required to warm the furnace, and then regular charges of 5 cubic feet, or 3 bushels, are made. In "blowing in," the quantity of mixed ore and flux added to this charge of coal is, at first, 56 pounds; then 112 pounds; and when the furnace is thoroughly hot the full charge of 203 pounds, which is the constant burden, to 3 bushels of charcoal. This is usually reached in the first twenty-four hours. Four hours after the first charge of ore and flux, the blast is turned on at first with a pressure of one-third inch and then one-half inch of mercury; or one-sixth and one-quarter of a pound to the square inch. After eight hours the slag begins to flow. The furnace is, of course, worked with a black throat.

The first matte forms 40 to 45 per cent. of the charge, the difference between this proportion and the 25 to 30 per cent. afforded by the ore being made up by roasted matte from the previous year. Its average composition is—

Iron 55. 1
Copper 4. 3
Zinc.. 3. 7
Lead...........‥.. 2. 1
Nickel, cobalt, arsenic, and antimony 4. 5
Sulphur.. 27. 9

                                                              ———
                                                              97. 6
                                                              ====

It contains 30 to 40 ounces, troy, of auriferous silver to the ton of 2,000 pounds; or, in American valuation, $100 to $150 in coin. From the fact that the ore is unroasted and the metals are so well "covered" by sulphur, the loss amounts to only 0.25 of 1 per· cent. About 38 bushels of charcoal are used to the ton of charge, and 9.75 tons are smelted in twenty-four hours.

The second fusion.—The first matte is roasted three times in stalls containing 28 tons, the roasting not being thorough, but carried only so far as to leave about 40 per cent. of unroasted matte. It is then re-smelted with quartz and siliceous slag; and to avoid the use of too much of the flux, a basic slag is made containing about 22 per cent. of silica. This requires very great care in managing the furnace, for the least irregularity of working causes the formation of sows. To secure proper working, whenever the furnace is tapped the hearth is examined by means of a bent bar. If lumps are felt, the front wall is broken out and they are removed; if the sole is slippery, the presence of reduced iron is indicated. A rough, hard, even sole is the proper one.

The pressure of blast is now reduced to one-sixth of an inch, or one-twelfth of a pound to the square inch; the hearth is made 10 to 12 inches larger in diameter than before, and the charge is increased to 222 pounds to 3 bushels of charcoal. These changes have for their object not only the prevention of iron sows, but also of speiss, a compound of arsenic with all the other metals present, and very difficult to utilize. The same precautions are used in blowing in as before. About 30 bushels of charcoal are used to the ton of ore and flux, and 13.5 tons are smelted in twenty-four hours. The second matte contains 52.67 ounces of auriferous silver to the ton, and is worth about $200.

Fusion with lead.—The second matte is roasted as before, but now 50 to 60 per cent. of unroasted matte is left. A stronger ·roasting would so enrich it that two fusions with lead, instead of one, would be necessary. The slag is again basic, and to keep the heat as low as possible, the pressure of blast is reduced to one and a half lines of mercury, while the charge is increased to 277 pounds of matte and flux to 4 bushels of charcoal. In order to keep the lead in contact with the matte as long as possible, as well as to decrease the heat, the crucible is made a foot deeper than before. The new slag has an average composition of—

Silica	27. 45
Oxide of iron	56. 52
Lime	10. 19
Magnesia	3. 48
Alumina	1. 25

The loss will not exceed 2.5 per cent. of the lead. When the hearth is full of melted matte it is tapped, the products running into a basin where they are well stirred with poles. The matte is then partially taken off, the lead remaining until 600 to 700 pounds have collected.

For a perfect extraction of the silver it is necessary to charge 120 to 130 pounds of lead for each pound of silver and gold. With this proportion 75 per cent. of these metals is extracted in one operation, and the matte ought not to contain more than 0.75 per cent. of lead. The extraction of 75 per cent. of auriferous silver, means that more than 90 per cent. of the gold and 73 per cent. of the silver have been obtained. A second operation removes so much more that, including amalgamation where the loss is very great, more than 90 per cent. of the silver and 96 per cent. of the gold is obtained. This second operation takes place only when the matte is worked for copper. At other times the gold and silver are obtained by charging the matte back in the first operation. The absolute loss in smelting is but 0.10 of one per cent. From 14 to 16 tons of matte and flux are smelted in twenty-four hours. A certain amount of lead-matte is obtained, and is charged back in the same operation. If the third matte is rich enough it now undergoes a second fusion with lead, but usually it is so poor that it is treated at once for copper. If, however, it contains less than 35 per cent. of copper, it is roasted and returned as a flux to the first fusion for raw matte. At Lend the conditions are such that this takes place every other year, copper being made one year, and only matte the next.

Cupellation is performed in a German furnace, with movable hood, made very low so that the heat from the fuel is thoroughly utilized. Inasmuch as none of the side products are sold, and there is no need of having them in great purity, there is, beside the fire-bridge, only one opening to the hearth, through which abzug, abstrich, litharge, and smoke, alike escape. From 6,000 to 7,000 pounds of lead are charged at once, and more is gradually added until about 21,000 pounds (the entire make of a year) have been melted. The blast is slow, and the litharge consequently flows rather cold. Refining follows the brightening of the silver, and metal of .985 to .995 is produced. Usually the loss of lead falls between 4 and 6 per cent., while that of silver and gold seldom reaches 0.10 of 1 per cent. About 3 tons are cupelled in twenty-four hours.

TABLES OF THE OPERATIONS.

The following tables will give at a glance all the foregoing particulars, and also exhibit the amount of material handled. The two fusions without lead are combined in one table.

Table of the first and second fusions, 1866.

	Weight in tons.	Ounces of gold.	Ounces of silver.	Per cent. of copper.
Charge:				
Ore...	81.15	21.00942	1373.4058	....
Matte and rich scraps...........................	89.61	19.29186	2216.6406	....
Flux, { basic 7.41 tons	58.29	....	....	....
{ siliceous 50.88 tons .... }				
Total	229.05	40.30128	3550.0464	....
Products:				
First matte...................................	61.38	18.71928	1612.7324	5
Second matte..................................	30.87	20.47698	1806.7302	10
Scraps..	10.25	.08228	95.0922	....
Total	....	39.87764	3514.5548	....

Labor: Thirty-nine 12-hour shifts, 5 men to each shift=195 days.
Charcoal: For warming furnace, bushels................. 291
Charcoal: For smelting, bushels......................... 6,820

7,111

Labor: Per ton of ore,* days............................ 1.8
Charcoal: Per ton of ore,* bushels...................... 65.2

Table of third fusion, 1866.

	Tons.	Lead, pounds.	Copper, per cent.	Gold, troy ounces.	Silver, troy ounces.
Charge:					
Containing gold and silver:					
Rich quartzose-ore..........................	1.96	....	....	0.9610	59.5404
Roasted second matte	30.87	....	10	20.4769	1806.7202
Scraps	2.09	....	....	0.1474	23.0778
Containing lead:					
Lead-matte.................................	3.71	741	6	1.9440	175.9600
Litharge	10.81	17,722	....	....	66.1500
Hearth....................................	3.30	3,304	....	....	42.4700
Flux:					
Scoria from first fusion....................	10.99	....	....	....	....
Quartz	2.04	....	....	....	....
Total....................................	66.77	....	....	23.5294	2175.9184
Products:					
Lead.......................................	10.51	21,030	....	21.5442	1281.8320
Third matte	15.59	....	....	....	624.5200
Lead-matte.................................	3.70	741	20	1.9440	174.9600
Scraps and flue-dust	4.06	....	10	0.2430	46.9900
Total....................................	....	....	....	23.7312	2130.2920

Labor: Ten 12-hour shifts, 5 men in each shift=50 days.
Charcoal: For warming furnace, bushels..................... 100
Charcoal: For smelting, bushels........................... 1,710

1,810

Labor: Per ton of ore,* days............................. 0.46
Charcoal: Per ton of ore, bushels........................ 16.6
Lead charged per ton of ore, pounds..................... 218

* In calculating this it is to be remembered that 28 tons of matte from the previous year were smelted, which must be counted as ore in calculating the expense of charcoal and labor.

Table of cupellation, 1866.

	Quantity.	Lead, per cent.	Gold, ounces.	Silver, ounces.
Charged: Lead.................................tons..	10. 06	100	21. 5442	1283. 8320
Products: Fine silver.......................ounces..	1208. 58	....	21. 5292	1186. 9560
Litharge ....tons..	10. 12	82	....	64. 9080
Hearth.............................tons..	2. 53	50	....	31. 9800

Loss in gold..	0. 1476
Gain in silver..	0. 0180
Labor : 26 days=per ton of ore,* days.....................	0.24
Wood : 7.52 cords=per ton of ore, cords...................	0.69
Charcoal : 40 bushels=per ton of ore, bushels..	0.37

Table of cost per ton of ore in units of labor and material.

	Labor, days.	Charcoal, bushels.	Wood, cords.	Lead, pounds.
First and second fusions	1. 8	65. 2	....	....
Third fusion	0. 46	16. 6	....	8. 0
Cupellation	0. 24	0. 37	0. 69	9. 7
Total	2. 50	82. 17	0. 69	17. 7

To this must be added a small quantity of wood, or refuse charcoal, and labor used in roasting the matte. The above is the cost for ores of the richness above given. With richer ores there is more matte to treat, and the expense of fuel, labor, and lead is therefore greater, and the cost per ton is more ; but *proportionately* richer ores are cheaper to treat than poor. The following table gives the relative cost for various ores, the poorest being taken as unity :

Auriferous silver in 2, 000 pounds.	Value in American coin.	Proportionate cost, poorest ore=unity.
0 to 14. 5 ounces	$0 to $61	1. 00
14. 5 to 29 ounces	$61 to 122	1. 10
29 to 58 ounces	$122 to 244	1. 31
58 to 116 ounces	$244 to 488	1. 73

The Lead ore falls under the first class. The milling ore of Colorado is worth from $15 to $30 a ton, and comes under the same category. The Colorado "smelting ore" so called is probably mostly in the second and third ranks.

Losses.—By reference to the above tables it will be found that the following is the loss and gain of the year :

	Loss.		Gain.	
	Gold.	Silver.	Gold.	Silver.
First and second fusions	4. 24 oz. = 1 p. ct.	35. 50 oz. = 1 p. ct.	....	....
Third fusion	....	45. 63 oz. = 2. 1 p. ct.	2. 02 oz. = 0. 86 p. ct.	....
Cupellation	0. 15 oz. = 0. 07 p. ct.	....	....	0. 02 oz. = 0. 0015 p. c.
Loss	4. 39 oz. = 1. 07 p. ct.	81. 13 oz. = 3. 1 p. ct.	....	....
Less gain	2. 02 oz. = 0. 86 p. ct.	0. 02 oz. = 0. 0015 p. ct.	....	....
Leaving loss	2. 37 oz. = 0. 21 p. ct.	81. 11 oz. = 3. 1 p. ct.	....	....

* In calculating this it is to be remembered that 28 tons of matte from the previous year were smelted, which must be counted as ore in calculating the expense of charcoal and labor.

These amounts are, however, so small that it is impossible to say whether the assayers' errors do not amount to more than the reported loss and gain. Dr. Turner's opinion, founded upon years of experience, and comparing the analyses of the ore with the yield by amalgamation and fusion through several years, was, as I have said, that he could count upon extracting more than ninety per cent. of the silver and ninety-six per cent. of the gold by the two processes of amalgamation and fusion. The loss of lead was nine per cent. of the amount charged.

The cost of all the operations at Lend, in 1866, was $883.88, and the balance-sheet shows a profit of $1,355. The expense was proportioned as follows: Labor, 17, materials, 43, direction, 40; total, 100.

I have dwelt thus particularly upon the minutiæ of each operation in order to indicate the means by which such excellent results are obtained. In our own country the losses in working silver-ores by fusion are so great (frequently from twenty to thirty per cent. in the West) that we can ascribe them only to very rude working. But even in works more pretentious in expense than the somewhat incomplete establishments to be found in the Territories, and which base upon long experience a claim to skillful treatment, we find such reckless application of heat and careless handling of valuable ore as must and does cause great loss. We see ore, worth one or more hundred dollars a ton, thrown in the state almost of powder into furnaces through which flames are roaring almost as violently as in a puddling-furnace. We see alloys of silver, lead, and zinc subjected to distillation in anthracite-fires at a heat far greater than that which has caused the rejection in Europe of all furnace methods of treating these alloys.

At the works which I present for consideration all avoidable causes of loss have been eliminated, or their operation reduced, with the greatest care. Two analyses a year determine the proportions of the charges and the composition of the scoria. Larger establishments would require more analytical work, but there is no reason why the largest works should not be conducted with equal care. The cost of the laboratory would not be more than $250, and the work would consume only a few days in each month.

Great care is necessary at Lend, because, with so small a quantity of ore, any disregard of proper precautions would hazard the profits of the works. In 1866 only 83 tons of ore, worth less than $6,400 in gold and silver, and containing a ton and a half of copper, were treated. And yet this small quantity, together with the ore which is treated by amalgamation in the mills, keeps alive two mining districts and a smelting-works. Beside the miners, an engineer, two smelters, and four assistants have to be supported for the whole year, though the work of smelting occupies only twenty-seven days of twenty-four hours. Of course, such a state of things can be maintained only by low prices, and we find the Austrian workmen paid at rates varying from $27\frac{1}{2}$ to 22 cents (coin) a day. Charcoal is $3\frac{1}{4}$ cents a bushel, and wood $1.17 a cord. In this country we have larger supplies of ore, sufficient to carry on the largest works on a correspondingly economical scale. The nature and higher value of our ores would enable us to work with less expenditure of labor and material to the Troy pound of silver and gold than at Lend.

In considering the results given in this paper for guidance in using a similar process at the West, it is evident that the American ores contain nothing to prevent the application of this method. Antimony, arsenic, and zinc, the bug-bears of the smelter, are, with the exception, perhaps, of zinc, quite as prevalent at Lend as in Colorado. Our ores contain more pyrites than those we have been considering, and there would be

II. Ex. 211——27

no necessity of a fusion for raw matte—an operation which has no object but to remove the gangue. Whether there ought to be a fusion for concentration depends upon the richness of the ore and its adaptability to concentration by machinery. A mixture of rich "smelting ore" and concentrated tailings, such as is now worked up by the smelters, could be roasted and immediately fused with lead. One more fusion with a fresh quantity of lead, if there were silver enough left in the matte to pay for the work, and cupellation, would complete the process. We should then have a process divided as follows:

1. Concentration of poor ore.
2. Roasting of concentrated and rich ore.
3. Fusion of roasted ore with lead.
4. Roasting of matte.
5. Fusion of matte with lead.
6. Cupellation.

The present imperfect concentration of tailings in Colorado is said to cost $6 a ton. A perfect concentration would cost no more. The other expenses would be—

	Days' labor.	Charcoal.	Wood.
			Cords.
Roasting in piles..	0.4	....	0.029
First and second fusions...	1.8	65.2	....
Roasting matte..	0.2	....	0.001
Third fusion..	0.46	16.6	....
Cupellation...	0.24	0.37	0.69
Total...	3.10	82.17	0.72

Mr. Hague says the millers expect to get 1 ton of concentrated pyrites from 6 tons of tailings, which seems to indicate a pretty heavy loss. At that basis, however, the theoretical expense would be—

Concentrating 6 tons to 1.....................................	$6 00
Smelting 1 ton, 3.10 days' labor, at $3........................	9 30
Smelting 1 ton, 82.17 bushels charcoal, at 25 cents............	20 54
Smelting 1 ton, 0.72 cords wood, at $8..	6 00
Total......	41 84
Mining at $10...	60 00
Total cost of treatment, 6 tons........................	101 84
Cost of one ton...	16 94

The expense of charcoal ought to be somewhat less than this, for in consequence of the small quantity of material treated at Lend, no less than 2.5 bushels per ton of ore are expended in heating the furnace. If we add one-half more for loss in blowing out, we have the very large proportion of 3.7 bushels—a quantity which would be lessened to 1 bushel if 500 tons of ore were smelted in one campaign. With proper management this could be very much exceeded, so that the expense of charcoal for blowing in and blowing out would be too little per ton to be worth reckoning.

It now remains to consider the adaptability of this process to western ores, and I will take those of Colorado as an example, for the reason that Mr. Hague's report on the mines of that Territory offers the best data for the calculation. He gives commercial assays of ores from various lodes, which prove their value to be as follows:

	Gold, ounces.	Silver, ounces.
First-class ore:		
Consolidated Gregory	5.6	20
Illinois	4	20
Gardner	3.5	11.5
California	3	18
Burroughs	6	12
Average	4.42	16.3
Milling ore:		
Burroughs, (1,340 tons)	1	4.5

The coin value of the first-class ore is therefore $91.36 for the gold and $21.03 for the silver; total, $112.13. By roasting the ore so as to leave one-third raw matte, and smelting with 180 to 195 pounds of lead to the ton, we ought to extract 90 per cent. of the gold,[*] or 4.05 ounces, worth $83.71; and 73 per cent. of the silver, or 11.90 ounces, worth $15.35, or $99.06 in all. The cost of this would be about as follows:

Mining one ton of ore $10 00
Roasting: 0.04 day's labor, at $3................ $1 20
        0.029 cord wood, at $8............... 23
        8 months' interest on $10, at 12 per
        cent............................ 80
                         — $2 23
Smelting: 1.5 day's labor...................... 4 50
        8 pounds lead, at 5 cents.............. 40
        46 bushels charcoal, at 25 cents........ 11 50
                         — 16 40

Total for roasting and smelting........................ 18 63

Total for mining, roasting, and smelting............... 28 63

If our ore contains no copper, and the matte will not pay for further treatment, and we proceed at once to cupellation, we have in addition:

Cupellation: 0.24 day's labor, at $3.................... $0 75
        0.37 bushels coal, at 25 cents.............. 09
        0.69 cord wood, at $8...... .... 5 52
        9 pounds lead, say at 5 cents.............. 45
                         — 6 81

Total for mining, roasting, smelting, and cupellation,.... 35 44

Profit, $99.06—$35.44=$63.62.

We have remaining a matte containing $13.07, and probably a certain amount of copper. Let us see whether this will pay to work by itself. · The cost will be:

---

[*] It will be observed by reference to the table of the third fusion that *all* the gold was extracted by one operation at Lend in 1866. I have, however, adhered to Dr. Turner's general estimate in making the above calculations.

Roasting: 0.30 day's labor,* at $3 .... $0 90
       0.02 cord wood, at $8 18
                                                    $1 08

Smelting: 0.60 day's labor, at $3 .... $1 80
       27 bushels coal, at 25 cents 6 75
       5 pounds lead, at 5 cents 25
                                                       9 80

                                                       10 88

This would cause a loss; for calculating 4 per cent. loss on gold and 10 per cent. on silver in the original ore, we have only $7.50 which can be extracted from the matte. The loss would therefore be $3.38, and unless the matte were worked for other products, as for copper, it probably could not be utilized at present, though in many cases it would be required as a basic flux in the first fusion. Considering the present state of the West and proportion of copper in the ore, the process would probably consist of three operations—first, roasting the ore; second, fusion; and third, cupellation, the copper matte being sold.

Accepting the Burroughs milling ore as an average of the second-class ore, we have for this, one ounce gold, worth $20.67, and 4.5 ounces silver, worth $5.81; total, $25.48. The cost of treating it would be:

Mining 6 tons $60 00
Concentrating 6 tons to 1 .... 6 00
Roasting and smelting 1 ton 18 63
Cupellation 6 81
Add for roasting† 2 00

Cost of treating 6 tons 93 44
Cost of treating 1 ton 15 57

Yield at 90 per cent. of the gold $18 60
Yield at 73 per cent. of the silver 4 24

      Total yield 22 84
Cost 15 57

Profit $7 27

This would leave a matte containing $1.86 in gold and silver, and perhaps some copper.

The above calculations are of course, theoretical, so far as they relate to works which have never yet been established in the Territory. There may be errors in the prices assumed for labor and materials; but there is no reason why the amount of labor and material expended per ton should be more than at Lend; that part of the calculations is not theoretical. Undoubtedly in establishing such works some difficulties would be experienced, but with a railroad to the foot of the mountains, and the improved facilities for communication, the difficulties in the way

---

* A certain correction has to be applied, because the amount of matte is taken as larger than at Lend. I have assumed it to be 50 per cent. more.
† Fine ore requires a more expensive roasting than coarse, for which reason I have added 50 per cent. to the cost. Roasting in furnaces, Mr. Hague says, costs $5 in Colorado.

cannot compare with those which have been overcome in establishing the milling system.

A chief drawback to extracting the precious metals in the Territory, instead of concentrating them in a matte to be exported, is thought to be the lack of lead-ores, since the Georgetown mines have not fulfilled their promise as lead mines. Let us see how much is required for works treating 25 tons of ore a day, a capacity which is considered to be quite respectable for a mill; and it is to be remembered that these 25 tons are concentrated ore representing several times that quantity of ore as it came from the mine. The loss amounts to about 17.7 pounds of lead per ton, or less than one per cent. Of galena-ore yielding, say, 70 per cent. lead, one ton daily suffices for, say, 70 tons pyrites, or 21,000 tons yearly; two tons daily suffice for, say, 140 tons pyrites, or 42,000 tons yearly; five tons daily suffice for, say, 350 tons pyrites, or 105,000 tons yearly.

If each ton of smelting ore represents 6 tons of ore from the mine, we have more than 600,000 tons of ore treated with 1,500 tons of galena-ore. Even if the mines of Georgetown and Argenta are unable to supply this amount, it could easily be bought in, and brought from Utah, at rates which would at least pay its own cost. My object, however, is not to urge any process upon the attention of western miners, or prove by full figures its applicability. I offer the Lead process as one which deals with ores precisely similar to those of Colorado, and leave it to those who are interested in the mines of that territory to work out its adaptability.

CHAPTER XII.

THE AMALGAMATION OF GOLD-ORES.

This chapter constitutes a supplement to the preceding one on economical results in the treatment of gold and silver ores by fusion, and was likewise furnished to me by John A. Church, E. M., of New York. I give the chapter without change or comment; but I do not fully concur in the theory of amalgamation which it presents.

It is commonly supposed that mercury takes up gold by reason of an affinity which causes the union of the metals whenever they are brought in contact, and in the use of amalgamated copper plates for catching the gold, the Americans have trusted the success of their gold-mills entirely to this action. In Austria they proceed on a different basis. There they acknowledge the affinity of gold for mercury, but confine it within small limits. The gold which is dissolved by the mercury, and which passes with it through the filter, is that which has a chemical union with the mercury; while that which remains in the filter, and after distillation forms the "retort," is merely particles of gold which have mechanically sunk into the mercury by force of gravity. Their surfaces are attacked by the fluid metal, which acts as a cement to bind them together; but in no sense do they form a definite amalgam. I will not discuss this point thoroughly here, but merely point out some facts in relation to Colorado ores which, on this hypothesis, give a ready explanation for the poor yield of those ores in the mill.

*The principles on which the separation of gold from its ores is effected by mechanical means are easily explained. If we have a substance composed of two elements, one having a specific gravity of 10 and the other of 5, it is clear that if we can provide a liquid having a density of, say, 7, the former can sink in it, and the latter cannot. To accomplish the separation of the two we have only to crush the substance to a certain fineness and place it in a bath of the liquid. As soon as each particle of gravity 10 comes in contact with the fluid, it sinks, and we have only to agitate the sand and bring every particle in contact with it to produce perfect separation. We have then the two elements, one at the bottom and the other on the top of the liquid.

This is precisely what takes place in the so-called amalgamation of gold-ores. Gold has a specific gravity of 19.33, and mercury of 13.60. The iron pyrites in which the gold of Colorado is found has a gravity of about 5, and quartz, another constituent of those ores, has a gravity of 2.6. It would appear, then, that in a mixture composed of gold, specific gravity 19.33, and pyrites, specific gravity 5, there should be no difficulty in effecting the separation when the gold in a finely divided state is passed over mercury in which the gold can and the pyrites cannot sink. The Austrian gold-mill was devised to satisfy these conditions, and it works perfectly. In it mechanical contact between the gold and mercury is effected in the most perfect way, and the mercury lying in a bath 1 to 1½ inches deep, is in a condition to act either by affinity or merely as a fluid of medium density. And yet this apparatus fails to extract the gold from most of its ores, and the tailings are sent to the smelting-

---

*What follows is partly taken from an article by me in the *Scientific American* of October 7, 1871. In that article an error was made in putting the "normal alloy" at 35 silver and 65 gold. It should have been the reverse, or 35 gold and 65 silver. The error, however, leaves the argument unaffected.

works, if they can be made to pay the cost of treatment. In some cases, as for instance at Zell, spoken of in the beginning of this paper, the ore, worth only $2 and less a ton, is unable to bear any expense but amalgamation, and it could not bear even this were it not for the fact that its gold is fine and contains little silver.

There is a difficulty in treating gold-ores with mercury, in the explanation of which we may perhaps account for the trouble experienced in Colorado. Native gold is rarely or never pure. It is alloyed with silver, which has a specific gravity of 10.56. An alloy of the two metals, therefore, has a specific gravity between 19.33 and 10.56, depending upon the proportion of the two metals. With gold 35 and silver 65 parts, the specific gravity of the alloy is about the same as that of mercury, and it cannot sink in that fluid; that is, it will not "amalgamate." The question is then, do the ores of Colorado contain more than 65 of silver to 35 of gold? Let us calculate the assays given above, and we have the following table, the 35 gold and 65 silver being taken as the normal alloy:

	Gold, ounces.	Silver, ounces.	Gold.	Silver.
Normal alloy..	....	....	35	65
Consolidated Gregory ore.............................	5.6	20	22	78
Illinois lode..	4	20	16	84
Gardner..	3.5	11.5	23	77
California...	3	18	15	85
Burroughs..	6	12	33	67
Various mines, (2,056 tons)..........................	4.5	11	30	70
Average..	4.43	15.41	23	77
Milling ore, Burroughs, (1,340 tons).................	1	4.5	18	82

These are fair specimens of Colorado ores, and we see that the gold they yield will not sink in mercury. And yet those who adhere to the milling process say it *does* amalgamate. That is true to a certain extent. Part of it amalgamates, and in that respect it exactly resembles the Lead ores, in which part of the gold amalgamates and part will not. The explanation is that Colorado ore contains 1 free gold, 2 gold alloyed with silver, and perhaps 3 silver not alloyed with gold. Mr. Hague thinks that the mills extract about 55 per cent. of the gold in the first operation, and 15 per cent. more by a repetition. If we construct a table for Colorado ores such as I have given for the Lead ores we shall have something like this:

	AMALGAMATED.			NOT AMALGAMATED.	
	Free gold.	Gold.	Silver.	Gold.	Silver.
	Ounces.	Ounces.	Ounces.	Ounces.	Ounces.
Gregory..	2.77	0.46	0.69	2.37	19.31
Illinois...	1.08	.33	.49	1.09	19.51
Gardner.......................................	1.73	.29	.43	1.48	11.07
California......................................	1.48	.24	.37	1.28	17.63
Burroughs.....................................	3.46	.58	.36	1.96	11.64
Average................................	11.42 2.28	1.90 0.32	2.34 0.46	8.78 1.76	70.16 15.80
		0.78		17.50	

	Free gold.	Alloy.	
Average................................	2.29	18.37	

Thus we see from this table that of the above Colorado ores only 57.7 per cent. of the gold and 2.9 per cent of the silver is extracted by amalgamation. These proportions are, of course, hypothetical; but we may regard them as near the truth.

The Burroughs milling ore contains 1 ounce gold and 4.5 ounces silver. At the same rate of yield the proportions would be:

	AMALGAMATED.			NOT AMALGA-MATED.	
	Free gold.	Gold.	Silver.	Gold.	Silver.
	Ounces.	Ounces.	Ounces.	Ounces.	Ounces.
Burroughs Milling	0.495	0.085	0.127	0.42	4.373

The value of the Burroughs milling ore is therefore $12.15 in gold that will amalgamate, and $14.42 alloy that will not amalgamate; or 45.5 of the former and 54.5 of the latter in 100 of value. Thus we see that to its great fault of not extracting more than 70 per cent. of the gold amalgamation adds the loss of nearly all the silver; so that the real saving, even by the best work, including a repetition of the milling, is under 60 per cent. of the value.

I judge that the Colorado ores contain silver not alloyed with gold, from the fact that, although a great deal of the gold has been removed by the mercury, which leaves nearly all the silver, the tailings show no proportionate increase of silver to gold. Silver has therefore been removed as well as gold, *and in about the same proportion*. In three tables, giving assays of tailings, which Mr. Hague publishes, we have the following proportions:

	Gold. Ounces.	Silver. Ounces.
1.	1.05	4.32
2.	0.66	2.63
3.	2.34	3.87
Average	1.35	3.61
Proportion	27	73

Compared with 23 gold and 77 silver, which is the average of the ores, these figures show that both silver and gold have disappeared, and about equally, in the process of milling. Though hardly necessary, I will say that nothing in the bullion explains this fact, for that is composed of 845 gold to 155 silver, on an average. The cause of this loss is undoubtedly defective concentration. The ores probably contain proper silverminerals, which are very brittle, reduce to a fine powder in crushing, and are easily carried off on the stream. It may be, too, that the small proportion of galena found in the ore is highly argentiferous, but contains little or no gold. This would partly account for the loss of silver, for when galena is stamped through a mesh of 25 to the inch, and then concentrated in a buddle we may be sure that very much of it goes in the water.

It must not be supposed that the above table, in which the gold of Colorado ores is divided into free gold and auriferous silver, is correct in its proportions. In the ore there are probably an unknown number of distinct alloys, and the gold we obtain comes (1) from fine gold, (2) from those alloys which contain more than 35 per cent. of gold. We know from the bullion that about one-sixth of the amounts which I have put down to fine gold is really silver. The great fact remains that, if we accept the Austrian explanation, the Colorado ores ought not to amalgamate well; and when we examine the results of practice, we find that they do not. This may be only a coincidence, but if so, it is one sufficiently remarkable to make us reconsider the determination to force those ores to amalgamate, to which we have so stubbornly held for ten or twelve years.

Before leaving this subject it may be well to inquire how it is that gold is amalgamated on copper plates in Colorado, where, of course, there is no mercury-bath. All experimenters, I believe, agree that more than half the amalgam obtained is made in the battery, and the plates placed there collect the larger part of the gold-sand. Probably still more comes in contact with the mercury within the battery, and issues from the screen with the surfaces of the gold-particles covered with mercury, or a true amalgam of gold and mercury, which gives them the power of adhering to the coat of amalgam on the plates. One of the signs for which the amalgamator constantly watches is, in fact, the appearance of hard, dry particles of "amalgam," which he knows by experience are apt to pass over the plate without adhering. He adds mercury, to the battery and the particles then come out with a softer coat and readily fix themselves to the plate. The success of the plates as amalgamators is also greater when there is a thick coat of soft amalgam on them, and all these facts point to the supposition that the gold is retained on the plates by virtue of the cementing properties of the mercury with which it becomes covered in the battery.

Another method of amalgamating gold, in use in this country, is the Washoe pan-amalgamation. In the pans, it is well known, there is no bath of mercury, but this metal is distributed through the pulp in small drops, the object being to secure not only thorough contact of the mercury with the silver, but also to maintain this contact long enough to have chemical action set in. To run sulphide or even chloride of silver over mercury, as gold is run over a bath of that metal in the Austrian mill, would not answer. But the Comstock ore contains gold as well as silver, and the Washoe secures a very good proportion of it; as much, probably, as any amalgamation will extract of gold that is not positively fine. The following results of numerous bullion assays taken from Mr. Hague's book represent in all 133,844 tons of third-class ore, and 5,105 tons of second-class ore: for third class, 32.4 gold and 67.6 silver; for second class, 36.7 gold and 63.3 silver. So that gold is really saved in the Washoe pans, and they seem to work, in fact, better than the Colorado mills, for they extract no less than 81.1 per cent. of the gold and 64.6 per cent. of the silver, estimated on the mill samples. How is it that gold can be taken up when the mercury is in fine drops all through the pulp, and there is no opportunity for its mechanical action as a fluid of medium density? Though no examination of this subject has been made, I have no doubt the bulk of the gold is obtained, either in the pan where the pulp is thinned, or in the settler where it undergoes still greater thinning, and the conditions are in fact extremely favorable for collecting the gold at the bottom, and entirely by mechanical means. There it meets with mercury and follows it in its subsequent movements. Undoubtedly some gold is taken up by the mercury while still distributed through the pulp, but the conditions under which mercury separates this metal from its ores teach us that this cannot be a very large proportion of the whole. It is, indeed, a very small proportion, and fortunately there is an analysis which proves this. In the chapter on the chemistry of the Washoe process we have the following analysis of some crystals of Washoe amalgam given:

Mercury.. 75.04
Silver.. 24.18
Gold.. .77

The proportions of silver=96.5 to gold=3.5, are here, it will be observed, very different from those found in the bullion of the third-class ore spoken of above, and which was, silver 67.6, and gold 32.4.

In working the small pan used in making the experiments upon which the chapter upon the chemistry of the pan-process is based, it is to be noted that no such proportion of silver to gold was obtained as that which occurs in the regular returns of the Savage Company. In trying the first-class Savage ore, the bullion contained silver 934.7, to gold 49.5; or, in 100 parts, silver 94.9 and gold 5.1; and yet the yield of silver was only 32.55 per cent., while that of the gold was 70.6. Had the yield of silver equaled that of gold, the proportion of the latter would have been very much smaller. Probably in working in the large way gold collects in the bottom of the agitator and is taken up by the mercury of the *next* charge of pulp, a state of things that could not exist in the experiments, because the apparatus was there thoroughly cleaned after each pan.

CHAPTER XIII.

THE AMALGAMATION OF SILVER-ORES IN PANS, WITH THE AID OF CHEMICALS.

The attention of the mining public has, within late years, been largely attracted to the existence of numerous veins and deposits of silver-ore, the beneficiation of which was, however, apparently impracticable, as the ores they furnish cannot be worked with any degree of success by the ordinary amalgamation-process, and are not sufficiently rich to bear the expense of roasting, even after the improved and cheaper methods now in vogue. An over-confidence in the adaptability of the ordinary Washoe process to the working of ores of the most different characters has led to numerous failures, while, on the other hand, the caution bred of experience causes many properties to lie idle which would prove remunerative under a method of treatment equal in thoroughness to the ordinary process, without greatly exceeding it in expense. The demand for such a process has within the last three years been practically filled, but the suspicion with which all innovations are regarded, and the uninvestigative spirit of the "practical" millman, have prevented this new method from being more generally known and adopted. The facts now set forth should demonstrate its usefulness beyond a question. A test of its merits can be readily made. Simple as it appears, it can be made to solve many metallurgical and financial problems in the way of working ores to a fair percentage of their value, and with profit, which hitherto have, by their apparent rebelliousness, proven only sources of loss.

The first successful attempt at amalgamating rebellious silver-minerals in pans, and without roasting, was made on the Comstock slimes. By far the greater portion of these slimes was in former years allowed to run to waste. This loss was probably looked upon as an evil unavoidable in wet-crushing mills, and was partially excused by the fact that not more than 35 per cent. of their gross value could be extracted by the then known methods of treatment. There was consequently but little induce-ment to millmen to save slimes assaying $40 per ton, or thereabout, as they could barely be worked to a profit. Thousands of tons were thus lost, which at the present day would yield an enormous fortune to their owner.

Some three years ago a series of experiments were made, which clearly proved that these slimes could be worked to a high percentage at a cost which, though greater than that of ordinary milling, still left a good margin for profit. Mill-owners were induced to save and sell what they had hitherto allowed to run to waste, and, shortly after, the working of slimes became a distinct feature of milling on the Comstock.

All this was effected simply by adapting to the Washoe pan-process certain features of the Mexican patio, namely, the use of sulphate of copper and salt in sufficient quantities to decompose the rebellious silver-minerals and leave them in a favorable state for amalgamation. The successful application of these chemicals to the ordinary Washoe process was entirely novel. It is true that in former years many millmen had tried these chemicals among a host of others, but their experiments were invariably carried on with such a want of knowledge that failure was inevitable. Some used bluestone without salt, and others salt with-ont bluestone. When the two were used simultaneously, the amount was so small that the effect could not be appreciated. At the time referred to it was a maxim among millmen that no benefit was to be derived from the use of chemicals. Mr. Küstel, in his book on the Washoe process, after giving a list of different chemicals employed or recommended—none of which were probably ever largely adopted—expresses strong doubts as to the efficacy of any or all chemicals; and in a later article, after admitting the extraordinary effects obtained on slimes, states that the same process would not be applicable to sands. So much to vindicate the practical originality of this process. It will be shown further on that it can be applied with equal advantage to the working of sands and tailings. There is no certainty as to the formulæ in obedience to which the chloride of copper—which is formed by the mutual action of salt and sulphate of copper—effects its purpose. Chemists are not agreed as to the various re-actions, and the subject is still under dispute. Its theoretical comprehension is, however, not essential to the practical operations which will be described.

We will first consider the application of chloride of copper to the working of slimes, and, later, its extension to the treatment of sands or pulp.

Slimes constitute the clayey portion of the ore which flows from the battery in wet-crushing mills, and is imperfectly caught or settled in the tanks constructed for the purpose of catching the crushed ore. It is needless to say that with a sufficient number of tanks these slimes could be saved, but lack of space generally renders this impracticable, and the usual plan is to allow them to flow into reservoirs outside of the mill, from which they are afterward dug out, allowed to dry, and hauled to the mill.

The mechanical treatment of slimes varies but little from that of ordinary ore. Their finely-divided condition renders grinding unnecessary, and calls merely for a sufficiently violent motion to thoroughly incorporate the quicksilver with the pulp.

The length of time devoted to the amalgamation of a charge of ore is influenced to a great extent by the amalgamating capacity of the mill, and by various reasons of economy. It often becomes advisable to sacrifice a certain percentage of the value for the sake of working a greater number of tons per diem. To insure good results a charge should not remain in the pan, subject to the action of quicksilver and chemicals, less than six to seven hours.

In working ordinary ores the quicksilver is not added until the pulp has been thoroughly ground. This is to prevent the flouring of the quicksilver; but since slimes require no grinding, one might suppose that it would be more advantageous to put the quicksilver into the pan on charging, thus allowing more time to the amalgamation.

Experience has, however, shown that better results are obtained by putting in the chemicals, i. e., sulphate of copper and salt, first, and adding the quicksilver from two to three hours later, thus allowing the intervening length of time for the decomposition or "preparation" of the refractory minerals. The reason of this is not immediately apparent, as the quicksilver charged simultaneously with the chemicals should attack the chloridized (?) silver-minerals in statu nascenti. The amount per ton of chemicals employed varies with the richness of the slimes. The quantity is always largely in excess of that called for by the chemical equivalents of the minerals to be acted upon, but in very large pans the relative proportions may be diminished. Taking ordinary slimes, assaying thirty dollars per ton, as a basis, the average quantities of sulphate of copper and salt employed per ton are 10 to 12 pounds of the former and 20 to 25 pounds of the latter.

These quantities may be advantageously increased on slimes of higher grade. The exact amounts must be determined by conscientious assays. In working slimes of uniform character, the varying fineness of bullion furnishes an excellent empirical test of the amount of chemicals to be employed. The baseness of the bullion increases with the quantity of sulphate of copper used, owing to the precipitation of copper by the iron of the pan, and its consequent amalgamation. The millman soon discovers that beyond a certain point an excess of sulphate of copper produces no adequate results besides reducing the fineness of his bullion. Having once determined the average fineness of his bullion when working under the most favorable circumstances, he increases or diminishes the amount of sulphate of copper as the bullion becomes finer or baser than his standard.

The same amount of chemicals with richer slimes will produce finer amalgam, which is due to the fact that a larger amount of silver is amalgamated while the same percentage of copper is precipitated. Should the value of the slimes and consequent fineness of the bullion increase materially, it becomes necessary to determine by assay the most advantageous proportion, and to adopt a new standard of fineness. A source of great expense in the working of slimes is the excessive and apparently unavoidable loss of quicksilver. Working with all the advantages of settlers and agitators, this loss seldom falls below four pounds per ton, whereas in quartz-mills it does not as a rule exceed $1\frac{1}{4}$ to $1\frac{1}{2}$ pounds per ton of ore. This is undoubtedly due to the nature of the slimes, which have the same effect upon the quicksilver as oil or grease, forming a film or coating over the surface of the metal, and preventing the globules from uniting. The minute particles settle with difficulty, and to a great extent flow off with the tailings. The excessive loss in working slimes cannot with justice be ascribed to the action of chloride of copper, as in working the tailings by the same process the loss of quicksilver per ton is not greater than in ordinary quartz-mills, thus proving sufficiently that the deleterious action of chloride of copper upon quicksilver is not appreciable upon so large a scale.

Many attempts have been made to effect a saving of quicksilver in milling, but so far only mechanical appliances have met with success. Although sodium, sodium amalgam, cyanide of potassium, &c., are very effective in temporarily "enlivening" the quicksilver, their action speedily dies out after the latter has been transferred to the pan, and experience has sufficiently demonstrated that when employed in practicable quantities they have no effect in aiding the amalgamation. The only thorough method of cleansing quicksilver is to retort it, and the use of proper straining-sacks will render a recourse to the method unnecessary.

The following table will show results obtained in working on a large scale two different lots of slime:

1. Slimes from various mills.
2. Slimes from Savage ore.

Number.	Assay value per ton.			Yield per ton.			Per cent. extracted.			Fineness of bullion.		
	Gold.	Silver.	Total.	Gold.	Silver.	Total.	Gold.	Silver.	Total.	Gold.	Silver.	Total.
1	$7.28	$21.44	$29.72	$4.01	$20.20	$24.21	55.08	90.04	81.46	005	401	406
2	6.03	35.70	41.73	4.60	31.08	35.68	76.23	87.06	85.50	005	537	542

The results obtained in the working of lot No. 2 are somewhat above the average, which can be estimated at about 80 per cent. of the assay value.

The method of treatment now in vogue is susceptible of modification and improvement. The yield can be increased by charging the chemicals at two different periods, say two-thirds at first and the remaining third two or three hours later. This is due to the fact that the action of the chloride of copper becomes more feeble as time elapses, and is revivified by the addition of this second portion.

A saving in the amount of salt consumed can be effected by dissolving it together with the sulphate of copper in vats in proportions approximating to their chemical equivalents, and charging the resulting chloride of copper in the liquid state. The plan of putting the sulphate of copper and salt into the pan separately, necessitates a large excess of the latter to insure the complete decomposition of the former. An excess of salt was originally deemed advantageous on the supposition that it would act beneficially in holding the chloride of silver in solution, thus presenting more surfaces of contact to the action of the quicksilver, and, moreover, would convert the Hg Cl formed to Hg Cl², in which state it would be precipitated by the iron of the pans in the metallic form. It is, however, improbable that such re-actions should occur on a large scale where the re-agents represent so small a proportion of the entire mass.

The discovery of a cheap method of working slimes led to important results. The same process was soon applied to the working of low-grade tailings, of which vast quantities had accumulated on the plains bordering on the Carson River, at Dayton, and at other points. Several mills, with a capacity of working from 250 to 300 tons per diem each, are now in operation in Dayton and vicinity. The tailings, which were formerly allowed to run into the Carson River, are now carefully saved, and in fact to such an extent has the re-action set in against the wasteful practices of the past, that many mill-owners save the tailings after they have passed through the pans, with a view to reworking them at some future time. The tailings being mostly very poor, it becomes necessary to work a large amount daily in order to obtain a fair interest on the capital invested. This is effected by using very large pans and by rapid amalgamation. Economy in the time devoted to amalgamation cannot, however, be effected without sacrificing a portion of the precious metals which might otherwise be saved. It is, however, considered policy to suffer this loss. The method of treatment is practically the same as in working slimes, with the exception that a smaller proportion of sulphate of copper and salt is used, and that in other respects less attention is paid to the extraction of a high percentage, the value of the tailings not justifying too careful a manipulation. For the same reason the extra expense of grinding the tailings, which would undoubtedly increase the yield, is dispensed with.

As a rule the bullion produced is very base, not exceeding on the average .450. and often going as low as .350.

In the mill of Birdsall & Co., in Dayton, where the crude bullion rarely exceeds in fineness the latter figure, an inexpensive process has been adopted for partially refining it.

As the bullion comes from the retort a great portion of the copper is found in a crust on the surface of the slab. This slab-crust, which contains some silver and gold, is separated from the finer metal beneath, and the latter is cast into bars 500 to 600 fine. The cupreous crust is roasted in a reverberatory furnace and crushed, and the copper-oxide separated by means of sulphuric acid. The bullion from other mills is sold in the crude state, and suffers a discount proportionate to its baseness.

Even with the use of sulphate of copper and salt the percentage extracted from low-grade tailings rarely exceeds 50 to 55 per cent. This is due principally, as above stated, to the necessity of a cheap, rapid, and consequently imperfect treatment. The same tailings, however, worked without chemicals, but otherwise in the same manner, would not yield over 20 to 25 per cent. of their value, and in some cases not so much as that. Richer tailings give better results, and are also treated more carefully.

In a mill of large capacity, and in good working order, the cost of working does not exceed $3 to $4 per ton. Hence there is still a margin for profit on tailings assaying only $9 to $10 per ton.

The pans now almost universally in use in the Washoe tailing-mills have a capacity of from 6 to 8 tons, and the time of working the charge generally does not exceed four hours. More than half an hour, however, is consumed in charging and discharging, in heating the pulps, &c. The actual time of treatment, therefore, is less than three and a half hours.

The following table will show the amount extracted from one half hour to one half hour, from a charge of tailings assaying $15.31 per ton, worked in a pan of eight tons' capacity.

Value of tailings, $15.31.

	Silver.	Yield per ton.
Assay value of sample from pan, end of first half hour................	$12.40	2.91
Assay value of sample from pan, end of second half hour.............	11.70	.70
Assay value of sample from pan, end of third half hour..............	10.83	.87
Assay value of sample from pan, end of fourth half hour.............	9.58	1.25
Assay value of sample from pan, end of fifth half hour..............	9.02	.56
Assay value of sample from pan, end of sixth half hour.............	8.09	.93
Assay value of sample from pan, end of seventh half hour...........	7.85	.24
Assay value of sample from pan, end of eighth half hour............	7.77	.08
		7.54

Total percentage extracted, 49.30 per cent.

The falling off in the yield in the seventh and eighth half hours is due to the insufficiency of chemicals used, their action having spent itself in the earliest part of the treatment.

Until the introduction of the use of the sulphate of copper and salt, the working of tailings was confined to rich lots, where a small percentage of the value would still prove remunerative, and to blanket-concentrations. The richness of the latter consisted to a great extent in amalgam, and its successful extraction was consequently an easy matter.

Before dismissing the subject of slimes and tailings, it is necessary to refer to certain classes of these ores which do not yield to the action of sulphate of copper and salt to the extent claimed in the foregoing pages. The precise cause of this rebelliousness has never been accurately ascertained, but from the fact that the bullion produced in working them is always of exceeding fineness ($\frac{990}{1000}$ and over) even when a large excess of sulphate of copper is used, it is evident that the evil consists in the presence of some substance which decomposes the sulphate of copper, thus producing the same effect as if none were used. This injurious agent is probably an alkali or alkaline earth. It is to be regretted that we have no knowledge of any cheap chemical re agent producing practically the same effect as chloride of copper, and which would not be subject to this decomposition.

The following table will show to what extent the treatment of slimes is affected by the elimination of the copper-salt. Five hundred and sixty-two and one-quarter tons of slime from Chollar ore were worked in charges of 2,500 pounds each, with 15 pounds sulphate of copper and 30 pounds salt to the charge. Total amount worked, 562¼ tons.

	Gold.	Silver.	Total.
Value per ton	$12 60	$30 91	$43 51
Total values	7,084 35	17,379 14	4,463 49
Yield	1,651 16	6,200 93	7,852 09
Percentage extracted	23.30	35.68	32.97

	Gold.	Silver.	Total.
Fineness of bullion	⁵⁵⁄₁₀₀₀	²⁸⁰⁄₁₀₀₀	³³⁵⁄₁₀₀₀

Within half an hour of putting the sulphate of copper into the pan, chemical tests could not detect the presence of copper in solution in the pulp, thus proving that it had been almost immediately decomposed. The same phenomenon has presented itself in the working of tailings.

Such cases, however, are exceptional, but prove why in certain instances the use of sulphate of copper and salt may not produce the effects claimed for them. Apart from these exceptional cases the beneficial effects of chloride of copper being thoroughly established, the question naturally arises why its use is principally limited to the working of slimes and tailings, and why it is not more generally employed in quartz-mills. There is not the slightest doubt that an intelligent use of chemicals would prove highly beneficial in many cases where the old method of treatment without chemicals is adhered to. The ores of the Comstock ledge are particularly docile under their influence. It is true that many mill-men claim to have investigated the matter, and have pronounced against the use of chemicals, but, as has already been stated, their experiments were carried on in such an unfair manner that failure was inevitable. It could not be expected that a few ounces, or even a few pounds, of sulphate of copper and salt would produce appreciable results upon a ton of ore, and yet mill-men were deterred from using larger quantities through the fear of producing baser bullion. It is a common superstition that fine bullion is a guarantee of skillful amalgamation, than which nothing is less true. Until this prejudice is eradicated it is useless to look for the employment of sulphate of copper in quartz-mills in quantities sufficient to establish its merits. Another cause of indifference on the part of Washoe mill-men is the fact that they generally find no difficulty in returning by their present method of working 65 per cent. of the value of the ore, the amount universally guaranteed by custom mills, and are, naturally enough, not disposed to increase the cost of working without corresponding benefit to themselves. The mine-owners also appear to be satisfied with these results, but their apparent indifference to their own interests does not admit of so simple an explanation.

The first quartz-mill to adopt the use of sulphate of copper and salt was that of the Meadow Valley Company, near Pioche, Nevada. This mill commenced operations in the summer of 1870. Owing to a misunderstanding only a small supply of sulphate of copper was on hand at the time, and when this was exhausted the difference in percentage extracted was immediately apparent. During the first week's run the yield was equal to 80 per cent. of the value of the ore. When the chemicals were exhausted the yield fell below 40 per cent., only to recover on the receipt of a fresh supply of sulphate of copper. The following table will show the difference in percentage extracted while working with and without chemicals:

July and August, 1870—Percentage extracted, 54.40, working three-quarters of the time without chemicals.

September—64.39, working one-half the time without chemicals.

October—87.90, working with chemicals.

November—82.5, working with chemicals.

December—81.8, working with chemicals.

January—76.7, working with chemicals.

February—77.6, working with chemicals.

March—68, working with chemicals.

April—67, working with chemicals.

May—74.4, working with chemicals.

June—74, working with chemicals.

July—73, working with chemicals.

Average assay value of the ore for the year, $143.21.

The ores operated upon were very rich. In such cases the policy of working ores with chemicals in preference to roasting may be questioned. The matter must be decided by a careful comparison of the expense and results of the respective methods. It must furthermore be taken into consideration that roasting involves dry crushing, and consequently a decrease in the working capacity of a mill of a given number of stamps. This again necessitates a larger outlay for the erection of more mills in order to crush a sufficient number of tons per diem to render possible the payment of any considerable amount in dividends. The expense of building furnaces, &c., must also be taken into consideration. Should, however, the ores prove of so rebellious a nature that the difference in percentage extracted by the roasting process be sufficient to overrule these considerations, then it is evident that the wet process must give way to dry-crushing and roasting.

At the Nevada Butte Mill, where this method of working with chemicals was introduced during the past summer, it was absolutely necessary to find some means of reducing the ore without roasting, as it was not sufficiently rich to admit of the latter method of treatment. Here also it was found that whereas only 35 per cent. of the value could be extracted by quicksilver alone, the use of sulphate of copper and salt increased the value to 75 per cent. and upwards. These ores contained still more lead than those of the Meadow Valley mine, and finally became so base as practically to belong to the class of smelting ores, causing an enormous loss of quicksilver, and rendering their beneficiation by the amalgamation process an impossibility. The enterprise had therefore to be abandoned.

The only chemicals thus far known to be applicable to the amalgamation process are salts of copper. A great benefit would be conferred upon the mining public by the discovery of other chemicals equally effective, but less expensive, and not so subject to decomposition by the iron of the pan or by the ore itself.

H. Ex. 211——28

CHAPTER XIV.

THE TREATMENT OF ORES OF NATIVE SILVER IN CHI-HUAHUA.

The following account was written by H. B. Cornwall, E. M., of the School of Mines of Columbia College, New York, and is to be published in April in the columns of the Engineering and Mining Journal. It furnishes an interesting view of a remarkable industry.

Several districts in Mexico yield considerable quantities of native silver, but nowhere does this class of ore occur so abundantly as at Batopilas, in southern Chihuahua, and the neighboring country. Batopilas especially has become famous for its silver-ore, and the object of this article is to describe the mining and reduction of the native silver-ores of Batopilas, as exemplified by the actual workings of an American company, which has now been in eminently successful operation there for several years. In a future article, the treatment of the sulphureted and other combined ores of silver will be given, with some remarks on mining in Northern Mexico in general.

Batopilas is situated in a deep and very narrow valley, or *barranca*, among the western ranges of the Sierra Madre, in southeastern Chihuahua, about eight days' journey, by mules, from the nearest port on the Gulf of California. The neighboring mountain-ranges show different formations; sometimes the trails lead over trachytic rock, then over granite or diorite, and again over conglomerate and porphyritic formations. All of these may be met within a six hours' ride from Batopilas. The silver-bearing veins, however, are confined to the diorite, and their universal vein-rock or gangue is carbonate of lime, in the white, crystalline form of calcite. Accompanying the native silver, as will be more minutely described presently, are black sulphuret of silver, (*plata negra*,) ruby silver, (*rosiclara*,) arsenical iron, (*fierro blanco*,) galena, (*plomo*,) and zinc-blende, (*copelilla ;*) all of which occur, however, in very small quantity. Through the valley runs a river, always supplying a great deal more water than would be necessary to run as many mines as could be worked, and during the rainy season swelling to a torrent. On both sides of the river rise steep mountain ranges, and the silver-veins occur in considerable number in both ranges. The particular mine to be described lies on the east side of the river, and is worked by a main tunnel, entering the mountain some sixty feet above high water, and but a few yards from the river-bank.

Three of the veins cut by this tunnel had been worked in old times, and from one of them it is reported that some ten millions of dollars were taken. Probably the report is little if at all exaggerated, for the immense waste-heaps and the size of the workings under ground show how extensively the vein was worked, and if the present richness of the ore extracted from another vein can be taken as a criterion, certainly an immense amount of silver must have been obtained. The veins have the general trend of the mountain-range, although they converge at different points, and some of them cross the valley. All are proved by the present tunnel to be true veins, as they are cut from 300 to 600 feet below the surface-level of the old workings, and show the same character below as above. As is always the case where the silver occurs pure, the ore is not uniformly distributed, but occurs in pockets, and sometimes a vein which has yielded a large *bonanza* at one time may be worked for several hundred feet without yielding more than a very moderate amount of silver. Still, although the indications of silver is not continuous throughout, yet, in the vein above alluded to, the old workings show a continuous body of ore, varying from 3 to 10 or 12 feet in width, and to a depth (these statements are made from memory, and not from notes) of over 200 feet. The present tunnel has cut eight large veins, all bearing silver, and by judicious working of these veins, following the indications of rich ore, new deposits may be constantly opened, so that once such a district is developed it may be as successfully worked as if there were but one vein with a continuous body of ore. The Mexicans were not prudent miners. If a large and rich deposit was opened they worked it out as rapidly as possible, not providing by dead work on other veins for the time when their *bonanza* should be exhausted.

It has been said that galena and zinc-blende accompanied the silver. These minerals, taken in connection with a *lively* appearance of the calcite, and the presence of arsenical iron, are the miner's guides. By their occurrence he judges where to look for the silver, and an experienced man can follow them up, until from a merely promising rock he proceeds to the silver itself. The country-rock is diorite, very hard and tough near the mouth of the tunnel, but becoming more tractable further in, and always changing decidedly when near a vein. Occasionally the diorite, in a somewhat altered state, and mixed with calcite, forms the vein-rock, as is the case in the largest and richest vein now worked in the tunnel; but invariably, wherever pay rock occurs, there the silver is found with the above accompanying minerals.

As regards the method and expense of mining and reducing the ore, the following facts are given, taken from notes furnished the writer, during a twelvemonths' stay in Batopilas, by the superintendent of the company, who is also the vice-president.

The cost of mining per ton, including all dead work, such as running the front of the tunnel, prospecting, &c., is $33; the actual cost of extracting the silver-ore, including necessary drifts, and the work on all the veins yielding silver in paying quantities, is $8 per ton. Hauling to the works, on donkeys, about half a mile, 62 cents per ton. In another article other details of mining expenses and methods will be given.

The ore is sorted into three classes: first class, value $2,500 and upwards per ton; second class, value $1,000 to $2,500; third class, all under $1,000, averaging perhaps $250.

The third-class ore is dumped at the stamps, the better ore is kept in a store-room and weighed out. All the ore is crushed in a battery of three small stamps, weighing about 300 pounds each, with a fall of 9 inches, and a capacity of 8 tons per twenty-four hours. The ore falls through a screen with five-eighth inch slits, and is then charged in the arrastra. The lumps of silver are separated by the screen, cleaned by hand, and, with the larger lumps of pure silver from the mine, refined with the retort silver. The stamps are run by a horizontal water-wheel, which will be described under the arrastra. This latter apparatus it is unnecessary to describe. Suffice it to say that it is a large Mexican arrastra, 9 feet in diameter, with two stone mullers or runners, weighing 600 to 800 pounds each. The wheel that runs it is, however, peculiar. The arrastra is built on the top of a pile of masonry in a deep pit. In the center of the arrastra rises a shaft, revolving on a pivot which rests in a plate raised a little above the bed of the arrastra, and from this shaft horizontal arms project beyond the rim of the arrastra. From these arms descend rods which support a horizontal wheel, that thus revolves around the arrastra a few inches above the bottom of the pit. In the periphery of this wheel, at intervals of 6 inches, are inserted rectangular floats, slightly concave, and set up edgewise, as if to receive the water from a tangential, horizontal chute. These floats are called *cucharas*, (spoons,) and hence these arrastras are called *arrastras de cuchara* as distinguished from the *arrastra de mula*. The water acts on this wheel solely by its momentum acquired while descending very rapidly through a tapering chute, having a fall of 8 feet, with a length of 12 to 15 feet. It is very evident that there is a great loss of power here, but as the works are supplied with a superabundance of water by a ditch, and the three arrastras are capable of reducing all the ore required, this makes little difference. These arrastras are universally employed in Mexico when water-power is at hand. Such a wheel, with a diameter of 20 feet, will carry the two runners of the arrastra as fast as four stout mules, which could not work more than eight hours per day, and it runs the battery of stamps as above stated.

From the stamps the ore is taken to the arrastras, into each of which a ton, more or less, according to the size, is charged at once. A few buckets of water are thrown in, just enough to give the mass a certain consistency, which is very essential to the proper conduct of the grinding process. If there is too little water, the ore is raised up and pushed forward by the mullers, without being ground; if there is too much water, it packs beneath the mullers. Water is from time to time added to preserve the proper consistency of the ore, and after the operation has been carried on about eight hours sufficient quicksilver is added to amalgamate all the silver in the ore. Generally the arrastra is charged with one ton per day of the third-class ore, requiring some 25 pounds of quicksilver, and after three days' run, or whenever the amalgamator thinks proper, rich ore is added, requiring proportionally more quicksilver, for the purpose of getting a suitable amount of amalgam collected in the arrastra, preparatory to cleaning up. Some hours after adding the quicksilver the amalgamator takes a portion of the charge out in a horn spoon, washes it, and thus judges whether there is the proper amount of quicksilver present. These assays are regularly made, but, after a little experience with any ore, he soon learns to gauge the amount of quicksilver very closely.

Every morning, after the silver appears to be thoroughly amalgamated, a large excess of water is added and the arrastra kept in motion for four to six hours; the heavier particles then settle, the amalgam separates from the fine ore, and after the machine has been at rest for a short time, the water is run off, carrying with it all of the finely ground and desilverized ore. The coarser grains of ore, not yet sufficiently reduced, remain and are ground with the next charge. The tailings thus obtained are very poor—so poor that the most experienced men in the place are unwilling to pay $3 per ton for them, with the object of extracting the silver on the patio. They contain nearly all of the galena, zinc-blende, and arsenical iron of the ore, a very little quicksilver and amalgam, and any arsenical silver (ruby silver) that may occur, with the exception of a trace that stays in the amalgam, either owing to its density or to native silver adhering to it. Most of the sulphuret of silver, being less brittle, and therefore not so easily reduced to powder, settles to the bottom of the arrastra and is taken out with the amalgam, in which it is plainly visible after washing. The rich tailings, removed after the rich silver-ore has been added, and just before a clean-up, are more valuable, and are saved for concentration or treatment on the patio.

As regards the quicksilver required in this part of the process, it is found that ore containing coarse silver needs less than ore with fine silver, in proportion to the amount of silver.

When the rich tailings have been run off, the top layer of coarsely ground ore is removed with iron scrapers and reserved for the next charge. Then the amalgam is scraped up and carried in wooden bowls (*bateas*) to the washing-tank. This amalgam seems, to the superficial observer, scarcely anything more than coarse sand and slime, but, on adding to a suitable amount of it, in a shallow wooden bowl, the proper quantity of quicksilver, and washing it, stirring and rubbing it constantly with the hand, the clean amalgam is obtained. The dirt thus removed from it is rich and is reserved for concentration by washing on the plane-table. About 10 per cent. of the quantity of quicksilver already employed in the arrastra is added to the amalgam in this process of cleaning. A small portion of amalgam from the finer-grained silver-ores is sometimes very carefully washed, the black silver (sulphuret) being removed by grinding the amalgam on a stone and washing it thoroughly, and the resulting very pure amalgam, after straining, is retorted carefully, and furnishes the bullion used in paying expenses. It is even purer than the refined silver cast into bars.

The clean amalgam is now strained in canvas cloths, and this is the most tedious part of the process. As the amalgam is at present retorted a very firm amalgam is required, so that it is not found sufficient to strain the amalgam in large bags by merely twisting them with a stick, but the quicksilver must be thoroughly pressed out from small balls, not over 2 to $2\frac{1}{2}$ inches in diameter, by squeezing and rubbing them in the canvas with the hands. The coarseness of the silver, which is frequently present in *nails*, renders the separation of the quicksilver impracticable by any other means yet tried; from the very fine-grained amalgam obtained on the patio the quicksilver is much more easily expressed. Probably if some other system of retorting were introduced this part of the process might be made less laborious. The strained amalgam is charged into quicksilver-flasks, from which the bottom has been removed. About 65 to 70 pounds of amalgam are introduced, and then the flasks are set aside to allow as much of the quicksilver to drain off as possible, and also to harden the amalgam. In a day or so, or as soon as four of these flasks are ready, they are removed to the retorting-furnace, where they are set on end over holes in a slab which forms the bottom of the furnace. There is nothing to prevent the amalgam from falling out of the flasks except four narrow strips of iron set into the mouth of the flask; the amalgam never runs out when the fire is properly managed. Other quicksilver-flasks, open at each end, are placed below the holes in the bottom of the furnace, their lower ends being beneath the surface of water in a tank which lies under the furnace. After luting the lower ends of the amalgam-flasks with clay and ashes there is thus no outlet for the quicksilver except into the water, where it condenses. A charcoal fire is slowly kindled around the flasks and they are thus retorted. This simple furnace, universal in that part of Mexico, has supplanted the old copper-bell apparatus, but is itself susceptible of great improvement.

We have now followed the silver as far as the refining process. It is evident that whatever loss of quicksilver may have resulted is purely mechanical. There is no chemical action in the arrastra, nor is any needed. Consequently the loss of quicksilver is small. A careful account is kept of it, and the result shows only three-quarters of an ounce lost per marc (eight pounds) of silver produced.

The retort-silver is refined in a small reverberatory furnace, built of adobes and fed with wood, which can receive a charge of 600 pounds of crude bullion. This charge is worked off in four hours. A little litharge and lead are added to remove the impurities, sulphur, arsenic, lead, iron, and, possibly, a very little zinc which is present as zinc-blende, and carbonate of soda and borax are also used as fluxes. The loss is 7 per cent. on the crude bullion, and consists, to some extent, of silver and quicksilver; the amount has not been exactly ascertained, but is, doubtless, very small, as the silver is ladled out very rapidly when refined. It is cast into bars weighing about 70 pounds each, which assay .988 of silver on the average.

The slags from the refining-furnace, with the concentrated tailings or slimes from the tanks for washing amalgam, and sometimes other secondary products, are occasionally smelted in a small shaft-furnace, with addition of galena, and the resulting lead is used in refining the retort-silver.

An experienced amalgamator, who was working up some of the ordinary tailings from the every-day run of the arrastras, furnished the writer with the following facts relative to their treatment on the patio. The Mexicans sometimes make the operation pay, especially when the ores contain other ores than native silver, because they employ their *peons*, whose labor costs very little.

The tailings were made into a heap (*torta*) containing 100 *cargas*, (about 16 tons,) and to this were added 720 pounds of salt, which was thoroughly mixed in, with water enough to keep the whole at a proper consistency. The next day 50 pounds of *magistral* (in this case sulphate of copper) were added and thoroughly incorporated. The third day 100 pounds of quicksilver were added and the whole left standing one day. Then every other day the mass was thoroughly mixed by driving mules about in it, in the usual way, and this operation, called the *repaso*, repeated until, by assays in the horn spoon, it was shown that the heap had been properly amalgamated, or was *rendida*. This requires more or less time, according to the temperature, the size of the heap, the nature of the ore, &c., and the whole operation requires great experience and care. When it was found that the heap was ready for washing, 25 pounds of quicksilver were added, with plenty of water, and the whole thoroughly mixed. In this operation of the patio there is a chemical action; the loss of quicksilver is necessarily large, and in this particular case amounted to 25 per cent. of the original amount charged. The resulting silver from the retort showed .990 fine. The amount obtained from the heap of 16 tons of tailings was only $145, leaving a loss of $5 on the expenses, not counting the amalgamator's time, and showing conclusively that simple amalgamation of native silver-ores, in the arrastra, is as effective and cheap a treatment as it is possible to employ, in the absence of an important proportion of combined ores, as sulphuret, chloride, or arsenical ores, and in a country so difficult of access as the interior and mountainous portion of Mexico.

CHAPTER XV.

THE REDUCTION OF SILVER-ORES IN CHILI.

This chapter was prepared at my request by Mr. James Douglas, jr., of Quebec, a gentleman whose high professional standing, and personal acquaintance with the subject, entitle his statements to respect.

The silver mines of Chili extend from its northern far toward its southern limits, the two last-discovered *minerals* happening to be those of Caracoles, in the debatable ground between Chili and Bolivia, and San Carlos in south latitude 36°. But the most prominent mines heretofore worked have been in the Department of Copiopó, and there are to be found the only extensive and scientifically managed establishments for the reduction of silver-ores. All through the central provinces, however, but especially in the neighborhood of Arqueros, the landscape is often enlivened by a pretty little mill, consisting of a single light stone *trapeze*, turned by a rude turbine attached to the vertical shaft beneath the mortar, and a single Freiberg barrel or open tub amalgamator. They work up two or three cwt. a day of tractable ores, but never touch the more refractory. In Copiopó, on the other hand, the most difficult ores are treated; the machinery is very perfect; and the extraction of the silver as thorough as in any mills in the world.

There are seventeen establishments on or near the Copiopó River which work up the ores from the three *minerals* of Chañareillo, Lomas Bayas, and Tres Puntas; but the three mills owned by Messrs. Escobar & Ossa, in which the patent process of Herr Krähnke is used, do more proportionately than any of the others.

In these establishments calcination is not employed. The different ores are mixed accurately in given proportions—the chlorides and native silvers of Chañareillo, with the polybasite of Lomas Bayas and the base metals of Tres Puntas. The refractory ores, however, largely preponderate, as the pure silver-ores are yearly becoming scarcer. The broken ore is sampled by falling from a hopper, accurately placed above the apex of a pyramidal cone, from the angles of which protrude partitions. The stuff which collects in each compartment is resampled separately in like manner, till by repeating the act a perfectly accurate sample is obtained. As the hopper and the pyramid are carefully protected from wind by being incased in canvas, no dust escapes, and thus the error, which formerly resulted from the difficulty of always taking up proper portions of dust and coarse ore in sampling with the shovel, has been remedied. This error was found to be so great (for the ores there are always absolutely dry) that more silver has, at the end of the year, been obtained from the furnace than was supposed to have been put into the mill.

The grinding is done altogether in the *trapezes* or Chilian mills; but these have been perfected in all their details. In Messrs. Escobar & Ossa's mill in the town Copiopó there are three double *trapezes*. Each wheel weighs 60 cwt. and is of metal. An automatic feed delivers the ore from a hopper, filled twice in the twenty-four hours. The three *trapezes* reduce to impalpable powder 12 tons in the twenty-four hours.

The pulp is received in slime-pits, whence it is carefully shoveled and allowed to dry by exposure to the air, every precaution being taken to prevent the lumps from breaking up. When perfectly dry, the cakes are ready for the barrels; of these there are five, arranged in a row, and driven by the same line of shafting. They differ in size and in mode of gearing from the common Freiberg barrel. Each barrel is 8 feet long and 5 feet diameter, and revolves on spindles, which form the centers of heavy spiders covering the barrel-heads, and bolted to one another by strong iron bars. These form a cage, within which the barrel lies firmly secured. The barrel is made to revolve by a pinion playing into a toothed hoop, and it can be raised out of and lowered into gear with the greatest ease by means of an ingenious mechanism. The barrels are charged with 80 cwt. of ore, a 10 per cent. solution of dichloride of copper in brine, mercury, and metallic lead or zinc; but of course upon the accurate proportioning of the quantity and quality of the ore to the reducing reagents depends the whole success of the operation; and as the establishments of Messrs. Escobar & Ossa, which are under the immediate supervision of Mr. Krähnke alone, command the necessary skill, there alone this delicate process is worked with satisfactory results. In a notice appearing some months ago in Dingler's Journal, the active agent is said to be protochloride of copper. This is incorrect, as the equation afterward proposed in explanation of the reaction shows.*

The dichloride plays the same rôle in reducing the silver-sulphide directly to the metallic state as M. Laur, in his recent articles in the *Annales des Mines*, ascribes to it in the *patio* process :

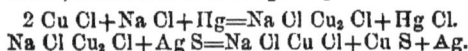

$$2 \text{ Cu Cl} + \text{Na Cl} + \text{Hg} = \text{Na Cl Cu}_2 \text{ Cl} + \text{Hg Cl.}$$
$$\text{Na Cl Cu}_2 \text{ Cl} + \text{Ag S} = \text{Na Cl Cu Cl} + \text{Cu S} + \text{Ag.}$$

Six hours suffice to effect the amalgamation; but one charge only is put into each barrel daily.

The separation of the amalgam from the sand is effected in a series of tanks provided with agitators.

There is a peculiarity in the subsequent treatment of the amalgam worthy, perhaps, of imitation. After being filtered in the usual way it is still further freed of mercury by being dried in a centrifugal machine, such as are employed in sugar-houses. The amalgam comes from the machine as fine sand, more uniformly deprived of free mercury than it can be in the filter. The *plata piña* is obtained as in Mexico, and then smelted in a small reverberatory to a fineness of .890. loss of silver by the draught being prevented by covering it with a very fusible slag.

The ores treated contain on an average 50 marcs to the *cajon* of 64 cwt. Ores with less than 20 marcs to the *cajon* are smelted with copper and gold ores at the works of the same firm at Nautoko, whence a rich argentiferous and auriferous matte is shipped to England and Germany.

---

* I cannot agree with Mr. Douglas here. The reactions as given in Dingler's Journal and in the *Berg und Hüttenmannische Zeitung* (quoted September 26, 1871, in the Engineering and Mining Journal) are represented by the equation—

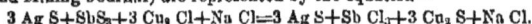

$$3 \text{ Ag S} + \text{SbS}_3 + 3 \text{ Cu}_2 \text{ Cl} + \text{Na Cl} = 3 \text{ Ag S} + \text{Sb Cl}_3 + 3 \text{ Cu}_2 \text{ S} + \text{Na Cl.}$$

If the argentic sulphide thus obtained is again treated in a hot solution with cupric subchloride and sodium-chloride, and zinc is added, metallic silver is almost instantaneously formed. The reactions are—

$$\text{Ag S} + \text{Cu}_2 \text{ Cl} + \text{Na Cl} + \text{Zn} = \text{Ag} + \text{Cu}_2 \text{ S} + \text{Na Cl} + \text{Zn Cl.}$$

The zinc probably acts as electro-positive metal, predisposing the atoms of argentic sulphide and cupric subchloride to a mutual exchange, so that the cupric subsulphide and argentic chloride are formed, which last is decomposed in a nascent state by the zinc, with the formation of zinc-chloride and silver. This may not be the correct theory; but the equations do not, in my opinion, bear evidence of its incorrectness.

R. W. R.

CHAPTER XVI.

THE METALLURGICAL VALUE OF THE LIGNITES OF THE WEST.

This chapter was prepared by my deputy, Mr. A. Eilers, after thorough personal examination and inquiry.

No one who has visited our western mining districts and studied the economical relations of the beneficiation of their ores, can underrate the importance of the question of fuel.

By far the larger number of the districts which contain smelting-ores, *i. e.*, argentiferous and auriferous lead or copper ores, are situated in the Great Basin, that great plateau between the Rocky Mountains on the east and the Sierra Nevada on the west, almost the whole of which is comprised at present in the boundaries of Nevada, Utah, and part of Arizona. This region is essentially a barren country. The extreme dryness of the atmosphere permits but a very scanty vegetation in the plains; and even in the detached mountain-chains running through it—generally from north to south, or from northwest to southeast—there are no trees found, except dwarf-pines and mahogany, at the head of sheltered ravines, and a few cottonwoods and willows, which fringe the insignificant streams, before the water sinks into the arid plains. Nearly all the mountain-chains in this region are rich in silver-ores. That class of these ores which is adapted to amalgamation, and rich in silver, has been worked with profit for more than ten years. But, before the advent of the transcontinental railways, mining was restricted to these ores alone, and the consumption of fuel could be met with the scanty supply of forest-trees in the immediate vicinity of the mining districts. Since, however, the Union and Central Pacific Railroads have brought the Great Basin nearer to the commercial centers of the East and the Pacific coast, thus reducing the expenses of freight and labor materially, other silver-deposits, containing poorer ores in greater abundance, have been rapidly taken up and worked. During the last year this industry has so expanded, that the State of Nevada alone has been able to show a production of over $22,000,000 in silver. But not alone are the poorer grades of amalgamating ores now worked profitably, aided, as the metallurgical process is, by such excellent inventions as that of the Stetefeldt and the Brückner roasting-furnaces, but the working of smelting ores has also been largely entered into. If I say "largely," I do not only mean to say that smelting-works are now scattered widely over the Great Basin, but that some of these conduct their operations on a really grand scale. In Eureka, Nevada, for instance, there are twelve furnaces in operation, which produced, during the last year, 5,665.5 tons of base bullion, worth $2,035,588, although only a small part of them ran regularly. Four of the Eureka furnaces have each a capacity of from 35 to 40 tons of ore per day. Three of these belong to the Eureka Consolidated Company, who have also two smaller furnaces. Nearly throughout the year this company have kept four furnaces running at a time, and one idle, and the daily consumption was 120 to 140 tons of ore and 4,000 bushels of charcoal. At this rate of smelting, the wood for ten miles around Eureka has been used up in a little over a year, which is not a strange statement, when we consider what I said before, that there is very little wood in those regions, any way. Thus the question of fuel becomes, at once, a very important one, for the price of 33 cents per bushel of coal, which is now paid at the works, cannot rise much without threatening the very life of the industry.

In Utah, where over twenty furnaces were built, and had been partly in operation, in the fall of last year, some of the works have been compelled to pay as high as 30 cents per bushel for their charcoal, and very few are so favorably located as to get their coal for less than 18 cents per bushel. Many more smelting-works have been erected since the time spoken of, and the addition of every one of them must inevitably tend to raise the price of fuel. Even the most fortunate ones—those located high up in the mountains, where timber is comparatively plenty—cannot hope to escape in the next few years the danger of an enormous rise in the cost of wood and charcoal. And almost every smelting and amalgamating works in the Great Basin finds itself in precisely the same position. While the masses of poor ores are growing on their hands, fuel has a continual upward tendency.

Now there are two means by the combination of which this threatening danger can be averted. The first is the building up of a net-work of narrow-gauge railroads along the principal valleys, which will connect the mining districts with the Central Pacific Railroad; and the second is the employment of the vast stores of lignites occurring in the Rocky Mountain region, for metallurgical purposes. The utilization of this coal for the purpose named has not yet been attempted successfully, and I propose, therefore, to-night, to say a few words on this subject.

According to a late lecture of Professor Newberry, these lignites underlie not less than 50,000 square miles in the Great Basin and along both flanks of the Rocky Mountains. The principal beds now open and wrought I have had the good fortune to visit during the last summer. The mines are located at Carbon, Rock Springs, and Evanston, all three stations on the Union Pacific Railroad, and along the eastern slope of the Rocky Mountains, in Colorado. The coal in these localities, though from different beds, hardly varies in external appearance, but analysis has established a somewhat differing composition. It has a black color, shining luster, a brown streak, and is very compact, the wood-structure, which is found intact in so many lignites, being almost totally obliterated.

The Carbon seam, one hundred and forty miles west of Cheyenne, is 8 to 10 feet thick, and had been extensively worked for over a year when the unfortunate fire broke out, in the latter part of 1870, which caused the whole mine to cave in. At the time of my visit, in the summer of 1871, work was progressing rapidly to re-open the mine, and regular operations have since been resumed. The coal in this bed is distinguished from that in the other beds by many small patches of resinous matter, very similar in appearance to amber. An analysis of this coal, furnished me by Mr. Wardell, the superintendent of the Wyoming Coal and Mining Company, gives—water, 6.80; ash, 8.00; volatile matter, 35.48; fixed carbon, 49.72.

The Rock Springs seam is opened in the midst of the Bitter Creek Desert. It is 10 to 12 feet thick, and a smaller seam lies close above it. This coal contains also some resinous matter, but not as much as the foregoing. The analysis shows—water, 7.00; ash, 1.73; volatile, 36.81; fixed carbon, 54.46.

The Evanston seam is by far the largest. It is from 22 to 26 feet thick, but the coal is not as good as that of the last locality. According to analysis it contains—water, 8.58; ash, 6.30; volatile matter, 35.22;

and carbon, 49.90. This bed presents also the great disadvantage in mining it that innumerable joints run through it at right angles to the strike and dip, undoubtedly resulting from great pressure, and that the coal is very hard and brittle, so that in undermining only slow head way can be made, and a very large proportion of waste results.

In regard to the Colorado beds now opened I cannot give any details, as I was prevented from visiting the mines.

The coal mines along the Union Pacific Railroad have furnished a considerable product since they were first opened in 1868, viz:

There were mined by the Wyoming Coal and Mining Company—

	Tons.
In 1868	8,755
In 1869	54,723
In 1870	84,356
In 1871	93,348
Total	241,182

By the Rocky Mountain Coal and Iron Company—

	Tons.
In 1869	2,473
In 1870	18,187
In 1871	53,869
Total	74,529

Altogether by these two companies, up to the end of 1871 315,711

The capacity of the mines of the Rocky Mountain Coal and Iron Company has been much increased lately, so that in the first three months of 1872 this company has been able to mine and ship 24,933 tons.

Almost all this coal has been used up by the two great railroad companies, the Union Pacific and the Central Pacific, the quantities shipped to San Francisco and other points being insignificant.

Here, then, is an almost inexhaustible source of supply for the pressing wants of the metallurgical works of the Great Basin and the Pacific States and Territories generally.

But if you suggest the use of these lignites for metallurgical purposes to the superintendents of works in those regions, you receive the unanimous answer that they are not fit to be employed for the production of high temperatures. You are told that the main difficulty in using this coal is the fact that it breaks into small pieces as soon as it is exposed to the heat; that in the fire-box of the reverberatory the draught cannot after that penetrate it, and that in the frequent stirrings which are necessary the small pieces fall through the grate half burned, while on account of the frequent opening of the fire-doors for the purpose of stirring the fire a great part of the heat produced is lost. In the blast-furnace, it is claimed, the blast cannot penetrate the fine coal and ore, and thus the necessary temperature is unattainable.

Such and similar opinions in regard to this coal are held by almost every one connected with mining and reduction in the far West. It is considered a settled affair that this coal cannot be used to advantage in metallurgical operations.

Now let us see whether this is really the case; and to do this, we must first examine the experiments by means of which people have arrived at such a conclusion.

As to the experiments for the use of this coal in reverberatories there are two unsuccessful ones on record, one in Colorado, the other in Utah. In both cases the grate used in the common fire-box was the horizontal grate, and the supply of air was provided by the draught of the chimneys only. In both cases the coal broke up into small pieces, and could not be burned rapidly enough to produce the required temperatures.

In blast-furnaces these lignites have been frequently tried in different localities in the West. But no smelting temperatures could be attained and the furnaces would come near chilling. This effect was also rightly attributed to the cracking and breaking up of the coal, and its use in blast-furnaces in the raw state is now virtually given up. I should mention here that the blowing-engines used in the West are ventilators, with which you can produce no pressure, and Root's blower, with which you can reach a very slight one. But then it was proposed to first coke the coal. To look at the analyses of these coals there appears to be no good reason why it should not be possible to make good coke of them. But it is the unanimous verdict of everybody, who has tried the experiment, that no serviceable coke for smelting purposes can be produced from them. Specimens which I saw last summer at various places along the Union Pacific Railroad are certainly not calculated to encourage the idea that the existence of the lignites in this region is a guarantee for the perpetuity of the mining industry in that barren country. The coke is not at all coherent, in fact so soft that a slight pressure of the hand crumbles it into a thousand fragments. How could such material resist the pressure of the superincumbent mass in a blast-furnace? It is evident that it could not be used at all, for the powdered mass would give the blast less chance to penetrate than the raw coal. It would seem, then, at first sight, that the existence of these lignites brings no relief to a threatened industry. At least this appears to be the conviction of the majority out West, and we do not now hear of further experiments. Yet, what have those already made proved? They have proved that under the conditions given in the various trials the Rocky Mountain lignites cannot be used to advantage in metallurgy, and nothing more.

But there are a great number of devices in modern metallurgy by which this fuel can be made to do effectual duty. I do not intend to discuss these at length, but I wish to point out a few ways in which, I am confident, the desired end may be easily reached. As to using this lignite in its raw state, in the common fire-box and on the common horizontal grate, with natural draught only, it might have been expected that a material containing 8 per cent. of hygroscopic, and certainly from 12 to 20 per cent. of chemically bound water, would fall to pieces and thus render the production of a high heat impossible; especially as so much heat is inevitably consumed in converting the water into steam. On the locomotives of the railroads, where no very high temperature is necessary, a sufficiently rapid combustion cannot be reached except through the increased draught by means of the exhaust; and even with this improvement the engineers on the Union and Central Pacific Railroads complain continually about the difficulty of keeping up steam.

But this whole difficulty can be overcome, as far as reverberatories are concerned, by using this coal in gas-generators, instead of in the common fire-place, and by doing the metallurgical work with gaseous fuel instead of the solid. I could adduce numerous examples, where lignites far inferior to those of the Rocky Mountains are used to great advantage in this way, and some, where even the high temperature necessary in iron-works are thus produced. According to Tunner, gases from good lignites are capable of producing a temperature as high as 2,600° C.

The lignites of the West are eminently fitted for use in gas-generators; for the very fact that they break up into small pieces, when exposed to the heat, is an advantage, because it would be much the easiest, this way, to convert *all* the carbonic acid formed in the lower part of the generator into carbonic oxide, as a very large surface of glowing carbon is thus presented. They are *not* bituminous, and their contents in ash are so small that they will not interfere. It may be, indeed, necessary, and it is certainly highly advantageous, to use a blast under the grate in order to further a rapid development of the gases, but this has also the advantage that the danger of explosions will be lessened. It is my opinion that generators with stair-grates and under-blast will be found the most advantageous; and if still higher temperatures than can be produced by this means should be required, an increase can easily be obtained by using hot wind, both under the grate and for the combustion of the gases.

But the use of the lignites in blast-furnaces is of far higher importance to the western mining districts than that in reverberatories. Experiments so far have proved unsuccessful, principally, I am sure, because with the blowing-engines in use the required pressure could not be attained. To burn that material in the blast-furnace, cylinder-blasts are required, and perhaps it would also be necessary to close the tops of the furnaces in order to smelt under a high pressure, which may be regulated by the damper in the flue. The extraordinary results thus attained in producing high temperatures by Bessemer are too new to require recalling. Nothing of this kind has, however, yet been tried in the West, but I hope that during the present year this subject will be thoroughly investigated.

The coke produced from the lignites by the simple method employed is, as I have said before, not fit for the blast-furnace. But the coal used was, as far as I am aware, of the inferior kind occurring in Colorado and in the Wasatch near Coalville. The Rock Springs coal, which is by far the best lignite, has not been tried. And if, instead of trying to coke this material in imperfectly constructed bee-hive ovens and in pits, more perfect apparatus, like the Belgian oven or Appolt's oven, had been used, I think the result, even with the poorer qualities of Rocky Mountain lignite, would have been more encouraging. The Rock Springs coal, I am confident, will make coke in good apparatus, and if it should not be quite as firm as required for the blast-furnace, its hardness might be increased according to experiments which I learn were made in the West several years ago, by coking it under pressure. To produce this pressure in the coking-ovens the escape of the gases need only be regulated; and the ovens themselves must be constructed with the special view of resisting a pressure from within. Success in this direction would of course be of the utmost moment; for even if we assume as a settled fact that the lignites can be used in the blast-furnace with the proper blowing-machinery, in a raw state, their high percentage in water will always be fatal to the production of very high temperatures and their maintenance. It is, besides, much more agreeable and economical to use coarse fuel than fine stuff, as every smelter well knows.

Finally, I wish to draw attention to the importance which these lignite-beds have in regard to the vast magnetic-iron-ore deposits near Laramie, and the hematites of Rawlins. The latter are very pure, and rich in iron, and the former also contain nothing deleterious except a little sulphur, the precise amount of which I have forgotten. If a method is found in which good coke can be made from the coal, there is of course nothing in the way of the railroad companies making their own rails, but if this should not be the case, it seems to me highly desirable that the late experiments of Siemens and Pousard, for the purpose of making wrought iron and steel directly from the ore, and so avoiding the blast-furnace, should be continued with a special view to the utilization of the iron-ores and lignites of the far West. It is true that the respective means employed by these two gentlemen, though technically successful, have not been so economically. There are, indeed, at the present time experiments going on in this country with apparatus different from those used by the English and French engineers, which are very likely to solve this problem favorably, it being the special object of these experiments to produce large quantities of iron in a given time, and with the greatest possible economy in fuel.

CHAPTER XVII.

THE METALLURGY OF NATIVE SULPHUR.

The discovery of large quantities of native sulphur, mixed more or less with earthy matter, in Nevada and Utah, and quite recently in the so-called Yellowstone region in Wyoming and Montana, induces me to say a few words in regard to the above subject.

All those of the above sulphur-deposits, which I have examined personally, owe their origin undoubtedly to the condensation of sulphurous vapors in the overlying colder layers. They are situated in volcanic regions, in some of which the subterranean forces are still active, the deposition of sulphur going on continually at the present time.

To determine the amount of sulphur which can be extracted profitably by the methods of beneficiation now in use, on a large scale, several simple tests are employed, one or two of which will be briefly mentioned. According to Authon,* two grams of coarsely pulverized ore are heated in a glass tube of 10 to 16 inches in length and four lines in width, which is closed on one side, and into the open end of which another tube, also closed on one side, is introduced up to within 3 inches of the ore. When no more sulphur issues from the ore, that piece of the latter tube in which the sulphur has been condensed, is cut off and weighed. The sulphur is then removed, and the empty tube weighed again, the difference of the two weights giving the available amount of sulphur in the ore. To make the test on a larger scale, one or two pounds of the crushed ore are introduced into a good clay retort, which is put into a wind-furnace, so that its neck protrudes about 15 centimeters. To this a porcelain tube is luted, one end of which just dips into water. The retort is now heated to a strong red heat; the sulphur-vapors are condensed in the porcelain tube and the liquid sulphur drops into the water. When there is no more sulphur in the ore, the tube is taken from the retort, heated strongly over the water, and the sulphur, which has remained in it in solid form, will also be collected in the water. The whole product is then taken out, dried carefully, and weighed.

For the utilization of the sulphur from the class of ores here under consideration various methods are in use, which can be classed under two main heads:

1. Eliquation of the sulphur in entirely open or partly closed apparatus.

2. Eliquation, sublimation or distillation of the sulphur in closed furnaces.

The methods coming under the second head require considerable outlay of capital for apparatus, and greater expense for labor. They also require fuel. In their favor, however, is the more perfect extraction and utilization of sulphur which they effect; but the gain by these methods is not great enough to overbalance in our western Territories the increased expense of securing the product. In some regions the absolute want of a cheap fuel precludes their employment altogether. For these reasons I shall not dwell on them here, but rather present a description of a few of the methods belonging under the first head.

a. Melting of the sulphur in cast-iron kettles.—This method can only be employed with profit in working the richest ores, containing over 70 per

---

* Dingler's Polytechnisches Journal, vol. 161, page 115.

cent. of sulphur, because with poorer ores the unavoidable retention of sulphur in the dross would render the percentage actually saved proportionately too small, and the process would be unprofitable. Rich ores are treated in cast-iron kettles, of not over two cubic meters contents, which are heated by means of a separate fire-place. The heat maintained is over 111° Centigrade, and must not rise over 150° Centigrade. The nearer the temperature can be kept to the melting-point of sulphur, (109° Centigrade = 228° Fahrenheit,) the better is the result, because at such a temperature sulphur is most liquid and does not burn. The kettles are filled with ore, which is melted down, and occasional additions of raw ore are made, until the kettle is filled with the liquid mass. Meanwhile all the earthy parts which can be reached are taken out with perforated iron ladles. After the kettle is full, the mass is permitted to settle for a short time. The scum on top is then taken off, and the clear sulphur cast into molds, until the sediment at the bottom of the kettle is reached. A new quantity of ore is then introduced, and the process is repeated. After several operations the sediment is taken out of the bottom of the kettle, and either thrown aside or used with poorer ores in pits or furnaces.

b. *Eliquation of sulphur in furnaces or pits.*—Formerly the sulphur was extracted from the ores of Sicily by means of shaft-furnaces, not over 4 to 5 feet high and 7 to 15 feet wide. They had an inclined bottom, at the lowest point of which a canal communicated with the outside. The largest pieces of ore were put on a bench on the inside of the furnace, near the bottom, and upon these as a base an arch was built, a small hole only being left in the center. Upon this arch smaller ore was thrown, until a small pyramid was formed protruding above the furnace-walls. This was finally covered with fine ore, upon which straw was thrown and ignited. The fire communicated to the sulphur and traveled from the outside toward the inside. After eight or ten hours the liquid sulphur had collected at the bottom, and was tapped into moistened molds or into water. This process furnished only from 40 to 50 per cent. of the sulphur in the ore, and is now nearly everywhere abandoned.

At present pits, or rather stalls, called calcaroni, are almost universally used in Sicily, Spain, and elsewhere; the yield in these being, according to Professor B. Kerl, 67 per cent, of the sulphur in the ore.

Mr. H. Sewell, who has had considerable experience with sulphur-ores in Spain, describes this method in the Engineering and Mining Journal as follows:

The governing principle in this method is the working of large masses of ore at low cost. Each calcarone works up per month from 800 to 1,000 tons of ore, the apparatus being constructed of common stone and plaster, and costing $300 apiece. No fuel is required, as one-seventh of the ore is used as combustible for reducing the rest; so that if the ore contained 23 per cent. of sulphur, 20 per cent. net would be produced.[*]

The dimensions of a calcarone differ much, according to the percentage of the ores; that is, the poorer the ore, the larger must be the furnace. When I commenced to use them in Spain, I found that stalls about 15 feet in diameter were the most successfully managed by workmen not versed in the process; but I found, also, that for economy, and a greater production in the liquation, a larger diameter, say 33 feet, gave the best results, and this is the size of the stall in the accompanying drawings.

The height at X, on the front or tapping door, varies from 6 to 8 and 12 feet, (though seldom the latter,) and that at L, the aperture for loading, is about 4 feet. At X X, also, in the ground and vertical plan, an aperture reaching from the bottom to the top of the stall exists. This is also used for loading; but after that operation

* This yield, as claimed by Mr. Sewell, is very high, and at variance with the statements of other authorities. Mr. Sewell seems to allow nothing for the sulphur retained in the ore after treating it.

has been concluded, the aperture is closed with a cast of plaster of Paris, (or pieces put together,) the thickness being only 2 inches. This thin door is built up new every time, and destroyed for discharging. It is used as a pyrometer, the heat easily piercing it, and indicates to the smelter how far the sulphur has *sweated* down. The ore is placed in large bowlders, just as it comes out of the mines, from the middle to the bottom of the furnace, which has a declivity of about 15 to 20 degrees, such being necessary at the end of the operation, in order that the last remnants of melted sulphur shall run toward the tapping-door at point M. In loading the stall, all the smaller-sized ore is reserved for filling near the top, where it is piled into the shape of a cone, as at F F F; and chimneys are left at points D D D, about 2 feet deep. These

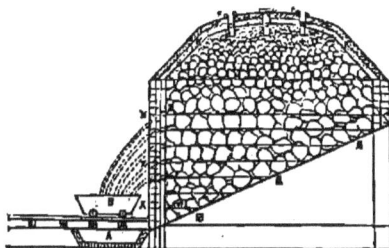

Calcarone—Vertical section.

are filled with brush-wood, and in this way the ore is made to ignite. These hollows are left while piling the ore and building the cone. The object of placing the small ore at the top is simply to prevent any of the earth and sifted stone from falling through the large crevices that would be left, if large bowlders were placed at the top. The earth and sifted stone or gravel play an important part in the manipulation. At K K K K we have, immediately in contact with the small ore, a stratum of about 6 inches of small sifted gravel, about the size of a nut, and on this again, at N N N N, we have a coating of earth; this is to make the interior of the stall as impermeable as possible to the oxidizing action of the air, and this coating is increased or decreased according to the amount of heat required, which in turn depends on the strength of the winds and their direction. The brushwood ignited, the ore commences to burn, and the chimneys are kept open for about twenty hours, at the end of which period the ore has ignited all over the surface of the heap, and to the depth of, say, some 15 or 20 inches. The chimneys

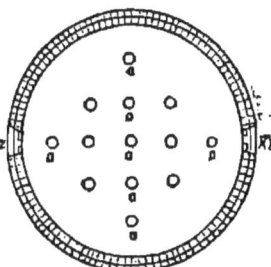

Calcarone—Horizontal section.

are then all closed as follows: bricks are placed over them, as at point P; and should the burning of the ore be too rapid, earth is then thrown over the bricks; but these chimneys are opened toward the middle of the operation, to increase the heat, and closed again, according to circumstances, to decrease it. After forty-eight hours, the melted sulphur begins to trickle down through the interstices of the stone, and congeals, forming, as it were, a conglomerate with the ore; the heat also travels downwards, and so we have remelting and congealing consecutively, till the sulphur arrives at the bottom of the furnace, forming a massive conglomerate of sulphur and ore; for it fills the interstices up to the point where the first tap-hole is drilled, through the thin door of plaster at point Z. The lines across the stall denote the lines or levels of tapping; and this commences naturally at Z, and so downwards, as the congealed sulphur is remelted with the descending heat. Every twenty-four hours a fresh tapping takes place, the former holes are plastered up, and a new hole drilled lower down, and so on till we get down to the lowest point or bottom of the furnace. At the end of the operation, that is, during the last three days, nearly all the chimneys are left open, so that the air shall descend to the lower part of the furnace, and aid the combustion of the ore. The jet of sulphur is received into wooden molds, as at point B. These have been soaked in water, to prevent the sulphur sticking to the wood, and are shaped wheelbarrow fashion, in order that the block of sulphur may easily fall out, without breaking. During the carrying away of a mold that has been filled, and the bringing of an empty one to be filled, the jet runs into a reservoir made for the purpose at A. One of the principal reasons for placing large blocks of ore, from the middle of the furnace downwards, is to leave sufficient interstices for receiving the sulphur, otherwise the first tap-hole would be too high, and near the ignited ore, thus setting fire to the stream of sulphur.

Two of the principal things to be guarded against are overheating the apparatus, and, on the other hand, carrying on the process so slowly, by the complete closing of the chimneys, that the operation would take two months instead of four weeks from the commencement. In the former case, instead of the sulphur congealing between the interstices, it would all be in a melted state from the top tap-hole to the bottom, thus not only consuming an unnecessary amount of sulphur in keeping up the heat, but likewise giving, by overheating, a bad chocolate-brown color to the sulphur. This quality would hardly be salable, even for sulphuric acid. Many stalls or heaps, say ten, after having been loaded, can be attended by two men, one in the daytime, the other at night. As soon as the operation is over, which takes about a month, both apertures are opened, to allow a current of air to pass through the apparatus. Otherwise it would not cool for a month; but by this precaution it can be discharged in a few days.

A modification of this process is the following: Before loading the stall, a number of iron bars are set obliquely from the inclined bottom, against the front wall, in which a single tap-hole is located at the lowest point of the pit. To force the liquid sulphur to run to this point, the bottom of the furnace is inclined, from both sides, toward a central line, and from the back toward the front, thus making a sort of trough, dipping forward. These bars form a complete grate, the space underneath remaining empty when the stall is filled with ore. The cone above the walls of the stall is, in this case, made much higher than in the method described above. The smelted sulphur collects continually on the clean bottom beneath the grate, and is from time to time tapped into wet molds, or into a basin with water.

The crude sulphur obtained by any of the above methods must of course be refined, if intended for other use than that of its manufacture into sulphuric acid. But as it is not likely that refined sulphur can be profitably made in our western districts for years to come, I omit treating the subject in this report.

PART III.

MISCELLANEOUS.

CHAPTER XVIII.

THE MINING LAW.

The following is the text of the proposed new mining law.* It embodies much that I have advocated in former reports, and I think it will be approved by the large body of practical miners in the United States, who, whatever criticisms they may make upon particular provisions, must agree in commending the tone which mining legislation has assumed, and the character of the protection offered to their property.

This law aims to offer means for a fair adjustment of thousands of claims upon all kinds of mining property, and lying between men of every class and nationality. Probably the most eager curiosity in reading it for the first time will be directed, not to the sections which prescribe how mines may hereafter be taken up, but to the inquiry, how the law affects present interests, and the disputed points of the past. The first section will be distasteful to some. None but citizens, and those persons who declare their intention to become such, can have any ownership in the virgin mines of the public domain, except by purchase of United States patents from citizens. If rumor is true, this will be a blow to some of the Mormon miners who have thought to air their supposed independence of our Government by refusing to become citizens. This provision, which is of course most natural and proper, is continued from the former law, where it has already made trouble for those who were anxious to reap without being willing to sow. Probably a large proportion of the Chinese will also be debarred from ownership, as they are not citizens. But this provision does not prevent any one who is willing to become a citizen, from taking up mines. If I read the civil-rights bill correctly, any person in the world may become a citizen of the United States, all sectional or State laws to the contrary notwithstanding.

The section giving absolute title to a certain surface and all veins "topping" within vertical lines drawn from the boundaries of that surface-claim, is necessary to prevent special litigation.

On the subject of tunnels the law follows a course contrary to the views I have held, and I still feel confidence enough in the strength of my position to believe that this section will be annulled by the practice of miners and eventually by Congress. Let us take the case of a hill in which a very promising mine is discovered. Instantly some sharper claps down a tunnel-claim, and by that act he invalidates or threatens to invalidate every discovery made on that hill within 3,000 feet of his location. The prohibition to work the hill continues six months at least, and longer if he begins work. I call it prohibition, because it is such if the law is effective. But I look upon this section with less alarm than I otherwise should feel, because I know that it will be very limited in its action. Very few of the innumerable tunnel-claims are ever prosecuted 10 feet into the hill. Others, and perhaps most of them, even if carried on, would come to nothing. Take the Emma Hill in Little Cottonwood. The tunnels there must run about parallel to the veins, and a tunnel that would intersect more than one would have to be very long.

---

* Since passed and signed by the President. The text here given is that of the final form of the law.

But the Emma Hill offers an example showing how hurtful this section of the law could be if it were effective. The mines in Little Cottonwood depend for their future upon pockets, bulges, or other enlargements of veins, and these enlargements show no sign at the surface. A tunnel might strike one of these and draw out the whole wealth of a mine, while the true owners would be working patiently down, unable to immediately prove that their narrow seam had any connection with the immense mass which was making the tunnel-owners rich. The Emma itself could have been worked for two years through a tunnel before the real owners found their way into the *bonanza*.

I cannot agree with the provision in section 5 to allow owners who have abandoned a mine to resume work without relocation. Something is needed to prevent a man whose whole time is spent in Utah from holding claims in California, Oregon, Montana, and all the other Territories. If the recorders cannot make yearly inspection the owners should be required to swear each year to having performed the work required. I am sorry to see, too, that there is no provision for the use of timber, and but a very incomplete settlement of the important questions of water-power and drainage.

Nevertheless it is certain that the present law is a great advance on anything we have had. Our legislators have no more puzzling task than to adjust the claims of miners. If their measures sometimes bear hardly on those who cherish vague but golden dreams of wealth, it must be remembered that their object is not to deny riches to any, but by early prescription of legal means to prevent conflict. The law, indeed, is made in the interest of the disputing claimant as much as in that of the actual holder, for in exact terms he finds clearly indicated what he can or cannot do; and if he is wise and honest he may save himself expense.

The following is the text of the new mining law:

Be it enacted, &c., That all valuable mineral deposits in lands belonging to the United States, both surveyed and unsurveyed, are hereby declared to be free and open to exploration and purchase, and the lands in which they are found to occupation and purchase, by the citizens of the United States and those who have declared their intention to become such, under regulations prescribed by law; and according to the local customs or rules of miners, in the several mining districts, so far as the same are applicable and not inconsistent with the laws of the United States.

SEC. 2. That mining claims upon veins or lodes of quartz or other rock in place bearing gold, silver, cinnabar, lead, tin, copper, or other valuable deposits, heretofore located, shall be governed as to length along the vein or lode by the customs, regulations, and laws in force at the date of their location. A mining claim located after the passage of this act, whether located by one or more persons, may equal, but shall not exceed, fifteen hundred feet in length along the vein or lode; but no location of a mining claim shall be made until the discovery of the vein or lode within the limits of the claim located. No claim shall extend more than three hundred feet on each side of the middle of the vein at the surface, nor shall any claim be limited by any mining regulation to less than twenty-five feet on each side of the middle of the vein at the surface, except where adverse rights existing at the passage of this act shall render such limitation necessary. The end lines of each claim shall be parallel to each other.

SEC. 3. That the locators of all mining locations heretofore made, or which shall hereafter be made, on any mineral vein, lode, or ledge, situated on the public domain, their heirs and assigns, where no adverse claim exists at the passage of this act, so long as they comply with the laws of the United States, and with State, territorial, and local regulations not in conflict with said laws of the United States governing their possessory title, shall have the exclusive right of possession and enjoyment of all the surface included within the lines of their locations, and of all veins, lodes, and ledges throughout their entire depth, the top or apex of which lies inside of such surface lines extended downward vertically, although such veins, lodes, or ledges may so far depart from a perpendicular in their course downward as to extend outside the vertical side lines of said surface locations: *Provided*, That their right of possession to such outside parts of said veins or ledges shall be confined to such portions thereof as lie between vertical planes drawn downward as aforesaid, through the end lines of their locations, so continued in their own direction that such planes will intersect such exterior parts of said veins or ledges : *And provided further*, That nothing in this section shall authorize the locator or possessor of a vein or lode which extends in its downward course beyond the vertical lines of his claim to enter upon the surface of a claim owned or possessed by another.

SEC. 4. That where a tunnel is run for the development of a vein or lode or for the discovery of mines, the owners of such tunnel shall have the right of possession of all veins or lodes within three thousand feet from the face of such tunnel on the line thereof, not previously known to exist, discovered in such tunnel, to the same extent as if discovered from the surface; and locations on the line of such tunnel of veins or lodes not appearing on the surface, made by other parties after the commencement of the tunnel, and while the same is being prosecuted with reasonable diligence, shall be invalid; but failure to prosecute the work on the tunnel for six months shall be considered as an abandonment of the right to all undiscovered veins on the line of the said tunnel.

SEC. 5. That the miners of each mining district may make rules and regulations not in conflict with the laws of the United States, or with the laws of the State or Territory in which the district is situated, governing the location, manner of recording, amount of work necessary to hold possession of a mining claim, subject to the following requirements : the location must be distinctly marked on the ground, so that its boundaries can be readily traced. All records of mining claims hereafter made shall contain the name or names of the locators, the date of the location, and such a description of the claim or claims located by reference to some natural object or permanent monument as will identify the claim. On each claim located after the passage of this act, and until a patent shall have been issued therefor, not less than one hundred dollars' worth of labor shall be performed or improvements made during each year. On all claims located prior to the passage of this act ten dollars' worth of labor shall be performed or improvements made for each one hundred feet in length along the vein until a patent shall have been issued therefor; but where such claims are held in common such expenditure may be made upon any one claim; and upon a failure to comply with these conditions, the claim or mine upon which such failure occurred shall be open to relocation in the same manner as if no location of the same had ever been made : *Provided*, That the original locators, their heirs, assigns, or legal representatives, have not resumed work upon the claim after such failure and before such location. Upon the failure of any one of several co-owners to contribute his proportion of the expenditures required by this act, the co-owners who have performed the labor or made the improvements may, at the expiration of the year, give such delinquent co-owner personal notice in writing, or notice by publication in the newspaper published nearest the claim, for at least once a week for ninety days, and if at the expiration of ninety days after such notice in writing or by publication such delinquent should fail or refuse to contribute his proportion to comply with this act, his interest in the claim shall become the property of his co-owners who have made the required expenditure.

SEC. 6. That a patent for any land claimed and located for valuable deposits may be obtained in the following manner: any person, association, or corporation authorized to locate a claim under this act, having claimed and located a piece of land for such purposes, who has, or have, complied with the terms of this act, may file in the proper land-office an application for a patent, under oath, showing such compliance, together with a plat and field-notes of the claim or claims in common, made by or under the direction of the United States surveyor-general, showing accurately the boundaries of the claim or claims, which shall be distinctly marked by monuments on the ground, and shall post a copy of such plat, together with a notice of such application for a patent, in a conspicuous place on the land embraced in such plat previous to the filing of the application for a patent, and shall file an affidavit of at least two persons that such notice has been duly posted as aforesaid, and shall file a copy of said notice in such land-office, and shall thereupon be entitled to a patent for said land, in the manner following: the register of the land-office, upon the filing of such application, plat, field-notes, notices, and affidavits, shall publish a notice that such application has been made, for the period of sixty days, in a newspaper to be by him designated as published nearest to said claim; and he shall also post such notice in his office for the same period. The claimant at the time of filing this application, or at any time thereafter, within the sixty days of publication, shall file with the register a certificate of the United States surveyor-general that $500 worth of labor has been expended or improvements made upon the claim by himself or grantors, that the plat is correct, with such further description by such reference to natural objects or permanent monuments as shall identify the claim, and furnish an accurate description, to be incorporated in the patent. At the expiration of the sixty days of publication the claimant shall file his affidavit that the plat and notice have been posted in a conspicuous place on the claim during said period of publication. If no adverse claim shall have been filed with the register and the receiver of the proper land-office at the expiration of the sixty days of publication, it shall be assumed that the applicant is entitled to the patent, upon the payment to the proper officer of five dollars per acre, and that no adverse claims exist; and thereafter no objection from third parties to the issuance of a patent shall be heard, except it be shown that the applicant failed to comply with this act.

SEC. 7. That where an adverse claim shall be filed during the period of publication, it shall be upon oath of the person or persons making the same, and shall show the nature, boundaries, and extent of such adverse claim, and all proceedings, except the publication of notice and making and filing of the affidavit thereof, shall be stayed until the controversy shall have been settled or decided by a court of competent jurisdiction, or the adverse claim waived. It shall be the duty of the adverse claimant, within thirty days after filing his claim, to commence proceedings in a court of competent jurisdiction, to determine the question of the right of possession, and prosecute the same with reasonable diligence to final judgment; and a failure so to do shall be a waiver of his adverse claim. After such judgment shall have been rendered, the party entitled to the possession of the claim, or any portion thereof, may, without giving further notice, file a certified copy of the judgment-roll with the register of the land-office, together with the certificate of the surveyor-general that the requisite amount of labor has been expended, or improvements made thereon, and the description required in other cases, and shall pay to the receiver five dollars per acre for his claim, together with the proper fees, whereupon the whole proceedings and the judgment-roll shall be certified by the register to the Commissioner of the General Land-Office, and a patent shall issue thereon for the claim, or such portion thereof as the applicant shall appear, from the decision of the court, to rightfully possess. If it shall appear from the decision of the court that several parties are entitled to separate and different portions of the claim, each party may pay for his portion of the claim, with the proper fees, and file the certificate and description by the surveyor-general, whereupon the register shall certify the proceedings and judgment-roll to the Commissioner of the General Land-Office, as in the preceding case, and patents shall issue to the several parties according to their respective rights. Proof of citizenship under this act, or the acts of July 26, 1866, and July 9, 1870, in the case of an individual, may consist of his own affidavit thereof, and in case of an association of persons unincorporated, of the affidavit of their authorized agent, made on his own knowledge or upon information and belief, and in case of a corporation organized under the laws of the United States, or of any State or Territory of the United States, by the filing of a certified copy of their charter or certificate of incorporation; and nothing herein contained shall be construed to prevent the alienation of the title conveyed by a patent for a mining claim to any person whatever.

SEC. 8. That the description of vein or lode claims, upon surveyed lands, shall designate the location of the claim with reference to the lines of the public surveys, but need not conform therewith; but where a patent shall be issued as aforesaid for claims upon unsurveyed lands, the surveyor-general, in extending the surveys, shall adjust the same to the boundaries of such patented claim, according to the plat or description thereof, but so as in no case to interfere with or change the location of any such patented claim.

SEC. 9. That sections one, two, three, four, and six of an act entitled "An act granting the right of way to ditch and canal owners over the public lands, and for other purposes," approved July 26, 1866, are hereby repealed, but such repeal shall not affect existing rights. Applications for patents for mining-claims now pending may be prosecuted to a final decision in the General Land-Office; but in such cases, where adverse rights are not affected thereby, patents may issue in pursuance of the provisions of this act; and all patents for mining-claims heretofore issued under the act of July 26, 1866, shall convey all the rights and privileges conferred by this act where no adverse rights exist at the time of the passage of this act.

SEC. 10. That the act entitled "An act to amend an act granting the right of way to ditch and canal owners over the public lands, and for other purposes," approved July 9, 1870, shall be and remain in full force, except as to the proceedings to obtain a patent, which shall be similar to the proceedings prescribed by sections six and seven of this act for obtaining patents to vein or lode claims; but where said placer-claims shall be upon surveyed lands, and conform to legal subdivisions, no further survey or plat shall be required. And all placer-mining claims hereafter located shall conform as near as practicable with the United States system of public-land surveys and the rectangular subdivisions of such surveys, and no such location shall include more than twenty acres for each individual claimant; but where placer-claims cannot be conformed to legal subdivisions, survey and plat shall be made as on unsurveyed lands: *Provided*, That proceedings now pending may be prosecuted to their final determination under existing laws; but the provisions of this act, when not in conflict with existing laws, shall apply in such cases: *Provided also*, That where by the segregation of mineral lands in any legal subdivision a quantity of agricultural land less than forty acres remains, said fractional portion of agricultural land may be entered by any party qualified by law for homestead or pre-emption purposes.

SEC. 11. That where the same person, association, or corporation is in possession of a placer-claim, and also a vein or lode included within the boundaries thereof, application shall be made for a patent for the placer-claim, with the statement that it includes such vein or lode, and in such case (subject to the provisions of this act and the act entitled "An act to amend an act granting the right of way to ditch and canal owners over the public lands, and for other purposes," approved July 9, 1870) a patent shall issue for the placer-claim, including such vein or lode, upon the payment of five dollars per acre for such vein or lode claim, and twenty-five feet of surface on each side thereof. The remainder of the placer-claim, or any placer-claim not embracing any vein or lode claim, shall be paid for at the rate of $2.50 per acre, together with all costs of proceedings; and where a vein or lode, such as is described in the second section of this act, is known to exist within the boundaries of a placer-claim, an application for a patent for such placer-claim which does not include an application for the vein or lode claim shall be construed as a conclusive declaration that the claimant of the placer-claim has no right of possession of the vein or lode claim; but where the existence of a vein or lode in a placer-claim is not known, a patent for the placer-claim shall convey all valuable mineral and other deposits within the boundaries thereof.

SEC. 12. That the surveyor-general of the United States may appoint in each land-district containing mineral lands as many competent surveyors as shall apply for appointment to survey mining-claims. The expenses of the survey of vein or lode claims, and the survey and subdivision of placer-claims into smaller quantities than one hundred and sixty acres, together with the cost of publication of notices, shall be paid by the applicants, and they shall be at liberty to obtain the same at the most reasonable rates, and they shall also be at liberty to employ any United States deputy surveyor to make the survey. The Commissioner of the General Land-Office shall also have power to establish the maximum charges for surveys and publication of notices under this act; and, in case of excessive charges for publication, he may designate any newspaper published in a land-district where mines are situated for the publication of mining notices in such district, and fix the rates to be charged by such paper; and, to the end that the Commissioner may be fully informed on the subject, each applicant shall file with the register a sworn statement of all charges and fees paid by said applicant for publication and survey, together with all fees and money paid the register and the receiver of the land-office, which statement shall be transmitted, with the other papers in the case, to the Commissioner of the General Land-Office. The fees of the register and the receiver shall be five dollars each for filing and acting upon each application for patent or adverse claim filed, and they shall be allowed the amount fixed by law for reducing testimony to writing, when done in the land-office, such fees and allowances to be paid by the respective parties; and no other fees shall be charged by them in such cases. Nothing in this act shall be construed to enlarge or affect the rights of either party in regard to any property in controversy at the time of the passage of this act, or the act entitled "An act granting the right of way to ditch and canal owners over the public lands, and for other purposes," approved July 26, 1866, nor shall this act affect any right acquired under said act; and nothing in this act shall be construed to repeal, impair, or in any way affect the provisions of the act entitled "An act granting to A. Sutro the right of way and other privileges to aid in the construction of a draining and exploring tunnel to the Comstock lode, in the State of Nevada," approved July 25, 1866.

SEC. 13. That all affidavits required to be made under this act, or the act of which it is amendatory, may be verified before any officer authorized to administer oaths within the land-district where the claims may be situated, and all testimony and proofs may be taken before any such officer, and when duly certified by the officer taking the same, shall have the same force and effect as if taken before the register and the receiver of the land-office. In cases of contest as to the mineral or agricultural character of land, the testimony and proofs may be taken as herein provided on personal notice of at least ten days to the opposing party; or if said party cannot be found, then by publication of at least once a week for thirty days in a newspaper, to be designated by the register of the land-office as published nearest to the location of such land; and the register shall require proof that such notice has been given.

SEC. 14. That where two or more veins intersect or cross each other, priority of title shall govern, and such prior location shall be entitled to all ore or mineral contained within the space of intersection: *Provided, however*, That the subsequent location shall have the right of way through said space of intersection for the purposes of the convenient working of the said mine: *And provided also*, That where two or more veins unite, the oldest or prior location shall take the vein below the point of union, including all the space of intersection.

SEC. 15. That where non-mineral land not contiguous to the vein or lode is used or occupied by the proprietor of such vein or lode for mining or milling purposes, such non-adjacent surface-ground may be embraced and included in an application for a patent for such vein or lode, and the same may be patented therewith, subject to the same preliminary requirements as to survey and notice as are applicable under this act to veins or lodes: *Provided*, That no location hereafter made of such non-adjacent land shall exceed five acres, and payment for the same must be made at the same rate as fixed by this act for the superficies of the lode. The owner of a quartz-mill or reduction-works, not owning a mine in connection therewith, may also receive a patent for his mill-site, as provided in this section.

SEC. 16. That all acts and parts of acts inconsistent herewith are hereby repealed: *Provided*, That nothing contained in this act shall be construed to impair, in any way, rights or interests in mining property acquired under existing laws.

CHAPTER XIX.

AMERICAN SCHOOLS OF MINING AND METALLURGY.

Since the publication of my first report, in 1869, in which the subject
of professional education was treated at some length, a great advance
has been made in the facilities afforded by American schools and colleges, though nothing has been done by the Government, I regret to say,
toward the establishing of a national school of mines. There are now
some thirty institutions in this country, in the plans of which room is
made for instruction in mining and metallurgy. Of course this department is not organized with equal thoroughness or furnished with equal
liberality in all these cases; in too many of them trustees have added to the
old curriculum merely a nominal course, because it was the fashion, and
in order to attract students. But it is gratifying to know that a considerable number of these mining and technological schools mean business, and not show. Without intending to slight any which I omit, I
have collected full information concerning some of the principal institutions east of the Rocky Mountains. There is an inchoate school in Colorado, and there is a promising department for this subject in the
University of California; but these have had no chance, as yet, to show
what they can do.

The schools to which I shall refer are, the Massachusetts Institute of
Technology, (Boston;) the School of Mines of Columbia College, (New
York;) the Sheffield Scientific School, (New Haven;) the Stevens Insti-
tute of Technology, (Hoboken;) the Pardee Scientific Department of
Lafayette College, (Easton;) the School of Mining and Metallurgy of
Lehigh University, (Bethlehem;) the School of Mining and Practical
Geology, of Harvard University, (Cambridge;) the Rensselaer Poly-
technic Institute, (Troy;) the Scientific Department of the University
of Pennsylvania, (Philadelphia;) the Missouri School of Mines and
Metallurgy, (Rolla;) and the Polytechnic Department of Washington
University, (Saint Louis.)

THE MASSACHUSETTS INSTITUTE OF TECHNOLOGY.

Officers of instruction.—John D. Runkle, Ph. D., LL. D., President; John D. Runkle,
Ph. D., LL. D., Walker Professor of Mathematics and Mechanics; William Watson, Ph.
D., Professor of Mechanical Engineering; John B. Henck, A. M., Hayward Professor of
Civil and Topographical Engineering; William R. Ware, S. B., Professor of Architec-
ture; William P. Atkinson, A. M., Professor of English and History; George A. Osborne,
S. B., Professor of Mathematics, Astronomy, and Navigation; Alfred P. Rockwell, A. M.,
Professor of Mining Engineering; Edward C. Pickering, S. B. Thayer Professor of
Physics; Samuel Kneeland, A. M., M. D., Professor of Zoölogy and Physiology; John
M. Ordway, A. M.,* Professor of Metallurgy and Industrial Chemistry; James M. Crafts,
S. B., Professor of Analytical and Organic Chemistry; Robert H. Richards, graduate
of the Institute, Professor of Mineralogy and Assaying, in charge of the Mining and
Metallurgical Laboratory; Thomas Sterry Hunt, LL. D., Professor of Geology; George
H. Howison, A. M., Professor of Logic and the Philosophy of Science; S. Edward War-
ren, C. E., Professor of Descriptive Geometry, Stereotomy, and drawing; ————,
Professor of Modern Languages; Henry L. Whiting, United States Coast Survey,
Professor of Topography; Henry Mitchell, A.M., United States Coast Survey, Professor
of Physical Hydrography; Alpheus Hyatt, S. B., Custodian of the Boston Society of
Natural History, Professor of Palæontology; Lewis B. Monroe, Professor of Vocal Cul-

* The instruction in botany is at present given by Professor Ordway.

ture and Elocution; Willliam H. Niles, Ph. B., A. M., Professor of Physical Geology and
Geography; William Ripley Nichols, graduate of the Institute, Assistant Professor of
General Chemistry; Charles R. Cross, graduate of the Institute, Assistant Professor of
Physics; Ernest Schubert, Instructor in Free-Hand and Machine Drawing; Eugene
Letang, Assistant in Architecture; John A. Whipple, Instructor in Photography; Wil-
liam E. Hoyt, graduate of the Institute, Instructor in Civil Engineering and Drawing;
Jules Lévy, Instructor in French, Spanish, and Italian; E. C. F. Kraus, Instructor in Ger-
man; Edward K. Clark, graduate of the Institute, Instructor in Mechanical Drawing;
Gaetano Lanza, S. B., C. E., Instructor in Mathematics; Foster E. L. Beal, graduate of
the Institute, Instructor in Mathematics; G. Russell Lincoln, graduate of the Institute,
Instructor in General Chemistry and Qualitative Analysis; Charles F. Stone, graduate
of the Institute, Instructor in Quantitative Analysis; Hobart Moore, Instructor in
Military Tactics.

Summary of students.—Resident graduates, 3; fourth year, 17; third year, 33; second
year, 39; first year, 91; students not candidates for a degree, 81; total, 264, for the
year 1871-72.

Courses of study.—The regular course in the department of geology and mine engineering extends over four years, and the range of studies pursued is indicated by the
following scheme of instruction:

FIRST AND SECOND YEARS.

Mathematics.—Algebra; solid geometry; mensuration; plane trigonometry and spher-
ical trigonometry; analytic geometry; elements of the calculus.
Surveying.—Field-work; plotting surveys; computing areas; plans.
Physical and industrial geography.
Physics.—Mechanics of solids, liquids, and gases; sound; light; heat; magnetism;
electricity.
Chemistry.—Qualitative analysis; chemistry, organic and inorganic.
French; German; English; descriptive geometry; mechanical and free-hand drawing.

THIRD YEAR.

Civil engineering.—Survey and construction of roads and railways; measurement of
earth-work and masonry; field practice.
Mathematics.—Differential and integral calculus; analytic mechanics.
Applied mechanics.—Stress; stability; strength and stiffness.
Geology.—General descriptive, and theoretical geology.
Zoölogy and palæontology.
Mining.—Ore-deposits; prospecting; boring; sinking shafts, &c.; methods of
mining.
Mineralogy.—Descriptive and determinative; crystallography; use of the blow-pipe.
Chemistry.—Lectures and laboratory practice in quantitative analysis.
Assaying.—Wet and dry ways.
Metallurgy.—Metallurgical processes; constructions and implements.
Physics.—Laboratory practice.
English and constitutional history; French or Spanish; drawing.

FOURTH YEAR.

Mining.—Ventilation; winding machinery; underground transport; pumps; dressing
and concentration of ores; practice in mining-laboratory in ore-dressing.
Economic geology.—Detailed description of American ore deposits and mines.
Strength of materials and hydraulics.
Machinery and motors.—Hand-machinery; water-wheels; boilers; steam-engines.
Chemistry.—Lectures and laboratory practice; synthetic experiments; quantitative
analysis.
Geology.—American geology; lithological, stratographical, palæontological.
Chemical geology.—Origin of rocks, vein-stones, ore-deposits, coal, petroleum, &c.
Metallurgy, practical lithology, and building-materials; physics.
Drawing.—Geological maps and sections; plans of mines; mining-machinery and
implements.
English literature; political economy; French or Italian; German.
The four years' course is so arranged as to secure to the student a liberal mental development and general culture, as well as the strictly technical education, which is his
chief object. The studies of the first and second years are somewhat general in character, but are regarded as a necessary foundation for the more special studies of the
two succeeding years. The special professional studies peculiar to this department commence with the third year. Instruction is given by lectures and recitations, and by
practical exercises in the field, the laboratories, and the drawing-rooms. In most of the

subjects problems are given the students to be worked outside the lecture-room. A high value is set upon the educational effect of these practical exercises.

The space devoted to laboratories and the prominence given to laboratory work, in physics, chemistry, assaying, blow-pipe analysis, metallurgy, and ore-dressing, is a marked feature in the scheme of instruction of the institute. It is believed that this school offers unusual facilities in this regard. The chemical laboratories cover 4,000 square feet; the mining, metallurgical, and assay laboratories, 2,000 square feet; the blow-pipe laboratory, 550 square feet; the physical laboratories, 3,500 square feet; and the drawing-rooms 8,556 square feet.

A course of thirty lectures on physical geology and geography is given to the students of the second year by Professor Niles. The study of the surface of the earth, of its external features, their origin and modifications, is essentially the subject of this course. A proper knowledge of the surface includes necessarily a corresponding acquaintance with the arrangement of rock-masses, in so far as they have determined the character of the surface features, and especially the geological agencies which are constantly producing the changes of the surface. The aim of the instruction, therefore, is to present clearly the most important relations between surface features and underlying geological formations and to show the action of the great dynamical forces, or, in other words, to teach physical geography and physical geology in their natural relations. The knowledge of these relations becomes of great practical value in determining the extent or even probable occurrence of certain ore-bearing rocks and of coal-beds in certain districts, since, where the rocks are completely covered by soil, the topographical features may be the only guide in "prospecting."

Descriptive and Theoretical Geology—30 lectures by Professor Hunt. American Geology—30 lectures by Professor Hunt. Practical Lithology and Building-Stones—15 lectures by Professor Hunt. Chemical Geology—15 lectures by Professor Hunt. Mining—70 lectures by Professor Rockwell. Economic Geology—20 lectures by Professor Rockwell. Palæontology—50 lectures by Professor Hyatt. Metallurgy—40 lectures by Professor Ordway. Industrial Chemistry—40 lectures by Professor Ordway. Quantitative Analysis—40 lectures by Professor Crafts. Chemical Laboratory Practice—10 to 15 hours a week, by Professor Crafts. Assaying, dry way—10 exercises (2¼ hours each) by Professor Richards. Blow-pipe and Determinate Mineralogy—45 exercises (1 to 2 hours each) by Professor Richards. Descriptive Mineralogy—15 lectures by Professor Richards. Mining and Metallurgical Laboratory Practice—10 hours a week. Physical Laboratory Practice—3 hours a week, 1 year, Professor Pickering. Calculus—50 lessons by President Runkle. Mechanics—50 lessons by President Runkle. Civil Engineering—40 lessons by Professor Henck. Strength of Materials and Hydraulics—40 lessons by Professor Watson. Machinery and Motors—25 lessons by Professor Watson. Natural History—25 lectures by Professor Kneeland. French—2 hours a week. German—2 hours a week. English—2 hours a week. Drawing—3 to 6 hours a week.

Geology, Lithology, &c.—The instruction in geology and certain related subjects is given by Professor Hunt, in four courses, delivered yearly to students of the third and fourth years. The first is a yearly course of thirty lectures on descriptive and theoretical geology. This embraces the classification of the related sciences; scope of geological studies; nature of rocks, or lithology; stratigraphy; succession of formations; zoological history; geological dynamics; chemical and physical forces; aqueous and igneous agencies; currents; sedimentation; elevation and subsidence; geographical distribution of formations; nature and origin of mountains; volcanic action. The second is a yearly course of thirty lectures on American geology, comprising introduction; geological history; geology of North America, considered lithologically, stratigraphically, and palæontologically; comparative geognosy. The third is a yearly course of fifteen lectures on practical lithology, comprising mineralogical composition of rocks; building-stones, their cohesion, porosity; granites, marbles, limestones, sandstones, slates, &c.; limes, cements, and mortars; ornamental stones and gems. The fourth is a yearly course of fifteen lectures on chemical geology, or the chemical history of the globe; comprising the origin of rocks, both stratified and unstratified; the history of veinstones and ore-deposits; the formation of coal and petroleum; the chemistry of salt-deposits and of mineral-waters; the seat and origin of volcanic and earthquake phenomena.

Mining and Economic Geology.—The instruction in mining and in economic geology is given by Professor Rockwell, in two yearly courses, delivered to students of the third and fourth years. The first is a yearly course of seventy lectures on mining. The student is made acquainted with the general character of the various deposits of the useful minerals, and with the theory and practice of mining operations, such as the methods of search or "prospecting;" boring for oil, coal, or water; the sinking of shafts, with the timbering, walling or tubing of the same; the driving of levels; the different methods of working lodes, coal-beds, &c.; the underground transportation of the mineral; hoisting, pumping, ventilation, and lighting, together with the machinery and other appliances connected with these and other operations; in short, the great variety of operations comprised under the general term "exploitation." Ore-dressing, or the mechanical separation of ores from their gangues, is discussed somewhat at length, and the machines described by means of which this concentration is most economically effected. The practical course of ore-dressing and smelting in the mining and metallurgical laboratory affords the student opportunities for acquiring a familiar knowledge of the treatment of ores, such as can be got under ordinary circumstances only at the best mines. The second is a yearly course of twenty lectures on economic geology, mainly devoted to a detailed description of the coal and ore deposits of North America, especially such as are most extensively worked.

The student who is a candidate for the degree of the institute is expected to spend a portion of his vacations in some one of the principal mining districts in the study of the local peculiarities of the ore-deposit, and the details of actual working, and to submit a full report upon the same, with drawings. Those who intend to become metallurgists may take smelting-works instead. Through the kindness of several owners, certain mines in different regions have been made accessible to students for the purpose of systematic study.

Metallurgy and Industrial Chemistry.—The instruction in metallurgy and in industrial chemistry is given by Professor Ordway, in two two-years' courses, delivered to students of the third and fourth years. The first is a two years' course of forty lectures on metallurgy. The subjects discussed are fuels, fluxes, slags, furnace-construction, and the roasting, smelting, and refining of the various metals. The second is a two years' course of forty lectures on industrial chemistry. The manufacture of acids, alkalies, salts, pottery, glass, and organic products, and the arts of dyeing and printing, are the principal subjects treated. In connection with these lectures excursions are made to manufactories and metallurgical works, and practical exercises are given in the laboratories. The students are required to make drawings and designs of apparatus used or to be used in large operations.

Mining and Metallurgical Laboratory.—The purpose of this laboratory is to furnish the means of studying experimentally the various processes of ore dressing and smelting. Ores of all kinds are here subjected to precisely the same treatment, and by the same machinery and other appliances that are in use at the best mines and metallurgical works of this and other countries. The laboratory has already in successful operation the most approved ore-dressing and mill machinery for gold and silver ores now in use in California and Nevada, consisting of a *five-stamp battery*, an *amalgamating-pan*, a *separator*, and a *concentrator*, complete in every respect and capable of treating half a ton of ore a day. These were obtained the past summer in San Francisco. There will be added during the present year an ore-crusher, a hydraulic jigger, a Rittinger shaking-table, and all other appliances necessary for the treatment of every kind of ore. The machinery is driven by a steam-engine of upward of 15 horse-power. For metallurgical treatment the laboratory contains at present a reverberatory roasting-furnace, crucible and assay furnaces, and a blacksmith's forge; and there are now being erected reverberatory and blast smelting-furnaces capable of working 400 pounds of ore per day, and a cupelling-furnace sufficient for working 50 pounds of lead at once. To these will be added retort and other smaller furnaces for various uses. All of these will be ready for use by October of the present year.

The experimental work of this laboratory is carried on by the students under the immediate supervision of Professor Richards. A sufficiently large quantity of ore is assigned to each student, who first samples it, and determines its character and value by analysis and assays, and makes such other preliminary examinations as serve to indicate in a general way the proper method of treatment. He then treats the given quantity, makes a careful examination of the products at each step of the process, ascertains the amount of power, water, chemicals, fuel, and labor expended. In this way the same ore is subjected to several methods of treatment, and by a comparison of the results obtained, the student learns the relative effectiveness and economy of different methods as applied to the same ore. It is believed that the experiments conducted in this way and upon such a scale, will prove of direct practical value not only to the student, but to the mining interest at large, by showing how existing methods of treatment may be advantageously modified to meet the requirements of new complex ores. The institute has now on hand eleven tons of gold and silver ores, representing over seventy different mines in Colorado and Utah, which were collected by the institute party of professors and students during their recent trip to these Territories. These ores will be worked and reports of the results sent to those who so generously contributed them; and it is hoped that by such co-operation the laboratory will continue to receive the necessary amount and variety of ores.

Palæontology.—The instruction in palæontology is given in a yearly course of sixty lectures, by Professor Hyatt, half of which are delivered to the third year and half to the fourth year students. Palæontology, or the history of ancient animal life, and stratigraphical palæontology, or the study of the distinctive and characteristic fossils of the different formations, are taught as a necessary foundation for the further study of geology. The aim of the course is to give the student a practical acquaintance with the structure of the characteristic families and orders of living and extinct animals, and by a judicious selection of examples to familiarize him to some extent with the genera which characterize various formations. The handling and drawing of specimens by the students is an essential feature of the method of instruction. The lecture of the instructor is devoted largely to explanatory demonstrations of the specimens, which the students are at the same time drawing. The success attending this mode of teaching palæontology has shown its value.

Mineralogy and blow-pipe practice.—The instruction in mineralogy and the use of the blow-pipe is given by Professor Richards, in two courses to students of the third year. The first is a course of forty-five exercises, (one to two hours each,) in which the student is taught determinative mineralogy by the study of crystalline forms, and the physical properties of minerals. He is instructed in the use of the blow-pipe in the qualitative determination of minerals, and in the quantitative assay of silver and copper ores. The second is a course of fifteen lectures on descriptive mineralogy, accompanied by a critical examination and handling of specimens on the part of the student.

Assaying.—The instruction in assaying is given by Professor Richards in a course of ten exercises, (two to three hours each,) in which the student learns to perform the ordinary dry assays of gold, silver, lead, and other ores. Instruction in wet assaying is given by Professor Crafts.

Chemical course for mining engineers.—The chemical instruction of the mining students extends through the four years. The course in general chemistry occupies the first year; and during this time the students work two hours each week in the chemical laboratory. Each student performs for himself a great variety of experiments designed to illustrate the properties of the various chemical elements, and of their more important compounds; he also prepares a number of such simple and compound substances as are of use in the arts or serve to illustrate the laws of chemical change. The knowledge thus acquired by the practical work of the laboratory is supplemented and enforced by lectures, recitations, and frequent examinations. During the second year more particular attention is given to the theory of chemistry and to qualitative analysis. The latter branch of the subject is taught by laboratory exercises, each student working four hours a week. As during the first year's course, these laboratory exercises are accompanied by recitations and examinations. During the third year the mining students take a systematic course of quantitative analyses, occupying six hours per week in the laboratory, and attend a weekly lecture, or exercise, in which methods of analysis are discussed, and the results of investigations too recent to be found in text-books are presented to them. Mining students may take in addition, as a voluntary exercise, a course of special analytical methods, reciting from German text-books. In the fourth year the students are engaged in laboratory work during the hours between 9 a. m. and 5 p. m., which are not devoted to recitations and problems in drawing, metallurgy, &c., which more nearly concern their professional studies. They all get about three hours daily for chemical work, and many are able to spend more time in the laboratory without neglect of their other studies. They accomplish a tolerably thorough analytical course, comprising the analyses of salts, the more common minerals, and particularly of ores and metallurgical products, and extending as far as the more difficult analyses, such as the determination of all the constituents of cast iron or steel, so that a student in his professional work as a mining engineer may be independent of the assistance of a chemist, and competent to deal with all ordinary investigations. Some students, who show a special aptitude for chemistry, are encouraged to take up special investigations connected with metallurgy, and all accomplish work which may be considered a sufficient preparation for their professional career.

Physics.—The instruction in physics, extending through the first three years of the course, is given by Professor Pickering. During the first two years the whole subject is thoroughly discussed mathematically and experimentally in lectures illustrated from the extensive collection of physical apparatus of the institute.

In the third year the students enter the physical laboratory and learn to use the different instruments and to perform a variety of experiments. Special attention is paid to the testing of physical laws, by comparing the observed and computed results.

They further carry on systematic investigations of particular subjects during the fourth year, and pursue such courses of experimentation as have a direct bearing on their professional studies.

Collections.—The geological collection of the late Professor Henry D. Rogers, of the University of Glasgow, presented to the Institute by Mrs. Rogers, is made up chiefly of fossils and rock-specimens from American localities, and in certain branches is peculiarly valuable for instruction. The collection of ores and veinstones is already large and varied, and is constantly receiving additions from the various mining regions.

A typical set of models of mining-machinery, chiefly from Freiburg, Saxony, is used in the course of instruction. They are designed mainly to illustrate the principles of the various processes of mining and ore-dressing, but combine also the latest improvements in machines. They show in detail the methods of working underground by underhand and overhand stoping, the timbering and walling of shafts and levels, the arrangements of pumps, man-engines, ladder-ways, hoisting-ways, the sinking of shafts, &c. The machines for ventilation, as well as those for ore-dressing, are working models. The latter illustrate all the stages of the concentration of ores.

The collections and library of the Boston Society of Natural History are, by an agreement between the society and the institute, freely open to the students. These collections rank among the first in the country for extent and value, and in many departments are unsurpassed. The library is rich in works on geology and natural science, and embraces the leading American and European journals and periodicals on those subjects. The instruction in certain subjects is given by the professors of the institute in the lecture-room of the Natural History Society, whose building is upon the same square. The private collections of some of the professors, especially that of Professor Hunt, are available for purposes of instruction. The professors and students of the institute are allowed the full use of the extensive and valuable libraries of the Boston Public Library and the library of the Boston Society of Natural History.

Other courses, degrees, fees, &c.—This school is divided into seven courses, namely, mechanical engineering, civil and topographical engineering, geology and mining engineering, building and architecture, chemistry, science and literature, and natural history. The foregoing description refers to the third course only; but the course of study for the first two years is the same in all. A degree is given in each course, the title being "Graduate of the Massachusetts Institute of Technology in the Department of ———." To be entitled to either of these degrees, the student must pass a satisfactory examination in all the studies and exercises prescribed for his department in the courses of the third and fourth years; and in all the studies of the previous years in which he has not already passed a satisfactory examination. He must, moreover, prepare a dissertation on some subject included in the course of study, or an account of some research made by himself, or an original report upon some machine, work of engineering, industrial works, mines, or mineral survey, or an original architectural design accompanied by an explanatory memoir. This thesis or design must be approved by the faculty. He will be required, also, to have sufficient familiarity with French and German to be able to read without difficulty works in these languages relating to science and the arts. The examinations for degrees are held in the month of May, and are partly oral and partly in writing. Certificates of attainment in special subjects will be given to such students as on examination are found to have attained the required proficiency in them.

The catalogue shows eleven students in the third and fourth years taking the full mining course, and one pursuing a special course in that department. Eight out of the seventeen in the fourth year are mining students.

The school-year begins on the first Monday in October, and ends on the Saturday preceding the first Monday in June. On legal holidays the exercises of the school are suspended. As the exercises of the school begin at 9 o'clock in the morning, and end at half-past 4 or 5 o'clock in the afternoon, students may conveniently live in any of the neighboring cities or towns on the lines of the various railroads, if they prefer to do so. The cost of board and rooms in Boston, and the neighboring cities and towns, need not exceed, on the average, from six to eight dollars a week; and the cost of books, drawing-instruments, and paper, from twenty-five to thirty dollars a year. The regular fee for each year is $150, payable, by students who have given bonds, $100 at the beginning, and $50 at the middle (first Monday in February) of the school-year. For one-half or any less fraction of the school-year, the fee is $100. Students who pursue a partial course pay, in general, the full fee. The fees for special students vary, according to the character of the study chosen, and cannot be specified, except for such special courses as from time to time may be advertised.

THE SCHOOL OF MINES OF COLUMBIA COLLEGE.

Officers of instruction.—Frederick A. P. Barnard, S. T. D., LL. D., President; Thomas Egleston, jr., A. M., E. M., Professor of Mineralogy and Metallurgy; Francis L. Vinton, E. M., Professor of Mining Engineering; Charles F. Chandler, Ph. D., Professor of Analytical and Applied Chemistry, and Dean of the Faculty; John Torrey, M. D., LL. D., Lecturer on Botany; Charles A. Joy, Ph. D., Professor of General Chemistry; William G. Peck, LL. D., Professor of Mechanics; John H. Van Amringe, A. M., Professor of Mathematics; Ogden N. Rood, A. M., Professor of Physics; John S. Newberry, M. D., LL. D., Professor of Geology and Palæontology; Frederick Steugel, Instructor in German; Jules E. Loiseau, Instructor in French; Alexis A. Julien, A. M., Assistant in Analytical Chemistry; Paul Schweitzer, Ph. D., Assistant in Analytical Chemistry; Elwyn Waller, A. M., E. M., Assistant in Analytical Chemistry; Thomas M. Blossom, A. M., E. M., Assistant in Assaying; Pierre De P. Ricketts, E. M., Assistant in Mineralogy; William Pistor, E. M., Assistant in Drawing; Henry Newton, A. B., E. M., Assistant in Geology ; Henry B. Cornwall, A. M., E. M., Assistant in Metallurgy ; Edward J. Hallock, A. B., Assistant in General Chemistry; Gracie S. Roberts, E. M., Honorary Assistant in Civil Engineering ; Edward C. II. Day, Librarian and Registrar.

Summary of students.—Third year, 8 ; second year, 5 ; first year, 19 ; preparatory, 25 ; special students, 58 ; total, 115.

General description.—The School of Mines was established in 1864, for the purpose of furnishing to students the means of acquiring a thorough knowledge of those branches of science which form the basis of the industrial pursuits that are to play the most important part in the development of the resources of our country. The system of instruction followed in the school includes five parallel courses of study, viz :

I. Civil engineering.
II. Mining engineering.
III. Metallurgy.
IV. Geology and natural history.
V. Analytical and applied chemistry.

The school is provided with fine mineralogical and geological collections; physical, mechanical, engineering, and mathematical instruments and models; chemical and physical apparatus; chemical and metallurgical laboratories ; and a scientific library and reading-room.

These are all sustained by liberal annual appropriations, which enable the professors to rapidly increase these important means of illustration and practical instruction.

Communication has been established with kindred institutions in Europe, and very valuable additions to the cabinets and library have already been received from France, Belgium, Germany, and Russia.

The success of the School of Mines has surpassed the most sanguine expectations of its projectors. The average number of pupils for the past three years has been about one hundred, of whom a large number are college-graduates.

Although the school has been in existence but seven years, it has already sent forth fifty-five graduates, most of whom have already been appointed to responsible positions as mining engineers, metallurgists, geologists, chemists, or professors.

The officers of the school, most of whom were educated in Europe, are satisfied that the school now offers to American students every facility necessary to enable them to prepare themselves for any of the professions which involve the practical application of the branches of science therein taught ; and that it is no longer necessary for young men to visit Europe to study applied science ; in fact, that they can be better fitted here for this field of labor, which is characterized by peculiar conditions of labor, transportation, &c.

Plan of instruction.—The plan of instruction pursued in the school includes lectures and recitations in the several departments of study ; practice in the chemical and metallurgical laboratories ; projects, estimates and drawings for the establishment of mines, and for the construction of metallurgical and chemical works; reports on mines, industrial establishments, and field geology.

The course of instruction occupies three years. Those who complete it receive the degree of civil engineer, engineer of mines, or bachelor of philosophy.

For candidates not qualified to enter the first year, a preparatory year has been added.

The year is divided into two sessions. The first commences on the first Monday in October ; the second, on the first Thursday in February. The lectures close on the first Friday in June. The annual examinations are then held on all the studies of the year.

The method of instruction is such that every pupil may acquire a thorough theoretical knowledge of each branch, of which he is required to give evidence at the close of the session by written and oral examinations. At the commencement of the following year he is required to show, from reports of works visited, that he not only understands the theoretical principles of the subjects treated, but also their practical application.

SYNOPSIS OF STUDIES.

First year.—First session : analytical geometry,* descriptive geometry ; inorganic chemistry,* qualitative analysis, crystallography, qualitative blow-pipe analysis, botany, French, German, drawing. Second session : calculus,† descriptive geometry, organic chemistry,† qualitative analysis, blow-pipe analysis, zoology, French, German, stoichiometry, drawing, memoir and journal of travel during the summer vacation.

Second year.—I. For civil-engineering students : mechanics, civil engineering, spherical projections, geodesy, quantitative analysis, metallurgy, geology, mineralogy, physics, drawing. II. For mining-engineering students : mechanics, mining engineering, quantitative analysis, metallurgy, geology, mineralogy, mathematical physics, drawing.

* Optional for students of the geological and chemical courses.
† Optional for students of the mining engineering course.

III. For students of metallurgy: quantitative analysis, metallurgy, geology, mineralogy, quantitative blow-pipe analysis, drawing. IV. For students in geology and natural history: quantitative analysis, metallurgy, geology, mineralogy, drawing. V. For students in analytical and applied chemistry: quantitative analysis, metallurgy, geology, applied chemistry, drawing, memoir and journal of travel during the summer vacation.

Third year.—I. For civil-engineering students: mechanics, constructions, economic geology, drawing, projét. II. For students of mining engineering: mining engineering, assaying, economic geology, metallurgy, quantitative analysis, drawing, projét. III. For students in metallurgy: assaying, economic geology, metallurgy, quantitative analysis, lithology, drawing, projét. IV. For students of geology and natural history: economic geology, lithology, palæontology, drawing, dissertation. V. For students of analytical and applied chemistry: assaying, economic geology, metallurgy, quantitative analysis, applied chemistry, drawing, dissertation.

Preparatory year.—First session: geometry, physics, chemistry, French, German, drawing. Second session: algebra and trigonometry, physics, chemistry, French, German, drawing.

Mathematics.—The course of mathematics in the preparatory year embraces algebra, so far as to include the general theory of equations, geometry, plane, volumetric and spherical; trigonometry, plane, analytical, and spherical; mensuration of surfaces and of volumes. In the first year, analytical geometry of two and three dimensions; differential and integral calculus; differentials of algebraic and transcendental functions; successive differentials; maxima and minima; transcendental curves; curvature; integration of regularly formed differentials; integration by series; integration of fractions; special methods of integration; rectification of curves; quadrature of surfaces; cubature of volumes; applications to mechanics and astronomy.

Physics.—The students of the preparatory year are occupied during the first term with the subject of heat, including the steam-engine, while the second term is employed in the study of voltaic electricity, magnetism, and electro-magnetism. These courses of lectures are fully illustrated by appropriate experiments. The instruction is conveyed by lectures and recitations, practical problems being occasionally proposed for solution. During the second year courses of lectures are delivered on the laws of electro-dynamics, on the mechanical theory of heat, on mathematical optics, and on the undulatory theory of light. Portions of these courses are accompanied by experimental demonstrations. The cabinet of physical apparatus will rank with the best on this continent, and extensive additions are made to it each year.

Mechanics.—This subject is taught during the second year. The course of instruction embraces the following subjects: composition and equilibrium of forces; center of gravity and stability; elements of machinery; hurtful resistances; rectilinear and periodic motion; moment of inertia; curvilinear and rotary motion; mechanics of liquids; mechanics of gases and vapors; hydraulic and pneumatic machines.

Drawing and descriptive geometry.—During the first session of the preparatory year the student is taught to execute topographical maps. He is first instructed in the use of the pen to delineate lines of level, shaded with lines of declivity, and completed with the conventional signs of different features, such as water, forests, marshes, cultivated ground, outcrops of veins, &c.; subsequently he is taught to represent the same in shading of India ink or sepia, with the application of the conventional signs and colors used by our Government and civil engineers. During the second session, the course of instruction includes sketching in pencil from plane models, and from nature; afterward colored sketches or landscape drawing in water-colors.

During the first year descriptive geometry is taught. The course of instruction includes the study of Davies's treatise on this subject, with lectures and blackboard exercises, illustrated by Olivier, and other models, showing the more difficult problems of intersections, and the generation of warped surfaces.

The instruction of drawing includes the use of mathematical instruments in constructing on paper the problems of descriptive geometry.

During the second session graphics are taught, including the study of Davies's Shades and Shadows, and Perspective; and Mahan's Stone-Cutting, with explanatory lectures; the exhibition of models; and the solution of various new problems of shades and shadows.

The course in drawing includes instruction and practice in the use of instruments; the pen and brush, with India ink, in drawing mathematical forms in projection and perspective; shading them; casting their shadows, and washing them. This is followed by an application of the principles learned to the execution of a drawing of a machine, or the section of a furnace, wherein the shadows are accurately calculated and washed, and the drawing is appropriately colored.

In the second year the course includes, during the first session, the drawing of machines, mills, furnaces, &c., from plane models. These are shaded, their shadows calculated and cast, and the whole properly colored. The dimensions are also quoted, so that these drawings serve as types of working drawings.

During the second session the students draw from various models in relief, chiefly furnaces and machines. They first make a free-hand sketch from the relief, and upon it place the dimensions, which they measure; subsequently they draw the finished representation in the academy to a proper scale, with shades, shadows, colors, and dimensions. This practice is of benefit in accustoming the student to take rapid sketches of established works, upon which he may be required to report, or by which he may wish to inform himself.

Modern languages.—The design in this department is to teach the student how to read French and German scientific books with facility.

Instruction is given for two hours a week in each of these languages, during two years; and as the text-books employed in the class-room are altogether works on science, the students can acquire a sufficient vocabulary to enable them to use French and German authors in all the departments of the school.

No attempt is made to produce accomplished scholars in all branches of German and French literature, but attention is concentrated upon the immediate wants of the young men. In this way no time is lost, and the instruction becomes thoroughly practical.

General chemistry.—The preparatory class attend three exercises a week in general chemistry throughout the year. It is intended to lay the foundation of a thorough knowledge of the theory of the subject preliminary to the practical instruction in the chemical laboratory. For this purpose the class is drilled upon the lectures, with free use of the best text-books. The students are expected to write out full notes, which must be exhibited to the professor at the close of each session. At the end of the year the class must pass a rigid examination before they can be admitted to a higher grade.

The first year students also attend three times a week, during the year, in general chemistry. The text-book for reference in this department is Roscoe's Chemistry, English edition, 1869; and the notation adopted is in accordance with the unitary atomic system.

Analytical chemistry.—There are two laboratories devoted to qualitative analysis, and one of larger size to quantitative analysis, besides the assay laboratory. These laboratories are provided with all the necessary apparatus and fixtures, and each is under the special charge of a competent assistant. Each student is provided with a convenient table, with drawers and cupboards, and is supplied with a complete outfit of apparatus and chemical reagents.

During the first year qualitative analysis is taught by lectures and blackboard exercises, and the student is required to repeat all the experiments at his table in the laboratory. Having acquired a thorough experimental knowledge of the reactions of a group of bases or acids, single members of the group or mixtures are submitted to him for identification. He thus proceeds from simple to complex cases till he is able to determine the composition of the most difficult mixtures. Constant use is made of the spectroscope in these investigations.

When the student shows on written or experimental examination that he is sufficiently familiar with qualitative analysis, he is allowed to enter the quantitative laboratory.

During the second and third years quantitative analysis is taught by lectures and blackboard exercises, and the student is required to execute in the laboratory, in a satisfactory manner, a certain number of analyses. He first analyzes substances of known composition, such as crystallized salts, that the accuracy of his work may be tested by a comparison of his results with the true percentages. These analyses are repeated till he has acquired sufficient skill to insure accurate results. He is then required to make analyses of more complex substances, such as coals, limestones, ores of copper, iron, nickel, and zinc, pig-iron, slags, technical products, &c.; cases in which the accuracy of the work is determined by duplicating the analyses, and by comparing the results of different analysts.

Volumetric methods are employed whenever they are more accurate or more expeditious than the gravimetric methods. In this way each student acquires practical experience in the chemical analysis of the ores and products which he is most likely to meet in practice.

Stoichiometry.—Stoichiometry, the arithmetic of chemistry, is taught in a special course of lectures and blackboard exercises, during the second session of the first year.

Assaying.—During the third year the student is admitted to the assay laboratory, where he is provided with a suitable table and a set of assay apparatus, and where he has access to crucible and muffle furnaces, and to volumetric apparatus for bullion assay by the wet process. The general principles as well as the special methods of assaying are explained in the lecture-room, and at the same time the ores of the various metals are exhibited and described. The student is then supplied with suitable material, ores of known composition, and is required to make assays himself. He first receives ores of lead, the sulphuret, carbonate, and phosphate, which he mixes with the proper fluxes, and heats in the furnace, obtaining a button of lead which he carefully weighs, thus determining the percentage of metal in the ore. He then determines by cupellation the amount of silver in the lead. Silver-ores are next given to him, at first those which are most easily assayed, such as mixtures of chloride of silver with quartz; afterward more complex ores, such as galena, ruby-silver ore, mispickel, fahlerz, &c. These he is required to assay both in the crucible and in the scorifier. Ores of gold are next supplied, auriferous quartz, slates, pyrites, blende, &c., which are assayed by the most reliable methods.

To facilitate the assay of ores of the precious metals a system of weights has been introduced, by which the weight of silver or gold globules obtained in the assay shows at once, without calculation, the number of troy ounces in a ton of ore.

The student then passes on to the assay of silver and gold bullion, the former by Gay-Lussac's volumetric method, the latter by "quartation," or "parting." Ores of tin, antimony, and iron are then assayed in the dry way, when the course is completed. Each student thus executes two or three hundred assays himself, under the immediate supervision of the instructor.

Applied chemistry.—The instruction in applied chemistry extends through the second and third years, and consists of lectures, illustrated by experiments, diagrams, and specimens. The subjects discussed are:

I. Chemical manufactures, acids, alkalies, and salts.

II. Glass, porcelain, and pottery.

III. Limes, mortars, and cements.

IV. Fuel and its applications.

V. Artificial illumination, candles, oils and lamps, petroleum, gas and its products.

VI. Food and drink, bread, water, milk, tea, coffee, sugar, fermentation, wines, beer, spirits, vinegar, preservation of food, &c.

VII. Clothing, textile fabrics, bleaching, dyeing, calico-printing, paper-tanning, glue, India rubber, gutta-percha, &c.

VIII. Artificial fertilizers, guano, superphosphates, poudrettes, &c.

IX. Disinfectants, antiseptics, preservation of wood, &c.

Mineralogy.—The studies in the department of mineralogy continue through two years. In the first year the students are instructed in crystallography and the use of the blow-pipe. The lectures on crystallography are illustrated by models, which the students are required to determine under the eye of the professor. A collection of glass models, and of models in wood, illustrating all of the important actual and theoretical forms, is always accessible to the students. The exercises in blow-pipe determination are entirely practical; known mixtures are first given to the student to examine, and when he is sufficiently familiar with them, unknown mixtures are determined. In the second year the lectures are illustrated by conferences, where the student is required to determine minerals by their physical and blow-pipe characters. The mineralogical cabinet contains about ten thousand specimens, which are labeled, and open to the public. Besides this, there is a collection of about two thousand specimens, to which the students have an unrestricted access.

Geology.—The course of instruction in this department is as follows: First year.—Botany and zoology as an introduction to palæontology; lectures throughout the year. Second year.—Lithology: minerals which form rocks, and rock-masses of the different classes; lectures and practical exercises. Geology: cosmical, physiographic, and historical; lectures throughout the year. Third year.—Economic geology: theory of mineral-veins, ores, deposits, and distribution of iron, copper, lead, zinc, gold, silver, mercury, and other metals; graphite, coal, lignite, peat, asphalt, petroleum, salt, clay, limestone, cements, building and ornamental stones, &c. Palæontology: systematic review of recent and fossil forms of life; lectures throughout the year.

Metallurgy.—The metallurgical course includes lectures on the preparation of fuels, construction of furnaces, the manufacture of metals, projects and estimates for the erection of metallurgical works. The lectures cover a period of two years, and discuss in detail the methods in use in the best establishments in this country, and in Europe, for the working of ores, with practical details of charges, labor, and cost of erection, obtained from the most authentic sources. Special attention is given to ores of this country which are difficult to treat, and to the solution of practical problems which are likely to occur. The lectures are illustrated by models, drawings of furnaces, and collections of metallurgical products. The projects assigned to the students familiarize them with the method of making plans and estimates for the erection of works. The ore to be worked and the various conditions which are required are given to the student at the close of the second year. During the summer vacation he is expected to visit works, and to ascertain what the practical requirements are. During the third year the drawings, estimates, and descriptions of the processes are completed and submitted for inspection and approval.

Mining engineering.—Mining engineering is taught during the second year. The instruction comprises a course of lectures illustrating the theory and practice of mining operations at home and abroad; giving the general principles of reconnoitering and surveying mineral property and mines; the attack, development, and administration of mines, and the mechanical preparation of ores, with the exhibition and use of all necessary reconnoitering and surveying instruments, particularly the mining theodolite, and the exhibition of various models.

In surveying, the student is taught to make surface surveys of the limited extent he needs, and subterranean surveys to direct and adjust his works; also, the solution of some problems of underground surveying by descriptive geometry, and many special examples of determining lines on the surface corresponding to given lines below, &c.

Attack describes the miner's methods, the use of drills, picks, powder, nitro-glycerine, compressed air, &c.; the proper location and construction of tunnels, slopes, shafts, wells for sounding, artesian wells, salt and oil wells, preceded by a theory and description of the most typical veins, true or irregular, and other deposits of ore, salt, coal, and oil, exemplified at home and abroad.

Development includes the best methods for laying out subterranean works for production and conservation in the present and future; for proper and economic ventilation, transportation, hoisting, pumping, or draining, distribution of workmen, &c.

Administration includes a review of the foregoing, with regard to a concentration of ideas and a general comparison of production cost to market price of untreated ore. Here the student is taught to forecast the expense of the establishments he must make, their annual cost, the cost of miners, employés, machines, material, &c., and offset these with the result of production, so endeavoring to solve the problem of making a given mine pay in given circumstances, by scientific attack, distribution, and general rational economy.

Mechanical preparation describes the various accepted methods of reducing massive ores to a condition either yielding metal or fitting the material for metallurgical processes. Models of stamps, crushers, shaking-tables, sluices, &c., are exhibited with plans and sections of mills and coal-breakers.

Machines.—The course on machines, which is inseparable from that of mining engineering, is given during the third year. It teaches the theory of the machines used in mining-works. It is the application of mechanics to the construction of water-wheels, turbines, windmills, steam and hot-air engines, pumps, and ventilators, transmission of force by compressed air, and the formulæ, with their theory, for the resistance of materials. Models of water-wheels, steam-cylinders, steam-engines, blowing-machines, &c., are exhibited.

In the resistance of materials the calculations are shown for the sections of different parts of machines, the fly-wheel, pump-rods, connecting-rods, &c.; also, for such constructions as retaining-walls, arches, timbering, supports, &c. The course of the third year also includes a plan of drawing and estimates from some projected work of mining, or the construction of a machine for some of the uses of mining.

This system of projects is to the young engineer a real practical application of all his three years' study, by which he is made to investigate prices, compare theories, models, methods, and dispositions, and, in competing with his class, to take pains to furnish the best arguments, illustrations, and calculations he can, in order to support his views.

Library and collections.—A special scientific library and reading-room have been provided for the use of the students of the school, which already numbers two thousand volumes, and which is rapidly increasing. Seventy of the best foreign and American scientific journals are regularly received. Collections of specimens and models illustrating all the subjects taught in the school are accessible to the student, including crystal models, minerals, ores, and metallurgical products, models of furnaces, collections illustrating applied chemistry, fossils, economic minerals, rocks, Olivier's models of descriptive geometry, models of mining-machines, models of mining-tools.

The lectures on crystallography are illustrated by a collection of one hundred and fifty models in glass, which show the axes of the crystals, and the relation of the derived to the primitive form. This suite is completed by three hundred and fifty models in wood, showing most of the actual and theoretical forms.

The collection of minerals comprises about ten thousand specimens, arranged in table cases. The minerals are accompanied by a large collection of models in wood, showing the crystalline form of each. Arranged in wall-cases are large specimens, showing the association of minerals.

A collection of metallurgical products, illustrating the different stages of the type process in use in the extraction of each metal, is accessible to the students. This collection is constantly increasing. Most of the specimens have been analyzed and assayed.

An extensive collection of models of furnaces has been imported from Europe. A very large number of working drawings of furnaces and machines used in the different processes are always accessible to the students; and several thousand specimens of materials and products illustrating applied chemistry have already been collected.

The geological collection consists of over sixty thousand specimens, including systematic series of rocks, fossils, and useful minerals. In this series is to be found the largest collection of fossil plants in the world, including many remarkably large and fine specimens, and over two hundred new species, of which representatives are not known to exist elsewhere. Also, the most extensive series of fossil fishes in the country, including, among many new and remarkable forms, the only specimens known of the gigantic dinichthys; a suite of Ward's casts of extinct saurians and mammals; a fine skeleton of the great Irish elk, &c.

Requirements for admission.—Candidates for admission to the first year of the school must not be less than eighteen years of age. They must pass a satisfactory examination in algebra, geometry, and plane, analytical, and spherical trigonometry, physics, and general chemistry.

Candidates for the preparatory year must be seventeen years of age, and must pass a satisfactory examination in arithmetic, including the metric system of weights, measures, and moneys, and in portions of algebra and geometry. Those who are not candidates for a degree may pursue any of the branches taught in the school.

During the vacation each student is expected to visit mines, metallurgical and chemical establishments, and to hand in, on his return, a journal of his travels, and a memoir on some subject assigned him. He is also required to bring collections, illustrating his journal and memoir, which collections are placed in the museum, reserved as a medium of exchange, or made use of in the laboratories. For pupils who have been proficient, and who desire to devote special attention to any one branch, application will be made for permission to work in particular mines or manufactories. This will be done only as the highest reward of merit that the institution can give. Prizes are awarded to students who pass the best examination in mineralogy, qualitative and quantitative analysis, and assaying, &c. At the close of the course are conferred degrees of Civil Engineer, Engineer of Mines, and Bachelor of Philosophy. The fee for the full course is $200 per annum.

Special students in chemistry pay $200 per annum. Special students in assaying are admitted for two months for a fee of $50 in advance. The fees for single courses of lectures vary from $10 to $30. Students unable to meet the expenses of the school are instructed gratuitously.

THE SHEFFIELD SCIENTIFIC SCHOOL OF YALE COLLEGE, NEW HAVEN, CONNECTICUT.

Officers of instruction.—William A. Norton, civil engineering and mathematics; Chester S. Lyman, physics and astronomy; William D. Whitney, linguistics and German; William P. Trowbridge, dynamic engineering; Samuel W. Johnson, agricultural and analytical chemistry; George J. Brush, metallurgy and mineralogy; William H. Brewer, agriculture; Daniel C. Gilman, physical geography and history; Daniel C. Eaton, botany; Othniel C. Marsh, palæontology; Addison E. Verrill, zoology and geology; Eugene C. Delfosse, French; Louis Bail, drawing; Mark Bailey, elocution; Oscar D. Allen, metallurgy and assaying; Daniel H. Wells, analytical and descriptive geometry; Thomas R. Lounsbury, English; William G. Mixter, elementary chemistry; Sidney I. Smith, zoology; Albert B. Hill, surveying and mechanics; Russell W. Davenport, assistant in chemistry; Charles S. Hastings, assistant in physics.

The chief instructor and their specialties may be thus grouped:

I. *Engineering, &c.*—Mathematics and civil engineering, W. A. Norton; mechanical or dynamic engineering, W. P. Trowbridge; astronomy, theoretical and practical, C. S. Lyman; analytical and descriptive geometry, D. H. Wells; land surveying, A. B. Hill; drawing, mathematical and free-hand, L. Bail.

II. *Chemistry, &c.*—Theoretical and analytical chemistry, S. W. Johnson; metallurgy and assaying, G. J. Brush and O. D. Allen; elementary chemistry, W. G. Mixter; agricultural chemistry, S. W. Johnson; agriculture, W. H. Brewer; laboratory practice, W. G. Mixter and O. D. Allen; physics, C. S. Lyman.

III. *Natural history, &c.*—Mineralogy, G. J. Brush; botany, D. C. Eaton; zoology, A. E. Verrill and S. I. Smith; palæontology, O. C. Marsh; geology, A. E. Verrill; physical geography, D. C. Gilman.

IV. *Language, &c.*—German, W. D. Whitney; French, E. C. Delfosse; English, T. R. Lounsbury; elocution, M. Bailey; linguistics, W. D. Whitney; modern history and political economy, D. C. Gilman.

Summary of students.—Graduates, 27; seniors, 21; juniors, 35; freshmen, 55; special students, 8; total, 146.

Relations to Yale College—The relations of the scientific department to the classical department of Yale College may be thus stated: The instructors, terms of admission, courses of study, and methods of instruction in the two departments are different; but both institutions are harmoniously organized under one board of trustees, and consequently the students have in common certain university privileges, and are alike entitled to become graduates of Yale College.

It is this union and this individuality which give the Sheffield School at New Haven the steadiness of a firm and well-tried institution, with the freedom of a new foundation. The combination has been in many respects highly advantageous to the new department, and is probably not without some influence for good upon the old and well-known classical department.

Plan of instruction.—This institution, which is partly co-ordinate with a classical college and partly with professional schools, receives three classes of students:

1. Those who wish to pursue a three years' course of training,* in accordance with a prescribed curriculum, largely based upon mathematical, physical, and natural science, with instruction in German, French, and English.

2. Those who have already graduated in some college or school of science, and desire to pursue advanced courses of scientific study.

3. Those who desire under peculiar circumstances to attend for a short time instructions in special branches.

These three classes are known as under-graduates, graduates, and special students.

Instruction for graduate students.—The degree of doctor of philosophy will be bestowed by the corporation of Yale College on young men who have already taken a bachelor's degree, and who here pursue for two or three years advanced special studies, passing satisfactory examinations, and submitting a graduation thesis as evidence of their attainments. Great freedom in the choice of work is permitted to such students; and all the resources of the institution in teachers, apparatus, laboratories, collections, &c., are at the service of those who need them. Persons desirous of availing themselves of opportunities to pursue higher studies are invited to state their special requirements or wishes to any of the instructors, and thus to become acquainted with the facilities which the institution affords. As examples of what may be done it may be mentioned that in mathematics, Professors Norton, Trowbridge, and Lyman, with the co-operation of Professor H. A. Newton, will direct the studies of those who wish to avail themselves of the class instructions in the calculus, in analytical and descriptive geometry, mathematical drawing, practical astronomy, &c. The Hillhouse Mathematical Library, open for consultation daily, and the astronomical instruments belonging to the school, may be freely used by advanced students. The higher course in engineering leads to the degree of civil engineer. The chemical laboratory is fitted for the instruction of those who wish to become proficients in practical analysis, either in preparation for professorships, technical pursuits, the medical profession, or other purposes. Instruction in natural history may be received in the zoological laboratory, where the collection, description, and classification of specimens are continually in progress; or, by private arrangement, in the herbarium of Professor Eaton. The public and private collections of minerals, ores, fossils, &c., afford special facilities for the study of mineralogy and geology.

Instruction for undergraduates.—The courses of study for undergraduates occupy three years; it is hoped that they will be soon extended to four years. The requirements for admission and the first year's work are the same as for all this class of students: during the last two years the courses are to some extent coincident, but are chiefly special and technical.

For admission the student must pass a thorough examination in Davies's Bourdon's Algebra as far as the general theory of equations, or in its equivalent; in geometry, in the nine books of Davies's Legendre, or their equivalent; and in plane trigonometry, analytical trigonometry inclusive; and also in arithmetic, including "the metrical system," geography, United States history and English grammar, including spelling. An acquaintance with the Latin language is also required, sufficient to read and construe some classical author, and Allen's Latin Grammar is commended as exhibiting the amount of grammatical study deemed important. Practice in drawing, if it can be obtained before entrance, will be of great advantage to the scholar.

The studies of the freshman year are: In the mathematics, analytical and descriptive geometry, spherical trigonometry, and surveying, (with practical field-work;) in chemistry, recitations and laboratory practice; in physics, recitations, with experimental illustrations of the subjects taken up; in language, the commencement of German, and lessons in respect to the use of English, with practice in writing and in elocution; in botany, recitations, excursions, and lectures; in drawing, Binn's First Course of Orthographic Projection; perspective and free-hand drawing.

At the close of the freshman year, the students distribute themselves into various sections with reference to special lines of work for the senior and junior years; but in all these sections the study of German is continued; the study of French is pursued; and practice in writing English is required. Drawing also occupies a part of the time. At the close of three years, every candidate for a bachelor's degree presents a thesis as evidence of his powers of investigation and his capacity as a writer.

In each section, the students attend to some of the studies appropriate to other sections; thus geology is taught to all the scholars; zoology to the students in chemistry as well as to those in natural history and in the select course; and so on.

---

* Soon to be made a four years' course.

The special courses most distinctly marked out are the following:

(a.) In chemistry and metallurgy;
(b.) In civil engineering;
(c.) In mechanical or dynamic engineering;
(d.) In agriculture;
(e.) In natural history;
(f.) In studies preparatory to medical studies;
(g.) In studies preparatory to mining;
(h.) In select studies preparatory to other higher pursuits, to business, &c.

(a.) For chemistry and metallurgy the Sheffield laboratory is fitted up in a complete and convenient manner, is provided with all the requisite apparatus and instruments of research, possesses a considerable collection of chemical preparations, and has a consulting-library of the best treatises on chemistry and the chemical arts. It is open for chemical practice seven hours daily, for five days of the week, but is closed on Saturday. The student works through a course of qualitative and quantitative analysis, which is varied according to his capacity and the object he has in view. Each pupil proceeds by himself independently of the others, under the constant guidance of the instructors. The regular students in chemistry are prepared for chemical work by their practical exercises in the laboratory during freshman year. In the junior and senior years they are required to occupy four to six hours in the laboratory each working-day. Special students who have not had adequate instruction in inorganic chemistry are required to join the freshman class in Eliot and Storer's Manual. Junior students have recitations in analytical chemistry and lectures on theoretical and organic chemistry. Senior students have recitations and lectures on agricultural chemistry and metallurgy. Mineralogy is taught in the junior year by lectures, which are fully illustrated with hand-specimens and models, and by weekly exercises throughout the senior year in the identification of minerals from physical and chemical characters. Instruction is also given in metallurgy, and especial attention is devoted to assaying and the investigation of ores and furnace-products. The student in agriculture has opportunity to acquaint himself with the modes of research employed in agricultural chemistry. The applications of the science to other branches of industry are taught as occasion requires. To advanced students, whether belonging to the regular classes or not, who desire to give attention to particular branches of chemistry, or to pursue original investigations, every facility is accorded. The private libraries of the professors, containing the chemical journals and the recent foreign literature of chemistry and mineralogy, the large collections of ores, furnace-products, &c., belonging to the school, and the extensive private cabinet of the professor of mineralogy, are freely used as aids in instruction.

(b.) The special course of civil engineering comprises the following departments of study: 1. The higher mathematics, consisting of spherical trigonometry, higher analytical geometry, differential and integral calculus, descriptive geometry, and co-ordinate branches of study, &c. 2. Applied mathematics, which include all the field-operations and plotting comprised in the various branches of practical surveying. 3. A course of drawing, comprising Binn's Course of Orthographic Projection, with application to mechanical and engineering drawing; shading and tinting; linear perspective; free-hand drawing; isometrical, topographical, architectural, and structural drawing. 4. Theoretical mechanics; and mechanics applied to engineering in the construction and operation of machines, the utilization of water-power, the employment of prime movers, including hydraulic motors, and the steam-engine, &c. 5. Field-engineering, which embraces the laying out of curves, and all the field-operations necessary in locating a line of road, establishing the grade, and determining the amount of excavation and embankment, &c. 6. Civil engineering, proper, or the science of construction, in its various departments, including, among many other topics the strength of materials, the establishment of foundations, the construction and stability of walls and arches, the theory and detail of the construction of bridges, roof-trusses, &c., in wood and iron, and the graphics of stone-cutting.

Students who pursue a higher course in engineering, for one year after graduating as bachelors, may receive the degree of civil engineer.

(c.) The course in dynamic engineering comprehends in its various branches of study and preparation all that have an immediate bearing on industrial pursuits, requiring the use of: 1. Instrumental drawing. Beginning with the elements of drawing, the students receive continuous instruction in all the conventional modes and practices of representing objects, machines, or structures, from the study of the objects, by plans, elevations, sections, shading and coloring, while at the same time, and by graphical representation, they learn the detailed construction of all classes of machinery, the application of mechanical movements, and the modes of transmitting motion and power. To these ends a large collection of standard drawings, models, and machines has been obtained and arranged for ready reference. 2. The higher mathematics. Spherical trigonometry, analytical geometry of three dimensions, differential and integral calculus, and descriptive geometry. 3. Applied mathematics and analytical mechanics.

The principles of thermodynamics, or the application of mathematics to the investigation of the laws of heat, the principles of cinematics or the comparison of motions; the theory of mechanism. 4. Applied mechanics and thermodynamics. The application of mechanics, cinematics, and thermodynamics to the construction of boilers, or steam-generators, the construction of steam or heat engines, the construction of water-wheels, shafting, gearing, and the construction and use of tools and machines for performing all kinds of useful work, the construction of iron bridges and structures of iron, the properties of materials as regards resistances to strains, or stresses, elasticity, durability, chemical reactions, friction, &c. 5. Dynamic engineering. The application of the principles of mathematics, mechanics, cinematics, thermodynamics, mechanism, and properties of materials to industrial operation, steamships, railway motive-power, manufactures, mills, forges, fabrication of materials, heating and ventilation, the utilization of water-power, draining, and irrigation, windmills, &c. 6. For students desiring to take a degree of dynamic engineer, two additional years will be required, during which the application of the foregoing studies will be continued in connection with the examination of existing works of industry in the various branches, and the exercises will be extended to the planning of such works, and the original designing of the various kinds of machinery applicable to them.

(d.) Students of agriculture, in addition to those general studies needed for mental discipline or general knowledge and culture, receive instruction in agricultural and analytical chemistry, vegetable physiology and botany, zoology, entomology, geology, the culture of our staple crops, the principles of stock-breeding and rearing, and rural economy. These instructions are given partly by lectures and partly by recitations. In the coming year, the lectures on stock-breeding, rural economy, and the cultivation of crops, will be given during the full term only.

(e.) Either geology, mineralogy, zoology, or botany may be made the principal study in natural history, some attention in each case being directed to the other three branches of natural history. In botany the extended course begins with structural and physiological botany, taught by text-books, lectures, and practical work with the microscope. Excursions and practice in identification of species and proper preservation of specimens follow. Familiarity with standard botanical literature is encouraged, and, lastly, students are taught to record their observations in scientific language, and to contribute, if possible, something to botanical science. In geology the instruction consists of recitations in Dana's Manual, illustrated by specimens of minerals, rocks, and fossils. Excursions are made to interesting localities to illustrate certain principles of the science which can be best studied in the field. Special students in geology pursue the practical study of fossils in the zoological laboratory, and of minerals and rocks in the chemical laboratory.

The instruction in zoology includes courses of lectures on systematic zoology, comparative anatomy, and the geographical distribution of animals, illustrated by specimens and a large number of diagrams; excursions for the purpose of studying the habits of living animals and collecting specimens; and practical instruction in the zoological laboratory, in comparative anatomy, embryology, and the identification, description, and classification of animals, together with their preservation and arrangement. The purpose is, in every case, to induce habits of close observation and accurate generalization, and, finally, to lead the student to make original investigations upon the objects of his study.

In mineralogy a course of lectures on elementary crystallography, and the physical properties of minerals, their chemical composition, classification, and the detailed description of mineral species, illustrated by constant reference to the mineral cabinets. Also a course of practical exercises in blow-pipe and determinative mineralogy.

(f.) During one year the work of the medical course will be chiefly under the direction of the instructors in chemistry; during the second year under that of the instructors in zoology and botany. In chemistry especial attention will be given to the examination of urine and the testing of drugs and poisons; in zoology to comparative anatomy, reproduction, embryology, the laws of hereditary descent and human parasites; and in botany to a general knowledge of structural and physiological botany, and to medicinal, food-producing, and poisonous plants.

(g.) Young men desiring to become mining engineers, can pursue the regular course in civil or mechanical engineering, and at its close can spend a fourth year in the study of metallurgy, mineralogy, &c. Should there be a sufficient number of students desiring it, a course of lectures on the subject of mining will also be provided.

(h.) In accordance with a demand for systematic instruction in scientific studies, without reference to technical pursuits, and with a just regard to intellectual culture, a course is arranged as a basis for higher scientific pursuits, for teachers, business men, those designing to engage in editorial work, and others. This course, in addition to the instructions in German, French, and English, common to all departments of the school, includes instruction from Professor Whitney in the general principles of language, and from Mr. Lounsbury in the critical study of the English language, in its structure, history, and literature. Constant practice in writing is also required. Stu-

PLATE I.

The Stevens Institute of Technology.

dents desiring to pursue the study of Latin, or of other languages, can easily make arrangements for doing so, if their time permits. The course also provides systematic instruction in the physical geography of the globe; in the special physical and historical geography of Europe and the United States; in the outlines of modern history, and in political economy. The students in this course receive from the various professors instruction in agricultural chemistry, botany, zoology, geology, and mineralogy. They attend the lectures on agriculture, rural economy, stock-breeding, &c., and those on general and theoretical chemistry. Their mathematical studies are continued in astronomy. They are expected to keep up the practice of drawing, especially of free-hand drawing. So far as it does not interfere with appointments in the school, students in this course are permitted to attend the lectures of the academical department.

For the benefit of those who desire to pursue some particular studies, without reference to a college degree, most of the various instructors are willing to receive special students for a longer or shorter time. Only persons of mature minds are received. For example, in agriculture the instruction is so arranged that by attendance during the ensuing autumn term the scholar may bear the various lectures, and receive as much technical instruction in this one branch as by remaining through the winter. In the chemical laboratory, students properly qualified are received for short periods of work. In the various departments of natural history special lessons will also be given. Instruction may also be received in practical astronomy and the use of instruments. These opportunities are not offered to persons who are incompetent to go on with regular courses, but are designed to aid those who have been educated elsewhere to increase their proficiency in special branches.

Building and apparatus.—Sheffield Hall, bearing the name of the donor, Mr. Joseph E. Sheffield, of New Haven, is a large and well-arranged building, containing recitation and lecture rooms for all the classes, a hall for public assemblies and lectures, laboratories for chemical and metallurgical investigations, a photographical room, an astronomical observatory, museums, a library and reading-room, besides studies for some of the professors, where their private technical libraries are kept.

The following is a summary statement of the collections belonging to the school:

1. Laboratories and apparatus in chemistry, metallurgy, mechanics, photography, and zoology.
2. Metallurgical museum of ores, furnace-products, &c.
3. Agricultural museum of soils, fertilizers, useful and injurious insects, &c.
4. Collections in zoology.
5. Astronomical observatory, with an equatorial telescope by Clark and Son of Cambridge, a meridian circle, &c.
6. Library and reading-room, containing the Hillhouse Mathematical Library, books of reference, and a selection of German, French, English, and American scientific journals.
7. A collection of physical apparatus, constituting the Collier cabinet, recently bought by Professor Lyman.
8. Models in architecture, civil engineering, and mechanics, and diagrams adapted to public lectures.
9. Maps and charts, topographical, hydrographical, geological, &c.

The mineralogical cabinet of Professor Brush, the herbarium of Professor Brewer, the collection of native birds of Professor Whitney, and the astronomical instruments of Professor Lyman, are all deposited in the building. Professor Eaton's herbarium, near at hand, is freely accessible.

Students in this department are also admitted to the college and society libraries, the college reading-room, the cabinet of minerals and fossils, the school of the fine arts, and the gymnasium for physical exercise.

The instructions of this institution are given chiefly in small class-rooms, by recitations or familiar lectures, illustrated by all the apparatus at the command of the various teachers. A public course of lectures is given every winter on topics of popular interest. On Sunday evenings during a portion of the year lectures are given by resident clergymen of different denominations, and by members of the theological and other college faculties.

Tuition charges.—The tuition charge is $150 per year. Besides this there is a charge of $5 annually for the use of the academical reading-room and gymnasium. Freshmen pay $5 for chemicals; and the special students in the chemical laboratory are likewise charged $25 per term for the materials they use—besides breakage. The graduation fee is $10.

Vacations correspond with those of the academical department.

THE STEVENS INSTITUTE OF TECHNOLOGY, HOBOKEN, NEW JERSEY.

Officers of instruction.—Henry Morton, Ph. D., President; Alfred H. Mayer, Ph. D., Professor of Physics; Robert H. Thurston, C. E., Professor of Mechanical Engineering;

Lieutenant-Colonel H. A. Hascall, Professor of Mathematics; C. W. MacCord, A. M., Professor of Mechanical Drawing; Albert R. Leeds, A. M., Professor of Chemistry; Charles F. Kroeh, A. M., Professor of Languages; Rev. Edward Wall, A. M., Professor of Belles-Lettres.

Summary of students.—First class, 16; second class, 3; third class, 2; total, 21.

Foundation.—This institution was founded in accordance with the will of Mr. Edwin A. Stevens, of Hoboken, who bequeathed for the purpose a large block of land in that city, and a sum amounting, at the discretion of the trustees, to $650,000.

Plan and buildings.—The Stevens Institute is especially a school of mechanical engineering; but the fact that chemistry, metallurgy, and mineralogy, as well as the whole science of machinery, so important to mining engineers of the present day, are taught here, and the magnificent completeness of the buildings and apparatus of instruction, justify me in including it in the present chapter. My principal object is to present a description of the building, which will be highly interesting and useful to those who have to deal with the arrangement of such institutions. The following description and plates are extracted from the New York Enginering and Mining Journal of April 16, 1872, and have been inserted at this place since this report was transmitted to Congress.

The building is situated in the pleasantest portion of the city, its windows commanding a beautiful view of the surrounding country, as well as of our harbor and bay, and the edifice itself presenting a fine appearance when viewed from the deck of the ferry-boat as we cross the river to visit it. (Plate I.)

The building is very substantially built of blue trap-rock, with brown-stone trimmings, from designs by Upjohn. It extends from street to street, and has two wings in the rear. It is three stories in height, and has a dry and roomy basement. (Plate II.)

In the basement is a work-shop, occupying the whole of the right-hand wing, and containing tools for working in both wood and metal, together with the steam engine which is to drive them.

Here, also, are gas-holders for oxygen and hydrogen, and from them pipes are led to the several lecture-rooms, where the lecturers may have occasion to use the lime-light or the oxyhydrogen blow-pipe.

At the opposite end of the basement are the boilers for heating the building, and for supplying steam to the engines, and at this end are the furnaces for metallurgical work, and under the wing are the assay-room and the rooms of the janitor.

On the first floor, (Plate IV,) we find at the right a splendidly lighted, high and airy hall, fitted up as a physical laboratory, and stocked with numerous ingenious and delicate forms of apparatus, such as were made use of by Faraday in his splendid researches in electricity, by Regnault, and by Tyndal, and by Melloni in their investigations of the nature and laws of heat, and by other physicists in other almost equally classical labors.

In the large room at the rear of the main building, which is the public lecture-hall of the Institute, we find seats for six hundred persons. The stage is fitted with all needed appurtenances. A trap-door being raised, pipes are discovered bringing water and gas from the street-mains, oxygen and hydrogen from the tanks in the basement, and steam from the main boilers. Heavy copper wires connect with the large electric battery, and these, as well as the oxygen and hydrogen pipes, are also led under the floor to the different points in the room, (marked O H in the plate,) where they may be required for the magic lantern, or for other purposes.

The large room at the extreme left, No. 3, in the main building, is the library and model-room. It is of the same size as the physical laboratory, and is also a beautifully-proportioned and well-lighted room. Here are kept the books which form the germ of what is intended to be a fine technical library, and the models and apparatus which are not needed in the lecture-rooms in illustration of the regular courses of instruction. This is also the reading-room, and the tables are furnished with a well-selected list of periodicals, some of which are contributed by the publishers.

The room in the wing at the left is the chemical laboratory, which, although not lofty, is well ventilated, and well fitted up with the best of modern apparatus. The balance-room is immediately adjacent, and contains some fine apparatus.

The second floor (Plate V) is occupied by the several lecture-rooms. At the right is the lecture-room used by the president when it becomes necessary for him to take part in instructing advanced classes, and at other times in special researches.

The little room off the stair-landing is also used by the president, as a work-room.

The lecture-room of the professor of physics is next to the preceding—a pleasant, well-arranged room, fitted up with every imaginable convenience, including all that were noticed in the larger lecture-hall, and also a pneumatic trough, and a set of Bunsen air-pumps.

Immediately over the main entrance is a room containing the principal part of the optical collection of the Institute, which is said to be the finest in the world, and con-

Stevens Institute of Technology.

PLATE II.

B'. Boilers, 70 Horse Power.—Harrison.
B. Battery Room.
C. Closets.
C'. Crucible Furnace
E. Entrances.
E'. Engine, Steam, 25 Horse Power.
F. Forge.
F'. File Bench.
G. Gas Furnace for Oxygen.
H. Hydrogen Reservoir.
L. Lift.
L'. Lathe Screw Cutting.
M. Milling Machine.

M'. Magneto-Electric Machine for Electric
    Light.
O. Oxygen Reservoir.
P. Planer.
R. Reverberatory Furnace.
S. Sinks.
V. Ventilating Shafts.
W. Wash Rooms.
W'. Work Benches for Carpentry.

JANITOR'S ROOM

JANITOR'S ROOM

ASSAY ROOM

COAL.

COAL.

GRINDING

TABLE

BASEMENT.

PLATE IV.

Stevens Institute of Technology.

No. 20. Private Laboratory of Prof. of Chem.
" 21. Study of Prof. of Chem.
" 22. Cabinet of Minerals.
" 23. Wash Room.
" 24. Drawing Department.
" 25. "
" 26. "
" 27. Stairway to Attic and Tower.
" 28. Prof. of Belles Lettres.
" 29. Cabinet of Instruments for Electrical
   Measurement.
" 30. Store Room.
" 31. Photographic Room with Dark Closet.
" 32. Prof. of Languages.

No. 33. Photometric Room.
" 34. Workshop of Messrs. Hawkins &
   Wale, Apparatus Makers to the
   Institute.

C. Closets or Cases.
L. Lifts.
S. Sinks.
S'. Skylights.
V. Ventilating Shafts.
W. Wash Basins.

THIRD FLOOR.

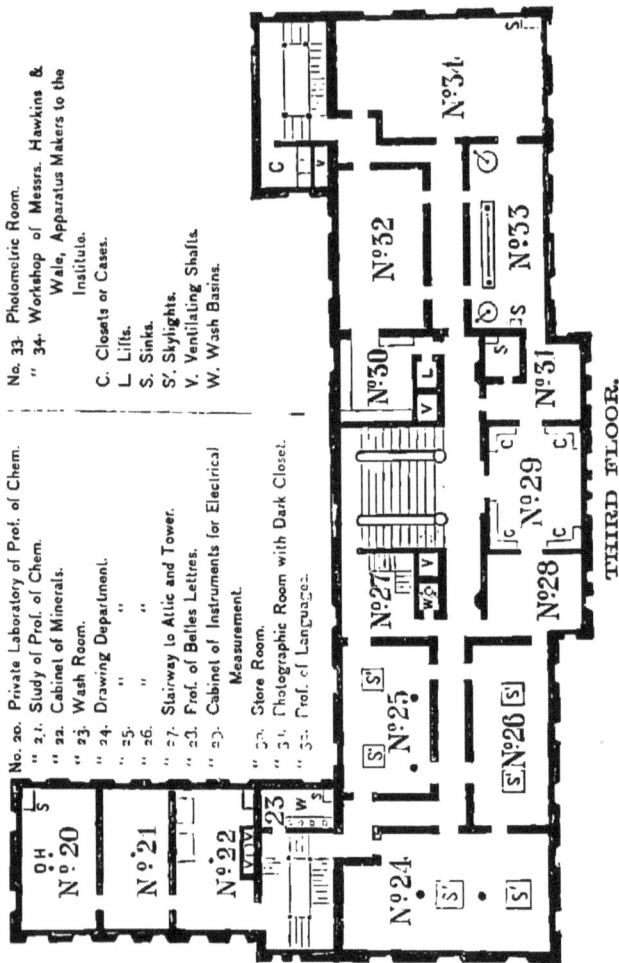

tains some exceedingly interesting pieces of apparatus. One of the largest instruments was made for M. Arago, the celebrated French philosopher.

On each side of this room is a pleasant apartment; the one at the right being the study of the professor of physics, and that at the left the study of the professor of mechanical engineering.

The large room next the room just mentioned is the lecture-room of the professor of mechanical engineering. This room is probably thirty-five feet square, sixteen feet to the ceiling, and well lighted, and thoroughly provided with means of heating and ventilation.

A large model-case nearly covers the farther side of the room, and contains a fine collection of apparatus acquired partly by purchase and partly by contribution from friends of the school and of the enterprise. Among the latter we are pleased to find many of our best-known and most enterprising as well as liberal manufacturers. The purchased apparatus is from the German makers of models at Darmstadt and Frankfort, and the French makers at Paris. The Germans seem to make them most substantially, and the French by far the most elegantly, and as a natural consequence the latter are the most expensive. Some few pieces are from London, and others are American, while still others were made at the Institute.

Among the most noticeable are the German models of gearing, several French sectional models of steam-engines, one of them being on quite a large scale, well proportioned, and very neatly made; a set of apparatus of different kinds for measuring the velocity of flow of water, dynamometers, cranes, and windlasses, pumps, water-wheels, turbines, and steam-engines of various kinds.

The most beautiful model in the collection is a copy of the English oscillating engine with feathering paddles as made by Penn & Co. We hope that some one will have enough of enterprise and public spirit to make the Institute the possessor of an equally perfect model of the American steamboat-engine.

We find in this collection models of the engine described by Hero of Alexandria over two thousand years ago, a modern copy of the old devices of DoCaus and Savery, models of the engines of Newcomen and of Watt, and a little horizontal stationary engine of 1½-inch cylinder and 4½ inches stroke, with a drop cut-off which is adjusted by a little fly-ball regulator. It is supplied with steam by a copper boiler fed by a separate steam-pump. It was made, we understand, by the head of this department when himself a boy at school.

At one end are boiler-models, and, at the other, roof and bridge trusses, and scattered through the case are many pieces that would be unusually interesting if they were not surrounded by so many others of still greater interest. A model of the Fontaine turbine by Sallerou, of Paris, is the most perfect specimen of model-making imaginable, and hardly less beautifully made is the small Giffard injector, cut open to exhibit its internal construction. This was made and contributed by Sellers & Co., of Philadelphia. Blake & Brothers, of New Haven, contributed a powerful little stone-crusher, that will crumble between its jaws any mineral that can be inserted there. The professor of chemistry often finds this a convenient substitute for the mortar. William D. Andrews & Brother, of New York, have given a neat little centrifugal pump; A. K. Rider, a model of his steam-engine; D. P. Davis and the Recording Steam-Gauge Company each have presented a finely finished specimen of the recording steam-gauge. A. L. Holley furnished model rolls and specimen tuyeres as used in the Bessemer-steel manufacture, and other manufacturers have exhibited equally active and helpful sympathy in this important enterprise by presenting other models and samples, and have already placed in this department a valuable set of drawings which occupy a good proportion of the space allowed in the set of drawers, which extend along the side of the room opposite the model-case.

Mr. George B. Whiting, chief draughtsman of the Bureau of Steam Engineering of the Navy Department, has placed here a large proportion of his own private collection of drawings, embracing some extremely valuable complete sets of drawings of steam-machinery.

At the end of the room, at the right of the entrance, is a case which is beginning to fill up with samples of useful ores, minerals, and metals. A set of specimens from the Pennsylvania Steel Company illustrate the Bessemer-steel manufacture by exhibiting samples of the steel rail and of all the material, ores, irons, spiegeleisen, ganister, &c., &c., that are used in its manufacture.

The Chrome-Steel Company contributed samples of their remarkable metal; the Pembroke irons and the iron of Catasauqua and elsewhere are shown, with the ores from which they are made.

The Manhattan Oil Company, of New York, present samples of all commercial, animal, and vegetable oils; and the Dover Company, of mineral oils.

Professor Thurston is gradually collecting a very interesting set of engineering relics.

In the library and museum on the first floor we saw the identical high-pressure, direct-acting engine and tubular boiler, and the remains of *the screw* used by John

Stevens on the Hudson River in 1804, the little steamer attaining a speed of eight miles an hour at times. With these are the twin screws used by that great engineer in 1805.

In the lecture-room we find the patent issued from the English patent-office for this *tubular* boiler in 1805, to John Cox Stevens, the oldest son of its inventor; and in the model-case is a model of the same boiler, which exhibits a strong resemblance to some of the safety tubular boilers of the present day, and was evidently quite as efficient.[*]

Here is to be seen a drawing of the engine of Fulton's first boat, the Clermont, drawn by Fulton's own hand, an autograph letter from Robert Fulton to Mr. Stevens, an autograph letter from Robert Stephenson, the distinguished son of the even more distinguished George Stephenson, which contains the assurance that *then*, 1835, the tendency in Great Britain was toward heavier rails and more powerful locomotives, and that the latest of his own design weighed *nine* tons, and could draw one hundred tons at the rate of sixteen or eighteen miles an hour on a level. A sketch accompanies the description.

There are other things of interest to be seen here, but space will not allow of further description of this, to us, most interesting of the many interesting departments of the Stevens Institute of Technology.

We would like, had we space, to describe the collections of physical apparatus, the apartments, with their apparatus and fittings, belonging to the department of chemistry, to which is devoted all available space in the whole west wing, the pleasant drawing-rooms and recitation-rooms and the work-shops of Hawkins and Wales, the instrument-makers to the Institute.

Instruction, during the first two years of the course, which is four years in length, is similar to that pursued in other colleges, except that the classics are not taught, all the available time being spent upon mathematics, English and foreign languages and literature, and the usual courses in science, and this constitutes a course preparatory to entering upon the technical and professional work of the last two years.

During the last two years the student enters the laboratories and work-rooms and pursues his professional studies with the intention of securing a practical and immediately useful knowledge of the several branches. In the physical laboratory he makes for himself, and with his own hands, the experiments that the student usually in our college courses merely witnesses from his seat at a distance from the lecture-table, and when he has acquired some familiarity with the adjustments and uses of the apparatus, he enters upon a final course of independent research, the results of which, when new and valuable, are at once published.

In mechanical engineering, the course commences with the study of the nature of materials used by the engineer, the methods adopted in obtaining them and preparing them for the market and for use, and the best methods of preserving them from decay. The course is illustrated by specimens which, thanks to the great interest taken in the school by all who have visited it, are continually coming in.

The course continues with the investigation of the facts and laws governing the strength of materials, by means of the apparatus of the Institute. In investigating tensile strengths, the use of the excellent and powerful testing-machine of the Camden and Amboy repair-shops is generously allowed by Mr. Francis B. Stevens.

Instruction in the use of tools and in designing machinery follows, partly in this course and partly in that of the professor of drawing, which is really almost as much a department of engineering as that which is so called. The course closes with the study from text-book and lecture of the principal prime movers.

The departments of engineering and drawing work together from beginning to end, and the time given them is in the aggregate fully commensurate with their importance as leading departments.

Occasionally, in response to the many invitations received, the students are given opportunity to do useful work outside the regular course, and to visit manufacturing establishments and places of interest.

During the past year students of the Institute attended the competitive trial of steam-boilers at the fair of the American Institute, keeping the logs with commendable accuracy, and exhibiting a professional interest in the work. They have engaged in at least one test of the performance of a newly-designed steam-engine, have visited, among other places of interest, the Allen Engine-Works, the Chrome-Steel Works, the caissons of the East River bridge, the machine-department of the Brooklyn navy-yard, and the iron-clad Dictator.

The cost of tuition is fixed at a minimum figure, and, in special cases, is remitted entirely if the student, proving pecuniarily deficient, exhibits unusual attention to duties. The number admitted is, however, limited, and when the number of applicants capable of passing the preliminary examination exceeds this limit, the requisite number is obtained by selecting the most worthy.

Students who, after studying two years, exhibit special fondness for science, are

---

[*] A sketch of this boiler is given in the Journal of the Franklin Institute for September, 1871, and in London Engineering, January 5, 1872.

Stevens Institute of Technology.

PLATE V.

SECOND FLOOR.

No. 10 Chemistry, (Lecture Room.)
" 10' Cabinet of Chemical Apparatus.
" 11 Mathematics.
" 12 Mechanical Engineering.
" 13 Studio of Prof. of Engineering.
" 14 Cabinet of Optical Instruments.
" 15 Studio of Prof. of Physics.
" 16 Physics, (Lecture Room.)
" 17 Cabinet of Physical Apparatus.
" 18 Mechanics, (Lecture Room.)
" 19 Studio of Prof. of Mechanics

C Cases for Apparatus, &c.
E Evaporating Closet.
L Lifts or Elevators.
O H Oxygen and Hydrogen Outlets.
P & P' Passage Ways and Cloak Rooms.
S Sinks or Pneumatic Troughs
T Lecture Table.
V Ventilating Shafts.
W Wash Basins.

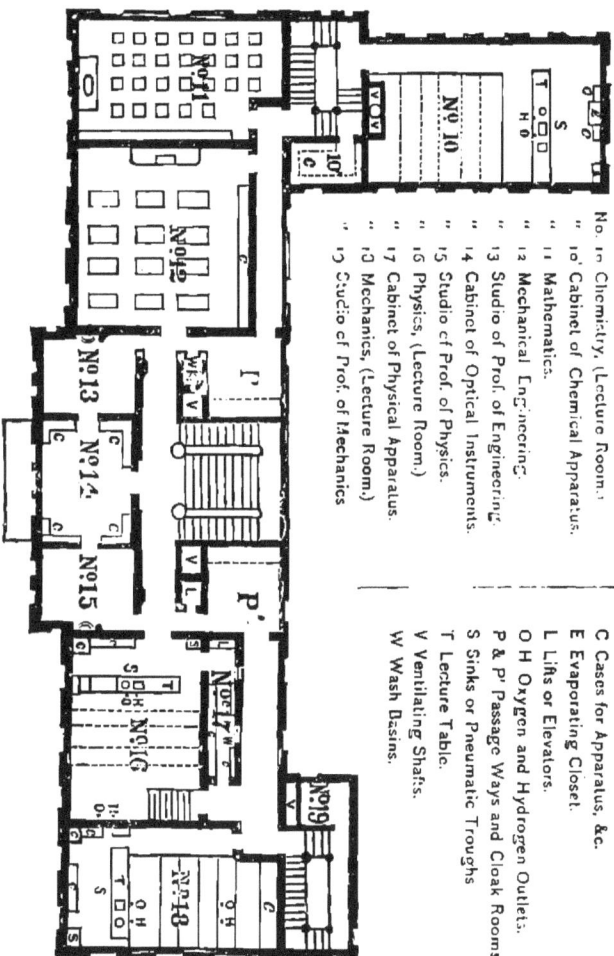

No. 11 No. 12 No. 10 No. 13 No. 14 No. 15 No. 16 No. 17 No. 18 No. 19

allowed to devote themselves to science during the remainder of the course, and are given, at graduation, the degree of doctor of philosophy, but the students are usually expected to take the course and to graduate as mechanical engineers, and students who have a special fondness for that branch are the class most desired.

Lectures on scientific subjects are delivered during the winter and spring in the great hall of the Institute.

THE PARDEE SCIENTIFIC DEPARTMENT OF LAFAYETTE COLLEGE, EASTON, PENNSYLVANIA.

Officers of instruction.—Rev. William C. Cattell, D. D., President and Professor of Mental and Moral Philosophy; Traill Green, M. D., LL.D., Adamson Professor of General and Applied Chemistry; James Henry Coffin, LL.D., Professor of Mathematics and Astronomy; Rev. John Leaman, A. M., M. D., Professor of Human Physiology and Anatomy; Rev. Lyman Coleman, D. D., Professor of Latin and of Biblical and Physical Geography; Rev. Thomas C. Porter, D. D., Professor of Botany and Zoology; Augustus A. Bloombergh, A. M., Professor of Modern Languages; Henry Francis Walling, C. E., Professor of Civil and Topographical Engineering; Frederick Prime, jr., A. M., Professor of Metallurgy and Mineralogy; E. Hubbard Barlow, A. M., Professor of Rhetoric, Elocution, and of Physical Culture; Rossiter W. Raymond, Ph. D., Lecturer on Mining Geology; Selden Jennings Coffin, A. M., Adjunct Professor of Mathematics; James W. Moore, A. M., M. D., Adjunct Professor of Mechanics and Experimental Philosophy; Edward S. Moffat, A. M., M. E., Lecturer in the Department of Mining; Justus M. Silliman, M. E., Adjunct Professor of Mining Engineering and Graphics; Theodore F. Tillinghast, C. E., Adjunct Professor of Civil Engineering; Charles McIntire, A. M., Assistant in Chemistry; John B. Grier, A. M., Tutor in Modern Languages; Joseph Johnston Hardy, A. B., Tutor in Mathematics; Alexander Hamilton Sherrerd, B. S., Assistant in Chemistry.

This list does not include those members of the college faculty of instruction who are exclusively occupied with the classical and literary courses.

History and plan.—This department was added to the classical course of the college, 1866, to carry into effect the conditions of a donation from Mr. A. Pardee, of Hazleton, Pennsylvania. In July, 1867, in response to the growing wants of the department, the original donation was increased to $200,000, on condition that other friends of the college should add the same sum to its general endowment. The donations for that purpose (completing more than half a million of dollars lately added to the college funds) having been completed in 1868, Mr. Pardee made an additional donation of $200,000 for the erection of a building designed for the departments of engineering, metallurgy, and chemistry. This building is now in course of erection, and will be finished in 1873. Meanwhile the technical studies of the department are carried on in the other buildings of the college, the laboratories of Jenks Hall and West College offering special facilities for applied chemistry and the metallurgical processes.

The situation of Lafayette College in the great manufacturing and mining region of the Middle States offers peculiar advantages for combining theoretical studies with actual practice. Every process used in the mining and working of the various ores of iron, and in the manufacture of iron into the thousand forms in which it is used, is going on almost within sight. Near by are the coal mines which supply the markets of Philadelphia and New York. Mineral wealth abounds on all sides. The expert is continually called on to examine new tracts of land, to analyze new ores, and to devise new ways of working and handling them. Here every resource of engineering is displayed in the works connected with the preparation and transport of lumber, and the carrying of railroads and canals through the mountains and over the rivers. Those who wish to prepare themselves to be working engineers in any of these departments, come from all parts of the country to observe and study these works, and it is most desirable that adequate means should be provided for the prosecution of scientific studies in the midst of them.

In addition, therefore, to the classical and the general scientific courses, which are designed to lay a substantial basis of knowledge and scholarly culture, courses of four years each have been arranged for studies essentially practical and technical, viz:

I. *Engineering, civil and mechanical.*—This course is designed to give professional preparation for the location, construction, and superintendence of railways, canals, and other public works; chemical works and pneumatic works; the design and construction of bridges; the trigonometrical and topographical survey of States, counties, &c., the survey of rivers, lakes, harbors, &c., and the direction of their improvement; the design, construction, and use of steam-engines and other motors, and of machines in general; and the construction of geometrical, topographical, and machine drawings.

II. *Mining engineering and metallurgy.*—This course offers the means of special preparation for exploring undeveloped mineral resources, and for taking charge of mining or metallurgical works. It includes instruction in engineering as connected with the survey, exploitation, and construction of mines, with the construction and adjustment of furnaces and machines, and with machine drawings; also, instruction in chemistry and assaying, as applied to the manipulation of minerals.

III. *Chemistry.*—This course includes text-book study, lectures, and laboratory practice, every facility for which is found in the laboratories of Jenks Chemical Hall. Particular attention is given to the chemistry of agriculture, medicine, metallurgy, and the manufacturing processes. Provision is made for students who may wish to make original researches, or to fit themselves to take charge of mines or manufactories, or to explore and work up the mineral resources of our own and other countries.

Resident graduates, and others having suitable preparation, may pursue the special studies of these departments in a post-graduate course, under the direction and instruction of the professors, and have the use of the laboratories, apparatus, collections, libraries, &c., while prosecuting researches in any department.

Students who have completed the studies in the full course of four years, and have passed satisfactory examinations upon the same, receive the degree usually given to graduates in these departments. Those who have gone over a part of the course, or who have pursued special studies to the satisfaction of the faculty, receive a certificate to that effect.

This department is still young, (though Lafayette is one of the old, established American colleges,) but its fine endowment, judicious plan, and excellent.location, will rapidly give it a high position among the schools of mining and metallurgy.

LEHIGH UNIVERSITY, BETHLEHEM, PENNSYLVANIA.

Officers of instruction.—Henry Coppee, LL.D., President and Professor of History and English Literature; Hiero B. Herr, esq., Professor of Mathematics and Astronomy; Major Lorenzo Lorain, U. S. A., Professor of Physics and Mechanics; Charles McMillan, C. E., Professor of Civil and Mechanical Engineering; William H. Chandler, Professor of Chemistry; Benjamin W. Frazier, A. M., Professor of Mining and Metallurgy; Richard P. Rothwell, C. E., Demonstrator of Mining and Metallurgy; Waldron Shapleigh, A. C., Instructor and Assistant in Chemistry; William A. Lamberton, A. M., Instructor in Latin and Greek; Frank Lourent Clerc, C. E., Instructor in Mathematics; S. Ringer, esq., Instructor in French and German; Spencer V. Rice, C. E., Instructor in Graphics and Field-work.

History.—During the year 1865 the Hon. Asa Packer, of Mauch Chunk, announced, unsolicited, to the bishop of the diocese, the Right Reverend William B. Stevens, D. D., LL.D., his intention to appropriate the sum of $500,000, and an eligible spot in South Bethlehem, containing fifty-six acres, (since enlarged by the donation of seven acres by Charles Brodhead, esq.,) for the purpose of founding an educational institution in the beautiful valley of the Lehigh, which should bear the name of the Lehigh University. The bishop was appointed president of the board of trustees.

The purpose of the founder in making this munificent endowment was to provide the means for imparting to young men of the valley, of the State, and of the country, a complete professional education, which should not only supply their general wants, but also fit them to take an immediate and active part in the practical and professional duties of the time. The system determined upon proposes to discard only what has been proved to be useless in the former systems, and to introduce those important branches which have been heretofore more or less neglected in what purports to be a liberal education, and especially those industrial pursuits which tend to develop the resources of the country—pursuits, the paramount claims and inter-relations of which natural science is daily displaying, such as engineering, civil, mechanical, and mining; chemistry, metallurgy, architecture and construction.

Courses of study are devoted by all regular students to the study of those elementary branches in which every young man should be instructed, for whatever profession or business in life he may be placed, viz, mathematics, languages, elementary physics, chemistry, drawing, history, rhetoric, logic, declamation, and composition.

At the end of two years, having acquired this necessary knowledge, the student, following the bent of his own mind, and aided by his parents and professors, will be ready to select some special professional course, to which all his studies and efforts will be directed. To enable him to do this there are several technical schools, which branch off from the end of the common course. In each, the term of study is two additional years, and the student, at his graduation in any one of them, receives a special degree. By this means a young man is relieved from the overpowering and confusing study of those branches for which he has no taste, and pursues with cheerfulness the special course which he has selected, and for which he is suited by inclination and intelligence. The students in the first two classes are called *First* and *Second Classmen*. Those in the schools are called *Junior* and *Senior Schoolmen*.

The schools at present provided for are: 1. General literature; 2. Civil engineering; 3. Mechanical engineering; 4. Mining and metallurgy; 5. Analytical chemistry.

In the studies of the school of mining and metallurgy are included mineralogy and geology; metallurgy, with the modes of extracting all metals from ores; the methods of mining for various ores, with special instructions as to iron, coal, zinc, lead, copper, gold, and silver. The aim will be to fit the student for immediate service in the rapidly developing mines of these metals in many parts of our country. The students in this school will be taken to the mines for ocular instruction. The graduate in this school will receive the degree of E. M., (engineer of mines.)

The course, after the first two years, is as follows: *For Junior Schoolmen.* Mathematics.—Differential and integral calculus; Mechanics.—Mathematical theory of motion, science of motion in general, statics, dynamics and equilibrium of bodies, inertia, statics of fluids; Moral Philosophy.—Whewell; Physics.—Optics and acoustics; Chemistry.—Work in laboratory, qualitative and quantitative analysis; Mining.—Modes of occurrence of the useful minerals, rules for research of mineral deposits, and examination of mining properties, boring artesian and oil wells, miners' tools, blasting, drilling, and coal-cutting machines, tunneling and sinking shafts, timbering and walling of tunnels and shafts, tubing of shafts, construction of dams, &c., methods of exploitation, open-air mining, hydraulic mining, ore-mining in veins, beds, and irregular deposits, coal-mining, salt-mining, examples from different mining districts, underground transportation, hoisting, (engines, ropes, cages, cars, safety-catches, man-engines, &c.,) visits to neighboring coal, iron, and zinc mines; Geodesy.—Use and adjustment of field-instruments, leveling, triangulation, topographical surveying, leveling with the barometer, mine surveying; General metallurgy.—Classification of metallurgical processes, furnaces, classification and modes of construction, natural and artificial refractory building-materials, manufacture of fire-bricks, crucibles, retorts, &c., nature of combustion, and conditions favorable to it, draught, natural and artificial, chimneys, fans, blowing-engines, &c., smoke-consuming processes, gas furnaces, Siemens' regenerating-furnace, calorific power of fuels, methods of computing quantity and intensity of heat, coal, lignite, peat, wood, manufacture of charcoal, coke, and patent fuel, drying peat and wood; Metallurgy of iron.—Physical and pyrochemical qualities of iron, description of iron-ores, preparation of ores for blast-furnaces, methods of working, influence of temperature and pressure of blast, form of furnace, &c., chemical reactions in the blast-furnace, gases, slags, &c., hot-blast stoves, hoists, charging-apparatus, &c., casting in iron, preparation of molds, remelting in cupola and reverberatory furnaces, manufacture of wrought iron from pig-iron, forges, puddling and reheating furnaces, hammers, rolling-mills, &c., manufacture of wrought iron directly from the ore, bloomeries, &c., manufacture of steel directly from the ore, manufacture of steel from cast iron in forges and puddling-furnaces, cementation, casting steel, Bessemer's process, &c.; Mineralogy.—Crystallography, exercises in drawing crystals and determining crystalline forms in models and minerals, descriptive mineralogy, exercises in determining minerals, practical instruction in the use of the blow-pipe, access to the mineralogical cabinet; French and German.—Throughout the year; Drawing.—Problems in descriptive geometry, crystals, plans and sections of mines and mining-machinery, furnaces, apparatus, and machinery of smelting-works, plans of trigonometrical surveys, contour maps, geological charts. The students in this department will be required to execute plans or projects for the establishment and working of mines and smelting-works, under given conditions, with drawings and written memoirs.

For senior schoolmen. Applied Mechanics.—Elasticity and strength of materials, including forms of uniform strength, stability of structures, theory of the arch, elementary machines, practical hydraulics, theory of trussed frames, general theory of machines; Christian Evidences.—Lectures; French and German.—Throughout the year; Drawing.—Plans, sections, and elevations of mines, furnaces, machinery, &c., as in the junior year; Astronomy.—Loomis's Manual, observatory; Mining.—Pumps and pumping-machinery, nature of gases found in mines, natural and artificial ventilation, furnaces, mechanical ventilators, distribution of air in mines, measurement of ventilation and work done by ventilators, lights, safety-lamps, precautionary measures, means of preventing and extinguishing fires, mechanical preparation of ores, stamps, mills, screens, jigging-machines, percussion-tables, &c., washing and dressing of coal, coal-breakers, &c.; Metallurgy.—Metallurgy of zinc, pyrochemical properties of zinc, ores of zinc, English, Belgian, and Silesian processes of extraction, manufacture of oxide, properties and ores of tin, preparation of ores, German and Cornish methods of extraction, properties and ores of copper, reduction of oxidized ores, Swedish, English, and mixed methods of treatment of sulphurous ores, methods of extraction from poor ores by the wet way, properties and ores of lead, American, Carinthian, English, and Belgian processes of extracting lead "by reaction," processes of reduction of roasted ores in the blast-furnaces, processes by precipitation, mixed processes, extraction of silver from lead, Pattinson's and Parkes's process, German and English cupellation, properties and ores of silver, American and European amalgamation, amalgamating-pans, smelting with lead, methods of extraction by wet way, properties and ores of gold, washing, amalgamation, smelting with lead, extraction by the wet way, separation from silver, metallurgy of platinum, aluminum, mercury, arsenic, antimony, bismuth, nickel, and cobalt; Chemistry.—Quantitative analysis, volumetric analysis of ores, dry assaying; Geology.—(Dana's Manual,) physical geography, lithology with practical exercises, stratigraphical and dynamic geology; Plans for the establishment of mining and smelting works, as in the junior year.

Fees and expenses.—Through the generosity of the founder, and by a resolution of the trustees, passed in July, 1871, tuition was declared to be *free* in all branches and classes. The following are the expenses approximately stated:

Board, (40 weeks, at about $5)	$200
Books	20
Washing	25
Total	245

Books, materials, paper, pencils, chemical materials used in the analytical laboratory, and instruments, are furnished by the student.

SCHOOL OF MINING AND PRACTICAL GEOLOGY OF HARVARD UNIVERSITY, CAMBRIDGE, MASSACHUSETTS.

Officers of instruction.—Charles W. Eliot, LL.D., president; Josiah D. Whitney, LL.D., dean, and professor of geology; Asa Gray, LL.D., professor of natural history; Henry L. Eustis, A. M., professor of engineering; Wolcott Gibbs, M. D., professor of physics; Joseph Winlock, A. M., professor of astronomy; Josiah P. Cooke, A. M., professor of chemistry; Charles F. Hoffmann, professor of topographical engineering; Raphael Pumpelly, professor of mining; William H. Pettee, A. M., assistant professor of mining; Nathaniel S. Shaler, S. B., professor of palæontology.

There are five students at present.

This school has for its object the instruction of students in practical geology, the art of mining, and kindred branches. The full course occupies four years, and on those who pass through it, and sustain the necessary examinations, the degree of mining engineer will be conferred.

The full course, prescribed for candidates for the degree of mining engineer, occupies four years, the first three of which are identical, as regards the subjects of instruction and the order thereof, with the first three years of the engineering course in the Lawrence Scientific School. The subjects of instruction during the fourth year of the course are as follows: economical geology and the phenomena of veins; mining-machinery and the exploitation of mines; general and practical metallurgy; assaying; working up, plotting, and writing out notes of summer excursions.

From time to time opportunities will be offered to the students, by excursions with the professors, of becoming practically acquainted with astronomical and geodetic work, as also with the method of making geological surveys, and with mining and metallurgical operations.

The terms of admission on examination are the same as those of the engineering course in the Lawrence Scientific School.

Graduates of colleges will be admitted without previous examination, and those who have taken the mathematical and scientific studies of the elective courses in Harvard College, or their equivalents in other institutions, should be able to enter at the commencement of the second year.

Persons properly qualified, and able to pass the necessary examination, will be admitted to any part of the course, at the beginning of any half year, but no later than the beginning of the second half of the third year.

Any person, however, who is not desirous of being considered as a candidate for a degree, may attend any special branch taught in the school, or any course of lectures, at his own pleasure, on paying such proportion of the fees for instruction as may be fixed by the professor to whose department he desires to be attached.

The tuition-fee for the academic year is $150; for half or any smaller fraction of a year, $75; for any fraction of a year greater than one-half, the fee of the whole year is charged. The other expenses of a student for an academic year may be estimated as follows: Room, from $50 to $100; board for thirty-eight weeks, from $152 to $304; books, from $20 to $25; fuel and lights, from $15 to $35; washing, from $10 to $38; total, from $236 to $502.

RENSSELAER POLYTECHNIC INSTITUTE, TROY,* NEW YORK.

Officers of instruction.—Hon. James Forsyth, president; Charles Drowne, C. E., A. M., director, and professor of theoretical and practical mechanics; James Hall, LL.D.,

* I have included this institution in the present chapter, partly on account of its age and past reputation as a school for civil engineers, and partly because I find in its last catalogue the names of seven students "in mining engineering," and the statement that metallurgy and an increased amount of geology are to be, hereafter, parts of the regular civil-engineering course. But the present list of instructors shows no adequate preparation for the fulfillment of such a pledge. Mining and metallurgy cannot be properly taught by throwing the work upon professors in other departments, already overburdened with their own specialties; and the Rensselaer Institute, by discontinuing its only professorship in mining and metallurgy, has practically (I trust only temporarily) retired from the ranks of the schools of those professions. The fact is to be much regretted.

New York State palæontologist, professor of theoretical, practical, and mining geology; Dascom Greene, C. E., professor of mathematics and astronomy; S. Edward Warren, C. E., Professor of descriptive geometry and stereotomy; Henry B. Nason, A. M., Ph. D., professor of chemistry and natural science; Charles McMillan, C. E., professor of Geodesy, road engineering, and topographical drawing; R. Halsted Ward, A. M., M. D., professor of botany; J. H. C. Lajoie De Marceleau, A. B., professor of French; Alexander G. Johnson, instructor in the English language and literature; Arthur W. Bower, C. E., instructor in mathematics and mechanics; Edward Nichols, B. S., assistant in analytical chemistry.

THE UNIVERSITY OF PENNSYLVANIA, PHILADELPHIA, PENNSYLVANIA.

In the scientific department of this institution it is intended to provide complete theoretical and practical training in mining and metallurgy. As the courses are not yet organized, the professors not appointed, and the handsome building not completed, little can be said of the details of this plan. The following is a description of the plans of the new building of the collegiate and scientific departments, by T. W. Richards, architect:

The building has a front on Locust street (between Thirty-fourth and Thirty-sixth streets) of 254 feet, by 102 feet 4 inches in depth, exclusive of towers, bay windows, buttresses, &c., with an additional projection of the center 21 feet 10 inches beyond the wings.

The cellar is arranged for the storage of coal, and an apartment in connection, outside of the building, is provided for the boilers of the steam-heating apparatus.

The basement on the sides and rear is entirely above ground, 15 feet high. There is an entrance in the rear for students to the assembly-room, 44 feet by 50 feet, and entrances on the east and west ends to a wide corridor, which extends the whole length of the building, in all the stories. The eastern wing contains: Laboratory, 30 by 45 feet, apparatus and store-room, 24 by 29 feet, metallurgical laboratory, 30 by 50 feet, a fire-proof furnace-room, 24 by 34 feet, balance-room, 14 by 17 feet, as well as smaller rooms for silver and gold assaying. The western wing contains: Laboratory, 30 by 45 feet, and apparatus and diagram room, 24 by 29 feet, for the chemical-lecture room on first floor, one laboratory, 30 by 50 feet, and one, 24 by 34 feet, for the physical department. Apartments for janitor and assistants are arranged on this floor, and for machinery, storage, dumb-waiters, water-closets, &c.

The first or principal floor is 16 feet high. On the eastern side of the main entrance is the faculty-room of the scientific department, 13 by 22 feet, professors' laboratory, 19 by 45 feet, preparing laboratory, 21 by 24 feet, qualitative laboratory, 30 by 45 feet, quantitative laboratory, 30 by 50 feet, laboratory for organic analysis, 24 by 34 feet, two balance-rooms and two assistants' rooms. On the western side is the reception and secretary's room, 13 by 22 feet, trustees' and faculty room, 19 by 37 feet, provost's recitation-room, 24 by 33 feet, and private room, 14 by 18 feet, chemical-lecture room, 30 by 45 feet, physical-lecture room, 30 by 50 feet, and apparatus-room, 24 by 34 feet. The library in center of rear, 44 by 50 feet, is entered from a hall 34 by 40 feet. This part of the building is fire-proof.

The second floor is 15 feet 6 inches high. The chapel, 50 by 80 feet, occupies the front of center building, and is 23 feet high. The eastern side contains lecture-rooms for civil engineering, mining and metallurgy, and mineralogy, and a large museum for these departments.

The western side and center of rear is divided into six large recitation-rooms, with adjoining private rooms.

The third floor is 14 feet high, and contains three large recitation-rooms, lecture and model rooms for mechanical engineering, three large rooms for the study and practice of drawing in the departments of civil and mechanical engineering, architecture, &c. A large examination-hall is in the rear of the chapel.

The fourth floor, over the chapel, has two society-rooms for students, each with an adjoining library.

The design is in the collegiate gothic style; the material to be used is Lieperville stone, for the basement, with base course of Hummelstown brownstone. The walls above are to be serpentine marble, with cornices, gables, arches, &c., of Ohio stone. The entrance-porch is to be of Franklin stone, with arch supported on polished red-granite columns, with enriched capitals of Ohio stone. The windows of chapel and gables are decorated with geometrical tracery.

The space devoted to the sciences more immediately connected with mining and metallurgy is one-half of the building. The chemical departments were planned by the late Prof. Wetherell, of Bethlehem.

MISSOURI SCHOOL OF MINES AND METALLURGY ROLLA, MISSOURI.

The legislative act " to locate and dispose of the congressional land-grant of July 2, 1862, to endow, support, and maintain a school of agriculture and the mechanic arts, and a school of mines and metallurgy, and to promote the liberal education of the industrial classes in the several pursuits and professions of life," provided that three-fourths of the proceeds from the sale or lease of the 330,000 acres of land granted to Missouri should be for the benefit of an agricultural and mechanical college, and the remaining one-fourth for the support of a school of mines and metallurgy. It also provided that the latter should be located in the mineral district of Southeast Missouri, and that the county therein, having mines, which might offer the highest bid of money and land for building and other purposes, should secure the location, provided it should offer not less than $20,000 in cash, and 20 acres of land for a building-site, in addition to suitable lots of mineral lands "for practical and experimental mining." The same act empowered the county courts of such county to issue bonds to raise the money and purchase the lands, providing that said bonds should not run longer than twenty years, and should bear not more than 10 per cent. interest annually, payable half yearly.

Phelps County and its county-seat, Rolla, having offered $75,000 in twenty-year 10 per cent. bonds, together with 7,879 acres of land, including a building-site of 130 acres, the locating committee of the board of curators of the State University reported in favor of locating the school at that point. At the regular meeting of the board of curators, in June, 1871, the report of the committee was adopted, and the school of mines and metallurgy was then definitely located at Rolla.

The plans and specifications have been prepared, and the contract let for the erection of a building 130 feet by 65 feet, three stories high, at a cost of $85,000, exclusive of the completion of the third story. The first floor contains three working laboratories, besides balance and apparatus rooms, and private laboratories for professors and assistants. The second floor is devoted to the large lecture-room, library, and recitation-rooms, while the third floor, when completed, is designed to contain additional class-rooms, the collections and cabinets, and a large well-lighted room for drawing purposes. In the mean time the school occupies ample and comfortable quarters in a new public-school building, recently completed, at a cost of $30,000. In this have been fitted up a laboratory, (24 by 55 feet,) lecture-room, cabinet and apparatus and recitation rooms.

The school was formally opened on November 23, 1871, the first session having begun on the 6th of the same month. The regular course is designed to extend through three years, though a preparatory year's course is provided for students not sufficiently advanced for admission to the school proper. The collegiate year begins in September, and ends in June following, and is divided into two sessions, with no intermediate vacation. The first semester of each collegiate year ends early in February. Technical excursions and field-practice are designed to fill up most of the vacation. The laboratory is accessible to students of the regular or special departments at all working hours of the day, not otherwise employed, and at least fifteen recitations, or their equivalents, exclusive of laboratory and drawing-room work and military drill are exacted for each week from each student. French and German are optional studies.

In the preparatory year are taught arithmetic, metrical system of weights and measures, algebra to quadratic equations, rhetoric, and English composition, physical geography, (two lectures weekly,) and the elements of physics and chemistry, (by recitation from text-books and illustrative experiments.)

The first year proper: Algebra, from quadratic equations; geometry and trigonometry, mineralogy, (descriptive,) laboratory work, blow-pipe analysis, and determinative mineralogy; drawing.

Second and third years: Pure and applied mathematics, organic chemistry and parts of chemical technology, geology and mining, metallurgy and assaying, physics, drawing, (free-hand and mechanical;) laboratory work, qualitative and quantitative analysis.

In accordance with the provisions of the congressional act, a military bias is given to the organization of the school. Special students are admitted.

Tuition: Forty dollars per year; chemical apparatus furnished at cost prices by the school, and the value of so much of it as may be returned is refunded,

Total number of students, 31.

In preparatory classes 20—average age, 18.2 years; in first year, regular, 8—average age, 20 years; in analytical chemistry and assaying, 3.

The faculty is not yet organized, but will be, partially, during the coming summer. Instruction to the first-year students is at present given by the director of the school, Professor C. P. Williams, assisted by Mr. N. W. Allen, instructor in mathematics, and Mr. William Cooch, as laboratory assistant.

POLYTECHNIC DEPARTMENT OF THE WASHINGTON UNIVERSITY, SAINT LOUIS, MISSOURI.

Officers of instruction.—W. G. Eliot, D. D., chancellor; Abram Litton, M. D., Eliot professor of chemistry; —— ——, Wayman Crow professor of physics; George H. Stone, A. M., professor of rhetoric; Calvin M. Woodward, A. M., Thayer professor of mathematics and applied mechanics, and dean of the department; Marshall S. Snow,

A. M., professor of belles-lettres; Leopold Noa, professor of ancient and modern languages; Henry Pomeroy, A. M., professor of astronomy and mathematics; William Eimbeck, U. S. C. S., professor of practical astronomy; William B. Potter, A. M., E. M., Allen professor of mining and metallurgy; F. William Raeder, S. B., professor of architecture; Denham Arnold, A. M., assistant professor of physics; Charles A. Smith, C. E., assistant professor of civil and mechanical engineering; Frederick M. Crandon, A. B., instructor in mathematics and elocution; J. W. Pattison, teacher of drawing.

Summary of students.—Seniors, 3; juniors, 6; sophomores, 10; freshmen, 13; students not candidates for degree, 4; total, 36.

Courses of study.—These are five in number, viz: I. Civil engineering; II. Mechanical engineering; III. Chemistry; IV. Mining and metallurgy; V. Building and architecture.

The studies are the same for all the courses during the freshman and sophomore years, but during the junior and senior years they diverge more or less, though certain branches still remain common. Students not proposing to become professionals are not required to adhere strictly to either course, but, with the approval of the faculty, may select such studies as will constitute a general course, the completion of which will entitle the student to the degree of bachelor of science. Special students will be received in any of the courses, if it is made clear that such arrangements are the best for the students, and not prejudicial to the interests of the department.

The course in mining and metallurgy.—This was established during the past year by the appointment of Professor Potter. It is evident that Saint Louis possesses great advantages for instruction in these branches, being a large and growing commercial and manufacturing center, within easy access of nearly all varieties of mining and metallurgical operations.

The studies during the first two years are somewhat general in character, preparatory to the special work of the course on mining and metallurgy, to which the remaining two years are devoted. The full details are omitted here, as they closely resemble the schedules of other institutions already quoted. The plan of instruction includes lectures and recitations on the various subjects pertaining to the course; practical work in the physical, chemical, and metallurgical laboratories; field-work in geology, &c.; projects, estimates, and plans for the establishment of mines and metallurgical works; examination of and reports on mines and manufacturing establishments.

Collections have already been made and are constantly being added to, embracing models of crystals and specimens illustrating the various minerals and rocks and their association; ores, coals, petroleum, fire-clays, building-materials, &c., from many parts of this country and Europe; characteristic fossils of the different geological ages; metallurgical products illustrating the various operations in the treatment of ores by the wet and dry methods. Models of furnaces and mine constructions will in time be added, together with sets of mining-tools and instruments. These collections are used to illustrate lectures, &c., and are at all times accessible to the students, so that they may become thoroughly familiar with the character and modes of occurrence of the minerals, rocks, and ores they are likely to meet with in the field, and the various products in metallurgical operations.

Assay-laboratories will, before the opening of the next term, be completely furnished with crucible, scorification, and cupelling furnaces, and everything necessary for practical work in the assay of ores of lead, silver, gold, iron, tin, &c., and with volumetric apparatus for the assay of silver coin and bullion by the wet methods. The general principles as well as the special methods of assaying are explained in the lecture-rooms, and at the same time ores of the various metals exhibited and described. From a large stock of these ores from various parts of the country the students are required to make a large number of assays themselves, under the immediate supervision of the instructor. In the chemical laboratories a practical course is pursued in connection with lectures on qualitative and quantitative analysis, the students being required to make tests and full analyses of coals, limestones, ores of iron, copper, lead, zinc, nickel, pig-iron, clays, technical products, &c., that they may acquire a practical experience in the chemical examination of the materials and products liable to be met with in practice.

Every opportunity is afforded the students through the term for visiting and examining the various mines, smelting and manufacturing establishments in the vicinity. During the summer vacations they are required to visit some mining or metallurgical district, and, at the opening of the following term, to hand in a journal of travels, with a report of the operations conducted there, illustrated with drawings. Before receiving the degree of engineer of mines, they will be required to execute plans or projects for the establishment and working of mines or smelting-works under given conditions, with drawings, estimates, and written memoirs. An endeavor is thus made to combine thorough practical with theoretical instruction in this course, and to fit the student for the successful practice of his profession hereafter, and for a field of usefulness in the country at large.

The chemical building above mentioned contains three work-rooms, besides a lecture-room, the professors' room, and two rooms for storage and apparatus. Besides, two large rooms in the basement of the new wing will soon be fitted up for assaying and industrial chemistry. Until the present accommodations are crowded, the large room, 43 by 41 feet, on the first floor of the large building, will be appropriated to the State geological cabinet. Students who propose to become professional chemists will spend almost their entire time during their third and fourth years in the laboratories. This institution is the headquarters of the Missouri State geological survey.

CHAPTER XX.

THE BURLEIGH DRILL.

Since the preceding pages contain (see chapter on California) a tolerably full account of the late operations of the diamond drill, it is but fair that some attention should be given in this report to its principal competitor, the Burleigh, now the most prominent representative* of the percussion machine-drills. In point of fact the Burleigh drill has never been successfully pushed in the West. One of them is about to be introduced at the Yellow Jacket mine, and another has been, it is said, contracted for by the Sutro Tunnel Company. But their actual use is confined at present to quarrying and tunneling operations in the East, and in mining, so far as I know, to the Lake Superior region.

I give, as the best statement of the performances of the machine, the following extracts from a letter addressed to me at my request by Mr. H. A. Willis, of Fitchburgh, Massachusetts, the treasurer of the Burleigh Rock-Drill Company:

We have not much to add to what has heretofore been published in regard to our machinery, except to state that time has fully proved the value of the machinery, and its economy over hand-work wherever any considerable amount of rock is to be removed.

The drills have been introduced into nearly every State of the Union, and are at work in Canada and South America. They are also largely used in England, being manufactured at Manchester. They are just about being started in a tunnel in Italy, and we are about shipping three of our largest compressors there to run them, as they have not yet made many compressors at Manchester, and are in immediate want of them. The Lake Superior copper mines are using them largely, the Copper Falls, Allouez, and Central Companies being already equipped, and the Calumet and Hecla Company awaiting at present our completion of their order for eighteen drills. These companies expect to do all their drilling by machinery.

We inclose a reprint of letter to London Mining Journal from Lake Superior, giving comparisions of hand and machine labor; also, a copy of Mr. Steele's article on Neaquehoning tunnel; also, letters of Messrs. Shanly & Co., showing the value of the machinery at Hoosac tunnel. With these data you will be able to say something about the machinery. You can use either or all of the cuts now in your possession to illustrate with, if desired. I would suggest that the cut you had engraved for us is the best representation of the drill as at present constructed, it having been entirely built over since your former report was made, and so strengthened that no important part of the machine ever breaks. Of course, the ratchets and springs wear out.

Yours truly,

H. A. WILLIS, *Treasurer.*

HOOSAC-TUNNEL CONTRACT,
North Adams, Massachusetts, February 23, 1872.

DEAR SIR: I do not find a copy of the letter of 2d April, 1870, as published on page 19 of the Burleigh Rock-Drill Company's pamphlets, and which was written for our firm by my brother. I can, however, testify to the correctness of the facts stated in that letter. The compressors we have been using ("No. 2") have been doing their work very well, "driving from two to three *tunnel-size* machines" each. As respects the difference in rapidity and cost between drilling in rock with these machines and by hand, we could not say without going into figures what it may actually be; but this we can say, that without the "Burleigh drills" we would not undertake such a work as the Hoosac tunnel on almost any terms.

W. SHANLY,
For F. Shanly & Co.

Hon. GEO. E. TOWNE,
President Burleigh Rock-Drill Company, Fitchburgh.

---

* The Gardner drill, a machine resembling the Burleigh, is in satisfactory use at various places in the East. I cannot enter into the merits of the controversy between the proprietors of these two machines as to the patent-right.

HOOSAC-TUNNEL CONTRACT,
North Adams, Massachusetts, February 26, 1872.

GENTLEMEN: In reply to your letter of the 20th instant, asking some particulars of our experience in the use of the Burleigh rock-drill, the machines we have been using are those known as the "tunnel-drill," having, in the terms of your inquiry, "their operating-end made as axial continuation of the piston-rods." We have some sixty of these machines in service, and they have given great satisfaction, working under an atmospheric pressure of from 55 to 60 pounds on the square inch, and making upward of 200 strokes per minute. We estimate the saving in expense, as compared with hand-drilling, at about 33¼ per cent., and in point of time there is a gain of fully 50 per cent.; in other words, effecting a saving of at least five years in the finishing of the Hoosac tunnel.

Yours, respectfully,

F. SHANLY & CO.

Messrs. CROSBY & GOULD, *&c., &c.,*
Boston, Massachusetts.

[From the London Morning Journal, November, 1871.]

SIR: Regarding the introduction of drilling-machinery into mines as a very important subject, and as I happen to be familiar with the results obtained from the working of the Burleigh drill, on Lake Superior—where, by the way, it is by no means common—perhaps you will allow me some space for the accompanying remarks.

Doering's machine was tried in Tincroft and in Dolcoath mines, Cornwall, and thrown out, I believe, because it would not pay. I was never fortunate enough to learn the results obtained from working it; but it seems to me that somebody ought to have been sufficiently interested in this machine to find out what work it did, as well as what work it could do, and make it public. I saw a statement made that the machine drifted a given number of feet more in a month than six good miners could do; but, as its use has been discontinued, I infer that it cost more to break the ground than by hand-labor. I was underground in this country with Mr. Nobel, when he was making efforts to introduce nitro-glycerine; he, of course, was praising the compound, and remarking on the success attending his endeavors to get it into use; "but," said he, "I could not succeed in Cornwall—they are prejudiced there against everything new." I felt my "Cornish" get up, at the time, and was inclined to dispute the assertion made, but, on reflection, it seems to me that there is a deal of truth in what was said. I believe the putting in of the man-engine at Tresavean mine was due as much to the efforts of the Polytechnic Society as to those running the mine. One of the deepest and best-managed mines in the Camborne district was a long time seeing the propriety of using skips, and how many now stick to the kibble! Ten years ago the wheelbarrow was as common as the tram-wagon. I have yet to learn that it is gone out of fashion. It is only of late that any attention has been given to increasing the stamping duty in mines; and when Messrs. Harvey & Co. set up and tried the pneumatic stamps, in their very laudable efforts to reduce the cost of stamping, if I remember aright the tenor of the remarks made by the "astute" manager of a very rich tin mine was to the effect that "we will let somebody else try them, and in that way learn if they are a success."

There is a difference in starting a drilling-machine in a mine, with the authorities interested in, or indifferent to, its success; the men commonly regard an innovation with disfavor; and I would defy any inventor to succeed in working a machine by Cornish miners if they considered it was against their interest that it should succeed, unless he personally supervised it, or had a competent person in his interest to do so. Cornishmen are good miners, and good mine managers—they ought to be—but they are just as apt as others to conclude that what they do not know is not worth knowing.

I am not going to draw the inference from the foregoing that the Doering machine did not get a fair show, nor would I for a moment suppose that the authorities in the mines where it was tried had prejudged it; even if they had, they would exert no undue influence against it. Still, if they were not in favor of it, I would certainly venture the opinion that the Doering machine did not do its very best. I am ready, however, to drop the Doering as a failure, and will try to tell you what I know of the Burleigh drill.

The first machine of the kind brought into the copper region of Lake Superior was tried at the Pewabic five years ago. The Red Jacket mine used one for a short time just afterward in sinking a perpendicular shaft from surface. The motive-power applied was steam in both instances. I cannot conceive that a hot-drilling machine could be a success. The next trial—and the first with air-compressors—was made at the Aztec mine, Ontonagon County; this was a disgraceful failure.

The Central Mining Company next procured a Burleigh, about two years ago, to work in an incline shaft which they have been sinking for several years. The said shaft is being sunk in the country 14 by 8 feet, at an angle of 30° from horizontal; this machine is still at work. In last year's report of the mine the mining captain stated that by the use of the drill they had increased the rate of sinking 50 per cent. This was the first machine of the kind I saw at work; and it very forcibly struck me that the machine could drill more ground in an hour than three of the best miners could in a day. After that at the Central mine had been working some months the Copper Falls Company decided on trying one on what they term the Ashbed, a lode of amygdaloidal character, varying in width from 7 to 10 feet, and dipping at an angle of 26° from horizontal. The lode is known here as a "stamps lode;" the proportion of copper contained therein is about 1 per cent. of mineral, or ₁/₁₀ per cent. of ingot-copper. The copper varies in size from the finest particles to pieces of 1 pound weight; rarely larger. The lode forms an integral part of the formation; the over and under-lying belts of trap protrude irregularly into it, consequently there is no regular or de-fined foot or hanging wall. Another feature is the almost entire absence of "slips," or "breast-heads." The ground cannot be called hard, but is "short" to "break," re-quiring more than ordinary care in planning holes. Four good men can drive from 18 to 23 feet per month in an ordinary-sized level; the same number can stope from 10 to 12 fathoms in the same time. For the past two years, instead of letting the stope to the miner per fathom, he has been paid so much per foot to drill holes, under the direc-tion of a competent person. A more trying place for a drilling-machine cannot be found, the inclination of the lode being a serious disadvantage in carrying a wide breast on a level. After getting fairly under way, it was found that three men and one boy in a shift, or six men and two boys with the machine, could drift from 40 to 44 feet per month, carrying a breast 18 by 8 feet; this was doing the work of 16 men, but at no reduction of cost. It was then decided to try what could be effected by stoping; and after a carriage was constructed for the purpose, work was commenced. The car-riage and machine weigh about 1½ tons. To move them up over the foot-wall a pair of common blocks and a small crab-winch are used. The mode of working is to set the carriage in the level, and commence cutting in for a stope, which is carried toward the bottom of the level over the stope worked out, lower the carriage down, and com-mence another. In working this way less drilling is performed with the machine, be-cause more time is occupied in moving it; but it pays best. Early this summer three drills were started, two No. 1 compressors supplying motive-power; these last cannot be relied on to do good duty without hinderances; very commonly the pressure of air being insufficient to work with. To obviate this, a No. 3 compressor has been set up and was started two weeks ago. This gives ample air to run three, or even four drills, going from 60 to 70 revolutions per minute. The gauge shows a pressure of from 45 to 55 pounds per inch, varying, of course, with the number of drills running at the time. Since starting this an increase of duty has been effected, as well as a material saving in fuel.

I have been fortunate enough to obtain the results of last month's running with the three drills now in use; these figures may be taken as the result of running three ma-chines, with two No. 1 compressors supplying air:

Number of machine.	Number of party.	Shifts worked.	Days worked.	Holes drilled, each shift.	Holes drilled, each machine.	Number of feet drilled, each shift.	Number of feet drilled, each machine.	Holes, per shift.	Holes, per day.	Feet, per shift.	Feet, per machine, per day.
1	1	21	......	186	......	982.2	....	8.85	......	46.76	......
1	4	21	21	188	374	961.1	1,943.3	9.0	17.85	45.76	92.52
3	2	21.5	......	175	......	905.9	....	8.13	......	42.13	......
4	5	23	22.25	184	350	930.8	1,842.7	8.0	16.13	40.73	82.86
3	3	21	......	189	......	868.9	....	8.57	......	41.28	......
3	0	21	21	183	363	843.1	1,711.1	8.71	17.28	40.14	81.42

No. 1 machine is the improved tunnel-drill; No. 2, the small machine, as constructed five years ago; No. 3 is same as No. 1, but worked irregularly, frequent stoppages being necessary to blast. The timing an average day's work with No. 1 machine be-fore and after starting the new compressor gave the following figures, (time is given in minutes.) Men leave the "dry" at 7 o'clock; quit work at 6 o'clock:

	Two No. 1 compressors working.	No. 3 compressors working.
Men going to and returning from work	15	30
Moving carriage	69	94
Shifting, elevating, and fastening machine	126	128
Cutting collars for holes	38	23
Changing drills, 14 and 11 times respectively	55	31
Dinner-time	13	14
Blasting	55	67
Compressor idle	11	21
Drilling-time	278	252
Number of holes drilled	10	11
Number of feet drilled	45.9	55
Fastest drilling rate per minute, in inches	3.26	3.86
Slowest drilling rate per minute, in inches	94	1.54
Average drilling rate per minute, in inches	1.93	2.66

The diameter of holes varies from 2 inches to 2¼ inches, none less than 2 inches. The heaviest day's work, or rather the heaviest shift's work, performed so far has been the drilling of 13 holes, or 64 feet of ground. Some shifts, when the machine is employed in drilling "dry holes" in the back, only about half that amount of work is performed. Copper commonly offers a serious impediment to the drill; but for this it would be easy to drill 60 feet per shift on an average. In the day's work given above one hole required 67 minutes of drilling-time to sink it 5.5 feet deep, when, but for the presence of copper, the same work could have been done in 22 minutes. The rock broken in the mine last month was at the fourth level, by hand-drilling exclusively, at the fifth level by Nos. 1 and 2 machines, at the sixth level by No. 3 machine and hand-labor combined. The rock from each level is carefully reckoned; that from fourth level amounted to 1,035 tons, from fifth level 1,941 tons. This is sufficient to show comparisons regarding cost, which, at fourth level, was as follows:

Drilling holes, 3,035.7 feet, at 26 cents	$789 28	
Man in charge	65 00	
		$854 28
Supply—Candles, 16 pounds, at 20 cents	3 20	
Powder, 46 kegs, at $4	184 00	
Fuse, 2,850 feet, at $10	28 50	
Powder-cans, three, at 50 cents	1 50	
		217 20
Cost of breaking 1,035 tons of rock, at $1.035 per ton		1,071 48

The cost of running Nos. 1 and 2 machines at fifth level was as follows:

Four foremen, with machines, at $65	$260 00	
Four engineers, with machines, at $60	240 00	
Four assistant miners, with machines, at $55	220 00	
Three boys, carrying water, tamping, &c., at $21	63 00	
		$783 00
Supply—270 pounds candles, at 20 cents	54 00	
102 kegs powder, at $4	408 00	
4,550 feet fuse, at $10	45 00	
Two powder-cans	1 00	
		508 50
Fuel for compressors, 45 tons coal, at $8	360 00	
Engineers for compressors, two at $45	90 00	
Oil, &c., (say)	50 00	
	500 00	
Deduct ⅓ for sixth-level machine	166 66	
		333 34
Cost of breaking 1,940 tons of rock, at 83.7 cents		1,624 84

There is nothing charged for repairs, which for the month were trifling, and could be covered for a cent per ton. This answers the question whether the Burleigh drill will pay or not; and I have no hesitation in saying that better figures than these can be attained. These two machines broke, with twelve men and three boys, as much rock as could be obtained from thirty good miners. Better work can be done in a shaft where the ground is moderately hard, because a great deal more working time can be got out of the machine. Very much depends on the facilities for handling the machine; and it will require thought, experience, and time to decide what appliances are best. The mechanic puts into the miner's hands a machine that will drill 2-inch or 3-inch holes in diameter, from 40 to 60 feet in the shift, and he ought surely to have brains enough to handle that power to the best advantage. There surely can be no reason why a charge of powder in a machine-drilled hole cannot break the same amount of rock as if exploded in a hole drilled by hand-labor. Going back to the time when the United mines, Gwennap, were at work, I remember that over £100 per fathom was paid to sixteen men for cross-cutting toward the "hot lode," when, but for the excessive heat, £10 would have been a good price. What would have been the value of cold-compressed air and the Burleigh drill there? How many deep and hot engine-shafts are now being sunk, where the rate of sinking is nearer 6 feet than 12 feet per month, and where the sinking could be doubled, or even quadrupled, by using a drill-ing-machine?

I am not writing in the interest of the manufacturer, who, by the way, could improve the machines by putting in better material, but simply as one who firmly believes that machinery will, in less than ten years, very generally supersede hand-labor in mines.

MINER.

KEWEENAW COUNTY, *Michigan, October 9.*

To these statements I add the following, from a paper read before the American Society of Civil Engineers, on the Nesquehoning tunnel, in Pennsylvania, in the construction of which the Burleigh drill was employed. The paper contains valuable statistics of its economy:

Nesquehoning tunnel, in Carbon County, Pennsylvania, is a work of the Lehigh Coal and Navigation Company. It pierces Locust Mountain, and will connect their railroad in Nesquehoning Valley with their extensive coal operations in the valley of Panther Creek. At present this coal finds its way to market by that interesting system of inclined planes and gravity-roads known as the "Switch-backs of Mauch Chunk," which has commanded the admiration of travelers for more than forty years, not only on account of the beautiful scenery which the route displays, but also from its early and admirable adaptation to the purpose for which it was designed. It has, however, become worked up to its capacity, and in arranging to extend their coal-mining operations, the company have wisely determined to avail themselves of the locomotive, which has had its practical development since they were the pioneers in railway enterprise.

It passes through the base of the mountain at an elevation of some 15 feet above the water on either side, and 564 feet below the crest, and cuts the strata at right angles, where they have a south dip of about 45°. Its length is 3,800 feet, of which 1,300 feet are through the coal-measures, with all their various strata of coal, coal-shale, sandstone, and conglomerate; 1,200 feet through the conglomerate formation, with its occasional strata of coal-slates and sandstone; 1,000 feet through the red shale, with occasional strata of sandstone, and 300 feet at the north end through the *débris*, and soft and decomposed red shale which is found overlying the red shale formation. It has encountered in its progress as hard and as soft material as is often met with in tunneling.

After mature investigation it was determined to use the Burleigh drills, driven by compressed air. With the advantage of the experience at Mont Cenis and Hoosac before us, we should, and it is believed we have, obtained better results, as to cost and progress, than attended either of those works in their early stages, and I may here state that I believe no other known process is capable of penetrating this conglomerate formation with that economy and rapidity which are necessary to meet the present demands of capital. This whole work has been done with 6 of the "two-drill" compressors, made at Fitchburgh, Massachusetts, and with sixteen-drill engines, and we have averaged as much as one-half of the drill-engines constantly in operation, and sometimes two-thirds.

The explosive used was gunpowder, ignited by the electic spark; but the requirements of ventilation and the hardness of the rock demanded powder of the highest Government standard. Some doubts which existed as to the economy in the use of the more powerful explosives, when the cost of drilling was reduced by machinery, and their greater danger, with the existing knowledge of workmen of their use, caused them to be rejected, and the result, in the freedom from serious accident, has been satisfactory, as we have not, thus far, lost a life from premature explosions.

American steel has been used. Several of our own makers produce a better and cheaper article for the purpose than can be obtained from abroad, and the best we have had is from the William Butcher Steel-Works at Philadelphia.

The headings are driven at the bottom, 8 feet high by 16 feet wide, and where arching is required, the full width for a double track is taken out, that the tunnel may hereafter be enlarged without disturbing the arches. At this date both headings are in the red shale and about 500 feet apart; they will be joined in August, and, until the tunnel is finished, full details of the work cannot be given; but the accompanying statement of Thomas C. Steele, chief assistant engineer, of the operations at the south end, up to June 1, may be of some interest.

The heading to which the tabular statement refers has been twelve months in the conglomerate, and two months in the red shale; the progress in the conglomerate has been about 100 feet per full month's work, and in the red shale 160 feet. The holes drilled per cubic yard of rock removed have been in the conglomerate about 11 feet, and in the red shale about 6¼ feet. The powder used per cubic yard has been in the conglomerate about 6 pounds, and in the red shale about 3½ pounds, though a bad lot of powder ran the consumption in the conglomerate up to 7¼ pounds for two months.

The operation in the enlargement to which the statement refers has been eight months in the coal-measures, and two months in the conglomerate; its average monthly progress has been 166 feet; its average holes drilled per cubic yard of rock removed, 3⁷⁄₁₀ feet; and its average powder used 2⁷⁄₁₀ pounds per cubic yard.

In this enlargement a portion of hand-drilling is included, which extended over the operations of one month, and it increased both the holes drilled and the powder consumed, showing that men do not use better judgment in directing hand than machine drilling.

Statement of the workings of the south end of Nesquehoning tunnel.

HEADING.

Month.	Feet of holes drilled.	Pounds of powder used.	Feet of progress.	Cubic yards of progress.	Feet of holes per cubic yard.	Pounds of powder per cubic yard.
1870.						
April	2,740	1,123	73	413	6.6	2.7
May	6,698	2,725	139	781	8.6	3.5
June	6,779	4,200	104	590	11.5	7.1¹
July	5,862	4,275	97.5	553.5	10.5	7.7*
August	5,170	2,150	82	464.7	11.1	6.0
September	4,920	2,695	72.5	410.8	11.9	6.4
October	6,052	2,525	91	520	11.6	4.9
November	5,054	2,700	103.5	586.5	8.6	4.0
December	4,638	2,400	78	442	10.5	5.4
1871.						
January	5,438	2,909	101	496	11.0	5.8
February	6,161	2,893	93.5	530	11.6	5.3
March	6,043	3,400	104	589.5	10.2	5.8
April	6,114	3,030	157	890	6.8	3.4
May	5,793	3,860	164	928	6.3	3.9
Total	77,471	41,050	1,459	8,194	9.4	5.0

Average, 104.   * Bad powder.

ENLARGEMENT.

Month.	Feet of holes drilled.	Pounds of powder used.	Feet of progress.	Cubic yards of progress.	Feet of holes per cubic yard.	Pounds of powder per cubic yard.
1870.						
August	2,223	1,450	125	1,000.9	2.2	1.4
September	2,384	2,125	121.5	905.7	2.0	2.3
October	3,491	2,050	153.5	1,237	2.8	2.1
November	4,408	2,700	120	1,276.5	3.4	2.1
December	4,410	5,775	156	1,009	4.4	5.7*
1871.						
January	3,667	3,075	164.5	1,158	3.2	2.7
February	3,432	3,900	205	1,701	2.0	2.3
March	5,316	3,900	215.5	1,359.5	3.9	2.9
April	2,633	1,900	195.5	666	4.0	2.9
May	4,438	2,600	145.5	873	5.0	2.9
Total	36,602	30,075	1,604	11,179.6	3.3	2.7

Average, 166.   * Partly hand-drilling.

CHAPTER XXI.

STAPFF'S CONTINUOUS JIG.

I find no space in the present volume for a discussion at length of the principles of the concentration and separation of ores, which I hope to publish next year. Meanwhile, I desire to describe and recommend a continuous jig, designed by Dr. F. M. Stapff, M. E., a Swedish engineer of skill and experience, who, having tested its efficiency in practice, offers it for trial and the *free* use of the American mining community, at the same time preventing its being patented by publishing this description. He does not claim the entire invention of an apparatus which is a combination of parts frequently and successfully used in jigs of different construction, it being easy enough for experts to recognize those new and essential features in arrangement and construction by which his machine surpasses similar ones of older date.

The cut subjoined represents a machine built in full accordance with Dr. Stapff's designs, which was constructed at Bethlehem, Pennsylvania, for the Dolores lead mine, Mexico. The ores from this mine, galena and black carbonate of lead, are associated with small quantities of zinc-blende, copper and iron pyrites; they contain calcspar as essential gangue, and are easily dressed. But in Sweden good use has been made of quite similar jigging-apparatus for dressing copper-ores, containing copper pyrites, iron pyrites, and blende in a gangue of quartz, hornblende, and other silicates.

Sizing.—It should be understood that the jigger is constructed for the treatment of sized stuff. We cannot here enter into a description of sizing-machines, and will only mention that the size of meshes in the sizing-screens, and the number of screens subsequently used, must be made dependent upon the specific gravity of the minerals to be separated by the jigging process. For the separation of calcspar and zinc-blende from galena, (Dolores,) the width of meshes in successive sizing-screens should be about 1.00, 0.64, 0.41, 0.26, 0.17, 0.11, 0.07, 0.04 inches, when perforated plates, with round holes, or, 0.70, 0.44, 0.29, 0.18, 0.12, 0.07, 0.05 inches, when wire-gauze with square holes is used. Stuff passing through 0.04-inch meshes is too fine, and stuff remaining upon ¾ to 1 inch meshes is too coarse for proper treatment by this jig, which, however, by some slight alterations in constructive details, can be made fit also for the working of coarser or finer stuff. All material fed on the jig should be free from dust.

General arrangement and modus operandi.—The main box contains six compartments, viz: A, open for the circulating water; B' and B", containing the pump-pistons E and E"; C', and C", receptacles for the jigged products; D, a space for filtration of the water from the refuse. The raising pistons press water through the sieve-beds, F' and F", while they draw water from A through the valves, e' and e", to the chambers B' and B". A consequence of the water's rising in B' and B", and of its sinking at the same time in A, is a current from left to right along the sieve-beds to D, and thence through the holes, k' and k", back to A, by which the medium water-level in the whole vessel is restored. The valves, d", d", allow the water to pass through the sinking pistons, so that there is no suction against the sieve-beds, and no current from right to left is effected by the back-stroke of pistons. Sized stuff fed upon the sieve-bed F', from hopper G, is jigged by the thrusts of water from below; but at the same time being exposed to the horizontal water-current, its lighter particles are carried across the ridge o', and after being exposed to a new jigging operation on sieve-bed F'', the refuse is carried across the ridge c'' to the chamber D. The heaviest parts of the jigged ore move close along the sieve-beds F' and F'', and enter the receptacles C'' and C'', through the gates g' and g'', respectively. Screens of woven wire, attached before the holes k' and k'', prevent the refuse from being carried to chamber A. Along the inclined screen l'', the refuse is led to a discharge-opening, m''. Continual feeding from hopper G is regulated by moving the plate f, which allows more or less stuff to be

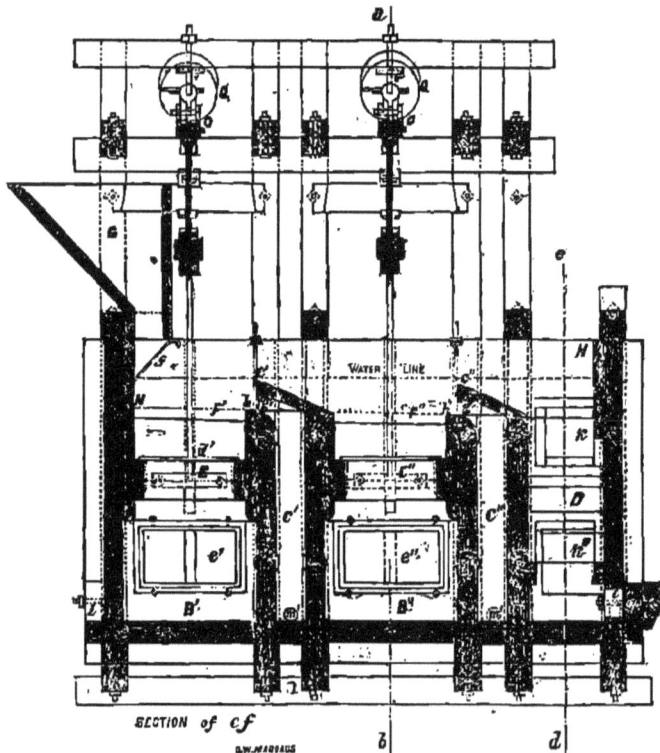

SECTION of c f

drawn from the hopper by the fluctuations of the water. The regular horizontal movement of the ore from left to right is promoted by a slope of sieve-beds of 1 to 36. By slides k' and h'', in front of the openings g' and g'', the quantity of products entering the receptacles C' and C'', is so regulated that the upper surface of the ore in process of treatment is about level with the ridges C' and C''. If plenty of water (one cubic foot per second) can be disposed of, the discharge through the holes m', m'', m''', from the chambers C', C'', and D, should be continual; but if water is scarce, the receptacles C' and C'' are emptied periodically. The refuse-hole m''' should never be totally closed, in order to allow the refuse to escape continually with a certain quantity of water, which has to be replaced from above.

The pistons, by through-rods and cross-heads connected with levers O, receive their movement from cams L, working against the lever. It being favorable for the jigging process that the pistons sink slowly, but rise rapidly, the down-stroke is caued directly by cam, and the up-stroke effected by a counterpoise Q, on the end of lever O. By alteration of the weight of this counterpoise, the velocity of the rising pistons may be changed at pleasure. Fine ore must be jigged by short strokes, coarse ore by long ones. For changing the length of stroke at pleasure, a set-screw, t, is placed over the guide-bar v of the lever O. By this means the back-arm of lever O can be raised so much that the cam L does not catch the head of the other lever-arm at all. Then the piston will rest, though the cam-axle is rotating; or it can be lowered so much as to allow the cam to catch the lever-head 3 to 4 inches above the releas-

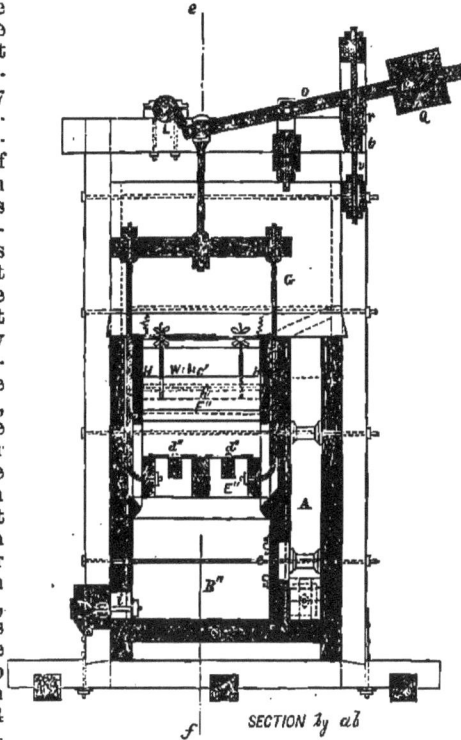

SECTION *by* ab

ing points, and then the stroke will be about 3 inches. The set-screw can be moved along the guide-bar while the machine is in full motion. An elastic cushion of rubber, r, is applied above the set-screw, to moderate the shocks of the falling lever.

All other circumstances unchanged, the quantity of ore jigged in a certain time depends essentially upon the number of piston-strokes. But this number must not be increased so much that the ore on the bed has not time enough to settle between two strokes, consequently the jig should be run slowly, (about 60 strokes a minute,) if coarse ore is jigged by long strokes, and fast, (about 180 strokes a minute,) if the finest ore is jigged by very short strokes. The most favorable number of strokes in any case can easily be produced by running the driving-belt on one of the three pulleys of different diameter on the same cam-shaft.

From the preceding it is easy enough to see the wide applicability of this jig. If the hopper is regularly filled, and everything else regulated in accordance with the nature of the ore jigged, viz, supply of water, number and length of strokes, feeding (by regulator f,) discharge, (by regulators, g' g'',) outlet of water, products and refuse, (by holes, m' m'' m''',) the receptacle C' will receive the heaviest ore, receptacle C'' an intermediate class of ore, box D gangue, or rock fit for the stamp-mill. The products of the jigging process, and the further operations they have to pass through, vary in accordance with the mineralogical character and the size of the treated material. Fine Dolores ore will give galena and lead carbonate in receptacle C', blende in receptacle C'', and calcspar in box D; coarse Dolores ore, galena and carbonate in C', blende, mixed with galena, carbonate and spar in C'', spar, with little blende, and traces of lead-ore in D. In this case it is necessary to submit the crushed products from C'' and D to further dressing operations.

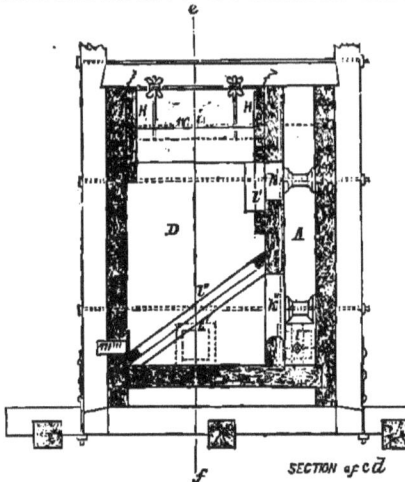

SECTION of c d.

The power necessary to drive this jigger depends upon the area of pistons, and upon the number and length of piston-strokes. Half a horse-power is in all cases sufficient to work a jigger of 18 by 18 inches piston area. The quantity of material worked in a certain time is greatest if the stuff is rich and of middle size. Of poor copper-ore, 6 to 7 cubic feet are worked in an hour.

Some details of construction.—Most of the constructive details of a jigger built of wood can be seen from the cut, without further explanation. If acid water is to be used for the jigging operation, wood is the best material; and, besides, it is the cheapest. Leaks in a well-constructed wooden jigger-box are usually calked by dirt after the jigger has been used for some time, and, besides, it is of no practical account if a few drops of water leak from a vessel through which ¼ to 1 cubic foot is run per second. The outside walls of the jigger-box should be at least 3 inches thick, the interior partition-walls 2 to 2½ inches. The planks forming these walls should be united by hard-wood wedges filling grooves. It must be remembered that soft and dry wood expands transversely about ¾ inch per foot by soaking. A wooden box constructed in accordance with the cut remains not only tight enough, but it can also be easily taken apart, and put together. Grooves in the side walls hold the bottom, which is stiffened by transverse rails; the end walls are fastened in the bottom and side walls, the long partition in the bottom and the end walls, the short partitions in the bottom, long partition, and front walls. Bottom rails and posts form a frame around the box, and by pieces between the rear wall and long partition-wall the whole construction is secured. The discharge-holes m' m'' m''' should not be closed otherwise than by wooden plugs or exterior trap-doors, which by weights working on knee-levers are pressed against nozzles. Close above the bottom are gates i for cleaning the chambers A B' B'' D' when they become obstructed by dirt.

The pistons do not slide directly on the walls of partitions B' and B'', but on hard-wood linings, which can be replaced, piece by piece, if necessary. The main valves c' and c'', of rubber, are stiffened by thin sheet-covers; their wooden valve-seats are kept in place by buttons. Each piston is covered by four light rubber valves.

The meshes in sieve-beds F' F'' and in screens l' l'' must be fine enough to prevent the ore from passing through. But it is not necessary to use as many different sets of sieve-beds and filtrating-screens as there are sizes of ore. Two sets answer all practical wants. Wire gauze, with very fine meshes, which has to be used if the finest stuff is worked, should always be protected against too speedy abrasion by wrappers of coarser gauze. The frames covered with wire cloth (F' F'', l' l'') rest loosely upon and behind wooden strips, and are kept in position by the large frame H, which is common for all compartments in the front part of the jigger. This frame contains the ridges c' and c'', and the gates f, h', h'', which move in grooves; f is kept in position by friction only, h' and h'' by wing-screws besides.

Behind frame H and sieve-beds F' and F'' move the piston-rods in spacious grooves. The lever-heads receiving the motion from cams L must be covered with steel; the counter-weights Q are made of disks of metal, kept in place by eye-bolts. By replacing one or more of those metal disks by wooden disks, and by moving them nearer to or farther from the fulcrum o, it is easy to change the velocity of the rising pistons at pleasure.

H. Ex. 211——32

CHAPTER XXII.

WIRE-ROPE TRANSPORTATION.

In my report, rendered 1870, for the year 1869, will be found (pages 579 to 581) some remarks on wire ropes, to which I take pleasure in adding the following notes upon steel cables, kindly furnished me by Mr. A. S. Hallidie, of San Francisco.

Steel-wire flat ropes are in general use throughout California and Nevada, more particularly at Gold Hill and Virginia City. The durability of steel rope, as compared to iron ropes, is as 6 to 5. The weight (for equal strength) of steel rope, as compared with iron rope, is as 6 to 10. The life of a steel rope working in the Virginia City mines, hoisting at the rate of 1,000 feet per minute, 6 to 8 hoists per hour, for a vertical height of 1,000 feet, is, on the average, about two years. The steel wire of which the rope is composed is especially manufactured for this purpose, ordinary steel wire being unsuitable. The best wire does not become brittle after two years' wear, but possesses still the admirable quality of being tough as lead and hard as steel.

The great tensile strength of steel wire recommends it strongly for the purpose of hoisting from great depths. A 13-gauge steel wire sustains a breaking-strain of 1,400 pounds; whereas the same size best charcoal bright wire sustains a breaking-strain of but 770 pounds; and the steel wire will bear, without breaking, two turns over its own part.

The importance of using ropes of high tensile strength, such as steel alone affords, may be illustrated by the case of a rope manufactured by Mr. Hallidie about the first of January. This rope was 2,000 feet long, 5 inches wide, ½ inch thick, 9,360 pounds in weight, and made of iron wire. The breaking-strain of this rope was estimated at 72,000 pounds, and the working load would be one-sixth of this, or 12,000 pounds. Subtracting from the latter amount the weight of the rope itself, we have 2,840 pounds as the weight of cage and ore that could be safely hoisted; that is, 22 per cent. of working load.

Now, a steel rope of the same capacity would weigh only 4,800 pounds, and the weight of cage and ore that could be safely hoisted would be 7,200 pounds, or 60 per cent. of the working load; and there would be a saving in dead work equal to the difference in weight of the ropes, or 4,560,000 foot-pounds at each hoist. For rough work, moreover, the steel rope has an advantage, inasmuch as it stands abrasion much better than iron.

In round wire ropes steel again shows its superiority over iron, both in its life and useful effects. The life of a round steel wire rope varies according to the character of the hoisting-machinery. In many cases such ropes have lasted three and four years. As a rule, the drum and pulleys should be 100 times the size of the rope.

Endless wire-rope tramways.—The use of endless wire ropes for above-ground transportation, which was alluded to in my report of 1870, (page 568,) has been perfected on a somewhat different principle, already mentioned in Chapter I of this report, and now to be more fully described and illustrated.

In the rough mountainous portions of the gold and silver mining regions of the Pacific coast, there is an immense amount of ore which

HALLIDIE'S SYSTEM OF CABLE TRANSPORTATION.

would be of great value if it were not for the cost of hauling or transporting it to the mill or furnaces to be worked; in many cases this is
done by packing on mules' backs, hauling by teams, or sliding down
chutes, as the case may require, and the cost is from 50 cents to $10 per
ton per mile, according to circumstances.

During the winter months, when the snow falls to great depths in the
Sierra Nevadas, it is not practicable to transport ores, except at intervals, on sleds, and consequently work has to be suspended.

The importance of a cheap and regular mode of transporting the ores
from the mine to its reducing-works has called forth many ingenious
arrangements by those interested in mining.

In Europe, where the Hodgson wire tramway is in use, which was

Fig. 1.

referred to in the chapter on
Mining Appliances, report for
1870, some success seems to
have attended these experiments. So far, the only patents
granted to American citizens in the United States have been issued to
Mr. A. S. Hallidie, of San Francisco, California, for various improvements and inventions
in endless-rope ways
for transporting ores
and other material
over mountainous
and difficult roads.

In the application of
this system, the route
to be followed having
been determined, and
in the selection of
which it is better to
make sharp horizon-
tal curves than verti-
cal ones, a peculiar
pulley called a "grip-
pulley" is placed hori-
zontally at each end
of the line, or at what-
ever point the motive-
power is obtained.

The grip-pulley has
already been referred
to in the last report,
(page 564,) its office
being to receive the
rope in its groove, and
by the pressure of the
rope on the clips in
the circumference of

Fig. 2.

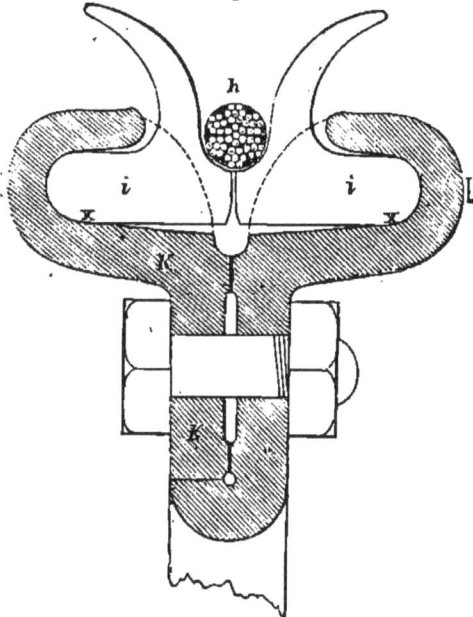

the pulley, to grip the rope and prevent it from slipping in the groove of
the pulley. By referring to Figs. 1 and 2, it will be seen how this is accomplished: h, rope; $i\ i$, clips working in recesses cast in the circumference
of wheel $l\ l$, and on fulcrums XX. The part K is cast separately and
bolted on to the wheel after the clips are fitted.

On the line of the route, at distances of about 250 feet, but regulated by
the configuration of the country, are erected strong posts with horizontal

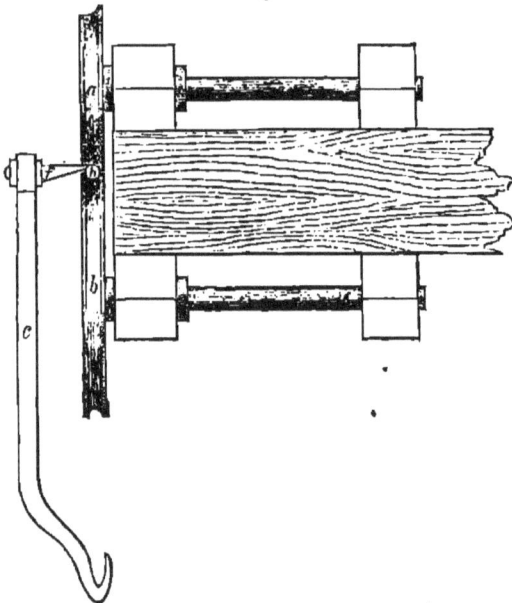

cross-arms, sufficiently high above the ground to clear obstructions, &c. On each end of the horizontal arm is a bearing-pulley, the groove of which is semicircular, and of sufficient size to allow the rope to run in it, and covering half its circumference. Immediately over each of the bearing-pulleys is another pulley, smaller in diameter, the groove of which is a quarter circle, covering one-fourth of the circumference of the rope. Fig. 3 shows the pulleys in position; *a* the upper pulley, *b* the lower pulley, *h* the rope.

Fig. 3.

A steel-wire rope of three-fourths inch diameter is stretched along the route around the end or grip-pulleys, and in the grooves of the bearing-pulleys which are attached to the horizontal arms of the posts. The upper pulley, *a*, is placed over the bearing-pulley and rope, as shown in Fig. 3, the circumference of the two pulleys running in close contact, but having an open space sufficiently large to allow the carrier *f* to pass between the pulleys on their outer sides.

The ends of the steel-wire rope (made from spring-steel wire) are spliced together, forming an endless rope; and motion being imparted to it through the grip-pulley, it will travel in the direction actuated, supported at intervals by the bearing-pulleys, and retained in position between the pulleys on the horizontal arms of the posts, as shown in Fig. 3.

Fig. 4.

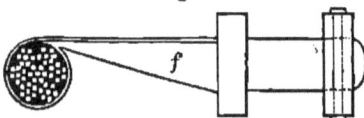

About 50 feet apart there are attached to the rope, by means of thin steel clasps, projecting arms also of steel, about four inches long, and of a form as shown in Fig. 4, the outer end of which is fitted with a journal and collars so as to take a suspension-bar, which hangs vertically and being at right angles to the arm, keeps it in horizontal position. *c* and *f*, Fig. 3, show this arrangement complete. It is designated as the " carrier."

For conveying an ore-sack or box holding about 150 pounds, one of these carriers is used, having a hook at the lower end of a curved suspension-rod; but when it is necessary to convey a car or self-dumping buckets, or a load greater than 200 pounds, the number of these carriers

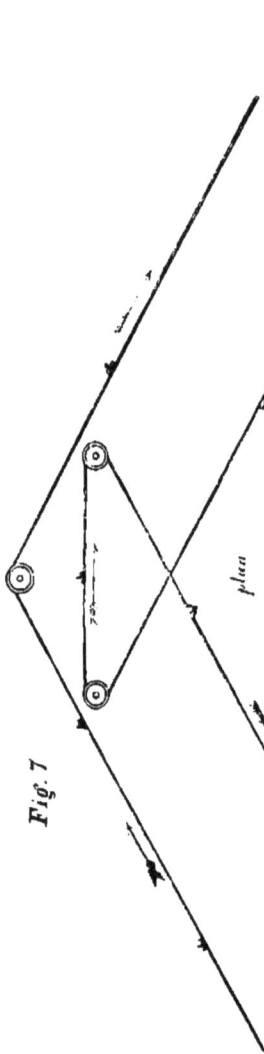

Fig. 7

plan

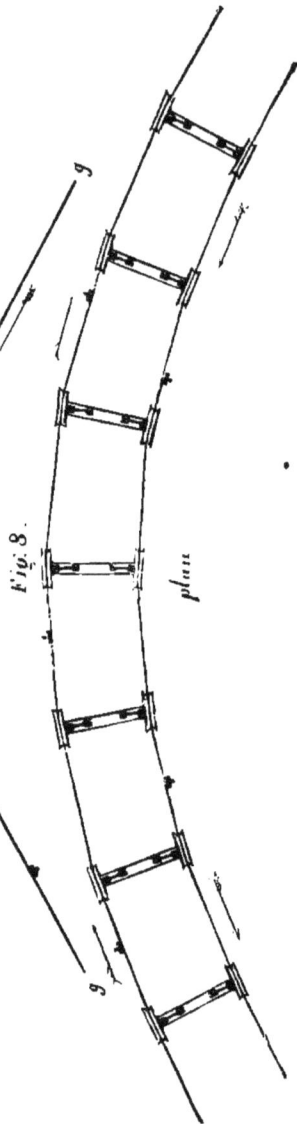

Fig. 8.

plan

is increased, and there is attached to the lower end of the suspension bars a suitable frame, which, in combination with the " carrier," is furnished with joints, enabling all necessary angles and curves to be passed.

Figs. 5 and 6 show the double carrier for conveying a self-dumping bucket or car holding 500 pounds of ore. By increasing the number of

Fig. 5.   Fig. 6.

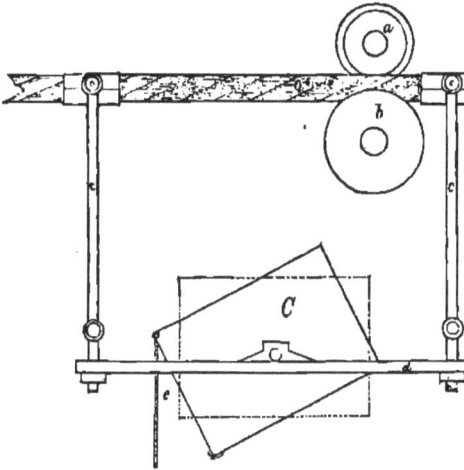 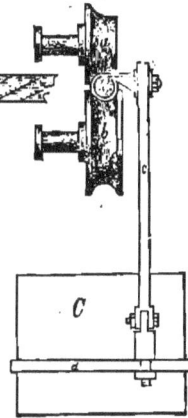

carriers, and thus distributing the load along the rope, the load or weight to be conveyed can be also increased.

The rope being set in motion, carries with it the various loads of ore or whatever material is transported; passing without impediment all the bearing and guide-pulleys, as well as the end grip-pulleys. On account of the arrangement of the bearing and guide pulleys a and b, Fig. 3, the rope cannot jump out from the groove of the pulleys under any circumstances, while they permit the rope and its load to pass over any mountain or through any valley.

The curves are passed in two ways. For a very acute angle two horizontal grooved pulleys are employed, the rope of the interior angle passing around in manner shown in plan, Fig. 7; and for ordinary curves a series of pulleys are so placed that the rope always leads fair on to the next pulley, being deflected in passing off at an angle not to exceed 11°, until the curve is completed. This manner is shown in Fig. 8. Fig. 9 shows another mode of conveying the ore. The car being on an

Fig. 9.

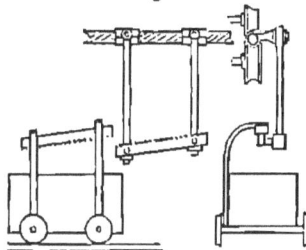

inclined track, is taken up by the carrier, which has attached to the lower end an inclined bar fitted with notches.

The speed of the rope is usually 200 feet per minute. If the ore-sacks contain 150 pounds, and are suspended every 50 feet, 36,000 pounds are delivered per hour, at a cost, including interest, wear and tear, &c., estimated by the patentee at from 20 to 40 cents per ton per mile.

CHAPTER XXIII.

ELECTRICITY AND ROCKS.

This chapter was read by me as a paper before the American Institute of Mining Engineers, at their Troy meeting, in November, 1871. There is much vague theorizing about the connection between electrical currents or discharges and the formation of mineral-deposits, and those who substitute the word "magnetism" for "electricity" mean probably the same thing. There is no need of being exact when one is explaining things on a grand scale, and without reference to the details, that is to say, the facts! In a brief treatise on mineral-deposits, contained in my report of 1870 on mines and mining, I have intimated my view of the limits to which electrical theories of vein phenomena should be confined, namely, those of chemical reactions, either taking place in vapors or solutions of mineral substances, and resulting in precipitation, or occurring in the permeable contents of mineral-deposits already once mechanically or chemically precipitated, and resulting in varied metamorphosis. In the latter case, as well as in the former, the process, strictly speaking, involves the presence of vapors or liquids, since this is a condition of all chemical reactions. The effects produced, aside from such as I have described, by the mere transmission of magnetic or electrical currents through solid rocks, I believe to be trivial and rare. Mr. Darwin, in his "Voyage of the Beagle," describes the vitrified siliceous tubes of La Plata, caused by lightning entering loose sand. In the second volume of the "Geological Transactions," referred to by the same author, there is an account of the similar formations, called fulgurites, occurring at Drigg, in Cornwall; and another case is described by Ribbentrop, in Germany. I quote a part of Darwin's description:

"Four sets entered the sand perpendicularly; by working with my hands I traced one of them two feet deep; and some fragments, which evidently had belonged to the same tube, when added to the other part, measured five feet three inches. The diameter of the whole tube was nearly equal, and therefore we must suppose that originally it extended to a much greater depth. These dimensions are, however, small, compared to those of the tubes from Drigg, one of which was traced to a depth of not less than thirty feet.

"The internal surface is completely vitrified, glossy, and smooth. A small fragment exhibited under the microscope appeared, from the number of minute entangled air, or perhaps steam-bubbles, like an assay fused before the blow-pipe. The sand is entirely, or in greater part, siliceous; but some points are of a black color, and from their glossy surface possess a metallic luster. The thickness of the wall of the tube varies from a thirtieth to a twentieth of an inch, and occasionally even equals a tenth. On the outside the grains of sand are rounded, and have a slightly glazed appearance. I could not distinguish any signs of crystallization. In a similar manner to that described in the Geological Transactions, the tubes are generally compressed, and have deep longitudinal furrows, so as closely to resemble a shriveled vegetable stalk, or the bark of the elm or cork tree. Their circumference is about two inches, but in some fragments, which are cylindrical and without any furrows, it is as much as four inches. The compression from the surrounding loose sand, acting while the tube was still softened, has evidently caused the creases or furrows. Judging from the uncompressed fragments, the measure or bore of the lightning (if such a term may be used) must have been about one inch and a quarter. At Paris, M. Hachet and M. Beudant succeeded in making tubes, in most respects similar to these fulgurites, by passing very strong shocks of galvanism through finely-powdered glass; when salt was added, so as to increase its fusibility, the tubes were larger in every dimension. They failed both with powdered feldspar and quartz. One tube, formed with powdered glass, was very nearly an inch long, namely, 0.982, and had an internal diameter of 0.919 of an inch. When we hear that the strongest battery in Paris was used, and that its power on a substance of such easy fusibility as glass was to form tubes so diminutive, we must feel greatly astonished at the force of a shock of lightning, which, striking the sand in several places, has formed cylinders, in one instance at least, thirty feet long, and having an internal bore, when not compressed, of full an inch and a half; and this in a material so extraordinarily refractory as quartz!"

It is unnecessary to dwell upon the power of electricity thus manifested. It is manifested to us in several instances where houses are struck by lightning, and metallic objects are instantly melted by the surcharging current. What I wish to point out is the comparative rarity of such electrical effects in nature, and particularly in ore-deposits. We do not find in these, as a general rule, any traces of vitrifying fusion, and we may fairly conclude that they are not particularly liable to this form of electrical action. It has, however, occurred to me that some puzzling cases in mineralogy might be due to this cause. Every mineralogist now and then encounters specimens sincerely alleged to be native, i. e., in a natural state, but which he recognizes as the products of more or less perfect fusion. In most instances, no doubt, the fusion has been artificial, and the specimens are really forge or furnace products. Sometimes, however, they may be really native, and vitrified by lightning. Mr. Daubré, in a paper on meteorites in the *Annales des Mines* for 1868, remarks that lightning produces on the rocks of the earth a varnish which is not without analogy to that of meteorites. It occasions, namely, on certain rocks, particularly toward the summit of mountains, the formation of little drops, or of a glaze, to which De Saussure first called attention. It was on account of this resemblance that the savants to whom certain meteorites were submitted, which fell at Lucè in 1768, expressed their opinion that they were merely terrestrial stones vitrified by lightning.

But the rarity and comparatively insignificant extent of such phenomena, and the fact that nothing of the kind is observed as normal to mineral-deposits, even at their outcrops, warrants me in saying that electrical discharges of this character cannot be considered as active agents in the formation, filling, or metamorphosis of veins.

I pass to consider another class of electrical phenomena, namely, those connected with the electric resonance or *boudonnement* of mountains.

Mr. George S. Dwight, of Montclair, New Jersey, has communicated to me a recent case of this kind, personally observed by him on Gray's Peak, in Colorado, the highest summit of the Rocky Mountains. He ascended this peak with a party about the 10th of June last, reaching the top at 2 p. m. Clouds had been gathering for an hour or two, and storms were in progress on the adjoining ranges, principally to the north and east, with heavy, rumbling thunder at brief intervals. "A strong wind from the west," says Mr. Dwight, "drove us to shelter behind a pile of stones some four feet high, which former visitors had erected as a screen. Behind this we crouched for some time, resting and viewing the gorges below. Presently one of the party arose to a standing position, and the instant his head and shoulders were elevated above the protecting line of wall, a hissing sound was heard by all of us. Our friend, with a possible suspicion of snakes, turned about in a bewildered manner to ascertain whence the noise came, and in a moment exclaimed, 'Why, it is me!' His hair stood out, and the gold spectacles he wore, about which there was doubtless some small amount of steel, crackled. At first we were disposed to laugh heartily at his experience, but, as one by one we rose, (there were four of us,) and encountered the same phenomena, we thought best to beat a hasty retreat from a spot which might prove dangerous, and discuss the theory on a lower plane. Each of us experienced the sensation the moment we entered the draught or current from the west. The sound was as loud as that produced by the effervescence of ale from a partially uncorked bottle, and similar in character, though a trifle more whistling in tone. It was accompanied by a strong smell, as of sulphuric or muriatic acid fumes. We all felt the prickling sensation in our fingers also, and a certain exhilaration. These passed away as we descended, and I should say left us entirely within 250 feet of the summit."

Similar experiences are described by M. Fournet, in the *Comptes Rendus de l'Académie des Sciences* of 1867, and also by M. Henri de Saussure, in an article translated for the Smithsonian Report of 1868. The cases enumerated are seven in number, and the circumstances strongly resemble those above related, except that the odor perceived by Mr. Dwight and his companions, possibly due to the presence of ozone, is not mentioned in any of them, though, in a number of these instances, the electric tension appears to have been very great, and, in almost all, there was a crackling of the soil and rocks themselves, and a peculiar vibration of the staffs or alpenstocks of the last observers, called the *chant des bâtons*. Invariably, according to the authorities quoted, there was an attendant shower of hail or sleet at the summit of the mountain. M. Fournet mentions also an instance of nocturnal luminosity on the *Grands Mulets* (Mont Blanc) as referable to a similar electrical condition.

De Saussure draws from the observations discussed by him the following conclusions:

1. The efflux of electricity from the culminant rocks of mountains is produced under a clouded sky, charged with low clouds, enveloping the summits, or passing at a small distance above them, but without the occurrence of electric discharges above the place whence the continuous efflux is proceeding. It would seem, therefore, that when this efflux takes place, it sufficiently relieves the electric tension to prevent lightning from being formed.

2. The continuous efflux of electricity from the ground toward the clouds is not unconnected with the formation of vapor, and probably also with that of the hail.

These electrical phenomena seem not to be rare in high regions, though they are by no means frequent. Many persons accustomed to climb mountains, such as guides and hunters, have never observed the electric resonance; others have heard it but once or twice in their lives. But, as De Saussure acutely remarks, it is precisely on those days when menacing skies repel adventurers from the highest altitudes that the phenomenon manifests itself.

If we now inquire what are the permanent physical traces left by this electric tension or efflux, we find nothing at all. Gray's Peak is a locality within my personal acquaintance, and it bears the marks of far other agencies than this. The whole mountain, for some hundreds of feet below the summit, appears to be a heap of broken fragments, sometimes erroneously called boulders. These have undoubtedly been produced from the exposed crags and ledges, and chiefly through the agency of frost. There is good reason to believe that this and many other summits have been frozen through, and that the summer thaws do not penetrate into their solid portions, except so far as disintegration may be still advancing year by year. I have been informed that the tunnel of the Baker mine, which is above timber-line on the Kelso Mountain, adjoining Gray's Peak, did not, by penetrating 200 feet, get beyond frozen ground. But neither the Baker mine nor the Stevens mine, which is about at an equal altitude on the McClellan Mountain opposite, nor any other of the numerous mines in our western districts, situated at great elevations, presents, so far as my observation goes, peculiar appearances referable to electricity as the cause.

In reply, then, to the assertion of electric theories of vein-formation, it may be said that they lack the basis of direct proof, and that the indirect evidence of analogy is against them. We are acquainted with certain effects of electricity upon rocks; these effects we do not find in ore-deposits; and what we do find there is referable to other causes. The prudent theorist will be content, for the present, with electrochemical, not electro-physical, action, and confine himself to the study, in this department, of the possible existence and effects of galvanic currents in vein-contents, depending upon chemical reactions.

APPENDIX.

THE BULLION PRODUCT.

Estimates of the bullion product of the country are as vague and variable as ever. In my last report I discussed at length the different methods by which it has been attempted to ascertain our production of gold and silver, and vindicated the estimates at which I had arrived by laborious and careful comparisons. I shall not here repeat the argument, but merely recall the fact that I showed the insufficiency of the data obtained by adding together the amount of domestic gold and silver deposited for coinage at the Mint and branches, and the reported amount of uncoined gold and silver bullion exported through the custom-house.

I shall give these figures for the year 1871 presently; but first I will quote a statement courteously furnished me by Mr. John J. Valentine, general superintendent at San Francisco, of Wells, Fargo & Co.'s Express. Mr. Valentine says:

With a view to ascertaining as accurately as possible the product of precious metals for 1871, in the States and Territories west of the Missouri River, I have caused statements to be carefully prepared at each of the company's offices showing the amount shipped monthly during the year named, viz, 1871. The results are:

Territory or State.	Silver bullion or amalgam.	Gold bullion, amalgam, or dust.
Arizona		$163,739 93
British Columbia		1,349,580 83
Colorado	$441,235 82	2,605,681 50
California	231,870 84	16,167,484 05
Idaho	936,934 37	1,471,067 21
Kansas		
Montana	29,417 00	4,031,502 00
Nevada	22,477,045 75	
Nebraska		
Oregon		1,693,602 15
Utah	130,175 81	221,262 14
Washington		320,107 09
Wyoming		
Total	24,246,680 59	28,024,026 90

I submit the following as an approximately correct statement and estimate combined of the total yield of precious metals for the States and Territories of the United States west of the Missouri River, excepting New Mexico, for which I have no data, viz:

Arizona shipments	$163,739 03
Estimate like amount forwarded by other routes and conveyances	163,739 93
British Columbia shipments	1,340,580 83
Estimate 20 per cent. by other conveyances	269,916 16
Colorado—excessive if any variation	3,046,917 32
California	16,309,354 89
Estimate 20 per cent. for undervaluation and by other conveyances	3,279,870 77
Idaho	2,408,001 58
Estimate 20 per cent. by other conveyances	481,600 31
Montana	4,060,929 00
Estimate 20 per cent. by other conveyances	812,185 80

Nevada, full	$22,477,045	75
Oregon, full	1,693,602	15
Utah	357,437	95
Estimate ores and pig-metal by weight	1,000,000	00
Washington, full	320,107	09
Total yield for 1871	58,284,029	66

I am confident that the allowance of 20 per cent. for, we may say, undervaluation and other conveyances, is a liberal concession, and that the total product did not exceed the above amount.

I take leave to differ widely in many points from these estimates, and for most of my corrections of them I have positive evidence. Mr. Valentine's addition of 20 per cent. for undervaluations and private shipments may be sufficient for California, but it seems far too little for Idaho and Montana, while for Colorado he makes no such allowance at all, but transfers the exact amount of the express shipments from his first to his second table, with the enigmatical comment, "excessive, if any variation." If this means anything, it means that the Colorado shipments of bullion are overvalued, and that the amounts upon which express charges are paid exceed the total amount produced! By comparing his figures with those given in my chapter on Colorado it will be seen that he ignores $923,000 shipped in matte, $500,000 shipped in ores, and $100,000 used by manufacturers, and that the product of the Territory is consequently about $1,523,000 more than he calculates.

With regard to Utah this statement is equally imperfect. He estimates the shipments of ores and base bullion at $1,000,000. This is a mere guess, and not a successful one. The shipments of ore from Salt Lake City in 1871 amounted to 10,806 tons, averaging at least $150 silver per ton, and the shipments of base bullion amounted to 2,378 tons, averaging $175 silver per ton. This gives us $1,620,900 as the value of the ores and $316,150 as the value of the base bullion, to which should be added $500,000 for the lead contained in ores and bars. The total of these items is $2,437,050, against $1,000,000 in Mr. Valentine's estimate. Considering that he makes apparently but $6,000 allowance for undervaluations and private shipments, it is quite within bounds to say that the product of Utah for 1871 was about $2,800,000, instead of $1,357,437, as he has it.

The express shipments from Arizona he doubles to obtain the total yield. I have direct evidence that this result is too small, and though I cannot say precisely how much too small it is, I believe my estimate is near the truth.

The British Columbia shipments are omitted from my table.

The product of New Mexico, omitted by Mr. Valentine, was about $500,000, and the product of Wyoming, also omitted by him, was about $100,000.

In calculating for the whole country, east and west, I add, under the head of "other sources," $200,000 to cover the product of the southern States, and the extraction of silver from lead-ores not otherwise taken into the calculation.

My estimate of the gold and silver production of the United States for the year 1871 is as follows, compared with former years:

State or Territory.	1869.	1870.	1871.
Arizona	$1,000,000	$800,000	$800,000
California	22,500,000	25,000,000	20,000,000
Colorado	*4,000,000	3,675,000	4,663,000
Idaho	7,000,000	6,000,000	5,000,000
Montana	9,000,000	9,100,000	8,050,000
Nevada	14,000,000	16,000,000	22,500,000
New Mexico	500,000	500,000	500,000
Oregon and Washington	3,000,000	3,000,000	2,500,000
Wyoming		100,000	100,000
Utah		1,300,000	2,300,000
Other sources	†500,000	525,000	250,000
Total	61,500,000	66,000,000	66,663,000

* Including Wyoming.   † Including Utah.

I exclude from the statement for 1871 the product of the smelting-works at Wyandotte, Michigan, which is believed to amount to $800,000, because the ores there reduced are obtained from Silver Islet, on the north coast of Lake Superior, and outside of the United States.

Further comment upon the above figures is unnecessary. It is evident that the product from placer-mining has continued to fall off, and that there has been a great advance in those districts which are chiefly occupied with quartz mining. In California the placer and hydraulic mines have continued to suffer from lack of water; and the reduced product of that State is probably not to be taken as a measure of actual decline in these branches of mining.

The amount of gold and silver coined at the mints of the United States during the year ended December 31, 1871, is shown by the following tables furnished by the chief coiner:

Statement of deposits and coinage at the Mint of the United States, and branches, during the year ended December 31, 1871.

DEPOSITS.

Mint and branches.	Gold-deposits.	Silver-deposits.	Total deposits.
United States Mint, Philadelphia	$2,884,645 61	$2,124,924 26	$5,009,569 87
Branch mint, San Francisco	24,960,122 61	1,347,567 05	26,307,689 66
Branch mint, Carson City	2,515,132 79	2,944,465 17	5,459,597 96
Branch mint, Denver	1,020,223 37	2,937 67	1,023,161 04
Assay office, Charlotte	16,122 37	164 52	16,286 89
Total	31,396,246 75	6,320,058 67	37,716,305 42

GOLD COINAGE

Mint and branches	Double eagles		Eagles		Half eagles		Three dollars		Quarter eagles		Dollars		Total	
	Pieces.	Value.	Pieces.	Value.	Pieces.	Value.	Pieces.	Value.	Pieces.	Value.	Pieces.	Value.	Pieces.	Value.
United States Mint, Philadelphia..	80,150	$1,603,000	1,790	$17,800	3,520	$16,150	1,330	$3,990	5,350	$13,375	3,930	$3,930	95,770	$1,658,245
Branch mint, San Francisco..	928,000	18,560,000	14,500	145,000	25,000	125,000	......	......	22,000	55,000	......	......	991,500	18,905,000
Branch mint, Carson City..	12,207	244,140	7,185	71,850	17,410	87,050	......	......	......	......	......	......	30,802	403,040
Total......	1,020,357	20,407,140	25,463	254,650	45,640	228,200	1,330	3,990	27,350	68,375	3,930	3,930	1,124,079	20,966,285

SILVER COINAGE

Mint and branches	Dollars		Half dollars		Quarter dollars		Dimes		Half dimes		Three cents		Total	
	Pieces.	Value.	Pieces.	Value.	Pieces.	Value.	Pieces.	Value.	Pieces.	Value.	Pieces.	Value.	Pieces.	Value.
United States Mint, Philadelphia..	1,115,760	$1,115,760	1,134,560	$567,280	119,239	$29,809 00	907,610	$90,761	1,748,860	$87,443	4,290	$127 80	5,030,282	$1,891,179 60
Branch mint, San Francisco..	......	......	2,178,000	1,089,000	30,900	7,725 00	320,000	32,000	161,000	8,050	......	......	2,689,900	1,136,775 00
Branch mint, Carson City..	1,376	1,376	139,950	69,975	10,890	2,722 50	20,100	2,010	......	......	......	......	172,316	76,063 50
Total......	1,117,136	1,117,136	3,452,510	1,726,255	161,029	40,255 50	1,247,710	124,771	1,909,860	95,493	4,290	127 80	7,892,496	3,104,038 30

BASE COINAGE.

Mint.	NICKEL.				BRONZE.				Total.	
	Five cents		Three cents		Two cents		One cent			
	Pieces.	Value.	Pieces.	Value.	Pieces.	Value.	Pieces.	Value.	Pieces.	Value.
United States Mint, Philadelphia......	561,000	$28,050	604,000	$18,120	721,250	$14,425	3,929,500	$39,295	5,815,750	$99,890

Gold coinage.		Silver coinage.		Base coinage.		Total.	
Pieces.	Value.	Pieces.	Value.	Pieces.	Value.	Pieces.	Value.
1,124,072	$20,906,285 00	7,892,498	$3,104,008 30	5,815,730	$99,890 00	14,832,320	$24,170,213 30

A. LOUDON SNOWDEN, *Chief Coiner.*

I have not been able to obtain the exact figures of the domestic gold and silver deposited for coinage. The following is the amount, as given by the Alta California, of refined gold and silver deposited in the United States branch mint during 1871 by the San Francisco Assaying and Refining Works:

Month.	Gold.		Silver.	
	Ounces.	Value.	Ounces.	Value.
January	50,953. 18	$1,166,144 83	18,074. 00	$24,254 93
February	47,932. 67	982,408 50	9,384. 15	12,580 25
March	45,183. 42	918,516 99	22,814. 50	30,587 00
April	96,574. 03	1,978,479 79	5,584. 05	7,502 08
May	101,370. 17	2,075,571 83	8,981. 75	12,054 94
June	84,000. 25	1,720,279 01	11,956. 55	16,043 45
July	92,943. 53	1,903,463 60	23,661. 40	34,133 00
August	99,648. 30	2,041,517 97	101,714. 35	136,502 09
September	80,583. 85	1,648,633 19	85,404. 15	114,575 48
October	77,245. 29	1,682,775 19	5,611. 00	7,529 29
November	66,457. 60	1,361,280 00	67,159. 80	90,128 27
December	46,762. 13	958,274 24	40,398. 60	54,197 03
Total deposited	895,664. 41	18,337,175 20	400,754. 90	537,668 04
Fine gold sold in market	38,228. 71	783,222 37		
Total product	933,893. 12	19,120,397 57		

This gives the large amount of $19,120,397.57 refined, and $783,222.37 was sold for Japan, China, and elsewhere.

The following, from the Commercial Herald of San Francisco, gives a comparative view of the coinage at the branch mint in that city for the years 1868, 1869, 1870, and 1871, as follows:

Month.	1868.	1869.	1870.	1871.
January	$97,000	$467,000	$1,600,000	$1,570,000
February	640,000	165,000	985,000	1,171,725
March	575,000	743,000	2,155,000	965,000
April	710,000	1,579,000	1,330,000	1,850,000
May	714,000	965,000	2,083,000	2,178,050
June	622,000	1,345,000	2,106,000	884,000
July	2,355,000	1,040,000	120,000	2,700,000
August	1,465,000	689,500	2,370,000	1,900,000
September	2,455,000	2,530,000	2,030,000	2,210,000
October	2,415,000	1,609,300	1,875,000	1,689,000
November	2,595,000	1,648,000	1,965,000	1,684,800
December	2,442,000	1,450,750	1,676,000	1,218,000
Total	17,365,000	14,363,550	20,335,000	20,020,775

The coinage of the mints is undoubtedly in excess of the domestic deposits of bullion for coinage, since deposits of United States coin, &c., are also recoined.

From the reports of the Bureau of Statistics of the Treasury Department I have compiled the following statement of the imports, exports, and re-exports of the precious metals during the year ending December 31, 1871:

Description.	Imports.	Exports.	Re-exports.
Bullion:			
Gold	$1,335,196	$6,068,173	$4,780
Silver	147,682	20,165,739	91,342
Total	1,482,878	26,233,912	96,122
Coin:			
Gold	4,506,752	37,293,426	1,549,596
Silver	10,779,785	1,904,004	10,363,410
Total	15,286,537	39,197,430	11,913,006

Exports of gold and silver bullion for five years.

Year.	Gold.	Silver.	Total.
1867	$19,192,299	$15,503,527	$34,695,826
1868	17,402,625	13,987,210	31,389,835
1869	13,681,934	12,748,315	26,430,200
1870	15,599,680	13,171,419	28,771,299
1871	6,068,173	20,165,739	26,233,912
Total for five years	71,944,961	75,576,210	147,521,171

Adding now the total coinage to the total exports of bullion, we have:

For the year ending December 31, 1871.	Gold.	Silver.	Total.
Coinage	$20,966,285	$3,104,038	$24,070,323
Exports	6,068,173	20,165,739	26,233,912
Total	27,034,458	23,269,777	50,304,235

It thus appears that even when all the coinage of the United States mints is added to all the exports of bullion, the aggregate is far below even Mr. Valentine's estimate, which I have demonstrated to be too small. This is an additional proof (if one were needed) of the futility of attempting to calculate the gold and silver product from the mint and custom-house returns, making no allowance for ores and mattes or base bullion shipped abroad and the large amount consumed (not as coin) by manufacturers. In my last report I discussed this subject fully.

I add some statistics of general interest concerning principally the commerce of San Francisco.

[From the San Francisco Commercial Herald.]

TREASURE-PRODUCT, IMPORTS, ETC.

The receipts of treasure from all sources, through Wells, Fargo & Co.'s express during the past twelve months, as compared with the same period in 1870, have been as follows:

	1870.	1871.
From northern and southern mines	$38,402,152	$35,608,385
Coastwise, north and south	4,472,594	3,215,431
Imports, foreign	5,466,883	4,108,724
Totals	48,341,629	42,932,540

RECEIPTS OF TREASURE.

The following table comprises the receipts of treasure in this city, through Wells, Fargo & Co.'s express, during the year 1871:

From the northern and southern mines.

187L	Silver bullion.	Gold dust.	Coin.	Total.
January	$1,212,104	$1,048,572	$808,738	$3,069,464
February	1,198,295	985,047	657,774	2,841,116
March	1,245,340	1,363,130	699,912	3,308,382
April	1,331,934	1,314,122	429,089	3,075,145
May	1,390,895	1,616,931	781,584	3,793,410
June	1,375,711	1,375,602	492,026	3,243,342
July	1,235,873	1,211,742	730,253	3,177,868
August	1,014,820	1,143,134	438,734	2,596,688
September	1,104,074	900,238	476,622	2,480,934
October	1,120,902	1,037,640	617,079	2,775,661
November	1,195,511	926,318	570,601	2,692,430
December	1,178,227	950,172	423,526	2,551,925
Total, 1871	14,609,809	13,872,648	7,125,928	35,608,385
Total, 1870	14,152,984	17,762,131	6,487,037	38,402,152
Total, 1869	not separated.	not separated.	11,572,594	44,045,445
Total, 1868	not separated.	not separated.	6,620,897	45,932,040
Total, 1867	not separated.	not separated.	4,812,787	45,404,770

From the northern coast.

1871.	Silver bullion.	Gold dust.	Coin.	Total.
January	$4,865	$147,004	$100,317	$252,186
February		86,597	43,835	130,432
March	4,920	92,425	51,728	149,073
April		197,828	25,170	223,058
May		232,599	18,524	251,123
June		153,749	61,393	215,102
July		264,155	23,181	287,336
August		354,525	29,676	384,211
September		297,961	118,813	416,774
October		217,971	86,818	304,789
November		199,974	94,800	294,834
December		307,850	53,781	361,631
Total, 1871	9,785	2,552,668	708,096	3,270,549
Total, 1870		3,380,566	532,901	3,913,467
Total, 1869	not separated.	not separated.	300,397	2,938,458
Total, 1868	not separated.	not separated.	728,651	2,930,955
Total, 1867	not separated.	not separated.	1,396,429	3,801,489

From the southern coast.

1871.	Silver bullion.	Gold dust.	Coin.	Total.
January		$14,073	$48,491	$62,504
February		31,119	47,325	78,444
March	$5,750	22,791	43,097	71,238
April		25,414	44,137	69,551
May		44,987	47,464	92,451
June		38,386	29,174	77,560
July		29,925	56,478	86,403
August		39,667	58,885	98,552
September		26,460	53,827	80,313
October		18,430	28,078	46,508
November		27,904	40,411	68,315
December		28,845	44,046	72,891
Total, 1871	5,750	347,627	551,413	904,790
Total, 1870		399,888	844,548	1,244,436
Total, 1869	not separated.	not separated.	227,000	2,282,571
Total, 1868	not separated.	not separated.	557,050	2,304,060
Total, 1867	not separated.	not separated.	1,096,440	2,391,341

The receipts from the northern and southern mines include the amounts sent East from the Virginia office, viz:

1871.	Silver bars.	1871.	Silver bars.
January	$638,966	July	$604,905
February	708,885	August	554,711
March	611,226	September	542,676
April	713,450	October	667,179
May	826,607	November	723,190
June	741,467	December	767,516
Total			8,100,778

COMBINED EXPORTS.

The combined exports, treasure, and merchandise, for 1871, as compared with the same time in 1869 and 1870, were as follows:

	1869.	1870.	1871.
Treasure exports	$37,287,117	$32,983,140	$17,253,347
Merchandise exports	20,888,991	17,848,160	13,951,149
Total	58,176,108	50,831,300	31,204,496

MOVEMENT OF COIN IN THE INTERIOR.

The following has been the circulation of coin through Wells, Fargo & Co.'s express, during 1871:

	To interior.	From interior and coastwise.
January	$869,311	$957,536
February	945,397	748,934
March	687,722	794,737
April	1,353,710	498,396
May	1,528,431	847,572
June	1,394,225	592,593
July	1,543,891	809,912
August	1,924,787	527,295
September	1,807,413	649,262
October	1,864,941	731,975
November	1,582,925	705,872
December	1,687,129	521,353
In 1871	17,389,882	8,385,437
In 1870	18,632,438	9,599,947
Decrease	1,242,556	1,214,510

Statement of the amount of treasure exported from San Francisco, through public channels, to eastern domestic and foreign ports during the year 1871, exclusive of shipments through United States mail.

To New York:
In January	$804,436 37
In February	1,141,165 01
In March	1,091,048 40
In April	674,052 04
In May	852,453 26
In June	930,101 70
In July	617,852 23
In August	535,371 88
In September	342,471 60
In October	399,285 50
In November	276,008 15
In December	384,942 22
	$8,057,279 33

To England:
In January	114,243 53
In February	143,273 41
In March	154,677 52
In April	335,761 45
In May	338,776 80
In June	301,849 18
In July	305,833 00
In August	392,656 92

In September	$216,074 60
In October	367,658 20
In November	273,740 02
In December	239,396 33
	$3,184,841 74

To China:
In January	
In February	381,021 26
In March	312,116 65
In April	327,656 50
In May	177,338 85
In June	361,811 26
In July	275,561 93
In August	270,316 70
In September	581,938 10
In October	
In November	327,629 68
In December	427,825 79
	3,443,208 72

To Japan:
In January	
In February	15,950 00
In March	7,918 00

H. Ex. 211——33

In April.....	$1,000 00	In July	$14,110 00
In May.................	60,000 00	In August	5,150 00
In June	26,011 29	In September	12,100 00
In July	63,500 00	In October	
In August		In November........	20,184 94
In September	107,956 95	In December	40,113 00
In October			
In November..........	202,920 91	To Montevideo:	$133,368 54
In December....	129,249 52	In June	13,091 62
	$738,412 67		13,091 62
To Panama:		To Callao:	
In January............	10,000 00	In August	500,000 00
In February...........	10,000 00	In September	1,000,000 00
In March..............	10,000 00		1,500,000 00
In April	10,000 00	To Tahiti:	
In May	10,000 00	In August.............	26,000 00
In June...............	12,779 00		26,000 00
In July	11,058 71	To Honolulu:	
In August.............	10,000 00	In November........	30,000 00
In September	10,000 00		30,000 00
In October	10,000 00	To Mexico:	
In November..........	6,308 78	In December	10,000 00
In December	5,000 00		10,000 00
	115,146 49		
To Central America:		Total for 1871.............	17,253,347 11
In January............	20,000 00	Total for 1870.............	32,983,140 04
In February	20,000 00		
In April	2,800 00	Decrease for this year..........	15,720,792 93
In June	908 60		

TREASURE EXPORTS.

(*Another statement from the Alta California.*)

The export and destination of treasure during the years 1870 and 1871, respectively, were as follows:

	1870.	1871.
To China	$6,055,080 49	$3,364,529 99
To Central American ports	284,475 08	229,008 55
To England..	9,788,310 20	3,010,584 45
To France...	190,410 24	....
To Japan	855,975 42	747,627 67
To Tahiti	....	26,000 00
To New York......................................	14,107,800 21	7,737,180 60
by post-office....................................		12,287,291 21
To Peru	2,130,084 70	1,500,000 00
To Sandwich Islands..............................	25,500 00	30,000 00
To Mexico	....	10,000 00
To Brazil	....	13,091 00
Total	33,566,898 39	28,953,813 45
Add duties..	4,901,150 31	7,378,270 42
Net	38,468,048 70	36,332,083 87

Exports of treasure via Panama.

Date.	Gold bars.	Silver bars.	Gold coin.	Total.
January 3			$5,570 50	$5,570 50
January 18	$28,713 61	$84,959 42	35,000 00	138,673 03
February 3	43,788 05	97,407 69	27,077 67	168,273 41
February 18			5,000 00	5,000 00
March 4		100,511 33	5,880 10	115,391 43
March 18	21,603 03	22,683 06	5,000 00	49,286 09
April 3	56,619 95	89,114 18	9,388 00	155,122 13
April 17	77,020 59	111,415 73	5,000 00	193,439 32
May 3	83,627 89	119,945 75	7,000 00	210,573 64
May 17	47,866 37	85,336 79	5,000 00	138,203 16
June 3	50,712 97	128,982 24	6,687 06	185,382 27
June 17	36,474 89	87,699 00	19,091 62	143,265 51
July 3		204,775 53	20,168 71	224,943 24
July 17	36,141 31	64,916 16	5,000 00	106,037 47
August 3	70,585 48	98,263 74	10,959 00	179,809 22
August 17	141,960 63	81,037 05	505,000 00	727,997 70
September 2	63,135 82	41,793 15	518,283 00	623,211 97
September 17	19,396 64	91,466 06	505,000 00	615,862 72
October 3	57,026 36	216,782 22	5,000 00	278,808 58
October 17	23,937 61	69,912 10	5,000 00	98,849 71
November 3	24,000 37	119,179 50	21,493 72	164,676 59
December 2	14,887 30	91,242 63	45,113 00	151,242 99
December 16	24,111 39	139,154 95	10,000 00	143,266 34
Total	921,613 33	2,123,554 31	3,168,168 01	6,213,340 56

NOTE.—The total exports of treasure, as above, include $45,000 in silver coin.

Railroad.

Date.	Gold bars.	Silver bars.	Gold coin.	Total.
January	$288,356 85	$492,136 22	$34,098 00	$814,591 07
February	332,530 43	579,337 69		911,408 12
March	504,589 65	550,136 11	19,699 00	1,074,424 76
April	270,584 19	305,657 95		576,242 14
May	445,757 58	504,525 60		950,283 18
June	335,240 14	529,408 08	7,205 10	871,853 33
July	122,020 77	650,475 65	5,191 80	784,288 22
August	600 00	435,674 35		436,274 75
September		338,263 39		338,263 89
October	29,055 07	370,240 43		389,295 50
November	10,002 18	264,345 97		275,008 13
December	112,494 62	165,703 93		178,198 55
Total	2,410,431 46	5,171,900 37	66,193 90	7,712,180 60

Exports to China, &c., per steamers and sail-vessels.

Date.	Gold bars.	Mexican dollars.	Gold coin.	Total.
To China:				
February 1	$14,900 00	$334,193 76	$31,929 50	$381,023 26
March 1	200 00	190,387 15	124,447 50	315,034 65
April	5,740 00	184,338 00	137,045 50	327,123 50
May	21,445 85	87,073 00	58,820 00	117,338 35
June	14,074 50	286,032 00	72,915 00	357,511 26
July	42,003 38	164,638 00	71,567 55	278,268 93
August	163,551 70	16,065 00	90,700 00	270,316 70
September	25,427 00	212,274 00	77,942 50	315,643 50
September 30	94,544 69	78,404 00	76,681 50	250,210 60
November 1	78,429 78	130,945 00	115,909 90	325,248 68
November 8			30,000 00	30,000 00
December 1	111,343 91	36,105 20	214,387 75	361,833 75
December 1	(silver bars.)			65,991 94
To Japan:				
February 1		$15,590 00		$15,590 00
March 1		5,000 00		5,000 00
April		4,000 00		4,000 00
May		60,000 00		60,000 00
June	(silver bars.)	24,911 29	$2,000 00	26,911 29
July		63,500 00		63,500 00
September	$59,460 34	101,000 00		160,460 34
September 30	5,996 61	14,000 00		19,996 61
November 1	202,926 91			202,926 91
December 1	129,249 52			129,249 52
Total	1,029,354 05	2,008,456 41	1,102,346 70	4,112,157 66

Total gold received in San Francisco according to the Alta California:

Sent to mint for coinage	$18,337,175
Exported by Panama steamers	$921,613
Exported East by railroad	2,410,431
Exported to China, Japan, &c.	1,029,354
	4,361,398
Total in San Francisco	22,698,573

Tabulated statement of the dividends disbursed to stockholders in each month of the year 1871 by the mining companies of California, Nevada, and Idaho, whose offices are located in San Francisco.

[From the Alta California.]

Name of mining company.	January.	February.	March.	April.	May.	June.	July.
Amador, (California)				$14,800	$9,250		
Chollar-Potosi;	$280,000	$280,000	$280,000	280,000	280,000	$56,000	$56,000
Crown Point						120,000	120,000
Eureka, (California)	40,000	40,000	40,000	40,000	40,000	20,000	20,000
Eureka Consolidated		50,000	37,500	37,500			
Golden Chariot, (Idaho)	40,000	60,000	70,000				
Greenville, (California)						4,000	
Hale and Norcross	40,000	40,000	40,000	40,000			
Keystone Quartz							
Meadow Valley	60,000	60,000					60,000
North Star, (California)	9,000	9,000	9,000	9,000	12,000	12,000	6,000
Pioche							
Raymond and Ely			30,000	30,000		30,000	30,000
Redington Quicksilver				6,300	6,300	6,300	6,300
Sierra Nevada	20,000						
Succor Mill							
Yellow Jacket	48,000	48,000	48,000	60,000	60,000	60,000	60,000
Yale Gravel							5,000
Total	537,000	587,000	554,500	517,600	407,550	308,300	363,300

Name of mining company.	August.	September.	October.	November.	December.	Total.
Amador, (California)						$24,050
Chollar-Potosi	$28,000	$28,000	$28,000	$28,000	$28,000	1,652,000
Crown Point	120,000	120,000				480,000
Eureka, (California)						240,000
Eureka Consolidated	50,000	50,000	50,000			275,000
Golden Chariot, (Idaho)						170,000
Greenville, (California)						4,000
Hale and Norcross						160,000
Keystone Quartz		5,000		7,500	7,500	20,000
Meadow Valley	60,000	60,000	90,000	90,000	90,000	570,000
North Star, (California)						66,000
Pioche		20,000				20,000
Raymond and Ely	30,000	45,000	120,000	150,000	150,000	615,000
Redington Quicksilver	6,300					31,500
Sierra Nevada						20,000
Succor Mill		11,400				11,400
Yellow Jacket	60,000					444,000
Yule Gravel	5,000	5,000	10,000	5,000	5,000	35,000
Total	359,300	344,400	298,000	280,500	280,500	4,837,950

The foregoing disbursements compare as follows with the amounts paid by the same and other mining incorporations for the previous year:

	1870.	1871.
Amador	$155,400	$24,050
Argenta	21,000	.....
Chollar-Potosi	658,000	1,652,000
Crown Point	.....	480,000
Eureka	430,000	340,000
Eureka Consolidated	.....	275,000
Golden Chariot	75,000	170,000
Golden Rule	3,000	.....
Greenville	.....	4,000
Gould and Curry	48,000	.....

	1870.	1871.
Hale and Norcross	$504,000	$130,000
Ida Elmore	20,000	.....
Kentuck	30,000	.....
Keystone	.....	20,000
Meadow Valley	150,000	570,000
Metropolitan Mill	10,000	.....
North Star	16,500	66,000
Original Hidden Treasure	32,000	.....
Piocho	.....	20,000
Raymond and Ely	.....	615,000
Redington Quicksilver	.....	31,500
Sierra Nevada	37,500	20,000
Succor Mill	.....	11,400
Union	30,000	.....
Yellow Jacket	.....	444,000
Yule Gravel	.....	35,000
Wheeler	6,000	.....,
Total	**2,226,400**	**4,837,950**

Tabular statement of the number and amounts of assessments levied and dividends disbursed and other information pertaining to the leading mines of California, Nevada, and Idaho, dealt in at the San Francisco stock-board, being an extract from the tabular statement published by R. Wheeler, editor of the San Francisco Stock Report, December 13, 1871.

Companies.	Number of assessments.	Number of feet in mine.	Number of shares in mine.	Number of dividends.	Total amount of assessments levied.	Total amount of dividends disbursed.	Amount of assessment per share.	Amount of dividend per share.
CALIFORNIA.								
Amador	...	1,850	3,700	33	.....	$860,050	.....	$232 50
Eureka	..7	1,680	20,000	62	.....	1,734,000	.....	86 70
Oriental	6	1,800	18,000	.....	$54,000	.....	$3 00	.....
Union Mill and Mining Company	2	.....	5,000	44	12,500	134,250	1 00	.....
Maxwell	19	.....	4,000	.....	53,880	.....	13 47	.....
St. Patrick Gold Mining Company	3	1,800	5,000	.....	55,000	.....	11 00	.....
Independent Gold Mining Co ...	.....	1,800	25,000	.....	24,000	.....	3 00	.....
Bellevue	4	20,000	8,000	.....	.....	.....	....	.....
Total					199,380	2,708,300		
NEVADA.								
Washoe district:								
Alpha Consolidated	5	330	6,000	.....	132,000	.....	22 00	.....
Belcher	8	1,040	10,400	.....	660,800	421,200	63 50	40 50
Bullion	43	2,500	5,000	.....	1,744,500	.....	215 50	.....
Buckeye	3	.....	16,000	.....	40,000	.....	.....	2 50
Chollar-Potosi	3	2,800	28,000	38	462,000	3,024,000	16 50	108 00
Confidence	9	130	1,560	6	218,860	78,000	140,00	50 00
Consolidated Virginia	10	1,160	11,600	.....	174,000	.....	15 00	.....
Crown Point	21	600	12,000	20	623,370	1,338,000	51 94	111 50
Daney	37	2,000	8,000	2	480,000	56,000	60 00	7 00
Empire Mill	7	75	1,200	21	104,400	513,600	87 00	428 00
Exchequer	8	400	8,000	.....	128,000	.....	16 00	.....
Flowery	1	3,600	12,000	.....	12,000	.....	1 00	.....
Gold Hill Quartz	4	134	500	8	35,000	41,250	70 00	82 50
Gould and Curry	11	1,200	4,800	30	705,600	3,826,800	147 00	797 25
Hale and Norcross	35	400	8,000	30	770,000	1,598,000	96 25	199 75
Imperial	12	184	4,000	30	530,000	1,067,500	132 50	266 87
Julia	30	2,000	10,000	.....	111,200	.....	11 12	.....
Justice	11	.....	.....	.....	.....	.....	.....	.....
Kentuck	5	95	2,000	32	90,000	1,250,000	45 00	625 00
Occidental	5	.....	10,000	1	165,000	20,000	16 50	2 00
Ophir	19	1,400	16,800	22	1,144,000	1,394,400	68 33	82 97
Overman	19	1,200	12,800	.....	798,628	.....	62 39	.....
Savage	5	800	16,000	52	468,000	4,288,000	29 25	268 00
Segregated Belcher	14	160	6,400	.....	212,800	.....	33 25	.....
Sierra Nevada	36	.....	20,000	11	500,000	102,500	25 00	5 12
Succor Mill and Mining Company	1	7,600	22,800	1	.....	22,800	.....	1 00
Yellow Jacket	14	1,200	24,000	25	1,518,000	2,184,000	63 25	91 00
Total					11,917,238	21,527,550		

Tabular statement of the number and amounts of assessments, &c.—Continued.

Companies.	Number of assessments.	Number of feet in mine.	Number of shares in mine.	Number of dividends.	Total amount of assessments levied.	Total amount of dividends disbursed.	Amount of assessment per share.	Amount of dividend per share.
White Pine district:								
Consolidated Silver Wedge			20,000					
Consolidated Chloride	3		50,000		250,000		5 00	
Hidden Treasure Consolidated	2	600	12,000		9,000		75	
Mammoth	8	1,809	36,000		59,400		1 65	
Noonday	9	1,000	20,000		44,000		2 20	
Metropolitan Mill and Mining Co.	4	4,000	10,000	2	60,000	10,000	4 50	1 00
Original Hidden Treasure	5		21,333		168,664	31,999	8 00	1 50
Silver Wave	6	1,600	20,000	1	122,000		6 10	
Silver Vault	4	3,000	30,000		6,000		20	
Virginia	4	800	21,333		41,265		1 95	
General Lee	3	1,000	20,000		6,000		30	
Total					757,298	41,999		
Battle Mountain district:								
Nevada Butte	3	4,200	20,000		60,000		3 00	
Ely district:								
Meadow Valley	7		60,000	10	210,000	720,000	3 50	12 00
Raymond and Ely		4,000	30,000	8		465,000		15 50
Meadow Valley West'n Extension	5	200	6,000		18,000		3 00	1 00
Pioche		1,000	20,000			20,000		
Washington and Creole	2	200	30,000		30,000		1 00	
Lillian Hall	1	1,000	15,000		7,500		50	
Total					265,500	1,185,000		
Eureka district:								
Eureka Consolidated			50,000	4		275,000		5 50
Jackson	2		50,000		25,000		50	
Mineral Hill			50,000					
Phenix			50,000		87,500		1 75	
Total					112,500	275,000		
Grand total					13,224,916	26,147,849		
IDAHO.								
Golden Chariot	2	750	10,000	9	80,000	350,000	8 00	35 00
Ida Elmore	3		10,000	6	95,000	60,000	9 50	6 00
Mahogany	5	440	8,000		54,000		9 00	
Rising Star	7	1,200	12,000		364,000		32 00	
Total					613,000	410,000		

The first statement of this kind ever published was issued by Mr. Wheeler early in the spring of 1871, and was just in time to be incorporated in my last report, while it was in the hands of the Public Printer. Since this first issue of April 8 (says Mr. Wheeler) the dividends have increased from $22,797,849 to $26,147,849, and the assessments from $11,327,237 to $13,199,916. The above figures exhibit a splendid showing, which will, however, be further improved upon the coming year. Since November 15 Daney levied an assessment of $8,000; Overman, $25,600; Original Hidden Treasure, $31,499; General Lee, $2,000; Golden Chariot, $30,000; Washington and Creole, $15,000; Lillian Hall, $7,500; Jackson, $12,500, and Phenix, $25,000, the whole aggregating $157,099. The dividends disbursed within the same period are as follows: Chollar, $28,000; Raymond and Ely, $150,000, and Meadow Valley, $90,000, being a total of $268,000, exceeding the assessments by $110,901.

There are some slight clerical errors in the foregoing table, all of which I am not able to correct by official statements, and therefore I leave the table as it stands. Mr. Wheeler's total of dividends up to April 8, 1871, $22,797,849, is $844,800 larger than the total given in my extract from his table of that date, (see my last report, page 111,) viz, $21,953,049. The difference is due to the inclusion in his table of the Idaho mines, which I then omitted, but now retain, (on account of his totals as now given,) and to some other completions or corrections of the first table which I cannot trace, but which amount to $434,800 at least.

In the present table I have noted the following discrepancies: The Amador dividends, at $232.50 per share, amount to $860,250; the total Eureka dividends are given in the company's official report at $1,694,000; the total for California dividends should be $2,728,300, according to the items given. Raymond and Ely has made (including the December dividend) $615,000 dividends, instead of $465,000. Mr. Wheeler has simply omitted the December dividend, doubtless declared after the preparation of his table.

Highest and lowest prices of mining stocks for the past twelve months.

[From the San Francisco Weekly Stock Report.]

Name of company.	DECEMBER. Highest.	Lowest.	JANUARY. Highest.	Lowest.	FEBRUARY. Highest.	Lowest.	MARCH. Highest.	Lowest.	APRIL. Highest.	Lowest.	MAY. Highest.	Lowest.
Alpha	$8 00	$5 00	$5 50	$3 00	$4 80	$4 00	$18 00	$5 00	$12 00	$8 00	$10 00	$6 50
Amador	245 00	245 00	325 00	287 00	327 00	303 00	345 00	325 00	390 00	350 00	366 00	290 00
Buckeye			4 00	3 00	3 00	1 67	2 50	1 00	3 50	2 25	2 87	2 00
Belcher	8 00	4 00	14 00	6 50	20 50	9 50	87 50	17 00	75 00	30 00	142 00	63 00
Bullion	5 00	5 00										
Chollar-Potosi	91 00	68 00	79 00	65 75	76 00	69 50	82 00	64 00	90 00	71 00	85 00	45 00
Crown Point	18 00	10 00	41 00	15 75	55 00	31 00	160 00	35 00	195 00	132 00	310 00	183 00
Consolidated Virginia	6 75	3 00	3 00	3 00	10 50	1 62	12 50	5 00	18 37	8 25	11 50	8 12
Confidence	5 00	5 00					25 00	10 00			15 00	15 00
Consolidated Chloride												
Consolidated Silver Wedge							1 85	80	1 00		95	
Daney	7 75	4 00	6 00	2 50	6 25	3 00	7 25	5 00	7 50	5 00	10 50	6 00
Eureka	380 00	375 00	365 00	73 00	78 00	73 00	95 00	82 50	90 00	81 00	90 00	80 00
Eureka Consolidated	18 75	15 50	16 62	7 25	14 50	9 50	13 25	10 00	11 62	9 87	12 75	9 50
Empire Mill	6 00	5 00							5 00	1 00	25 00	12 00
Exchequer	5 00	5 00	3 00	3 00	3 00	3 00	6 00	4 00	6 00	5 00	7 00	6 00
Flowery												
Gould and Curry	90 00	45 00	51 50	41 00	60 00	37 00	70 00	40 00	67 00	50 00	115 00	65 00
Gold Hill Quartz												
General Lee							80	70	1 00		90	
Golden Chariot	93 50	67 50	83 25	64 00	85 00	71 00	80 00	37 50	43 00	32 50	46 00	37 00
Hale and Norcross	121 00	102 00	107 00	97 00	103 00	86 00	95 00	76 50	82 50	51 00	65 00	52 00
Hidden Treasure Consolidated												
Ida Elmore	17 00	12 00	12 00	7 00	12 00	7 00	14 50	8 00	18 00	12 50	16 00	14 00
Imperial	22 00	11 00	22 00	10 50	10 50	4 00	48 50	6 00	42 00	18 00	85 00	24 00
Independent											3 50	3 50
Jackson			6 25	2 50	3 69	1 50	2 25	1 00	1 62	62	1 12	87
Julia	6 00	2 00	2 37	2 50	2 50	2 25	2 25	2 25				
Justice												
Kentuck	41 00	30 00	40 00	31 00	50 00	27 00	120 00	46 00	90 00	64 00	112 00	65 00
Lady Bryan												
Mammoth			40	34	30	22	80	35	70	40	77	45
Mahogany			19 50	5 00	15 00	8 00	11 50	8 25	10 00	7 50	8 25	5 00
Maxwell			9 00	8 50								
Meadow Valley	28 00	23 00	35 00	23 50	30 00	22 00	26 50	13 50	19 50	15 00	18 75	14 00
Meadow Valley Western Extension												
Mineral Hill												
Metropolitan Mill and Mining Company												
Noonday			40	20	30	30	1 00	30	80	45	1 50	45
Nevada Butte												
Occidental	50	50										
Ophir	6 00	3 00	5 75	3 12	7 25	3 25	19 00	5 87	16 75	6 00	10 00	6 25
Oriental												
Original Hidden Treasure			6 00	2 87	3 25	3 00	7 00	3 25	9 50	6 75	11 25	7 00
Overman	6 00	3 00	6 00	2 00	6 75	2 00	5 62	2 50	6 25	3 50	7 50	2 25
Phenix			2 12	1 00	1 37	60	1 50	80	1 80	1 12	2 00	1 60
Raymond and Ely			21 25	15 00	25 00	16 50	25 00	21 00	23 25	20 00	22 00	16 50
Rising Star												
Savage	52 00	30 00	56 75	45 50	52 00	34 50	80 00	36 75	70 00	35 00	54 00	46 00
Segregated Belcher	3 00	1 00	2 75	1 50	3 69	1 50	11 00	3 75	16 00	6 00	14 50	8 75
Sierra Nevada	21 00	18 00	22 00	13 00	15 00	12 00	15 50	10 00	16 75	12 00	18 00	12 50
Silver Wave			4 00	1 00			6 50	3 25	7 00	5 25		
Silver Vault												
St. Patrick Gold Mining Co.											55 00	55 00
Succor			7 50	4 00	6 00	3 50	10 00	5 50	6 50	3 87	4 00	2 50
Union Mill and Mining Co.												
Virginia			3 50	1 50	3 50	3 00	6 00	2 00	5 00	4 00		
Yellow Jacket	48 00	35 00	48 00	37 00	45 00	29 75	80 00	42 00	71 00	62 50	99 00	64 00
Pioche									12 00	12 00	10 87	8 00
Washington and Creole												

Highest and lowest prices of mining stocks, &c.—Continued.

Name of company.	JUNE. Highest.	JUNE. Lowest.	JULY. Highest.	JULY. Lowest.	AUGUST. Highest.	AUGUST. Lowest.	SEPTEMBER. Highest.	SEPTEMBER. Lowest.	OCTOBER. Highest.	OCTOBER. Lowest.	NOVEMBER. Highest.	NOVEMBER. Lowest.
Alpha	$12 00	$7 00	$10 00	$8 00	$10 00	$8 00	$20 00	$9 50	$20 00	$9 00	$10 00	$8 50
Amador	315 00	275 00	300 00	270 00	275 00	265 00	290 00	290 00	290 00	287 00	290 00	270 00
Buckeye	4 12	2 75	3 75	1 50	2 50	1 50	4 00	1 87	6 00	3 00	4 00	2 50
Belcher	245 00	125 00	242 00	140 00	240 00	196 00	405 00	255 00	405 00	305 00	400 00	330 00
Bullion												
Chollar-Potosi	60 00	40 00	52 00	31 00	35 50	28 50	39 50	27 00	35 00	28 50	34 00	29 25
Crown Point	340 00	267 00	330 00	290 00	310 00	274 00	340 00	295 00	316 00	280 00	350 00	290 00
Consolidated Virginia	10 50	8 75	9 75	6 00	10 00	6 54	11 50	7 00	10 50	8 00	9 50	8 00
Confidence					20 00	20 00						
Consolidated Chloride												
Consolidated Silver Wedge												
Daney	8 50	4 00	6 50	3 50	5 00	3 50	12 50	4 00	8 00	6 00	7 00	5 75
Eureka	68 00	48 00	52 00	12 50	20 00	11 00	25 50	17 50	19 50	12 00	25 00	19 00
Eureka Consolidated	15 00	12 75	15 00	13 75	16 62	14 25			19 00	9 00	29 00	20 75
Empire Mill												
Exchequer	9 00	6 00	6 00	6 00			18 50	18 00	14 00	6 00	7 00	6 00
Flowery												
Gould and Curry	178 00	82 00	125 00	102 00	124 00	95 00	127 00	93 00	120 00	94 00	113 00	90 00
Gold Hill Quartz							17 00	17 00				
General Lee										23		20
Golden Chariot	56 00	41 50	51 50	36 00	31 50	17 00	29 00	13 00	15 50	10 00	23 00	4 50
Hale and Norcross	72 00	58 00	145 60	58 00	122 00	72 00	111 00	82 50	130 00	90 00	119 00	95 00
Hidden Treasure Consolidated												
Ida Elmore	15 75	8 87	9 00	6 25	7 00	4 75	5 75	3 00	5 00	2 50	4 00	2 00
Imperial	78 00	33 00	45 00	35 00	4 00	4 00	5 50	3 50	5 75	5 75	37 00	35 00
Independent					4 00	4 00	5 50	3 50	5 75	5 75	5 50	3 50
Jackson	1 00	50	80	60	66	52	95	50	1 75	75	2 00	67
Julia	1 50	1 50			30	30						
Justice												
Kentuck	191 00	105 00	130 00	100 00	148 00	100 00	177 00	134 00	172 00	125 00	157 00	125 00
Lady Bryan												
Mammoth			60	45	50	40	30	40	40	80	40	65
Mahogany	8 25	4 00	12 50	7 50	15 00	8 00	10 50	7 75	9 75	7 87	10 00	6 00
Maxwell												
Meadow Valley	19 50	17 75	20 00	18 50	23 25	20 12	33 50	20 00	35 50	31 00	36 50	31 50
Meadow Valley Western Extension												
Mineral Hill												
Metropolitan Mill and Mining Company												
Noonday	1 50	40	40	50	65	2	50 00	35 00	1 00	40	63	65
Nevada Butte												
Occidental												
Ophir	10 75	6 00	10 25	6 25	28 00	9 00	25 00	21 00	28 00	21 00	27 00	20 00
Oriental												
Original Hidden Treasure	12 12	5 62	9 00	5 25	9 50	7 00	8 25	7 62	7 50	5 23	8 50	5 50
Overman	13 00	8 25	11 00	6 00	7 50	5 25	41 00	5 50	38 00	14 00	28 00	17 50
Phenix	4 37	1 67	3 75	3 25	3 12	1 62	4 12	3 00	4 12	3 25	4 62	3 00
Raymond and Ely	21 00	18 87	18 50	15 60	21 00	18 73	92 00	24 75	150 00	77 00	97 00	64 00
Rising Star												
Savage	50 50	43 00	45 00	34 00	44 25	33 00	43 75	37 00	47 00	40 00	48 50	42 50
Segregated Belcher	11 50	8 50	10 00	7 50	13 00	7 00	41 00	12 75	45 00	20 00	28 00	19 00
Sierra Nevada	17 25	15 00	18 00	13 87	9 25	9 00	18 50	17 50	32 00	20 00	22 00	20 00
Silver Wave												
Silver Vault												
St. Patrick Gold Mining Co	50 00	50 00	45 00	42 50	27 00	27 00	21 00	20 00	41 00	20 00	25 00	22 50
Succor	5 25	3 50	7 00	4 62	7 50	5 50	7 75	6 00	7 00	4 50	5 00	3 50
Union Mill and Mining Co										7 00		
Virginia												
Yellow Jacket	77 00	66 00	74 00	44 50	75 00	41 50	68 50	53 00	62 50	54 50	60 00	55 00
Pioche			9 62	9 50	12 75	11 00	28 25	13 00	20 75	9 00	9 87	7 00
Washington and Creole							7 25	5 00	6 62	4 75	6 87	4 00

San Francisco coal trade for 1870 and 1871.

[From the Commercial Herald of January 12, 1872.]

Imports.	1870.	1871.
Anthracite, tons	21,320	7,321
Australian, tons	83,982	38,942
Bellingham Bay, tons	14,355	20,284
Cumberland, tons	9,322	6,060
Coos Bay, tons	20,567	28,690
Chili, tons	7,350	4,164
English, tons	31,196	54,191
Mount Diablo, tons	129,761	133,485
Vancouver Island, tons	12,640	15,621
Queen Charlotte Island, tons	....	565
Sitka, tons	....	18
Rocky Mountain, tons	....	1,025
Seattle, tons	....	4,918
Total	**330,493**	**315,194**

Specified on the way from domestic Atlantic ports, December 31:

	1870.	1871.
Tons	2,464	3,679
Casks	690	885

The business of the past year in this leading article of consumption has been characterized by great firmness of all varieties, with some extreme prices. The largely increased demand for bituminous coal, for gas purposes, and the disposition of our ocean steamship companies to employ it extensively in the place of anthracite, heretofore used, have produced some marked changes in the trade. This latter article, so long the leading feature of the market, has been gradually falling off for some years, and has now touched the lowest point ever reached; while the market now takes with firmness an amount of foreign, which a few years since would have entirely prostrated it. It will be observed that while the aggregate of our foreign varieties is somewhat under the imports of 1870, (which were excessive, and greatly depressed prices,) yet the English varieties have greatly exceeded any former year, and yet the market has remained remarkably firm under these heavy arrivals, and nearly all varieties were wanted at full prices at the close of the year. Indeed, many of the vessels arriving with English coal, failing to get return cargoes of grain, have been dispatched under charters to Australia, to return with coal, and our supplies from this source, through the coming spring, will be quite large, but being all ordered by parties here, little or none of it will probably come upon the market, and we may not look for much change in prices until the arrival of the full fleet. In the mean time our domestic varieties arrive freely, and control the market within the sphere of their special adaptability, and show a gratifying increase on last year's production. In addition to the amounts reported, it is understood that the Central Pacific Railroad Company have brought over, for its own consumption, more or less Rocky Mountain coal, which would probably swell the figures of our consumption of domestic varieties to something over 200,000 tons for the year. Below will be found our usual five-year comparative statement of the various varieties received:

Variety.	1867.	1868.	1869.	1870.	1871.	Total.
	Tons.	*Tons.*	*Tons.*	*Tons.*	*Tons.*	*Tons.*
Foreign	64,000	93,000	109,000	135,108	113,483	514,051
Eastern	62,500	32,700	38,600	30,890	13,291	177,911
Domestic	124,500	157,000	184,100	167,183	188,420	821,203
Total	**251,000**	**282,700**	**331,700**	**333,171**	**315,194**	**1,513,705**

In regard to markets at date, recent imports continue light, and business much restricted for want of needed supplies. The last sale of Chili was at $13.25. There is no Nanaimo on the market. Seattle sells to the trade at $10; Coos Bay at $9.75; Bellingham Bay, $8.50; Mount Diablo, $6.25 to $8.25, for fine and coarse respectively. Anthracite is scarce and high, say $25 to $35 for the driblets to be had. Cumberland is also out of first hands, jobbing at $22.50 to 27.50. West Hartley is held at $13.50; Scotch and English steam, $12.50 to $13; Australian is wanted at $13 to $13.50.

Product of quicksilver in 1869, 1870, and 1871, and exports from 1852 to 1871.

[From the Commercial Herald of January 12, 1872.]

The following is the quicksilver produced in 1869, 1870, and 1871:

Mine.	1869.	1870.	1871.
	Flasks.	*Flasks.*	*Flasks.*
New Almaden mine	17,000	14,000	18,703
New Idria mine	10,450	10,000	9,227
Redington mine	5,000	4,546	2,128
Sundry other mines	1,150	1,000	1,703
Total	33,600	29,546	31,881

The exports to the different countries for 1871, and the three previous years, were as follows:

To—	1868.	1869.	1870.	1871.
	Flasks.	*Flasks.*	*Flasks.*	*Flasks.*
New York	4,500	1,500	1,000	800
Great Britain	3,500			
China	17,785	11,600	4,050	7,900
Mexico	14,120	8,060	7,088	3,081
South America	2,500	2,900	1,300	2,200
Australia	1,580	300	300	1,100
British Columbia	20	4	9	6
Other countries	501	51	41	118
Total	44,506	24,415	13,788	15,205

And our exports previously have been—

	Flasks.			*Flasks.*
In 1867	28,653		In 1859	3,399
In 1866	30,287		In 1858	24,142
In 1865	42,469		In 1857	27,262
In 1864	36,997		In 1856	23,740
In 1863	26,014		In 1855	27,165
In 1862	33,747		In 1854	20,963
In 1861	35,965		In 1853	12,737
In 1860	9,448		In 1852	900

The Redington company has produced nothing since October 31, and the Phenix mine has produced, during 1871, 763 flasks, from a partial working of the mine. The entire product of all the mines on this coast for the year 1871 aggregates 31,881 flasks, against 29,546 the year previous. High prices under the monopoly rule have been kept up for three years past. The three-year contract purchase expires on the 1st of April next, when a complete change in the programme may be looked for. Already shipments from London and New York are *en route* to this coast, and lower prices may be looked for at any moment. Present nominal prices, 82½ to 85 cents.

INDEX OF MINES.

A.

Page.

H.

534   INDEX OF MINES.

INDEX OF MINES. 537

INDEX OF MINES. 537

Page.

Martha Washington, Tooele County, Utah ... 316, 316
Martin, White Pine County, Nevada ... 197
Mary Ellen, Wyoming ... 374
Maryland, White Pine County, Nevada ... 197, 198
Maxon, Amador County, California ... 89
Maxwell, Amador County, California ... 45, 88, 517, 520, 521
Mayflower, Humboldt County, Nevada ... 221
Mazeppa, White Pine County, Nevada ... 184, 195, 198, 199
McBride, White Pine County, Nevada ... 197
McCormick, White Pine County, Nevada ... 197, 198
McDonald & Co., El Dorado County, California ... 100
McElroy's, Calaveras County, California ... 77
McFadden, William, Summit County, Colorado ... 363
McMahon Ledge, White Pine County, Nevada ... 200
McMechan's, San Diego County, California ... 91
Meade, Gilpin County, Colorado ... 341
McNevin, White Pine County, Nevada ... 197
Meadow Valley, Lincoln County, Nevada ... 224, 432, 433, 516, 517, 518, 520, 521
Meadow Valley Mining Company, Lincoln County, Nevada ... 223, 249
Meadow Valley, Western Ex., Lincoln County, Nevada ... 518, 520, 521
Medeon, Amador County, California ... 89
Morrill's Claim, Beaver Head County, Montana ... 262
Metropolitan, Tooele County, Utah ... 305
Metropolitan Mill, White Pine County, Nevada ... 198, 199, 517, 518, 520, 521
Mexican, Storey County, Nevada ... 143
Michigan Bluff, Placer County, California ... 114
Miller, Salt Lake County, Utah ... 405
Miller, White Pine County, Nevada ... 197
Milton, Beaver Head County, Montana ... 267
Mineral Hill, Tooele County, Utah ... 305
Mineral Hill Silver Mining Company, Lander County, Nevada ... 180, 518, 520, 521
Mineral Point, White Pine County, Nevada ... 195, 197
Miner's Delight, Wyoming ... 374, 375
Miner's Dream, Calaveras County, California ... 70
Miner's Dream, White Pine County, Nevada ... 185
Minneapolis, Clear Creek County, Colorado ... 353, 357, 358
Miser's Dream, White Pine County, Nevada ... 195, 197, 401
Miwanotack, Beaver Head County, Montana ... 270
Moffat Shock, Summit County, Colorado ... 362
Moffit, White Pine County, Nevada ... 190
Mollie Stark, White Pine County, Nevada ... 195, 197, 401
Monarch, Tooele County, Utah ... 308, 309
Monitor, Humboldt County, Nevada ... 219
Monitor, Nye County, Nevada ... 182
Monitor Ledge, Lander County, Nevada ... 170
Monitor and Magnet, Salt Lake County, Utah ... 323
Monitor and Northwestern, Alpine County, California ... 93
Mono, White Pine County, Nevada ... 197
Monroe, Nevada ... 36
Monte Cinto, Nevada County, California ... 117
Montezuma, Colfax County, New Mexico ... 337
Montezuma, Salt Lake County, Utah ... 319, 321, 324, 325
Montezuma, White Pine County, Nevada ... 196
Montgomery, White Pine County, Nevada ... 195, 199
Montrose, White Pine County, Nevada ... 187
Moorhouse, Nevada County, California ... 45
Moose, Park County, Colorado ... 365, 366
Morgan Ground, Calaveras County, California ... 75
Mormon Chief, Tooele County, Utah ... 313, 314
Morning Star, Alpine County, California ... 95
Morning Star, Iron Rod District, Montana ... 286
Morning Star, Placer County, California ... 115
Morning Star, Radersburgh, Montana ... 295, 297
Morning Star, Salt Lake County, Utah ... 321, 323
Mountain Boy, Humboldt County, Nevada ... 223
Mountain Chief, White Pine County, Nevada ... 198
Mountain Gem, Tooele County, Utah ... 305
Mountain King, Humboldt County, Nevada ... 213, 214
Mountain Lion, Clear Creek County, Colorado ... 353, 358

R.

S.

U.

V.

W.

INDEX OF COUNTIES, DISTRICTS, ETC.

A.

B.

J.

K.

L.

M.

INDEX OF SUBJECTS.

A.

T.

○

www.ingramcontent.com/pod-product-compliance
Lightning Source LLC
Chambersburg PA
CBHW020854210326
41598CB00018B/1661